Wolf Prize in Mathematics

Volume 1

Wolf Prize in Mathematics

Volume 1

Edited by

S S Chern

University of California, Berkeley, USA &
Mathematical Sciences Research Institute, USA

F Hirzebruch

Universität Bonn &
Max-Planck-Institut für Mathematik, Germany

World Scientific
Singapore • New Jersey • London • Hong Kong

Published by

World Scientific Publishing Co. Pte. Ltd.

P O Box 128, Farrer Road, Singapore 912805

USA office: Suite 1B, 1060 Main Street, River Edge, NJ 07661

UK office: 57 Shelton Street, Covent Garden, London WC2H 9HE

Library of Congress Cataloging-in-Publication Data

Wolf prize in mathematics / edited by S.S. Chern, F. Hirzebruch.
p. cm.
Includes bibliographical references.
ISBN 9810239459 (v. 1)
1. Mathematicians--Biography. 2. Mathematics. 3. Wolf Foundation Prizes.
I. Chern, Shiing-Shen, 1911- . II. Hirzebruch, Friedrich, 1927–

QA28.W65 2000
510.92'2--dc21
[B] 00-031994

British Library Cataloguing-in-Publication Data

A catalogue record for this book is available from the British Library.

The editors and publisher would like to thank the following organisations and publishers of the various journals and books for their assistance and permission to reproduce the selected reprints found in this volume:

Academic Press
Almqvist & Wiksell
American Institute of Physics
American Mathematical Society
Gauthier-Villars
International Press
Israel Mathematical Society
The Johns Hopkins University Press
Mathematical Association of America
Mathematical Society of Japan
Nagoya University Press
National Academy of Sciences (USA)
Princeton University Press
The Royal Swedish Academy of Sciences
Société Mathématique de France
Springer-Verlag
University of Illinois Press
Vandenhoeck & Ruprecht
The Wolf Foundation

While every effort has been made to contact the publishers of reprinted papers prior to publication, we have not been successful in some cases. Where we could not contact the publishers, we have acknowledged the source of the material. Proper credit will be accorded to these publishers in future editions of this work after permission is granted.

Printed in Singapore by Uto-Print

List of Wolf Prize Winners

1978	Izrail M. Gelfand	1913
	Carl L. Siegel	1896–1981
1979	Jean Leray	1906–1998
	André Weil	1906–1998
1980	Henri Cartan	1904
	Andrei N. Kolmogorov	1903–1987
1981	Lars V. Ahlfors	1907–1996
	Oscar Zariski	1899–1986
1982	Hassler Whitney	1907–1989
	Mark Grigor'evich Krein	1907–1989
1983/4	Shiing-Shen Chern	1911
	Paul Erdös	1913–1996
1984/5	Kunihiko Kodaira	1915–1997
	Hans Lewy	1904–1988
1986	Samuel Eilenberg	1913–1998
	Atle Selberg	1917
1987	Kiyosi Itô	1915
	Peter D. Lax	1926
1988	Friedrich Hirzebruch	1927
	Lars Hörmander	1931
1989	Alberto P. Calderon	1920–1998
	John W. Milnor	1931
1990	Ennio De Giorgi	1928–1996
	Ilya Piatetski-Shapiro	1929
1992	Lennart A. E. Carleson	1928
	John G. Thompson	1932
1993	Mikhael Gromov	1943
	Jacques Tits	1930
1994/5	Jürgen K. Moser	1928–1999
1995/6	Robert Langlands	1936
	Andrew J. Wiles	1953
1996/7	Joseph B. Keller	1923
	Yakov G. Sinai	1935
1999	László Lovász	1948
	Elias M. Stein	1931
2000	Raoul Bott	1923
	Jean-Pierre Serre	1926

Preface

There is no Nobel prize in mathematics. Perhaps this is a good thing. Nobel prizes create so much public attention that mathematicians would lose their concentration to work.

There are several other prizes for mathematicians. There is the Fields medal (only for mathematicians). This medal is awarded to mathematicians who are at most 40 in the year of the International Congress of Mathematicians where the medals are presented. Thus it honours outstanding work and encourages further efforts. The Fields medal is perhaps best known and is often called the Nobel prize in mathematics. World Scientific has published a book of the Fields medallists' lectures.

Then there is the Wolf prize. The Wolf foundation describes the prize as follows: "The WOLF FOUNDATION began its activities in 1976, with an initial endowment fund of 10 million U.S. dollars donated in its entirety by the Wolf family. The main founders were Dr. Riccardo Subirana Lobo Wolf and his wife Francisca .... Since 1978 five or six annual prizes are awarded to outstanding scientists and artists, irrespective of nationality, race, colour, religion, sex or political view, for achievements in the interest of mankind and friendly relations among people. In Science, the fields are: AGRICULTURE; CHEMISTRY; MATHEMATICS; MEDICINE; PHYSICS, and in ARTS, the prize rotates annually among Music, Painting, Sculpture and Architecture .... The official presentation of the prizes takes place at the Knesset building (Israel's parliament) and the winners are handed their awards by the President of the State of Israel at a special ceremony .... ."

The Fields medal goes to young people, and indeed many mathematicians do their best work in the early years of their life. The Wolf prize often honours the achievements of a whole life. But it may also honour the work of young people.

The first Wolf prize winners in mathematics were Izrail M. Gelfand and Carl L. Siegel (1978). Siegel was born in 1896 and Gelfand in 1913. Gelfand is still active at Rutgers University. Several prize winners were born before 1910. Thus the achievements of the prize winners cover much of the twentieth century.

The documents collected in these two volumes characterize the Wolf prize winners in a form not available up to now: bibliographies and curricula vitae, autobiographical accounts, early papers or especially important papers, lectures and speeches, for example at International Congresses, as well as reports on the work of the prize winners by others. Since the work of the Wolf laureates covers a wide spectrum, a large part of contemporary mathematics comes to life in these books.

Mathematical prize winners are usually quite modest. They know that the selection committee had to choose from a large list of excellent candidates and that not only merit is needed to receive a prize, but also much luck. Quite different sets of mathematicians could illustrate with their work just as well the development of mathematics in the period covered by the Wolf prizes.

The volumes are also a symbol of thanks to the donors who made the Wolf foundation possible and to all who worked, and work, for the success of the foundation. The editors also thank all mathematicians who prepared the material for deceased Wolf prize winners. Without their generous help the two volumes would be very incomplete.

The editors want to express their thanks to World Scientific for a splendid cooperation. They especially profited from the work and valuable advice of Dr. Sen Hu of World Scientific office in the USA and thank him most cordially.

S. S. Chern
F. Hirzebruch

Contents

Wolf Prize in Mathematics

Volume 1

Lars Valerian Ahlfors
Curriculum Vitae

Born

April 18, 1907

Degrees

Fil.-Kand.	University of Helsinki	1928
Fil.-Lic.	University of Helsinki	1930
Fil. Mag. - Fil. Dr.	University of Helsinki	1932
M.A. (complimentary)	Harvard University	1945
Ll.D. (h.c.)	Boston University	1953
Fil.-Dr. (h.c.)	Abo Adademi	1970
Ph.Dr. (Fakultat II, h.c.)	University of Zürich	1977
Sc.D. (h.c.)	University of London	1978

Positions Held

Abo Adademi	Lecturer	1929–1933
University of Helsinki	Adjunct	1933–1935
Harvard University	Lecturer	1935–1938
University of Helsinki	Professor	1938–1944
University of Zürich	E.O. Professor	1945–1946
Harvard University	Professor	1946–1977
Harvard University	Emeritus Professor	1977–

Visiting Appointments

Mittag-Leffler Institute	1972
Columbia University	Spring 1978
University of Michigan	Fall 1979

Prizes

Fields' Medal	1936
International Prize (Finland)	1968
Wolf Prize in Mathematics	1981

Memberships

Societas Scientiarum Fennica
Academia Scientiarum Fennica
National Academy of Sciences
Swedish Royal Academy of Science
Danish Royal Academy of Science

PUBLICATIONS OF LARS V. AHLFORS

1929 [1] Sur le nombre des valeurs asymptotiques d'une fonction entière d'ordre fini. *C. R. Acad. Sci. Paris* 188: 688-689.

[2] Über die asymptotischen Werte der ganzen Funktionen endlicher Ordnung. *Ann. Acad. Sci. Fenn.*, Ser. A., 32, 6: 1-15

1930 [3] Sur quelques propriétés des fonctions méromorphes. *C. R. Acad. Sci. Paris* 189: 720-722.

[4] Beiträge zur Theorie der meromorphen Funktionen. *7th Scand. Math. Congr.*, Oslo, pp. 84-87.

[5] Untersuchungen zur Theorie der konformen Abbildung und der ganzen Funktionen. *Acta Soc. Sci. Fenn.*, N. S. A1, pp. 1-40.

1931 [6] Zur Bestimmung des Typus einer Riemannschen Fläche. *Comm. Math. Helv.* 3: 173-177.

[7] Ein Satz von Henri Cartan und seine Anwendung auf die Theorie der meromorphen Funktionen. *Comm. Phys. Math. Soc. Sci. Fenn.* 5, 16: 1-19.

1932 [8] Sur une généralisation du théorème de Picard. *C. R. Acad. Sci. Paris* 194: 245-247.

[9] Sur les fonctions inverses des fonctions méromorphes. *C. R. Acad. Sci.* 194: 1145-1147.

[10] Über die asymptotischen Werte der meromorphen Funktionen endlicher Ordnung. *Acta Acad. Aboensis* 6: 1-8.

[11] Über eine in der neueren Wertverteilungstheorie betrachtete Klasse transzendenter Funktionen. *Acta Mathematica*, pp. 375-406.

[12] Ein Satz über die charakteristische Funktion und den Maximalmodul einer meromorphen Funktion. *Soc. Sci. Fenn. Comm. Math. Phys.* 6, no. 9: 1-4.

[13] Quelques propriétés des surfaces de Riemann correspondant aux fonctions méromorphes. *Bull. Soc. Math. France* 60: 197-207.

1933 [14] Über, die Kreise die von einer Riemannschen Fläche schlicht überdeckt werden. *Comm. Math. Helv.* 5: 28-38.

[15] Sur les domaines dans lesquels une fonction méromorphe prend des valeurs appartenant à une région donnée. *Acta Soc. Sci. Fenn.*, N. S. 2: 1-17.

1935 [16] Über eine Methode in der Theorie der meromorphen Funktionen. *Soc. Sci. Fenn. Comm. Phys.-Math.* 8, 10: 1-14.

[17] Sur le type d'une surface de Riemann. *C. R. Acad. Sci. Paris* 201: 30-32.

[18] Über die konforme Abbildung von Überlagerungsflächen. *8th Scand. Congr. of Math.* (Stockholm, 1934), pp. 299-305.

[19] Zur Theorie der Überlagerungsflächen. *Acta Math.* 65: 157-194.

Reprinted from Lars V. Ahlfors: Collected Papers, Vol. 1, pp. xv–xix.

1936 [20] Über eine Klasse von Riemannschen Flächen. *Soc. Sci. Fenn. Comm. Phys.-Math.* 9: 6.

1937 [21] On Phragmén-Lindelöf's Principle. *Trans. Amer. Math. Soc.* 41, no. 1: 1-8.

[22] Über die Anwendung differentialgeometrischer Methoden zur Untersuchung von Überlagerungsflächen. *Acta Soc. Sci. Fenn.*, N. S. A, Tom II (6) pp. 1-17.

[23] (With H. Grunsky) Über die Blochsche Konstante. *Math Zeitschr.* 42: 671-673.

[24] Geometrie der Riemannschen Flächen. *Comptes rendus du congrès international des mathématiciens*, Oslo (1936), pp. 239-248.

1938 [25] An Extension of Schwarz's Lemma. *Trans. Amer. Math. Soc.* 43, no. 3: 359-364.

1939 [26] Zur Uniformisierungstheorie. *Comptes rendus du congrès des mathématiciens scandinaves à Helsinki* (1938), pp. 235-248.

1941 [27] The Theory of Meromorphic Curves. *Acta Soc. Sci. Fenn.* N. S. A, Tom III (4) pp. 1-31.

1943 [28] Die Begründung des Dirichletschen Prinzips. *Soc. Sci. Fenn. Comm. Phys.-Math.* 11, 15: 1-15.

1946 [29] (With A. Beurling) Invariants conformes et problèmes extrémaux. *Comptes rendus du 10me congrès des mathématiciens scandinaves*, Copenhagen, pp. 341-351.

1947 [30] Bounded analytic functions. *Duke Mathematical Journal* 14, 1: 1-11.

[31] Normalintegrale auf offenen Riemannschen Flächen. *Annales Acad. Scient. Fenn.*, Ser. A1, no. 35.

[32] Das Dirichletsche Prinzip. *Math. Annalen* 120: 36-42.

1948 [33] Book review: Topological methods in the theory of functions of a complex variable by Marston Morse. *Bull. Amer. Math. Soc.* 54, no. 5: 489–491.

1949 [34] Questions of regularity connected with the Phragmén-Lindelöf principle. *Annals of Math.* 50, no. 2: 341-346.

1950 [35] (With A. Beurling) Conformal invariants and function-theoretic null-sets. *Acta Mathematica* 83: 100–129.

[36] Open Riemann surfaces and extremal problems on compact subregions. *Commentarii Mathematici Helvetici* 24: 100–129.

1951 [37] Book review: Coefficient regions for schlicht functions by A. C. Schaeffer and D. C. Spencer. *Bull. Amer. Math. Soc.* 57: 328-331.

[38] Remarks on the classification of open Riemann surfaces. *Ann. Acad. Sci. Fenn.*, Ser. A1, 87 pp. 1-8.

1952 [39] Remarks on the Neumann-Poincaré integral equation. *Pacific J. Math.* 2: 271-280.

[40] On the characterization of hyperbolic Riemann surfaces. *Ann. Acad. Sci. Fenn.*, Ser. A1, Math.-Phys. 125 p. 1925.

[41] (With A. Beurling) Conformal invariants. Construction and applications of con-

formal maps. Proceedings of a symposium, pp. 243-245. National Bureau of Standards, Appl. Math. Ser. no. 18, U.S. Government Printing Office.

[42] (With H. L. Royden) A counterexample in the classification of open Riemann surfaces. *Ann. Acad. Sci. Fenn.*, Ser. A1, Math-Phys. 120: (5).

1953 [43] Development of the theory of conformal mapping and Riemann surfaces through a century. Contributions to the theory of Riemann surfaces. *Annals of Math. Studies* 30: 3-13. Princeton, N.J.: Princeton University Press.

1954 [44] On quasiconformal mappings. *Jour. d'Analyse Math* 3: 1-58.

1955 [45] Conformality with respect to Riemannian metrics. *Ann. Acad. Sci. Fenn.*, Ser. A1, no. 206: 22.

[46] Two numerical methods in conformal mapping. Experiments in the computation of conformal maps, pp. 45-52. National Bureau of Standards Appl. Math. Series, U.S. Government Printing Office.

[47] Remarks on Riemann surfaces. *Lectures on functions of a complex variable*, pp. 45-48. Ann Arbor; University of Michigan Press.

1956 [48] Square integrable differentials on open Riemann surfaces. *Proc. Nat. Acad. Sci.* 42: 758-760.

[49] (With A. Beurling) The boundary correspondence under quasiconformal mappings. *Acta Math.* 96: 125-142.

1958 [50] Extremalprobleme in der Funktionentheorie. *Ann. Acad. Sci. Fenn.*, Ser. A1, no. 249/1: 9.

[51] The method of orthogonal decomposition for differentials on open Riemann surfaces. *Ann. Acad. Sci. Fenn.*, Ser. A1, no. 249/7: 15.

[52] Abel's theorem for open Riemann surfaces. *Seminars on Analytic Functions*, vol. II, pp. 7-19. Institute for Advanced Studies, 1957.

1960 [53] (With L. Bers) Riemann's mapping theorem for variable metrics. *Annals of Math.* 72 pp. 385-404.

[54] The complex analytic structure of the space of closed Riemann surfaces. *Analytic Functions*, pp. 45-66. Princeton, N.J.: Princeton University Press.

1961 [55] Classical and contemporary analysis. *SIAM Reviews.* 3: 1-9.

[56] (With L. Bers) Spaces of Riemann surfaces and quasiconformal mappings. *Iz dat. Inostr. Lit. Moscow*, 177.

[57] Some remarks on Teichmüller's space of Riemann surfaces. *Ann. of Math.* 74: 171-191.

1962 [58] Curvature properties of Teichmüller's space. *Jour. d'Analyse Math.* 9: 161-176.

[59] Geodesic curvature and area. *Studies in mathematical analysis and related topics.* Stanford, Calif.: Stanford University Press.

[60] (With G. Weill) A uniqueness theorem for Beltrami equations. *Proc. Amer. Math. Soc.* 13: 975-978.

1963 [61] Teichmüller spaces. *Proc. International Congr. Math.* (1962) pp. 3-9. Inst. Mittag-Leffler.

[62] The use of quasiconformal deformations for the study of Kleinian groups. Outlines Joint Symp. Partial Differential Equations (Novosibirsk 1963), pp. 10-13. Acad. Sci., USSR, Siberian Branch, Moscow.

[63] Quasiconformal reflections. *Acta Math.* 109: 291-301.

1964 [64] Extension of quasiconformal maps from two to three dimensions. *Proc. Nat. Acad. Sci.* 51: 768-771.

[65] Finitely generated Kleinian groups. *Am. J. Math.* 86 pp. 413-429 and 87 p. 759.

[66] Eine Bemerkung über Fuchssche Gruppen. *Math. Z.* 84: 244-245.

[67] Kleinian groups. *Scripta Math.* 27: 97-103.

[68] Quasiconformal mappings and their applications. *Lectures on Modern Mathematics,* Vol. II, pp. 161-164. New York: Wiley.

1965 [69] The modular function and geometric properties of quasi-conformal maps. *Proc. Conf. Complex Analysis* (Minneapolis, 1964), pp. 296-300. Berlin: Springer.

[70] Some remarks on Kleinian groups. *Proc. conf. on quasiconformal mappings, moduli, and discontinuous groups* (Tulane), pp. 7-13.

1966 [71] Fundamental polyhedrons and limit point sets of Kleinian groups. *Proc. Nat. Acad. Sci.* 55: 183-187.

[72] Remarks on Carleman's formula for functions in a half-plane. *SIAM J. Numer. Anal.* 3: 183-187.

[73] Kleinsche Gruppen in der Ebene und in Raum. *Festband zum 70. Geburtstag R. Nevanlinna,* pp. 7-15, Springer: Berlin.

1967 [74] Hyperbolic motions. *Nagoya Math. J.* 29: 163-165.

1968 [75] Eichler integrals and the area theorem of Bers. *Michigan Math. J.* 15: 257-263.

1969 [76] The structure of a finitely generated Kleinian group. *Acta Math.* 122: 1-17.

1971 [77] Two lectures on Kleinian groups. Proc. Romanian-Finnish Seminar on Teichmüller spaces and quasiconformal mappings, Brasov, Romania, 1969.

[78] Remarks on the limit point set of a finitely generated Kleinian group. *Advances in the theory of Riemann surfaces* (Stony Brook 1969). Princeton, N.J.: Princeton University Press.

1973 [79] Sufficient conditions for quasiconformal extension. Proc. conf. University of Maryland.

1974 [80] Conditions for quasiconformal deformations in several variables. *Contributions to analysis.* New York: Academic Press.

[81] A remark on schlicht functions with quasiconformal extensions. Proc. Symp. Complex Analysis, Kent University.

1975 [82] Invariant operators and integral representations in hyperbolic space. *Math. Scand.* 36: 27-43.

1976 [83] Quasiconformal deformations and mappings in R^n. *Jour. d'Analyse Math.* 30: 74-97.

[84] Das mathematische Schaffen Rolf Nevanlinnas. *Ann. Acad. Sci. Fenn.*, Ser. A1, 2: 1-15.

[85] On a class of quasiconformal mappings. *Österreich. Akad. Wiss. Math.-Naturw.*, Kl. S. BII 185, no. 1-3: 5-10.

[86] An inequality between the coefficients a_3 and a_4 of a univalent function. *Amer. Math. Soc. Translations* 2, 104.

1977 [87] A somewhat new approach to quasiconformal mappings in R^n. Complex Analysis, Kentucky, *Lecture Notes in Mathematics*, New York: Springer Verlag. pp. 1-6.

[88] A singular operator in hyperbolic space. *Akad. Nauk SSSR*, Proc. of the Steklov Institute, pp. 40-44.

1978 [89] A singular integral equation connected with quasiconformal mappings in space. *L'Enseignement Mathématique*, IIme Série, Fascicule 3-4.

[90] Quasiconformal mappings, Teichmüller spaces, and Kleinian groups. Proc. of the international congress of mathematicians, Helsinki, 1978.

1979 [91] The Hölder continuity of quasiconformal deformations. *Amer. Journ. of Math.* 101: 1-9.

[92] Ergodic properties of groups of Möbius transformations. Proc. of a conference in Kozubnik, Poland, 1979. *Lecture Notes in Mathematics* 798: 1-8.

1982 [92] Rolf Nevanlinna och riemannska vtor. Arkhimedes, Rolf Nevanlinna-erikoisnumero, pp. 30–33.

[93] Riemann surfaces and small point sets. *Ann. Ac. Sci. Fenn., Series A. I. Math.* 7: 49–57.

1984 [94] Old and new in Möbius groups. *Ann. Ac. Sci. Fenn. Series, A. I. Math.* 9: 93–105.

1985 [95] Möbius transformations and Clifford numbers. Dedicated to the memory of Harry Ernest Rauch. Berlin, Heidelberg, Springer-Verlag.

1986 [96] On the fixed points of Möbius transformations in R^n. *Ann. Ac. Sci. Fenn. Series A. I. Math.* 10: 15–27.

[97] Möbius transformations in R^n expressed through 2×2 matrices of Clifford numbers. *Complex Variables* 5: 215–224.

1988 [98] Cross-ratios and Schwarzian derivatives in R^n. *Complex Analysis*, Vol. Dedicated to Professor Albert Pfluger. Basel, Birkhäuser Verlag.

1989 [99] (with Pertti Loanesto), Some remarks on Clifford algebras. *Complex Variables* 12: 201–209.

1991 [100] The binomial theorem in the algebra A^r. In Constantin Carathéodorg: An International Tribute. World Scientific.

Books

[101] *Complex Analysis*, New York: McGraw-Hill. 1st ed. (1953), 2nd ed. (1966), 3rd ed. (1979).

[102] *Riemann Surfaces* (with L. Sario), Princeton, N.J.: Princeton University Press (1960).

[103] *Lectures on Quasiconformal Mappings*, New York: D. Van Nostrand Company (1966).

[104] *Conformal Invariants*, New York: McGraw-Hill (1973).

Ahlfors' Preface to his Collected Papers

When first confronted with the prospect of having my collected papers published, I felt both awe and confusion, but I calmed down when I realized that the purpose was not to honor the author, but to be of service to the mathematical community. If young scholars of a future generation should desire to find out what some mathematicians of the twentieth century were up to, they would indeed have reason to be thankful if spared the need to seek this information from a multitude of sources.

As an introduction it seems polite and useful to begin with a brief outline of my life, especially as related to my professional activity. I was born the eighteenth of April 1907 in Helsingfors, Finland. My father was a professor of mechanical engineering at the Polytechnical Institute. My mother died in childbirth when I was born.

At the time of my early childhood Finland was under Russian sovereignty, but with a certain degree of autonomy, sometimes observed and sometimes disregarded by the czar who was, by today's standards, a relatively benevolent despot. Civil servants, including professors, were able to enjoy a fairly high standard of living, a condition that was to change radically during World War I and the Russian revolution that followed.

As a child I was fascinated by mathematics without understanding what it was about, but I was by no means a child prodigy. As a matter of fact I had no access to any mathematical literature except in the highest grades. Having seen many prodigies spoiled by ambitious parents, I can only be thankful to my father for his restraints. The high school curriculum did not include any calculus, but I finally managed to learn some on my own, thanks to clandestine visits to my father's engineering library.

I entered Helsingfors University in 1924 and soon realized how very fortunate I was to have two truly outstanding mathematicians as teachers, Ernst Lindelöf and Rolf Nevanlinna. At that time the university was still run on the system of one professor for each subject. Fortunately, the need for more professors had become acute and Nevanlinna soon occupied the second chair in mathematics. The elementary teaching was in the hands of two "adjunct professors" and several part-time teachers with the title of "docent." There were essentially no courses on the graduate level; advanced reading was done under the supervision of Lindelöf.

Ernst Lindelöf is rightly considered the father of mathematics in Finland. In the 1920s *all* Finnish mathematicians were his students. He was essentially self-taught and found much of his inspiration in the works of Cauchy. His worldwide reputation as a leading complex analyst was well founded, but when I knew him he had given up research in favor of teaching, at which he was a master. I still remember many Saturday mornings when I had to visit him in his home at 8 A.M. to be praised or scolded – as the case may have been.

In the spring of 1928 I earned my degree of "fil. kand," a title that was to be changed to the equivalent "fil. mag." at a public "promotion." In the fall term of 1928 Hermann Weyl was on leave from ETH in Zürich and Rolf Nevanlinna had been invited to take his place. At the urging of Lindelöf, my father agreed to let me go along to Zürich, no doubt at a nontrivial sacrifice. He may have been moved by the fact that he himself spent some time at ETH as a young man. This was my first trip abroad, and I found myself suddenly transported from the periphery to the center of Europe.

Reprinted from Lars V. Ahlfors: Collected Papers, Vol. 1, pp. xi–xiv.

It was also my first exposure to live mathematics. Nevanlinna was a young man of 33 who had already won widespread acclaim, and I was a very immature 21. The course covered contemporary function theory, including the main parts of Nevanlinna's theory of meromorphic functions, and was essentially a forerunner of his famous "Eindeutige analytische Funktionen." Among other things, Nevanlinna introduced the class to Denjoy's conjecture on the number of asymptotic values of an entire function, including Carleman's partial proof. I had the incredible luck of hitting upon a new approach, based on conformal mapping, which with very considerable help from Nevanlinna and Pólya led to a proof of the full conjecture. With unparalleled generosity they forbade me to mention the part they had played, and Pólya, who rightly did not trust my French, wrote the Comptes Rendus note. For my part I have tried to repay my debt by never accepting to appear as coauthor with a student.

With a small grant from a student organization I was able to follow Nevanlinna to Paris for three more months. There I discovered a geometric interpretation of the Nevanlinna characteristic, which, as it turned out, had been found independently by Shimizu in Japan. Nevertheless, this was the upbeat to an intense involvement with meromorphic functions.

On my return to Finland I entered my first teaching assignment as lecturer (lektor) at Åbo Akademi, the Swedish-language university in Åbo (Turku). At the same time I began work on my thesis, which I defended in the spring of 1930. For formal reasons my degree of Ph.D. was delayed until 1932.

During 1930-1932 I made several trips to continental Europe, including a longer stay in Paris, with a grant from the Rockefeller Institute. My name was becoming known, and I met many of the leading mathematicians. In 1933 I was able to return to Helsingfors as adjunct professor. That same year I married Erna Lehnert, a girl from Vienna, who with her parents had settled first in Sweden and then in Finland. This was the happiest and most important event in my life.

Our life in Helsingfors was pleasant, but uneventful. Quite unexpectedly I received a letter from Harvard University, which to me was hardly more than a name. It turned out that the Mathematics Department was looking for a young mathematician, and Carathéodory, whom I had met in Munich, had recommended me warmly. I was hesitant at first, but after persuasion by my sponsor W. C. Graustein I agreed to a trial period of three years beginning in the fall of 1935. We found life in Cambridge very rewarding, although the social life seemed somewhat Victorian. However, the friendliness of my colleagues was absolutely disarming. I also found that my knowledge of mathematics was still rather spotty, and I learned a lot during these three years.

I was in for the surprise of my life when in 1936, at the International Congress in Oslo, I was told only hours before the ceremony that I was to receive one of the first two Fields medals ever awarded. The prestige was perhaps not yet the same as it is now, but in any case I felt singled out and greatly honored. The citation by Carathéodory mentions explicitly my paper "Zur Theorie der Überlagerungsflächen," which threw some new light on Nevanlinna's theory of meromorphic functions. The award contributed in great measure to the confidence I felt in my work.

In the spring of 1938 I had to decide whether to stay at Harvard or return to Finland, where I was offered a professorship at the University of Helsinki. Lindelöf, who was already retired, urged me to return home as a "patriotic duty." In the end I think it was plain homesickness that decided my return, and back in Finland we enjoyed a very happy year together with our firstborn child.

Alas, the happiness did not last. The war broke out, and it became clear that Finland would not be spared. My wife and children – the second a newborn infant – were evacuated and found refuge with relatives in Sweden. The university was closed for lack of male students, but otherwise life went on, partly in air-raid shelters. Because of an earlier physical condition, I had never been called to military duty, and my only part in the war effort was as an insignificant link in a communications setup.

Soon after the end of the Winter War, my family was able to return home and resume a seemingly normal life. Politics in Finland took an unfortunate turn, however, and when Hitler attacked the Soviet Union in 1941 Finland was his ally. When the Russians were finally able to repulse the attack, they could also intensify the war in Finland with foreseeable results. The Finnish-Russian war ended with a separate armistice in September 1944, whereupon Finland was able to expulse the German troops stationed there. The harsh terms of the armistice left Finland in a very difficult position.

Although the uncertainty of the future and the suffering of the bereaved were on everybody's mind, the wartime was not a complete loss. It unified the nation and paved the way for a return to relatively stable conditions. Paradoxically, I was myself able to do a lot of work during the war, although without the benefit of accessible libraries.

During the summer of 1944, I received an offer from the University of Zürich. Because I saw this as my only opportunity to be reunited with my family and in view of the bleak future in Finland, I accepted in principle although for the moment I saw no physical possibility of following through the invitation. My health had declined, and, because I had no military duties, I was allowed to go to Sweden to recuperate. The problem was now to get from Sweden to Switzerland. By that time Finland was at war with Germany and, at least on paper, also with England. An appeal to the German legation in Stockholm proved fruitless, but the British were willing to let us pass through Great Britain if an opportunity arose. The Swedes had organized some semiregular "stratospheric flights" on moonless nights from Stockholm to Prestwick, Scotland. With the help of diplomatic channels we managed to be placed on the list of potential passengers. Obtaining permission from the British was still necessary and depended on the military situation. Unfortunately, when our low priority had made us eligible, the Battle of the Bulge put a temporary stop to our hopes. But, finally, one day in March 1945 we were told to be ready to leave, weather permitting. It is difficult to forget that flight. The plane was a reconditioned Flying Fortress, with perhaps a dozen passengers. It was not pressurized, and breathing was accomplished by individual oxygen masks. Swim vests were worn by all. Our children, ages 5 and 6, were quite capable of understanding the implications of danger.

We left Sweden with feelings of deep gratitude. Virtually penniless, we had been taken care of not only by close relatives, but also by mathematical colleagues, who made it possible to stay in Uppsala for months. I am forever indebted to Arne Beurling, who showed what true friendship can be.

An arduous train ride took us from Glasgow to London, where we had to wait several days for the Channel to be cleared for a ferry to Dieppe. Much of the time was spent in the London zoo, but the frequent explosions of V2 rockets made us wonder if this was wise.

To cut a long story short, we were finally shipped to Paris, where the Swiss legation was unforgettably helpful in securing lodgings in a luxury hotel – and in providing Swiss cigars to the station master. On arriving at the Swiss border, somewhat disheveled and again penniless, we were met by the Red Cross, who lent us Swiss money. Although there was some rationing even in Switzerland, we were overwhelmed by the chocolates, cakes, and other foods, the likes of which we had not seen for years.

The University of Zürich was ready to begin the summer term. I was met by the Director of the

Mathematics Institute, Professor R. Fueter, in the single room that served as office for the institute and all the professors. My first disappointment came when I learned that I would be responsible for Descriptive Geometry, a subject that for some reason had survived in the Swiss high school and undergraduate curriculum. My second shock was that it was to be taught from 7 to 9 o'clock in the morning.

Nevertheless, I slowly adjusted to my work, which even included some serious, although not very advanced, mathematics. Professor Fueter and his colleague Professor Finsler were getting on in years, and it became clear that the reason for inviting me was that no competent native successor was in sight. I took over a class of students in their formative years, and I am happy to say that many have remained my friends and are now important mathematicians in their own right.

I cannot honestly say that I was happy in Zürich. The postwar era was not a good time for a stranger to take root in Switzerland. The whole nation, although spared from war, was in a state of suspension, with, understandably, quite a bit of xenophobia in the lower classes. My wife and I did not feel welcome outside the circle of our immediate colleagues.

I was therefore very pleased when in 1946 I was asked if I would like to return to Harvard. My Swiss colleagues and the "Erziehungsdirektion des Kantons Zürich" did their best to persuade me to stay, but I was convinced that for the sake of my mathematical career I should go to America, and I was consoled when I learned that my teacher Rolf Nevanlinna had agreed to become my successor. This arrangement turned out to be a great success both for Nevanlinna and for the university.

In the fall of 1946 I took up my duties at Harvard, where I was to stay until my retirement in 1977 and as emeritus to this day. My relationship with Harvard, with both the administration and my friends in the department, has been singularly happy. I have enjoyed the many excellent students Harvard has had to offer, many of whom I have watched become leaders in my own or some other field. Harvard has also offered me an optimal milieu for my research, and it has been a source of great satisfaction that the Mathematics Department has been able to maintain the high standards that have always been its hallmark.

The annotations to my papers will serve as a running commentary to my scientific activity. I take this opportunity to thank Birkhäuser Boston, Inc., for their willingness to publish these collected works, and above all my friend Gian-Carlo Rota for taking the initiative and supervising the editing of this book.

Carathéodory's Report on Ahlfors' work for Fields Medal at ICM in Oslo 1936

Lars Valerian Ahlfors wurde in Helsingfors am 18. April 1907 als Sohn des Professors für Maschinenbau an der dortigen Technischen Hochschule Axel Ahlfors geboren. Nachdem er die »Nya svenska samskolan« absolviert hatte, wurde er im Mai 1924 an der Universität Helsingfors immatrikuliert, wo er am 5. März 1928 »Filosofie kandidat« und am 20. November 1930 »Filosofie licentiat« wurde. Bei der solemnen Promotion an derselben Universität wurde er am 31. Mai 1932 zum »Filosofie magister« und »Filosofie doktor« promoviert.

Von 1929—1933 war er Lektor der Mathematik an der Åbo Akademie und seit Juni 1933 ist er Adjunkt der Mathematik an der Universität Helsingfors. Im akademischen Jahre 1935—1936 ist er beurlaubt worden, um an der Universität Harvard Vorlesungen zu halten. Im Jahre 1936 wurde er Mitglied der Societas Scientiarum Fennica.

Wiederholt haben ihn Studienreisen ins Ausland geführt: im Winter 1928—1929 war er in Zürich, im Frühjahr 1929 in Paris, im Sommer 1931 in Göttingen und im Sommer 1934 in München.

Das akademische Lehrjahr 1931—1932 verbrachte er in Paris als »International Research Council Fellow« der »Rockefeller Foundation«.

Ahlfors ist einer der glänzendsten Vertreter der berühmten Finnländischen Funktionentheoretischen Schule, die von Ernst Lindelöf gegründet, seit dreißig Jahren so viele wichtige Beiträge der Wissenschaft geschenkt und so zahlreiche große Mathematiker hervorgebracht hat. Er ist Schüler von Ernst Lindelöf und von Rolf Nevanlinna. Unter den Auspizien des letzteren ist seine Dissertation entstanden und die Ideen und Theorien von R. Nevanlinna haben weiterhin seine ganze Entwickelung beeinflußt.

Ausgegangen ist er von der Theorie der meromorphen Funktionen, die er von Anbeginn mit originellen Methoden befruchtet hat. So konnte er einmal einen wichtigen Satz von Henri Cartan für seine Zwecke benutzen, und seine Dissertation »Untersuchungen zur Theorie der konformen Abbildungen und der ganzen Funktionen« (1930) enthält eine Methode für die Behandlung der konformen Abbildung des verallgemeinerten Streifens mit beliebig gekrümmten Rändern, die sofort bei ihrem Erscheinen Aufsehen erregt hat.

Die reifste Frucht, die das Studium der meromorphen Funktionen durch Ahlfors gezeitigt hat, findet man in seiner Arbeit »Über eine Methode in der Theorie der meromorphen Funktionen« (1935). Wenn man diese Arbeit liest, so weiß man nicht, was mehr zu bewundern ist: die Kunst, mit der Ahlfors die ganze, große Nevanlinnasche Theorie mit wunderbarer Klarheit auf nur 14 Seiten auseinanderzusetzen versteht, oder die geniale Intuition von Rolf Nevanlinna, der zu einer Zeit, wo die geometrischen Zusammenhänge noch ganz versteckt lagen, nur solche Begriffe entdeckt, ausgebildet und benutzt hat, die später die einfachste geometrische Deutung erfahren sollten.

Bei dem Entschluß des Komitees, eine der Fields-Medaillen Lars Ahlfors zuzuerkennen, ist aber vor allen eine andere Richtung seiner Arbeiten ins Gewicht gefallen. Das Studium der Riemannschen Flächen der Umkehrfunktionen von ganzen oder meromorphen Funktionen hat Ahlfors dazu geführt, allgemeinere Eigenschaften von Überlagerungsflächen zu betrachten. Es stellte sich allmählich heraus, daß die fertige, unberandete Überlagerungsfläche gewisse ganz allgemeine Eigenschaften besitzt, die man nur beobachten kann, wenn man sie gewissermaßen in statu nascendi als Grenze von berandeten, immer weiter um sich greifenden Überlagerungsflächen betrachtet. Bisher war diese Entstehungsweise einer Riemannschen Fläche, die Hermann Amandus Schwarz die Methode des Ölflecks nannte, nur für die Zwecke der konformen Abbildung und der Uniformisierungstheorie benutzt worden. Es blieb Ahlfors vorbehalten, asymptotische Gesetze aufzustellen, in welche der Inhalt der behandeten Approximationsflächen, die Länge des Randes dieser Flächen und die charakteristische Zahl der Grundfläche eingehen.

Die Arbeit »Zur Theorie der Überlagerungsflächen« (1935), die diesen Untersuchungen gewidmet ist, kann als das erste Kapitel eines neuen Zweiges der Analysis angesehen werden, für welchen vielleicht der Name »metrische Topologie« der geeignete ist. Die Überlagerungsflächen sollen nämlich alle auf einer Grundfläche liegen, die gewisse metrische Eigenschaften hat.

Den Gebieten auf der Grundfläche und ihren Rändern werden in allgemeinster Weise gewisse Zahlen zugeordnet, die, wenn man sie als Inhalt und Länge deutet, einer isoperimetrischen Ungleichheit genügen sollen. Diese scheinbar so wenig sagende Bedingung ist aber hinreichend, um die Ahlforsche Theorie abzuleiten.

Die schönsten Anwendungen dieser Resultate beziehen sich auf die Theorie der quasikonformen Abbildungen. Es zeigt sich z. B., daß der Picardsche Satz schon für quasikonforme Abbildungen der dreifach punktierten Kugel gilt, also eigentlich als Satz der metrischen Topologie angesehen werden muß. Wenn nämlich eine Überlagerungsfläche der Kugel auf die ganze euklidische Ebene quasikonform abgebildet wird, so gibt es in der Ebene immer Folgen von konzentrischen Kreisen, deren

Bilder auf der Kugel folgende Eigenschaften haben: die Länge des Randes dieser sphärischen mehrblättrigen Fläche dividiert durch ihren Gesamtflächeninhalt, konvergiert gegen Null. Nach der Ahlforschen Theorie ist es aber unmöglich, diese Bedingung zu befriedigen wenn drei Punkte der Kugel von der Figur nicht überdeckt werden dürfen.

Noch merkwürdiger ist der Ahlforsche Scheibensatz, von dem ein partikulärer Fall folgendermaßen ausgesprochen werden kann: Die ganze xy-Ebene werde auf eine Riemannsche Fläche der uv-Ebene quasikonform irgendwie abgebildet. Man wähle drei beliebige einfach zusammenhängende schlichte Gebiete der uv-Ebene, die außerhalb einander liegen. Dann hat die Riemannsche Fläche die Eigenschaft, daß mindestens das eine ihrer Blätter auch das eine dieser drei beliebigen, aber fest gewählten Gebiete enthalten muß. Oder mit anderen Worten: es gibt in der xy-Ebene mindestens ein Gebiet, das eineindeutig auf das eine oder andere der drei Gebiete der uv-Ebene abgebildet wird. Daß man mit zwei Gebieten keinen derartigen Satz aufstellen kann, zeigt schon die konforme Abbildung, die durch die Gleichung $w = \sin z$ hervorgerufen wird.

Est ist leider unmöglich, die ganze Tragweite der Ahlforschen Theorie in wenigen Worten verständlich zu machen; nach diesen wenigen Proben sieht man aber schon, daß es sich um eine ganz erstklassige Leistung handelt.

Ahlfors' Address as Honorary President at opening of ICM in Berkeley 1986

I accept this great honor with a good conscience because I consider myself a link between this International Congress and the one in 1936, fifty years ago, the occasion on which the Fields Medals were given for the first time. I understand that my only duty here will be the pleasure of handing out the Fields Medals and the Nevanlinna Prize.

At that time the circumstances were quite different; the idea of the medals had been approved in Zurich in 1932, but there had been no publicity about it and when I arrived in Oslo I did not know that the Medal had become a reality, and if I had known it I would not have considered myself the right candidate. As a matter of fact, I had not been told anything officially until I entered the room where the opening ceremony would take place, but there I was shown a place somewhere in front, and I may have had my suspicions. Well, I had more than that. I had been warned beforehand by somebody who by mistake congratulated me a day before. But up to that point it had been a secret at least officially, even to myself. There was no tradition to go by and no protocol to follow. As was mentioned here, two medals were given, one to me and one to Jesse Douglas, who was then at MIT while I happened to be a visiting lecturer at Harvard. In that way it so happened that both medals went to Cambridge, Massachusetts. Unfortunately Douglas could not accept his medal in person because according to the Congress record he was too tired. I don't know, but maybe he had good reason to be tired after a long and strenuous journey. I would not expect that to happen today. His medal was then accepted by Norbert Wiener as representative of MIT.

There are two traditions that go back to the very beginning. In the first place, the Committee to select the winners should consist of the top brass of contempory mathematics. In 1936 the members of that Committee were G.D. Birkhoff, Caratheodory, Elie Cartan, Severi, and Takagi. Truly I would call that a panel of Olympian heroes. And I think that this tradition has been continued at subsequent Congresses. The other tradition is that the works of the winners should be commented on by prominent persons in the field. In 1936 both prizes were explained by Caratheodory.

As was mentioned there was no Congress until 1950, fourteen years later. On that occasion, which took place at Harvard, the medals were given to Atle Selberg and Laurent Schwartz, both known and admired by all mathematicians. From then on the Fields Medals have become more and more prestigious and it is a safe bet that many dream of getting it. Whether true or not that the existence of the medal has contributed to the phenomenal growth of mathematics both in quantity and quality during the last fifty years must remain anybody's guess.

Today it is safe to congratulate the winners in advance and I use this occasion to offer them my sincerest compliments to their success. I share their feeling of pride and accomplishment and I know that their continued success is guaranteed. I also share the disappointment of the many who may feel that they have been passed by. I wish them better luck next time or, if there is not a next time, that posterity will prove them right and the Committee wrong. Thank you.

AN EXTENSION OF SCHWARZ'S LEMMA*

BY
LARS V. AHLFORS

I. The fundamental inequality

1. To every neighborhood on a Riemann surface there is given a map onto a region of the complex plane. For any two overlapping neighborhoods the corresponding maps are directly conformal.† We agree to denote points on the surface by $\mathfrak{w}$, corresponding values of the local complex parameter by w.

We introduce a Riemannian metric of the form

$$ds = \lambda |dw|, \tag{1}$$

where the positive function λ is supposed to depend on the particular parameter chosen, in such a way that ds becomes invariant. The metric is regular if λ is of class C_2. In this paper we shall, without mentioning it further, allow λ to become zero, although such points are of course singularities of the metric.

It is well known that the Gaussian curvature of the metric (1) is given by

$$K = -\lambda^{-2} \Delta \log \lambda, \tag{2}$$

and that this expression remains invariant under conformal mappings of the w-plane. We are interested in the case of a metric with negative curvature, bounded away from zero. It is convenient to choose the upper bound of the curvature equal to -4. From (2) it follows that the corresponding λ satisfies the condition

$$\Delta \log \lambda \geqq 4\lambda^2. \tag{3}$$

When we set $u = \log \lambda$ this is equivalent to

$$\Delta u \geqq 4e^{2u}. \tag{4}$$

The hyperbolic metric of the unit circle $|z| < 1$ is defined by

$$d\sigma = (1 - |z|^2)^{-1} |dz| \tag{5}$$

and has the constant curvature -4.

2. Consider now an analytic function $\mathfrak{w} = f(z)$ from the circle $|z| < 1$ to a Riemann surface W. The analyticity is expressed by the fact that every local parameter w is an analytic function of z. To a differential element dz corresponds an element dw whose length does not depend on the direction of dz. The corresponding value of $ds = \lambda |dw| = \lambda_z |dz|$ is therefore uniquely de-

* Presented to the Society, September 8, 1937; received by the editors April 1, 1937.

† For the definition of a Riemann surface see T. Radó, *Über den Begriff der Riemannschen Fläche*, Acta Szeged, vol. 2 (1925).

Reprinted from Trans. Amer. Math. Soc., Vol. 43, No. 3 (1938), pp. 359–364.

termined, and we have $\lambda_s = \lambda |w'(z)|$. It is also seen that $u = \log \lambda_s$ satisfies the condition (4) whenever the given metric has a curvature $\leqq -4$. An exception has to be made for the possible zeros of λ_s, corresponding to the zeros of λ and $w'(z)$.

THEOREM A. *If the function* $w = f(z)$ *is analytic in* $|z| < 1$, *and if the metric* (1) *of* W *has a negative curvature* $\leqq -4$ *at every point, then the inequality*

$$ds \leqq d\sigma \tag{6}$$

will hold throughout the circle.

Proof: Choose an arbitrary $R < 1$ and set $v = \log R(R^2 - |z|^2)^{-1}$ for $|z| < R$. We note that $\Delta v = 4e^{2v}$ and consequently

$$\Delta(u - v) \geqq 4(e^{2u} - e^{2v}). \tag{7}$$

Let us denote by E the open point set in $|z| < R$ for which $u > v$. It is clear that E cannot contain any zeros of λ_s. Hence (7) is valid and shows that $u - v$ is subharmonic in E. It follows that $u - v$ can have no maximum in E and must approach its least upper bound on a sequence tending to the boundary of E. But E can have no boundary points on $|z| = R$, for v becomes positively infinite as z tends to that circle, and at interior boundary points we must have $u - v = 0$, by continuity. A contradiction is thus obtained, unless E is vacuous. The inequality $u \leqq v$ consequently subsists for all points with $|z| < R$, and letting R tend to 1 we find $u \leqq -\log (1 - |z|^2)$ at all points. This is equivalent to (6).

If W is the unit circle and ds its hyperbolic metric, Theorem A is simply the differential form of Schwarz's lemma given by Pick.*

3. Several generalizations of the theorem just proved suggest themselves at once. Since the only thing we need is to prevent the function $u - v$ from having a maximum in E, it is obvious that the assumptions on λ can be considerably weakened, without affecting the validity of the argument. We shall give below two such generalizations which are found to be particularly useful for the applications.

THEOREM A1. *Let* λ *be continuous and such that at every point, either* (a) *the second derivatives of* $u = \log \lambda$ *are continuous and satisfy* (4), *or* (b) *it is possible to find two opposite directions* n', n'' *for which* $\partial u/\partial n' + \partial u/\partial n'' > 0$. *Then the statement of the previous theorem is still true.*

Opposite directions in the w-plane correspond to opposite directions in the z-plane. At a maximum of $u - v$ we have $\partial u/\partial n \leqq \partial v/\partial n$ in any direction, when-

* An account of all questions related to Schwarz's lemma will be found in R. Nevanlinna, *Eindeutige analytische Funktionen*, Springer, 1936, pp. 45–58.

ever the directional derivative exists. For opposite directions $\partial v/\partial n' + \partial v/\partial n'' = 0$; hence $\partial u/\partial n' + \partial u/\partial n'' \leqq 0$ in case of a maximum. It follows that no maximum can be attained in points satisfying condition (b).

We shall call $ds' = \lambda' |dw|$ a supporting metric of $ds = \lambda |dw|$ at the point w_0 if: (1) $\lambda' = \lambda$ at w_0, (2) λ' is defined and $\leqq \lambda$ in a neighborhood of w_0.

THEOREM A2. *Suppose that λ is continuous, and that it is possible to find a supporting metric, satisfying* (4), *at every point of W. Then the inequality* (6) *still holds.*

If $u - v > 0$ at z_0, then $u' - v$ will also be positive, and consequently subharmonic, in a neighborhood of z_0.* A maximum of $u - v$ will *a fortiori* be a maximum of $u' - v$. Hence $u - v$ can have no maximum in E.

II. SCHOTTKY'S THEOREM

4. As a first application we prove Schottky's theorem with definite numerical bounds.

THEOREM B. *If $f(z)$ is analytic and different from* 0 *and* 1 *in* $|z| < 1$, *then*

$$\log |f(z)| < \frac{1+\theta}{1-\theta}\left(7 + \overset{+}{\log} |f(0)|\right) \tag{8}$$

for $|z| \leqq \theta < 1$.†

Let $\zeta_1 = \zeta_1(w)$ map the region outside of the segment $(0, 1)$ onto the exterior of the unit circle, so that $w = \infty$ corresponds to $\zeta_1 = \infty$, $w = 1$ to $\zeta_1 = 1$, and $w = 0$ to $\zeta_1 = -1$. We also set $\zeta_2(w) = \zeta_1(w^{-1})$ and $\zeta_3(w) = \zeta_2(1 - w)$. Clearly these functions define similar maps of the regions outside of the segments $(1, \infty)$ and $(-\infty, 0)$. Explicitly, $\zeta_1(w)$ is obtained from the equation

$$\zeta_1 + \zeta_1^{-1} = 4w - 2. \tag{9}$$

We introduce the coordinates $\rho_1 = |w|$, $\rho_2 = |w - 1|$ and divide the plane into regions Ω_1: $\rho_1 \geqq 1$, $\rho_2 \geqq 1$; Ω_2: $\rho_1 \leqq 1$, $\rho_1 \leqq \rho_2$; Ω_3: $\rho_2 \leqq 1$, $\rho_2 \leqq \rho_1$. The metric

$$ds_i = \frac{|d \log \zeta_i|}{2(4 + \log |\zeta_i|)} = \lambda_i |dw| \tag{10}$$

---

* u' corresponds to λ' as u to λ.

† Schottky's original theorem was purely qualitative. Numerical relations have been studied at great length, notably by Ostrowski (*Studien über den Schottky'schen Satz*, Basel, 1931, and *Asymptotische Abschätzung des absoluten Betrags einer Funktion, die die Werte* 0 *und* 1 *nicht annimmt*, Commentarii Mathematici Helvetici, vol. 5 (1933)), but no simple inequality comparable with (8) has ever been proved.

Added in proof: Numerical bounds of the same order of magnitude are found by A. Pfluger, *Über numerische Schranken im Schottky'schen Satz*, Commentarii Mathematici Helvetici, vol. 7 (1935). His proof depends on the use of modular functions, while ours is strictly elementary.

is readily recognized as the hyperbolic metric of a half-plane with the constant curvature -4. Computing the derivatives $\zeta_i'(w)$ we find

$$\begin{aligned}\lambda_1^{-1} &= 2(\rho_1\rho_2)^{1/2}(4 + \log |\zeta_1|),\\ \lambda_2^{-1} &= 2\rho_1\rho_2^{1/2}(4 + \log |\zeta_2|),\\ \lambda_3^{-1} &= 2\rho_2\rho_1^{1/2}(4 + \log |\zeta_3|).\end{aligned} \tag{11}$$

We now set $ds = \lambda|dw|$ with $\lambda = \lambda_i$ in Ω_i. This metric is regular and satisfies condition (3) except at the singular points 0, 1, ∞ and on the lines separating the regions Ω_i. On these lines λ is still continuous, as seen from (11) and the relations between ζ_1, ζ_2, and ζ_3.

Next we wish to show that condition (b) in Theorem A1 holds on the singular lines. We consider the arc $\rho_1 = 1$, $\rho_2 > 1$ and choose n', n'' as the outer and inner normals of the circle. The required condition is

$$\frac{\partial}{\partial n'}\log\lambda_1 + \frac{\partial}{\partial n''}\log\lambda_2 = \frac{\partial}{\partial n'}\log\frac{\lambda_1}{\lambda_2} > 0.$$

From (11) we obtain

$$\frac{\partial}{\partial n'}\log\frac{\lambda_1}{\lambda_2} = \frac{1}{2} - \frac{\dfrac{\partial}{\partial n'}\log\left|\dfrac{\zeta_1}{\zeta_2}\right|}{4 + \log|\zeta_1|},$$

which is also equal to

$$\frac{1}{2} - 2(4 + \log|\zeta_1|)^{-1}\frac{\partial\Phi_1}{\partial\phi},$$

where $\Phi_1 = \arg\zeta_1$, $\phi = \arg w$. For Φ_1 we have the simple relation $\cos\Phi_1 = \rho_1 - \rho_2$, which for $\rho_1 = 1$ becomes $\cos\Phi_1 = 1 - 2\sin\phi/2$. Differentiating we find

$$\frac{\partial\Phi_1}{\partial\phi} = \frac{1}{2}\left(1 + \csc\frac{\phi}{2}\right)^{1/2},$$

and by use of the inequalities $\pi/3 \leqq \phi \leqq 5\pi/3$, $|\zeta_1| > 1$, we are finally led to the desired result,

$$\frac{\partial}{\partial n'}\log\frac{\lambda_1}{\lambda_2} > \frac{1}{2} - \frac{3^{1/2}}{4} > 0.$$

By symmetry, the same must be true for the arc $\rho_2 = 1$, $\rho_1 > 1$. The transformation $w' = (1 - w)^{-1}$ takes Ω_1 into Ω_2 and Ω_2 into Ω_3. Since the function λ is invariant under the transformation we conclude at once that condition (b) will hold also on the line separating Ω_2 and Ω_3.

From Theorem A1 we can now conclude that $w = f(z)$ satisfies the differential inequality $\lambda|dw| \leqq (1-|z|^2)^{-1}|dz|$. Integrating, we find that the shortest distance between the points $f(0)$ and $f(z)$, $|z| = \theta$, measured in the metric $ds = \lambda|dw|$, cannot exceed $[\log (1+\theta)/(1-\theta)]/2$.

The shortest path between the circles $\rho_1 = m$ and $\rho_1 = M$, where $M > m \geqq 2$, is a segment of the negative real axis, whose length is found to be

$$\frac{1}{2} \log \frac{4 + \log |\zeta_1(-M)|}{4 + \log |\zeta_1(-m)|}.$$

To simplify we introduce the lower and upper bounds $|\zeta_1(-M)| \geqq 4M$, $|\zeta_1(-m)| \leqq 5m$. Setting $M = |f(z)|$ and m equal to the greater of the numbers $|f(0)|$ and 2 we obtain

$$4 + \log 4M \leqq \frac{1+\theta}{1-\theta}(4 + \log 5m).$$

Here $\log 5m \leqq \log 10 + \overset{+}{\log} |f(0)| < 3 + \overset{+}{\log} |f(0)|$ and we find

$$4 + \log 4M < \frac{1+\theta}{1-\theta}(7 + \overset{+}{\log} |f(0)|)$$

which is stronger than (8).

III. Bloch's theorem

5. Let $w = f(z)$ be analytic in $|z| < 1$ with $|f'(0)| = 1$. Let $B' = B'(f)$ be the l.u.b. of the radii of all simple (*schlicht*) circles contained in the Riemann surface W generated by $f(z)$. Bloch's theorem is $B = \min B' > 0$. Landau has proved $B > .396$.* Grunsky and Ahlfors proved in a recent paper $B < .472$.†

We show that the method developed in this paper gives an immediate proof of Bloch's theorem with a better lower bound for B. For an arbitrary point $\mathfrak{w}$ on W let $\rho(\mathfrak{w})$ denote the radius of the largest simple circle of center $\mathfrak{w}$ contained in W. It is clear that $\rho(\mathfrak{w})$ is continuous, and equal to zero only at the branch-points. We introduce the metric $ds = \lambda|dw|$ with

$$\lambda = \frac{A}{2\rho^{1/2}(A^2 - \rho)} \qquad (\rho = \rho(\mathfrak{w})) \tag{12}$$

and w denoting the variable of the function plane (not the uniformizing variable). A is a constant satisfying the preliminary condition $A^2 > B'$.

In the neighborhood of a branch-point $\mathfrak{a}$ we have $\rho = |w - a|$. Let n be the multiplicity of $\mathfrak{a}$; then $w_1 = (w-a)^{1/n}$ is a uniformizing variable, and

* E. Landau, *Über den Blochschen Satz und zwei verwandte Weltkonstanten*, Mathematische Zeitschrift, vol. 30 (1929).

† L. V. Ahlfors and H. Grunsky, *Über die Blochsche Konstante*, Mathematische Zeitschrift, vol. 42 (1937). The result was found independently by R. M. Robertson.

the corresponding λ_1 is determined from $\lambda_1|dw_1| = \lambda|dw|$. We obtain $\lambda_1 = n\rho^{1/2-1/n}/2(A^2-\rho)$, and it is seen at once that the metric is regular in case $n=2$ and that λ_1 becomes zero in case $n>2$.

We wish to apply Theorem A2 and therefore look for a supporting metric satisfying the requirements of that theorem. For a regular point $\mathfrak{w}_0$ the surrounding circle of radius $\rho(\mathfrak{w}_0)$ must pass through at least one singularity $\mathfrak{b}$ which is either a branch-point or a boundary point for the surface. We set $\rho' = |w-b|$ and define $\lambda' = A/[2\rho'^{1/2}(A^2-\rho')]$. This metric has the curvature -4 for it is obtained from the hyperbolic metric of a circle by means of the transformation $w' = w^{1/2}$. In all points of our circle we have $\rho \leqq \rho'$ by the definition of ρ. The inequality $\lambda' \leqq \lambda$ is therefore satisfied in a neighborhood of $\mathfrak{w}_0$ if the function $t^{1/2}(A^2-t)$ increases for $t \leqq \rho(\mathfrak{w}_0)$. Under this condition λ' will be a supporting function of λ, for at the center w_0 we have $\lambda' = \lambda$. The function $t^{1/2}(A^2-t)$ is increasing as long as $t < A^2/3$. Consequently all the conditions in Theorem A2 are fulfilled if we suppose that $A^2 > 3B'$.

Apply the theorem with $z=0$. Using the condition $|dw/dz|_{z=0} = 1$ we get

$$A \leqq 2\rho_0^{1/2}(A^2 - \rho_0), \tag{13}$$

where ρ_0 is the radius of the largest simple circle with center at the image of $z=0$. The function in the right member of (13) is increasing, and we can replace ρ_0 by B' obtaining $A \leqq 2B'^{1/2}(A^2-B')$. Letting A tend to $(3B')^{1/2}$ we finally get $B' \geqq 3^{1/2}/4$. This implies that Bloch's constant $B \geqq 3^{1/2}/4 > .433$.

On the other side, if we insert $A^2 = (3B')^{1/2}$ in (13), lower and upper bounds for ρ_0 in terms of B' can be found.

6. Landau has considered a closely related constant L. Let $L' = L'(f)$ be the l.u.b. of the radii of all circles in the w-plane contained in the projection of W, that is, whose values are taken by the function $w=f(z)$, $|f'(0)|=1$. L is defined as the minimum of all such L'. Clearly, $L \geqq B$.

The method employed above is immediately applicable if we choose $\lambda = (2\rho \log C/\rho)^{-1}$. This metric is regular at all branch-points, and when we replace ρ by the distance ρ' from a fixed boundary point, the curvature becomes -4. In order that the function λ' thus obtained be a supporting function it is sufficient that $t \log C/t$ is increasing. This is true for $t < Ce^{-1}$. We therefore choose $C > eL'$, obtaining the inequality $1 \leqq 2L' \log C/L'$ as above. Letting C tend to eL' we find $L' \geqq 1/2$ and hence $L \geqq 1/2$.

This lower bound is the best known. It shows in particular that $L > B$.*

HARVARD UNIVERSITY,
CAMBRIDGE, MASS.

* In the other direction R. M. Robinson has proved $L < .544$. This result has not been published.

QUASICONFORMAL REFLECTIONS

BY

LARS V. AHLFORS

Harvard University, Cambridge, Mass., U.S.A.(¹)

Let L be a Jordan curve on the Riemann sphere, and denote its completmentary components by Ω, Ω^*. Suppose that there exists a sense-reversing quasiconformal mapping λ of the sphere onto itselfs which maps Ω on Ω^* and keeps every point on L fixed. Such mappings are called quasiconformal reflections. Our purpose is to study curves L which permit quasiconformal reflections.

Let U denote the upper and U^* the lower halfplane. Consider a conformal mapping f of U on Ω and a conformal mapping f^* of U^* on Ω^*. Evidently, $f^{*-1}\lambda f$ defines a quasiconformal mapping of U on U^* which induces a monotone mapping $h = f^{*-1}f$ of the real axis on itself. It is not quite unique, for we may replace f by fS and f^* by f^*S^* where S and S^* are linear transformations with real coefficients and possitive determinant. This replaces h by $S^{*-1}hS$ which we shall say is equivalent to h. Observe that h, or rather its equivalence class, does not depend on λ. It is also unchanged if we replace the triple (Ω, L, Ω^*) by a conformally equivalent triple $(T\Omega, TL, T\Omega^*)$ where T is a linear transformation.

The mapping f of U has a quasiconformal extension to the whole plane, namely by the mapping with values $\lambda f(\bar{z})$ for $z \in U^*$. It is known that quasiconformal mappings carry nullsets into nullsets. Therefore L has necessarily zero area.

From this we may deduce that h determines Ω uniquely up to conformal equivalence. In fact, let f_1, f_1^* be another pair of conformal mappings on complementary regions and suppose that $f_1^{*-1}f_1 = f^{*-1}f$ on the real axis. For a moment, let us write F for the mapping given by $f(z)$ in U and by $\lambda f(\bar{z})$ in U^*, and let F_1 have the corresponding meaning. The mapping $H = F_1^{-1}f_1^*f^{*-1}F$ is defined in U^* and reduces to the identity on the real axis. We extend it to the whole plane by setting $H(z) = z$

(¹) This work was supported by the Air Force Office of Scientific Research.

Reprinted from Acta Math., Vol. 109 (1963), pp. 291–301.

in U. Then F_1HF^{-1} is a quasiconformal mapping. It reduces to f_1f^{-1} in Ω and to $f_1^*f^{*-1}$ in Ω^*. It is thus conformal, except perhaps on L. But a quasiconformal mapping which is conformal almost everywhere is conformal. Hence $f_1 = Tf$ where T is a linear transformation.

What are the properties of h? A necessary condition is that h can be extended to a quasiconformal mapping of U on U^*, namely to $f^{*-1}\lambda f$. This condition is also sufficient. To prove it, let g be a quasiconformal mapping of U on U^* with boundary values h. The function $g^*(z) = g(\bar{z})$, defined in U^*, has weak derivatives which satisfy an equation

$$g^*_{\bar{z}} = \mu g^*_z$$

with $|\mu| \leq k < 1$ (k constant). Set $\mu = 0$ in U. Consider the equation

$$F_{\bar{z}} = \mu F_z$$

for the extended μ. An important theorem (see [1]), sometimes referred to as the generalized Riemann mapping theorem, asserts the existence of a solution F which is a homeomorphic mapping of the sphere. Because z is a solution in U and g^* a solution in U^* it is possible to write $F = f$ in U, $F = f^*g^*$ in U^*, where f and f^* are conformal mappings. Clearly, $\Omega = f(U)$ and $\Omega^* = f^*(U^*)$ are quasiconformal reflections of each other.

To sum up, we have established a correspondence between equivalence classes of boundary correspondences h, conformal mappings f, and curves L which permit a quasiconformal reflection. It is a natural program to try to characterize the possible h, f and L in a more direct way. For boundary correspondences h this problem has been solved; we shall have occasion to recall the solution.

In Part I we solve the corresponding problem for L. It turns out that the curves which permit a quasiconformal reflection can be characterized by a surprisingly simple geometric property. (Partial results in this direction have been obtained by M. Tienari whose paper [7] came to my attention only when this article was already written.)

We have been less successful with the mappings f, but in Part II we show, at any rate, that the mappings f form an open set. To understand the meaning of this, we observe that the mappings equivalent to f are of the form TfS. To account for T we replace f by its Schwarzian derivative $\varphi = \{f, z\}$. The Schwarzian of fS is $\varphi(S)S'^2$, and to eliminate S it is indicated to consider φdz^2 in its role of quadratic differential.

If f is schlicht in U, Nehari [6] has shown that $|\varphi| y^2 \leq \frac{3}{2}$. We take the least upper bound of $|\varphi| y^2$ to be a norm of φ. In the linear space of quadratic differentials with finite norm, let Δ be the set of all φ whose corresponding f is schlicht and has a quasiconformal extension. We are going to show that Δ is an open set. For the significance of this result in the theory of Teichmüller spaces we refer to the companion article of L. Bers [4] in the next issue of this journal.

Part I

1. In 1956 A. Beurling and the author derived a neccessary and sufficient condition for a boundary h to be the restriction of a quasiconformal mapping of U on itself (or on its reflection U^*). This work is an essential preliminary for what follows.

We recall the main result. Without loss of generality it may be assumed that $h(\infty) = \infty$. Then h admits a quasiconformal extension if and only if it satisfies a ϱ-condition, namely an inequality

$$\varrho^{-1} \leqslant \frac{h(x+t) - h(x)}{h(x) - h(x-t)} \leqslant \varrho, \tag{1}$$

which is to be fulfilled for all real x, t and with a constant $\varrho \neq 0, \infty$. More precisely, if h has a K-quasiconformal extension, then (1) holds with a $\varrho(K)$ that depends only on K, and if (1) holds, then h has a $K(\varrho)$-quasiconformal extension.

The necessity follows from the simple observation that the quadruple $(x-t, x, x+t, \infty)$ with cross-ratio 1 must be mapped on a quadruple with bounded cross-ratio. The sufficiency requires an explicit construction. We set $w(z) = u + iv$ with

$$\left.\begin{aligned} u(z) &= \int_0^1 [h(x+ty) + h(x-ty)]\,dt, \\ v(z) &= \int_0^1 [h(x+ty) - h(x-ty)]\,dt. \end{aligned}\right\} \tag{2}$$

It is proved in [2] that $w(z)$ is $K(\varrho)$-quasiconformal.

I am indebted to Beurling for the very important observation that the mapping (2) is also quasi-isometric, in the sense that corresponding noneuclidean elements of length have a bounded ratio. This condition can be expressed by

$$\left.\begin{aligned} |w_z| &\leqslant C(\varrho)\frac{v}{y}, \\ |w_{\bar z}| &\geqslant C(\varrho)^{-1}\frac{v}{y}, \end{aligned}\right\} \tag{3}$$

where $C(\varrho)$ depends only on ϱ. The proof is an immediate verification based on the estimate given in Lemma 6.5 of the cited paper.

2. Let L be a Jordan curve through ∞ which admits a quasiconformal reflection. The complementary regions determined by L are denoted by Ω, Ω^*, and the reflection is written as $z \to z^*$. We assume that the reflection is K-quasiconformal.

Constants which depend only on K will be denoted by $C(K)$, with or without subscripts. In different connections $C(K)$ may have different values. We emphasize that $C(K)$ is not allowed to depend on L.

The shortest distance from a point z to L will be denoted by $\delta(z)$.

LEMMA 1. *The following estimates hold for all z in the plane and all z_0 on L:*

(a)
$$C(K)^{-1} \leqslant \left|\frac{z^* - z_0}{z - z_0}\right| \leqslant C(K)$$

(b)
$$\frac{|z - z^*|}{\delta(z)} \leqslant C(K)$$

(c)
$$C(K)^{-1} \leqslant \frac{\delta(z^*)}{\delta(z)} \leqslant C(K).$$

Proof. If the cross-ratio of a quadruple has absolute value $\leqslant 1$, then the cross-ratio of the image points under a K-quasiconformal mapping has an absolute value $\leqslant C(K)$. This assertion is contained in [1], Lemma 16. It is a rather elementary result.

If $|z^* - z_0| \leqslant |z - z_0|$ we can apply the above remark to (z^*, z, z_0, ∞) and conclude that $|z - z_0| \leqslant C(K)|z^* - z_0|$. Symmetrically, $|z - z_0| \leqslant |z^* - z_0|$ implies $|z^* - z_0| \leqslant C(K)|z - z_0|$. In all circumstances (a) follows.

From (a) we obtain

$$|z - z^*| \leqslant (C(K)+1)|z - z_0| = C_1(K)|z - z_0|$$

and (b) follows when $|z - z_0| = \delta(z)$. Since $\delta(z^*) \leqslant |z - z^*|$ the second inequality (c) follows from (b), and the first is true by symmetry.

2. We introduce now the noneuclidean metrics $ds = \varrho(z)|dz|$ in Ω and Ω^*. Explicitly, if $z = z(\zeta)$ is a conformal map of $|\zeta| < 1$ on Ω we set

$$\varrho(z)|dz| = (1 - |\zeta|^2)^{-1}|d\zeta|.$$

The classical estimates

$$\delta(z) \leqslant \varrho(z)^{-1} \leqslant 4\delta(z) \tag{4}$$

follow by use of Schwarz' lemma and Koebe's one-quarter theorem.

Lemma 2. *If L passes through ∞ and permits a K-quasiconformal reflection, then it also permits a $C(K)$-quasiconformal reflection with the additional property that corresponding euclidean line elements satisfy*

$$C_1(K)^{-1}|dz| \leqslant |dz^*| \leqslant C_1(K)|dz|. \tag{5}$$

Proof. As shown in the introduction, the given K-quasiconformal reflection induces a K-quasiconformal mapping of U on U^* with a boundary correspondence h. This h must satisfy a $\varrho(K)$-condition of type (1). The Beurling-Ahlfors construction permits us to replace the mapping of U on U^* with a $C(K)$-quasiconformal mapping with the same boundary values, in such a way that it satisfies condition (3). It follows that the corresponding reflection about L is $C(K)$-quasiconformal and satisfies

$$C_1(K)^{-1}\varrho(z)|dz| \leqslant \varrho(z^*)|dz^*| \leqslant C_1(K)\varrho(z)|dz|.$$

Use of (4) and (c) leads to the desired inequality (5).

3. We are now ready to characterize the curves L in purely geometric form:

Theorem 1. *A Jordan curve L through ∞ permits a quasiconformal reflection if and only if there exists a constant C such that*

$$\overline{P_1P_2} \leqslant C \cdot \overline{P_1P_3} \tag{6}$$

for any three points P_1, P_2, P_3 on L which follow each other in this order.

Again, there is a more precise statement to the effect that C depends only on the K of the reflection, and vice versa. If L does not pass through ∞ condition (6) must be replaced by

$$\overline{P_1P_2} : \overline{P_1P_3} \leqslant C(\overline{P_4P_2} : \overline{P_4P_3}),$$

where (P_1, P_3) separates (P_2, P_4).

4. *Proof of the necessity.* We follow the segment P_1P_3 from P_1 to its last intersection with the subarc $P_2P_1\infty$ of L, and from there to the first intersection P_3' with the arc $P_2P_3\infty$. If $P_1P_2 > P_1P_3$ it is geometrically evident that

$$\overline{P_1P_2} : \overline{P_1P_3} \leqslant \overline{P_1'P_2} : \overline{P_1'P_3'}.$$

Therefore we may assume from the beginning that $\overline{P_1P_3}$ has only its endpoints on L. For definiteness, we suppose that the inner points lie in Ω.

By Lemma 2 there exists a quasiconformal reflection which multiplies lengths at most by a factor $C(K)$. Hence P_1 and P_3 can be joined in Ω^* by an arc γ^* of length $\leq C(K)\cdot\overline{P_1P_3}$. The Jordan curve formed by $\overline{P_1P_3}$ and γ^* separates P_2 from ∞. Hence γ^* intersects the extension of $\overline{P_1P_2}$ over P_2 and we conclude that $\overline{P_1P_2}\leq$ length of $\gamma^*\leq C(K)\cdot\overline{P_1P_3}$.

5. *Proof of the sufficiency*. We shall use the notations

$$\alpha=\text{arc } P_2P_3,\quad \alpha'=\text{arc } P_1P_2,$$
$$\beta=\text{arc } P_1\infty,\quad \beta'=\text{arc } P_3\infty.$$

Denote by $d(\alpha,\beta)$ and $d^*(\alpha,\beta)$ the extremal distances of α and β with respect to Ω and Ω^* respectively. With similar notations for α',β' one has the relations

$$d(\alpha,\beta)\,d(\alpha',\beta')=d^*(\alpha,\beta)\,d^*(\alpha',\beta')=1.$$

In a conformal mapping of Ω on the halfplane U with ∞ corresponding to ∞, let P_1, P_2, P_3 be mapped on x_1, x_2, x_3. It is evident that $d(\alpha,\beta)=1$ if and only if $x_3-x_2=x_2-x_1$. Furthermore, the ratio $|x_3-x_2|:|x_2-x_1|$ is bounded away from 0 and ∞ if and only if this is true of $d(\alpha,\beta)$. Consequently, in order to prove that the boundary correspondence induced by L satisfies (1) it is sufficient to show that $d(\alpha,\beta)=1$ implies $K(C)^{-1}\leq d^*(\alpha,\beta)\leq K(C)$.

Two elementary estimates are needed. We show first that $d(\alpha,\beta)=1$ implies

$$\overline{P_1P_2}:\overline{P_2P_3}\leq C^2e^{2\pi}. \tag{7}$$

Indeed, it follows from (6) that the points of β are at distance $\geq C^{-1}\cdot\overline{P_1P_2}$ from P_2 while the points of α have distance $\leq C\cdot\overline{P_2P_3}$ from P_2. If (7) were not true, α and β would be separated by a circular annulus whose radii have the ratio $e^{2\pi}$. In such an annulus the extremal distance between the circles is 1, and the comparison principle for extremal lengths would yield $d(\alpha,\beta)>1$, contrary to hypothesis. Hence (7) must hold. If P_1 and P_3 are interchanged we have in the same way

$$\overline{P_2P_3}:\overline{P_1P_2}\leq C^2e^{2\pi}. \tag{8}$$

Consider points $Q_1\in\alpha$, $Q_2\in\beta$. By repeated application of (6)

$$\overline{Q_1Q_2}\geq C^{-1}\,\overline{Q_1P_1}\geq C^{-2}\,\overline{P_1P_2}$$

and with the help of (8) we conclude that the shortest distance between α and β is $\geq C^{-4}e^{-2\pi}\,\overline{P_2P_3}$. To simplify notations, write $d=\overline{P_2P_3}$, $M_1=Cd$, $M_2=C^{-4}e^{-2\pi}d$. Because of (6), all points on α are within distance M_1 from P_2.

We recall that the definition of extremal length implies

$$d^*(\alpha,\beta) \geqslant \frac{(\inf \int_\gamma \varrho |dz|)^2}{\iint_{\Omega^*} \varrho^2 dx dy},$$

where the infimum is with respect to all arcs γ that join α and β within Ω^*, and ϱ is any positive function for which the right-hand side has a meaning. We choose $\varrho = 1$ in a circular disk with center P_2 and radius $M_1 + M_2$, $\varrho = 0$ outside of that disk. Then $\int_\gamma \varrho |dz| \geqslant M_2$ for all curves γ. Indeed, this is so whether γ stays within the disk or contains a point on its circumference. We conclude that

$$d^*(\alpha,\beta) \geqslant \frac{1}{\pi}\left(\frac{M_2}{M_1+M_2}\right)^2 = \pi^{-1}(1+C^6 e^{2\eta})^{-2}.$$

The same inequality, applied to α', β', yields an upper bound for $d^*(\alpha,\beta)$, and our proof of Theorem 1 is complete.

Part II

1. In the introduction we saw that the boundary correspondences h give rise to conformal mappings f, and with these we associated their Schwarzian derivatives $\varphi = \{f, z\}$. The set of all such φ was denoted by Δ. We formulate a precise definition:

The set Δ consists of all functions φ, holomorphic in U, such that the equation $\{f, z\} = \varphi$ has a solution f which can be extended to a schlicht quasiconformal mapping of the whole plane.

Our purpose is to prove:

Theorem 2. *Δ is an open subset of the Banach space of holomorphic functions with norm $\|\varphi\| = \sup |\varphi(z)| y^2$.*

We know already that all $\varphi \in \Delta$ have norm $\leqslant \frac{3}{2}$. It will follow that the norms are in fact strictly less than $\frac{3}{2}$.

2. It is a known result that Δ contains a neighborhood of the origin ([3], [5]). As an illustration of the method we shall follow it is nevertheless useful to include a proof.

Lemma 3. *Δ contains all functions φ with $\|\varphi\| < \frac{1}{2}$.*

Proof. Let η_1 and η_2 be linearly independent solutions of the differential equation

$$\eta'' = -\tfrac{1}{2}\varphi\eta. \tag{9}$$

normalized by $\eta_1'\eta_2 - \eta_2'\eta_1 = 1$. It is well known that $f = \eta_1/\eta_2$ satisfies $\{f, z\} = \varphi$. Observe that f may be meromorphic with simple poles, and that $f' \neq 0$ at all other points!

It is to be shown that f is schlicht and has a quasiconformal extension. To construct the extension we form

$$F(z)=\frac{\eta_1(z)+(\bar{z}-z)\eta_1'(z)}{\eta_2(z)+(\bar{z}-z)\eta_2'(z)}\quad (z\in U). \tag{10}$$

Because $\eta_1'\eta_2-\eta_2'\eta_1=1$ the numerator and denominator cannot vanish simultaneously. If the denominator vanishes we set $F=\infty$, and local assertions about F will apply to $1/F$.

A simple computation which makes use of (9) gives

$$F_z/F_{\bar{z}}=\tfrac{1}{2}(z-\bar{z})^2\varphi(z).$$

Under the assumption $\|\varphi\|<\tfrac{1}{2}$ we conclude that F is quasiconformal and sense-reversing. The mapping $z\to F(\bar{z})$ is quasiconformal and sense-preserving in U^*.

Our intention is to show that

$$f(z)=\begin{cases} f(z) \text{ in } U\\ F(\bar{z}) \text{ in } U^*\end{cases} \tag{11}$$

gives the desired extension. To see this it is sufficient to know that f can be extended to the real axis by continuity, that the extended function is locally schlicht at points of the real axis, and that it tends to a limit for $z\to\infty$. Indeed, f will then be locally schlicht everywhere, and by a familiar reasoning is must be globally schlicht.

The missing information is easy to supply under strong additional conditions. We suppose that φ is analytic on the real axis, including ∞, where φ shall have a zero of order $\geqslant 4$ (this means that the quadratic differential φdz^2 is regular at ∞). It is immediate that f and F agree on the real axis, and that they are real-analytic in the closed half-planes. It follows easily that f is locally schlicht. At ∞ the assumption implies that equation (9) has solutions whose power series expansions begin with 1 and z respectively. Hence

$$\eta_1=a_1z+b_1+O(|z|^{-1}),$$
$$\eta_2=a_2z+b_2+O(|z|^{-1})$$

with $a_1b_2-a_2b_1=1$. Substitution in (10) shows that

$$F(z)=\frac{a_1\bar{z}+b_1+O(|z|^{-1})}{a_2\bar{z}+b_2+O(|z|^{-1})}$$

and therefore f and F have the same limit a_1/a_2 as $z\to\infty$.

To prove the lemma without additional assumptions we use an approximation method. We can find a sequence of linear transformations S_n such that the closure of $S_n U$ is contained in U and $S_n z \to z$ for $n \to \infty$. Take $\varphi_n(z) = \varphi(S_n z)\, S_n'(z)^2$. It follows by Schwarz' lemma that $\|\varphi_n\| < \|\varphi\|$. Moreover, φ_n is analytic on the real axis and has at least a fourth order zero at ∞. Consequently, there exist quasiconformal mappings f_n, holomorphic with $\{f_n, z\} = \varphi_n$ in U, with uniformly bounded dilatation. A subsequence of the f_n converges to a limit function f which is itself schlicht and quasiconformal, and which satisfies $\{f, z\} = \varphi$ in U. This completes the proof.

With suitable normalizations it is possible to arrange that $f_n \to f$, the mapping defined by (11).

3. The method of the preceding proof can be carried over to the general case, although with some significant modifications.

Suppose that $\varphi_0 \in \Delta$ and $\{f_0, z\} = \varphi_0$. We may assume that f_0 maps U on a region Ω whose boundary L passes through ∞, and we know that L admits a quasiconformal reflection $w \to w^* = \lambda(w)$. We choose λ in accordance with Lemma 2.

If $\|\varphi - \varphi_0\| < \varepsilon$ and $\{f, z\} = \varphi$ the identity

$$\{f, z\} = \{f, f_0\}\, f_0'^2 + \{f_0, z\}$$

yields

$$|\{f, f_0\}|\, |f_0'|^2 y^2 < \varepsilon.$$

The non-euclidean metric in Ω is given by

$$\varrho(w)\,|dw| = \frac{|dz|}{2y},$$

and if we write $\tilde{f} = f f_0^{-1}$ we obtain

$$|\{\tilde{f}, w\}| < 4\,\varepsilon\,\varrho(w)^2 \tag{12}$$

If ε is sufficiently small it is to be proved that $\tilde{f}$ is schlicht and has a quasiconformal extension.

We set $\tilde{\varphi} = \{\tilde{f}, w\}$ and $\tilde{f} = \eta_1/\eta_2$ where η_1, η_2 are normalized solutions of

$$\eta'' = -\tfrac{1}{2}\tilde{\varphi}\eta.$$

In close analogy with (10) we form

$$F(w) = \frac{\eta_1(w) + (w^* - w)\,\eta_1'(w)}{\eta_2(w) + (w^* - w)\,\eta_2'(w)},$$

where $w \in \Omega$ and $w^* = \lambda(w)$. Computation gives

$$\frac{F_w}{F_{\bar{w}}} = \frac{\lambda_w}{\lambda_{\bar{w}}} + \frac{\bar{\varphi}(w - w^*)^2}{2\lambda_{\bar{w}}}. \tag{13}$$

Here $|\lambda_w/\lambda_{\bar{w}}| \leqslant k < 1$ because λ is quasiconformal. To estimate the second term we have first, by (12), Lemma 1 (*b*) and (4),

$$|\bar{\varphi}||w - w^*|^2 < 4\varepsilon C^2.$$

On the other hand, $|\lambda_{\bar{w}}|$ stays away from 0, for Lemma 2 gives

$$C^{-1}|dw| \leqslant |dw^*| \leqslant 2|\lambda_{\bar{w}}||dw|.$$

We conclude that $|F_w/F_{\bar{w}}| \lesssim k' < 1$ provided that ε is sufficiently small.

4. We wish to show that

$$f = \begin{cases} f(w) & \text{in } \Omega \\ F(w^*) & \text{in } \Omega^* \end{cases}$$

is schlicht and quasiconformal. Again, the proof is easy under strong assumptions. This time we assume that L is an analytic curve, that $\bar{\varphi}$ is analytic on L and that it has a fourth order zero at ∞. It is clear that we can prove f to be a quasiconformal homeomorphism exactly as in the proof of Lemma 3.

To complete the proof, let $\zeta = \omega(w)$ be a conformal mapping of Ω on $|\zeta| < 1$. Let Ω_n be the part of Ω that corresponds to $|\zeta| < r_n$, L_n its boundary. Here $\{r_n\}$ is a sequence which converges to 1.

A quasiconformal reflection λ_n across L_n can be constructed as follows: If $r_n^2 < |\omega(w)| < r_n$ we define $\lambda_n(w)$ so that $\omega(w)$ and $\omega(\lambda_n(w))$ are mirror images with respect to $|\zeta| = r_n$. If $|\omega(w)| \leqslant r_n^2$ we find w_n so that $\omega(w_n) = r_n^{-2}\omega(w)$ and choose $\lambda_n(w) = \lambda(w_n)$. The definitions agree when $|\omega(w)| = r_n^2$, and L_n is kept fixed. The dilatation of λ_n is no greater than the maximum dilatation of λ.

After a harmless linear transformation which throws a point on L_n to ∞ the part of the theorem that has already been proved can be applied to Ω_n. It is to be observed that $\varrho_n \geqslant \varrho$ where $\varrho_n|dw|$ is the noneuclidean metric in Ω_n. Therefore $\bar{\varphi}$ satisfies

$$|\bar{\varphi}| < 4\varepsilon\varrho_n(w)^2$$

with the same ε as before. Hence there exists a quasiconformal mapping f_n of the whole plane which agrees with f on Ω_n and whose dilatation lies under a fixed bound. A subsequence of the f_n tends to a limit mapping f which is schlicht, quasiconformal, and equal to f in Ω. The theorem is proved.

References

[1]. L. AHLFORS & L. BERS, Riemann's mapping theorem for variable metrics. *Ann. of Math.* 72 (1960), 385–404.

[2]. L. AHLFORS & A. BEURLING, The boundary correspondence under quasiconformal mappings. *Acta Math.*, 56 (1956), 125–142.

[3]. L. AHLFORS & G. WEILL, A uniqueness theorem for Beltrami equations. *Proc. Amer. Math. Soc.*, 13 (1962), 975-978.

[4]. L. BERS (to appear).

[5]. ——, Correction to "Spaces of Riemann surfaces as bounded domains". *Bull. Amer. Math. Soc.*, 67 (1961), 465–466.

[6]. Z. NEHARI, The Schwarzian derivative and schlicht functions. *Bull. Amer. Math. Soc.* 66 (1949), 545–551.

[7]. M. TIENARI, Fortsetzung einer quasikonformen Abbildung über einen Jordanbogen. *Ann. Acad. Sci. Fenn., Ser. A*, I, 321 (1962).

Received February 18, 1963

FINITELY GENERATED KLEINIAN GROUPS.*

By Lars V. Ahlfors.

1. Introduction. 1.1. Let Γ be a group of linear transformations

$$Az = \frac{az+b}{cz+d},\ ad-bc=1$$

of the extended complex plane. A point z_0 is called a *limit point* of Γ if it is an accumulation point of points Az_1, $A \in \Gamma$, some some z_1. The set of limit points will be denoted by $\Sigma(\Gamma)$, or by Σ when it is clear to what group we are referring.

$\Sigma(\Gamma)$ is a closed set, and invariant under Γ; we denote its complement by $\Omega = \Omega(\Gamma)$. As soon as $\Omega(\Gamma)$ is not empty we say that Γ is discontinuous, and $\Omega(\Gamma)$ is its *set of discontinuity*. It is easy to classify all groups for which Σ is void, consists of a single point, or of two points. All other discontinuous groups will be called *Kleinian groups*. In other words, a Kleinian group is a discontinuous group with more than two limit points. In this paper Γ will always denote a Kleinian group.

For Kleinian groups the images Az_1, $A \in \Gamma$, of any point z_1 accumulate toward all of Σ. It is an immediate consequence that Σ is a nowhere dense perfect set, and that every nonvoid closed invariant set includes Σ.

A Kleinian group which leaves a circle invariant is said to be Fuchsian. The set of limit points lies on the invariant circle. If Σ is the whole circle the Fuchsian group is of the first kind. If not, Σ is nowhere dense on the invariant circle, and the group is of the second kind.

The fixpoints of non-elliptic transformations in Γ lie on Σ. There are always infinitely many hyperbolic or loxodromic transformations, and their fixpoints are dense on Σ. The fixpoints in Ω belong to elliptic transformations, and they are isolated.

1.2. In order to study the action of Γ on Ω we consider the quotient space $S = \Omega/\Gamma$. It has a natural complex structure such that the projection map $\pi: \Omega \to S$ is holomorphic. Thus, the components S_i of S are Riemann

Received October 30, 1963.
* This work was supported (in part) by the Air Force Office of Scientific Research.

Ahlfors, L., Finitely Generated Kleinian Groups. American Journal of Mathematics. [Vol. 86, 1964], pp. 413–429.

surfaces. We shall write $\Omega_i = \pi^{-1} S_i$. In general, the Ω_i are not connected, and we denote the components of Ω_i by Ω_{ij}.

With any Kleinian group Γ we have thus associated the decompositions $\Omega = \cup \Omega_i = \cup \Omega_{ij}$ and $S = \cup S_i$. Each Ω_i is invariant under the full group Γ, and the boundary of Ω_i is all of Σ. The components Ω_{ij} are ramified covering surfaces of S_i whose branch points are elliptic fixpoints.

The projections of the branch points will be denoted by p_k. They are isolated, and with each p_k there is associated an integer $m_k \geqq 2$, the order of the corresponding elliptic transformation. We say that S has a signature $I = \{p_k, m_k\}$. S together with its signature is denoted by $S(I)$, and $S_i(I)$ has a similar meaning.

We shall say that S is of *finite type* if there are finitely many components S_i and points p_k, and if each S_i can be obtained from a compact surface by omission of a finite number of points. When this is so it is convenient to let S_i denote the compact surface, and to regard the omitted points as points p_k with $m_k = \infty$. It will be clear from the context when S_i should be interpreted in this manner.

1.3. The main result to be proved in this article is the following statement:

If Γ is finitely generated, then S is of finite type.

This is known for Fuchsian groups, but to the author's knowledge there is no proof in the published literature. A proof is contained in a manuscript of Nielsen and Fenchel, where the result is obtained at the end of a penetrating study of Fuchsian groups. Recently, L. Bers has found a short proof by variational methods, the same that have proved so useful in the theory of Teichmüller spaces.[1]

My task has been to show that Bers' method can be extended to arbitrary Kleinian groups. The main difficulty was the correct formulation and proof of a crucial lemma, which is to be considered the core of this paper. We have formulated it in two parts, as Lemma 8 and Lemma 9.

In the last section we prove some results on functions groups, that is, groups with an invariant region. These were the results that gave the initial impetus to this study.

1.4. Perhaps of greater interest are the theorems I have not been able to prove. For instance, it seems very likely that the set Σ for any finitely generated group Γ has zero area, but I cannot prove it.

[1] Unpublished.

Any Kleinian group can be extended to a corresponding group of Möbius transformations in the Poincaré half-space. If Γ is finitely generated it seems plausible that the Poincaré group has a fundamental polyhedron with only a finite number of vertices on Σ. Again, I cannot prove this, but if this is so I am able to show that Σ has indeed zero measure.

2. The Poincaré metric. 2.1. Each component Ω_{ij} carries a Poincaré metric $ds = \rho \mid dz \mid$ with constant negative curvature equal to -1, and such that Ω_{ij} is complete in this metric. The metric is unique, and therefore

$$\rho(Az) \mid A'(z) \mid = \rho(z) \tag{2.1}$$

for all $A \in \Gamma$ which map Ω_{ij} onto itself. We shall consider ρ on all of Ω, defined in each Ω_{ij}. Then (2.1) holds for all $A \in \Gamma$.

The projection π induces a corresponding metric $\bar{\rho} \mid d\zeta \mid$ on S. Here ζ is a local parameter on S, and $\bar{\rho}$ is determined by $\bar{\rho} \mid d\zeta \mid = \rho \mid dz \mid$. At the branch points $d\zeta/dz = 0$, and we see that $\bar{\rho}$ becomes infinite at the points p_k.

If S_i is compact the Gauss-Bonnet formula yields

$$\chi(S_i) + \sum_{p_k \in S_i} (1 - \frac{1}{m_k}) = \frac{1}{2\pi} \int_{\Omega_i/\Gamma} \rho^2 \, dxdy, \tag{2.2}$$

where $\chi(S_i)$ is the Euler characteristic. The formula remains valid for any $S_i(I)$ of finite type, provided that $\chi(S_i)$ is the characteristic of the compact surface and the infinite signatures are included in the sum. To prove (2.2) for this case requires knowledge of the asymptotic behavior of $\bar{\rho}$ at the punctures. If the omitted point corresponds to $\zeta = 0$ this behavior is given by

$$\log \bar{\rho} = -\log \mid \zeta \mid - \log\log (1/\mid \zeta \mid) + O(1), \tag{2.3}$$

a classical result.

An immediate consequence of (2.2) is that

$$\chi(S_i) + \sum_{p_k \in S_i} (1 - \frac{1}{m_k}) > 0. \tag{2.4}$$

This rules out certain possibilities. Without insisting on the details, we conclude that S_i cannot be a sphere with certain low signatures, nor can it be a torus without signature.

A more careful analysis of the possible cases shows that (2.4) can be sharpened to

$$\chi(S_i) + \sum_{S_i} (1 - \frac{1}{m_k}) \geqq 1/42.$$

This is important because it shows that S *is of finite type if and only if it total Poincaré area is finite.* The conclusion uses the elementary fact that a single surface with finite Poincaré area is necessarily of finite type.

2.2. We show next that the points p_k with $m_k = \infty$ correspond to parabolic fixpoints. Since the result does not require S to be of finite type, let us merely assume that $S_i = \hat{S}_i - \{p_k\}$ where $\hat{S}_i$ is a Riemann surface and $p_k \in \hat{S}_i$.

LEMMA 1. *There exists a parabolic transformation $A_k \in \Gamma$ with the following property: If the linear transformation U is such that $U^{-1}A_kUz = z + 1$, then Ω_i contains a set $\{z \mid \operatorname{Im} Uz > c\}$ and $\pi(z) \to p_k$ when $z \in \Omega_i$ approaches the fixpoint $z_k = U\infty$ in such a way that $\operatorname{Im} Uz \to \infty$.*

We remark that A_k can of course be replaced by any conjugate transformation $B^{-1}A_kB$ with $B \in \Gamma$, in which case z_k is replaced by Bz_k. The set $\{z \mid \operatorname{Im} Uz > 0\}$ is a disk or a halfplane whose boundary passes through z_k. It follows that there are at most two p_k which correspond to the same z_k.

For the proof of the lemma we may regard p_k as the center of a punctured disk $\bar{\Delta} \subset S_i$, represented by $0 < |\zeta| < 1$, and we may assume that no point in $\bar{\Delta}$ has a signature. Let Δ be a component of $\pi^{-1}(\bar{\Delta})$. It is clear that the cover transformations of Δ over $\bar{\Delta}$ are generated by a single transformation $A \in \Gamma$ which is not elliptic.

If A is hyperbolic or loxodromic we can take it to be of the form $Az = kz$, $|k| \neq 1$. In terms of $s = \log \zeta$ the projection π induces a one-one correspondence $z = z(s)$ between the halfplane $\operatorname{Re} s < 0$ and Δ which satisfies $z(s + 2\pi i) = kz(s)$. Because 0 is a limit point it is not in Δ, and therefore the distance from $z(s)$ to the boundary of Δ is at least $|z(s)|$. With this information Koebe's one-quarter theorem easily yields

$$|z'(s)|/|z(s)| \leqq 2/|\operatorname{Re} s|.$$

Integration from s to $s + 2\pi i$ along a vertical line gives

$$|\log |k|| \leqq 4\pi/|\operatorname{Re} s|.$$

But this is absurd, for $|\operatorname{Re} s|$ can be chosen arbitrarily large.

We conclude that A is parabolic, for instance $Az = z + 1$. In this case $z(s + 2\pi i) = z(s) + 1$ and it is elementary to deduce that

$$\exp(2\pi iz) = \zeta + a_0 + a_1/\zeta + \cdots,$$

from which

$$z(s) = \frac{s}{2\pi i} + O(|\zeta|^{-1}) \tag{2.5}$$

and, for later use,

(2.6) $$dz/d\zeta = (2\pi i)^{-1} + O(|\zeta|^{-2}).$$

It is immediate from (2.5) that Ω_i contains a halfplane $y > c$, and that $y \to \infty$ implies $\pi(z) \to p_k$. This is the assertion of the lemma when U is the identity, and the general case follows by application to the group $U^{-1}\Gamma U$.

2.3. We shall use (2.6) to derive the asymptotic behavior of ρ in terms of the uniformizer ζ. Because $\rho\,|dz| = \bar{\rho}\,|d\zeta|$ we obtain, with the help of (2.3),

(2.7) $$\log \rho = -\log\log (1/|\zeta|) + O(1).$$

In the derivation of this formula we have assumed that U is the identity, but it is easy to see that it remains valid as soon as the fixpoint is at ∞.

In case of a finite fixpoint we make the transformation $z = Uw$. The Poincaré metric in $U^{-1}\Omega$ is given by $\rho_0(w)\,|dw| = \rho(z)\,|dz|$. It is for ρ_0 that (2.7) is valid, while (2.5) is applicable to w in the place of z. We have $\log\rho = \log\rho_0 - \log|U'(w)| = \log\rho_0 + 2\log|w| + O(1)$, for since $U\infty$ is finite, $w^2U'(w)$ has a finite limit when $w \to \infty$. It follows that

$$\log\rho = -\log\log(1/|\zeta|) + 2\log|\log\zeta| + O(1).$$

The expression on the right is not single-valued, but we may conclude that

(2.8) $$\log\rho = \log\log(1/|\zeta|) + O(1)$$

when z approaches the fixpoint in an interior angle.

2.4. The coefficient of the Poincaré metric decreases when the region increases (this is essentially Schwarz' lemma). Denote the euclidean distance from z to Σ by $\delta(z)$. If we compare Ω with the circle about z of radius $\delta(z)$ it follows at once that $\rho(z) \leqq 2\delta(z)^{-1}$. In case Ω_{ij} is simply connected and does not contain ∞ an opposite inequality can be obtained by use of Koebe's one-quarter theorem. One finds under these circumstances $\rho(z) \geqq \tfrac{1}{2}\delta(z)^{-1}$.

In the general case we use the majorization $\Omega \subset \Omega_0$, where Ω_0 is the sphere punctured at three points $z_0, z_1, z_2 \in \Sigma$. We have $\rho \geqq \rho_0$, and ρ_0 can be estimated by means of (2.3). It follows that

$$\log\rho(z) \geqq -\log|z - z_0| - \log\log(1/|z - z_0|) - O(1)$$

in a neighborhood of z_0. It is not hard to see that the remainder is uniformly bounded when z_0 varies but stays away from z_1, z_2. If Σ is assumed to be compact we conclude that

(2.9) $$\log\rho \geqq -\log\delta - \log\log(1/\delta) - O(1)$$

for sufficiently small δ.

3. Quadratic differentials. 3.1. In all that follows a dominant role is played by the quadratic differentials on Ω. A quadratic differential is a holomorphic function ϕ on Ω which satisfies the functional equations

$$\phi(Az)A'(z)^2 = \phi(z) \tag{3.1}$$

for all $A \in \Gamma$. More accurately, the quadratic differential is the invariant expression $\phi\, dz^2$, but it is convenient to use the same term for the coefficient ϕ.

The projection map induces a corresponding quadratic differential on S, namely by the relation $\phi\, dz^2 = \bar{\phi}\, d\zeta^2$. At a point p_k with finite signature, let the projection be given by $\zeta = z^m$ in local coordinates. Then $\bar{\phi} = O(|\zeta|^{-2+2/m})$, and since $\bar{\phi}$ is single-valued it has at most a simple pole at p_k. Conversely, if $\bar{\phi}$ has at most a simple pole, then ϕ is holomorphic.

As soon as S is not compact the class of all quadratic differentials on Ω is too extensive to be useful. We shall consider the following restriction:

$$\int_{\Omega/\Gamma} |\phi|\, dxdy < \infty, \tag{Q}$$

$$|\phi| = o(\rho^2). \tag{Q*}$$

Observe that the integral in (Q) is meaningful because of (3.1). The quadratic differentials which satisfy these conditions form linear spaces which we denote by $Q(\Gamma)$ and $Q^*(\Gamma)$ respectively.

In terms of $\bar{\phi}$ condition (Q) becomes

$$\int_S |\bar{\phi}|\, d\xi d\eta < \infty. \tag{3.2}$$

Similarly, (Q^*) yields $|\bar{\phi}| = O(\bar{\rho}^2)$. In both cases $\bar{\phi}$ has at most a simple pole at points with infinite signature. In fact, for Q this follows directly from (3.2), and for Q^* we may use (2.3) to deduce that $\bar{\phi} = O(|\zeta|^{-2})$ in terms of a local variable, and hence that the singularity is at most a simple pole.

We shall let $\bar{Q}(S)$ and $\bar{Q}^*(S)$ denote the spaces formed by all $\bar{\phi}$ for $\phi \in Q(\Gamma)$ and $\phi \in Q^*(\Gamma)$ respectively. If S is of finite type these spaces are identical, and we conclude:

LEMMA 2. *If S is of finite type the spaces $Q(\Gamma)$ and $Q^*(\Gamma)$ are identical and of finite dimension.*

The dimension over the complex numbers is $3g + n - 3k$ where g is the sum of the genera of the surfaces S_i, n is the total number of points with a signature, and k is the number of components. It follows from condition (2.4) that there are non-zero quadratic differentials on each S_i.

3.2. We have just proved one part of the following result, which is particularly important for our purposes:

THEOREM 1. *The space $Q(\Gamma)$ is finite dimensional if and only if S is of finite type.*

If there are infinitely many S_i, all of finite type, it follows from the remark above that $Q(\Gamma)$ has infinite dimension. Suppose now that one S_i is not of finite type. If S_i has infinite genus there exist infinitely many linearly independent square integrable first order differentials θ_j on S_i. If they are multiplied by a fixed one, θ_0, the $\theta_0\theta_j$ are linearly independent quadratic differentials, and they are integrable. Assume next that S_i has finite genus. Then it can be imbedded in a compact Riemann surface $\bar{S}_i$. If $\bar{S}_i - S_i$ is an infinite point set we can find infinitely many quadratic differentials on $\bar{S}_i$ with distinct simple poles on $\bar{S}_i - S_i$. Their restrictions to S_i are integrable and linearly independent. Finally, if $\bar{S}_i - S_i$ is a finite set there must be infinitely many p_k on S_i, and we can find infinitely many quadratic differentials with distinct poles among the p_k. Thus, whenever S is not of finite type, $Q(\Gamma)$ has infinite dimension.

4. Beltrami differentials. 4.1. We regard $Q(\Gamma)$ as a Banach space with the norm

$$\| \phi \| = \int_{\Omega/\Gamma} | \phi | \, dxdy$$

and $Q^*(\Gamma)$ as one with the norm

$$\| \phi \|_* = \sup | \phi | \rho^{-2}.$$

One of our aims is to determine the conjugate space of $Q(\Gamma)$. Several characterizations will be obtained, and their comparison will contribute to our knowledge of Kleinian groups.

4.2. Choose an arbitrary $\phi_0 \in Q(\Gamma)$, not identically zero, and define a measure on $S = \Omega/\Gamma$ by $dm = | \phi_0 | \, dxdy$. Evidently, a quadratic differential ϕ belongs to $Q(\Gamma)$ if and only if ϕ/ϕ_0 is of class L^1 with respect to this measure. Therefore, every linear functional on $Q(\Gamma)$ can be represented as

$$\int_{\Omega/\Gamma} \phi_0^{-1} | \phi_0 | \beta \, dxdy$$

where β is a bounded measurable function on S. We may define β as an automorphic function on Ω, and we find that $\nu = \phi_0^{-1} | \phi_0 | \beta$ satisfies the equations

$$\nu(Az)\bar{A}'(z)/A'(z) = \nu(z) \tag{4.1}$$

for all $A \in \Gamma$. Such functions are called *Beltrami differentials*; we emphasize that ν has to be bounded and measurable. We have proved that every linear functional on $Q(\Gamma)$ is of the form

$$\int_{\Omega/\Gamma} \phi\nu \, dxdy \tag{4.2}$$

where ν is a Beltrami differential.

4.3. Let $B(\Gamma)$ be the linear space of Beltrami differentials, normed by $\| \nu \|_\infty$, and let $N(\Gamma)$ denote the subspace of those ν which satisfy

$$\int_{\Omega/\Gamma} \phi\nu \, dxdy = 0 \tag{4.3}$$

for all $\phi \in Q(\Gamma)$. Our result can be stated as follows:

LEMMA 3. *The conjugate space of* $Q(\Gamma)$ *can be identified with* $B(\Gamma)/N(\Gamma)$.

Clearly, any characterization of $N(\Gamma)$ will therefore be a characterization of the conjugate space of $Q(\Gamma)$.

5. The Fuchsian case. 5.1. Many problems in the theory of Kleinian groups can be reduced to the corresponding problem for Fuchsian groups. We shall therefore begin with a study of the Fuchsian case.

To emphasize the distinction, Fuchsian groups will be denoted by the letter G. We assume that G is discontinuous and acts on the unit disk $D = \{z \mid |z| < 1\}$, but it need not be of the first kind. We shall constantly disregard the outside of D. Therefore, we shall set $S(G) = D/G$, and the notations $Q(G)$ and $Q^*(G)$ will refer to quadratic differentials in D. The corresponding remark applies to the spaces $B(G)$ and $N(G)$. Observe that $\rho(z) = 2(1-|z|^2)^{-1}$.

We shall make frequent use of the known representation formula[2]

$$\phi(\zeta) = 3\pi^{-1} \int_D \phi(z)(1-|z|^2)^2(1-\bar{z}\zeta)^{-4} \, dxdy \tag{5.1}$$

[2] L. Ahlfors, "Some remarks on Teichmüller's space of Riemann surfaces," *Annals of Mathematics*, vol. 74, No. 1, 1961, p. 176.

which is valid as soon as ϕ is holomorphic in D and

$$\int_D |\phi|(1-|z|^2)^2\,dxdy<\infty. \tag{5.2}$$

This condition is obviously fulfilled when $\phi\in Q^*(G)$.

The identity

$$\int_D |1-\bar z\zeta|^{-4}\,dxdy=\pi(1-|\zeta|^2)^{-2} \tag{5.3}$$

will also be needed.

5.2. With any Beltrami differential $\nu\in B(G)$ we associate

$$T\nu(\zeta)=3\pi^{-1}\int_D \bar\nu(z)\,(1-\bar z\zeta)^{-4}\,dxdy. \tag{5.4}$$

A simple computation shows that $T\nu$ is a quadratic differential, and by use of (5.3) we conclude that $T\nu\in Q^*(G)$. In the opposite direction, $\phi\in Q^*(G)$ can be mapped on $T^*\phi=\bar\phi(1-|z|^2)^2=4\bar\phi\rho^{-2}\in B(G)$. Formula (5.1) takes the form $TT^*\phi=\phi$, or $TT^*=I$. The operator T^* is of course meaningful also when $\phi\in Q(G)$, but then we can no longer assert that $T^*\phi$ belongs to $B(G)$.

As an abbreviation we introduce

$$\langle\phi,\nu\rangle=\int_{D/G}\phi\nu\,dxdy, \tag{5.5}$$

and we use this notation as soon as the integral on the right is absolutely convergent. One verifies at once that

$$\langle\phi_1,T^*\phi_2\rangle=\langle\phi_2,T^*\phi_1\rangle^- \tag{5.6}$$

as soon as $\phi_1\in Q(G)$, $\phi_2\in Q^*(G)$ or vice versa (the bar stands for the complex conjugate).

Even more important is the following identity:

LEMMA 4. *$\langle\phi,T^*T\nu\rangle=\langle\phi,\nu\rangle$ for $\phi\in Q(G)$, $\nu\in B(G)$.*

Both sides of the identity represent absolutely convergent integrals, and by (5.6) the left hand side can be replaced by $\langle T\nu,T^*\phi\rangle^-$. By virtue of (5.1) the equality $\langle T\nu,T^*\phi\rangle^-=\langle\phi,\nu\rangle$ is equivalent to

$$\int_{D/G}\phi(\zeta)(1-|\zeta|^2)^2\,d\xi d\eta\int_D \nu(z)\,(1-\bar z\zeta)^{-4}\,dxdy$$

$$=\int_D \phi(\zeta)\,(1-|\zeta|^2)^2\,d\xi d\eta\int_{D/G}\nu(z)\,(1-\bar z\zeta)^{-4}\,dxdy,$$

an identity which is proved by subdivision into "fundamental regions" and subsequent changes of the integration variables. The absolute convergence justifies the procedure.

5.3. Let $A(D)$ be the space of holomorphic functions which are integrable over D. For $F \in A(D)$ the Poincaré series

$$\Theta F(z) = \sum_{A \in G} F(Az) A'(z)^2 \tag{5.7}$$

converges and represents an element of $Q(G)$. In fact, one obtains

$$\| \Theta F \| \leqq \int_D | F | \, dxdy. \tag{5.8}$$

It is also easy to verify that

$$\langle \Theta F, \nu \rangle = \int_D F\nu \, dxdy. \tag{5.9}$$

5.4. We now deduce the most important result in this direction, which is due to L. Bers:[3]

THEOREM 2. *The mapping $T: B(G) \to Q^*(G)$ is a surjection with kernel $N(G)$. Therefore, the conjugate space of $Q(G)$ can be identified with $Q^*(G)$.*

That the mapping is onto follows from $TT^* = I$. Another way of expressing Lemma 4 is to say that $\nu - T^*T\nu \in N(G)$. Therefore, $T\nu = 0$ implies $\nu \in N(G)$. To prove the converse, let F be the function $3\pi^{-1}(1 - z\bar{\zeta})^{-4}$ with fixed $\zeta \in D$. The definition (5.4) of $T\nu$ together with (5.9) gives

$$T\nu(\zeta)^- = \langle \Theta F, \nu \rangle,$$

and hence $\nu \in N(G)$ yields $T\nu = 0$ as asserted.

Specifically, we have shown that every coset of $N(G)$ in $B(G)$ contains an element $T^*\phi$ with $\phi \in Q^*(G)$, namely for $\phi = T\nu$. This is of course the only element with this property, for $T^*\phi \in N(G)$ implies $\phi = TT^*\phi = 0$.

5.5. As an immediate corollary we obtain:

LEMMA 5. *Every $\phi \in Q(G)$ can be written in the form ΘF with $F \in A(D)$.*

Indeed, it follows from (5.8) that Θ maps $A(D)$ onto a closed subspace of $Q(G)$. If it were not all of $Q(G)$ we could find $\phi \in Q^*(G)$, not identically zero, such that $\langle \Theta F, T^*\phi \rangle = 0$ for all $F \in A(D)$. By (5.9) this gives

[3] Oral communication.

$$\int_D FT^*\phi\, dxdy = 0.$$

We choose again $F(z) = 3\pi^{-1}(1 - z\bar{\zeta})^{-4}$ and obtain $\phi(\zeta) = 0$, contrary to assumption.

5.6. There is a generalization of Lemma 5 which will be of importance when we pass to the case of Kleinian groups. Let G_0 be an arbitrary subgroup of G. For any $\phi_0 \in Q(G)$ we construct

$$\Theta_0\phi_0(z) = \sum \phi_0(Az)A'(z)^2 \tag{5.10}$$

where the summation is now over a set of representatives of the left cosets of G_0, one from each coset. Because ϕ_0 is a quadratic differential with respect to G_0 the choice of representatives is irrelevant, and one finds that $\| \Theta_0\phi_0 \| \leqq \| \phi_0 \|$. Equation (5.9) is replaced by

$$\langle \Theta_0\phi_0, \nu \rangle = \langle \phi_0, \nu \rangle, \tag{5.11}$$

valid for all $\nu \in B(G)$. It is to be observed that the first inner product is over D/G, the second over D/G_0.

Lemma 6. *Every $\phi \in Q(G)$ can be written as $\Theta_0\phi_0$ with $\phi_0 \in Q(G_0)$.*

Again, if this were not true we could find $\phi \in Q^*(G)$, $\phi \neq 0$, such that $\langle \Theta_0\phi_0, T^*\phi \rangle = 0$ for all $\phi_0 \in Q(G)$. According to (5.11) this gives $\langle \phi_0, T^*\phi \rangle = 0$ for all ϕ_0, and hence $\phi = 0$, a contradiction.

6. **The general case.** 6.1. We return to the case of an arbitrary Kleinian group Γ, except that we shall assume ∞ to be a limit point. This minor restriction has the advantage that the Poincaré series

$$\Theta F(z) = \sum_{A \in \Gamma} F(Az)A'(z)^2 \tag{6.1}$$

remains holomorphic for any $F \in A(\Omega)$, that is, whenever F is holomorphic and integrable in Ω. It is clear that ΘF enjoys the same properties as in the Fuchsian case.

We shall show that the conclusion in Theorem 2 as well as Lemma 5 remain valid. As far as the conjugate space is concerned it is quite clear that we may restrict our attention to a single component Ω_{ij} of Ω and replace Γ by the subgroup Γ_{ij} which leaves Ω_{ij} invariant. It amounts to the same thing if we assume from the beginning that Ω is an invariant region with respect to Γ, but not necessarily the whole set of discontinuity.

Let $z = \omega(\zeta)$ be a conformal mapping of $D = \{\zeta \mid |\zeta| < 1\}$ onto the universal covering surface of Ω. We introduce a Fuchsian group G as follows: $B \in G$ if and only if there exists $A \in \Gamma$ such that

$$\omega(B\zeta) = A\omega(\zeta). \tag{6.2}$$

It is evident that A is uniquely determined, and the mapping $A \to B$ is a surjective homomorphism whose kernel we denote by G_0.

The equation

$$\hat{\phi}(\zeta) = \phi(\omega(\zeta))\omega'(\zeta)^2 \tag{6.3}$$

sets up a one-one correspondence between quadratic differentials with respect to G and Γ. One verifies that $Q(\Gamma)$ corresponds to $Q(G)$ and $Q^*(\Gamma)$ to $Q^*(G)$. A similar correspondence between $B(\Gamma)$ and $B(G)$ is defined by

$$\hat{\nu}(\zeta) = \nu(\omega(\zeta))\overline{\omega'}(\zeta)/\omega'(\zeta). \tag{6.4}$$

The identity

$$\int_{\Omega/\Gamma} \phi\nu\, dxdy = \int_{D/G} \hat{\phi}\hat{\nu}\, d\xi d\eta \tag{6.5}$$

prevails. We conclude:

THEOREM 3. *The conjugate space of $Q(\Gamma)$ has a canonical identification with $Q^*(\Gamma)$.*

6.2. For the counterpart of Lemma 5 it is also true that it suffices to consider a component Ω_{ij} and the corresponding subgroup Γ_{ij}. To see this, let Θ_{ij} denote the theta-operator with respect to Γ_{ij}. If $\phi = \Theta_{ij}F$ in Ω_{ij} with $F \in A(\Omega_{ij})$, extend F to all of Ω by setting $F = 0$ outside of Θ_{ij}. At the same time ϕ is extended to all of Ω_i as a quadratic differential. It follows that $\phi = \Theta F$ in Ω_i, and this is what we wish to prove. It is therefore no restriction to assume again that Ω is an invariant region for Γ.

LEMMA 7. *Every $\phi \in Q(G)$ can be written as ΘF with $F \in A(\Omega)$.*

We use the same notations as before. By Lemma 6 it is possible to write $\phi = \Theta_0\phi_0$ with $\phi_0 \in Q(G_0)$. The relation

$$\phi_0(\zeta) = F(\omega(\zeta))\omega'(\zeta)^2 \tag{6.6}$$

defines F uniquely, and one verifies that $\Theta_0\phi_0(\zeta) = \Theta F(\omega(\zeta))\omega'(\zeta)^2$, and this makes $\phi = \Theta F$.

7. Direct characterization of $N(\Gamma)$. 7.1. In this section we shall derive some results which, as far as we are able to judge, cannot be obtained by reduction to the Fuchsian case. It will be convenient to assume, in this connection, that the points 0, 1, ∞ belong to Σ. All invariantly formulated results are of course independent of this hypothesis.

The method we shall use is intimately connected with infinitesimal deformations of conformal structure. However, there will be no need to make this connection explicit.

Given $\nu \in B(\Gamma)$ we construct

$$(7.1) \qquad f(\zeta) = -\pi^{-1} \int_\Omega \nu(z) \left[\frac{1}{z-\zeta} - \frac{\zeta}{z-1} - \frac{1-\zeta}{z}\right] dxdy.$$

The integral converges because the rational function in brackets has a third order zero at ∞ and only simple poles at ζ, 0, and 1. It represents a continuous function in the finite plane. What is more, it satisfies a uniform Hölder-type condition

$$(7.2) \qquad |f(\zeta_1) - f(\zeta_2)| \leqq C\,|\zeta_1 - \zeta_2| \log 1/|\zeta_1 - \zeta_2|,$$

for instance for $|\zeta_1 - \zeta_2| \leqq 1/2$. At ∞ it is of order $O(|\zeta| \log |\zeta|)$.

It is well known that f has locally square integrable distributional derivatives, and that

$$(7.3) \qquad f_{\bar{z}} = \tfrac{1}{2}(f_x - if_y) = \nu$$

almost everywhere (provided that we set $\nu = 0$ on Σ).

7.2. We shall prove:

Lemma 8. *$\nu \in N(\Gamma)$ if and only if $f = 0$ on Σ.*

The necessity is immediate. Indeed, for any fixed $\zeta \in \Sigma$ the bracketed expression in (7.1) is a holomorphic and integrable function F on Ω. Therefore, $\nu \in N(\Gamma)$ implies

$$\int_\Omega \nu F \, dxdy = \langle \Theta F, \nu \rangle = 0,$$

and hence $f(\zeta) = 0$.

To prove the sufficiency, let λ be a C^∞ function which vanishes in a neighborhood of Σ. In addition we require that λ and $f\lambda_{\bar{z}}$ be bounded. Under these conditions, if $F \in A(\Omega)$ Stokes' formula yields, with the help of (7.3),

$$(7.4) \qquad \int_\Omega \lambda F \nu \, dxdy = \int_\Omega f_{\bar{z}} \lambda F \, dxdy = -\int_\Omega f \lambda_{\bar{z}} F \, dxdy.$$

If we can choose λ, depending on a parameter, in such a way that λ tends boundedly to 1 and $f\lambda_{\bar{z}}$ tends boundedly to 0, it will follow that $\langle \Theta F, \nu \rangle = 0$, and hence, by Lemma 7, that $\nu \in N(\Gamma)$.

We are assuming that $f = 0$ on Σ. As before, $\delta(z)$ denotes the distance from z to Σ. We deduce from (7.2) that

$$|f| \leqq C\delta \log 1/\delta \tag{7.5}$$

for $\delta \leqq \delta_0$, say. We choose $\delta_0 \leqq e^{-1}$.

For the moment, let us ignore the condition that λ be of class C^∞. Given $\epsilon > 0$, let $\Lambda(t)$ be 0 for $0 \leqq t \leqq \epsilon$, 1 for $t \geqq 2\epsilon$, and linear between ϵ and 2ϵ. We shall set

$$\lambda(z) = \Lambda[(\log\log 1/\delta(z))^{-1}] \tag{7.6}$$

if $\delta < \delta_0$ and $\lambda(z) = 1$ if $\delta \geqq \delta_0$. For ϵ small enough, λ will be continuous and identically zero near Σ.

If $\delta(z)$ were differentiable, the obvious inequality $|\delta(z_1) - \delta(z_2)| \leqq |z_1 - z_2|$ would yield $|\delta_z| \leqq 1$, and hence

$$|\lambda_z| \leqq \epsilon^{-1}\delta^{-1}(\log 1/\delta)^{-1}(\log\log 1/\delta)^{-2}$$

in the part where λ is not constant. Since $(\log\log 1/\delta)^{-1} \leqq 2\epsilon$ for these values it follows that

$$|\lambda_z| \leqq 4\epsilon\delta^{-1}(\log 1/\delta)^{-1},$$

and in view of (7.5) we obtain $|\lambda_z f| \leqq 4C\epsilon$ throughout Ω. The desired conclusion follows.

The reasoning is completed by an obvious smoothing process, which we need not explain in detail. First, we replace δ in (7.6) by a smooth mean value δ^* obtained by convolution. In the relevant part of the plane δ^* will satisfy the same Lipschitz condition as δ. Secondly, Λ has to be replaced by a smooth function whose derivative is nowhere much greater than ϵ^{-1}. With these modifications the argument goes through exactly as before.

7.3. An alternative characterization of $N(\Gamma)$ is the following:

LEMMA 9. *$\nu \in N(\Gamma)$ if and only if $f(Az) = f(z)A'(z)$ for all $A \in \Gamma$.*

The condition means that f/dz is invariant, in which case f is said to be an inverse differential.

The definition (7.1) of f shows that $f(0) = 0$. If the condition in the lemma is fulfilled it follows that $f(A0) = 0$ for all $A \in \Gamma$. But the points $A0$ are dense on Σ and f is continuous. Therefore f vanishes on Σ, and Lemma 8 shows that $\nu \in N(\Gamma)$.

Consider now the function

$$P_A(z) = f(Az)A'(z)^{-1} - f(z). \tag{7.7}$$

By use of (7.3) we obtain

$$(P_A)_{\bar z} = \nu(Az)\bar{A}'(z)A'(z)^{-1} - \nu(z) = 0 \tag{7.8}$$

almost everywhere; in particular, also on Σ. Furthermore, the derivatives of P_A are locally square integrable (except perhaps at $A^{-1}\infty$). It follows that P_A is analytic, and if it vanishes on Σ it must be identically zero. This proves the lemma.

A more detailed investigation of the singularities at ∞ and $A^{-1}\infty$ shows that P_A is always a quadratic polynomial. It is of some interest to note the formula

$$P_{AB}(z) = P_A(Bz)B'(z)^{-1} + P_B(z). \tag{7.9}$$

7.4. The last lemma leads immediately to the principal result:

THEOREM 4. *If Γ is finitely generated, then $S = \Omega/\Gamma$ is of finite type.*

Indeed, all relations $f(Az) = f(z)A'(z)$ are consequences of the corresponding relations for the generators. On the other hand, each relation is equivalent to the vanishing of the coefficients of P_A, and thus to three linear conditions on ν. For a finitely generated group the space $B(\Gamma)/N(\Gamma)$ has thus finite dimension. By Lemma 3 and Theorem 3 this means that $Q^*(\Gamma)$ is finite dimensional, and hence the same is true of $Q(\Gamma)$. We apply Theorem 1 to conclude that S is of finite type.

8. Invariant regions. 8.1. It is possible to obtain more precise information by adding the hypothesis that one or several Ω_i are connected, and hence invariant under the full group Γ. Such groups have been called function groups.[4]

The fact that all invariant regions Ω_i have the same boundary Σ does not by itself preclude the existence of any number of such regions. However, the existence of nonelliptic fixpoints does impose a severe restriction. The following theorem was communicated to me by R. Accola.

THEOREM 5. *There are at most two invariant regions Ω_i, and if there are two they are simply connected.*

For the simple proof, which is purely topological, we refer to a forthcoming article by Accola.

8.2. If Γ is finitely generated we are able to prove a more precise result:

[4] L. Ford, Automorphic Functions (2nd ed., Chelsea, 1951), p. 64.

THEOREM 6. *Suppose that Ω_1 and Ω_2 are invariant regions for the finitely generated group Γ. Then $S = \Omega/\Gamma$ has only the components $S_1 = \Omega_1/\Gamma$ and $S_2 = \Omega_2/\Gamma$. Moreover, S_1 and S_2 are homeomorphic with isomorphic signatures.*

Since Γ is finitely generated we need not distinguish between $Q(\Gamma)$ and $Q^*(\Gamma)$. We shall denote by Q_i the subspace of quadratic differentials that vanish outside of Ω_i. The main part of the theorem is an easy consequence of

LEMMA 10. *If Ω_1 is connected, then* $\dim Q \leqq 2 \dim Q_1$.

Indeed, under the hypothesis of the theorem the lemma yields $\dim Q_1 + \dim Q_2 \leqq \dim Q \leqq 2 \dim Q_1$. Hence $\dim Q_2 \leqq \dim Q_1$, and by symmetry $\dim Q_1 = \dim Q_2$. Furthermore, $\dim Q = \dim Q_1 + \dim Q_2$, and this shows that there is no room for a third component. If we can show that S_1 and S_2 have the same signature it will follow, by the formula for the number of linearly independent quadratic differentials, that they have the same genus.

The points with finite signature arise from elliptic fixpoints, and it is easy to see that each elliptic transformation must have one fixpoint in Ω_1 and one in Ω_2 (the simplest way to see this is by conformal mapping of Ω_1 and Ω_2 on the unit disk). It is therefore evident that the points with finite signature can be matched in pairs, in such a way that corresponding points have the same signature.

The facts for points with infinite signature are quite similar. We know by Lemma 1 that any such point, whether on S_1 or S_2, determines a class of conjugate parabolic transformations. In the present situation the converse is also true. Let $A \in \Gamma$ be parabolic. We map Ω_1 conformally on the unit disk D and use the construction in 6.1 to determine an isomorphic mapping of Γ on a Fuchsian group G. The image B of A is necessarily parabolic. In the elementary case of a Fuchsian group it is quite evident that B corresponds to a point on S_1 with infinite signature (M. Heins has proved this for arbitrary Fuchsian groups).[5] Moreover, non-conjugate transformations correspond to different points. It follows that the points with infinite signature are indeed in one-one correspondence.

8.3. It remains to prove Lemma 10. Let us denote by Q'_1 the space of quadratic differentials which vanish on Ω_1. Given $\phi \in Q'_1$ we form $\nu = \bar{\phi}\rho^2$, which is a Beltrami differential, and construct

[5] M. Heins, "On Fuchsoid groups that contain parabolic transformations" (*Contributions to function theory*), Tata Institute, Bombay, 1960.

$$S\phi(\zeta)=\int_{\Omega-\Omega_1}\nu(z)(z-\zeta)^{-4}\,dxdy \tag{8.1}$$

for $\zeta\in\Omega_1$. One verifies that $S\phi$ is a quadratic differential. Because ν is bounded the integral is majorized by

$$\int_{|z-\zeta|>\delta(\zeta)}|z-\zeta|^{-4}\,dxdy$$

which is a multiple of $\delta(\zeta)^{-2}$. Since Ω_1 is simply connected $\delta(\zeta)^{-1}$ is comparable with $\rho(\zeta)$ (we are assuming that $\infty\in\Sigma$). It follows that $S\phi\in Q_1$.

We shall show that the antilinear mapping $S\colon Q'_1\to Q_1$ is one to one. Assuming that $0,1,\infty\in\Sigma$ we construct, as in 7.1,

$$f(\zeta)=-\pi^{-1}\int_{\Omega-\Omega_1}\nu(z)\left[\frac{1}{z-\zeta}-\frac{\zeta}{z-1}-\frac{1-\zeta}{z}\right]dxdy. \tag{8.2}$$

If $S\phi=0$ we see by comparison with (8.1) that $f'''(\zeta)=0$ in Ω_1. Hence $f(\zeta)$ is a quadratic polynomial in Ω_1. Its behavior at ∞ shows that it is in fact a linear polynomial, and because it vanishes at 0 and 1 it reduces identically to zero. By continuity $f=0$ on Σ, and we conclude by Lemma 8 that $\nu\in N(\Gamma)$. Since $\nu=\bar{\phi}\rho^2$ this gives $\phi=0$, and we have proved that S is one to one.

We emphasize that the connectedness of Ω_1 is essential for the proof. If Ω_1 were not connected we could merely conclude that f is a quadratic polynomial in each component, but not necessarily the same in all components.

HARVARD UNIVERSITY.

Henri Cartan
Curriculum Vitae

CARTAN Henri Paul, né à Nancy (France) le 8 juillet 1904
1923–26 Elève à l'Ecole Normale Supérieure
1926 Agrégé de mathématiques
décembre 1928 Docteur és Sciences mathématiques (Université de Paris)
1928–29 Professeur au Lycée Malherbe à CAEN
1929–31 Chargé de cours à la Faculté des Sciences de LILLE
1931–35 Chargé de cours, puis maître de conférences à la Faculté des Sciences de STRASBOURG
1936–40 Professeur à la Faculté des Sciences de STRASBOURG
1940–49 Professeur sans chaire à la Faculté des Sciences de PARIS
1945–47 Détaché à la Faculté des Sciences de STRASBOURG
1948 Professeur invité à l'Université HARVARD
1949–69 Professeur à la Faculté des Sciences de PARIS
1940–65 Responsable de l'enseignemùent des mathématiques à l'Ecole Normale Supérieure
1966–67 Invité à l'Institute for Advances Study de Princeton
1969–75 Professeur à l'Université Paris-Sud (ORSAY)

1967–70 Président de l'Union mathématique internationale
1957–75 Président de l'Association Européenne des Enseignants (section française)
1974–85 Président du Mouvement Fédéraliste Européen (section française)

Foreign Honorary Member of the American Academy (Boston), 1950
Foreign Honorary Member of the London Mathematical Society, 1959
Membre de l'Académie Royale des Sciences et des Lettres du Danamark, 1962
Membre correspondant de l'Académie des Sciences (Institut de France), 1965
Honorary Member of the Cambridge Philosophical Society, 1969
Foreign Member of the Royal Society of London, 1971
Membre correspondant de l'Académie des Sciences de Göttingen, 1971
Membre correspondant de l'Académie des Sciences de Madrid, 1971
Foreign Associate of the National Academy of Sciences (Washington), 1972
Membre de l'Académie des Sciences (Institut de France), 1974
Membre correspondant de l'Académie bavaroise des Sciences (Munich), 1974
Membre associé de l'Académie Royale de Belgique, 1978
Foreign Honorary Member of the Japan Academy, 1979
Membre étranger de l'Académie des Sciences et Lettres de Finlande, 1979
Membre étranger de l'Académie Royale de Suède, 1981
Associé étranger de l'Académie polonaise des Sciences, 1985
Membre d'honneur de la Deutsche Mathematiker Vereinigung, 1994
Membre d'honneur de la Société mathématique de Moscou, 1995

Docteur honoris causa de l'Ecole Polytechnique Fédérale de Zürich, 1955
Docteur honoris causa des Universités de:

Münster, 1952
Oslo, 1961
Sussex, 1969
Cambridge, 1969
Stockholm, 1978
Oxford, 1980
Saragosse, 1985
Athènes, 1992

Médaille d'Or du Centre National de la Recherche Scientifique (Paris), 1976
Prix WOLF de mathématiques, 1980

Brève analyse des travaux*

I. Fonctions analytiques

1) Fonctions d'une variable complexe

C'est à elles que sont consacrés mes tout premiers travaux. Quelques Notes aux Comptes Rendus se rapportent à la fonction de croissance de Nevanlinna et à la répartition des valeurs des fonctions méromorphes. Dans ma Thèse [3], j'ai réussi à prouver, en la précisant, une inégalité conjecturée par André BLOCH: pour tout nombre réel $h>0$, les points du plan complexe où un polynôme unitaire de degré n est, en valeur absolue, au plus égal à h^n peuvent être enfermés dans des disques dont la somme des rayons est au plus égale à $2\,eh$ ($e =$ base des logarithmes népériens). J'ai montré de plus que l'on peut considérablement généraliser ce résultat; cette généralisation a été ensuite reprise et utilisée par Ahlfors. L'inégalité de Bloch s'est révélée un instrument précieux dans l'étude de la répartition des valeurs d'une fonction analytique.

Dans [25], j'ai étudié la croissance d'un système de fonctions holomorphes, c'est-à-dire, en fait, d'une application holomorphe dans un espace projectif, généralisant à cette situation les théorèmes de NEVANLINNA. Cette étude a été reprise, d'une façon indépendante, par Hermann et Joachim WEYL.

C'est dans ma Thèse [3] que j'ai étudié les familles normales d'applications holomorphes d'un disque dans l'espace projectif $P_n(\mathbb{C})$ privé de $n+2$ hyperplans en position générique. Ce sujet semble redevenu d'actualité à la suite de quelques travaux récents (notamment de P. KIERNAN et S. KOBAYASHI, Nagoya Math. J. 1973).

2) Problèmes d'itération et de limite pour les fonctions holomorphes de plusieurs variables complexes ([14], [24], [29])

J'ai notamment prouvé le résultat suivant: soit D un domaine borné de $\mathbb{C}^n$, et soit f une application holomorphe $D \to D$. Si, dans l'adhérence de la suite des itérées f^k, il existe une transformation dont le Jacobien n'est pas identiquement nul, f est nécessairement un *automorphisme* de D. Ce résultat est susceptible de nombreuses applications; M. HERVÉ l'a utilisé avec succès à diverses occasions. En voici une application immédiate [24]: pour $n=1$, s'il existe un point a du plan complexe $\mathbb{C}$, hors de D, et une courbe fermée de D dont l'indice par rapport

* écrite par H. Cartan en 1973.

Reprinted from Oeuvres, Collected Works (Springer-Verlag, 1979), Vol. 1, pp. ix–xix.

à *a* soit non nul, si de plus *f* transforme cette courbe en une courbe dont l'indice est non nul, alors *f* est nécessairement un automorphisme de D. Autre application: pour *n* quelconque, si *f*: D→D possède un point fixe en lequel le Jacobien est de valeur absolue égale à 1, *f* est un automorphisme de D.

3) Automorphismes des domaines bornés ([13], [20], [33])

Que peut-on dire du groupe de tous les automorphismes holomorphes d'un domaine *borné* D de $\mathbb{C}^n$? (Cf. aussi *4)* ci-dessous). Soit G(*a*) le groupe d'isotropie d'un point *a* ∈ D, c'est-à-dire le sous-groupe formé des automorphismes qui laissent fixe le point *a*. Un premier résultat est le suivant: l'application qui, à chaque élément de G(*a*), associe la transformation linéaire tangente en *a*, est un isomorphisme de G(*a*) sur un sous-groupe (compact) du groupe linéaire GL(*n*, ℂ). J'ai prouvé cela à partir d'un lemme très simple, qui dit que si une transformation holomorphe *f* de D dans D (non supposée bijective) laisse fixe un point *a* ∈ D et est tangente à l'identité en *a*, c'est l'application identique. Ce lemme est aussi valable pour les groupes formels (cf. le livre classique de Bochner et Martin). Il a aussi l'avantage de pouvoir s'appliquer tel quel aux fonctions holomorphes dans un espace de Banach complexe de dimension infinie, beaucoup étudiées aujourd'hui.

Le résultat précédent m'a conduit à une démonstration très simple du théorème suivant: soient D et D′ deux domaines *cerclés* dont l'un au moins est supposé borné (un domaine D est dit cerclé s'il est stable par toute homothétie de rapport λ tel que $|\lambda| = 1$ et s'il contient l'origine); alors tout isomorphisme holomorphe *f*: D→D′ qui transforme l'origine en l'origine est nécessairement *linéaire*. Ce théorème était auparavant connu dans des cas particuliers, ou sous des hypothèses restrictives relatives à la frontière (Behnke). Il est, lui aussi, valable dans un espace de Banach.

L'article [13] contient beaucoup d'autres résultats, notamment sur l'existence de développements en séries de types particuliers.

La détermination du groupe de *tous* les automorphismes d'un domaine cerclé borné a été faite complètement pour le cas de deux variables dans [20]. A part quelques types spéciaux de domaines cerclés (qui sont explicités), le groupe de tous les automorphismes se réduit au groupe d'isotropie de l'origine.

4) Groupes de transformations holomorphes en général

Le groupe des automorphismes holomorphes d'un domaine borné D de $\mathbb{C}^n$ est *localement compact:* c'est un résultat nullement évident que j'ai prouvé dans [24]. La question se posait ensuite de savoir si c'est un *groupe de Lie*. Ce problème ne doit pas être confondu avec le fameux cinquième problème de Hilbert, qui du reste n'était pas encore résolu à l'époque (1935). Dans [32], j'ai démontré le théorème fondamental suivant: tout «noyau» compact de groupe de transformations holomorphes, dans $\mathbb{C}^n$, est un noyau de groupe de Lie. Il en résulte d'une part que le groupe des automorphismes holomorphes d'un domaine borné est un groupe de Lie (à paramètres réels); d'autre part que le groupe des automorphismes d'une variété analytique complexe *compacte* est un groupe de Lie, comme BOCHNER l'a montré plus tard. Quant au théorème fondamental ci-dessus, publié en 1935, il fut retrouvé huit ans plus tard par MONTGOMERY sous une forme plus générale, valable pour les groupes de transformations différentiables; la méthode de Montgomery est essentiellement la même, mais en utilisant le théorème de Baire il réussit à l'appliquer au cas différentiable.

5) Domaines d'holomorphie et convexité ([16], [23])

La notion de «domaine d'holomorphie» est bien connue aujourd' hui. Dans l'article [16], j'ai pour la première fois montré qu'un domaine d'holomorphie possède certaines propriétés de «convexité» par rapport aux fonctions holomorphes. Cette notion de «convexité» s'est, depuis lors, montrée féconde et elle est devenue classique. Dans [16], j'ai prouvé que la «convexité» est non seulement nécessaire pour que D soit un domaine d'holomorphie, mais qu'elle est suffisante pour certains domaines d'un type particulier (par exemple les domaines cerclés). Qu'elle soit suffisante dans le cas général a été démontré peu après par P. THULLEN. En mettant en commun nos idées, Thullen et moi avons écrit le mémoire [23] consacré à la théorie des domaines d'holomorphie. La notion de convexité holomorphe s'introduit aussi dans les problèmes d'approximation.

6) Problèmes de Cousin

Le premier problème de Cousin (ou problème *additif* de Cousin) consiste à trouver une fonction méromorphe dont on se donne les parties principales (polaires). Le deuxième problème de Cousin (ou problème *multiplicatif*) consiste à trouver une fonction méromorphe admettant un «diviseur» donné (variété des zéros et des pôles avec leurs ordres de multiplicité). On sait aujourd'hui que le problème additif est toujours résoluble pour un domaine d'holomorphie, et plus généralement pour une «variété de Stein». Ce résultat a été prouvé pour la première fois par K. OKA. Avant Oka, j'avais vu (cf. [31]) que le problème additif pouvait se résoudre en utilisant l'intégrale d'André WEIL, mais comme à cette époque il manquait certaines techniques permettant d'appliquer l'intégrale de Weil au cas général des domaines d'holomorphie, je renonçai à publier ma démonstration. Par ailleurs, je savais que, dans le cas de deux variables, le premier problème de Cousin n'a pas toujours de solution pour un domaine qui n'est pas un domaine d'holomorphie. En revanche, pour trois variables, j'ai donné le premier exemple (cf. [34]) d'ouvert qui n'est pas domaine d'holomorphie et dans lequel cependant le problème additif de Cousin est toujours résoluble; il s'agit de $\mathbb{C}^3$ privé de l'origine. Ma méthode de démonstration pour ce cas particulier (utilisation des séries de Laurent) a été utilisée plusieurs fois depuis dans des cas plus généraux, notamment par Frenkel dans sa Thèse.

Aujourd'hui, les problèmes de Cousin trouvent leur solution naturelle dans le cadre de la théorie des faisceaux analytiques cohérents (voir ci-dessous, *7)*).

7) Théorie des faisceaux sur une variété analytique complexe

L'étude des problèmes globaux relatifs aux idéaux et modules de fonctions holomorphes m'a occupé plusieurs années, en partant des travaux d'Oka. Dès 1940, j'avais vu qu'un certain lemme sur les matrices holomorphes inversibles joue un rôle décisif dans ces questions. Ce lemme est énoncé et démontré en 1940 dans [35]; dans ce même travail, j'en fais diverses applications, et je prouve notamment que si des fonctions f_i (en nombre fini), holomorphes dans un domaine d'holomorphie D, n'ont aucun zéro commun dans D, il existe une relation $\Sigma c_i f_i = 1$ à coefficients c_i holomorphes dans D. Dans [36], j'introduis la notion de «cohérence» d'un système d'idéaux et je tente de démontrer les théorèmes fondamentaux de ce qui deviendra la théorie des faisceaux analytiques cohérents sur une variété de Stein; mais je n'y parviens pas dans le cas le plus général, faute de réussir à prouver une conjecture que K. Oka démontrera plus tard (1950) et qui, en langage d'aujourd'hui, exprime que le faisceau des germes de fonctions holomorphes est *cohérent.* Sitôt que j'eus connaissance de ce théorème d'Oka (publié avec beaucoup d'autres dans le volume 78 du Bulletin de la Société mathématique de France), je repris l'ensemble de la question dans [38], en introduisant systématiquement la notion de *faisceau* (introduite alors par Leray en Topologie) et celle de faisceau cohérent (mais pas encore dans le sens plus général et définitif qui sera celui de mon Séminaire 1951–52). Il s'agit essentiellement de ce qu'on appelle aujourd'hui les «théorèmes A et B». Cependant, la formulation cohomologique générale du théorème B ne viendra que dans le Séminaire cité, à la suite de discussions avec J.-P. Serre. La conférence [41] est consacrée à une exposition d'ensemble de ces questions (sans démonstrations), avec indications sur les diverses applications qui en découlent pour la théorie globale des variétés de Stein, et en particulier pour les problèmes de Cousin.

8) Un théorème de finitude pour la cohomologie

Il s'agit du résultat suivant, obtenu en collaboration avec J.-P. Serre (cf. [42], ainsi que mon Séminaire 1953–54): si X est une variété analytique complexe *compacte,* et F un faisceau analytique cohérent, les espaces de cohomologie $H^q(X,F)$ sont des $\mathbb{C}$-espaces vectoriels de dimension finie. Le même résultat vaut, plus généralement, si X est un espace analytique compact.

Ce théorème n'est aujourd'hui que le point de départ du fameux théorème de Grauert qui dit que les images directes d'un faisceau analytique cohérent par une application holomorphe et propre sont des faisceaux cohérents.

9) La notion générale d'espace analytique

C'est après 1950 qu'apparaît la nécessité de généraliser la notion de variété analytique complexe, pour y inclure des singularités d'un type particulier, comme on le fait en Géométrie algébrique. Par exemple, le quotient d'une variété analytique complexe par un groupe proprement discontinu d'automorphismes n'est pas une variété analytique en général (s'il y a des points fixes), mais c'est un espace analytique (cf. [43]). Dès 1951, BEHNKE et STEIN tentaient d'introduire une notion d'espace analytique en prenant comme modèles locaux des «revêtements ramifiés» d'ouverts de $\mathbb{C}^n$; mais leur définition était assez peu maniable. Ma première tentative date de mon Séminaire 1951–52 (Exposé XIII); j'ai repris cette définition des espaces analytiques dans mon Séminaire de 1953–54 en introduisant la notion générale d'*espace annelé,* qui a ensuite été popularisée par SERRE, puis par GRAUERT et GROTHENDIECK. En 1953–54, ma définition conduisait aux espaces analytiques *normaux* (c'est-à-dire tels que l'anneau associé à chaque point soit intégralement clos). C'est SERRE qui, le premier, attira l'attention sur l'utilité d'abandonner la condition restrictive de normalité. Ensuite GRAUERT puis GROTHENDIECK introduisirent la catégorie plus générale des espaces annelés dans lesquels l'anneau attaché à un point n'est plus nécessairement un anneau de germes de fonctions mais peut admettre des éléments nilpotents.

J'ai démontré dans [48] un théorème de «prolongement» des espaces analytiques normaux, suggéré par des travaux de W. L. BAILY, et qui s'applique à la compactification de SATAKE dans la théorie des fonctions automorphes.

10) Quotients d'espaces analytiques ([43], [51], et Séminaire 1953–54)

Tout quotient d'un espace annelé X est canoniquement muni d'une structure d'espace annelé (ayant une propriété universelle aisée à formuler). Le problème suivant se pose: lorsque X est un espace analytique, trouver des critères permettant d'affirmer que l'espace annelé quotient est aussi un espace analytique. J'ai montré que lorsque la relation d'équivalence est définie par un groupe proprement discontinu d'automorphismes de X, le quotient est toujours un espace analytique. Puis, dans [51], j'ai donné un critère valable pour toutes les relations d'équivalence «propres» et j'ai étendu au cas des espaces analytiques généraux un théorème prouvé (par une autre méthode) par K. STEIN dans le cas des variétés sans singularités, et que voici: si f: X→Y est une application holomorphe, et si les composantes connexes des fibres de f sont compactes, le quotient de X par la relation d'équivalence dont les classes sont les composantes connexes des fibres est un espace analytique. D'autres applications du critère sont données dans [51].

11) Fonctions automorphes et plongements

Ayant défini le quotient d'un espace analytique X par un groupe G proprement discontinu d'automorphismes, il s'agissait de réaliser dans certains cas cet espace quotient comme sous-espace analytique d'espaces d'un type simple. Le premier cas que j'ai traité est celui où X est un ouvert borné de $\mathbb{C}^n$ et où X/G est compact: en m'appuyant sur des résultats de M. HERVÉ (repris dans [47]), j'ai prouvé dans [43] que les formes automorphes d'un poids convenable fournissent un plongement de X/G comme sous-espace analytique (fermé) d'un espace projectif. Donc X/G s'identifie à l'espace analytique sous-jacent à une «variété *algébrique* projective». Au même moment, ce résultat était démontré tout autrement par KODAIRA, mais seulement dans le cas où G opère sans point fixe (la variété algébrique étant alors sans singularité). C'est par ma méthode que, plus tard, W. L. BAILY prouva la possibilité de réaliser dans l'espace projectif le compactifié de SATAKE du quotient X/G dans le cas où G est le groupe modulaire de SIEGEL; X/G est alors isomorphe à un ouvert de ZARISKI d'une variété algébrique projective. J'ai moi-même repris la question dans mon Séminaire 1957–58 et prouvé la réalisation projective de X/G non seulement pour le groupe modulaire, mais pour tous les groupes qui lui sont «commensurables».

12) Fibrés holomorphes

Les premières indications relatives à l'utilisation de la théorie des faisceaux pour l'étude des fibrés holomorphes remontent à une conférence que j'ai faite au Séminaire BOURBAKI (décembre 1950). Ma contribution à la théorie a ensuite simplement consisté en une mise au point, au Colloque de Mexico (1956), des théorèmes fondamentaux de GRAUERT sur les espaces fibrés principaux dont la base est une variété de Stein, théorèmes dont la démonstration n'était pas encore publiée mais dont les grandes lignes m'avaient été communiquées par l'auteur. Dans la rédaction [49], j'ai donné des démonstrations complètes.

13) Variétés analytiques réelles ([44], [45], [46])

L'un des buts de [44] était de prouver l'analogue des théorèmes A et B pour les variétés analytiques réelles, dénombrables à l'infini. A cette époque le théorème de plongement de GRAUERT n'était pas encore connu; il a pour conséquence que les théorèmes que j'ai énoncés pour les variétés plongeables sont, en fait, toujours vrais. A partir de là on obtient, par les procédés usuels de passage du local au global, une série de résultats de caractère global; par exemple, une sous-variété analytique fermée d'une variété analytique réelle (dénombrable à l'infini) peut être définie globalement par un nombre fini d'équations analytiques. Toutefois, il est une propriété (d'ailleurs de caractère local) qui différencie le cas réel du cas complexe: le faisceau d'idéaux défini par un sous-ensemble analytique réel n'est pas toujours cohérent, contrairement à ce qui se passe dans le cas complexe; j'en donne des contre-exemples dans [44], et je donne aussi un exemple d'un sous-ensemble analytique A de $\mathbb{R}^3$, de codimension un, tel que toute fonction analytique dans $\mathbb{R}^3$ qui s'annule identiquement sur A soit identiquement nulle. D'autres situations pathologiques sont étudiées dans les Notes [45] et [46], écrites en collaboration avec F. Bruhat.

II. Topologie algébrique

1) Fibrés et groupes d'homotopie

Dans les Notes [89] et [90], en collaboration avec J.-P. Serre, nous introduisons l'opération qui consiste à «tuer» les groupes d'homotopie d'un espace X «par le bas», c'est-à-dire à construire un espace Y et une application f: Y→X de manière que les groupes d'homotopie $\pi_i(Y)$ soient nuls pour $i \leqq n$ (n entier donné), et que $\pi_i(Y) \to \pi_i(X)$ soit un isomorphisme pour $i > n$. L'on peut choisir pour f une application fibrée (en construisant avec Serre des espaces de chemins), et l'on a donc une suite spectrale reliant les homologies de X, de Y et de la fibre. Cette méthode permet le calcul (partiel) des groupes d'homotopie d'un espace à partir de ses groupes d'homologie.

2) Détermination des algèbres d'Eilenberg-MacLane $H_*(\Pi, n)$ ([91], [92], [93])

Rappelons que $K(\Pi,n)$ désigne un espace dont tous les groupes d'homotopie sont nuls, sauf π_n qui est isomorphe à une groupe abélien donné Π. Un tel espace est un espace de Hopf et par suite ses groupes d'homologie forment une algèbre graduée $H_*(\Pi,n)$. Le problème du calcul explicite de ces algèbres avait été posé par Eilenberg et MacLane. Je suis parvenu à ce calcul par des méthodes purement algébriques, basées sur la notion de «construction», et qui permettent un calcul explicite. Les résultats s'énoncent particulièrement bien lorsqu'on prend comme anneau de coefficients le corps $\mathbb{F}_p$ à *p éléments* (*p* premier). Le cas où $p = 2$ et où le groupe Π est cyclique avait été entièrement résolu par J.-P. Serre, par une méthode un peu différente. A l'occasion de ces calculs j'ai été amené à introduire la notion d'algèbre graduée *à puissances divisées;* l'algèbre d'Eilenberg-MacLane possède de telles «puissances divisées». C'est une notion qui s'est avérée utile dans d'autres domaines, et notamment dans la théorie des groupes formels (Dieudonné, Cartier).

3) Suite spectrale d'un espace où opère un groupe discret ([82], [83])

On considère un groupe G opérant sans point fixe, de façon proprement discontinue, dans un espace topologique X. Dans une Note commune, J. Leray et moi avions envisagé le cas où le groupe est fini. J'ai étudié ensuite le cas général, qui a de nombreuses applications. On trouve une exposition de cette question au Chapitre XVI de mon livre «Homological Algebra» écrit en collaboration avec S. Eilenberg.

4) Cohomologie des espaces homogènes de groupes de Lie ([86], [87])

Il s'agit de la cohomologie à coefficients réels d'un espace homogène G/g, G étant un groupe de Lie compact connexe et g un sous-groupe fermé connexe de G. La méthode utilisée est celle de l'«algèbre de Weil» d'une algèbre de Lie. J'obtiens pour la première fois une détermination complète de la cohomologie réelle de G/g; il suffit de connaître la «transgression» dans l'algèbre de Lie de G, et l'homomorphisme I(G)→I(g) (où I(G) désigne l'algèbre des polynômes sur l'algèbre de Lie de G, invariants par le groupe adjoint; de même pour I(g)). Ces résultats ont été ensuite repris par A. BOREL qui les a en partie étendus au cas plus difficile de la cohomologie à coefficients dans $\mathbb{F}_p$. A ce sujet, on peut consulter le rapport de BOREL dans le Bulletin de l'A.M.S. (vol. 61, 1955, p. 397–432).

5) Opérations de STEENROD

La première démonstration de la formule du produit pour les «carrés de Steenrod», improprement appelée «Cartan formula» puisque c'est WU-WEN-TSÜN qui m'avait proposé de prouver cette formule, se trouve donnée dans la Note [85]. Son seul mérite est d'avoir suggéré à STEENROD une démonstration de la formule analogue $\mathcal{P}^k_p(xy) = \sum_{i+j=k} \mathcal{P}^i_p(x)\, \mathcal{P}^j_p(y)$ pour les opérations de Steenrod modulo p (p premier impair). Aujourd'hui on a de meilleures démonstrations de ces relations.

Dans [94], je détermine explicitement les relations multiplicatives existant entre les générateurs St^i_p de l'algèbre de Steenrod pour p premier impair (le cas $p = 2$ avait été traité par J. ADEM; le cas où p est impair a ensuite été traité indépendamment par J. Adem au moyen d'une méthode différente de la mienne).

6) Cohomologie à coefficients dans un faisceau

Cette notion maintenant fondamentale, aussi bien en Topologie qu'en Analyse, avait été introduite par J. LERAY d'une façon relativement compliquée. Dans mon Séminaire de 1950–51 j'en donne la première exposition axiomatique, qui est aujourd'hui adoptée (voir par exemple le livre classique de R. GODEMENT). Cette présentation a permis ultérieurement de faire rentrer la théorie des faisceaux (de groupes abéliens) dans celle des «catégories abéliennes» et de lui appliquer les méthodes de l'Algèbre homologique (foncteurs dérivés, etc. ...). D'autre part, c'est dans le cadre de la cohomologie à valeurs dans un faisceau que j'ai placé le théorème de DE RHAM (relatif au calcul de la cohomologie réelle d'une variété différentiable au moyen des formes différentielles), ainsi que la «dualité» de POINCARÉ des variétés topologiques, triangulables ou non. Ces idées sont devenues courantes; elles ont permis à P. DOLBEAULT d'étudier le complexe de d''-cohomologie d'une variété analytique complexe.

III. Théorie du potentiel ([70], [71], [72], [73], [74], [75], [84])

C'est sous l'influence de M. BRELOT que je me suis intéressé pendant la guerre aux problèmes de la théorie du potentiel (potentiel newtonien et généralisations diverses). J'ai utilisé d'une manière systématique la notion d'*énergie*, en commençant par prouver le théorème suivant: l'espace des distributions positives d'énergie finie, muni de la norme déduite de l'énergie, est *complet*. Ce fut l'occasion d'employer la méthode de projection sur un sous-ensemble convexe et complet (dans un espace fonctionnel). Le théorème précédent suggéra à J. DENY d'introduire en théorie du potentiel les distributions de SCHWARTZ; il prouva que l'espace vectoriel de toutes les distributions d'énergie finie (et plus seulement les distributions positives) est complet.

J'ai aussi introduit la notion de *topologie fine* (la moins fine rendant continues les fonctions surharmoniques), qui s'est avérée utile notamment dans les questions d'effilement à la frontière, et, plus récemment, dans les nouveaux développements axiomatiques de la théorie du potentiel en relation avec les Probabilités.

J'ai donné la première démonstration d'un théorème que désirait BRELOT, et qui se formule ainsi: la limite d'une suite décroissante (ou, plus généralement, d'un ensemble filtrant décroissant) de fonctions surharmoniques, si elle n'est pas identiquement $-\infty$, ne diffère d'une fonction surharmonique que sur un ensemble de capacité extérieure nulle.

Enfin, je crois avoir été le premier à introduire une théorie du potentiel dans les espaces homogènes [71].

IV. Algèbre homologique

Ecrit entre 1950 et 1953, paru seulement en 1956, le livre «Homological Algebra» est dû à une longue collaboration avec Samuel EILENBERG. On y expose pour la première fois une théorie qui englobe diverses théories particulières (homologie des groupes, homologie des algèbres associatives, homologie des algèbres de Lie, syzygies de HILBERT, etc. ...), en les plaçant dans le cadre général des foncteurs additifs et de leurs foncteurs «dérivés». Les foncteurs $\mathrm{Tor}_n(A,B)$ (foncteurs dérivés gauches du produit tensoriel $A \otimes B$) sont introduits dans cet ouvrage, ainsi que les foncteurs $\mathrm{Ext}^n(A,B)$ (foncteurs dérivés droits du foncteur $\mathrm{Hom}(A,B)$). Auparavant, seul le foncteur $\mathrm{Ext}^1(A,B)$ avait été explicitement considéré dans la littérature (Eilenberg-MacLane). On montre notamment le rôle qu'ils jouent dans la «formule de Künneth», qui est pour la première fois énoncée en termes invariants.

Cet ouvrage de 400 pages semble avoir servi de catalyseur: il a été à l'origine de rapides développements tant en Algèbre pure qu'en Géométrie algébrique et en Géométrie analytique. Le terme lui-même d'«algèbre homologique», donné comme titre à notre livre, a fait fortune. Dans ce livre nous avions traité le cas des modules sur un anneau; mais l'exposition avait été conduite de telle sorte qu'elle pouvait immédiatement se transposer à d'autres cas, comme il était d'ailleurs indiqué dans l'Appendice à notre livre écrit par D. BUCHSBAUM. Il devait revenir à GROTHENDIECK d'introduire et d'étudier systématiquement les «catégories abéliennes», ce qui permit aussitôt, par exemple, d'intégrer dans l'Algèbre homologique la théorie de la cohomologie d'un espace à coefficients dans un faisceau de groupes abéliens. C'est aussi GROTHENDIECK qui, à la suite de SERRE, introduisit systématiquement l'Algèbre homologique comme un nouvel outil puissant en Géométrie algébrique et en Géométrie analytique. Faut-il mentionner, à ce sujet, l'immense ouvrage de DIEUDONNÉ et GROTHENDIECK, les fameux E.G.A. (Éléments de Géométrie Algébrique)? Les élèves de GROTHENDIECK (et, pour n'en citer qu'un, Pierre DELIGNE) ont montré tout le parti que l'on peut tirer des méthodes d'Algèbre homologique, non seulement pour explorer de nouveaux domaines, mais aussi pour résoudre des problèmes anciens et justement réputés difficiles.

V. Divers

1) Théorie des filtres

J'ai introduit en 1937 la notion de filtre dans deux Notes aux Comptes Rendus ([61], [62]). Cette notion est devenue d'un usage courant en Topologie générale, ainsi que celle d'ultrafiltre qui lui est liée. Cette dernière intervient aussi dans certaines théories logiques.

2) Théorie de Galois des corps non commutatifs ([79])

La théorie a ensuite été étendue aux anneaux simples, notamment par DIEUDONNÉ.

3) Analyse harmonique

Il s'agit d'un article écrit en collaboration avec R. GODEMENT [80]. C'est l'une des premières présentations «modernes» de la transformation de Fourier dans le cadre général des groupes abéliens localement compacts, sans faire appel à la théorie «classique».

4) Classes de fonctions indéfiniment dérivables ([63] à [68])

J'ai établi par voie élémentaire de nouvelles inégalités entre les dérivées successives d'une fonction d'une variable réelle. Puis, en collaboration avec S. MANDELBROJT, nous les avons appliquées à la solution définitive du problème de l'équivalence de deux classes de fonctions (chacune des classes étant définies par des majorations données des dérivées successives).

5) Extension et simplification d'un théorème de RADO ([40])

J'ai formulé ce théorème de la manière suivante: une fonction continue f qui est holomorphe en tout point z où $f(z) = 0$ est holomorphe aussi aux points où $f(z) = 0$. La démonstration que j'en ai donnée est très simple et basée sur la théorie du potentiel. De là on déduit le théorème de RADO sous sa forme usuelle (i.e.: une fonction holomorphe qui tend vers zéro à la frontière est identiquement nulle, sous des hypothèses convenables relatives à la frontière). De plus, sous la forme où je l'énonce, le théorème s'étend trivialement aux fonctions d'un nombre quelconque de variables, et même aux fonctions dans un ouvert d'un espace de Banach.

VI. Collaboration au Traité de N. BOURBAKI

Pendant vingt ans, de 1935 à 1954, j'ai participé au travail collectif d'élaboration des «Eléments de mathématique» de Nicolas BOURBAKI. Ceci doit être mentionné dans cette Notice, non pour évoquer ma contribution personnelle qu'il est d'ailleurs bien difficile d'évaluer, mais pour dire tout l'enrichissement que j'en ai retiré. Ce travail en commun avec des hommes de caractères très divers, à la forte personnalité, mus par une commune exigence de perfection, m'a beaucoup appris, et je dois à ces amis une grande partie de ma culture mathématique.

105.

Les Séminaires CARTAN

Allocution prononcée à l'occasion du Colloque Analyse et Topologie, Orsay, 17 Juin 1975

Chers amis,

J'ai fait la connaissance de Cartan en 1948, quand j'étais agrégatif à l'École Normale. Il revenait de Harvard, et nous faisait un séminaire d'Analyse Harmonique : transformation de Fourier, dualité, thèse de Godement. Je me souviens avoir été particulièrement impressionné par la facilité avec laquelle il avait démontré le "théorème de décomposition spectrale", épouvantail de nos cours de Sorbonne à l'époque !

Ce séminaire, n'ayant pas été rédigé, n'a pas eu l'influence considérable qu'ont eue les suivants, ceux que l'on appelle les "**Séminaires CARTAN**".

Ce sont ceux-là que je voudrais surtout évoquer :

Le premier Séminaire (48-49) était une introduction à la Topologie Algébrique. On y apprenait ce qu'est une suite exacte, et quelles belles conséquences on peut tirer de l'identité $d^2 = 0$; il y avait aussi des produits tensoriels, la formule de Künneth et l'homologie singulière, basée sur le "Singular Homology Theory" d'Eilenberg. Après ces préliminaires, Cartan est passé à une première version de la théorie des faisceaux, que Leray venait juste de créer (et d'appliquer avec le succès que l'on sait) ; le couronnement en était la dualité de Poincaré, sans hypothèses de triangulation, mais à grands coups de "carapaces". A vrai dire, cela nous passait un peu par-dessus la tête ; je me souviens en effet que, au début de l'année suivante, quand Cartan nous a demandé "Qu'est-ce qu'on fait dans le Séminaire ? ", certains ont suggéré "Si l'on reprenait le sujet de l'an dernier ? ". Cartan a fait semblant de ne pas entendre, et le Séminaire 49-50 a été consacré aux espaces fibrés et aux groupes d'homotopie. Borel, Wu Wen-Tsun y participaient. On s'est beaucoup servi d'un rapport secret de Weil pour Bourbaki sur les espaces fibrés. Quant aux groupes d'homotopie, on ne pouvait guère que les définir, et en donner des exemples — y compris $\pi_{n+2}(S_n)$ que l'on croyait nul à l'époque. A la fin de l'année, Cartan nous a exposé ses propres résultats sur l'homologie réelle des espaces fibrés et des espaces homogènes, l'algèbre de Weil, etc, prolongeant les résultats de Leray d'abord, et aussi de Hirsch, Koszul, Chevalley et Weil.

Topologie encore pour le séminaire 50-51 : cohomologie des groupes et des espaces fibrés, suites spectrales, et surtout théorie des faisceaux nouvelle manière, cette fois essentiellement définitive (c'est elle qui a été reproduite dans le livre de Godement,

Reprinted from Oeuvres, Collected Papers of J. P. Serre (Springer-Verlag, 1986), pp. 235–239.

puis axiomatisée par Grothendieck dans le "Tôhoku"). En même temps —et en dehors du Séminaire— Cartan revenait à ses premières amours, les fonctions de plusieurs variables complexes. (Comment trouvait-il le temps d'y travailler ? Il lui fallait préparer son Séminaire, distribuer les exposés, expliquer au conférencier ce qu'il aurait à dire, le critiquer pendant l'exposé et corriger sa rédaction ensuite . . . Heureusement, il y avait les vacances !). Il avait posé quelques années auparavant deux problèmes "de cohérence" comme il disait (et comme nous continuons à dire). L'un d'eux, celui de la cohérence du faisceau des fonctions analytiques, venait d'être résolu par Oka. Dès qu'il a eu connaissance de la solution d'Oka, Cartan a vu qu'une méthode semblable permettait de résoudre également le deuxième problème (cohérence du faisceau d'idéaux défini par un sous-ensemble analytique), et il a publié le tout au Bulletin de la S.M.F.

Rien d'étonnant, donc, à ce que le Séminaire suivant (51-52) ait été consacré aux fonctions de plusieurs variables complexes, d'autant plus que, grâce au Séminaire précédent, Cartan avait en main tous les outils topologiques nécessaires. Une fois les résultats de base débrouillés, et en particulier le "lemme sur les matrices holomorphes inversibles", il a pu attaquer la théorie de ce qu'il a appelé les "variétés de Stein". Immédiatement, il est apparu que les résultats de son article au Bulletin de la S.M.F. s'énonçaient bien mieux et se démontraient tout aussi bien dans le langage, mis au point l'année précédente, de la théorie des faisceaux. C'est ainsi que sont nés les fameux "théorèmes A et B" — terminologie peu suggestive, mais devenue classique.

En 52-53, retour à la Topologie : groupes d'homotopie. Toutefois, ce Séminaire, à la différence des quatorze autres, n'a pas été rédigé.

En 53-54, nouveau grand Séminaire sur les fonctions de plusieurs variables complexes. Et d'abord, un travail de "fondation" : Cartan a l'idée de définir la structure d'espace analytique (éventuellement à singularités) par un faisceau, le faisceau des fonctions holomorphes. Cette idée a eu un tel succès, elle a été transposée à tant de situations, qu'elle nous paraît maintenant naturelle, presque banale (bientôt, l'Enseignement Secondaire (*) fera réciter "Qu'est-ce qu'une fonction ? C'est une section du faisceau des germes de fonctions . . ."). A l'époque c'était une idée tout à fait originale, et qui a été immédiatement mise à profit dans la suite même du Séminaire : d'abord pour exposer (d'après Oka) la construction du "normalisé" d'un espace analytique, puis pour définir le quotient d'un espace analytique par un groupe discret opérant proprement, mais pouvant avoir des points fixes ; lorsque l'espace analytique est un domaine symétrique, et que son quotient par le groupe discret est compact, Cartan démontre que ce quotient est algébrique : plus précisément, on peut le plonger, au moyen de fonctions automorphes, dans un espace projectif.

(*) Primaire ! Primaire ! (interruption de J. Dieudonné)

C'est à la même époque (1953) que Cartan et Eilenberg achèvent leur "Homological Algebra" ; pieusement conservé dans les tiroirs de l'éditeur, l'ouvrage voit le jour en 1956. La nouveauté du sujet, ses nombreuses possibilités d'applications, en font un classique – tout comme (pour des sujets différents) les "Opérations Linéaires" de Banach, les "Variétés Abéliennes" de Weil ou les "Distributions" de Schwartz.

C'est aussi d'Algèbre Homologique que traite le Séminaire suivant (54-55), ou plutôt de ce que l'on appelle, avec Moore, "Differential Homological Algebra". Cartan y expose sa théorie des "constructions", résumée auparavant en deux Notes aux Proc. Nat. Acad. Sci. U.S.A. ; cette théorie permet le calcul de la cohomologie des complexes d'Eilenberg-MacLane, et du même coup la détermination de toutes les opérations cohomologiques primaires, ainsi que de leurs relations (le cas particulier des puissances de Steenrod est développé par Cartan dans un travail paru la même année aux Comm. Math. Helv.). Ce sont là des résultats dont il avait eu l'idée dès 1950-51, mais qu'il s'était abstenu de mettre au point et de publier pour ne pas gêner l'un de ses élèves qui préparait alors une thèse sur les espaces de lacets, les groupes d'homotopie et les complexes d'Eilenberg-MacLane. Tel est du moins le sentiment de l'élève en question . . .

Changement de direction avec le Séminaire suivant (55-56) : Géométrie Algébrique, en collaboration avec Chevalley. Il s'agissait surtout de mettre au point les techniques de base : anneaux locaux, points simples, etc. Le cadre adopté n'est plus guère utilisé à l'heure actuelle : inconvénient mineur, le changement de cadre étant justement un excellent exercice mathématique.

Après un bref Séminaire de Topologie (56-57), arrive un grand Séminaire (57-58) sur les fonctions de plusieurs variables complexes et les fonctions automorphes ; Weil, Godement, Satake et Shimura y participent. J'hésite à parler de ce Séminaire, que je connais mal. Je sais seulement qu'il contient nombre de résultats originaux, souvent cités, et qui n'ont pas été reproduits ailleurs (ceux de Godement, notamment). La dernière partie du Séminaire, par Satake et Cartan, traite d'un sujet d'un grand intérêt pour les arithméticiens : la compactification des quotients de domaines symétriques par des groupes discrets tels que le groupe modulaire et ses généralisations. En dimension 1, pour les groupes fuchsiens, c'est la classique "adjonction des pointes". Le cas général est plus difficile ; tout d'abord, que doit-on ajouter à l'infini ? Pour les géomètres du siècle dernier, il semble que chaque espace possédait une compactification "naturelle", au point qu'ils se sont parfois disputés pour savoir laquelle était "la vraie" (ainsi, en géométrie élémentaire, on peut adjoindre au plan affine, soit une droite à l'infini, soit un point à l'infini, suivant qu'on s'intéresse à la géométrie projective, ou à la géométrie conforme). La théorie des espaces analytiques de Cartan clarifie la question : elle permet de dire ce qu'est **une** compactification. Encore faut-il prouver qu'il y en a, et de raisonnables ; c'est ce que font Cartan et Satake. D'autres compactifications, plus "grosses" mais plus "lisses", ont été ensuite construites par Igusa, Hirzebruch, Mumford, . . . ; le sujet est en pleine activité en ce moment.

Viennent ensuite deux Séminaires de Topologie (58-59) et (59-60), le dernier en collaboration avec J.C. Moore. Chacun est consacré à la démonstration d'un résultat : celui de 58-59, au théorème d'Adams suivant lequel il n'y a pas d'application entre sphères d'invariant de Hopf 1 (en dehors de celles que connaissait Hopf) ; celui de 59-60, au théorème de périodicité de Bott. Le théorème d'Adams est démontré, comme l'avait fait Adams lui-même, au moyen de constructions cohomologiques secondaires ; on n'en connaissait pas alors la démonstration si rapide et si élégante qui utilise la K-théorie (et repose donc, en définitive, sur le sujet du Séminaire suivant !). Quant à la périodicité de Bott, Cartan et Moore la démontrent sans théorie de Morse, uniquement par voie homologique ; un beau tour de force !

Retour à l'analytique complexe dans le Séminaire suivant (60-61). Le point de départ était la théorie des déformations de Kodaira-Spencer, et son application à l'espace des modules des surfaces de Riemann compactes, autrement dit à l'espace de Teichmüller. Au bout d'un certain nombre d'exposés, le Séminaire a été pris en charge par Grothendieck, et nous avons eu droit à un exposé systématique, d'ailleurs fort intéressant, des fondements de la "Géométrie Analytique" d'un point de vue algébrique. (Il y aurait beaucoup à dire sur les échanges entre "analytique" et "algébrique" à cette époque. C'est du côté analytique que vient la théorie des faisceaux cohérents, créée par Cartan, et transposée ensuite au cadre algébrique. Inversement, les propriétés purement algébriques des anneaux locaux (régularité, profondeur, etc) ont été utilisées avec succès pour simplifier certains résultats délicats de Géométrie Analytique, tel le théorème de normalisation dont je parlais tout à l'heure).

Les Séminaires suivants (61-62 et 62-63) sont consacrés à des questions diverses de Topologie Différentielle : théorie de Smale (sans oublier les variétés "à coins" et l'arrondissement de ceux-ci), travaux de Cerf sur π_0(Diff S_3), et théorème de préparation différentiable de Malgrange.

La formule de l'index d'Atiyah-Singer est le sujet du dernier Séminaire (63-64), en collaboration avec Schwartz. C'est vraiment le Séminaire qui correspond le mieux au titre de ce Colloque "Analyse et Topologie", vu qu'il s'agit de démontrer que A = B, et que A, c'est la Topologie qui le définit, et que B, c'est l'Analyse !

Telle est la liste des "Séminaires CARTAN". Le bref résumé que je viens d'en faire ne peut pas donner une idée de l'influence qu'ils ont eue sur l'Analyse et la Topologie, tant en France qu'à l'étranger ; heureusement, certaines conférences du Colloque s'en chargeront. Également grande a été l'influence que Cartan a exercée par (et à travers) ses élèves, qui furent nombreux ; parmi ces "thésitifs" figurent : Deny, Godement, Koszul, Thom, moi-même, Dolbeault, Cerf, Shih, Douady, Morin, Karoubi, . . . et j'en oublie. CARTAN s'en occupait comme il sait le faire – orthographe comprise !

L'influence d'un homme ne tient pas seulement à ce qu'il fait, mais aussi à la façon dont il le fait, à son style. Je crois que le style de Cartan est ce qu'on peut trouver de mieux en mathématiques.

41.

Variétés analytiques complexes et cohomologie

Colloque sur les fonctions de plusieurs variables, Bruxelles 41–55 (1953)

La théorie globale des idéaux de fonctions analytiques, due à K. Oka [**10**] et H. Cartan [**2**, **3**, **4**], vaut non seulement pour les domaines d'holomorphie, mais pour une classe plus vaste de variétés analytiques complexes introduite par K. Stein [**11**], et qui comprend notamment toutes les sous-variétés analytiques sans singularité, de dimension quelconque p, de l'espace numérique complexe de dimension quelconque $n > p$.

Les théorèmes fondamentaux de cette théorie se formulent bien dans le langage de la cohomologie, qui suggère des généralisations et fournit un outil commode en vue de l'exploitation des résultats. Dans cette conférence, nous exposerons d'abord les notions de base : celle de *faisceau*, et celle de *cohomologie* à coefficients dans un faisceau. Puis nous énoncerons les théorèmes fondamentaux. Enfin, nous ferons des applications à des problèmes globaux concernant les variétés de Stein; d'autres applications seront données dans la conférence de J.-P. Serre.

1. Faisceaux sur un espace topologique [1]

Soit X un espace topologique. Un *faisceau de groupes abéliens* sur X, ou simplement *faisceau*, est défini par la donnée :

[1] La notion de *faisceau* a été introduite par J. Leray à l'occasion de l'étude des propriétés homologiques d'une application continue. Voir J. Leray, *Journ. de Math. pures et appliquées*, **29**, 1950, pp. 1-139; c'est dans cet ouvrage que l'on trouve (bas de la page 75) une définition de la cohomologie à coefficients dans un faisceau, limitée à vrai dire au cas d'un espace X localement compact (et il s'agissait de la cohomologie « à supports compacts »). La définition des faisceaux adoptée ici est un peu différente; elle est due à Lazard et a été exposée dans mon *Séminaire polycopié* de l'E. N. S. 1950-1951, où la théorie de la cohomologie à coefficients dans un faisceau a été développée (exposés XIV à XX).

Reprinted from Oeuvres, Collected Works (Springer-Verlag, 1979), Vol. II, pp. 669–683.

1° D'une fonction $x \longrightarrow \mathcal{F}_x$ qui, à chaque point $x \in X$, associe un groupe abélien $\mathcal{F}_x$ (qu'on notera additivement);

2° D'une topologie (non nécessairement séparée) dans la réunion $\mathcal{F}$ des ensembles $\mathcal{F}_x$.

Avant de formuler les axiomes auxquels ces données sont astreintes, notons p l'application de $\mathcal{F}$ sur X qui, à chaque $\alpha \in \mathcal{F}$, associe le point x tel que $\alpha \in \mathcal{F}_x$. On pose les deux axiomes :

(F_I) l'application $\alpha \longrightarrow -\alpha$ qui, à chaque $\alpha \in \mathcal{F}$, associe l'opposé de α dans le groupe $\mathcal{F}_{p(\alpha)}$, est une application continue de $\mathcal{F}$ dans $\mathcal{F}$. L'application $(\alpha, \beta) \longrightarrow \alpha + \beta$, définie sur l'ensemble $\mathcal{G}$ des couples (α, β) tels que $p(\alpha) = p(\beta)$, et qui associe à un tel couple la somme $\alpha + \beta$ dans le groupe $\mathcal{F}_{p(\alpha)}$, est une application continue de la partie $\mathcal{G}$ de $\mathcal{F} \times \mathcal{F}$ dans $\mathcal{F}$.

(F_{II}) l'application p est un *homéomorphisme local*, i.e.: tout élément $\alpha \in \mathcal{F}$ possède un voisinage ouvert V tel que la restriction de p à V soit un homéomorphisme de V sur un ouvert de X.

Si U est une partie de X, la collection des $\mathcal{F}_x$ pour $x \in U$, munie de la topologie induite par celle de $\mathcal{F}$, est évidemment un faisceau sur U ; on le notera $\mathcal{F}(U)$, et on l'appellera le faisceau *induit* par $\mathcal{F}$ sur U.

Exemple. — Soit G un groupe abélien. Soit $\mathcal{F}$ le produit $G \times X$, muni de la topologie-produit (G étant muni de la topologie discrète). Le sous-ensemble $\mathcal{F}_x$ des couples (g, x), où g parcourt G, est évidemment muni d'une structure de groupe abélien (isomorphe à G). On vérifie aussitôt les axiomes (F_I) et (F_{II}). Ce faisceau s'appelle le « faisceau constant » défini par G, et se note aussi G.

Soit $\mathcal{F}$ un faisceau quelconque sur X. On appelle *section* de $\mathcal{F}$ au-dessus d'un ouvert $U \subset X$ une application *continue* $s : U \longrightarrow \mathcal{F}$ telle que $p \circ s$ soit l'identité ; s est alors un homéomorphisme de U sur son image $s(U)$. L'axiome (F_{II}) implique ceci : si deux sections sont égales en un point $x \in U$, elles sont égales en tous points d'un voisinage de x. L'application s qui, à chaque $x \in U$, associe l'élément neutre $0_x \in \mathcal{F}_x$, est une section [en vertu de (F_I)] ; on l'appelle la section nulle. L'ensemble $\Gamma(U, \mathcal{F})$ des sections de $\mathcal{F}$ au-dessus de U est muni d'une structure de groupe abélien, grâce à (F_I), et la section nulle est l'élément neutre de ce groupe.

Si U et V sont deux ouverts tels que $V \subset U$, toute section au-dessus de U induit une section au-dessus de V. Le groupe $\mathcal{F}_x$ est la limite inductive des groupes $\Gamma(U, \mathcal{F})$ relatifs aux voisinages ouverts U de x.

Dans la pratique, un faisceau sur X est souvent défini de la manière suivante : on se donne des groupes abéliens $\mathcal{F}_U$ attachés à certains ouverts $U \subset X$, formant un système fondamental d'ouverts de la topologie de X ; et, pour tout couple (U, V) tel que $V \subset U$, on se donne un homomorphisme $f_{VU} : \mathcal{F}_U \longrightarrow \mathcal{F}_V$, de manière que, pour $W \subset V \subset U$, on ait $f_{WU} = f_{WV} \circ f_{VU}$. On prend alors pour $\mathcal{F}_x$ la limite inductive des $\mathcal{F}_U$ pour les ouverts U contenant x ; et, sur la réunion $\mathcal{F}$ des $\mathcal{F}_x$, on définit la topologie $\mathcal{C}$ que voici : pour tout ouvert U et tout $\alpha \in \mathcal{F}_U$, soit $[\alpha]$ l'ensemble des images de α dans les $\mathcal{F}_x$ associés aux points $x \in U$; par définition, les sous-ensembles $[\alpha]$ de $\mathcal{F}$ constituent un système fondamental d'ouverts de la topologie $\mathcal{C}$. Les axiomes (F_I) et (F_{II}) sont satisfaits. On a un homomorphisme évident : $\mathcal{F}_U \longrightarrow \Gamma(U, \mathcal{F})$, mais ce n'est pas nécessairement un isomorphisme. Deux modes de définition distincts peuvent ainsi définir un même faisceau. Pour que l'homomorphisme $\mathcal{F}_U \longrightarrow \Gamma(U, \mathcal{F})$ soit un *isomorphisme*, il faut et il suffit que, pour tout système d'ouverts U_i de réunion U, et tout système d'éléments $\alpha_i \in \mathcal{F}_{U_i}$ tels que α_i et α_j aient même image dans $\mathcal{F}_{U_i \cap U_j}$, il existe un $\alpha \in \mathcal{F}_U$ et un seul, tel que $f_{U_iU}(\alpha) = \alpha_i$ pour tout i.

Exemple. — Soit $\mathcal{F}_U$ le groupe additif des fonctions numériques réelles (resp. complexes) définies et continues dans U. Si $V \subset U$, l'homomorphisme $\mathcal{F}_U \longrightarrow \mathcal{F}_V$ sera celui qui associe à une fonction définie sur U sa restriction à V. Alors $\mathcal{F}_x$ est le groupe additif des *germes de fonctions continues* au point x; et $\mathcal{F}$ s'appelle le faisceau des germes de fonctions continues réelles (resp. complexes). $\mathcal{F}_U$ est isomorphe au groupe des sections de $\mathcal{F}$ au-dessus de U.

Remarque. — On a défini des faisceaux de groupes abéliens; mais il est clair que des définitions analogues peuvent être données pour n'importe quelle structure algébrique.

2. Sous-faisceau, homomorphisme, faisceau-quotient

Soit $\mathcal{F}$ un faisceau sur X. Soit $\mathcal{G}$ un sous-ensemble de $\mathcal{F}$, tel que, pour tout $x \in X$, $\mathcal{G} \cap \mathcal{F}_x = \mathcal{G}_x$ soit un *sous-groupe* de $\mathcal{F}_x$. Pour que $\mathcal{G}$, pour la topologie induite par celle de $\mathcal{F}$, soit un faisceau, il faut et il suffit que $\mathcal{G}$ soit *ouvert* dans $\mathcal{F}$ [voir la condition (F_{II})]. On dit alors que $\mathcal{G}$ est un *sous-faisceau* de $\mathcal{F}$.

Soient $\mathcal{F}$ et $\mathcal{F}'$ deux faisceaux sur un même espace X. On appelle homomorphisme de $\mathcal{F}$ dans $\mathcal{F}'$ une application *continue* f de $\mathcal{F}$ dans $\mathcal{F}'$ telle que la restriction f_x de f à $\mathcal{F}_x$ soit un homomorphisme du groupe $\mathcal{F}_x$ dans le groupe $\mathcal{F}_x'$. L'image réciproque de la section nulle de $\mathcal{F}'$ (section qui est un ouvert de $\mathcal{F}'$) est un sous-faisceau $\mathcal{G}$ de $\mathcal{F}$, appelé *noyau* de l'homomorphisme f; pour chaque $x \in X$, $\mathcal{G}_x$ est le noyau de f_x. D'autre part, f est une application ouverte; donc l'image de $\mathcal{F}$ dans $\mathcal{F}'$ est un sous-faisceau $\mathcal{G}'$ de $\mathcal{F}'$, appelé l'*image* de l'homomorphisme f; $\mathcal{G}_x'$ est l'image de f_x.

Il est clair que tout homomorphisme $f: \mathcal{F} \longrightarrow \mathcal{F}'$ définit, pour chaque ouvert $U \subset X$, un homomorphisme des groupes de sections $\Gamma(U, \mathcal{F}) \longrightarrow \Gamma(U, \mathcal{F}')$.

Soit $\mathcal{G}$ un sous-faisceau d'un faisceau $\mathcal{F}$. Définissons un *faisceau-quotient* comme suit : soit $\mathcal{H}_x$ le groupe quotient $\mathcal{F}_x/\mathcal{G}_x$. Si $\mathcal{H}$ désigne la réunion des $\mathcal{H}_x$, les applications $\mathcal{F}_x \longrightarrow \mathcal{H}_x$ définissent une application de $\mathcal{F}$ sur $\mathcal{H}$, qui identifie $\mathcal{H}$ à un quotient de $\mathcal{F}$. On munit $\mathcal{H}$ de la topologie quotient, ce qui définit $\mathcal{H}$ comme faisceau. Ce faisceau est noté $\mathcal{F}/\mathcal{G}$. L'application $\mathcal{F} \longrightarrow \mathcal{F}/\mathcal{G}$ est un homomorphisme, dont le noyau est le faisceau $\mathcal{G}$. Cherchons les sections du faisceau-quotient $\mathcal{F}/\mathcal{G}$ au-dessus de X : si $s \in \Gamma(X, \mathcal{F}/\mathcal{G})$, tout point $x \in X$ possède un voisinage ouvert U tel que la section induite par s dans U soit l'image d'un élément de $\Gamma(U, \mathcal{F})$. Ainsi X peut être recouvert par des ouverts U_i, et dans chaque U_i on a un élément $s_i \in \Gamma(U_i, \mathcal{F})$ de manière que, dans $U_i \cap U_j$, $s_i - s_j$ soit une section du sous-faisceau $\mathcal{G}$. En général, une section de $\mathcal{F}/\mathcal{G}$ au-dessus de X n'est l'image d'aucune section de $\mathcal{F}$ au-dessus de X. Ainsi, la suite de groupes et d'homomorphismes

$$0 \to \Gamma(X, \mathcal{G}) \to \Gamma(X, \mathcal{F}) \xrightarrow{\varphi} \Gamma(X, \mathcal{F}/\mathcal{G})$$

est évidemment une *suite exacte* (i.e.: l'image de chaque homomorphisme est le noyau de l'homomorphisme suivant), mais l'homomorphisme φ n'est pas un épimorphisme [2], en général.

3. Cohomologie à coefficients dans un faisceau

Soit $\mathcal{F}$ un faisceau sur un espace topologique X. On va définir, pour tout entier $q \geqslant 0$, un groupe de cohomologie $H^q(X, \mathcal{F})$. Pour tout recouvrement $\mathcal{R}$ de X par des ouverts

[2] Un homomorphisme $\varphi: A \longrightarrow B$ de groupes abéliens (ou plus généralement de modules) s'appelle un *épimorphisme* si φ *applique* A *sur* B. Rappelons aussi la définition du *conoyau* de φ : c'est le quotient de B par l'image $\varphi(A)$. Définitions analogues pour les homomorphismes de faisceaux.

U_i, considérons ce qui suit : pour tout entier $q \geqslant 0$, nous posons $q+1=p$, et nous associons à chaque suite de p indices $i_1, \ldots, i_p$ un élément $f_{i_1 \ldots i_p} \in \Gamma(U_{i_1} \cap \ldots \cap U_{i_p}, \mathcal{F})$, zéro si $U_{i_1} \cap \ldots \cap U_{i_p}$ est vide; et l'on suppose que $f_{i_1 \ldots i_p}$ est une fonction *alternée* des indices (en particulier, est nulle si les indices ne sont pas tous distincts). On obtient ainsi le groupe additif des « cochaînes alternées » du recouvrement $\mathcal{R}$, de degré $q=p-1$, relatives au faisceau $\mathcal{F}$. On définit dans ce groupe un opérateur « cobord » à la manière habituelle; il augmente le degré de 1. D'où un groupe de cohomologie $H^q(\mathcal{R}, \mathcal{F})$. Si un recouvrement $\mathcal{R}'$ est plus fin que $\mathcal{R}$, on a un homomorphisme naturel, unique, de $H^q(\mathcal{R}, \mathcal{F})$ dans $H^q(\mathcal{R}', \mathcal{F})$. On peut donc considérer la limite inductive des $H^q(\mathcal{R}, \mathcal{F})$ relatifs à tous les recouvrements ouverts $\mathcal{R}$; c'est par définition, le groupe $H^q(X, \mathcal{F})$. On notera que $H^0(X, \mathcal{F})$ s'identifie canoniquement au groupe des sections $\Gamma(X, \mathcal{F})$.

Tout homomorphisme de faisceaux $\mathcal{F} \longrightarrow \mathcal{F}'$ définit évidemment des homomorphismes $H^q(X, \mathcal{F}) \longrightarrow H^q(X, \mathcal{F}')$. De plus, soit $\mathcal{G}$ un sous-faisceau de $\mathcal{F}$; on peut définir des homomorphismes naturels

$$\delta^q : H^q(X, \mathcal{F}/\mathcal{G}) \longrightarrow H^{q+1}(H, \mathcal{G}),$$

tout au moins lorsque X est paracompact.

Donnons par exemple la définition de δ^0 : on a vu (§ 2) qu'un élément $\alpha \in H^0(X, \mathcal{F}/\mathcal{G})$ peut être défini par des sections $s_i \in H^0(U_i, \mathcal{F})$ au-dessus des ouverts U_i d'un recouvrement convenable $\mathcal{R}$ de X ; et que $s_i - s_j \in H^0(U_i \cap U_j, \mathcal{G})$. Posons alors $s_i - s_j = f_{ij}$. Les f_{ij} définissent un cocycle (alterné) de degré 1, donc un élément de $H^1(\mathcal{R}, \mathcal{G})$, donc un élément $\beta \in H^1(X, \mathcal{G})$. On vérifie facilement que cet élément β est univoquement déterminé par α, c'est-à-dire est indépendant du choix du recouvrement $\mathcal{R}$ et des sections s_i. Par définition, $\delta^0(\alpha) = \beta$. Pour $q > 0$, la définition de δ^q est analogue, mais un peu plus compliquée.

La propriété fondamentale de la cohomologie est la suivante (³): l'espace X étant supposé paracompact, si $\mathcal{G}$ est un sous-faisceau d'un faisceau $\mathcal{F}$, *la suite illimitée de groupes abéliens et d'homomorphismes*

$$0 \to H^0(X, \mathcal{G}) \to H^0(X, \mathcal{F}) \to H^0(X, \mathcal{F}/\mathcal{G}) \xrightarrow{\delta^0} H^1(X, \mathcal{G}) \to \ldots$$

$$\to H^q(X, \mathcal{G}) \to H^q(X, \mathcal{F}) \to H^q(X, \mathcal{F}/\mathcal{G}) \xrightarrow{\delta^q} H^{q+1}(X, \mathcal{G}) \to \ldots$$

est une suite exacte.

(³) Voir mon *Séminaire* 1950-1951 cité en note 1, où la propriété de « suite exacte » était posée comme l'un des axiomes d'une théorie axiomatique de la cohomologie à coefficients dans un faisceau.

Le début de cette suite coïncide avec la suite

$$0 \longrightarrow \Gamma(X, \mathcal{G}) \longrightarrow \Gamma(X, \mathcal{F}) \longrightarrow \Gamma(X, \mathcal{F}/\mathcal{G})$$

considérée à la fin du § 2. Il en résulte ceci : pour que $\Gamma(X, \mathcal{F}) \longrightarrow \Gamma(X, \mathcal{F}/\mathcal{G})$ soit un épimorphisme [2], il suffit que $H^1(X, \mathcal{G})=0$.

4. Variétés analytiques complexes ; problème additif de Cousin

La définition bien connue d'une variété analytique-complexe de dimension (complexe) n, (donc de dimension réelle $2n$), peut se formuler en termes de faisceaux, comme suit : sur l'espace X, supposé séparé, on se donne un sous-faisceau $\mathcal{O}$ du faisceau $\mathcal{F}$ des germes de fonctions continues complexes (§ 1), et on lui impose l'axiome suivant :

(VA) pour chaque point $x \in X$, il existe un ouvert U contenant x, et n sections f_i de $\mathcal{O}$ au-dessus de U, nulles au points x, telles que : 1° les f_i définissent un homéomorphisme de U sur un ouvert de l'espace numérique complexe C^n, de dimension (complexe) n ; 2° les éléments de $\mathcal{O}_x$ soient exactement les fonctions composées $F(f_1, \ldots, f_n)$, où F est holomorphe à l'origine (dans C^n).

Les systèmes de n sections f_i jouissant de ces propriétés s'appellent systèmes de *coordonnées locales* au point x. Le faisceau $\mathcal{O}$ s'appelle le faisceau des *germes de fonctions holomorphes* ; les sections de $\mathcal{O}$ au-dessus d'un ouvert U sont les *fonctions holomorphes* dans U. On note $\mathcal{O}(U)$ le faisceau induit sur un ouvert U. On observe que $\mathcal{O}_x$ est un anneau d'intégrité.

Sur une variété analytique-complexe X, définissons le faisceau $\mathcal{M}$ des *germes de fonctions méromorphes* : $\mathcal{M}_x$ est défini comme le corps des fractions de l'anneau d'intégrité $\mathcal{O}_x$; sur la réunion $\mathcal{M}$ des $\mathcal{M}_x$, on définit la topologie suivante : pour tout couple de fonctions holomorphes f, g dans un ouvert U, tel que g n'induise la constante 0 dans aucun sous-ensemble ouvert non vide de U, soit $[f, g]$ l'ensemble des $f_x/g_x \in \mathcal{M}_x$ pour $x \in U$ (en notant f_x, resp. g_x, l'image de f, resp. g, dans $\mathcal{O}_x$); les ensembles $[f, g]$ constituent, par définition, une famille fondamentale d'ouverts de la topologie de $\mathcal{M}$. Une « fonction méromorphe » dans un ouvert U est, par définition, une section de $\mathcal{M}$ au-dessus de U.

Il est clair que $\mathcal{O}$ est un sous-faisceau de $\mathcal{M}$. Interprétons le faisceau-quotient $\mathcal{M}/\mathcal{O}$: si $m \in \mathcal{M}_x$, la classe de m dans $\mathcal{M}_x/\mathcal{O}_x$ s'appelle la *partie principale* de m. Une section de

$\mathcal{M}/\mathcal{O}$ s'appelle un *système de parties principales*. Considérons l'homomorphisme $\varphi : \Gamma(X, \mathcal{M}) \longrightarrow \Gamma(X, \mathcal{M}/\mathcal{O})$; à chaque fonction méromorphe dans X, φ associe un système de parties principales. Le classique *problème additif de Cousin* (ou premier problème de Cousin) ([4]) consiste à caractériser, parmi les systèmes de parties principales dans X, ceux qui proviennent d'une fonction méromorphe dans X ; autrement dit, à caractériser l'image de l'homomorphisme φ. Dire que le problème de Cousin est toujours résoluble, c'est dire que φ est un épimorphisme.

En vertu de la suite exacte de cohomologie (§ 3), la condition $H^1(X, \mathcal{O}) = 0$ est *suffisante* pour que le problème additif de Cousin soit toujours résoluble. Nous verrons qu'il en est notamment ainsi lorsque X est une « variété de Stein » (§ 7, théorème B). Avant d'énoncer des théorèmes généraux, affirmant que certains groupes de cohomologie $H^q(X, \mathcal{F})$ sont nuls dans certaines conditions, il nous faut définir une nouvelle notion : celle de « faisceau analytique cohérent ».

5. Faisceaux analytiques cohérents

Soit X une variété analytique-complexe. Un *faisceau analytique* est un faisceau $\mathcal{F}$ tel que, pour chaque point x, $\mathcal{F}_x$ soit muni d'une structure de *module* sur l'anneau $\mathcal{O}_x$ (anneau des germes de fonctions holomorphes au point x), et cela de manière que l'application $(f, \alpha) \longrightarrow f\alpha$, définie sur l'ensemble $\mathcal{G}$ des couples (f, α) tels qu'il existe $x \in X$ avec $f \in \mathcal{O}_x$ et $\alpha \in \mathcal{F}_x$, soit une application *continue* de $\mathcal{G} \subset \mathcal{O} \times \mathcal{F}$ dans $\mathcal{F}$.

Exemple. — Soit $\mathcal{O}^p$ la somme directe de p exemplaires du faisceau $\mathcal{O}$; un élément de $\mathcal{O}_x^p$ est une suite $(f_1, \ldots, f_p)$ de p germes de fonctions holomorphes au point x. Faisons opérer $\mathcal{O}_x$ dans $\mathcal{O}_x^p$ par la formule

$$f \cdot (f_1, \ldots, f_p) = (ff_1, \ldots, ff_p) ;$$

ceci définit $\mathcal{O}^p$ comme faisceau analytique.

Autre exemple. — Soit un sous-faisceau $\mathcal{J}$ de $\mathcal{O}$ tel que, pour chaque point x, $\mathcal{J}_x$ soit un *idéal* de l'anneau $\mathcal{O}_x$; alors $\mathcal{J}$ est un faisceau analytique.

Soient $\mathcal{F}$ et $\mathcal{F}'$ deux faisceaux analytiques sur X. Un homomorphisme de faisceaux $f : \mathcal{F} \longrightarrow \mathcal{F}'$ est dit *analytique* si, pour chaque point x, l'homomorphisme $f_x : \mathcal{F}_x \longrightarrow \mathcal{F}'_x$ est

([4]) Cf. [7]. Les problèmes de Cousin ont donné lieu à une abondante littérature; bornons-nous à renvoyer à [9].

compatible avec les opérations de $\mathcal{O}_x$. Le noyau, l'image et le conoyau de f sont alors des faisceaux analytiques.

Définition. — On dit qu'un faisceau analytique $\mathcal{F}$ sur X est *cohérent* [5] si chaque $x \in$ X possède un voisinage ouvert U tel que le faisceau analytique induit $\mathcal{F}$(U) soit isomorphe au conoyau d'un homomorphisme analytique $f: \mathcal{O}^p(U) \longrightarrow \mathcal{O}^q(U)$ (p et q entiers).

En particulier, nous dirons qu'un faisceau analytique est *localement libre* si chaque $x \in$ X possède un voisinage ouvert U tel que le faisceau induit $\mathcal{F}$(U) soit isomorphe à $\mathcal{O}^q(U)$ pour un entier q convenable. Alors, tout faisceau localement libre est cohérent.

On démontre ceci : *soient $\mathcal{F}$ et $\mathcal{F}'$ deux faisceaux analytiques cohérents, et f un homomorphisme analytique de $\mathcal{F}$ dans $\mathcal{F}'$. Alors le noyau, l'image et le conoyau de f sont des faisceaux cohérents.* La démonstration repose essentiellement sur le théorème d'Oka [6], qui affirme ceci : pour tout homomorphisme analytique $f: \mathcal{O}^p(X) \longrightarrow \mathcal{O}^q(X)$, chaque point $x \in$ X possède un voisinage ouvert U tel que le noyau de l'homomorphisme induit $\mathcal{O}^p(U) \longrightarrow \mathcal{O}^q(U)$ soit l'image d'un homomorphisme analytique $\mathcal{O}^r(U) \longrightarrow \mathcal{O}^p(U)$.

On notera une condition nécessaire et suffisante pour qu'un sous-faisceau analytique $\mathcal{G}$ d'un faisceau analytique cohérent $\mathcal{F}$ soit cohérent : c'est que, pour chaque $x \in$ X, il existe un voisinage ouvert U de x et un nombre fini de sections de $\mathcal{F}$ au-dessus de U, telles que, pour tout $y \in$ U, $\mathcal{G}_y$ soit le sous-$\mathcal{O}_y$-module de $\mathcal{F}_y$ engendré par ces sections. Ce critère s'applique notamment lorsque $\mathcal{F} = \mathcal{O}^q$, et plus particulièrement lorsque $\mathcal{F} = \mathcal{O}$; dans ce dernier cas, $\mathcal{G}_x$ est un idéal de $\mathcal{O}_x$, et l'on a un *faisceau cohérent d'idéaux.*

Exemple de faisceau cohérent d'idéaux. — Soit X une variété analytique-complexe. On appelle *sous-variété analytique* de X tout sous-ensemble fermé V de X tel que, pour chaque $x \in$ V, il existe un voisinage ouvert U de x et un nombre fini de f_i holomorphes dans U, telles que $y \in V \cap U$ soit équivalent à « $y \in$ U et $f_i(y) = 0$ ». Un point $x \in$ V est dit *régulier* s'il existe au voisinage de x, dans la variété ambiante X, un système de coordonnées locales $x_1, \ldots, x_n$ nulles au point x, et telles que V soit localement définie (au voisinage

(5) Cette définition est plus générale que celle donnée dans [3] et [4]; dans les cas plus particuliers envisagés dans [3] et [4], les définitions coïncident.

(6) Voir [10], pp. 18 et suiv.; voir aussi [4], p. 37, théorème 1; et [5], exposé XV, pp. 5-10.

de x) par l'annulation de certaines de ces coordonnées. Quand tous les points de V sont réguliers, on dit que V est *régulièrement plongée* dans X. Ces définitions étant posées, soit V une sous-variété analytique de X ; elle définit un faisceau d'idéaux sur X comme suit : en un point $x \in V$, on prend l'idéal $\mathcal{J}_x$ de $\mathcal{O}_x$ formé des germes qui s'annulent identiquement sur V au voisinage de x, et en un point $x \notin V$, on prend $\mathcal{J}_x = \mathcal{O}_x$. On démontre ([7]) que ce faisceau $\mathcal{J}$ (appelé le faisceau de la sous-variété V) est *cohérent* ; c'est évident lorsque V est régulièrement plongée, mais c'est vrai dans tous les cas.

6. Variétés de Stein

Une variété de Stein est, en gros, une variété analytique-complexe sur laquelle il y a suffisamment de fonctions holomorphes. D'une façon précise, c'est une variété analytique-complexe X (connexe ou non), réunion dénombrable de compacts, qui satisfait aux trois conditions suivantes :

(a) Si $x \in X$, $y \in X$ et $x \neq y$, il existe une fonction f holomorphe dans X, telle que $f(x) \neq f(y)$;

(b) Pour tout $x \in X$, il existe n fonctions holomorphes dans X qui induisent, dans l'anneau $\mathcal{O}_x$, un système de coordonnées locales au point x (n désigne la dimension complexe de X) ;

(c) L'*enveloppe* $\hat{K}$ de tout compact $K \subset X$ est compacte.

Rappelons la définition de l'enveloppe d'un compact K : c'est l'ensemble $\hat{K}$ des points $x \in X$ tels que

$$|f(x)| \leqq \sup_{y \in K} |f(y)|$$

pour toute f holomorphe dans X.

On montre (en utilisant le théorème de Baire) que la condition (c) équivaut à la suivante :

(c') Pour toute suite infinie S de points de X, sans point adhérent dans X, il existe une fonction holomorphe dans X et *non bornée* sur S.

Exemples

1. La condition (a) montre qu'une variété *compacte* X de dimension $n > 0$ n'est jamais une variété de Stein.

2. Pour $n = 1$, toute « surface de Riemann » connexe.

([7]) Voir [4], théorème 2 ; et [5], exposé XVI.

non compacte, est une variété de Stein : cela résulte d'un mémoire de Behnke et Stein [1].

3. Soit X un ouvert de l'espace numérique complexe C^n. Les conditions (*a*) et (*b*) sont trivialement vérifiées; la condition (*c*) exprime que X est un *domaine d'holomorphie* (cf. [6]).

4. Soit X un « domaine étalé » dans C^n, c'est-à-dire une variété munie d'une application φ dans C^n, φ étant un homéomorphisme local. La condition (*b*) est trivialement vérifiée; les conditions (*a*) et (*c*) entraînent que X est un domaine d'holomorphie. Réciproquement, tout domaine d'holomorphie (étalé dans C^n) est-il une variété de Stein ? La réponse est affirmative ([8]) lorsque X est « de type fini » (i.e.: s'il est impossible de trouver dans C^n un disque ouvert A et, dans X, une infinité d'ouverts que φ applique homéomorphiquement sur A). La question reste ouverte dans le cas général.

5. Le produit $X \times Y$ de deux variétés de Stein est évidemment une variété de Stein.

6. Soient X une variété de Stein, et V une sous-variété analytique régulièrement plongée dans X (cf. § 5). Alors V est une variété de Stein : cela résulte aussitôt des définitions. Ainsi, toute sous-variété analytique régulièrement plongée dans C^n est une variété de Stein ; plus particulièrement, toute variété algébrique affine est une variété de Stein.

7. Soient X une variété de Stein, et g une fonction holomorphe dans X, non identiquement nulle. L'ensemble Y des points $x \in X$ tels que $g(x) \neq 0$ est une variété de Stein : pour vérifier que Y satisfait à la condition (*c*), on observe que $1/g$ est holomorphe dans Y ([9]). Par exemple, la variété du groupe linéaire complexe à r variables est une variété de Stein; donc la variété de tout sous-groupe fermé du groupe linéaire complexe est une variété de Stein.

([8]) Cf. [6], et aussi [5], exposé IX.

([9]) On a un résultat plus général : si d'une variété de Stein on enlève un « diviseur » D (c'est-à-dire une sous-variété analytique de X qui, au voisinage de chaque point $x \in D$, peut être définie par une seule équation $g_x = 0$, g_x étant holomorphe au point x et $\not\equiv 0$), *la variété restante* $X - D$ *est une variété de Stein*. Tout revient à prouver ceci : pour tout $x \in D$ il existe une fonction méromorphe f dans X, holomorphe dans $X - D$, et telle que $g_x f$ soit holomorphe et $\neq 0$ au point x. L'existence d'une telle f résulte du théorème A (§ 7 ci-dessous), appliqué au sous-faisceau $\mathcal{F}$ du faisceau $\mathcal{M}$ des germes de fonctions méromorphes, défini comme suit : $\mathcal{F}_x = \mathcal{O}_x$ si $x \notin D$; si $x \in D$, $\mathcal{F}_x$ se compose des $\varphi \in \mathcal{M}_x$ telles que $g_x \varphi$ soit holomorphe. Ce faisceau est cohérent. La démonstration qui vient d'être esquissée est due à J.-P. Serre.

7. Enoncé des théorèmes fondamentaux

Théorème A. — *Soient* X *une variété de Stein, et* $\mathcal{F}$ *un faisceau analytique cohérent sur* X. *Alors, pour tout point* $x \in X$, *l'image de* $H^0(X, \mathcal{F})$ *dans* $\mathcal{F}_x$ *engendre* $\mathcal{F}_x$ *pour sa structure de* $\mathcal{O}_x$*-module.*

Théorème B. — *Soient* X *une variété de Stein, et* $\mathcal{F}$ *un faisceau analytique cohérent sur* X. *Alors, pour tout entier* $q > 0$, *les groupes de cohomologie* $H^q(X, \mathcal{F})$ *sont nuls.*

La démonstration est trop longue et délicate pour pouvoir être donnée ici ([10]). La démonstration du théorème A, et celle du théorème B pour le cas $q = 1$, constituaient en fait l'objectif essentiel du mémoire [**4**], au moins dans le cas où X est un domaine d'holomorphie de type fini, et $\mathcal{F}$ un sous-faisceau cohérent de $\mathcal{O}^q(X)$. Les raisonnements se transposent sans difficulté au cas général d'une variété de Stein. La formulation cohomologique du théorème B, et l'idée d'étudier non seulement le cas $q = 1$, mais le cas $q > 0$ quelconque, sont dues à J.-P. Serre.

8. Applications

Théorème 1 ([11]). — *Sur une variété de Stein* X, *le problème additif de Cousin est toujours résoluble.*

En effet, d'après le théorème B, on a $H^1(X, \mathcal{O}) = 0$.

Théorème 2. — *Soient* X *une variété de Stein,* V *une sous-variété analytique de* X, $\mathcal{J}$ *le faisceau d'idéaux défini par* V (§ 5). *Alors les fonctions holomorphes dans* X *qui s'annulent en tout point de* V, *engendrent, en chaque point* $x \in X$, *l'idéal* $\mathcal{J}_x$.

En effet, $\mathcal{J}$ est un faisceau cohérent, auquel on applique le théorème A.

Corollaire. — Si $x \notin V$, il existe une fonction f holomorphe dans X, nulle sur V, et telle que $f(x) \neq 0$. Autrement dit : la sous-variété V peut être globalement définie par des équations (obtenues en égalant à 0 des fonctions holomorphes dans X). De plus : pour tout ouvert U de X, relativement com-

([10]) Une démonstration complète a été donnée dans [**5**], exposé XIX.
([11]) Démontré pour la première fois par Oka [**9**] dans le cas où X est un domaine d'holomorphie univalent.

pact, il existe un nombre *fini* de f_i holomorphes dans X, nulles sur V, et n'ayant pas, dans U, d'autre zéro commun que les points de $V \in U$.

THÉORÈME 3. — *Soient* X *une variété de Stein, et* V *une sous-variété analytique régulièrement plongée dans* X. *Alors toute fonction holomorphe sur la variété analytique-complexe* V *est induite par une fonction holomorphe sur* X.

En effet, soit $\mathcal{J}$ le faisceau d'idéaux défini par V. D'après le théorème B, on a $H^1(X, \mathcal{J})=0$, donc

$$H^0(X, \mathcal{O}) \longrightarrow H^0(X, \mathcal{O}/\mathcal{J}) \qquad (1)$$

est un *épimorphisme*. Or $\mathcal{O}_x/\mathcal{J}_x$ est nul si $x \notin V$; si $x \in X$, $\mathcal{O}_x/\mathcal{J}_x$ s'identifie à l'anneau des germes de fonctions holomorphes sur V au point x, parce que le point x est régulier pour V. Donc $H^0(X, \mathcal{O}/\mathcal{J})$ s'identifie à l'anneau des fonctions holomorphes sur V, et (1) est l'homomorphisme qui, à chaque fonction holomorphe sur X, associe la fonction qu'elle induit sur V. D'où le théorème.

Considérons le cas particulier où V est un sous-ensemble infini discret de X : si X est une variété de Stein, le théorème 3 montre qu'il existe une f holomorphe sur X et *prenant des valeurs arbitrairement données aux points de* V. Cette propriété renforce évidemment les conditions (a) et (c') des variétés de Stein. Plus généralement :

A chaque point x d'un ensemble discret A, associons un entier $r(x)$. Pour $x \in A$, soit $\mathcal{J}_x$ l'idéal de $\mathcal{O}_x$ formé des germes dont le développement de Taylor, au point x, n'a pas de terme de degré $\leqslant r(x)$; pour $x \subset A$, prenons $\mathcal{J}_x = \mathcal{O}_x$. Il est immédiat que le faisceau $\mathcal{J}$ des $\mathcal{J}_x$ est cohérent. Si $x \notin A$, un élément de $\mathcal{O}_x/\mathcal{J}_x$ est un « développement limité d'ordre $r(x)$ ». On peut alors énoncer :

THÉORÈME 4. — *Si* X *est une variété de Stein, il existe une fonction holomorphe dans* X *et admettant, en chaque point* $x \in A$, *un développement limité arbitrairement donné* (et d'ordre $r(x)$ arbitraire).

En effet, se donner un tel système de développements limités, c'est se donner un élément de $H^0(X, \mathcal{O}/\mathcal{J})$. Or l'homomorphisme

$$H^0(X, \mathcal{O}) \longrightarrow H^0(X, \mathcal{O}/\mathcal{J})$$

est un épimorphisme, puisque $H^1(X, \mathcal{J})=0$ en vertu du théorème B.

Le théorème 4, appliqué au cas où A possède un seul point, renforce la propriété (b) de la définition des variétés de Stein.

Remarque (due à J.-P. Serre). — Pour qu'une variété analytique-complexe X, réunion dénombrable de compacts, soit une variété de Stein, il faut et il suffit que X satisfasse à la condition :

(S) pour tout faisceau cohérent d'idéaux $\mathcal{J}$, on a $H^1(X, \mathcal{J})=0$.

La condition est nécessaire, d'après le théorème B. Elle est suffisante, car (S) entraîne la validité de la conclusion du théorème 4, et ceci entraîne que X satisfait aux conditions (*a*), (*b*) et (*c'*).

THÉORÈME 5. — *Soient* X *une variété de Stein, et* $\mathcal{F}$ *un faisceau analytique cohérent. Si des sections* $u_i \in H^0(X, \mathcal{F})$ *en nombre fini sont telles que, pour tout* $x \in X$, *le* $\mathcal{O}_x$*-module* $\mathcal{F}_x$ *soit engendré par les images des* u_i *dans* $\mathcal{F}_x$*, alors les* u_i *engendrent* $H^0(X, \mathcal{F})$ *pour sa structure de module sur l'anneau* $H^0(X, \mathcal{O})$ *des fonctions holomorphes sur* X.

En effet, soit p le nombre des u_i ; les u_i définissent un homomorphisme analytique de faisceaux : $\mathcal{O}^p \longrightarrow \mathcal{F}$. Par hypothèse, c'est un épimorphisme. Or son noyau est un faisceau cohérent; donc, d'après le théorème B, $H^0(X, \mathcal{O}^p)\longrightarrow H^0(X, \mathcal{F})$ est un épimorphisme, et ceci démontre le théorème.

Exemple. — Prenons $\mathcal{F}=\mathcal{O}$. Dire que l'idéal de $\mathcal{O}_x$ engendré par les $u_i \in H^0(X, \mathcal{O})$ est $\mathcal{O}_x$, c'est dire que les fonctions holomorphes u_i n'ont aucun zéro commun dans X. Le théorème 5 affirme alors qu'il existe une identité de la forme

$$1=\sum c_i u_i \tag{2}$$

à coefficients c_i holomorphes dans X ([12]). Ce résultat vaut notamment quand X est un ouvert de C^n et est domaine d'holomorphie. Montrons qu'il est en défaut quand X, ouvert de C^n, *n'est pas* un domaine d'holomorphie : il existe alors un point a de la frontière de X, tel que toute fonction holomorphe dans X se prolonge en une fonction holomorphe dans un ouvert $Y \supset X$ tel que $a \in Y$. Prenons $u_i=x_i-a_i$ (x_i : coordonnées complexes d'un point de C^n ; a_i : coordonnées du point a). Les u_i n'ont pas de zéro commun dans X, mais il n'y a pas d'identité telle que (2), car les c_i, étant holomorphes dans X, le seraient au point a ; or la relation (2) ne peut être vérifiée au point a.

9. EXTENSIONS DIVERSES ; PROBLÈMES NON RÉSOLUS

Le problème additif de Cousin peut être résoluble sans que X soit nécessairement une variété de Stein. Par exemple,

([12]) Cf. [3], p. 189, pour le cas où X est un domaine d'holomorphie.

il est résoluble pour l'*espace projectif complexe* P, de dimension quelconque n ; en effet, *on a* $H^q(P, \mathcal{O}) = 0$ *pour tout* $q > 0$. La démonstration de ce résultat est toute différente de celle du théorème B : il résulte d'un théorème de Dolbeault ([13]) que si X est une variété kählérienne compacte, l'espace vectoriel (complexe) $H^q(X, \mathcal{O})$ est isomorphe à l'espace des formes harmoniques de type $(0, q)$; or, dans le cas où X est l'espace projectif P, toute forme harmonique non identiquement nulle est de degré pair $2p$ et de type (p, p), comme cela résulte de la structure multiplicative de l'anneau de cohomologie de P à coefficients complexes.

D'autre part, les théorèmes A et B du § 7 peuvent s'étendre au cas suivant : soit Y un sous-ensemble fermé d'une variété analytique-complexe X; la notion de faisceau analytique cohérent se définit d'une manière évidente pour les faisceaux sur l'espace Y. On démontre : si Y possède un système fondamental de voisinages ouverts dont chacun est une variété de Stein, les théorèmes A et B valent pour Y (et pour tout faisceau cohérent sur Y). Par exemple, prenons $X = C^n$, $Y = R^n$ (espace numérique réel plongé dans l'espace numérique complexe); on voit sans difficulté que l'on se trouve dans les conditions précédentes. On en déduit une extension de la théorie aux *sous-variétés analytiques-réelles* de l'espace R^n ; les théorèmes concernent les faisceaux analytiques-réels et cohérents (on se ramène au cas des faisceaux analytiques-complexes par extension du corps de base). On obtient par exemple le résultat suivant : si V, analytique-réelle, est régulièrement plongée dans R^n, toute fonction analytique-réelle, définie sur V, est induite par une fonction analytique-réelle de R^n.

Problème. — Au lieu de nous borner aux sous-variétés analytiques-réelles régulièrement plongées dans R^n, considérons, en général, les « variétés de Stein réelles », c'est-à-dire les variétés analytiques-réelles (abstraites) qui satisfont aux conditions (*a*), (*b*) et (*c*) du § 7, à cela près qu'on y remplace partout le mot « holomorphe » par « analytique-réelle ». Y a-t-il, pour les variétés de Stein réelles, des théorèmes analogues aux théorèmes A et B ?

Autre problème. — Soit X une variété de Stein (complexe). Étant donné un ouvert $U \subset X$, à quelles conditions U est-il une variété de Stein ? Une condition *nécessaire* est que tout point adhérent à U possède dans X un voisinage ouvert V tel que $V \cap U$ soit une variété de Stein (à ce propos, on notera que si U et V sont des ouverts de Stein, $U \cap V$ est un ouvert de

([13]) [8], théorème 1.

Stein). *Cette condition nécessaire est-elle suffisante?* Dans le cas particulier où X est un domaine d'holomorphie univalent de l'espace numérique C^n, la réponse est affirmative d'après un théorème de Oka ([14]), que celui-ci n'a du reste démontré que dans le cas $n = 2$.

Bibliographie

[1] Behnke, H. und Stein, K., *Entwicklung analytischer Funktionen auf Riemannschen Flächen* (*Math. Annalen*, **120**, 1948, pp. 430-461).

[2] Cartan, H., *Sur les matrices holomorphes de n variables complexes* (*Journal de Math. pures et appl.*, **19**, 1940, pp. 1-26).

[3] Cartan, H., *Idéaux de fonctions analytiques de n variables complexes* (*Ann. Ecole Normale Sup.*, **61**, 1944, pp. 149-197).

[4] Cartan, H., *Idéaux et modules de fonctions analytiques de variables complexes* (*Bull. Soc. Math. France*, **78**, 1950, pp. 28-64).

[5] Cartan, H., *Séminaire E. N. S.*, *1951-1952* (polycopié).

[6] Cartan, H. und Thullen, P., *Zur Theorie der Singularitäten der Funktionen mehrerer Veränderlichen : Regularitäts- und Konvergenzbereiche* (*Math. Annalen*, **106**, 1932, pp. 617-647).

[7] Cousin, P., *Sur les fonctions de n variables complexes* (*Acta math.*, **19**, 1895, pp. 1-62).

[8] Dolbeault, P., *Sur la cohomologie des variétés analytiques complexes* (*Comptes rendus, Paris*, **236**, 1953, pp. 175-177).

[9] Oka, K., *Sur les fonctions analytiques de plusieurs variables. II. Domaines d'holomorphie* (*Journ. Sci. Hiroshima*, Ser. A, **7**, 1937, pp. 115-130).

[10] Oka, K., *Sur les fonctions analytiques de plusieurs variables. VII. Sur quelques notions arithmétiques* (*Bull. Soc. Math. France*, **78**, 1950, pp. 1-27).

[11] Stein, K., *Analytische Funktionen mehrerer komplexer Veränderlichen zu vorgegebenen Periodizitätsmoduln und das zweite Cousinsche Problem* (*Math. Annalen*, **123**, 1951, pp. 201-222).

([14]) *Tohoku Math. Journal*, **49**, 1942, pp. 15-52.

89.

(avec J-P. Serre)

Espaces fibrés et groupes d'homotopie. I. Constructions générales

Comptes Rendus de l'Académie des Sciences de Paris 234, 288–290 (1952)

Construction d'espaces fibrés ([1]) permettant de « tuer » le groupe d'homotopie $\pi_n(X)$ d'un espace X dont les $\pi_i(X)$ sont nuls pour $i < n$. Cette méthode généralise celle qui consiste, pour $n = 1$, lorsque X est connexe, à « tuer » le groupe fondamental $\pi_1(X)$ en passant au revêtement universel de X.

1. Soient X un espace connexe par arcs, $x \in X$, $\mathfrak{S}(X)$ le complexe singulier de X. Pour tout entier $q \geq 1$, soit $\mathfrak{S}(X; x, q)$ le sous-complexe engendré par les simplexes dont les $(q-1)$-faces sont en x. Les groupes d'homologie (resp. cohomologie) de $\mathfrak{S}(X; x, q)$ à coefficients dans G sont les *groupes d'Eilenberg* ([2]) de l'espace X en x; on les notera $H_i(X; x, q, G)$, resp. $H^i(X; x, q, G)$. Ils forment des systèmes locaux. Rappelons ([2]) que $\pi_q(X; x) \approx H_q(X; x, q, Z)$ pour $q \geq 2$.

Définition. — Un espace Y, muni d'une application continue f de Y dans X, *tue* les groupes d'homotopie $\pi_i(X)$ pour $i \leq n (n \geq 1)$ si $\pi_i(Y) = 0$ pour $i \leq n$ et si f définit un isomorphisme de $\pi_i(Y)$ sur $\pi_i(X)$ pour $i > n$.

Théorème 1. — *Si un espace* Y *tue les* $\pi_i(X)$ *pour* $i \leq n$, *les groupes d'homologie* $H_j(Y)$ *sont isomorphes aux groupes d'Eilenberg* $H_j(X; x, n+1)$; *de même pour la cohomologie.*

Cela résulte du :

Lemme 1. — *Si une application f d'un* Y *dans un* X *applique* $y \in Y$ *en* $x \in X$ *et définit, pour tout* $i > n$, *un isomorphisme* $\pi_i(Y; y) \approx \pi_i(X; x)$, *l'homomorphisme* $\mathfrak{S}(Y; y, n+1) \to \mathfrak{S}(X; x, n+1)$ *défini par f est une chaîne-équivalence.* (*En considérant le « mapping cylinder » de f, on se ramène au cas où* Y *est plongé dans* X; *le lemme s'obtient alors par un procédé standard de déformation.*)

([1]) L'expression « espace fibré » est prise dans le sens général défini par Serre (*Ann. of Math.*, 54, 1951, p. 425-505). Ce Mémoire sera désigné par [S].

([2]) *Ann. of Math.*, 45, 1944, p. 407-447; *voir* § 32.

Reprinted from Oeuvres, Collected Works (Springer-Verlag, 1979), Vol. II, pp. 1294–1296.

(2)

Le théorème 1 justifie la notation $(X, n+1)$ pour n'importe quel espace qui tue les $\pi_i(X)$ pour $i \leqq n$.

2. Théorème 2. — *A tout X connexe par arcs, on peut associer une suite d'espaces* (X, n) [où $n = 1, 2, \ldots$ et $(X, 1) = X$] *et d'applications continues* f_n : $(X, n+1) \to (X, n)$, *de manière que* $(X, n+1)$ *tue les* $\pi_i(X, n)$ *pour* $i \leqq n$, *et que :*

(I) *l'application* f_n *munisse* $(X, n+1)$ *d'une structure d'espace fibré* ([1]) *de base* (X, n), *ayant pour fibre un espace* $\mathcal{K}[\pi_n(X), n-1]$([3]);

(II) *il existe un espace* X'_n *de même type d'homotopie que* (X, n), *et une fibration de* X'_n, *de fibre* $(X, n+1)$, *ayant pour base un* $\mathcal{K}[\pi_n(X), n]$.

Il suffira de dire comment $(X, n+1)$, f_n et X'_n se construisent à partir de (X, n). On utilise d'abord deux lemmes, déjà employés par certains auteurs ([4]) :

Lemme 2. — *Étant donné un espace connexe* A *et un entier* $k \geqq 1$, *on peut plonger* A *dans un espace* U *de manière que* $\pi_i(A) \to \pi_i(U)$ *soit un isomorphisme (sur) pour* $i < k$, *et* $\pi_k(U) = 0$. [*Pour tout* $\alpha \in \pi_k(A)$ *on choisit un représentant* g_α : $S_\alpha \to A$, *où* S_α *est une sphère de dimension* k, *frontière d'une boule* E_α *de dimension* $k+1$; *on « attache » à* A *les boules* E_α *au moyen des applications* g_α.]

Lemme 3. — *Étant donné un espace* A *tel que* $\pi_i(A) = 0$ *pour* $i < n$, *on peut plonger* A *dans un espace* V *de manière que* $\pi_n(A) \to \pi_n(V)$ *soit un isomorphisme sur, et* $\pi_i(V) = 0$ *pour* $i \neq n$. (*Se déduit du lemme* 2 *par itération, en prenant l'espace-réunion.*)

Constructions. Étant donné une application continue φ d'un espace A dans un $V = \mathcal{K}(\pi, n)$, soit A' l'espace des couples (a, ω) où $a \in A$ et ω est un chemin ([5]) de V d'extrémité $\varphi(a)$; A' se rétracte sur A, identifié à l'espace des couples (a, ω) tels que ω soit ponctuel en $\varphi(a)$. L'application g qui, à (a, ω), associe l'origine de ω, définit A' comme espace fibré de base V. Soit B la fibre au-dessus de $\varphi(a_0)$(a_0, point fixé de A); l'application f qui, à (a, ω), associe a, définit B comme espace fibré de base A, de fibre l'espace W des lacets sur V, qui est un $\mathcal{K}(\pi, n-1)$.

Appliquons ces constructions à l'espace $A = (X, n)$ supposé déjà obtenu, au groupe $\pi = \pi_n(X)$ et à l'injection φ de A dans V (lemme 3). La suite

([3]) Rappelons (*cf.* Eilenberg-MacLane, *Ann. of Math.*, 46, 1945, p. 480-509, § 17; *ibid.*, 51, 1950, p. 514-533) que si un espace V satisfait à $\pi_i(V) = 0$ pour tout $i \neq n$, $\pi_n(V) = \pi$, le complexe $\mathcal{S}(V)$ a même type d'homotopie qu'un complexe $K(\pi, n)$ explicité par ces auteurs, et qui dépend seulement de n et du groupe π (abélien si $n \geqq 2$). D'un tel espace V, nous dirons que c'est un espace $\mathcal{K}(\pi, n)$; ses groupes d'homologie $H_i(\pi; n)$ (resp. de cohomologie) sont les *groupes d'Eilenberg-MacLane* du groupe π, pour l'entier n.

([4]) *Voir*, par exemple, J. H. C. Whitehead, *Ann. of Math.*, 50, 1949, p. 261-263.

([5]) Pour tout ce qui concerne les espaces de chemins, *voir* [S], Chap. IV.

(3)

exacte d'homotopie des espaces fibrés montre que B tue les $\pi_i(A)$ pour $i \leq n$; on peut donc prendre $(X, n+1) = B$, $f_n = f$, $X'_n = A'$, et le théorème 2 est démontré.

3. *Utilisation.* — Chacune des fibrations (I) et (II) définit (pour chaque n) une suite spectrale ([6]). Dans la mesure où l'on connaît les groupes d'Eilenberg-MacLane d'un groupe π donné, on obtient une méthode de calcul (partiel) des groupes d'Eilenberg de X, et notamment des groupes d'homotopie de X.

La méthode utilisée par Hirsch ([7]) pour étudier $\pi_3(X)$ quand $\pi_1(X) = 0$ et que $\pi_2(X)$ est libre de base finie, rentre dans notre méthode générale; elle revient à prendre au-dessus de X une fibre $\mathcal{K}(\pi_2, 1)$ qui est ici un produit de cercles.

En vue des applications, la remarque suivante est utile : l'espace $W = \mathcal{K}[\pi_n(X), n-1]$ opère à gauche dans $B = (X, n+1)$, et par suite chaque $\alpha \in H_i[\pi_n(X), n-1]$ définit un endomorphisme λ_α de la suite spectrale d'homologie de la fibration (I); on démontre que λ_α commute avec toutes les différentielles de cette suite spectrale.

([6]) Il s'agit de la suite spectrale en homologie (resp. cohomologie) singulière; *voir* [S], Chap. I et II.

([7]) *Comptes rendus*, 228, 1949, p. 1920.

90.
(avec J-P. Serre)

Espaces fibrés et groupes d'homotopie. II. Applications

Comptes Rendus de l'Académie des Sciences de Paris 234, 393–395 (1952)

Applications de la méthode générale exposée dans une Note précédente ([1]). On retrouve la plupart des relations connues entre homologie et homotopie; les résultats nouveaux concernent notamment les groupes d'homotopie des groupes de Lie et des sphères.

Dans toute la suite X désignera un espace *connexe par arcs*.

Considérons la fibration (II) de la Note ([1]), pour $n \geqq 2$; en lui appliquant la Proposition 5 du Chapitre III de [S], on obtient :

PROPOSITION 1. — *Pour tout espace* X *et tout* $n \geqq 2$ ([2]), *on a une suite exacte :*

$$(1)\quad \left\{ \begin{array}{l} H_{2n}(X, n+1) \to H_{2n}(X, n) \to H_{2n}(\pi_n(X); n) \to H_{2n-1}(X, n+1) \to H_{2n-1}(X, n) \to \ldots \\ \ldots \to H_{n+2}(X, n+1) \to H_{n+2}(X, n) \to H_{n+2}(\pi_n(X); n) \to \pi_{n+1}(X) \to H_{n+1}(X, n) \to 0. \end{array} \right.$$

Compte tenu de ce que $H_{n+2}(\pi; n) = \pi/2\pi \,(n \geqq 3)$ et $H_{n+3}(\pi; n) = {}_2\pi \,(n \geqq 4)$, on retrouve des résultats de G. W. Whitehead ([3]).

COROLLAIRE 1. — *Les groupes d'homologie relatifs* $H_i[\mathcal{S}(X; x, n), \mathcal{S}(X; x, n+1)]$ (*où* x *est un point de* X) *sont isomorphes aux groupes d'Eilenberg-MacLane* $H_i(\pi_n(X); n)$ *pour* $1 \leqq i \leqq 2n$.

Ce résultat semble en rapport étroit avec une suite spectrale annoncée récemment par W. Massey et G. W. Whitehead (lorsque X est une sphère) ([4]).

COROLLAIRE 2. — *Si* $\pi_i(X) = 0$ *pour* $i < n$ *et* $H_j(X) = 0$ *pour* $n < j \leqq 2n$ (*en particulier si* X *est une sphère* $\mathbf{S}_n$), *on a des isomorphismes :*

$$H_j(X, n+1) \approx H_{j+1}(\pi_n(X); n) \qquad pour \quad n \leqq j \leqq 2n-1 \qquad (n \geqq 2).$$

([1]) *Comptes rendus*, **234**, 1952, p. 288. Nous renvoyons a cette Note dont nous conservons la terminologie et les notations.

([2]) Le cas $n = 1$ est spécial et n'apporte d'ailleurs rien de nouveau.

([3]) *Proc. Nat. Acad. Sc. USA*, 34, 1948, p. 207-211.

([4]) *Bull. Amer. Math. Soc.*, 57, 1951, Abstracts 544 et 545.

Reprinted from Oeuvres, Collected Works (Springer-Verlag, 1979), Vol. II, pp. 1297–1299.

(2)

On notera que, si $j < 2n - 1$, les groupes $H_{j+1}(\pi; n)$ sont « stables » et isomorphes aux groupes $A_{j-n+2}(\pi)$ introduits par Eilenberg-Mac Lane ([5]), ce qui fournit une interprétation géométrique de ces derniers groupes.

PROPOSITION 2. — *Si* $\pi_i(X) = 0$ *pour* $i < n$ *et* $n < i < m$ (*n et m étant deux entiers tels que* $0 < n < m$), *on a une suite exacte* :

$$H_{m+1}(X) \to H_{m+1}(\pi_n(X); n) \to \pi_m(X) \to H_m(X) \to H_m(\pi_n(X); n) \to 0.$$

Ceci se démontre au moyen de la fibration (II) et complète des résultats d'Eilenberg-MacLane ([6]) (à l'exception, toutefois, de ceux relatifs à l'invariant $\mathbf{k}$).

PROPOSITION 3. — *Supposons que* $\pi_1(X) = 0$, *que les nombres de Betti de* X *soient finis en toute dimension et que l'algèbre de cohomologie* $H^*(X, Q)$ (Q *désignant le corps des rationnels*) *soit le produit tensoriel d'une algèbre extérieure engendrée par des éléments de degrés impairs et d'une algèbre de polynomes engendrée par des éléments de degrés pairs; si* d_n *désigne le nombre des générateurs de degré n, on a*

$$\text{rang } ({}^7) \text{ de } \pi_n(X) = d_n \qquad \text{pour tout } n.$$

On utilise la fibration (I), et le calcul des algèbres de cohomologie d'Eilenberg-MacLane à coefficients dans Q; on montre par récurrence sur n que $H^*(X; n, Q)$ est l'algèbre quotient de $H^*(X, Q)$ par l'idéal engendré par les générateurs de degrés $< n$.

Remarques. — 1. La démonstration montre aussi que le noyau de l'homomorphisme $\pi_n(X) \to H_n(X)$ est un groupe de torsion.

2. La proposition subsiste même si $\pi_1(X) \neq 0$, pourvu que $\pi_1(X)$ soit abélien et opère trivialement dans $H^*(X; 2, Q)$.

3. La proposition 3 s'applique notamment : *a.* à une sphère de dimension impaire; *b.* à un espace de lacets sur un espace simplement connexe dont les nombres de Betti sont finis; *c.* à un groupe de Lie. En particulier, *les groupes d'homotopie d'un groupe de Lie sont finis en toute dimension où il n'y a pas d'élément « primitif »* (donc en toute dimension *paire*).

PROPOSITION 4. — *Soit* X *tel que* $\pi_1(X) = 0$, *et q un entier. Si* $H_i(X)$ *est un groupe de torsion pour* $1 < i < q$, *il en est de même du noyau et du conoyau* ([8]) *de l'homomorphisme* $\varphi_j : H_j(X, q) \to H_j(X)$ *pour tout j. Si en outre la compo-*

([5]) *Proc. Nat. Acad. Sc. USA*, 36, 1950, p. 657-663.

([6]) *Ann. of Math.*, 51, 1950, p. 514-533.

([7]) Le *rang* d'un groupe G est la dimension du Q-espace vectoriel $Q \otimes G$.

([8]) Le *conoyau* d'un homomorphisme $A \to B$ est le quotient de B par l'image de A.

(3)

sante p-primaire (p premier) de $H_i(X)$ *est nulle pour* $1 < i < q$, *il en est de même du noyau et du conoyau de* φ_j. *Ceci vaut notamment pour* $\varphi_q : \pi_q(X) \to H_q(X)$.

PROPOSITION 5. — *Les groupes d'homologie de la sphère* $\mathbf{S}_3$ *dont on a tué le troisième groupe d'homotopie sont les suivants :*

$$H_i(\mathbf{S}_3, 4) = 0 \quad \text{pour } i \text{ impair} \qquad \text{et} \qquad H_{2q}(\mathbf{S}_3, 4) = Z/qZ$$

(Les premiers groupes d'homologie sont donc : Z, 0, 0, 0, Z_2, 0, Z_3, 0, Z_4, ...).

COROLLAIRE. — *La composante p-primaire de* $\pi_{2p}(\mathbf{S}_3)$ *est* Z_p [9].

La proposition 5 permet de retrouver aisément les résultats connus sur les $\pi_i(\mathbf{S}_3)$, $i = 4, 5, 6$: pour $i = 4$, c'est évident; appliquant la suite (1) pour $n = 4$, et utilisant le fait que $H_7(Z_2; 4) = Z_2$, on obtient $\pi_5(\mathbf{S}_3) = Z_2$ et $H_6(\mathbf{S}_3, 5) = Z_6$; en appliquant la suite (1) pour $n = 5$ on obtient une suite exacte : $\pi_5(\mathbf{S}_3) \to \pi_6(\mathbf{S}_3) \to Z_6 \to 0$, qui montre que $\pi_6(\mathbf{S}_3)$ a 6 ou 12 éléments [4].

PROPOSITION 6. — *Les groupes* $\pi_7(\mathbf{S}_3)$ *et* $\pi_8(\mathbf{S}_3)$ *sont des groupes 2-primaires;* $\pi_9(\mathbf{S}_3)$ *est somme directe de* Z_3 *et d'un groupe 2-primaire.*

On utilise le fait que $H_i(Z_3; 5) = 0$ pour $i = 7$, 8, et $H_9(Z_3; 5) = Z_3$ [5].

Enfin, si l'on admet les résultats sur les groupes d'Eilenberg-MacLane obtenus par H. Cartan au moyen de calculs dont le fondement théorique n'a pas encore reçu de justification complète, on obtient les résultats suivants (que nous donnons donc comme *conjecturaux*) : *pour* n *impair* ≥ 3, *et* p *premier, la composante p-primaire de* $\pi_i(\mathbf{S}_n)$ *est* Z_p *si* $i = n + 2p - 3$, *nulle si* $n + 2p - 3 < i < n + 4p - 6$; *celle de* $\pi_{4p-3}(\mathbf{S}_3)$ *est* Z_p, *de même (si* $p \neq 2$) *que celle de* $\pi_{4p-2}(\mathbf{S}_3)$. *Par exemple,* $\pi_{10}(\mathbf{S}_3)$ *est somme directe de* Z_{15} *et d'un groupe 2-primaire.*

---

(9) Notre méthode montre également que l'homomorphisme $f_p : \pi_{2p}(\mathbf{S}_3) \to Z_p$ introduit par N. E. Steenrod est *sur*.

Médaille d'Or du CNRS

DISCOURS

DE

MONSIEUR HENRI CARTAN

prononcé le 1er février 1977 à l'occasion de la remise de la Médaille d'Or par Madame Alice Saunier-Séïté, Secrétaire d'Etat aux Universités

Madame le Ministre,

Pourquoi ne pas avouer ma confusion lorsque je lis la liste des savants à qui a été décernée la Médaille d'Or du CNRS depuis sa création en 1954 ? J'y vois entre autres les noms de deux mathématiciens illustres, Emile Borel en 1954 et Jacques Hadamard en 1956 : Emile Borel qui m'inspira toujours un respect un peu craintif malgré la grande amitié qui le liait à mon père, Hadamard dont la longue jeunesse et l'inlassable curiosité d'esprit m'ont toujours rempli d'admiration. La médaille qui lui fut remise alors qu'il avait dépassé l'âge de 90 ans n'ajoutait rien à sa gloire, mais elle honorait le CNRS.

Vous pouvez comprendre, Madame le Ministre, Monsieur le Directeur Général du CNRS, quels sont aujourd'hui ma confusion et mon embarras en constatant que je suis, après vingt ans d'interruption, le premier mathématicien à qui échoit cette médaille d'or. Mais j'éprouve aussi (pourquoi le cacher ?) une certaine satisfaction à la pensée qu'à travers ma personne c'est l'Ecole mathématique française que l'on a voulu honorer. Je me réjouis profondément de voir que son rayonnement mondial est ainsi reconnu en France même. Ce rayonnement n'a pas seulement sa source dans les grandes gloires du passé ; il s'explique par le maintien d'une tradition qui reste toujours vivante. Et en disant cela je ne pense pas seulement aux mathématiciens de ma génération, qui ont tant contribué à ma propre formation et dont quelques-uns auraient pu, aussi bien et même mieux que moi, être choisis pour l'honneur qui m'est fait aujourd'hui ; mais je songe aussi à ceux qui furent leurs élèves et qui, à leur tour, ont formé une nouvelle génération de brillants disciples.

Votre choix prouve aussi, me semble-t-il, que la mathématique est encore regardée dans notre pays comme une science, bien que l'on entende parfois exprimer l'opinion contraire ; et que la recherche mathématique peut et doit être considérée comme partie intégrante de la « recherche fondamentale », de cette recherche fondamentale dont on reconnaît à nou-

veau le rôle éminent parce qu'elle conditionne les applications techniques. Qu'il me soit permis de dire ici pourquoi la mathématique est, à mes yeux, la science fondamentale par excellence.

Il est vrai que certains prétendent que la mathématique n'est pas véritablement une science et qu'elle relève plutôt de la philosophie, pour la raison, dit-on, qu'en mathématique on n'a pas à se confronter avec le monde réel. Je ne sais si une telle vision implique un jugement flatteur sur la philosophie, mais je me réserve de discuter de son bien-fondé en ce qui concerne les mathématiques. L'on pourrait aussi défendre la thèse selon laquelle les mathématiques relèvent plutôt de l'art, et pour ma part je trouve qu'il y a une part de vérité dans cette façon de voir. Une théorie mathématique bien faite inspire en effet un sentiment esthétique, comme une belle construction en architecture ou en musique ; de plus il est certain que les qualités esthétiques d'une belle théorie en facilitent la diffusion et la rendent apte à une utilisation efficace. Cependant, vouloir réduire les mathématiques à une branche de la philosophie ou à une manifestation d'ordre purement artistique serait méconnaître leur véritable nature, et le mathématicien qui cherche sait bien que l'essentiel de sa démarche est d'un autre ordre.

Le mathématicien sent plus ou moins confusément qu'il est à la recherche d'une réalité cachée qui refuse de se dévoiler du premier coup. Il lui faut, sans se laisser décourager par des tentatives infructueuses, persévérer en se livrant à ce que j'appellerai des expériences variées, jusqu'au moment béni où il découvrira tout à coup ce qu'il cherchait, ou parfois ce à quoi il ne s'attendait pas du tout. Quelle est donc cette réalité que le mathématicien poursuit sans cesse de ses efforts et qu'il n'a jamais fini de découvrir entièrement ? La mathématique serait-elle une science expérimentale ?

Certains seront peut-être étonnés ou choqués de m'entendre parler de recherche expérimentale, alors qu'il est universellement admis que les mathématiques sont une construction qui ne repose que sur la logique, et qu'à partir du moment où les raisonnements sont corrects les conclusions sont assurées, du moins si l'on s'est préalablement mis d'accord sur le point de départ, c'est-à-dire sur les axiomes. Loin de moi l'idée de vouloir minimiser l'importance du raisonnement logique : il est l'outil indispensable et unique. Mais ce n'est qu'un outil. *Confondre cet outil avec l'objet même de la recherche mathématique serait, à mon avis, aussi erroné que de confondre, en physique, la recherche des lois de la nature avec la mise au point d'instruments plus ou moins perfectionnés. La construction des instruments n'est pas le but ultime du physicien ; ce n'est qu'un outil pour la découverte. Il en est de même en mathématique : on y fabrique, à l'aide de la pure logique, des outils de plus en plus perfectionnés, qui s'appellent des théories, dont le but ultime est d'aider à découvrir de nouveaux phénomènes, des lois nouvelles. Et de même que certaines lois de la biologie moléculaire ont pu être découvertes avec l'aide du microscope électronique (instrument qui lui-même n'a pu être conçu et réalisé que grâce à la connaissance de lois plus ou moins cachées de la physique), de même certaines lois de la théorie des nombres n'ont pu être découvertes, c'est-à-dire prouvées, qu'avec l'aide du puissant outil qui s'appelle, en mathématique, la théorie des faisceaux et de la cohomologie. Cet exemple montre en outre qu'un outil mathématique bien conçu peut avoir des applications dans un domaine des mathématiques tout différent de celui qui a permis de le concevoir. Il s'agit là d'un fait essentiel pour le développement de notre science : vouloir découper les mathématiques en morceaux séparés les uns des autres par des cloisons étanches ne peut conduire qu'à la stérilité. Et voilà pourquoi il faut sans cesse lutter contre la spécialisation excessive, voilà pourquoi nous aimons parler de l'unité de la mathématique.*

Mais je m'aperçois que je n'ai pas encore défini explicitement le véritable objet de la recherche mathématique. Je n'ai pas dit quelle est cette réalité, indépendante de nous, que nous nous efforçons de découvrir, et qui, lorsque nous en avons saisi quelques bribes, nous permet de mieux comprendre que chaque découverte soulève de nouveaux problèmes et que notre recherche est sans fin. Si je ne l'ai pas dit, c'est parce que je n'en suis point capable ; pour le pouvoir il faudrait être Dieu. La croyance en cette réalité n'est pas d'ordre logique, bien que nous y accédions par le raisonnement logique. Cette réalité que nous poursuivons, elle est pour le mathématicien le « monde réel » auquel il se trouve confronté et dont il n'est pas maître. Si cette réalité n'existait pas, si les mathématiques n'étaient qu'un jeu un peu vain et gratuit, comment pourrait-on expliquer qu'elles puissent servir avec efficacité dans les autres sciences ? La réalité mathématique est une inconnue dont nous découvrons des lambeaux ; ce qui nous est encore caché est imprévisible, ou du moins n'est pas prévisible à long terme.

C'est précisément le caractère imprévisible de la découverte en mathématique qui rend vain tout projet de planification de la recherche fondamentale dans ce domaine. Que les autorités responsables du CNRS veuillent bien ne pas s'offusquer d'une telle situation ! Le seul moyen de favoriser la recherche mathématique consiste à assurer aux chercheurs l'indispensable liberté d'esprit et la possibilité de contacts fréquents qui permettent l'échange des idées. Il n'y a pas de meilleure justification pour la création tant attendue du centre international de recherche de Luminy.

La pratique des mathématiques est une rude école de probité intellectuelle. Qu'il me soit permis de terminer sur une note plus personnelle. J'ai eu la chance, dès mon enfance, d'avoir quotidiennement sous les yeux un vivant exemple de probité intellectuelle et de désintéressement ; tous ceux qui ont connu mon père me comprennent. J'ai eu une autre chance, celle de trouver parmi les mathématiciens de ma génération, celle qui a suivi la guerre de 1914-18, des amis qui m'ont profondément influencé et qui ont contribué, par leur culture plus vaste que la mienne, à élargir mon horizon ; je pense à l'entreprise de Bourbaki, et en particulier à l'influence considérable qu'a exercée André Weil sur les mathématiciens de notre génération (pour ne parler que de ceux-là). J'ai eu enfin une troisième chance, celle d'avoir été en contact étroit, après 1940 et pendant plus de vingt ans, avec les promotions successives d'élèves de l'Ecole normale supérieure : j'ai trouvé là une tâche qui exige beaucoup de celui qui veut s'y consacrer, et j'en ai été largement récompensé par ce que les meilleurs de mes élèves m'ont appris.

Je souhaite que tous ceux qui m'ont aidé à devenir un mathématicien soient associés à l'honneur qui m'est fait aujourd'hui, et j'émets le vœu que les succès futurs de l'Ecole mathématique française justifient, dans un proche avenir, l'octroi d'une nouvelle médaille d'or à un mathématicien.

Lennart Carleson

CURRICULUM VITAE

Lennart Axel Edvard Carleson
Född 18 mars 1928 i Stockholm
Gift med *Sylvia* Edith Connie Elmstedt, född 1942
Barn: Caspar, född 1955; Beatrice, född 1958.

UTBILDNING

Studentexamen, Karlstad	1945
Fil. kand., Uppsala universitet	1947
Fil. lic., Uppsala universitet	1949
Fil. dr. och docent, Uppsala universitet	1950
Post-graduate-studier, Harvard University	1950–51

ANSTÄLLNINGAR

Docent, Uppsala universitet	1951–54
Professor, Stockholms högskola	1954–55
Professor, Uppsala universitet	1955–1993
Gästforskare vid MIT	hösten 1957
Gästprofessor vid Stanford University	1965–66
Gästprofessor vid MIT	1974–75

ANDRA BEFATTNINGAR

Member, Institute for Advanced Studies, Princeton	1961–62
Föreståndare för Mittag-Lefflers institut, Stockholm	1968–84
Redaktör, Acta Mathematica	1956–79
President, International Mathematical Union	1978–82
Medlem av kommittén för Salem-priset	1971–
Medlem av Vetenskapliga kommittén IHES, Paris	1983–
Inbjuden talare vid Internationella kongresserna	1962, 1966

MEDLEMSKAP I VETENSKAPLIGA SÄLLSKAP

Ledamot av Kungl. Vetenskapsakademin
Utländsk ledamot av
- American Academy, Boston
- Sovjetiska vetenskapsakademin, Moskva
- Danske Videnskabernes Selskab, Köpenhamn
- Norske Vetenskapsakademien, Oslo
- Finlands Vetenskapsakademi, Helsingfors

Reprinted from Acta Universitatis Upsaliensis, Section C, Vol. 58 (1995), p. 11.

Bibliografi

1. *On null-sets for continuous analytic functions.* Ark. Mat. **1** (1950), 311–318.
2. *On a class of meromorphic functions and its associated exceptional sets.* Diss. Uppsala, 1950.
3. *On Bernstein's approximation problem.* Proc. Am. Math. Soc. **2**:6 (1951), 953–961.
4. *Sets of uniqueness for functions regular in the unit circle.* Acta Mathematica **87** (1952), 325–345.
5. *On bounded analytic functions and closure problems.* Ark. Mat. **2** (1952), 283–291.
6. *On the zeros of functions with bounded Dirichlet integrals.* Math. Zeitschrift **56**:3 (1952), 289–295.
7. *On infinite differential equations with constant coefficients. I.* Math. Scand. **1** (1953), 31–38.
8. *On generators of normed rings.* Comptes rendus du Douzième Congrès des Mathématiciens Scandinaves (Lund, 1953), 16–17.
9. *A proof of an inequality of Carleman.* Proc. Am. Math. Soc **5**:6 (1954), 932–933.
10. *Random sequences and additive number theory.* Math. Scand. **4** (1956), 303–308.
11. *En matematisk modell för järnvägstrafik* (in Swedish). Nord. Mat. Tidskrift **5** (1957), 176–180.
12. *On the connection between Hausdorff measures and capacity.* Ark. Mat. **3** (1957), 403–406.
13. *Representations of continuous functions.* Math. Zeitschrift **66** (1957), 447–451.
14. *Two remarks on the basic theorems of information theory.* Math. Scand. **6** (1958), 175–180.
15. *An interpolation problem for bounded analytic functions.* Am. Journ. of Math. **80** (1958), 921–930.
16. with E.J. Akutowicz: *The analytic continuation of interpolatory functions.* Journ. d'anal. math. **7** (1959/60), 223–248.
17. *A representation formula for the Dirichlet integral.* Math. Zeitschrift **73** (1960), 190–196.
18. *A remark on Picard's theorem.* Bull. Am. Math. Soc. **67** (1961), 142–144.
19. *On universal moment problems.* Math. Scand. **9** (1961), 197–206.

Reprinted from Acta Universitatis Upsaliensis, Section C, Vol. 58 (1995), pp. 12–15.

20. *On the existence of boundary values for harmonic functions in several variables.* Ark. Mat, **4** (1961), 393–399.

21. *Interpolations by bounded analytic functions and the corona problem.* Ann. of Math. **76** (1962), 314–316.

22. *Interpolations by bounded analytic functions and the corona problem.* Proc. Internat. Congr. Math. (Stockholm, 1962), 314–316.

23. *Removable singularities of continuous harmonic funcitons in R^m.* Math. Scand. **12** (1963), 15–18.

24. *Mergelyan's theorem on uniform polynomial approximation.* Math. Scand. **15** (1964), 167–175.

25. *Maximal functions and capacities.* Ann. de l'Institut Fourier **15** (1965), 59–64.

26. *On convergence and growth of partial sums of Fourier series.* Acta Math. **116** (1966), 135–157.

27. *Convergence and summability of Fourier series.* Proc. Internat. Congr. Math. (Moscow, 1966), 83–88.

28. *On mappings, conformal at the boundary* (dédié au Jubilé Scientifique de Lars V. Ahlfors). Journ. d'anal. math. **19** (1967), 1–14.

29. "Selected problems in the theory of exceptional sets". Mathematical Studies, no. 13. D. Van Nostrand Co., Princeton 1967. 2nd ed., Wadsworth 1983. Russian edition: Избранные проблемы теории исключительных множествы. Translated by V. P. Havin. Edited by V. G. Maz'ja. Izdat. "Mir", Moscow (1971).

30. *The corona theorem.* Proc. of the Fifteenth Scandinavian Congr. (Oslo, 1968) in Lecture Notes (Springer) **118,** 121–132.

31. "Matematik för vår tid" (in Swedish). Bokförlaget Prisma, Stockholm 1968.

32. *On saddle point summability.* Mittag-Leffler report 1, 1971.

33. *On the Littlewood-Paley theorem.* Mittag leffler report 2, 1971.

34. *Some remarks on spectral synthesis on R^1 and a unified approach for special Cantor sets.* Mittag-Leffler report 3, 1971.

35. with Sigvard Jacobs: *Best uniform approximation by analytic functions.* Ark. Mat **10** (1972), 219–229.

36. with Per Sjölin: *Oscillatory integrals and a multiplier problem for the disc.* Studia Math. **44** (1972), 287–299.

37. *A moment problem and harmonic interpolation.* Mittag-Leffler report 5, 1972.

38. *Projections métriques de L^p sur H^p.* C. R. Acad. de Sci. **276** (1973), 1156–1160.

39. *On the distortion of sets on a Jordan curve under conformal mapping.* Duke Math. J. **40**:3 (1973), 547–559.

40. "The extension problem for quasiconformal mappings." Academic Press, N.Y., 1974.

41. *A counter example for measures bounded on H^p for the bi-disc.* Mittag-Leffler report 7, 1973.
42. with J. Garnett: *Interpolating sequences and separation properties.* J. Analyse Math. **28** (1975), 273–299.
43. *Two remarks on H^1 and BMO.* Advances in Math. **22** (1976), 269–277.
44. *Asymptotic paths for subharmonic functions in R^n.* Ann. Acad. Sci. Fenn. Ser. A I Math. **2** (1976), 35–39.
45. with G. Halász: *On an integral equation.* Mittag-Leffler report 10, 1978.
46. *A remark on Denjoy's inequality and Herman's theorem.* IHES, Publ. Math. **49** (1978), 235–241.
47. *The work of Charles Fefferman.* Proceedings of the International Congress of Mathematicians, Helsinki 1978.
48. *An example concerning analytic functions with finite Dirichlet integrals.* Investigations on linear operators and the theory of functions, IX. Zap. Nauchn. Sem. Leningrad Otdel. Mat. Inst. Steklov (LOMI) **49** (1979), 283–287.
49. *An explicit unconditional basis in H^1.* Bull. Sci. Math. **104** (1980), 405–416.
50. *Hacian's method in linear programming.* Mittag-Leffler report 3, 1980.
51. *Some analytic problems related to statistical mechanics.* Lecture notes in Math. **779**, Springer, Berlin, 1980.
52. with Peter Jones: *Weighted norm inequalities and a theorem of Koosis.* Mittag-Leffler report 2, 1981.
53. *BMO – 10 years' development.* Progress in Math. **11**, Boston, Mass., 1981.
54. *Estimates of harmonic measures.* Ann. Acad. Sci. Fenn. **7** (1982), 25–32.
55. with L. Gårding: *Obituary on Niels Erik Nørlund.* Acta Math. **148** (1982), 1.
56. *On H^∞ in multiply connected domains.* Anthony Zygmund honorary volume, Wadsworth 1983, 349–372.
57. with M. Benedicks: *On iterations of* $1-ax^2$ *on* $(-1,1)$. Mittag-Leffler report 3, 1983.
58. *On the support of harmonic measure for sets of Cantor type.* Ann. Acad. Sci. Fenn., Series A-I **10** ((1985), 113–123.
59. with S.-Y. A. Chang: *On the existence of an extremal funciton for an inequality of J. Moser.* Bull. Sci. Math **110** (1986), 113–127.
60. with Lars Ahlfors: *Arne Beurling in memoriam.* Acta Math. **161** (1988), 1–9.
61. with M. Benedicks: *On the Hénon attractor.* IXth Intern. Congr. Mathematical Physics, Swansea, 1988. Hilger, Bristol ((1989)), 498–500.
62. with C. J. Bishop, J. B. Garnett, P. W. Jones: *Harmonic measures supported on curves.* Pacific J. Math. **138** (1989), 233–236.
63. with M. Benedicks: *The dynamics of the Hénon map.* Ann. Math, **132** (1990), 629–725.

64. *The dynamics of non-uniformly hyperbolic systems in two variables.* Proc. of ICM, Kyoto 1990. Springer 1991.

65. *Stochastic behavior of deterministic systems.* Journ. Econ. Beh. and Org **16** (1991), 85–92.

66. *Stochastic models on some dynamical systems.* Lecture notes in Math. **1469**, Springer, Berlin (1991), 1–12.

67. with Peter W. Jones: *On coefficient problems for univalent functions and conformal dimension.* Duke Math. J. **66** (1992), 169–206.

68. with T. W. Gamelin: "Complex dynamics". University Tracts in Mathematics, Springer-Verlag, New York (1993), x+175 pp.

69. with N.G. Makarov: *Some results connected with Brennan's conjecture.*. Ark. Mat. **32** (1994), 33–62.

Lennart Carleson's work in Analysis

Peter W. Jones

0. Introduction

Lennart Carleson has made deep and fundamental contributions to the field of analysis during a period spanning more than forty years. I will attempt to give an overview of his work by focusing on some of the highlights. It is difficult in a short article to give a full report of the impact of Carleson's theorems and techniques, and I will only give the reader some glimpses which serve to indicate the influence this work has had on both analysis and other branches of mathematics.

I have rather arbitrarily divided Carleson's work into four areas: Complex Analysis, Fourier Analysis, Quasiconformal Mapping, and Harmonic Measure. Within these four categories the results are given in more or less chronological order.

While Carleson's work covers a wide area, one sees in his methodology a deep underlying unity. The problems he considers are always extremely concrete and the statements of his theorems are easily understood by any mathematician. But it is in the proofs where one sees the unmistakable stamp of Carleson. In virtually every major work there is an involved geometric construction carrying deep combinatorial structure of renormalizable (scale invariant) type. These methods form a massive toolchest where every analyst, from apprentice to master, must delve time and again.

1. Complex Analysis

Carleson's first paper [1] is based on his Ph.D. thesis from Uppsala and was published in Acta Math. in 1952. There he considers closed sets E in the unit circle and discusses when they are sets of uniqueness for certain classes of functions. By a set of uniqueness for a class $\mathcal{C}$ one means that if $f \in \mathcal{C}$ and f has radial limits zero on E,

$$f(\theta) = \lim_{r \nearrow 1} f(re^{i\theta}) = 0 \ , \ \theta \in E,$$

then f vanishes identically. For example, a classical result due to the brothers F. and M. Riesz asserts that E is a set of uniqueness for bounded holomorphic functions if and only if E has positive Lebesgue measure. Until Carleson's work, this was essentially all that was known.

The paper considers three classes of holomorphic functions:

Reprinted from Acta Universitatis Upsaliensis, Section C, Vol. 58 (1995), pp. 17–28.

1. Functions with high regularity (e.g. Lipschitz classes);
2. The Dirichlet space ($f' \in L^2(\mathbb{D})$);
3. Absolutely convergent Taylor series (A).

Theorem 1. *For both Lipschitz classes of order $\alpha > 0$ ($|f(z) - f(w)| \leq C|z - w|^\alpha$) and the spaces of functions $f = \Sigma a_n z^n$, $|a_n| = 0(n^{-p}), p > 1$, E is a set of uniqueness if and only if the open intervals I_j given by $\mathbb{T} \backslash E = UI_j$ have lengths ℓ_j satisfying*

$$\sum_{j=1}^{\infty} \ell_j \log \frac{1}{\ell_j} = +\infty. \tag{1.1}$$

Of course we also assume $|E| = 0$, for otherwise the F. and M. Riesz theorem applies. Beurling [Be] had earlier proved that condition (1.1) implies E is a set of uniqueness, and the nontrivial part of this result is to construct a non zero function when the sum in (1.1) converges.

For the Dirichlet space, Carleson studies sets E which satisfy a (now) so-called capacitary-density condition. let C_α denote capacity with respect to the kernel $|x - y|^{-\alpha}$. This capacity is a certain measure of size, and an excellent reference is Carleson's book [6]. it has the property that $C_\alpha(I) = |I|$ for intervals I (but not for general sets). E satisfies a capacitary-density condition for C_α if there is $\varepsilon > 0$ such that

$$C_\alpha(E \cap (x - t, x + t)) \geq \varepsilon t \tag{1.2}$$

whenever $x \in E, t > 0$. (E "looks thick.")

Theorem 2. *If E satisfies (1.2) for some $\alpha > 0$, then E is a set of uniqueness for the Dirichlet space if and only if (1.1) holds.*

For the third class of functions, A, Carleson first proved that there are sets of uniqueness having zero Lebesgue measure. He then considered the class $\Gamma = \Gamma_E$ of functions of bounded variation on $\mathbb{T}$ which are constant outside of E, and defined the subclass Γ_0 of functions normalized by $\int_E |d\mu(x)| = 1$. He then defined the *index of linear dependence* within E as

$$p(E) = \inf_{\mu \in \Gamma_0} \sup_{n \geq 0} \left| \int_E e^{inx} \, d\mu(x) \right|$$

so that $0 \leq p(E) \leq 1$. This definition turns out to be natural due to the arithmetic nature of the uniqueness problem for the class A.

Theorem 3. *If $p(E) > 0$, E is not a set of uniqueness for the class A.*

In fact, Carleson showed that if $p(E) > 0$, then every continuous function on E is the boundary function of an element in A.

In [2] Carleson solved the interpolation problem for H^∞ on the disk. This result is now understood to be one of the pillars of function theory, and it shows up in areas ranging from the corona problem to operator theory (and many places in between). Recall that a set $\{z_j\}$ in the disk $\mathbb{D}$ is an interpolating sequence for

$H^\infty = H^\infty(\mathbb{D})$ (the bounded analytic funcitons) if for every bounded sequence $\{\alpha_j\} \in \ell^\infty$ there is $f \in H^\infty$ such that

$$f(z_j) = \alpha_j \text{ for all } j.$$

It is elementary to see that this holds if and only if we can select f with a bound on its norm,

$$\|f\|_{H^\infty} \le C\|\{\alpha_j\}\|_{\ell^\infty}$$

for some constant C depending only on $\{z_j\}$.

Theorem 4. *The sequence $\{z_j\}$ is an interpolating sequence for H^∞ if and only if there is $\delta > 0$ such that*

$$\inf_j \prod_{\substack{k \\ k\neq j}} \left| \frac{z_j - z_k}{1 - z_j \bar{z}_k} \right| \ge \delta. \tag{1.3}$$

Notice that the factor $|(z_j - z_k)/(1 - z_j\bar{z}_k)|$ is the pseudo-hyperbolic distance $(\tilde{\rho}(z_j, z_k))$ between z_j and z_k, which equals the Gleason distance. The "only if" part of Theorem 4 is rather easy and is an argument with Blaschke products. The converse is the deep part. Carleson's solution involved a precursor of what are now called Carleson measures, i.e. measures μ on $\mathbb{D}$ with the property that there is $A < \infty$ such that the norm is $\le A$, i.e.

$$|\mu|(\{z \in \mathbb{D} : |z - e^{i\theta}| \le r\}) \le Ar$$

for all $e^{i\theta} \in \mathbb{T}$ and $r > 0$. This definition appears in Carleson's later paper [3]. These measures arise in the interpolation problem because (1.3) is equivalent to the existence of $\varepsilon > 0$ and $A < \infty$ such that

$$\inf_{j\neq k} \tilde{\rho}(z_j, z_k) \ge \varepsilon \tag{1.4}$$

and

$$\sum_j (1 - |z_j|)\delta_{z_j} \text{ is a Carleson measure of norm } A. \tag{1.5}$$

Carleson's proof of Theorem 4 used $H^1, L^\infty/H^\infty$ duality to reduce to a version of the next result (from [3]).

Theorem 5. *The measure μ is a Carleson measure if and only if the map*

$$H^1(\mathbb{T}) \ni f \to \int_{\mathbb{D}} |f(z)|\, d|\mu|(z)$$

is bounded on H^1, i.e. there is $B < \infty$ such that

$$\int_{\mathbb{D}} |f(z)|\, d|\mu|(z) \le B\|f\|_{H^1}.$$

Skipping ahead for a moment, we note this result (actually valid for $H^p, p > 0$) preshadows C. Fefferman's theorem on the duality of H^1 and BMO. The crucial step in any of the myriad proofs of duality is to use the fact that Theorem 5 holds (with suitable definitions) in any dimension. For example, if $\varphi(x)$ is a function on $\mathbb{R}^n$, with Poisson extension $\varphi(z)$ to $\mathbb{R}^{n+1}_+$, then $\varphi \in \mathrm{BMO}(\mathbb{R}^n)$ if and only if the gradient square of φ gives a weighted Carleson measure on $\mathbb{R}^{n+1}_+$,

$$\sup_{x\in\mathbb{R}^n,\, r>0} r^{-n} \iint_{B(x,r)} |\nabla\varphi(z)|^2 y \, d\mathrm{Vol}(z) < \infty. \tag{1.6}$$

Here $z = (x, y), x \in \mathbb{R}^n, y > 0$, and $B(x,r)$ is the ball centered at x of radius r.

One of the reasons Carleson measures play such a crucial role in function theory is that, when correctly interpreted, they are a conformal invariant. It is clear that the property of being an interpolating sequence is invariant under Möbius transformations that fix $\mathbb{D}$, so we should not be surprised that Carleson measures should have conformal invariance attached to them. Conversely, when one finds a Carleson measure, one knows there must be conformal invariance. To pick a (slightly misleading) example, consider (1.6). This should mean that BMO is a conformally invariant space, i.e.

$$\varphi(x) \in \mathrm{BMO}(\mathbb{R}^n) \Leftrightarrow \varphi\left(\frac{x}{|x|^2}\right) \in \mathrm{BMO}(\mathbb{R}^n).$$

This is indeed true, though it is easier to prove than (1.6).

To drive home this point, let us consider a seemingly unrelated problem: Cauchy integrals on curves. Let $\Gamma = \{x + iA(x) : x \in \mathbb{R}\}$ be a Lipschitz curve in $\mathbb{R}^2$, i.e.

$$|A(x) - A(y)| \leq C|x - y| \ , \ x, y \in \mathbb{R},$$

where C is the Lipschitz constant. Define

$$Cf(z) = \mathrm{p.v.} \int_\Gamma \frac{f(w)}{w - z}\, dw,$$

where *p.v.* denotes principal value. If $A \equiv 0$ we get (a multiple of) the Hilbert transform, which is bounded on L^2. A theorem (due to Calderón [C] for small C and Coifman, McIntosh, Meyer [C,M,M] for all $C < \infty$) asserts that C defines an operator bounded on $L^2(\Gamma)$ for any Lipschitz curve. The conformally invariant space here is L^∞ (because $A' \in L^\infty$), and the crucial step in any proof of L^2 boundedness uses a Carleson measure associated to A'. This line of reasoning is today very well developed.

In 1962 Carleson [3] published his celebrated proof of the corona theorem for the disk. Now $H^\infty(\mathbb{D})$ is a Banach algebra, and $\mathbb{D}$ is a subset of the maximal ideal space (under point evaluation). It thus makes sense to conjecture, as Kakutani did in the early 1940's, that the maximal ideal space is the closure of $\mathbb{D}$ (in what would be called today the Gelfand topology). The name "corona" comes from the following analogy. Let $\bar{\mathbb{D}}$ be the (Gelfand) closure of $\mathbb{D}$ in $\mathcal{M}$, the maximal ideal space. Let $\mathcal{M}$ be the sun and $\bar{\mathbb{D}}$ be the moon. If we consider an eclipse, the moon covers the sun, giving $\mathcal{M}\backslash\bar{\mathbb{D}}$. What is visible is the corona. Thus the corona conjecture asserted that there is no corona, $\mathcal{M}\backslash\bar{\mathbb{D}} = \emptyset$. To solve this problem, one must deal with a more concrete (but equivalent) one which is the following.

Theorem 6. *If* $\|f_j\|_{H^\infty} \le 1$, $1 \le j \le N$ *and there is* $\varepsilon > 0$ *such that*

$$\sup_j |f_j(z)| \ge \varepsilon \tag{1.7}$$

for all $z \in \mathbb{D}$, *then there are* $g_j \in H^\infty(\mathbb{D})$ *such that* $\|g_j\|_{H^\infty} \le C(\varepsilon, N), 1 \le j \le N$, *and*

$$\sum_j f_j g_j \equiv 1. \tag{1.8}$$

To appreciate the difficulties encountered, let us look at the case (considered in [2]) where the corona data (the f_j) consist of two Blaschke products B_1, B_2 with simple zeros, and suppose the zeros of B_1, $\{z_j\}$, form an interpolating sequence. If we can solve (1.8), it must be that $B_2(z_j)g_2(z_j) = 1$ for all j. Theorem 4 is exactly the statement that such a $g_2 \in H^\infty$ exists, because by (1.7), $|B_2(z_j)| \ge \varepsilon$ for all j. Now since $1 - B_2 g_2$ vanishes on $\{z_j\}$, elementary H^∞ theory tells us that

$$1 - B_2 g_2 = B_1 g_1$$

for some $g_1 \in H^\infty$, and $\|g_1\|_{H^\infty} = \|1 - B_2 g_2\|_{H^\infty}$. It therefore seems natural to prove the corona theorem for general data by using a strong version of Theorem 4, and this is exactly what Carleson did.

The crucial step in Carleson's proof was to introduce the "Carleson construction" of "Carleson contours." This is a method of separating the places where $F \in H^\infty$ is big from places where F is small, and doing it with control. More precisely, let $\|F\|_{H^\infty} = 1$ and let $\varepsilon > 0$. Then there is a $\delta = \delta(\varepsilon) > 0$ and a collection of closed contours $\{\Gamma_j\}$ in $\mathbb{D}$ such that

$$\{z : |F(z)| > \varepsilon\} \text{ is separated from } \{z : |F(z)| < \delta\}$$

by $\{\Gamma_j\}$ and if μ is arclength measure on $\bigcup_j \Gamma_j$,

$$\mu \text{ is a Carleson measure.}$$

This separation allows an interpolation argument to be used, and that is how the proof of Theorem 6 is then finished.

The corona construction is widely regarded as one of the most difficult arguments in modern function theory. Those who take the time to learn it are rewarded with one of the most malleable tools available. Many of the deepest arguments concerning hyperbolic manifolds are easily accessible to those who understand well the corona construction. An indication that this should be true is obtained by considering a Riemann surface S covered by the disk. Suppose $S \cong \mathbb{D}/\Gamma$, where Γ is a Fuchsian group, and let $\{z_j\}$ be the orbit of some point (say the origin) under Γ. Then S has Green's function if and only if the sequence $\{z_j\}$ satisfies (1.3), which in this case is equivalent to (1.5). Let us now consider the following result of P. Doyle [D]. Let G be a classical Schottky group acting on hyperbolic three space $\mathbb{H}$, and let $M = \mathbb{H}/G$ be the corresponding manifold.

Then the bass eigenvalue, λ, of M (the square of the fundamental frequency) is bounded away from zero,

$$\lambda \geq c_0 > 0,$$

where c_0 is independent of G. Doyle's argument uses an ingenious construction of corona type on a fundamental domain for M. The point of this perhaps idiosyncratic remark is not that Doyle's argument is a simple version of the corona construction (it is not!), but rather that the corona construction provides a framework, within which the reader can understand the technical difficulties arising in a wide variety of situations not obviously related to the corona problem.

Returning to the corona problem for other situations, we note that little is well understood beyond the case of the disk. The method of proof now usually given for the corona theorem (for $\mathbb{D}$) is a modification due to Hörmander (see [Hö]) and involves solutions of the $\bar{\partial}$ problem with L^∞ bounds. It is well understood that, at least in the disk, interpolation and L^∞ bounds on solutions of $\bar{\partial}$ are the same. T. Wolff has found a simpler proof for $\mathbb{D}$ (see [Gar]), and his proof much resembles certain proofs of H^1, BMO duality. No matter which proof one uses, Carleson measures appear explicitly or implicitly. B. Cole showed ([Gam]) the corona problem fails for certain Riemann surfaces, but the problem remains open for the ball or polydisk in dimension ≥ 2. The problem is also open for general planar domains, though there is some progress. In 1982, Carleson studied the following interesting case. Let $\Omega = \mathbb{C}\backslash E$, where $E \subset \mathbb{R}$ (a Denjoy domain) and suppose E satisfies

$$|E \cap (x-t, x+t)| \geq \varepsilon t \tag{1.9}$$

for all $x \in E$ and $t \leq$ diameter (E). Then the corona theorem is true for Ω ([11]). Notice that (1.9) is condition (1.2) for Lebesgue measure! The proof of this theorem uses a deep analysis of $\bar{\partial}$ operators on the universal covering surface, and actually proves much more than the corona theorem. This paper, and all the machinery of corona type constructions, was used by others to solve the corona problem for arbitrary Denjoy domains [G,J].

2. Fourier Analysis

Surely one of the most celebrated results in Fourier analysis is Carleson's 1966 solution [4] of the Lusin conjecture:

Theorem 7. *The Fourier series of an L^2 function on $\mathbb{T}$ converges almost everywhere.*

Before this result, very little was known except in the realm of counterexamples for L^1, produced by Kolmogorov. Since Carleson's result extensions have been given to $L^p, 1 < p < 2$, by R. Hunt [H], and by Carleson's student P. Sjölin to $L \log L \log\log L$ (see [Sj]). C. Fefferman [F2] has given an alternative approach to Carleson's theorem, but there is still no short proof. Indeed, Carleson's proof is one of the most complicated to be found in all of modern analysis. Suffice it to say that the main idea is to (repeatedly) renormalize the situation for a given f, keeping track of errors as the scale changes. The goal, of course, is to get a hold on the L^2 norm of the maximal partial sum operator

$$S_N^* f(x) = \sup_n \sum_{-n}^{n} \hat{f}(k) e^{2\pi i k x}.$$

Carleson's method is based on the premise that this operator must be understood at many scales.

After these sketchy remarks, it is amusing to focus on one of the small points of the proof of L^2 convergence, namely one of the error terms. Let $\{I_j\}$ be a collection of disjoint, open intervals in $[0,1]$ with centers x_j and lengths y_j. Let

$$\varphi_j(x) = \left(1 + \frac{|x - x_j|^2}{y_j^2}\right)^{-1}$$

be the scaled multiple of the Poisson kernel for the point $(x_j, y_j) \in \mathbb{R}_+^2$, and let $\varphi = \sum_j \varphi_j$. The function φ is one of the error terms arising in [4], and Carleson proves the estimate

$$|\{x : \varphi(x) > \lambda\}| \leq c_1 e^{-c_2 \lambda}. \tag{2.1}$$

This estimate (extended to $\mathbb{R}^n$ by Zygmund [Z]) is now understood to be the John-Nisenberg Theorem applied to $\varphi \in$ BMO. (Zygmund reinterpreted this result as a statement about the integral of Marcinkiewicz.) The modern proof that $\varphi \in$ BMO is that the Poisson balayage takes Carleson measures in $\mathbb{R}_+^{n+1}$ to BMO functions on $\mathbb{R}^n$. (The converse statement is also true, i.e. that BMO functions are Poisson balayages of Carleson measures because this is equivalent to H^1, BMO duality. Carleson gave in [10] a direct construction of the Carleson measure of which the given BMO function is a balayage.)

Carleson's second fundamental contribution to Fourier analysis is the solution of the Bochner-Riesz problem in two dimensions, the subject of a joint paper with P. Sjölin [7]. Let

$$m_\alpha(x) = (1 - |x|^2)_+^\alpha$$

be the multiplier (supported on the unit disk) associated to the operator T_α,

$$\widehat{T_\alpha f}(x) = m_\alpha(x) \hat{f}(x),$$

where $\wedge$ denotes Fourier transform. This operator is associated in a natural way with summation of Fourier series, and the problem is to see for which p it is true that for all $\alpha > 0$, T_α is a bounded operator on L^p.

Theorem 8. *If $\alpha > 0, T_\alpha$ is a bounded operator on $L^p(\mathbb{R}^2), 4/3 \leq p \leq 4$.*

The theorem is best possible in terms of p. In $\mathbb{R}^n$, $n \geq 3$, the corresponding result remains open, and it is conjectured that T_α is bounded (for all $\alpha > 0$) on $L^p(\mathbb{R}^n)$, where

$$\frac{2n}{n+1} \leq p \leq \frac{2n}{n-1}.$$

This is one of the deepest open problems in Fourier analysis. C. Fefferman [F1] proved that the multiplier for the disk in $\mathbb{R}^2$ (or ball in $\mathbb{R}^n$) defines a bounded operator only in the trivial case when $p = 2$. J. Bourgain has made some recent progress on the Bochner-Riesz problem, see [Bo].

3. Quasiconformal Mappings

Let $F : \mathbb{R}^2_+ \to \mathbb{R}^2_+$ be a quasiconformal homeomorphism, and suppose $F(\mathbb{R}) = \mathbb{R}$. Then F need not be absolutely continuous on $\mathbb{R}$. The reason is that the necessary and sufficient condition for $f : \mathbb{R} \to \mathbb{R}$ to be the boundary values of a quasiconformal F mapping $\mathbb{R}^2_+$ to itself (and with $F(\infty) = \infty$) is the Beurling-Ahlfors condition: There exists $M < \infty$ such that

$$\frac{1}{M} \leq \frac{|f(I)|}{|f(J)|} \leq M \tag{3.1}$$

for all intervals I and J such that I and J are adjacent and of equal length. (See [Be].) It is easy to construct f satisfying (3.1) but not absolutely continuous.

Carleson [5] was the first to find the natural condition under which f is absolutely continuous and this is expressed in terms of the Beltrami coefficient for F. Recall that on $\mathbb{R}^2_+$,

$$\bar{\partial} F = \mu \partial F$$

for some $\mu \in L^\infty(\mathbb{R}^2_+)$, $\|\mu\|_{L^\infty} < 1$.

Theorem 9. *Suppose there is a function $\varepsilon(y)$ such that ε is an increasing function of y and*

$$|\mu(x+iy)| \leq \varepsilon(y)\ , \ x \in \mathbb{R}, y > 0. \tag{3.2}$$

Then f is absolutely continuous provided that

$$\int_0^1 \varepsilon^2(y)\,\frac{dy}{y} < \infty. \tag{3.3}$$

If one inspects the proof one finds that, in more modern language, $\log |f'(x)| \in$ VMO, if one defines VMO as being with respect to intervals of length less than one. The theorem is essentially sharp in the sense that if $\varepsilon(y)$ has some regularity and the integral in (3.3) diverges, there is a corresponding f such that f' is singular to Lebesgue measure. S. Semmes [Se] found an extension of this theorem, where (3.1) and (3.2) are replaced by a quadratic Carleson measure condition on μ.

Taking a closer look at Carleson's proof of Theorem 9, one discovers that the hypothesis $F : \mathbb{R}^2_+ \to \mathbb{R}^2_+$ is actually irrelevant. If (3.2) and (3.3) hold, the restriction of F to $\mathbb{R}$ is absolutely continuous provided $F : \mathbb{R}^2 \to \mathbb{R}^2$ extends to be conformal on $\mathbb{R}^2_-$. Suppose $F(\mathbb{R}^2_+) = \Omega$. Then the Laplacian on Ω can be carried over to an operator of divergence form on $\mathbb{R}^2_+$, call it L. Carleson's theorem then asserts that the harmonic measure for L is absolutely continuous with respect to arclength on $\mathbb{R}$. This line of reasoning has now been extensively studied. See e.g. the paper of R. Fefferman, C. Kenig, and J. Pipher [F,K,P].

Another quasiconformal problem successfully attacked by Carleson is the so-called lifting problem. Suppose $f : \mathbb{R}^n \to \mathbb{R}^n$ is a quasiconformal homeomorphism (quasisymmetric when $n = 1$). Is f the restriction of a quasiconformal $F : \mathbb{R}^{n+1} \to \mathbb{R}^{n+1}$? When $n = 1$, this was proven by Beurling and Ahlfors [Be], and when $n = 2$ it is a theorem of Ahlfors [A]. The Ahlfors proof uses factorization to reduce the problem to $1 + \varepsilon$ quasiconformal maps, which is easily handled. The factorization problem is open in higher dimensions, but Carleson [9] was able to prove

Theorem 10. *A quasiconformal homeomorphism of* $\mathbb{R}^3$ *can be lifted (quasiconformally) to* $\mathbb{R}^4$.

To prove this it is sufficient to lift to $\mathbb{R}^4_+$. Carleson's idea was to make the extension (uniformly) piecewise linear (PL) on every Whitney cube in $\mathbb{R}^4_+$, in the sense of hyperbolic geometry. As in the Ahlfors case ($n = 2$), this is easy when F is $1+\varepsilon$ quasiconformal. The difficulty is thus when F has a large coefficient of quasiconformality. In $\mathbb{R}^3$, there is a reasonable amount of control over PL functions, and Carleson was able to lift that control to $\mathbb{R}^4_+$. When $n = 4$, PL functions are much much more difficult to control. However, it turns out that Carleson's ideas can be modified to solve the general lifting problem. This was carried out in a paper of P. Tukia and J. Väisälä [T,V].

4. Harmonic Measure

For $\mu \geq 0$ a Borel measure on $\mathbb{R}^n$ we can define the dimension of (the support of) μ by setting

$$\mathrm{Dim}(\mu) = \inf\{\dim(E) : \mu(E^c) = 0\},$$

where $\dim(E)$ denotes Hausdorff dimension. Carleson was the first to push the problem of evaluating $\mathrm{Dim}(\omega)$ for harmonic measure ω. Recall that if $\Omega \subset \mathbb{R}^2$ is a domain and $z \in \Omega$ there is a unique probability measure ω_z such that (for a suitable class of harmonic functions)

$$f(z) = \int_{\partial\Omega} f(x)\, d\omega_z(x).$$

The definition in $\mathbb{R}^n$ is similar. If $z, z' \in \Omega$, their respective harmonic measures are mutually absolutely continuous, so it makes sense to define $\mathrm{Dim}(\omega)$ without specifying the basepoint.

The Beurling theorem [Be] for simply connected domains states that for $z \in \Omega$ and $x \in \partial\Omega, r > 0$,

$$\omega_z(\{y \in \partial\Omega : |x-y| < r\}) \leq Cr^{1/2}, \tag{4.1}$$

and a moment's thought shows this implies $\mathrm{Dim}(\omega) \geq \frac{1}{2}$. Carleson made the first improvement on this in [8] where he proved

Theorem 11. *If* $\Omega \subset \mathbb{R}^2$ *is simply connected,* $\mathrm{Dim}(\omega) \geq \frac{1}{2} + c_0$.

In this result, $c_0 > 0$ does not depend on Ω. The proof uses a geometric argument to show there are not too many places where (very roughly speaking)

$$\omega_z(\{y \in \partial\Omega : |x-y| < r\}) \sim r^{1/2+\varepsilon}.$$

This was used by J. Brennan [Br] to get the best known results on what is now called the Brennan conjecture: If $F : \Omega \to \mathbb{D}$ is univalent and $\varepsilon > 0$,

$$\iint_{\Omega} |F'(z)|^{4-\varepsilon}\, dx\, dy < +\infty. \tag{4.2}$$

Carleson's next result on this subject concerns Cantor sets in the plane [12]. We state here a weak version of his theorem. Let $E \subset [0,1]$ be any Cantor set "formed from a pattern", and let

$$\Omega^c = E \times E.$$

Theorem 12. *For such domains Ω,* $\mathrm{Dim}(\omega) < 1$.

This is a remarkable result because $\partial\Omega$ could have any Hausdorff dimension less than two. His proof is also remarkable, as it introduces the use of entropy via the Shannon-McMillan-Breiman theorem. The point is that after suitable scaling,

$$\frac{1}{2\pi}\int_{\partial\Omega}\frac{\partial G}{\partial n}\log\frac{\partial G}{\partial n}\,ds = \Sigma_j G(z_j)$$

where the sum is taken over all critical points of Green's function. The left hand integral has a natural interpretation as an entropy, and the right hand sum is positive in a way amenable to calculation. This is why $\mathrm{Dim}(\omega)$ gets pushed below one.

Shortly after Carleson proved Theorem 11, N. Makarov [M] showed $\mathrm{Dim}(\omega) = 1$ for any simply connected planar domain (and much more). Carleson's approach was later used by J. and T. Wolff [J,W] to prove $\mathrm{Dim}(\omega) \leq 1$ for any planar domain. T. Wolff [W] has also used Carleson's approach to show there are domains in $\mathbb{R}^n, n \geq 3$, such that for some $\varepsilon_0 > 0$,

$$\mathrm{Dim}(\omega) \geq n - 1 + \varepsilon_0.$$

5. Selected Recent Work

Carleson's research continues unabated. I will take just a few examples to give a flavor of the directions this work has taken. In joint work [13] it is shown that certain coefficient problems are related to level lines of Green's function. Recall that Σ is the class of univalent functions in $\{|z| > 1\}$ with Laurent expansion

$$z + \sum_{n=0}^{\infty} b_n z^{-n}.$$

While the exact decay rate of b_n is unknown, it is proven in that paper that for fixed n

$$\sup_{\Sigma}|b_n| \sim \sup_{\Sigma} n^{-1}\int_{|z|=1+\frac{1}{n}}|f'(z)|\,ds(z). \tag{5.1}$$

The point here is that the constants in $\sim$ above do not depend on n. This reduces the coefficient problem to a geometric problem: How long can the level line $\{z \in \Omega : G(z,\infty) = \frac{1}{n}\}$ be? Here G is Green's function with pole at ∞ in an image domain Ω (under $f \in \Sigma$). For fixed Ω, the rate of growth of the level lines of G can be used to define a "conformal dimension" of $\partial\Omega$, and this is shown to be equivalent to a definition similar to those found in statistical mechanics. It is shown that near maximizers for length ($\{z : G(z) = \frac{1}{n}\}$) come from fractal domains, and computer studies indicate that Julia sets (as boundaries) may give near maximum values. It is interesting to note that one of the proofs in this paper requires the two dimensional version of estimate (2.1).

Carleson has also carried out research with N. Makarov. In [14] they prove some theorems closely related to Brennan's conjecture (4.2). In separate work they are studying properties of D.L.A., one of the most mysterious processes related to harmonic measure.

6. Some Remarks

Carleson's influence extends far beyond his research, a fact well known to the broad mathematical community. Besides his papers he has published an influential book on potential theory [6] and (with P. Malliavin, J. Neuberger, and J. Wermer) made accessible the unpublished work of Arne Beurling [Be]. (See Benedicks' article for Carleson's books on dynamical systems.) But Carleson's influence extends far beyond his publications. He has trained many Ph.D. students, and many more mathematicians who came from around the world to learn from him. As director of the Institut Mittag-Leffler he not only developed a world class research center, but molded an entire generation of analysts. His research in analysis is a series of towering and fundamental discoveries. His friends know well his generosity, encouragement, and selfless giving of himself.

Selected Works of Lennart Carleson

1. *Sets of uniqueness for functions analytic in the unit disc,* Acta Math. **87** (1952), 325–345.
2. *An interpolation problem for bounded analytic functions,* Amer. J. Math. **80** (1958), 921–930.
3. *Interpolation by bounded analytic functions and the corona problem,* Annals of Math. **76** (1962), 547–559.
4. *On convergence and growth of partial sums of Fourier series,* Acta Math. **116** (1966), 135–157.
5. *On mappings, conformal at the boundary,* J. Analyse Math. **19** (1967), 1–13.
6. *Selected Problems on Exceptional Sets,* Van Nostrand Co., 1967. (Reprinted by Wadsworth, 1983.).
7. (with P. Sjölin), *Oscillatory integrals and a multiplier problem for the disc,* Studia Math. **44** (1972), 287–299.
8. *On the distortion of sets on a Jordan curve under conformal mapping,* Duke Math. J. **40** (1973), 547–559.
9. *The extension problem for quasiconformal mappings,* . Contributions to Analysis, edited by L. Ahlfors et al., Academic Press, 1974, pp. 39–47.
10. *Two remarks on H^1 and* BMO, Advances in Math. **22** (1976), 269–277.
11. *On H^∞ in multiply connected domains,* Conference on Harmonic Analysis in Honor of Antoni Zygmund, Vol. 2, Wadsworth Inc., 1983, pp. 349–372.
12. *On the support of harmonic measure for sets of Cantor type,* Ann. Acad. Sci. Fenn. **10** (1985), 113–123.
13. (with P. Jones), *On coefficient problems for univalent functions and conformal dimension,* Duke Math. J. **66** (1992), 169–206.
14. (with N. Makarov), *to appear.*

References

[A] L. Ahlfors, *Extension of quasiconformal mappings from two to three dimensions,* Proc. Nat. Acad. Sci. U.S.A. **51** (1964), 768–771.

[Be] *The Collected Works of Arne Beurling, Vols 1, 2,* edited by L. Carleson et al., Birkhäuser, 1989.

[Bo] J. Bourgain, *Some new estimates on oscillatory integrals*, to appear in: Essays on Fourier Analysis in Honor of Elias M. Stein, Princeton University Press.

[Br] J. Brennan, *The integrability of the derivative in conformal mapping*, J. London Math. Soc. **18** (1978), 261–272.

[C] A. Calderón, *Cauchy integrals on Lipschitz graphs and related operators*, Proc. Nat. Acad. Sci. U.S.A. **74** (1977), 1324–1327.

[C,M,M] R. Coifman, A. McIntosh, and Y. Meyer, *L'intégrale de Cauchy définit un opérateur borné sur L^2 pour les courbes Lipschitziennes*, Annals of Math. **116** (1982), 361–387.

[D] P. Doyle, *On the bass note of a Schottky group*, Acta Math. **160** (1988), 249–284.

[F1] C. Fefferman, *The multiplier theorem for the ball*, Annals of Math. **94** (1971), 330–336.

[F2] ———, *Pointwise convergence of Fourier series*, Annals of Math. **98** (1973), 551–571.

[F,K,P] R. Fefferman, C. Kenig, and J. Pipher, *The theory of weights and the Dirichlet problem for elliptic equations*, Annals of Math. **134** (1991), 65–124.

[Gam] T. Gamelin, *Uniform Algebras and Jensen Measures*, London Math. Soc. Lecture Notes Series, No. 32, Cambridge University Press, 1979.

[Gar] J. Garnett, *Bounded Analytic Functions*, Academic Press, 1981.

[G,J] J. Garnett and P. Jones, *The corona theorem for Denjoy domains*, Acta Math. **155** (1985), 27–40.

[H] R. Hunt, *On the convergence of Fourier series*, (Edwardsville, Ill., 1967), Southern Illinois Univ. Press, Carbondale, Ill., 1968, Proc. Conf. Orthogonal Expansions and their Continuous Analogues, 235–255.

[Hö] L. Hörmander, *Generators for some rings of analytic functions*, Bull. Amer. math. Soc. **73** (1967), 943–949.

[J,W] P. Jones and T. Wolff, *Hausdorff dimension of harmonic measures in the plane*, Acta Math. **161** (1988), 131–144.

[Ma] N. Makarov, *On the distortion of boundary sets under conformal mappings*, Proc. London Math. Soc. **51** (1985), 369–384.

[Se] S. Semmes, *Quasiconformal mappings and chord-arc curves*, Trans. A.M.S. **306** (1988), 233–263.

[Sj] P. Sjölin, *An inequality of Paley and convergence a.e. of Walsh-Fourier series*, Ark. Mat. **7** (1969), 551–570.

[T,V] P. Tukia and J. Väisälä, *Quasiconformal extension from dimension n to $n+1$*, Annals of Math. **115** (1982), 331–348.

[W] T. Wolff, *Counterexamples with harmonic gradients in Euclidean space*, to appear in: Essays on Fourier Analysis in Honor of Elias M. Stein, Princeton University Press.

[Z] A. Zygmund, *On certain lemmas of Marcinkiewicz and Carleson*, J. Approx. Theory **2** (1969), 249–257.

PETER W. JONES,
MATHEMATICS DEPARTMENT, YALE UNIVERSITY,
NEW HAVEN, CT 06520-8283

Lennart Carleson's work in Statistical mechanics and Dynamical systems

Michael Benedicks

Lennart Carleson is, at least until recently, best known for his fundamental work in Fourier Analysis and Complex Analysis. In the field of dynamical systems his most appreciated work is on the Hénon attractor and chaotic behaviour in the one-dimensional quadratic family (Jakobson's theorem). Since there are already several surveys of this [11], [23], [21], [1], I would here like to emphasize other parts of his work, and that Lennart Carleson's interest in statistical mechanics and dynamical systems goes in fact back through his entire career.

I. Statistical Mechanics

We consider N particles $\{(q_i, p_i)\}_{i=1}^N$ evolving according to a Hamiltonian function $H = H(p,q) = H(p_1, \dots, p_N, q_1, \dots, q_N)$ by the classical equations of motion

$$\begin{cases} q_i = \dfrac{\partial H}{\partial p_i} \\ p_i = -\dfrac{\partial H}{\partial q_i} \end{cases} \qquad i = 1, 2, \dots, N.$$

The associated phase space is $\Omega \subset \mathbb{R}^{6N}$.

A typical form of the Hamiltonian function could be

$$H = \sum_{\nu=1}^{N} \frac{1}{2m_\nu} p_\nu^2 + \sum_{\nu \neq \mu} \Phi(q_\nu, q_\mu),$$

where Φ is a *pair potential* depending on the coordinates of two of the particles.

The starting point of the famous theory of Gibbs for the *canonical ensemble* is the following postulate:

(CE) The density of particles in phase space is given by

$$F = C_{N,\beta} e^{-\beta H},$$

where $C_{N,\beta}$ is a normalizing constant.

A consequence of this postulate is that the distribution of velocities for an ideal gas is given by the Maxwell-Boltzman distribution.

Reprinted from Acta Universitatis Upsaliensis, Section C, Vol. 58 (1995), pp. 29–40.

A fundamental problem in Statistical Mechanics is to verify this rigorously starting from basic principles. This is an area where Lennart Carleson has given several contributions.

A. Boltzmann's theory for gases.

In the Boltzmann model for gases the particle density is given by $F(x, v, t)$, $x \in \mathbb{R}^3$, $v \in \mathbb{R}^3$, $T \in \mathbb{R}$ and its time development is given by the equation

$$\frac{\partial F}{\partial t} + \sum v_i \frac{\partial F}{\partial x_j} + \sum X_i \frac{\partial F}{\partial v_i} = \iint \left[F(s^1)F(s_1^1) - F(s)F(s_1)\right] h(s, s_1, q)$$

or

$$D(F) = T(F).$$

If we assume no explicit dependence on the coordinates x_i we obtain the integrodifferential equation

$$\frac{\partial F}{\partial t} = T(F).$$

The existence, uniqueness and asymptotic properties of the solutions to this equation were treated in the famous book by T. Carleman "Problèmes mathématiques dans la théorie cinétique des gaz", [7].

At the time of Carleman's death in 1955 the book only existed as an incomplete manuscript, which was edited by Lennart Carleson and Otto Frostman. Lennart Carleson made essential contributions to complete the manuscript and fill in the gaps.

As an example we mention that the proof that

$$F(x, v, t) \to Ce^{-\beta v^2} \qquad \text{(Maxwell distr.)}$$

as $t \to \infty$ in Carleman's book depends on the following variational lemma due to Lennart Carleson.

Lemma. *Let μ be a positive measure and consider the set of functions*

$$\mathcal{E} = \left\{\varphi \colon \int_\Omega \varphi d\mu = A, \int_\Omega \varphi w^2 d\mu \leq B\right\}.$$

Suppose that there exists a function $\varphi_0 = ae^{-2w^2}$ so that equality holds in the conditions defining $\mathcal{E}$: $\int_\Omega \varphi d\mu = A$, $\int_\Omega \varphi w^2 d\mu = B$. Then for all $\varphi \in \mathcal{E}$

$$I(\varphi) = \int \varphi \log \varphi d\mu \geq I(\varphi_0),$$

and equality holds only if $\varphi \equiv \varphi_0$ on supp μ.

The proof of the convergence to the Maxwell distribution proceeds by first showing that the limiting function $\psi(v) = \lim_{t\to\infty} F(v, t)$ satisfies

$$\inf_{\varphi \in \mathcal{E}} I(\varphi) \geq I(\psi)$$

and then invoking the lemma.

B. Gibbs's theory for the harmonic oscillator

In his lectures in Maryland 1979, [10], Lennart Carleson again addressed the fundamental problems of deriving the Gibbs theory from first principles. The starting point is to consider the Hamiltonian for the harmonic oscilator

$$H = \sum_{\nu=1}^{N} \tfrac{1}{2} p_\nu^2 + \sum_{\nu \neq \mu} a_{\nu-\mu} q_\nu q_\mu .$$

The corresponding Newton equations are

$$(*) \qquad \begin{cases} \ddot{y}_\nu(t) = -2 \sum_{\nu \neq \mu} a_{\nu-\mu} y_\mu(t), \\ y_\nu(0) = 0, \\ y'_\nu(0) = b_\nu, \qquad \nu = 1, 2, \ldots, N. \end{cases}$$

Carleson's aim was to conclude that for a large set of initial conditions the velocity distribution was given by the Maxwell–Boltzman distribution.

Let us introduce the functions

$$\begin{cases} A(x) = \sum_{\nu=-\infty}^{\infty} a_\nu e^{i\nu x} \\ a(x) = \sqrt{A(x)} \\ b(x) = \sum_{\nu=-\infty}^{\infty} b_\nu e^{i\nu x} \end{cases}$$

Then there is an explicit formula that gives the solution of $(*)$:

$$y_\nu(t) = \frac{1}{2\pi} \int_{-\pi}^{\pi} b(x) \frac{\sin a(x) t}{a(x)} e^{-i\nu x} dx.$$

Carleson introduced the condition

$$A_z = \{a \colon a'(x) = z \text{ only on a set of measure } 0\}$$

and proved the following results.

Theorem. (a) *Assume A_z for all z and let $\{b_\nu\}$ be an independent identically distributed sequence of random variables with mean value zero and finite second moment. Then with probability 1 the limiting distribution is Maxwell-Boltzmann.*

(b) *Assume A_0 and that $\{b_\nu\}$ are independent normally and identically distributed random variables with zero expectation. Then with probability 1 the limiting distribution is Maxwell-Boltzmann.*

C. A statistical mechanical approach to Szegő's asymptotical formula for Toeplitz determinants (thesis of Kurt Johansson).

Let f be a function given on the unit circle $\mathbb{T}$ with complex Fourier coefficients $\{f_\nu\}_{\nu=-\infty}^{\infty}$. Then the Toeplitz determinant of order n with respect to the generating function f is given by

$$D_{n-1}(f) = \det(f_{\nu-\mu})_{0 \leq \nu, \mu \leq n-1}.$$

In 1952 G. Szegő proved the very precise asymptotic formula

$$D_{n-1}(f) = \exp\left\{nc_0 + \sum_{k=1}^{\infty} kc_k c_{-k} + o(1)\right\}$$

under the assumption $f > 0$, $f \in C^{1+\alpha}(\mathbb{T})$, $\alpha > 0$. Here $\{c_k\}$ are the complex Fourier coefficients of $\log f$. Later other proofs of Szegő's formula with various assumptions on the generating function f have appeared. Kurt Johansson (student of Lennart Carleson) proved in his thesis Szegő's formula for complex generating functions f under the assumption that $\sum_{k=1}^{\infty} k|c_k|^2 < \infty$. The starting point for Kurt Johansson's investigation was the following integral representation of a Toeplitz determinant

$$D_{n-1}(f) = \frac{1}{(2\pi)^n n!} \int_{[0,2\pi]^n} \exp\left\{-\sum_{0\le\mu<\nu\le n-1} \log\left|e^{i\theta_\mu} - e^{i\theta_\nu}\right|^{-1}\right\} \prod_{\mu=0}^{n-1} f\left(e^{i\theta_\mu}\right) d\theta,$$

which is an expectation in the canonical ensemble of a Coulomb gas on the circle.

D. Phase transitions in Statistical Mechanics.

Another topic treated in the Maryland lectures, [10], is van Hove's model for particle interactions. Carleson started with a continuously differentiable function Φ on the torus $\mathbb{T}$ of length 1. He assumed that Φ was even with mean value 0 and studied the asymptotic behaviour of the "state sum" integrals

$$e^{-Nf_N(\beta)} = \int_0^1 \cdots \int_0^1 \exp\left\{-\frac{\beta}{N}\sum \Phi(x_\nu - x_\mu)\right\} dx_1 \ldots dx_N.$$

He proved the existence of $f(\beta) = \lim_{N\to\infty} f_N(\beta)$ and showed that the limiting function has a relation to a variational problem

$$f(\beta) = \inf_{\psi}\left\{\int_0^1\int_0^1 \beta\Phi(x-y)\psi(x)\psi(y)\,dx\,dy + \int_0^1 \psi\log\psi\,dx\right\},$$

where $\psi \ge 0$ and $\int_0^1 \psi\,dx = 1$.

Moreover he proved the existence of a minimizing function $\psi = \psi_0$ and that this system shows a phase transition, i.e. the function f is not analytic everywhere. This was the starting point of the thesis of another student of Lennart Carleson, Martin Tamm. In his thesis "Subanalytic Sets in the Calculus of Variation" published in Acta Math. 1981, [20], Martin Tamm determined the singularities of f also in the case that f depends on several variables; e.g. if f depends on two variables, the singularity set is a finite union of analytic arcs.

II. Dynamical systems

A. Real dynamical systems.

1. M. Herman's work on Arnold's problem for Circle diffeomorphisms.

Let $f\colon \mathbb{T} \to \mathbb{T}$ be a diffeomorphism of the circle. Then f has a lifting to a mapping

F of the line such that $F(x+1) = F(x) + 1$. Poincaré introduced the rotation number of a circle homeomorphism as the limit

$$\rho(f) = \lim_{n\to\infty} \frac{F^n(x) - x}{n},$$

which always exists even for homeomorphisms f. The simplest possible circle diffeomorphism is the rigid rotation $R_\alpha(x) = x + \alpha$.

A natural question is: When is f topologically conjugate to a rotation? Denjoy proved that this is true if α is irrational and Df is of bounded variation and also gave a famous counterexample. Arnold proved that if α is Diophantine and f is a real analytic function sufficiently close to R_α, then f is real analytically conjugate to R_α (compare Siegel's theorem). He conjectured that this would be true for general f without the a priori assumption that f is close to R_α. M. Herman, [16], succeeded to prove Arnold's conjecture for a large class of α's.

In an appendix to M. Herman's paper Lennart Carleson [9] made improvements to the proof making the estimates more explicit. In particular Carleson was able to improve the estimate in the right side of Denjoy's inequality

$$\left| \frac{1}{q} \sum_{i=0}^{q-1} \varphi\left(f^i(x_0)\right) - \int_0^1 \varphi\left(f(x)\right) d\mu(x) \right| \leq \frac{C}{q} \operatorname{Var}(\varphi)$$

to $\leq C/(q_j\, j^\beta)$ in the case that $q = q_j$ — one of the convergents of the continued fractions of α. The advantage of this method is that it avoids the Hurewicz ergodic theorem (which in its turn is based on the Chacon-Ornsteen ergodic theorem) and thereby makes the proof constructive.

2. The Hénon attractor and chaotic dynamics of one-dimensional quadratic maps. Around 1981 Lennart Carleson and I were intrigued by the pictures of the Hénon attractor. This simple dynamical system was constructed by the astronomer M. Hénon, [15], to be the simplest possible two-dimensional model which exhibits a "strange attractor" of the type first studied by Lorenz for a three-dimensional system of non-linear ODE's.

Hénon studied the iterations of the following invertible polynomial mapping $f = T_{a,b}$ of the plane

$$f : \begin{pmatrix} x \\ y \end{pmatrix} \to \begin{pmatrix} 1 + y - ax^2 \\ bx \end{pmatrix}, \qquad (a, b) = (1.4, 0.3).$$

He concluded that there is an invariant open set $U = U_{a,b}$ with the property $f(\overline{U}) \subset U$.

The Hénon map has two fixed points one of which is a saddle point located in the first quadrant. We denote this fixed point by z_f. The numerical experiments seemed to indicate that, if one picked an initial point in U, the orbits were attracted to the unstable manifold $W = W^u$ of this fixed point and seemed to have chaotic behaviour, i.e. positive Lyapunov exponent. However this was doubted by several mathematicians in particular in view of the Newhouse phenomena:

arbitrarily close to a two-dimensional C^2 diffeomorphism with a homoclinic tangency, there are maps with infinitely many stable periodic orbits.

Initially Carleson and I believed that there was what is known as an invariant cone field (we didn't know the terminology at the time), i.e. at each point z of w there is a cone $C(z)$ with the property that $f(C(z)) \subset C(f(z))$. However the numerical experiments convinced us that there was an obstruction. There seemed to be points for which the tangent vectors of the unstable manifold W contracted indefinitely at an exponential rate. If we had interpreted the outcome of this numerical experiment correctly at this time maybe we could have shortened the process of understanding the Hénon problem.

Instead we turned to the simpler case of studying the corresponding problem for one-dimensional maps $f_a\colon x \mapsto 1 - ax^2, 0 < a \leq 2$, and gave a new proof for Jakobson's theorem, [2]. (In [2], we only proved a subexponential growth of the derivative at the critical value; the statement here depends on subsequent development [3], [5].)

Theorem. *There is a set E of parameters of positive Lebesgue measure such that for all $a \in E$.*

(a) $|Df_a^n(1)| \geq e^{cn}, \qquad \forall n \geq 1;$

(b) f_a *has an absolutely continuous invariant measure.*

The set of parameters is constructed as a subset of

$$E_1 = \{a \in [0,2]\colon |f_a^j(0)| \geq e^{-\alpha j}, \forall j \geq 1\},$$

where α is a fixed number, $\alpha = \frac{1}{100}$ say. Technically, the main idea is to prove that there is a constant $c > 0$ such that

$$|Df_a^n(1)| \geq e^{cn} \qquad \forall n \geq 0.$$

The basic idea in the proof is the following: Suppose that we have proved by induction that $|Df_a^{n-1}(1)| \geq e^{c(n-1)}$ and that $x_n(a) = f_a^n(0)$ is close to 0. By the chain rule

$$|Df_a^{n+j}(1)| = |D_a^{n-1}(1)| \cdot |2ax_n| \cdot |Df^j(1 - ax_n^2)|\,.$$

Now $|x_n| \geq e^{-\alpha n}$ is not too small compared with $|Df_a^{n-1}(1)|$ and the iterates $x_{n+1}, \ldots, x_{n+j}$ follow $x_1, \ldots, x_j$ closely for $j \leq j_0(x_n)$ and j can be chosen so that $|Df^j(1 - ax_n^2)|$ compensates the small factor $|2ax_n|$.

The proof has to a large extent the flavour of proofs in harmonic analysis (Calderon-Zygmund theory), where Whitney decompositions and crude estimates play an important rôle. In general the theory of dynamical systems contains a mixture of topological and analytical ideas. It was a main contribution, when S. Smale managed to isolate the topological ideas in Levinson's theory of van der Pol's equation in his famous horseshoe construction. However, this development lead to emphasize an axiomatic setting — the uniform hyperbolic theory — which does not apply to that many "real life" dynamical systems.

An important model problem which is outside the uniform hyperbolic setting is the Hénon map. Lennart Carleson and I managed to treat this problem when the $b = |\det(DT_{a,b})|$ is small. In this case the Hénon map can be viewed as a small pertubation from the one-dimensional map $x \mapsto 1 - ax^2$. However, many new problems related to the two-dimensional geometry appears. We managed to prove the following theorem ([BC2])

Theorem. *There is a constant $b_0 > 0$ such that for all b, $0 < b < b_0$, there is a set of a-parameters E_b of positive Lebesgue measure such that $\forall a \in E_b$ there is an attractor $\Lambda_{a,b} = \overline{W^u(z_f)}$, where $W^u(z_f)$ is the unstable manifold of the fixed point in first quadrant. The following properties hold*

(1) *There is an open set $U \supset \Lambda_{a,b}$ such that $\forall z \in \mathcal{U}$*

$$\text{dist}\left(T^j_{a,b}z, \Lambda_{a,b}\right) \to 0$$

as $j \to \infty$.

(2) *There is a point $z_0 \in \Lambda_{a,b}$ such that $\{T^j z_0\}^{\infty}_{jz_0}$ is dense in $\Lambda_{a,b}$ (topological transitivity).*

(3) *There is a constant $c > 0$ such that*

$$\left|DT^j_{a,b}(z_0)\binom{0}{1}\right| \geq e^{cj}, \qquad \forall j \geq 1.$$

The main idea is to isolate the non-uniformly hyperbolic behaviour in the critical points. There is a countable set $\mathcal{C}$ of points on the unstable manifold, discovered in the computer experiment described initially, with the property that the lengths of images of the tangent vectors decay exponentially. The set of parameters are chosen so that for $\forall z_0 \in \mathcal{C}$, dist $(T^j z_0, \mathcal{C}) \geq e^{-\alpha j}$, $j \geq 1$, in order to avoid too complicated interaction between the critical points. However, the critical points themselves are only defined for the final parameter set. This leads to a complicated induction procedure in which the critical points are successively constructed with better and better accuracy.

The papers on the one-dimensional quadratic family and the Hénon map were the starting point for a subsequent development. Let me mention the following:

(1) The presence of Hénon-like strange attractors for two-dimensional maps close to homoclinic bifurcations (Mora-Viana, [18]).
(2) The construction of Sinai-Bowen-Ruelle measures for the Hénon map with "good parameters" (Benedicks-Young, [5]).
(3) Decay of correlation and central limit theorems hold for the one-dimensional quadratic family (Young, [22]) and the Hénon family (Benedicks-Young, [4]).
(4) There is an open set D of parameters (a, b) for the Hénon family such that $\overline{D} \supset \bigcup E_b$ and such that $\forall (a, b) \in D$, the mapping $T_{a,b}$ has a stable periodic orbit. (Raul Ures, to appear in Erg. Theory Dynam. Sys.)

3. Sums of random Cantor sets (thesis of Per Larsson [17].) In connection with homoclinic bifurcations, compare Palis-Takens [19], Chapter 4, it is a natural problem to consider the algebraic difference $C_1 - C_2$ of dynamically defined one-dimensional Cantor sets, C_1 and C_2. It is important to find out whether $C_1 - C_2$ contains an interval and to compute the Hausdorff dimension of $C_1 - C_2$.

As suggested by Lennart Carleson, his student Per Larsson studied the problem in the case of random Cantor sets and obtained several interesting results. In particular Per Larsson gave sufficient conditions for $C_1 - C_2$ to contain an interval.

B. Complex Dynamical Systems.

1. Hans Brolin's thesis.

Hans Brolin was the first student of Lennart Carleson in dynamical systems. His thesis from 1965, [6], deals with iterations of rational functions. Now this field is one of the most active in mathematics. At that time there had been little activity in this field since the classical papers of Fatou and Julia from the turn of the century. One notable exception was the important paper of C.L. Siegel on analytic conjugation of univalent functions of the type $\lambda z + \sum_{n=2}^{\infty} a_n z^n$, $|\lambda| = 1$, to rotations. A large part of the thesis was dedicated to a modern exposition of the classical papers of Fatou and Julia and when the renewed interest in complex iterations arose in the 1980's Brolin's thesis was an important reference. This is now superseded — especially by Carleson-Gamelin, [12].

Let me mention a famous result from Brolin's thesis.

Theorem. *Let p be a d-degree polynomial and z_0 an unstable periodic point, i.e. $f^n(z_0) = z_0$, $|DF^n(z_0)| > 1$, where $f^n = f \circ \cdots \circ f$. Let $D_N = p^{-N}(z_0) = \{z: p^N(z) = z_0\}$ and*

$$\mu_N = d^{-N} \sum_{z \in D_N} \delta_z$$

Then $\mu_N \to \mu$ weak$-*$ where μ is an invariant measure, the equilibrium measure in the sense of potential theory on the Julia set of p.

This result is important also for the newly established theory of iterations in several complex variables, where basic methods like Montel's theorem are not available.

2. Coefficient problems for univalent functions and Julia sets

A famous class of problems in the theory of analytic functions are the problems on the properties of the coefficients of univalent functions on the unit disc U. The best known of these is the Bieberbach conjecture solved by de Branges. There are, however, other less well known coefficient problems, which are quite difficult. Some of these were considered by Lennart Carleson in a joint paper with Peter Jones, [14]. They considered the coefficient problems for the classes S_1, the class of univalent functions in the unit disk

$$S_1 = \left\{ f = \sum_{n=1}^{\infty} a_n z^n : \|f\|_\infty \leq 1, f \text{ univ. on } U \right\},$$

and the class S_2 of univalent functions in the complement of the closed unit disk

$$S_2 = \left\{ f = z + \sum_{n=1}^{\infty} a_n z^{-n} : f \text{ univ. on } U^* \right\}, \qquad U^* = \{|z| > 1\}.$$

Let $A_n = \sup_{f \in S_1} |a_n|$, $B_n = \sup_{f \in S_2} |a_n|$ and define

$$\begin{cases} \gamma_1 = -\lim_{n\to\infty} \dfrac{\log A_n}{\log n}, \\ \\ \gamma_2 = -\lim_{n\to\infty} \dfrac{\log B_n}{\log n}. \end{cases}$$

Lennart Carleson and Peter Jones prove that $\gamma_1 = \gamma_2$, they improve on a lower bound of $\gamma_1 = \gamma_2$ given by Littlewood and give an example in the class S_2 with slow decay of the Laurent coefficients. The example is the conformal map of the exterior of the unit disk onto the unbounded component of the Julia set of a polynomial of the form $P_c(z) = z^2 + c$, where c belongs to the boundary of the Mandelbrot set.

3. The geometry of the Julia sets

In a recent paper which in a remarkable way combines methods of classical function theory Lennart Carleson, Peter Jones and Jean-Cristophe Yoccoz [13] gave a geometric characterization of the Julia sets of polynomials the critical points of which satisfy a "Misiurewicz type" condition: the orbit of every critical point stays at a fixed distance from the critical point.

Recall that an unbounded domain Ω is a John domain if there is a constant C so that for each $z \in \partial\Omega$ there is a curve γ from ∞ to z so that $\forall \zeta \in \gamma$, $|\zeta - z| \leq C \operatorname{dist}(\zeta, \partial\Omega)$. We denote by $A_\infty = A_\infty(P)$ the immediate basin at ∞ of P.

More precisely they proved the following

Theorem. *The following statements are equivalent*

(1) (i) *For each critical point c of P*

$$|P^j(c) - c| \geq \delta, \qquad \forall j \geq 1.$$

(ii) *P has no parabolic periodic points.*

(2) *A_∞ is a John domain.*

(3) *A_∞ is a John domain and for every Fatou component $\mathcal{F}_j$, the boundary $\partial\mathcal{F}_j$ is a quasicircle.*

4. Complex dynamics (Book by L. Carleson and T.W. Gamelin, [12]).

Complex dynamics has recently been in the focus of interest. However, the subject has until recently lacked comprehensive introductions. There are excellent expository articles, in particular by P. Blanchard and M. Lyubich, but until recently there has been no treatments with complete proofs. This situation has now had its remedy by the appearance of the book by A. Beardon, the Stony Brook lecture notes by J. Milnor and in particular the book "Complex dynamics" by L. Carleson and T. Gamelin. A remark: The appearance of the monographs by Milnor and Carleson show the interdisciplinary rôle (within mathematics) of dynamical systems. Who would have believed 20 years ago, say, that Lennart Carleson and John Milnor would almost simultanously write books on the same topic. Of course the different backgrounds of the authors show up in the approach to the subject and what is emphasized. The book by Carleson and Gamelin is more comprehensive and contains a wealth of material not available elsewhere. For instance, it gives the necessary background in quasiconformal mappings and proceeds to give a complete proof of Sullivan's theorem on the classification of the components of the Fatou set of a rational map.

Of special interest is chapter VII on the local geometry of the Fatou set. Several interesting new results on the accessability of repulsive and parabolic periodic points by invariant curves in the Fatou set. The methods of proofs are related to those of the Ahlfors, Carleman, Denjoy theorem. There is also a discussion of the case when the Fatou components are John domains (cf. the paper of Carleson, Jones and Yoccoz, [13]). Another interesting result, not easily available elsewhere, is the theorem of Sibony on the regularity at the boundary of the Green function of the basin at ∞, in turn based on a theorem of Lennart Carleson from the book "Selected problems in exceptional sets", [8].

The final chapter on quadratic polynomials (Chapter VIII) also contains several innovative ideas. One such idea is in Theorem 4.2 where a criterion in terms of escape rates which would prove that the Mandelbrot set is locally connected. Another is the treatment of parabolic points on the Mandelbrot set where inverse Runge–Kutta type methods are introduced to find a flow corresponding to the discrete time dynamical system.

5. Diffusion limited aggregation.

One of the main current research interests of Lennart Carleson is the study of diffusion limited aggregation (DLA) in a joint project with N. Makarov. One version of the problem of DLA is the following: Introduce a *seed*, a sphere of radius ε at the origin. Then pick a disk of radius ε and centre z_0 of uniform probability distribution on $|z| = R$ and then start a Brownian walk $\{z_t\}$ at z_0. With probability 1, $B(z_t, \varepsilon)$ will hit the seed at some time $t_1 < \infty$. Then attach $B(z_{t_1}, \varepsilon)$ to the seed and repeat the procedure successively with new Brownian walks. This is a model for crystal growth. The mathematical problem is to understand the geometry of the growing object K_N. In particular, it is still not known how fast it grows. In particular for the exponents

$$\gamma_1 = \limsup_{N\to\infty} \frac{\log \operatorname{diam}(K_N)}{\log N}$$

$$\gamma_2 = \liminf_{N\to\infty} \frac{\log \operatorname{diam}(K_N)}{\log N}$$

so far one knows only the trivial estimates $\gamma_1 \le 1$ and $\gamma_2 \ge \frac{1}{2}$.[1] This problem is intimately related to problems on harmonic measure and fractal geometry, which have followed Lennart Carleson through his career, and offers a new challenge to his analytical power.

Acknowledgement. I wish to thank Peter Jones, L.S. Young and Gunnar Peters for valuable comments on drafts of this paper.

[1]Added in proof: Kesten now has a nontrivial upper bound for γ_1.

References

1. M. Benedicks, *New developments in the ergodic theory of nonlinear dynamical systems*, Phil. Trans. R. Soc. Lond. A **346** (1994), 145–157.
2. M. Benedicks and L. Carleson, *On iterations of* $1 - ax^2$ *on* $(-1, 1)$, Ann. Math. **122** (1985), 1–25.
3. M. Benedicks and L. Carleson, *The dynamics of the Hénon map*, Ann. Math. **133** (1991), 73–169.
4. M. Benedicks and L.-S. Young, *Decay of correlation for certain Hénon maps* (to appear).
5. M. Benedicks and L.-S. Young, *Sinai-Bowen-Ruelle measures for certain Hénon maps*, Invent. Math. **112** (1993), 541–576.
6. H. Brolin, *Invariant sets under iteration of rational functions*, Arkiv för Mat. **6** (1965), 103–144.
7. T. Carleman, *Problèmes mathématiques dans la théorie cinétique des gaz*, Almqvist & Wiksell Boktryckeri AB, Uppsala, 1957.
8. L. Carleson, *Selected problems in exceptional sets*, Van Nostrand Co., Princeton NJ, 1967.
9. L. Carleson, *A remark on Denjoy's inequality and Herman's theorem*, Publ. Math. IHES **49** (1979), 235–241.
10. L. Carleson, *Some analytical problems related to statistical mechanics*, Lecture notes in mathematics **779** (1979), Springer Verlag, Berlin, Heidelberg, New York.
11. L. Carleson, *The dynamics of non-uniformly hyperbolic systems in two variables*, Proceedings of the International Congress of Mathematicians (Kyoto, 1990), vol. II, The Mathematical Society of Japan, Springer-Verlag, Tokyo, Berlin and New York, 1991, pp. 1241–1247.
12. L. Carleson and T. W. Gamelin, *Complex Dynamics*, Springer-Verlag, New York, Heidelberg, Berlin, 1993.
13. L. Carleson, P. Jones and J.-C. Yoccoz, *Julia and John*, Bol. Soc. Brasil. Math. (NS) **25** (1994), 1–30.
14. L. Carleson and P. W. Jones, *On coefficient problems for univalent functions and conformal dimensions*, Duke Math. J. **66** (1992), 169–206.
15. M. Hénon, *A two-dimensional mapping with a strange attractor*, Commun. Math. Phys. **50** (1976), 69–79.
16. M. Herman, *Sur la conjugaison différentiable des difféomorphismes du circle a des rotations*, Publ. Math. IHES **49** (1979), 5–233.
17. P. Larsson, *L'ensemble différence de deux ensembles de Cantor aléatoires*, C.R.A.S. Paris **310** (1990), 735–738.
18. L. Mora and M. Viana, *Abundance of strange attractors*, Acta Math. **171** (1993), 1–71.
19. J. Palis and F. Takens, *Hyperbolicity & sensitive chaotic dynamics at homoclinic bifurcations*, Cambridge Unversity Press, 1993.
20. M. Tamm, *Subanalytic sets in the calculus of variation*, Acta Math. **146** (1981), 167–199.
21. J.-C. Yoccoz, *Polynômes quadratiques et attracteur de Hénon*, Séminaire Bourbaki, 43éme année, 1990–91, vol 734 (nov., 1990).
22. L.-S. Young, *Decay of correlation for certain quadratic maps*, Commun. Math. Phys. **146** (1992), 123–138.

23. L.-S. Young, *Ergodic theory of chaotic dynamical systems*, From topology to computations: proceedings of the Smalefest (M.W. Hirsch and J.E. Marsden and M. Shub, eds.), Springer Verlag, New York, 1993.

DEPARTMENT OF MATHEMATICS,
ROYAL INSTITUTE OF TECHNOLOGY,
S-100 44 STOCKHOLM, SWEDEN
E-mail address: michaelb@math.kth.se

ON CONVERGENCE AND GROWTH OF PARTIAL SUMS OF FOURIER SERIES

BY

LENNART CARLESON

Uppsala, Sweden, and Stanford, California, U.S.A. (1)

1. Introduction

In the present paper we shall introduce a new method to estimate partial sums of Fourier series. This will give quite precise results and will in particular enable us to solve the long open problem concerning convergence a.e. for functions in L^2. We denote by $s_n(x)$ the nth partial sum of a function $f(x) \in L^1(-\pi, \pi)$ and have the following theorem.

THEOREM. (a) *If for some* $\delta > 0$

$$\int_{-\pi}^{\pi} |f(x)| (\log^+ |f(x)|)^{1+\delta} dx < \infty, \tag{1.1}$$

then $$s_n(x) = o(\log \log n), \quad a.e.$$

(b) *If* $f(x) \in L^p$, $1 < p < 2$, *then*

$$s_n(x) = o(\log \log \log n), \quad a.e.$$

(c) *If* $f(x) \in L^2$, *then* $s_n(x)$ *converges a.e.*

Remarks. (a) This result should be compared with Kolmogorov's example of an a.e. divergent series in L^1. If we consider in detail the construction of Hardy–Rogosinski (see [1], pp. 306–308), we see that the following is actually true. Given $\varepsilon(n) \to 0$, $n \to \infty$, there is a function $f \in L^1$ such that

$$s_n(x) \neq O(\varepsilon(n) \log \log n), \quad a.e.$$

The best previous result in this case is $o(\log n)$.

(1) This research was partially supported by NSF grant 4079 at Stanford University.

Reprinted from Acta Math., Vol. 116 (1966), pp. 135–157.

(b) The best previous result here is the Littlewood–Paley theorem (see [2], p. 166) $s_n(x) = o((\log n)^{1/p})$, a.e. It is rather obvious from the proof of (c) that we actually have convergence a.e. in this case also, and the proof of (b) will only be sketched.

(c) This result was conjectured by Lusin. The best earlier result is the Kolmogorov–Seliverstov–Plessner theorem $s_n(x) = o((\log n)^{\frac{1}{2}})$, a.e.

The proof is quite technical and it is convenient to give an outline of the idea behind the proof here.

We assume f real and extend f periodically. We then consider the modified Dirichlet formula

$$s_n^*(x) = \int_{-4\pi}^{4\pi} \frac{e^{-int} f(t)}{x-t}\, dt, \qquad -\pi < x < \pi. \tag{1.2}$$

If ω is a subinterval of $(-4\pi, 4\pi)$, we let $E_\omega(f)$ denote the mean value of f over ω. We consider a suitable disjoint covering $\Omega = \{\omega_\nu\}$ of $(-4\pi, 4\pi)$. If $x \in \omega_\nu = \omega^*(x)$, we write

$$s_n^*(x) = \int_{\omega^*(x)} \frac{e^{-int} f(t)}{x-t}\, dt + \sum_{\mu \neq \nu} \int_{\omega_\mu} \frac{E_{\omega_\mu}(e^{-int} f)}{x-t}\, dt$$
$$+ \sum_{\mu \neq \nu} \int_{\omega_\mu} \frac{e^{-int} f(t) - E_{\omega_\mu}(e^{-int} f)}{x-t}\, dt. \tag{1.3}$$

If x is "strictly" inside ω^*, the first term gives the main contribution. If ω^* has length $2\pi \cdot 2^{-s}$, s an integer, we modify in this term n to the closest integer of the form $h \cdot 2^s$. This gives only a small change in the value of the integral. After the modification and a change of variables $x = 2^s \xi$, $t = 2^s \tau$, we have an integral of the same form as $s_n^*(x)$ but localized to ω^*. We can now repeat the argument.

To prove that the second term is small, we choose Ω so that the mean values $E_{\omega_\mu}(e^{-int} f)$ are all small, specifically of the same magnitude as $\int_{-\pi}^{\pi} f e^{-int}\, dt$.

In the third term, finally, we use the fact that the numerator has a vanishing integral over each ω_μ. In this way we can change the first order singularity into a second order singularity, which is easier to handle (Lemma 5). The situation should be compared with the trivial formulas

$$\int_\delta^1 \frac{dt}{t} = \log \frac{1}{\delta}; \qquad \delta \int_\delta^1 \frac{dt}{t^2} < 1.$$

For every combination of intervals $\omega^*(x)$ and integers n in formula (1.3) we get an exceptional set where the remainder terms are not small. In the proof of (a), which will first be given in Sections 2–5, we allow in a certain sense all combinations (n, ω^*).

The improvement that is needed to get (c) is a careful examination of which (n, ω^*)'s are necessary. The Parseval equation plays a fundamental role and to get the L^p-result, a sufficiently good substitute has to be found. In Section 12 we sketch how an interpolation argument gives (b). The author is indebted to L. Gårding, A. Garsia, L. Hörmander, J.-P. Kahane and I. Katznelson for many improvements in the presentation of the proof.

2. Some notations and lemmas

Let ω_{-1}^* be the interval $(-4\pi, 4\pi)$ and $\omega_{-10}, \omega_{00}, \omega_{10}, \omega_{20}$ be the intervals $(-4\pi, -2\pi), \dots, (2\pi, 4\pi)$. We shall restrict x to $(-\pi, \pi)$ and ω_{00} and ω_{10} will be our basic intervals. We subdivide $\omega_{00} \cup \omega_{10}$ into $2 \cdot 2^\nu$ equal intervals of lengths $2\pi \cdot 2^{-\nu}$, $\nu = 1, 2, \dots$. The resulting intervals are from left to right denoted $\omega_{j\nu}$, $j = -2^\nu + 1, \dots, 2^\nu$.

We further define

$$\omega_{j\nu}^* = \omega_{j\nu} \cup \omega_{j+1,\nu}, \quad -2^\nu + 1 \leqslant j \leqslant 2^\nu - 1. \tag{2.1}$$

For the length of an interval ω we use the notation $|\omega|$.

Let f be a given real function with period 2π such that

$$\int_{-\pi}^{\pi} |f(x)| (\log^+ |f|)^{1+\delta} dx = 1. \tag{2.2}$$

$\delta > 0$ is fixed and Const. will indicate numbers only depending on δ. For each $\omega =$ some $\omega_{j\nu}$, $\nu \geqslant 0$, we define the Fourier transform

$$c_\alpha(\omega) = \frac{1}{|\omega|} \int_\omega f(x) \exp\{-2^\nu \alpha i x\} dx, \tag{2.3}$$

where α is any real number. Together with the Fourier coefficients $c_n(\omega)$, n an integer, we consider the non-negative numbers $C_n(\omega)$ defined by

$$C_n(\omega) = \sum_{\mu=-\infty}^{\infty} |c_{n+\mu/3}(\omega)| (1 + \mu^2)^{-1}. \tag{2.4}$$

We shall also set $p = (n, \omega)$ and use the notation $C(p)$.

It will be necessary to have some estimates of c_n and C_n, as $n \to \infty$, for functions of type (1.1). The following form of the Hausdorff–Young inequality is sufficient.

LEMMA 1. *If*

$$\frac{1}{|\omega|} \int_\omega |f(x)| (\log^+ |f|)^{1+\delta} dx \leqslant \lambda, \tag{2.5}$$

then there are numbers $a(\lambda)$ and $A(\lambda) > 0$ only depending on λ and δ such that

$$\sum_{-\infty}^{\infty} \exp\{-a\, C_n(\omega)^{-1/(1+\delta)}\} \leqslant A. \tag{2.6}$$

Proof. Except for the fact that no constants are mentioned, (2.6) is given in [2], p. 158, for $|c_n|$ instead of C_n. It then also follows for $c_{n\pm 1/3}$ if we consider $e^{\pm i\beta x} f(x)$, $\beta = \frac{1}{3} 2^\nu$. To obtain (2.6) from this result we observe that

$$C_n \leqslant \text{Const.} \sup_\mu \frac{1}{(1+|\mu|)^{\frac{1}{2}}} |c_{n+\mu/3}|.$$

This means that, $q = 1/(1+\delta)$,

$$\exp\{-b C_n^{-q}\} \leqslant \sum_\mu \exp\{-a|c_{n+\mu/3}|^{-q}(1+|\mu|)^{\frac{1}{2}q}\} \tag{2.7}$$

for some other constant $b>0$. Summing (2.7) with respect to n and observing that for $|c| \leqslant 2\lambda$

$$\sum_{\mu=-\infty}^{\infty} \exp\{-a|c|^{-q}(1+|\mu|)^{\frac{1}{2}q}\} \leqslant A_1(\lambda) \exp\{-a|c|^{-q}\},$$

we obtain (2.6).

With the intervals $\omega_{j\nu}^*$, defined by (2.1), we associate the analogous numbers

$$C_n^*(\omega_{j\nu}^*) = \max_{\omega'} C_n(\omega') \tag{2.8}$$

where ω' ranges over the four subintervals $\omega_{k,\nu+1}$ of $\omega_{j\nu}^*$. If $\omega_{j\nu}^* = \omega_{-1}^*$, C_n^* is simply $C_n(\omega_{00})$. We also use the notations $p^* = (n, \omega^*)$ and $C^*(p^*)$.

Finally, given an integer n and an interval $\omega = \omega_{j\nu}$, we expand n to the basis 2

$$n = \sum_{i \geqslant 0} \varepsilon_i 2^i, \quad \varepsilon_i = 0, 1,$$

and define

$$n[\omega] = 2^{-\nu} \sum_{i \geqslant \nu} \varepsilon_i 2^i. \tag{2.9}$$

If ω' is related to ω^* as in (2.8), we also write $n[\omega^*]$ for $n[\omega']$.

The reason for considering $C_n(\omega)$ together with $c_n(\omega)$ is seen from the following lemma.

LEMMA 2. *For any integer n and any $\omega = \omega_{j\nu}$ we have the inequality*

$$|c_{n \cdot 2^{-\nu}}(\omega)| \leqslant \text{Const.}\, C_{n[\omega]}(\omega).$$

It will be convenient for future reference to collect the method of proof of Lemma 2 in a special lemma.

LEMMA 3. *Let $\varphi(t) \in C^2(\omega)$, $|\omega| = 2\pi \cdot 2^{-\nu}$. Then we can represent $\varphi(t)$*

$$\varphi(t) = \sum \gamma_\mu \exp\{-i 2^\nu 3^{-1} \mu t\}, \quad t \in \omega, \tag{2.10}$$

where

$$(1+\mu^2)|\gamma_\mu| \leqslant \text{Const.}\,(\max_\omega |\varphi| + 2^{-2\nu} \max_\omega |\varphi''(t)|). \tag{2.11}$$

Proof of Lemma 3. By a change of variables, $t=2^{-\nu}\cdot\tau$, we see that we may assume $\omega=(0,2\pi)$. If we determine a polynomial with the same derivatives of order $\leqslant 2$ as $\varphi(t)$ at $t=0$ and vanishing with these derivatives at $t=-2\pi$, and do analogously for $(2\pi, 4\pi)$, we realize thas we can extend φ to $(-2\pi, 4\pi)$ and then periodically so that $\max|\varphi|+\max|\varphi''|$ increases at most by a fixed factor. By partial integrations we see that (2.11) holds for the expansion of this function.

Proof of Lemma 2. We apply Lemma 3 to the function $e^{-i\beta t}$, $\beta=n-n[\omega]\,2^{\nu}$, $\omega=\omega_{j\nu}$, so that $|\gamma_\mu|\leqslant \text{Const.}\,(1+\mu^2)^{-1}$. We find

$$c_{n2^{-\nu}}(\omega)=\frac{1}{|\omega|}\int_\omega e^{-int}f(t)\,dt=\frac{1}{|\omega|}\int_\omega e^{-i\beta t}\,e^{-in[\omega]2^{\nu}t}f(t)\,dt=\sum \gamma_\mu c_{n[\omega]+\mu/3}(\omega).$$

The estimate of γ_μ and the definition (2.4) prove the lemma.

The estimations of the remainder terms of (1.3) depend on the following two lemmas, the first of which is of well-known type.

LEMMA 4. *Let $E(t)$ be defined on an interval ω^* and assume $|E(t)|\leqslant c$. Let σ_x denote subintervals of ω^* containing x inside their middle halves. We define (as principal values) the maximal Hilbert transform*

$$H^*(x)=\sup_{\sigma_x}\left|\int_{\sigma_x}\frac{E(t)}{x-t}\,dt\right|.$$

Let T be the set $$T=\{x\,|\,H^*(x)>a,\ x\in\omega^*\}.$$

Then $$mT\leqslant \text{Const.}\exp\left(-\text{Const.}\frac{a}{c}\right)|\omega^*|.$$

Proof. By a change of scale we may assume $\omega^*=(0,1)$ and $c=1$. Let $u(z)$ be harmonic in $y>0$, with boundary values $=E(x)$ on $(0,1)$ and $=0$ otherwise. Let v be the conjugate of u, $v(\infty)=0$. It is easy to see that (see [1], p. 103)

$$H^*(x)\leqslant \text{Const.}\sup_{y>0}|v(x+iy)|+\text{Const.}$$

For $k<\pi/2$ we have (see [1], p. 254)

$$\int_{-1}^{2} e^{k|v(x+iy)|}\,dx<\text{Const.},$$

and applying the Hardy–Littlewood maximal theorem ([1], p. 155) to $\exp\{\pm\tfrac{1}{2}ki(u+iv)\}$, which is in L^2, Lemma 4 follows.

LEMMA 5. *Let $\{\omega_k\}$ be a disjoint covering of an interval ω^* and let ω_k have midpoint t_k and length δ_k. We define the function*

$$\Delta(x) = \sum_k \frac{\delta_k^2}{(x-t_k)^2+\delta_k^2}, \quad x \in \omega^*,$$

and the set

$$U = \{x \mid \Delta(x) > M, \ x \in \omega^*\}.$$

Then

$$mU \leqslant \text{Const.} \exp(-\text{Const.}\, M)\,|\omega^*|.$$

Proof. By a change of scale we may again assume $\omega^* = (0,1)$. Let $g(x) \geqslant 0$ be integrable and have its support in $(0,1)$ and let $g(x+iy)$, $y>0$, be the corresponding harmonic function in $y>0$. Then

$$\int_{-\infty}^{\infty} \Delta(x)\, g(x)\, dx = \pi \sum_k \delta_k\, g(t_k + i\delta_k) \leqslant \text{Const.} \int_0^1 \sup_{y>0} g(x+iy)\, dx,$$

since by Harnack's inequality $\delta_k g(t_k + i\delta_k) \leqslant 3 \int_{\omega_k} g(t+i\delta_k)\, dt$. By the theorem on maximal functions, [1], p. 155, this last integral is bounded if

$$\int_0^1 g \log^+ g\, dx \leqslant 2. \tag{2.12}$$

Let μ be the measure of the set where $\Delta(x) > M$, $x \in (0,1)$, and define

$$g(x) = \left(\mu \log \frac{1}{\mu}\right)^{-1}$$

on this set and 0 otherwise. This function satisfies (2.12) and hence

$$\frac{M}{\log \dfrac{1}{\mu}} \leqslant \text{Const.},$$

which proves Lemma 5.

3. Construction of the exceptional set

We start from a large number λ and an integer N. Depending on λ, we shall determine a number λ_1 and a set $E_N(\lambda, \lambda_1)$ such that outside E_N the partial sums $s_n(x)$, $n \leqslant 2^N$, satisfy $|s_n(x)| \leqslant \text{Const.}\, \lambda_1 \lambda \log N$. In this section we shall construct E_N and in later sections show that E_N has the desired properties. E_N will consist of four different parts S, T, U, V, which will be constructed in the steps (A)–(E) below.

(A) We first consider the set of all $\omega=\omega_{j\nu}\subset(-2\pi,2\pi)$ such that

$$\int_{\omega}|f(x)|\,(\log^{+}|f|)^{1+\delta}\,dx>\lambda|\omega|. \tag{3.1}$$

We then define S to be the union of all intervals

$$S:\ \omega_{j+k,\nu},\ k=0,\pm1,\pm2,\pm3,\quad \omega_{j\nu}\ \text{satisfies (3.1)}.$$

Selecting successively suitable intervals $\omega_{j\nu}$ for $\nu=0,1,2,\ldots$, it is easy to see that

$$mS\leqslant 7\lambda^{-1}\int_{-2\pi}^{2\pi}|f|\,(\log^{+}|f|)^{1+\delta}\,dx\leqslant\frac{14}{\lambda}. \tag{3.2}$$

We observe certain properties of the set S.

(Aα) S does not depend on N.

(Aβ) If (3.1) does not hold, then $(\lambda>e)$ $|c_n(\omega)|<2\lambda$ and

$$C_n(\omega)\leqslant 2\lambda\sum_{\mu=-\infty}^{\infty}\frac{1}{1+\mu^2}<10\lambda. \tag{3.3}$$

(Aγ) Given (j_0,ν_0), $-1\leqslant\nu_0$, recall the definition of $\omega^*=\omega^*_{j_0\nu_0}$ in (2.1) and of $C^*_n(\omega^*)$ in (2.8). If $\omega_{j_0,\nu_0}\not\subset S$, (3.1) holds also for the neighboring intervals with $\nu=\nu_0+1$ and it follows from (3.3) that

$$C^*_n(\omega^*)<10\lambda.$$

(B) Again, given $\omega^*=\omega^*_{j_0,\nu_0}$, $-1\leqslant\nu_0\leqslant N-1$, and an arbitrary, nonnegative integer n, we shall define a partition $\Omega_n(\omega^*)$ of ω^*. To simplify our notations we assume $\omega^*=(-4\pi,4\pi)$, but the general case consists simply in a change of scale.

An interval $\omega=\omega_{j\nu}$, $\nu_0+1=0\leqslant\nu\leqslant N-1$, belongs to $\Omega_n(\omega^*)$ if

(Bα) $C_{n[\omega]}(\omega)\leqslant 2C^*_n(\omega^*)$;

(Bβ) the condition (Bα) holds with $\omega_{j\nu}$ replaced by any $\omega_{k\mu}$, $\omega_{j\nu}\subset\omega_{k\mu}\subset\omega^*$, but not for a certain $\omega_{l,\nu+1}\subset\omega_{j\nu}$.

(Bβ′) If $\omega_{j\nu}\subset\omega_{k\mu}\subset\omega^*$, $k<j$, then (Bα) and (Bβ) do not hold for $\omega_{k\mu}$.

(Bγ) $\Omega_n(\omega^*)$ contains all ω_{jN} not included in the $\omega_{j\nu}$ defined by B(α, β, β′).

Loosely speaking, the definition means that we subdivide ω^* into as small intervals as possible so that (Bα) holds and $\nu\leqslant N$.

(C) As a preparation for the construction of the sets T and U, we shall make a careful definition of the interval $\omega^*(x)$ in (1.3) so that x is "strictly" inside $\omega^*(x)$.

Let x belong to the middle half of ω^* and consider the set of intervals $\tilde\omega^*$ which are obtained by taking every $\omega_{j\nu}\in\Omega_n(\omega^*)$ and adjoining $\omega_{j-1,\nu}$ or $\omega_{j+1,\nu}$. Among these intervals $\tilde{\omega}^*$ there are those which contain x in their middle half. We then define $\omega^*(x)$, corresponding to $\Omega_n(\omega^*)$ and the point x as such an interval $\tilde{\omega}^*$ for which $|\tilde{\omega}^*|$ is as large as possible. We observe that

$$|\omega^*(x)| \leqslant \tfrac{1}{2}|\omega^*|. \tag{3.4}$$

Furthermore, $\omega^*(x)$ has the following properties:

(Cα) x belongs to the middle half of $\omega^*(x)$.

(Cβ) $\omega^*(x)$ is a union of intervals $\omega_{j\nu} \in \Omega_n(\omega^*)$ since $|\tilde{\omega}^*|$ was assumed maximal.

(Cγ) If $\omega^*(x) = \omega_{j\nu} \cup \omega_{j\pm 1, \nu}$, $\omega = \omega_{j\nu} \in \Omega_n(\omega^*)$, it follows from (B$\alpha$), (B$\beta$) and (C$\beta$) that

$$C_{n[\omega]}(\omega_{j\nu}), \qquad C_{n[\omega]}(\omega_{j\pm 1,\nu}) \leqslant 2C_n^*(\omega^*).$$

(Cδ) The complement of $\omega^*(x)$ with respect to ω^* is by (Cβ) the union of certain intervals in $\Omega_n(\omega^*)$. For each such interval σ, the distance from x to σ exceeds half the length of σ.

We now define
$$H_n(x) = \int_{\omega^* - \omega^*(x)} \frac{E_n(t)}{x-t}\,dt,$$

where (note $\nu_0 = -1$)

$$E_n(t) = \frac{1}{|\omega|}\int_\omega f(x)\,e^{-inx}\,dx, \quad t \in \omega \in \Omega_n(\omega^*). \tag{3.5}$$

By the construction of $\Omega_n(\omega^*)$ and Lemma 2, $|E_n(t)| \leqslant \text{Const.}\, C_n^*(\omega^*)$. As in Lemma 4, we define the maximal transform $H_n^*(x)$ of $E_n(t)$ and the set T_n of points $x \in \omega^*$ such that

$$T_n\colon\ H_n^*(x) > \lambda_1 C_n^*(\omega^*)^\varrho \log N,$$

where ϱ is a number, $0 < \varrho < \delta/(1+\delta)$, fixed from here on. By Lemma 4,

$$mT_n \leqslant \text{Const.} \exp\{-\text{Const.}\, \lambda_1 C_n^*(\omega^*)^{-1+\varrho} \log N\}\, |\omega^*|.$$

We also observe that $|H_n(x)| \leqslant 2H_n^*(x)$.

(D) With the same definition of $\omega^*(x)$ we now set

$$R_n(x) = R(x) = \int_{\omega^* - \omega^*(x)} \frac{e^{-int} f(t) - E_n(t)}{x-t}\,dt, \tag{3.6}$$

where again we have normalized the situation to $\omega^* = (-4\pi, 4\pi)$. If ω_k denote the intervals in $\Omega_n(\omega^*)$, $\omega^* - \omega^*(x)$ is by (Cδ) the union of a certain subset (x) of the ω_k's.

Denote by δ_k the lengths and by t_k the midpoints of the ω_k's and define $\Delta(x)$ as in Lemma 5. We rewrite the formula (3.6), using the fact that the numerator has vanishing integral over each ω_k,

$$R(x)=\sum_{(x)}\int_{\omega_k}\frac{t-t_k}{(x-t)(x-t_k)}e^{-int}f(t)\,dt-\sum_{(x)}\int_{\omega_k}\frac{t-t_k}{(x-t)(x-t_k)}E_n(t)\,dt. \tag{3.7}$$

Using (Cδ) and $|E_n(t)|\leqslant$ Const. $C_n^*(\omega^*)$, we see that the last sum is dominated by Const. $C_n^*(\omega^*)\,\Delta(x)$.

We shall now prove that also the first sum of (3.7) has this bound and shall use Lemma 3. We write, if $t\in\omega=\omega_k$, $|\omega|=2\pi\cdot 2^{-\nu}$,

$$e^{-int}\frac{t-t_k}{(x-t)(x-t_k)}=e^{-i2^\nu n[\omega]t}\varphi(t).$$

By Lemma 3, $\varphi(t)$ can be written

$$\varphi(t)=\sum_\mu \gamma_\mu \exp\{-2^\nu 3^{-1}i\mu t\},\quad t\in\omega, \tag{3.8}$$

where, again observing (Cδ),

$$|\gamma_\mu|\leqslant\frac{\text{Const.}\,\delta_k}{(1+\mu^2)\,[(x-t_k)^2+\delta_k^2]}.$$

We multiply (3.8) by $f(t)$ and integrate over ω_k. If we observe (Bα), we obtain after summation over *all* k the desired bound so that

$$|R(x)|\leqslant \text{Const.}\,C_n^*(\omega^*)\,\Delta(x). \tag{3.9}$$

Let us now define $U_n(\omega^*,\lambda_1,N)$ as the set where

$$U_n:\ \Delta(x)>\lambda_1 C_n^*(\omega^*)^{\varrho-1}\log N.$$

It follows from (3.9) and Lemma 5 that

$$mU_n\leqslant \text{Const.}\,|\omega^*|\exp\{-\text{Const.}\,\lambda_1 C_n^*(\omega^*)^{-1+\varrho}\log N\}. \tag{3.10}$$

Observe that outside $T_n\cup U_n$, $|H_n(x)|$ and $|R_n(x)|$ are less than Const. $\lambda_1 C_n^*(\omega^*)^\varrho\log N$.

(E) Finally, let V, not depending on N, be the set, where

$$V:\ \sup_{\sigma_x}\left|\int_{\sigma_x}\frac{f(t)}{x-t}\,dt\right|>\lambda_1,$$

σ_x being defined as in Lemma 4. As is well known $mV \leqslant \eta(\lambda_1)$, where η does not depend on f and $\eta \to 0$, $\lambda_1 \to +\infty$ (see [1], p. 279).

These definitions made, the exceptional set $E_N(\lambda, \lambda_1)$ is defined by

$$E_N = (S \cup V) \cup \bigcup_{n,\,\omega^*} (T_n(\omega^*) \cup U_n(\omega^*)), \tag{3.11}$$

where ω^* runs over all intervals $\omega^*_{j\nu} \not\subset S$ for which $-1 \leqslant \nu \leqslant N-1$.

4. Estimate of the exceptional set

We shall here estimate the measure of E_N and must now determine the relation between the numbers λ and λ_1.

Let us recall that by (Aγ) the numbers $C^*_n(\omega^*)$ associated with (3.11) are $<10\lambda$. Furthermore, by (2.8) $C^*_n(\omega^*) = C_n(\omega')$ for a certain subintervval ω' of ω^*. For a fixed ω', such a relation holds for at most two ω^*'s and $|\omega^*| = 4|\omega'|$. It therefore follows from (2.6) and the definition of S in (A) that

$$\sum_{(s)} \exp\{-a C^*_n(\omega^*)^{-q}\}\, 2\pi\, 2^{-s} \leqslant A(\lambda), \quad q = \frac{1}{1+\delta}, \tag{4.1}$$

where (s) indicates that the summation runs over all pairs (n, ω^*) used in the definition of E_N for which $|\omega^*| = 2\pi 2^{-s}$, $-2 \leqslant s \leqslant N-2$. Summing over s we get

$$\sum_{n,\,\omega^*} \exp\{-a C^*_n(\omega^*)^{-q}\}\, |\omega^*| \leqslant \text{Const.}\, A(\lambda)\, N. \tag{4.2}$$

In (4.2) we now only consider the set Q_0 of those pairs (n, ω^*) for which

$$Q_0\colon\ 1 \leqslant C^*_n(\omega^*) \quad (<10\lambda).$$

It follows that

$$\sum_{Q_0} |\omega^*| \leqslant \text{Const.}\, A(\lambda)\, N. \tag{4.3}$$

Disregarding the set $(S \cup V)$, it follows from the estimates of mT_n and mU_n above that only at most the fraction

$$\text{Const.} \exp\{-\text{Const.}\, \lambda_1 (10\lambda)^{-1} \log N\}$$

of each interval ω^* that corresponds to Q_0 can belong to E_N. Hence choosing λ_1 sufficiently large as a function of λ, we can make this fraction $<\text{Const.}\, N^{-3}$. (4.3) then implies that the corresponding part of E_N has measure $<\text{Const.}\, A(\lambda)\, N^{-2}$.

Similarly, let Q_{i+1} be the set of (n, ω^*) for which

$$2^{-i-1} \leqslant C^*_n(\omega^*) < 2^{-i}, \quad i = 0, 1, 2, \ldots.$$

As above, $$\sum_{Q_i+1} |\omega^*| \leqslant \text{Const.}\, A(\lambda) \exp\{2a\, 2^{qi}\}\, N. \tag{4.4}$$

The corresponding fractions are at most

$$\text{Const.} \exp\{-\text{Const.}\, \lambda_1 2^{(1-\varrho)i} \log N\}.$$

We observe that $1-\varrho > q$. Hence it follows that if we take the above fraction of each term in (4.4) and sum over i, the result will for $\lambda_1 \geqslant \lambda_1(\lambda)$ be less than N^{-2}. This implies

$$mE_N \leqslant m(S \cup V) + \text{Const.}\, A(\lambda)\, N^{-2}.$$

Hence, since S and V are independent of N, we obtain

$$m\left(\bigcup_{N=N_0}^{\infty} E_N\right) \leqslant \varepsilon(\lambda, \lambda_1) + \text{Const.}\, A(\lambda)\, N_0^{-1}, \tag{4.5}$$

so that the measure of $E = \bigcup_{N_0}^{\infty} E_N$ is less than a prescribed ε if λ, λ_1 and N_0 are larger than certain bounds, which only depend on ε and δ.

5. Proof of Theorem (a)

As in the introduction, we assume f real with the integral (1.1) equal 1 and extend f periodically. We first compare the Dirichlet formula for the nth partial sum,

$$s_n(x) = \frac{1}{2\pi} \int_{-\pi}^{\pi} \frac{\sin(n+\frac{1}{2})(x-t)}{\sin\dfrac{x-t}{2}} f(t)\, dt,$$

with $$t_n(x) = \frac{1}{2\pi i}\left(e^{inx} s_n^*(x) - e^{-inx} s_{-n}^*(x)\right),$$

where $s_n^*(x)$ is defined by (1.2). It is very easy to see that uniformly for $|x| < \pi$

$$|s_n(x)| \leqslant \text{Const.}\, |t_n(x)| + o(1), \quad n \to \infty. \tag{5.1}$$

We are going to prove that outside the set E of measure $< \varepsilon$ constructed in Section 4

$$|s_n^*(x)| \leqslant \text{Const.}\, \lambda_1 \lambda \log\log n, \quad n \geqslant 2^{N_0}. \tag{5.2}$$

By (5.1) this implies the same relation for $s_n(x)$. To obtain the $o(\log\log n)$ of the theorem, we simply consider $f(x) - T(x)$ which has a small integral (1.1), if the trigonometric polynomial T is suitable, and observe that the bounds on λ, λ_1 and N_0 only depend on δ and ε.

Let us now consider $0<n<2^N$, $N \geqslant N_0$, and a fixed x outside E_N. If we can prove $|s_n^*(x)| \leqslant \text{Const.}\, \lambda_1 \lambda \log N$, the proof is complete. In formula (1.3) we choose as Ω the covering $\Omega_n(-4\pi, 4\pi)$ defined by (Bα)–(Bγ) in Section 3. The interval $\omega^*(x)$ is defined in (C). Using the previous notations, (1.3) becomes

$$s_n^*(x) = \int_{\omega^*(x)} \frac{e^{-int} f(t)}{x-t}\, dt + H_n(x) + R_n(x). \tag{5.3}$$

Since $x \notin E_N$, the bounds $\text{Const.}\, \lambda_1 C_n^*(-4\pi, 4\pi)^\varrho \log N$ are valid for the two remainder terms.

In the integral (5.3) we are now going to replace n by the number $n_1 = n[\omega']\, 2^{\nu+1}$ if $\omega^*(x) = \omega_{j\nu}^*$ and $|\omega'| = 2\pi\, 2^{-\nu-1}$. Multiplying the integral in (5.3) by e^{imx}, $m = n - n_1$, and subtracting the integral with n replaced by n_1, we obtain the difference

$$\int_{\omega^*(x)} e^{-in_1 t} \frac{e^{im(x-t)} - 1}{x-t} f(t)\, dt. \tag{5.4}$$

If we observe that $C_{n[\omega]}(\omega) \leqslant 2 C_n^*(-4\pi, 4\pi)$ for the two intervals constituting $\omega^*(x)$, we see that we can estimate (5.4) by Lemma 3 applied to the function $u^{-1}(e^{imu}-1)$. We find $\gamma_\mu = O(2^\nu \mu^{-2})$ since $|m| < 2^{\nu+1}$, and for (5.4) we obtain the estimate

$$\text{Const.}\, C_n^*(-4\pi, 4\pi). \tag{5.5}$$

This is smaller than the previous bound if $N \geqslant N_1(\lambda)$ and can be included in that term.

The relation (5.3) now becomes

$$s_n^*(x) = e^{i\theta_0} \int_{\omega^*} \frac{e^{-i2^{\nu+1} n[\omega] t} f(t)}{x-t}\, dt + O(\lambda_1 \log N \cdot \alpha_0),$$

where $\omega^* = \omega^*(x) = \omega_{j\nu}^*$ for a certain $\omega_{j\nu}$ and where

$$\alpha_0 = C_n^*(-4\pi, 4\pi)^\varrho. \tag{5.6}$$

$O(\cdot)$ has the same dependence as the notion Const. Here, the integral over ω^* is of exactly the same type as the original integral with the fundamental difference that

$$\alpha_1 = C_{n[\omega]}^*(\omega^*)^\varrho > 2^\varrho \alpha_0$$

by the property (Bβ) of Ω_n and the definition (2.8) of C_m^*.

For the new interval ω^* we construct the covering $\Omega_{n[\omega]}(\omega^*)$, get a similar remainder term since x is so chosen that the estimates for the remainder terms are valid for *all* ω^* that come into account. We also get a new number $\alpha_2 \geqslant 2^{\varrho}\alpha_1$. The process cannot stop until we reach an interval $\omega^*(x)=I$ of length $2\pi \cdot 2^{-N+1}$, in which case $n[\omega']=0$ since $n<2^N$. Hence

$$s_n^*(x) = e^{i\theta}\int_I \frac{f(t)}{x-t}\,dt + \sum_{i=0}^{r} O(\lambda_1 \log N \cdot \alpha_i),$$

where
$$\frac{\alpha_{i+1}}{\alpha_i} > 2^{\varrho} \quad \text{and} \quad \alpha_r < 10\lambda. \tag{5.7}$$

The first integral is bounded by Const. λ_1 since $x \notin V$ and since x belongs to the middle half of I. Since α_i grow exponentially by (5.7) and are bounded above, we obtain the desired bound $O(\lambda_1 \lambda \log N)$, and the proof is complete.

6. Summary of proof of (c)

The result for L^2 is proved by analyzing carefully the weak point in the preceding proof. This is the fact that we have allowed all combinations $p^*=(n,\omega^*)$ when we estimate the size of the exceptional set in (4.2). The factor N that is introduced in this way must then be compensated by the factor $\log N$ in the exponent. However, in the proof itself only certain special combinations $(n, \omega^*(x))$ will occur. Furthermore, in changing n to the closest n_1, we obtain the very small error $C^*(n; (-4\pi, 4\pi))$. Obviously, we can allow a larger gap between n and n_1, i.e., restrict the choice of the new combination $(n_1, \omega^*(x))$. The basis for the construction of those pairs $p^*=(n,\omega^*)$ that may be used during the proof is certain trigonometric polynomials $P_k(x,\omega)$. They will be constructed in the next section. It will also be convenient to modify the definition of the coverings Ω, and we must add new exceptional points. This is done in Sections 7, 8 and 9, and the proof of (c) is completed in Sections 10 and 11.

7. The polynomials $P_k(x, \omega)$

Denote by b_k the numbers 2^{-2^k}, $k=0,1,\ldots$. Let $f(x)$ be real and periodic and assume

$$\varepsilon^2 = \int_{-\pi}^{\pi} |f(x)|^2\,dx \leqslant 1. \tag{7.1}$$

Denote by $a_{\mu 0}$ those integral Fourier coefficients c_n, $n=\lambda_{\mu 0}$, of f over ω_{j0} which are of absolute value $\geqslant b_k$. We define

$$P_k(x;\omega_{j0}) = \sum_{\mu} a_{\mu 0}\, e^{i\lambda_{\mu 0}x}, \quad j=-1,0,1,2.$$

On each of the two intervals ω_{l1} constituting ω_{j0}, we consider similarly those Fourier coefficients (now corresponding to even integers) of $f(x) - P_k(x; \omega_{j0})$ which are of modulus $\geqslant b_k$. They are denoted $a_{\mu 1}$, corresponding to $\lambda_{\mu 1}$, and will be called primitive for ω_{l1}. We define

$$P_k(x; \omega_{l1}) = P_k(x; \omega_{j0}) + \sum a_{\mu 1} e^{i\lambda_{\mu 1} x}.$$

It is clear how the construction proceeds, and we obtain the polynomials

$$P_k(x; \omega) = \sum_{(\omega)} a_\omega e^{i\lambda_\omega x}, \tag{7.2}$$

where the summation runs over a certain set (ω) of pairs $(a_\omega, \lambda_\omega)$. By the Parseval relation we have, summed over all ω, $|\omega| = 2\pi 2^{-\nu}$,

$$\sum_\omega \int_\omega |f(x) - P_k(x; \omega)|^2 dx + \sum_\omega \sum_{(\omega)} |a_\omega|^2 |\omega| = \int_{-4\pi}^{4\pi} |f(x)|^2 dx.$$

Hence if we define by an infinite sum where (P): a_ω primitive for ω, holds

$$A_k(x) = \sum_{x \in \omega} |a_\omega|^2,$$

it follows that
$$\sum_{(P)} |a_\omega|^2 |\omega| = \int_{-4\pi}^{4\pi} A_k(x)\, dx \leqslant 4\varepsilon^2 \leqslant 4. \tag{7.3}$$

Denote by X_k the set
$$X_k = \{x \mid A_k(x) > b_k^{-1}\}$$

so that $mX_k \leqslant 4b_k$. To every $\omega \subset X_k$ we associate its three right and left neighbors of equal length and define X_k^* as the union of all such intervals so that

$$mX_k^* \leqslant 7\, mX_k \leqslant 28 b_k. \tag{7.4}$$

If $\omega \not\subset X_k$, then $P_k(x; \omega)$ has at most b_k^{-3} terms and

$$|P_k(x; \omega)| \leqslant \sum_{(\omega)} |a_\omega| \leqslant b_k^{-2}. \tag{7.5}$$

In analogy with Section 3 we shall also consider the set S of points x included in an interval ω such that

$$\int_\omega |f(x)|^2 dx > \varepsilon |\omega|. \tag{7.6}$$

Such ω's together with their six neighbors are denoted by S^*. Then

$$mS^* \leqslant 28\varepsilon. \tag{7.7}$$

Denote by M_k the set of $\omega \not\subset (X_k \cup S)$.

8. Allowed pairs p^*

We first consider the set F_k of pairs $p=(n,\omega)$, $\omega\in M_k$, for which $P_k(x;\omega)$ contains a primitive term $ae^{i\lambda x}$, $\lambda[\omega]=n$. By (7.3)

$$\sum_{F_k}|\omega|\leqslant 4b_k^{-2}. \tag{8.1}$$

We are now going to define in ($\tilde{\text{F}}$a) and ($\tilde{\text{F}}$b) below a larger set $\tilde{F}_k$ by associating with each $p\in F_k$ a number of elements $\tilde{p}$. Let P defined in (7.2) be the k-polynomial corresponding to ω.

($\tilde{\text{F}}$a) If $p=(n,\omega)\in F_k$, then $\tilde{F}_k$ contains every $\tilde{p}=(\tilde{n},\tilde{\omega})$ such that $\tilde{\omega}\in M_k$ and

$$\tilde{\omega}\subset\omega,\quad |\tilde{\omega}|\geqslant b_k^{10}|\omega|,\quad |\tilde{n}-\lambda_\omega[\tilde{\omega}]|\leqslant b_k^{-10}, \tag{8.2}$$

where λ_ω is an arbitrary exponent in $P(x;\omega)$.

The condition means that we include all neighbors within b_k^{-10} of all exponents in P not only on ω but also on sufficiently large subintervals $\tilde{\omega}$ of ω.

The number of possibilities for $\tilde{n}$ is $\leqslant \text{Const.}\, b_k^{-3}b_k^{-10}\log b_k^{-1}$.

($\tilde{\text{F}}$b) Again let $p=(n,\omega)\in F_k$ and consider $\tilde{p}=(\tilde{n},\tilde{\omega})$, $\tilde{\omega}\in M_k$, $\tilde{\omega}\subset\omega$, and the polynomial $P(x,\omega)$. Then, by definition, $\tilde{p}\in\tilde{F}_k$ if there are two different exponents λ_ω and λ'_ω in P such that

$$b_k^{10}\leqslant|\lambda_\omega-\lambda'_\omega|\,|\tilde{\omega}|\leqslant b_k^{-10}\quad\text{and}\quad |\tilde{n}-\lambda_\omega[\tilde{\omega}]|<b_k^{-10}.$$

To estimate the number of such pairs $\tilde{p}$, we first observe that P contains $\leqslant b_k^{-3}$ exponents so that the number of pairs $(\lambda_\omega,\lambda'_\omega)$ is $<b_k^{-6}$. For each fixed pair, the first inequality holds for $\leqslant\text{Const.}\log b_k^{-1}$ choices of lengths of $\tilde{\omega}$ and the second for $<2b_k^{-10}$ different $\tilde{n}$'s. This implies that, for *fixed* $p\in F_k$,

$$\sum_{(\tilde{\text{F}}\text{b})}|\tilde{\omega}|\leqslant\text{Const.}\, b_k^{-17}|\omega|,$$

and the same inequality clearly also holds for ($\tilde{\text{F}}$a).

If we also observe (8.1), we find

$$\sum_{(\tilde{F}_k)}|\tilde{\omega}|\leqslant\text{Const.}\, b_k^{-19}. \tag{8.3}$$

The above construction has the following consequence: *if* $p=(n,\omega)\notin\tilde{F}_k$, *then*

$$P(x,\omega)=Q_0(x,\omega)+Q_1(x,\omega), \tag{8.4}$$

where the polynomials Q_i satisfy:

$$Q_0(x) = \varrho\, e^{i\lambda x} + O(b_k^8), \quad x \in \omega, \tag{8.5}$$

where ϱ is constant and $\lambda =$ some λ_ω; further $|\varrho| \leqslant b_k^{-2}$ by (7.5);

$$Q_1(x) \textit{ contains those } \lambda_\omega \textit{ for which } |n - \lambda_\omega[\omega]| \geqslant b_k^{-10}. \tag{8.6}$$

Since $p \notin \tilde{F}_k$, the exponents in Q_0 satisfy by ($\tilde{\text{F}}$b) $|\lambda_\omega - \lambda'_\omega||\tilde{\omega}| < b_k^{10}$, which gives (8.5). If Q_0 only contains one exponent, (8.5) is obvious.

With the primitive elements we also associate an exceptional set. If $p = (n, \omega)$ gives a primitive element, the set Y_k^* contains two intervals of lengths $2b_k^3|\omega|$, symmetric around the endpoints of ω. By (8.1)

$$mY_k^* \leqslant 16 b_k. \tag{8.7}$$

This set is of a purely technical nature and is introduced to secure the validity of Lemma 6 below.

With each $\tilde{p} = (\tilde{n}, \tilde{\omega})$ we associate the two intervals ω^*, $|\omega^*| = 4|\tilde{\omega}|$, which contain $\tilde{\omega}$. The set of such combinations $(n, \omega^*) = p^*$, $n = \tilde{n}$, for $\tilde{p} \in \tilde{F}_k$, is denoted F_k^* and the relation (8.3) becomes

$$\sum_{F_k^*} |\omega^*| \leqslant \text{Const.}\, b_k^{-19}. \tag{8.8}$$

Finally we shall need the following lemma.

LEMMA 6. *Let $p^* = (n, \omega^*)$ be given and assume $p^* \notin F_k^*$. Assume that for each $\omega' \subset \omega^*$, $4|\omega'| = |\omega^*|$, $\omega' \not\subset X_k \cup Y_k^*$. We further assume that for a certain choice ω_0' of ω' the corresponding polynomial P^0 contains an exponent λ^0 such that $|\lambda^0[\omega_0'] - n| < b_k^{-9}$. Then the four polynomials P corresponding to different choices of ω' are all identical.*

Proof. Let ω' be another choice and assume that the corresponding polynomial does not contain λ^0. This exponent was primitive for a certain $\omega_1 \supset \omega_0'$. By the construction ($\tilde{\text{F}}$a), $|\omega_1| \geqslant b_k^{-10}|\omega'|$ because the opposite inequality would imply $(n, \omega_0') \in \tilde{F}_k$, i.e., $p^* \in F_k^*$. Since $\omega_1 \not\supset \omega'$, it then follows that $\omega_0' \subset Y_k^*$ against our assumption.

In the same way it then also follows from ($\tilde{\text{F}}$a) that none of the four polynomials $P(x, \omega')$ can contain exponentials that are primitive for intervals $\omega_2 \supset \omega'$, $|\omega_2| < b_k^{-10}|\omega'|$, and that every primitive interval ω_2 for one ω' must contain the three others.

9. The coverings $\Omega(p^*; l)$ and the exceptional set

Consider a pair $p^* = (n, \omega^*)$. If the following conditon $\Omega(l)$

$$\Omega(l):\ p^* \in F^*_{l+3},\ C^*(p^*) < b_{l-1},$$

holds, we construct the partition $\Omega(p^*; l)$ in analogy with Section 3 (B). The condition (Bα) is simply changed to

$$(\mathrm{B}^*\alpha) \qquad C(n[\omega]; \omega) < b_{l-1}.$$

N is here arbitrary but fixed, and the main point in the proof is to make the estimates independent of N. We may take f as a trigonometric polynomial of degree N.

As in (C) and (D) of Section 3, we form the functions $H^*(x)$ and $\Delta(x)$ corresponding to the partition and define the exceptional sets

$$T^*(p^*):\ H^*(x) > b^{\frac{1}{2}}_{l-1}, \qquad U^*(p^*):\ \Delta(x) > b^{-\frac{1}{2}}_{l-1},$$

and as there

$$m(T^*(p^*) \cup U^*(p^*)) \leqslant \text{Const.} \exp\{-\text{Const.}\, b^{-\frac{1}{2}}_{l-1}\}\,|\omega^*|. \tag{9.1}$$

We next observe that for $\omega^* \not\subset S^*$, $C^*(p^*) < b_L$ will automatically hold where $L = L(\varepsilon) \to \infty$, $\varepsilon \to 0$. We shall therefore only consider $l \geqslant L$ and define

$$T^* = \bigcup_{l=L}^{\infty} \bigcup_{\Omega(l)} T^*(p^*)$$

and similarly for U^*. Since $\Omega(l)$ defines a subset of F^*_{l+3}, we have by (9.1) and (8.8)

$$m(T^* \cup U^*) \leqslant \text{Const.} \sum_{l=L}^{\infty} b_l^{-19\cdot 16} \exp\{-\text{Const.}\, b_l^{-\frac{1}{2}}\}. \tag{9.2}$$

Similarly, we define $\qquad X^* = \bigcup_{L}^{\infty} X^*_k, \qquad Y^* = \bigcup_{L}^{\infty} Y^*_k,$

and recall the definition of V in (E) of Section 3, here with $\lambda_1 = \varepsilon^{\frac{1}{2}}$. We then define the exceptional set

$$E = S^* \cup T^* \cup U^* \cup X^* \cup Y^* \cup V.$$

The results (7.7), (8.7) and (9.2) show that

$$mE \leqslant \delta(\varepsilon) \to 0, \quad \varepsilon \to 0,$$

and there is no reference to N in the estimate of E. It now remains to prove that every partial sum of order $< N$ is small outside E.

10. Proof of Theorem (c). Three propositions

To get a better organization of the proof, we isolate certain parts of it in propositions 1–3 in this section and complete the proof in Section 11. Proposition 1 is purely technical and is needed because we have used integers relatively each ω in the definition of the polynomials P. We must then deduce estimates for fractional Fourier transforms. Proposition 2 gives an estimate of the change in $s_n^*(x)$ when we move n to another position n_1 for which we have an estimate of the remainder terms. Proposition 3 finally shows how that estimate for a pair $p_0^* = (n_0, \omega_0^*)$ can be obtained from the corresponding expression for $(\bar{n}, \bar{\omega}^*)$ where $\bar{\omega}^* \supset \omega_0^*$. This fact is the crucial one; it is here essential that $H^*(x)$ is a maximal Hilbert transform and that $\Delta(x)$ has positive terms.

In the sequel we shall consider x fixed outside E.

PROPOSITION 1. *Let $g(t) \in L^2(\omega^*)$ and let n be given. We assume*

$$\int_{\omega^*} |g(t)|^2 \, dt \leqslant G^2 |\omega^*|$$

and for all $\omega' \subset \omega^*$, $|\omega^*| = 4|\omega'|$,

$$|c_m(\omega')| \leqslant \mu, \quad |m-n| < M.$$

Then

$$C_n^*(\omega^*; g) \leqslant \text{Const.} \left\{ \mu \log M + \frac{G}{\sqrt{M}} \right\}.$$

Proof. Take a fixed ω' and normalize to $\omega' = (0, 2\pi)$ and $n = 0$. Let

$$e^{i\alpha t} \sim \sum_{-\infty}^{\infty} \alpha_\nu e^{i\nu t}, \quad 0 < t < 2\pi.$$

Then

$$|\alpha_\nu| < \frac{2}{|\alpha - \nu| + 1}$$

and

$$c_\alpha(\omega') = \sum c_\nu \bar{\alpha}_\nu.$$

If $|\alpha| \leqslant \frac{1}{2}M$, we have

$$|c_\alpha(\omega')| \leqslant \text{Const.} \left\{ \mu \sum_{\nu=1}^{\frac{1}{2}M} \frac{1}{\nu} + \left\{ \sum_{-\infty}^{\infty} |c_\nu|^2 \right\}^{\frac{1}{2}} \left\{ \sum_{\frac{1}{2}M}^{\infty} \frac{1}{\nu^2} \right\}^{\frac{1}{2}} \right\} \leqslant \text{Const.} \left\{ \mu \log M + \frac{G}{\sqrt{M}} \right\} = q.$$

For $|\alpha| \geqslant \frac{1}{2}M$ we have $|c_\alpha| \leqslant 2G$. We find

$$C_n^* \leqslant q \sum_{-\infty}^{\infty} \frac{1}{1+\nu^2} + 4G \sum_{M}^{\infty} \frac{1}{1+\nu^2},$$

which proves the assertion.

For proposition 2, let us recall definition (1.2) of $s_n^*(x)$ for $\omega_{-1}^* = (-4\pi, 4\pi)$. The same integral over an arbitrary ω^* is denoted $s_n^*(x; \omega^*)$. We shall also assume ε sufficiently small.

PROPOSITION 2. *Assumptions. Let n_0 and $p_0^* = (n_0[\omega_0^*], \omega_0^*)$ be given and assume $\omega_0 \notin E$. Let l be defined by*

$$b_{l-1} > C^*(p_0^*) \geqslant b_l \quad (\text{so that } l \geqslant L(\varepsilon)) \tag{10.1}$$

and suppose $p_0^* \notin F_{l+3}^*$. Let x belong to the middle half of ω_0^*.

Assertions. There exist $\omega_1^* \supseteq \omega_0^*$, x belonging to the middle half of ω_1^*, and n_1 such that

$$p_1^* = (n_1[\omega_1^*], \omega_1^*) \in F_{l+3}^* \tag{10.2}$$

and

$$|n_1[\omega_0^*] - n_0[\omega_0^*]| \leqslant \frac{1}{b_l}. \tag{10.3}$$

Furthermore, if we set $p_{10}^* = (n_1[\omega_0^*], \omega_0^*)$, then

$$\left| |s_{n_1}^*(x; \omega_0^*)| - |s_n(x; \omega_0^*)| \right| \leqslant \text{Const.}\{C^*(p_{10}^*) + b_l\} \tag{10.4}$$

for all n such that

$$|n_1[\omega_0^*] - n[\omega_0^*]| < b_l^{-3/2} \tag{10.5}$$

and in particular for $n = n_0$.

Proof. Let ω_0' be the subinterval of ω_0^* for which $C_{n_0[\omega_0']}(\omega_0') = C^*(p_0^*)$ and let ω' be an arbitrary interval $\omega' \subset \omega^*$, $|\omega^*| = 4|\omega'|$. Let P_0 and P be the corresponding $(l+3)$-polynomials. The definition of $(l+3)$-polynomials implies for every such ω'

$$|c_m(\omega', f-P)| < b_{l+3}, \quad m \text{ an integer.}$$

Since $|P| \leqslant b_{l+3}^{-2}$, we can use Proposition 1 with

$$G = 3b_{l+3}^{-2}, \quad \mu = b_{l+3}, \quad M = b_{l+3}^{-10}, \quad n \text{ arbitrary.}$$

We obtain

$$C_n(\omega'; f-P) \leqslant b_{l+2} \quad (l \geqslant L(\varepsilon),\ \varepsilon < \varepsilon_0) \tag{10.6}$$

for all n.

In particular for $\omega' = \omega_0'$ and $n = n_0[\omega_0^*]$, (10.6) yields

$$C_n(\omega_0'; P_0) \geqslant C^*(p_0^*; f) - b_{l+2} \geqslant b_l - b_{l+2}.$$

Since $p_0^* \notin F_{l+3}^*$ by our assumption, we can use (8.4–6) with $k=l+3$ and find

$$\text{Const.}\frac{|\varrho|}{|\lambda[\omega_0'] - n_0[\omega_0']|} \geqslant b_l. \tag{10.7}$$

We now choose $n_1=\lambda$; $\lambda[\omega_0']$ is used in (10.6) and we obtain

$$|\varrho| \leqslant \text{Const.}\,(C^*(p_{10}^*)+b_{l+2})<1 \quad (\varepsilon<\varepsilon_1). \tag{10.8}$$

We finally insert this improved estimate of ϱ in (10.7) and have verified (10.3). The relation (10.2) also holds for $n_1=\lambda$ since λ is an exponent in P. ω_1' is the corresponding primitive interval or its left neighbor so that $\omega_1^* \supseteq \omega_0^*$ and contains x in its middle half. We also observe that by Lemma 6 the polynomials P corresponding to the four choices of ω' coincide and we have a unique P on ω_0^*.

To prove (10.4), we write

$$s_n^*(\cdot\,;f)=s_n^*(\cdot\,;f-P)+s_n^*(\cdot\,;P).$$

Since (10.6) holds for $f-P$, it follows from (5.4) and (5.5) used less than $b_l^{-3/2}$ times that

$$|e^{inx}s_n^*(\cdot\,;f-P)-e^{in_1x}s_{n_1}^*(\cdot\,;f-P)| \leqslant \text{Const.}\,b_l^{-3/2}b_{l+2},$$

if n satisfies (10.5). Since $P=Q_0+Q_1$ satisfies (8.5) and (8.6) and since P has at most b_{l+3}^{-3} terms an easy calculation shows that

$$|e^{inx}s_n^*(\cdot\,;P)-e^{in_1x}s_{n_1}^*(\cdot\,;P)| \leqslant \text{Const.}\,\{|\varrho|+b_l\}$$

for these values of n. The basic fact for this calculation is the relation

$$\int_{\omega_0^*}\frac{e^{-i\lambda t}}{x-t}\,dt=\pi i\,\text{sign}\,(\lambda)\,e^{-i\lambda x}+O((\lambda\,|\omega_0^*|)^{-1}).$$

The proof of proposition 2 is now complete.

PROPOSITION 3. *Assumptions as in Proposition 2.*

Assertions. There exist $\bar\omega$, x *belonging to the middle half of* $\bar\omega^*$, *an integer* m, $m\leqslant l$, *and an integer* $\bar n$ *such that*

$$|\bar n[\omega_0^*]-n_0[\omega_0^*]| \leqslant \text{Const.}\,b_l^{-1}, \tag{10.9}$$

$$\bar p^*=(\bar n[\bar\omega^*],\bar\omega^*)\in F_{m+3}^*, \tag{10.10}$$

$$C^*(\bar p^*)<b_{m-1}. \tag{10.11}$$

The relations (10.10) *and* (10.11) *imply that* $\Omega(\bar p^*;m)$ *is defined. For this partition it holds that*

$\bar{\omega}^*(x) \subset \omega_0^*$ *strictly and* $\omega_0^* - \bar{\omega}^*(x)$ *is a union of intervals belonging to* $\Omega(\bar{p}^*; m)$. (10.12)

Furthermore, if p_{10}^* *is given by Proposition 2, then*

$$C^*(p_{10}^*) < b_{m-1}. \tag{10.13}$$

Proof. Denote by Σ the set of triplets $(n; \omega^*; k)$ where n and k are integers and ω^* an interval $\omega_{j\nu}^*$ such that

$$\Sigma : \begin{cases} (1)\ \ \omega^* \supseteq \omega_0^* \text{ and } x \in \text{middle half of } \omega^*; \\ (2)\ \ k \leqslant l \text{ and } C^*(p_{10}^*), \text{ defined in Prop. 2, } < b_{k-1}; \\ (3)\ \ \text{if } n_1 \text{ is defined in Prop. 2, then } |n_1[\omega_0^*] - n[\omega_0^*]| \leqslant \sum_{j=k}^{l} b_j^{-1}; \\ (4)\ \ (n[\omega^*], \omega^*) \in F_{k+3}^*. \end{cases}$$

We first show that Σ is not empty. If $C^*(p_{10}^*) < b_{l-1}$, then $(n_1; \omega_1^*; l) \in \Sigma$. If $C^*(p_{10}^*) \geqslant b_{l-1}$, we define k by $b_k \leqslant C^*(p_{10}^*) < b_{k-1}$. If $p_{10}^* \in F_{k+3}^*$, then $(n_1; \omega_0^*; k) \in \Sigma$. If $p_{10}^* \notin F_{k+3}^*$, we use p_{10}^* as p_0^* in Proposition 2 to construct a new pair (n_2, ω_2^*). By (10.3)

$$|n_2[\omega_0^*] - n_1[\omega_0^*]| \leqslant b_k^{-1}$$

so that Σ (3) holds. It is now clear that $(n_2; \omega_2^*; k) \in \Sigma$.

Define $(\bar{n}; \bar{\omega}^*; m)$ of Proposition 3 as an element of Σ for which k is minimal. Then (10.9) holds by (3) in the definition of Σ since (10.3) holds and

$$\sum_{j=1}^{l} b_j^{-1} \leqslant \text{Const.}\, b_l^{-1}.$$

(10.10) follows from (4) and (10.13) is included in (2). It remains to prove (10.11) and (10.12).

Assume first $C^*(\bar{p}^*) \geqslant b_{m-1}$ and define k by $b_k \leqslant C^*(\bar{p}^*) < b_{k-1}$, $k < m$. Since m is minimal, $(\bar{n}; \bar{\omega}^*; k) \notin \Sigma$, i.e. $\bar{p}^* \notin F_{k+3}^*$. We then use $\bar{p}^*$ as p_0^* in Proposition 2 and obtain $\bar{n}_1$, $\bar{\omega}_1^*$ by that construction. Since

$$\begin{aligned} |\bar{n}_1[\omega_0^*] - n_1[\omega_0^*]| &\leqslant |\bar{n}_1[\omega_0^*] - \bar{n}[\omega_0^*]| + |\bar{n}[\omega_0^*] - n_1[\omega_0^*]| \\ &\leqslant |\bar{n}_1[\bar{\omega}^*] - \bar{n}[\bar{\omega}^*]| + \sum_{m}^{l} b_j^{-1} \\ &\leqslant \sum_{k}^{l} b_j^{-1}, \end{aligned}$$

$(\bar{n}_1; \bar{\omega}_1^*; k) \in \Sigma$, which again contradicts the minimality of m. We have thus proved (10.11).

We now know that we can use the construction $\Omega(m)$ on $\bar{p}^*$ and get $\bar{\omega}^*(x)=\bar{\omega}_1^*$ and a corresponding $\bar{p}_1^*$. Then $C^*(\bar{p}_1^*)\geqslant b_{m-1}$ and if $\bar{\omega}_1^*\supseteq\omega_0^*$, we could use the argument above to obtain a contradiction. Hence $\bar{\omega}_1^*\subset\omega_0^*$ strictly, and it is then easy to see that $\omega_0^*-\bar{\omega}^*(x)$ has the stated property.

11. Proof of Theorem (c), completed

Let us again consider formula (5.3) for $s_n^*(x)$, x fixed not in E. We have used the covering $\Omega((n,\omega_{-1}^*);k)$, $b_k\leqslant C_n^*(\omega_{-1}^*)<b_{k-1}$. Since all $c_n(\omega_{00})$, $|c_n|\geqslant b_{k+3}$, are coefficients in $P_{k+3}(x;\omega_{00})$, it follows from Proposition 1 and ($\tilde{\text{F}}$a) in Section 8 that $(n;\omega_{-1}^*)\in F_{k+3}^*$. The remainders in (5.3) therefore are $O(b_{k-1}^{\frac{1}{2}})$.

We write $\omega^*(x)=\omega_0^*$ and $p_0^*=(n[\omega_0^*],\omega_0^*)$. If $b_{l-1}>C^*(p_0^*)\geqslant b_l$, then $l<k$. If $p_0^*\in F_{l+3}^*$, we can make the construction $\Omega(p_0^*;l)$ and the remainders will be $O(b_{l-1}^{\frac{1}{2}})$. If $p_0^*\notin F_{l+3}^*$, we construct p_1^* and n_1 according to Proposition 2 and $\bar{n}$, $\bar{\omega}$ and m according to Proposition 3. We write, using $\Omega(\bar{p}^*;m)$ only on ω_0^*,

$$\int_{\omega_0^*}\frac{e^{-i\bar{n}t}f(t)}{x-t}\,dt=\int_{\bar{\omega}^*(x)}\frac{e^{-i\bar{n}t}f(t)}{x-t}\,dt+\bar{H}_{\bar{n}}(x)+\bar{R}_{\bar{n}}(x).$$

By the estimate (10.9) for $\bar{n}$ and (10.4) and (10.13), the left side integral differs in modulus from the corresponding integral in (5.3) by $O(b_{m-1})$. Since x belongs to the middle half of $\bar{\omega}^*$, ω_0^* and $\bar{\omega}^*(x)$, it follows that

$$|\bar{H}_{\bar{n}}(x)|\leqslant 2H_{\bar{n}}^*(x)<2b_{m-1}^{\frac{1}{2}}$$

since $x\notin T'^*$. A similar inequality holds for $\bar{R}_{\bar{n}}(x)$ since $\omega_0^*-\omega^*(x)$ satisfies (10.12) so that Const. $b_{m-1}\Delta(x)$ is a majorant also of $\bar{R}_{\bar{n}}(x)$. Finally we observe that unless $|\bar{\omega}^*(x)|=2\pi 2^{-N+1}$, which case is easy, since f is of degree N,

$$C^*(\bar{n}[\bar{\omega}^*(x)],\bar{\omega}^*(x))\geqslant b_{m-1}.$$

We can therefore repeat the argument and obtain, as in the proof of Theorem (a),

$$s_n^*(x)=\sum_{j=L(\varepsilon)}^{\infty}(O(b_j^{\frac{1}{2}})+O(b_j))+\sqrt{\varepsilon}\to 0,\quad \varepsilon\to 0,$$

for all $x\notin E$, $mE\to 0$, $\varepsilon\to 0$. Since no estimates depend on N, Theorem (c) is proved.

12. Theorem (b)

As mentioned before, it seems clear that $s_n(x)$ converges a.e. also if $f\in L^p$, $p>1$. We shall therefore only outline the proof of (b).

Let N be fixed and define in this case the exceptional sets as in (c) by the inequalities, $n \leqslant 2^N$,

$$T^*: \quad H^*(x) > b_{l-1}^{\frac{1}{2}};$$

respectively, $$U^*: \quad \Delta(x) > b_{l-1}^{-\frac{1}{2}}.$$

As in the proof of (a) we allow the construction $\Omega(p^*; l)$, $b_l \leqslant C^*(p^*) < b_{l-1}$ for *all* pairs p^* for which l is such that $b_{l-1} < (\log N)^{-k}$, where k is a suitable constant depending on p ($f \in L^p$). As in the proof of (a), it follows from the Hausdorff–Young inequality that the corresponding exceptional set has measure $O(N^{-2})$.

If $b_{l-1} \geqslant (\log N)^{-k}$, we have to select certain pairs $p = (n, \omega)$ for which we allow the construction. For ω_{00} we include all n such that $|c_n| \geqslant b_l$. The set $\{p\} = \Phi_l$, being defined for $|\omega| \geqslant 2\pi 2^{-\nu}$, we include $p' = (n', \omega')\, |\omega'| = 2\pi 2^{-\nu-1}$ if

$$|c(p'; f)| \geqslant b_l \quad \text{and} \quad \big||\omega| n' - |\omega'| n\big| > b_l^{-K} |\omega| \tag{12.1}$$

holds for every $(n, \omega) \in \Phi_l$ with $\omega \supset \omega'$; K is a suitable constant.

We now consider functions $\varphi(x)$, $|x| < 4\pi$, of the form

$$\varphi(x) = e^{i\theta} \cdot e^{\frac{2\pi n}{|\omega|} ix}, \quad x \in \omega, \ (n, \omega) \in \Phi_l,$$

where for fixed φ the different ω's do not intersect and are maximal with respect to this property. From the separation (12.1) it follows that this system is "almost" orthogonal. From this we obtain L^p estimates by a standard interpolation. It is then easy to define a set X^* for the set Φ_l as we did for F_l in (b).

By the above construction, we have for each $(n[\omega^*], \omega^*) = p^*$ with $C^*(p^*) \geqslant b_l$ a $\bar{p}^* \in \Phi_l$, as in Proposition 3, such that

$$|\bar{n}[\omega^*] - n[\omega^*]| < \text{Const.}\, b_l^{-K} = b.$$

Since $x \notin S^*$, $|s_{\bar{n}} - s_n| = O(\log 1/b)$ and summing this for $b_l \geqslant (\log N)^{-k}$, we obtain the desired estimate.

References

[1]. Zygmund, A., *Trigonometric series*, Vol. I, 1959.
[2]. Ibid., Vol. II.

Received January 31, 1966

Shiing-Shen Chern
CURRICULUM VITAE

Born: October 28, 1911, Kashing, Chekiang Province, China
Nationality: Naturalized U.S. citizen, 1961

Degrees

B.Sc., Nankai University, Tientsin, China. 1930
M.Sc., Tsinghua University, Peiping, China. 1934
D.Sc., University of Hamburg, Germany. 1936
LL.D. (honorary), The Chinese University of Hong Kong. 1969
D.Sc. (honorary), University of Chicago. 1969
D.Sc. (honorary), University of Hamburg. 1971
Dr.Math., Eidgenossische Technische Hochschule, Zurich. 1982
D.Sc. (honorary), State University of New York at Stony Brook. 1985
Honorary Dr., Nankai University. 1985
D.Sc. (honorary), University of Notre Dame, Indiana. 1994

Positions Held

China Foundation Postdoctoral Fellow, Sorbonne, Paris, France. 1936–37
Professor of Mathematics, Tsinghua University and Southwest Associated University, Kunming, China. 1937–43
Member, Inst. for Advanced Study, Princeton, New Jersey. 1943–45
Acting Director, Institute of Mathematics, Academia Sinica, Nanking, China. 1946–48
Professor of Mathematics, University of Chicago. 1949–60
Professor of Mathematics, University of California at Berkeley. 1960–79; Professor Emeritus, 1979–present
Vice President, American Mathematical Society, 1962–64
Director, Mathematical Sciences Research Institute, Berkeley, California. 1981–84; Director Emeritus, 1984–present
Director, Nankai Institute of Mathematics, Tianjin, China. 1984–92, Director Emeritus, 1992–present

Honorary Professorships

Beijing University 1978, Nankai University 1978, Institute of Systems Science. Academy of Sciences (China) 1980, Chinan University (Canton) 1980, Graduate School, Academy of Sciences (China) 1984, Nanjing University 1985, East China Normal University 1985, Chinese University of Science and Technology 1985, Beijing Normal University 1985, Chekiang University 1985, Hangchow University 1986, Fudan University 1986, Shanghai University of Technology 1986, Tianjin University 1987, Tohoku University (Japan) 1987

Visiting Professorships

Harvard University, Cambridge, Massachusetts. 1952
Eidgenossische Technische Hochschule, Zurich, Switzerland. 1953
Massachusetts Instituite of Technology, Cambridge, Massachusetts. 1957
Institute for Advanced Study, Princeton, New Jersey. 1964

Reprinted from A Mathematician and His Mathematical Work (World Scientific, 1996), pp. 695–696.

University of California at Los Angeles. 1966
Institut des Hautes Etudes Scientifiques, Paris, France. 1967
Instituto de Matematica Pura e Aplicada, Rio de Janeiro, Brazil. 1970
University of Warwick, Coventry, England. 1972
Rockefeller University, New York, New York. 1973
Eidgenossische Technische Hochschule, Zurich, Switzerland. 1981
Max Planck Institut für Mathematik, Bonn, Germany. 1982
University of Houston, Houston, Texas, 1988–

Memberships

Member, Academia Sinica, Nanking, China, 1948–
Honorary Member, Indian Mathematical Society, 1950–
Member, National Academy of Sciences, USA, 1961–
Fellow, American Academy of Arts and Sciences, 1963–
Corresponding Member, Brazilian Academy of Sciences, 1971–
Associate Founding Fellow, Third World Academy of Sciences, 1983–
Foreign Member, Royal Society of London, 1985–
Honorary Member, London Mathematical Society, 1986–
Corresponding Member, Academia Peloritana, Messina, Sicily, 1986–
Honorary Life Member, New York Academy of Sciences, 1987–
Foreign Member, Accademia dei Lincei, Rome, 1988–
Foreign Member, Académie des Sciences, Paris, 1989–
Member, American Philosophical Society, 1989–
Foreign Member, Chinese Academy of Sciences, 1994–

Prizes

Chauvenet Prize, Mathematical Association of America, 1970
National Medal of Science, 1975
Alexander von Humboldt Award, Germany, 1982
Steele Prize, American Mathematical Society, 1983
Wolf Prize, Israel, 1983–84

Bibliography of the Publications of S.S. Chern

I. *Books and Monographs*

1. *Topics in Differential Geometry* (mimeographed), Institute for Advanced Study, Princeton (1951), 106 pp.
2. *Differentiable Manifolds* (mimeographed), University of Chicago, Chicago (1953), 166 pp.
3. *Complex Manifolds*
   a. University of Chicago, Chicago (1956), 195 pp.
   b. University of Recife, Recife, Brazil (1959), 181 pp.
   c. Russian translation, Moscow (1961), 239 pp.
4. *Studies in Global Geometry and Analysis* (Editor), Mathematical Association of America (1967), 200 pp.
5. *Complex Manifolds without Potential Theory*, van Nostrand (1968), 92 pp. Second edition, revised, Springer-Verlag (1979), 152 pp, 2nd printing (1995), 160 pp.
6. *Minimal Submanifolds in a Riemannian Manifold* (mimeographed), University of Kansas, Lawrence (1968), 55 pp.
7. (with Wei-huan Chen) *Differential Geometry Notes*, in Chinese, Beijing University Press (1983), 321 pp.
8. *Studies in Global Differential Geometry* (Editor), Mathematical Association of America (1988), 350 pp.
9. (with R. Bryant, R. Gardner, H. Goldschmidt, P. Griffiths) *Exterior Differential Systems*, MSRI Series **18**, Springer Verlag (1991), 475 pp.

II. *Papers*

1932

[1] Pairs of plane curves with points in one-to-one correspondence. *Science Reports Nat. Tsing Hua Univ.* **1** (1932) 145–153.

1935

[2] Triads of rectilinear congruences with generators in correspondence. *Tohoku Math. J.* **40** (1935) 179–188.

[3] Associate quadratic complexes of a rectilinear congruence. *Tohoku Math. J.* **40** (1935) 293–316.

[4] Abzählungen für Gewebe. *Abh. Math. Sem. Univ. Hamburg* **11** (1935) 163–170.

1936

[5] Eine Invariantentheorie der Dreigewebe aus r-dimensionalen Mannigfaltigkeiten im R_{2r} *Abh. Math. Sem. Univ. Hamburg* **11** (1936) 333–358.

Reprinted from A Mathematician and His Mathematical Work (World Scientific, 1996), pp. 699–707.

BIBLIOGRAPHY OF THE PUBLICATIONS OF S.S. CHERN

1937

[6] Sur la géométrie d'une équation différentielle du troisième ordre. *C. R. Acad. Sci. Paris* **204** (1937) 1227–1229.

[7] Sur la possibilitè de plonger un espace à connexion projective donné dans un espace projectif. *Bull. Sci. Math.* **61** (1937) 234–243.

1938

[8] On projective normal coördinates. *Ann. of Math.* **39** (1938) 165–171.

[9] On two affine connections. *J. Univ. Yunnan* **1** (1938) 1–18.

1939

[10] Sur la géométrie d'un système d'équations différentielles du second ordre. *Bull. Sci. Math* **63** (1939) 206–212.

1940

[11] The geometry of higher path-spaces. *J. Chin. Math. Soc.* **2** (1940) 247–276.

[12] Sur les invariants intégraux en géométrie. *Science Reports Nat. Tsing Hua Univ.* **4** (1940) 85–95.

[13] The geometry of the differential equation $y''' = F(x, y', y'')$. *Science Reports Nat. Tsing Hua Univ.* **4** (1940) 97–111.

[14] Sur une généralisation d'une formule de Crofton. *C.R. Acad. Sci. Paris* **210** (1940) 757–758.

[15] (with C.T. Yen) Sulla formula principale cinematica dello spazio ad n dimensioni. *Boll. Un. Mat. Ital.* **2** (1940) 434–437.

[16] Generalization of a formula of Crofton. *Wuhan Univ. J. Sci.* **7** (1940) 1–16.

1941

[17] Sur les invariants de contact en géométrie projective différentielle. *Acta Pontif. Acad. Sci.* **5** (1941) 123–140.

1942

[18] On integral geometry in Klein spaces. *Ann. of Math.* **43** (1942) 178–189.

[19] On the invariants of contact of curves in a projective space of N dimensions and their geometrical interpretation. *Acad. Sinica Sci. Record* **1** (1942) 11–15.

[20] The geometry of isotropic surfaces. *Ann. of Math.* **43** (1942) 545–559.

[21] On a Weyl geometry defined from an $(n-1)$-parameter family of hypersurfaces in a space of n dimensions. *Acad. Sinica Sci. Record* **1** (1942) 7–10.

1943

[22] On the Euclidean connections in a Finsler space. *Proc. Nat. Acad. Sci. USA*, **29** (1943) 33–37.

[23] A generalization of the projective geometry of linear spaces. *Proc. Nat. Acad. Sci. USA*, **29** (1943) 38–43.

1944

[24] Laplace transforms of a class of higher dimensional varieties in a projective space of n dimensions. *Proc. Nat. Acad. Sci. USA*, **30** (1944) 95–97.

[25] A simple intrinsic proof of the Gauss–Bonnet formula for closed Riemannian manifolds. *Ann of Math.* **45** (1944) 747–752.

[26] Integral formulas for the characteristic classes of sphere bundles. *Proc. Nat. Acad. Sci. USA* **30** (1944) 269–273.

[27] On a theorem of algebra and its geometrical application. *J. Indian Math. Soc.* **8** (1944) 29–36.

BIBLIOGRAPHY OF THE PUBLICATIONS OF S.S. CHERN

1945

[28] On Grassmann and differential rings and their relations to the theory of multiple integrals. *Sankhya* **7** (1945) 2–8.

[29] Some new characterizations of the Euclidean sphere. *Duke Math. J.* **12** (1945) 279–290.

[30] On the curvature integra in a Riemannian manifold. *Ann. of Math.* **46** (1945) 674–684.

[31] On Riemannian manifolds of four dimensions. *Bull. Amer. Math. Soc.* **51** (1945) 964–971.

1946

[32] Some new viewpoints in the differential geometry in the large. *Bull. Amer. Math. Soc.* **52** (1946) 1–30.

[33] Characteristic classes of Hermitian manifolds. *Ann. of Math.* **47** (1946) 85–121.

1947

[34] (with H.C. Wang). Differential geometry in symplectic space I. *Science Report Nat. Tsing Hua Univ.* **4** (1947) 453–477.

[35] Sur une classe remarquable de variétés dans l'espace projectif à N dimensions. *Science Reports Nat. Tsing Hua Univ.* **4** (1947) 328–336.

[36] On the characteristic classes of Riemannian manifolds. *Proc. Nat. Acad. Sci USA*, **33** (1947) 78–82.

[37] Note of affinely connected manifolds. *Bull. Amer. Math. Soc.* **53** (1947) 820–823; correction ibid **54** (1948) 985–986.

[38] On the characteristic ring of a differentiable manifold. *Acad. Sinica. Sci. Record* **2** (1947) 1–5.

1948

[39] On the multiplication in the characteristic ring of a sphere bundle. *Ann. of Math.* **49** (1948) 362–372.

[40] Note on projective differential line geometry. *Acad. Sinica Sci. Record* **2** (1948) 137–139.

[41] (with Y.L. Jou) On the orientability of differentiable manifolds. *Science Reports Nat. Tsing Hua Univ.* **5** (1948) 13–17.

[42] Local equivalence and Euclidean connections in Finsler spaces. *Science Reports Nat. Tsing Hua Univ.* **5** (1948) 95–121.

1949

[43] (with Y.F. Sun). The imbedding theorem for fibre bundles. *Trans. Amer. Math. Soc* **67** (1949) 286–303.

[44] (with S.T. Hu) Parallelisability of principal fibre bundles. *Trans. Amer. Math. Soc.* **67** (1949) 304–309.

1950

[45] (with E. Spanier). The homology structure of sphere bundles. *Proc. Nat. Acad. Sci. USA*, **36** (1950) 248–255.

[46] Differential geometry of fiber bundles. *Proc. Int. Congr. Math.* (1950) **II** 397–411.

1951

[47] (with E. Spanier). A theorem on orientable surfaces in four-dimensional space. *Comm. Math. Helv.* **25** (1951) 205–209.

BIBLIOGRAPHY OF THE PUBLICATIONS OF S.S. CHERN

1952

[48] On the kinematic formula in the Euclidean space of N dimensions. *Amer. J. Math* **74** (1952) 227–236.

[49] (with C. Chevalley). Elie Cartan and his mathematical work. *Bull. Amer. Math. Soc.* **58** (1952) 217–250.

[50] (with N.H. Kuiper) Some theorems on the isometric imbedding of compact Riemann manifolds in Euclidean space. *Ann. of Math.* **56** (1952) 422–430.

1953

[51] On the characteristic classes of complex sphere bundles and algebraic varieties. *Amer. J. of Math.*, **75** (1953) 565--597.

[52] Some formulas in the theory of surfaces. *Boletin de la Sociedad Matematica Mexicana*, **10** (1953) 30–40.

[53] Relations between Riemannian and Hermitian geometries. *Duke Math. J.*, **20** (1953) 575–587.

1954

[54] Pseudo-groupes continus infinis *Colloque de Geom. Diff.* Strasbourg (1954) 119–136.

[55] (with P. Hartman and A. Wintner) On isothermic coordinates. *Comm. Math. Helv.* **28** (1954) 301–309.

1955

[56] La géométrie des sous-variétés d'un espace euclidien à plusieurs dimensions. *l'Ens. Math.*, **40** (1955) 26–46.

[57] An elementary proof of the existence of isothermal parameters on a surface. *Proc. Amer. Math. Soc.*, **6** (1955) 771–782.

[58] On special W-surfaces. *Proc. Amer. Math. Soc.*, **6** (1955) 783–786.

[59] On curvature and characteristic classes of a Riemann manifold. *Abh. Math. Sem. Univ. Hamburg* **20** (1955) 117–126.

1956

[60] Topology and differential geometry of complex manifolds. *Bull. Amer. Math. Soc.*, **62** (1956) 102–117.

1957

[61] On a generalization of Kähler geometry. *Lefschetz jubilee volume.* Princeton Univ. Press (1957) 103–121.

[62] (with R. Lashof) On the total curvature of immersed manifolds. *Amer. J. of Math.* **79** (1957) 306–318.

[63] (with F. Hirzebruch and J-P. Serre) On the index of a fibered manifold. *Proc. Amer. Math. Soc.*, **8** (1957) 587–596.

[64] A proof of the uniqueness of Minkowski's problem for convex surfaces. *Amer. J. of Math.*, **79** (1957) 949–950.

1958

[65] Geometry of submanifolds in complex projective space. *Symposium International de Topologia Algebraica* (1958) 87–96.

[66] (with R.K. Lashof) On the total curvature of immersed manifolds, II. *Michigan Math. J.* **5** (1958) 5–12.

[67] Differential geometry and integral geometry. *Proc. Int. Congr. Math. Edinburgh* (1958) 441–449.

BIBLIOGRAPHY OF THE PUBLICATIONS OF S.S. CHERN

1959

[68] Integral formulas for hypersurfaces in Euclidean space and their applications to uniqueness theorems. *J. of Math. and Mech.* **8** (1959) 947–956.

1960

[69] (with J. Hano and C.C. Hsiung) A uniqueness theorem on closed convex hypersurfaces in Euclidean space. *J. of Math. and Mech.* **9** (1960) 85–88.

[70] Complex analytic mappings of Riemann surfaces I. *Amer. J. of Math.* **82** (1960) 323–337.

[71] The integrated form of the first main theorem for complex analytic mappings in several complex variables. *Ann. of Math.* **71** (1960) 536–551.

[72] Geometrical structures on manifolds. *Amer. Math. Soc. Pub.* (1960) 1–31.

[73] La géométrie des hypersurfaces dans l'espace euclidean. *Seminaire Bourbaki*, **193** (1959–1960).

[74] Sur les métriques Riemanniens compatibles avec une reduction du groupe structural. *Séminaire Ehresmann*, January 1960.

1961

[75] Holomorphic mappings of complex manifolds. *L'Ens. Math.* **7** (1961) 179–187.

1962

[76] Geometry of quadratic differential form. *J. of SIAM* **10** (1962) 751–755.

1963

[77] (with C.C. Hsiung) On the isometry of compact submanifolds in Euclidean space. *Math. Annalen* **149** (1963) 278–285.

[78] Pseudo-Riemannian geometry and Gauss–Bonnet formula. *Academia Brasileira de Ciencias* **35** (1963) 17–26.

1965

[79] Minimal surfaces in an Euclidean space of N dimensions. *Differential and Combinatorial Topology*, Princeton Univ. Press (1965) 187–198.

[80] (with R. Bott) Hermitian vector bundles and the equidistribution of the zeroes of their holomorphic sections. *Acta. Math.* **114** (1965) 71–112.

[81] On the curvatures of a piece of hypersurface in Euclidean space. *Abh. Math. Sem. Univ. Hamburg* **29** (1965) 77–91.

[82] On the differential geometry of a piece of submanifold in Euclidean space. *Proc. of U.S.–Japan Seminar in Diff. Geom.* (1965) 17–21.

1966

[83] Geometry of G-structures. *Bull. Amer. Math. Soc.* **72** (1966) 167–219.

[84] On the kinematic formula in integral geometry. *J. of Math. and Mech.* **16** (1966) 101–118.

[85] Geometrical structures on manifolds and submanifolds. *Some Recent Advances in Basic Sciences*, Yeshiva Univ. Press (1966) 127–135.

1967

[86] (with R. Osserman) Complete minimal surfaces in Euclidean n-space. *J. de l'Analyse Math.* **19** (1967) 15–34.

[87] Einstein hypersurfaces in a Kählerian manifold of constant holomorphic curvature. *J. Diff. Geom.* **1** (1967) 21–31.

BIBLIOGRAPHY OF THE PUBLICATIONS OF S.S. CHERN

1968

[88] On holomorphic mappings of Hermitian manifolds of the same dimension. *Proc. Symp. Pure Math.* **11**. Entire Functions and Related Parts of Analysis (1968) 157–170.

1969

[89] Simple proofs of two theorems on minimal surfaces. *L'Ens. Math.* **15** (1969) 53–61.

1970

[90] (with H. Levine and L. Nirenberg) Intrinsic norms on a complex manifold. *Global analysis*, Princeton Univ. Press (1970) 119–139.

[91] (with M. do Carmo and S. Kobayashi) Minimal submanifolds of a sphere with second fundamental form of constant length. *Functional Analysis and Related Fields*, Springer-Verlag (1970) 59–75.

[92] (with R. Bott) Some formulas related to complex transgression. *Essays on Topology and Related Topics*, Springer-Verlag, (1970) 48–57.

[93] Holomorphic curves and minimal surfaces. *Carolina Conference Proceedings* (1970) 28 pp.

[94] On minimal spheres in the four-sphere, Studies and Essays Presented to Y. W. Chen, Taiwan, (1970) 137–150.

[95] Differential geometry: Its past and its future. *Actes Congrès Intern. Math.* (1970) **1**, 41–53.

[96] On the minimal immersions of the two-sphere in a space of constant curvature. *Problems in Analysis*, Princeton Univ. Press, (1970) 27–40.

1971

[97] Brief survey of minimal submanifolds. *Differentialgeometrie im Grossen.* W. Klingenberg (ed.), **4** (1971) 43–60.

[98] (with J. Simons) Some cohomology classes in principal fibre bundles and their application to Riemannian geometry. *Proc. Nat. Acad. Sci. USA*, **68** (1971) 791–794.

1972

[99] Holomorphic curves in the plane. *Diff. Geom., in honor of K. Yano*, (1972) 73–94.

[100] Geometry of characteristic classes. *Proc. 13th Biennial Sem. Canadian Math. Congress*, (1972) 1–40. Also pub. in Russian translation.

1973

[101] Meromorphic vector fields and characteristic numbers. *Scripta Math.* **29** (1973) 243–251.

[102] The mathematical works of Wilhelm Blaschke. *Abh. Math. Sem. Univ. Hamburg* **39** (1973) 1–9.

1974

[103] (with. J. Simons) Characteristic forms and geometrical invariants. *Ann. of Math.* **99** (1974) 48–69.

[104] (with M. Cowen, A. Vitter III) Frenet frames along holomorphic curves. *Proc. of Conf. on Value Distribution Theory*, Tulane Univ. (1974) 191–203.

[105] (with J. Moser) Real hypersurfaces in complex manifolds. *Acta. Math.* **133** (1974) 219–271.

1975

[106] (with S.I. Goldberg) On the volume decreasing property of a class of real harmonic mappings. *Amer. J. of Math.* **97** (1975) 133–147.

BIBLIOGRAPHY OF THE PUBLICATIONS OF S.S. CHERN

[107] On the projective structure of a real hypersurface in C_{n+1}. *Math. Scand.* **36** (1975) 74–82.

1976

[108] (with J. White) Duality properties of characteristic forms. *Inv. Math.* **35** (1976) 285–297.

1977

[109] Circle bundles. *Geometry and topology, III.* Latin Amer. School of Math, Lecture Notes in Math. Springer-Verlag, **597** (1977) 114–131.

[110] (with P.A. Griffiths) Linearization of webs of codimension one and maximum rank. *Proc. Int. Symp. on Algebraic Geometry, Kyoto* (1977) 85–91.

1978

[111] On projective connections and projective relativity. *Science of Matter*, dedicated to Ta-you Wu, (1978) 225–232.

[112] (with P.A. Griffiths) Abel's theorem and webs. *Jber. d. Dt. Math. Verein.* **80** (1978) 13–110.

[113] (with P.A. Griffiths) An inequality for the rank of a web and webs of maximum rank. *Annali Sc. Norm. Super.–Pisa, Serie IV*, **5** (1978) 539–557.

[114] Affine minimal hypersurfaces. *Minimal Submanifolds and Geodesics.* Kaigai Publications, Ltd. (1978) 1–14.

1979

[115] Herglotz's work on geometry. *Ges. Schriften Gustav Herglotz*, Göttingen (1979) xx–xxi.

[116] (with C.L. Terng) An analogue of Bäcklund's theorem in affine geometry. *Rocky Mountain J. Math.* **10** (1979) 105–124.

[117] From triangles to manifolds. *Amer. Math. Monthly* **86** (1979) 339–349.

[118] (with C.K. Peng) Lie groups and KdV equations. *Manuscripta Math.* **28** (1979) 207–217.

1980

[119] General relativity and differential geometry. *Some Strangeness in the Proportion: A Centennial Symp. to Celebrate the Achievements of Albert Einstein, Harry Woolf* (ed.), Addison-Wesley Publ. (1980) 271–287.

[120] (with W.M. Boothby and S.P. Wang) The mathematical work of H.C. Wang. *Bull. Inst. of Math*, **8** (1980) xiii–xxiv.

[121] Geometry and physics. *Math. Medley*, Singapore, **8** (1980) 1–6.

[122] (with R. Bryant and P.A. Griffiths) Exterior differential systems. *Proc. of 1980 Beijing DD-Symposium*, (1980) 219–338.

1981

[123] Geometrical interpretation of the sinh-Gordon equation. *Annales Polonici Mathematici* **39** (1981) 63–69.

[124] (with P.A. Griffiths) Corrections and addenda to our paper: "Abel's theorem and webs." *Jber. d. Dt. Math.–Verein.* **83** (1981) 78–83.

[125] (with R. Osserman) Remarks on the Riemannian metric of a minimal submanifold. *Geometry Symposium Utrecht 1980*, Lecture Notes in Math. Springer-Verlag **894** (1981) 49–90.

[126] (with J. Wolfson) A simple proof of Frobenius theorem. *Manifolds and Lie Groups, Papers in Honor of Y. Matsushima.* Birkhäuser (1981) 67–69.

[127] (with K. Tenenblat) Foliations on a surface of constant curvature and modified Korteweg–de Vries equations. *J. Diff. Geom.* **16** (1981) 347–349.

BIBLIOGRAPHY OF THE PUBLICATIONS OF S.S. CHERN

[128] (with C.K. Peng) On the Bäcklund transformations of KdV equations and modified KdV equations. *J. of China Univ. of Sci. and Tech.*, **11** (1981) 1–6.

1982

[129] Web geometry. *Proc. Symp. in Pure Math.* **39** (1983) 3–10.

[130] Projective geometry, contact transformations, and *CR*-structures. *Archiv der Math.* **38** (1982) 1–5.

1983

[131] (with J. Wolfson) Minimal surfaces by moving frames. *Amer. J. Math.* **105** (1983) 59–83.

[132] On surfaces of constant mean curvature in a three–dimensional space of constant curvature. *Geometric Dynamics*, Springer Lecture Notes **1007** (1983) 104–108.

1984

[133] Deformation of surfaces preserving principal curvatures, *Differential Geometry and Complex Analysis*, Volume in Memory of H. Rauch, Springer-Verlag (1984) 155–163.

1985

[134] (with R. Hamilton) On Riemannian metrics adapted to three-dimensional contact manifolds. *Arbeitstagung Bonn 1984* Springer Lecture Notes **1111** (1985) 279–308.

[135] (with J. Wolfson) Harmonic maps of S^2 into a complex Grassmann manifold. *Proc. Nat. Acad. Sci.* USA **82** (1985) 2217–2219.

[136] Moving frames, *Soc. Math. de France*, Astérisque, (1985) 67–77.

[137] Wilhelm Blaschke and web geometry, Wilhelm Blaschke—Gesammelte Werke. **5**, Thales Verlag, (1985) 25–27.

[138] The mathematical works of Wilhelm Blaschke—an update. Thales Verlag, (1985), 21–23.

1986

[139] (with K. Tenenblat) Pseudospherical surfaces and evolution equations. *Studies in Applied Math.* MIT **74** (1986) 55–83.

[140] On a conformal invariant of three-dimensional manifolds. *Aspects of Mathematics and Its Applications* Elsevier Science Publishers B.V. (1986) 245–252.

[141] (with P.A. Griffiths) Pfaffian systems in involution. *Proceedings of 1982 Changchun Symposium on Differential Geometry and Differential Equations*, Science Press, China, (1986) 233–256.

1987

[142] (with J. Wolfson) Harmonic maps of the two–sphere into a complex Grassmann manifold II. *Ann. of Math.* **125** (1987) 301–335.

[143] (with T. Cecil) Tautness and Lie Sphere geometry *Math. Annalen*, Volume Dedicated to F. Hirzebruch **278** (1987) 381–399.

1988

[144] Vector bundles with a connection. Studies in Global Differential Geometry, MAA, no. 27 (1988), 1–26.

1989

[145] (with T. Cecil) Dupin submanifolds in Lie sphere geometry, *Differential Geometry and Topology*, Springer Lecture Notes, No. 1369, 1–48.

1990

[146] Historical remarks on Gauss–Bonnet, *Analysis, et cetera*, Volume in Honor of Jurgen Moser, Academic Press (1990) 209–217.

[147] What is geometry? *Amer. Math. Monthly* **97** (1990) 678–686.

1991

[148] An introduction to Dupin submanifolds, *Differential Geometry, A Symposium in Honor of M. do Carmo*, Longman (1991) 95–102.

[149] Families of hypersurfaces under contact transformations in R^n, *International Symposium in Memory of Hua Loo Keng*, Springer (1991) 49–56.

[150] Surface theory with Darboux and Bianchi, *Miscellanea Mathematica*, Volume Dedicated to H. Gotze, Springer (1991) 59–69.

[151] Transgression in associated bundles, *Internat. J. Math.* **2** (1991) 383–393.

[152] Characteristic classes and characteristic forms, *Paul Halmos Celebrating 50 Years of Math.*, Springer, 169–177.

1992

[153] On Finsler geometry, *Comptes Rendus Sci.*, Paris **314** (1992) 757–761.

[154] Yang–Mills equations and Yang–Baxter equations, *Chinese J. of Phys.*, Yang Volume, Taiwan Physical Society **30** (1992) 949–953.

1993

[155] (with David Bao) On a notable connection in Finsler Geometry, *Houston J. Math.* **19** (1993) 135–180.

1994

[156] Characteristic classes as a geometric object, *Global Analysis in Modern Mathematics* (Palais Festival Volume), Publish or Perish, 221–226.

[157] Sophus Lie and Differential Geometry, The Sophus Lie Memorial Conference, Oslo 1992 Proceedings, 129–137.

1995

[158] (with S. Y. Ji) Projective geometry and Riemann's mapping problem, *Mathematische Annalen* **302** (1995) 581–600.

1996

[159] (with David Bao) A note on the Gauss-Bonnet theorem for Finsler spaces, *Ann. of Math.* **143** (1996) 233–252.

III. *Collected Works*

1. Shiing-Shen Chern Selected Papers, Springer Verlag (1978), 508 pp.
2. Shiing-Shen Chern Selected Papers, Springer Verlag, Vols. II, 444 pp., 504 pp., IV, 462 pp. (1989).
3. (in Chinese) Selected Essays by S S Chern, Science Press, Beijing (1989); Taipei (1993).
4. Chern — A Great Geometer of the Twentieth Century, International Press (1992), edited by S. T. Yau.

Reprinted from *Chern — A Great Geometer of the Twentieth Century*, International Press, 1992.

My Mathematical Education

Shiing-Shen Chern

1. Early Education in China

I entered Fulun Middle School in Tientsin in January 1923. It was a four-year high school and I was admitted to the second semester of the first year. The mathematical curriculum consisted of:

(1) First year, arithmetic, with textbook in Chinese;

(2) Second year, algebra, textbook by Hall and Knight;

(3) Third year, geometry, textbook by Wentworth and Smith;

(4) Fourth year, trigonometry and higher algebra, textbooks respectively by Wentworth-Smith and Hall-Knight.

My teachers were competent and devoted, and I did a large number of exercises. In the fourth year I was able to do many of the Cambridge-Tripos problems quoted in Hall-Knight's book.

I graduated from Fulun in 1926. In entering Nankai University the same year I skipped two years. As a result I never had a course on analytic geometry. It was worse than that: I had to take an entrance examination to Nankai whose mathematical part consisted principally of analytic geometry. For three weeks before the examination I studied by myself the book "Mathematical Analysis" by Young and Morgan.

If I remember correctly, I came out with the second best paper of the examination. The notion of "foci of conics" puzzled me for a long time until I learned projective geometry several years later.

After entering Nankai University I soon found that I was clumsy with experimental work and mathematics became my only choice. I was fortunate to find as teacher Professor Li-Fu Chiang, who received his Ph.D. from Harvard in 1918 with Julian Coolidge, with a thesis on the line-sphere contact transformation in a noneuclidean space. As a result I did a lot of study on geometry during my senior year. Among the books I studied were Coolidge's Non-euclidean Geometry, Geometry of the Circle and Sphere, Salmon's Conic Sections and Analytic Geometry of Three Dimensions, and Castelnuovo's Analytic and Projective Geometry, etc. I was particularly fascinated by Otto Staude's two-volume book on "Fadenkonstruktionen." The geometry of quadrics is a beautiful chapter of mathematics. I was glad to see it taken up by J. Moser in 1979 in his work on integrable Hamiltonian systems and spectral theory; cf.[3] of Bibliography. Even now a study of "Salmon" could be rewarding, and is in my opinion enjoyable.

After graduation from Nankai in 1930 I went to work with Professor Dan Sun of Tsing Hua University in Peiping. He was at that time the only mathematician in China publishing research papers on mathematics. His field was projective differential geometry, being a Ph.D. student of E.P. Lane at the University of Chicago. The subject was founded by E.J. Wilczynski in 1901 and was a natural outgrowth of projective geometry which had reigned over geometry for almost a century. I became familiar with the literature and wrote a few papers. Among them was my master's thesis on projective line geometry. Following Plücker and Klein line geometry had been a favorite topic of geometers. In fact, Klein's dissertation was on quadratic line complexes, i.e., line loci defined by quadratic equations in the Plücker coordinates. They have beautiful properties; a modern treatment can be found in the book of

Griffiths-Harris [1]. Line geometry is very much in the background of twistor theory.

My thesis studies line congruences, i.e., two-dimensional submanifolds of lines, and their osculation by quadratic line complexes.

Toward the end of my graduate years, i.e., around 1934, I began to realize the importance of global differential geometry, called differential geometry in the large at that time. It was generally considered to be a difficult subject, both in the mathematical breadth required and in the depth of the problems. My main inspiration came from Wilhelm Blaschke's books on differential geometry.

It was clear that algebraic topology was at the foundation of the whole area. But algebraic topology itself was then in a stage of development. Veblen's "analysis situs", published in 1922, introduced the "homology characters", i.e., Betti numbers and torsion coefficients, in terms of the incidence matrices. Lefschetz's Topology appeared in 1930, but it did not help the beginners to enter the subject. I had a course (1933-34) from Emanuel Sperner, then visiting at Peking University, where he gave, among other things, a careful and detailed account of Erhard Schmidt's proof of the Jordan curve theorem. I also had a course with Tsai-Han Kiang, a former student of Marston Morse and a former assistant to Lefschetz, on analysis situs, following Lefschetz's book. But I had the feeling that I was only at the door of the great edifice of algebraic topology. The situation changed greatly only with the appearance of the Seifert-Threlfall book in 1934 and the Alexandroff-Hopf book in 1935.

In the spring of 1932 Blaschke visited Peiping and gave a series of lectures on "topological questions in differential geometry". It was really local differential geometry where he took, instead of a Lie group as in the case of classical differential geometries, the pseudo-group of all diffeomorphisms and studied the local invariants. I was able to follow his lectures and to read many papers under the same general title published in the Hamburger Abhandlungen and other journals. The subject is now known as web geometry. With this contact and my previous knowledge of Blaschke's books on differential geometry, I decided to go to Hamburg as a student when a fellowship was made available to me in 1934.

2. European Student Life

I was in Hamburg in 1934-36, receiving my D.Sc. in 1936, and spent a postdoctoral year in Paris with Elie Cartan. The choice of Hamburg turned out to be a fortunate one. Hamburg had a strong Department, with professors Blaschke, Artin, and Hecke, and junior members including E. Kähler, H. Petersson, H. Zassenhaus.

Blaschke's mathematical interest was shifting from web geometry to integral geometry. When I first saw him in September 1934, he gave me a bunch of reprints on web geometry. I became interested in the notion of the rank of a web and webs of maximum rank. Recall that a d-web in R^n of codimension 1 consists of d foliations by hypersurfaces in general position. If $x_1, \cdots, x_n$ are the coordinates in R^n and the foliations given by the equations

$$u_i(x_1, \cdots, x_n) = \text{const}, 1 \leq i \leq d,$$

an equation of the form

$$\sum_{1 \leq i \leq d} f_i(u_i) = 0$$

is called an abelian equation. The maximum number of linearly independent abelian equations is called the rank of the web. If the d-web is defined by the hyperplanes of an algebraic curve of class d in R^n, it has abelian equations given by Abel's theorem applied to the abelian differentials. Hence its rank is at least the genus of the curve. In a short note I determined the maximum rank $\pi(d,n), n \leqq d-1$, of all codimension one d-webs in R^n. According to a theorem of Castelnuovo this integer is equal to the maximum genus of an algebraic curve of degree d in the projective space P^n of dimension n, which does not belong to a hyperplane, P^{n-1}. The remarkable fact is that not all maximum rank webs are given by maximum genus algebraic curves in the manner described above: there are exotic maximum rank webs, whose leaves are not all hyperplanes. The abelian equations are essentially functional equations, for in the classical cases they become addition theorems of well-known transcendental functions. In the plane ($n = 2$) a 5-web of curves has maximum rank 6 and there exists an exotic web (Bol's web) whose abelian equations involve the dilogarithm. Griffiths and I studied in 1978 the question of codimension one d-webs in R^n of maximum rank $\pi(d,n)$, but we did not reach the goal. I think the determination of such webs which are exotic is a problem of great interest and importance.

During 1934-35 my major effort was spent on Kähler's seminar. It was based on his famous booklet "Einführung in die Theorie der Systeme von Differentialgleichungen", which had just been published. The main result was later known as the Cartan-Kähler Theorem. At the first meeting all the people were present, including Blaschke, Artin, and Hecke, and everybody was given a copy of the book. The attendance decreased rapidly and I was one of very few who stayed till the end. I made an application of the theory to 3-webs of r-dimensional submanifolds in R^{2r}. Both Blaschke and Kähler thought this and my earlier result on maximum rank were enough for a thesis. So I had my thesis ready by the end of 1935.

Blaschke and his school were mainly concerned with integral geometry, on which he gave a course. The most beautiful results were found by L.A. Santalò. One result consists of expressing the isoperimetric defect of a plane convex curve as an infinite sum of positive terms each of which has a geometric meaning. Santalò went on to become a world leader on integral geometry. He was from Spain and immigrated to Argentina.

Another of my fellow students was the algebraic geometer Wei-Liang Chow. He came from Chicago to Göttingen in order to work with Hermann Weyl. Political developments in Göttingen and Germany made this impossible, and he went to work with van der Waerden in Leipzig. But for some reason he lived in Hamburg and occasionally took part in seminars. He was developing his "zugeordnete Formen" later known as Chow coordinates. Chow is an original mathematician. He made major contributions to algebraic geometry, including his theorem on compact subvarieties and his intersection theory. He came from a high mandarin family in China, which recognized early the need of westernization. As a result the family produced many distinguished people. Chow was a night worker. When he visited me, I lost some sleep but learned some mathematics.

I attended Artin's lectures whenever possible. During the two years they included: complex function theory, algebraic topology, relativity, and diophantine approximations. I also had a course with Hecke on algebraic number theory, following mostly his book. The scientific life in Hamburg was an ideal one, but political events did not allow it to last.

I had a postdoctoral year in 1936-37 and sought the opinion of Blaschke. He advised me either to stay on in Hamburg and work with Artin on number theory or to go to Paris to work with Elie Cartan. They were attractive alternatives, but Paris and Cartan won.

The timing was perfect. For in that year Cartan gave a course on exterior differential systems; the lecture notes later came out as a book. The "young" French mathematicians, who later became Bourbaki, began to be active. They organized a "Séminaire Julia", which met biweekly and was devoted to a topic to be chosen every year. The topic for 1936-37 was "Les travaux de M. Elie Cartan".

Cartan was a wonderful teacher. He suggested "little" problems, some of which became the subjects of my papers. Probably because of my responses to his questions he allowed me to visit him at his home, about once every two weeks. After the visit I usually received a letter from him the next day, which would say: "After your departure I thought more about your questions. ... It would be interesting ... ". It was an interesting and unforgettable year.

I also attended Montel's lectures on several complex variables and Hadamard's seminar at the Collége de France. At the end of a seminar Hadamard would give a summary, which was frequently more lucid and informative than the talk itself.

On July 10, 1937 I left Paris to return to China with a heavy heart, after learning the news that the Sino-Japanese war had broken out.

3. Mathematical Isolation

When I left Eruope for China in the summer of 1937, I was going to take up my position as professor of mathematics at Tsing Hua University in Peiping. Because of the Sino-Japanese war the goal was reached only ten years later. The University moved to Changsha and then to Kunming in 1938, until the end of the war in the summer of 1945.

Kunming is a beautiful city. With deprivations and uncertainties to be expected in a country at war, life was otherwise pleasant. Tsing Hua University joined with Peking University and Nankai University to form the Southwest Associated University, and Kunming immediately became the intellectual center of wartime China. My mathematics colleagues included Loo-keng Hua and Pao-lu Hsu. I gave classes and seminars on algebraic topology, Lie groups, sphere geometry, exterior differential systems, etc. and attracted a reasonable number of students. The great disadvantage was that the place was cut off from the outside: there was a period when even the "Burma Road" was closed and the only communication with the outside world was by air. I had a small personal library. At the beginning it was even fun to do some reading and thinking that I had wished and did not find the time. But frustrations came quickly and had to be overcome. I wrote to Elie Cartan telling him the situation and he sent me a large number of his reprints, including some old ones. I spent a great deal of time pondering over them and thinking about their implications and applications. This was definitely a gain. In the thirties people such as Weyl, Blaschke, and Kähler began to realize the importance of Cartan's work, but very few read his old papers (except those on Lie algebra). I was lucky to be forced to carry it out.

The Chinese ambassador in Washington, Dr. Hu Shih, sent by air mail a copy of the book by Hurewicz-Wallman on "Dimension theory". People now used to xerox may find it difficult to imagine that I copied the whole book by hand, with the exception of the last chapter, where they did "exact sequences" without the sequence and I found it difficult to follow. In fact, at that time it was common to take notes from the reading of a paper. In contrast to the xerox flood, it is not clear whether we have made progress.

I began to have students, among whom were Hsien-Chung Wang and Chih-ta Yen. Wang later made extensive contributions to topology, although he was best known for the Wang sequence. Yen was the first one to give the correct values of the Betti numbers of all the exceptional Lie groups.

Looking back I do not think I had a good idea of mathematics as a whole. I knew some of my deficiencies and was anxious to fill them. My mathematical strength lies in my ability in computation. Even now I do not mind doing lengthy computations, while years ago I could do them with relatively few errors. This is a training which is now relatively unpopular and has not been encouraged. It is still a great advantage in dealing with many problems.

I was fascinated by the Gauss-Bonnet formula and knew that the most conceptual proof is through the structural equation expressing the exterior derivative of the connection form. So when I went to Princeton in 1943, the ground was laid for a most satisfying piece of my mathematical work.

4. Princeton Sunshine

I arrived in Princeton in August 1943. The change of atmosphere was unforgettable. The Institute for Advanced Study was quiet during these days, as most people left for war work. Hermann Weyl was interested in my work. Before my visit he refereed a paper of mine on isotropic surfaces for the Annals of Mathematics and wrote a long favorable report, a fact he revealed to me personally. The report contained suggestions for improvements and showed that he went through the paper in detail. We had frequent conversations. Among his insights was the prediction that algebraic geometry was going to have a great future.

André Weil was at nearby Lehigh University. We soon met and had a lot to talk about. Weil had just published his paper with Allendoerfer on the Gauss-Bonnet formula, and it immediately entered into our discussions. From my understanding of the two-dimensional case I knew that the right proof had to be based on an idea that we now call transgression. There were two difficulties: 1) I did not know clearly the Poincaré-Hopf theorem on the singularities of a vector field; 2) the transgression has to be carried out in the unit tangent bundle and not in the principal bundle, which involves a non-trivial technical difficulty. These were overcome within a short time and the story had a happy ending. I still consider this my best piece of work.

After this it is natural to extend the result to the Stiefel-Whitney classes. That was the time when even in Princeton a talk on fiber bundles had to begin with a definition; there were no vector bundles, only sphere bundles. I noticed that the complex characteristic classes are simpler and admit a local curvature representation. The work was not difficult, but was not in the topology fashion at that time.

Although I was a member of the Institute, I spent a lot of time in Fine Hall of the University. Chevalley was writing his book on Lie groups. Lefschetz was opinionated and did not like the routine work on differential geometry then prevailing. When he asked me to referee a paper for the Annals and I recommended rejection, he made me an associate editor.

The surroundings and pace were most agreeable to me. I reached greater maturity in my mathematical outlook and I enjoyed the stay greatly. In recent years scientific competition has reached a proportion making the life of a scientist unpleasant, although the situation is much better with mathematics. I do not think there is a need for quick progress and I am not impressed by the discovery of electronic mail.

I left Princeton at the end of 1945, returning to China. Immediately on arrival I was given the task of organizing an Institute of Mathematics in in Academia Sinica, the Chinese National Academy. Although the second World War ended, China was torn by a civil war. I extended an invitation to Hermann Weyl to visit China. He responded favorably, but conditions in China made such a visit impossible.

At the end of 1948 the Nanking government was collapsing. I was grateful to the Institute for Advanced Study for taking the initiative to bring me out of China. I was at the Institute in the winter term of 1949. During the term I was the main speaker in Veblen's seminar on differential geometry. The notes were written up two years later and had a wide circulation; they are now published in volume IV of my "Selected Papers". The main result is the Weil homomorphism.

It is a generalization of the Chern classes from the unitary group to any Lie group. I knew the result while writing my paper on complex characteristic classes in 1944. I could not prove it, not being on top of Lie groups. Weil furnished a crucial idea, by considering a family of connections. I called the result the Weil homomorphism. My friends thought I should have a share of the credit, to which I naturally do not object.

5. Mathematics at the Midway

After the second world war Marshall Stone was called to reorganize the department of mathematics at the University of Chicago. He became chairman. As a proof of his insight in mathematics and the mathematical world his first two offers went respectively to Hassler Whitney and André Weil. Whitney declined, but after some negotiations Weil accepted.

When I was in China, Stone corresponded with me about a visiting appointment in Chicago. After I came in 1949, the Department decided to make me a permanent offer. I think the University of Chicago is the only American university whose main objective is the "advancement of knowledge" and not education. I had many friends in the Department and I joined it in summer of 1949. It turned out to be a very pleasant and profitable association.

In 1949-50 I gave a course entitled "differential geometry in the large" and I had a galaxy of brilliant students. I myself was finding my way; many of my errors and blunders were duly corrected by my students. It was an animated and interesting gathering. My memory goes to Arnold Shapiro, who led many of the discussions. Looking back, my knowledge of differential geometry was rudimentary. It is perhaps the strength of the subject that some issues are even now unsettled. For instance, what is a surface? Is it imbedded, immersed, or defined by equations with possible singularities? On the other hand, many of the topics touched in the class received extensive later developments.

I had a close association with Weil. He was always ready and available. Among the mathematicians I discussed mathematics with, and there were many, Weil was one of the few who grasped my ideas quickly and gave helpful comments. We took long walks along Lake Michigan when it was still safe.

I was also interested in algebraic topology, and occasionally taught a course. With Ed Spanier we did some joint work on sphere bundles. One of our results was to formulate Gysin's work as an exact sequence. René Thom did it in a cleaner way and the result is commonly known as the Thom isomorphism.

I found both Chicago and Hamburg very enjoyable. I think they are of the right size. Unfortunately developments in mathematics have forced everything to grow bigger.

6. Settlement on the West Coast

In 1960 I moved to Berkeley. The place was not unfamiliar to me. My teacher in China, Professor L.F. Chiang, received his B.S. from Berkeley. In 1946 and 1949 I stopped at Berkeley and spent some time with the Department. The Department was first-rate and was built up by G.C. Evans. On several occasions he asked about my interest in joining. His brother was the owner of the famous western bookstore in Tientsin, where I got some of my textbooks but was generally frightened by the prices.

Ironically the Berkeley offer became serious when Evans was retiring. It was true, as it was sometimes speculated, that I was attracted by the milder climate as I was getting older. But other factors, such as the expanding department and jet travel, making California less isolated, also favored the move.

Berkeley was improving her standing in the mathematical world and attracting excellent students. There were 31 students who got their Ph.D.'s with me, but my influence extended also to others. I began to write joint papers, with myself as the junior author, as in the case of Bott, Griffiths, Moser, Simons, etc. In such cases I had the feeling of a light responsibility. Life became more and more pleasant.

The colleagues with close scientific contact with me included Hans Lewy and Chuck Morrey, original and powerful analysts. Lewy and I spent some time on the problem of local isometric imbedding of a three-dimensional Riemannian metric in R^6. We were led to the cubic asymptotic cone and knew that it is hyperbolic, but stopped there.

The rôle of differentiation in mathematics is a mystery. One is inclined to think that the two pillars of mathematics are algebra and topology. But life is not that simple; Newton and Leibniz played a trick. This period saw the admission of differential geometry to the main stream of mathematics.

7. Something to Play With in My Eighties

My career is approaching an end and my only question is what to do. The answer is simple: I will continue to play with mathematics. I have never been alert in physical activities and it is now out of the question. Music I have found to be a waste of time. My occasional involvement is purely social. Fortunately global differential geometry still has many fundamental problems, although I most likely will be only a spectator in its development.

I think the restriction to smooth manifolds is dictated only by technical reasons and is unsatisfactory. Not only do non-smooth manifolds exist naturally. But also even if we start with a smooth manifold, geometrical constructions, such as the evolute, lead to non-smooth ones. Whitney introduced the notion of a stratified manifold, which allows singularities and the application of the infinitesimal calculus. Recent light was cast by the work of Robert McPherson. The Cheeger-Goresky-McPherson intersection homology and the McPherson Chern classes have given substance to the notion; cf [2].

It is also not clear to me whether the Riemannian structure is as basic as indicated by recent developments. After all, in his historical paper Riemann allowed the metric to be the fourth root of a quartic from, the general case being now called Finslerian. In a recent note [4] I showed that Finsler geometry can be developed simply, if the proper viewpoint is taken. Further developments are inevitable.

Due to my background I like algebraic manipulation, as Griffiths once observed. Local differential geometry calls for such work. But good local theorems are difficult to come by. The problem on maximum-rank webs discussed above is clearly an important problem, and will receive my attention.

My mathematical education goes on.

References

1. P. Griffiths and J. Harris, *Principles of Algebraic Geometry*, John Wiley, 1978.
2. Robert McPherson, *Global questions in the topology of singular spaces, Proc. ICM Warszawa*, vol. 1, 1983, pp. 213-235.
3. J. Moser, *Geometry of quadrics and spectral theory, Chern Symposium*, Springer-Verlag, 1979, pp. 147-188.
4. S. Chern, *On Finsler geometry*, Comptes Rendus, Académie des Sciences, Paris t. 314, 1992, p.p. 757-761.

Reprinted from *Shiing-Shen Chern Selected Papers*, Springer Verlag, 1978.

S. S. Chern as Geometer and Friend

by André Weil

The friends and colleagues of S. S. Chern who have planned this volume in his honor have asked me for a contribution. Such an invitation is also an honor and could not easily be declined. At the same time, while I have no doubt that future historians of differential geometry will rank Chern as the worthy successor of Elie Cartan in that field, I do not feel competent to give an assessment of his work, nor called upon to do so, since the best part of it, or at least a very representative selection from it, is reproduced in this volume and speaks for itself. All I can do on this occasion is to evoke memories from a friendship of long standing—a friendship which has been among the most valuable ones, personally and scientifically, that I have been privileged to experience.

I must confess that even Chern's name was unknown to me in 1942 when I was asked to review his *Annals of Mathematics* paper on integral geometry ([18] of his bibliography). As I found out later, I had met him briefly in Paris during the year that he spent there in 1936–1937; I was then on the Faculty in Strasbourg and came to Paris regularly to attend the Julia seminar, organized by my friends and myself, which met there every other week. The topic of the seminar for that year was Elie Cartan's work; of course it was of special interest to Chern. However, I was invited to spend the second term (January to April 1937) at the Institute for Advanced Study in Princeton, and did not come back until the fall. Thus I did see him in the fall of 1936, but formed no acquaintance with him; to me he was just an anonymous young man from China, soon lost sight of and forgotten.

Five years later Chern was hardly a beginner any more, but somehow none of his published papers had attracted my attention; in part they had appeared in journals which were not even accessible to me at the time. Having left France early in 1941, I was that year at Haverford and had just written, in collaboration with Carl Allendoerfer, a paper on the Gauss–Bonnet formula. My work on the Haar measure and invariant measures in homogeneous spaces, and the interest I was taking in de Rham's work, had brought me close to "integral geometry," which had been a favorite subject for Blaschke and his students in the thirties. This made it natural for *Mathematical Reviews*, then in its infancy, to send me Chern's article [18] for review.

As I duly mentioned, the paper had some weak points. Nevertheless, it lifted the whole subject at one stroke to a higher plane than where Blaschke's school had left it, and I was impressed by the unusual talent and depth of understanding that shone through it. I tried to indicate this in my review, and also pointed it out to Hermann Weyl. As it happened, Veblen was well aware of Chern's work on projective differential geometry, and he and Weyl were considering an invitation to Chern to come to the Institute in Princeton, in spite of the enormous practical problems which this involved; since Pearl Harbor, war was everywhere; a trip from China to America was more than adventurous, it was risky; merely to obtain the necessary visas and priorities on airplanes required setting the whole diplomacy of the USA in motion. Needless to say, none of this fell to my share; I was a helpless refugee myself, officially classified as "enemy alien." All I could do was to express to Hermann Weyl my warm approval of the whole plan, and it is a matter of no little self-satisfaction to me to think that thus, in a small way, I may have contributed to Chern's coming to Princeton in 1943.

When he reached America, I was still not too far from Princeton, and he soon came to visit me. As we found out at once, we had many interests in common. Both of us had been deeply impressed by Elie Cartan's work and by the masterly presentation that Kähler had given of part of that work in his *Einführung in die Theorie der Systeme von Differentialgleichungen*; both of us had known Kähler in Hamburg. We were both interested in the Gauss–Bonnet formula. We were both beginning to realize the major role which fibre-bundles were playing, still mostly behind the scenes, in all kinds of geometrical problems. Better still, we seemed to share a common attitude towards such subjects, or rather towards mathematics in general; we were both striving to strike at the root of each question while freeing our minds from preconceived notions about what others might have regarded as the right or the wrong way of dealing with it.

Chern and I had been particularly intrigued by the little which was then known about characteristic classes (for which no name had been devised yet). Some mystery seemed to hide behind the fact that some Stiefel–Whitney classes were only defined modulo 2. I was able to tell Chern about the "canonical classes" in algebraic geometry, as introduced in the work of Todd and Eger. Their resemblance with the Stiefel–Whitney classes was apparent, while they were free from the defect (if it was one) of being defined modulo 2; their status, however, was somewhat uncertain, since that work had been done in the spirit of Italian geometry and still rested on some unproved assumptions. As to Pontrjagin classes, they had not yet been heard of at that time.

Such were the topics which came up during Chern's first visit and on subsequent occasions, which we sought to renew as often as we could. Very soon, as every geometer knows, they were completely transformed at the hands of Chern, first with his proof of the Gauss–Bonnet formula and then with his fundamental discovery of the role played by complex or quasicomplex structures in global differential geometry. Any comment would be superfluous; I will merely point out what can now be realized in retrospect about Chern's proof for the Gauss–Bonnet formula, as compared with the one Allendoerfer and I had given in 1942, following in the footsteps of H. Weyl and other writers. The latter proof, resting on the consideration of "tubes," did depend (although this was not apparent at the time) on the construction of a sphere-bundle, but of a non-intrinsic one, viz., the transversal bundle for a given immersion in Euclidean space; Chern's proof operated explicitly for the first time with an intrinsic bundle, the bundle of tangent vectors of length 1, thus clarifying the whole subject once and for all.

Chern and I had then to part for a while; I left for Brazil at the end of 1944, while he had to wait until 1946 before being able to go home to his family which he had been compelled to leave behind when first coming to America. There was not much communication between us during those years. My own ideas about fibre-bundles in algebraic geometry were maturing slowly, under the influence of Chern's work on complex manifolds. I knew that he was organizing an Institute of Mathematics in Nanking; I was also watching the political and military developments in China, with increasing anxiety for his fate. In 1947 I came to Chicago, where Marshall Stone had thoroughly reorganized the department of mathematics; he wrote to Chern, offering a visiting appointment. In the Fall of 1948, the civil war was coming closer and closer to Nanking; Veblen and Weyl, obviously feeling as I did about Chern, sent him an invitation to the Institute, backed up, as he later told me, by a friendly cable from Oppenheimer. Chern realized that he had to act quickly; he sent two cables, one to me and one to Princeton, telling us that he was coming to the United States.

I have criticized, sometimes severely, the American system of higher education; but I have often quoted the episode of Chern's second coming to America as an example of the flexibility which is perhaps its best feature. When Chern's cable reached me, Stone was travelling in South America. A single exchange of cables with him was enough; on his suggestion and mine, the professors in the department voted to ask for Chern's immediate appointment as a full professor. In the following months, there was some difficulty with the administration; obviously they thought that Chern, as a refugee, could be acquired more cheaply; I knew this attitude, which I had personally experienced during the war. It took Stone's coming back, a threat of resignation from him, and a personal appeal to Robert Hutchins to clinch the matter. Hutchins was in bed with the flu; also, he was technically on leave from the University. But the appointment went through, to take effect in the summer of 1949. In the meanwhile, the Institute in Princeton had provided the funds for him to come over with his family and stay in Princeton until his appointment would start in Chicago.

So it came about, in January 1949, that I could welcome Chern in Chicago at the railway station (not the airport; one still had that option) when he stopped there on his way to Princeton. On that day I met his wife and children for the first time, and remember the occasion vividly. Chern, in his fur cap, looked very much the Manchurian general. But to me the most unforgettable sight was his daughter May, a small girl not yet two years old, all wrapped up in white furs; nothing more lovely could have been imagined.

Thus we became colleagues in Chicago, and remained so for the better part of the decade that followed. We were also close neighbors, housed in the same Faculty building; the University had just built it across the Midway. These were fruitful years scientifically, for him and for me. Fiber-bundles, complex manifolds, homogeneous spaces were prominent among our interests; we discussed them in our offices in Eckhart Hall, or at home, or, better still, on long or short walks through the neighborhood parks, where it was still possible to take a walk and come out alive. Relations with colleagues and graduate students were cordial;

visitors, American or foreign, succeeded one another in a steady flow, for longer or shorter stays. With Ed Spanier's appointment a true topologist was added to the team. A quick look through Chern's list of publications in those years, or through mine, will bear witness to the stimulating influence which this scientific atmosphere had on our work.

The time came when circumstances persuaded both Chern and myself to seek elsewhere, among other things, a better climate and more pleasant physical surroundings. As we had sometimes jokingly predicted, he found them by moving closer to China, and I to France. This did not slacken the bonds of friendship, but it is only natural that from then on we followed each other's work less closely, even though we did arrange to get together at not too distant intervals. It is entirely to him and to the ties he had kept up with his colleagues in China that I owed my invitation there in the fall of 1976–an unusual experience which left on me a deep impression. But, rather than commenting upon such personal matters, or upon Chern's work of the last fifteen years (which others would be more competent to discuss, and whose value is recognized by all), it is perhaps appropriate to conclude with a few words about the place of geometry in mathematics—the mathematics of today and presumably also the mathematics of tomorrow.

Obviously everything in differential geometry can be translated into the language of analysis, just as everything in algebraic geometry can be expressed in the language of algebra. Sometimes mathematicians, following their personal inclination or perhaps misled by a false sense of rigor, have turned their mind wholly to the translation and lost sight of the original text. It cannot be denied that this has led occasionally to work of great value; nevertheless, further progress has invariably involved going back to geometric concepts. The same has happened in our times with topology. Whether one considers analytic geometry at the hands of Lagrange, tensor calculus at those of Ricci, or more modern examples, it is always clear that a purely formal treatment of geometric topics would invariably have killed the subject if it had not been rescued by true geometers, Monge in one instance, Levi-Civita and above all Elie Cartan in another.

The psychological aspects of true geometric intuition will perhaps never be cleared up. At one time it implied primarily the power of visualization in three-dimensional space. Now that higher-dimensional spaces have mostly driven out the more elementary problems, visualization can at best be partial or symbolic. Some degree of tactile imagination seems also to be involved. Whatever the truth of the matter, mathematics in our century would not have made such impressive progress without the geometric sense of Elie Cartan, Heinz Hopf, Chern, and a very few more. It seems safe to predict that such men will always be needed if mathematics is to go on as before.

Reprinted from *Abh. Math. Sem. Univ. Hamburg* **11** (1935).

Topologische Fragen der Differentialgeometrie 60.

Abzählungen für Gewebe.

Von SHIING-SHEN CHERN in Hamburg.

In einer früheren Arbeit[1]) hat BLASCHKE einige Sätze bewiesen, die sich auf Abzählungen für Kurvengewebe der Ebene und Flächengewebe des Raumes beziehen. In der vorliegenden Arbeit wollen wir diese Sätze auf höhere Dimensionen verallgemeinern. Es wird also der „Höchstrang" für alle Hyperflächengewebe eines N-dimensionalen Raumes bestimmt.

§ 1. Der allgemeine Fall.

Es sei R_N ein N-dimensionaler Euklidischer Raum mit den Koordinaten $x_1, x_2, \cdots, x_N$. Wir sprechen von einem „n-Gewebe von Hyperflächen" in einem zusammenhängenden Gebiet G des Raumes R_N, wenn n Hyperflächenscharen

(1) $$t_i(x_1, x_2, \cdots, x_N) = \text{konst.} \qquad (i = 1, 2, \cdots, n)$$

sich dort so darstellen lassen, daß für alle ungleichen $i_1, i_2, \cdots, i_N$

(2) $$\frac{\partial(t_{i_1}, t_{i_2}, \cdots, t_{i_N})}{\partial(x_1, x_2, \cdots, x_N)} \neq 0$$

in G gilt. Um den trivialen Fall zu vermeiden, nehmen wir $n > N$ an. Wenn es nun m und nicht mehr linear unabhängige Identitäten in $x_1, x_2, \cdots, x_N$ von der Gestalt

(3) $$\sum_{i=1}^{n} f_i^{(k)}(t_i) = 0 \qquad (k = 1, 2, \cdots, m)$$

gibt (dabei bedeutet k einen Index), so sagen wir: das Hyperflächengewebe hat den „Rang" m. Linearkombination ist dabei mit festen Koeffizienten gemeint, und die vorkommenden Funktionen sollen regulär und analytisch in G sein.

Wir denken uns die Identitäten (3) abgeleitet und finden zwischen den „Pfaffschen Formen" dt_i die linearen Abhängigkeiten

(4) $$\sum_{i=1}^{n} \frac{df_i^{(k)}}{dt_i} dt_i = 0 \qquad (k = 1, 2, \cdots, m).$$

[1]) W. BLASCHKE, T_{40}, Hamb. Abhandl. 9 (1933), S. 299–312.

Wir deuten die $df_i^{(k)} : dt_i$ als homogene Koordinaten eines Punktes $\mathfrak{p}_i$ im linearen Raume R_{m-1}. Jeder solche Punkt

$$(5) \qquad (t_i) = \left\{ \frac{df_i^{(k)}}{dt_i} \right\}$$

beschreibt eine Kurve $\mathfrak{P}_i$ in diesem R_{m-1}. Dann können wir die Abhängigkeiten vektoriell so schreiben

$$(6) \qquad \sum_{i=1}^{n} \mathfrak{p}_i(t_i)\, dt_i = 0.$$

Führen wir einen Differentiator Δ im Raum R_N

$$(7) \qquad \Delta = \alpha_1 \frac{\partial}{\partial x_1} + \alpha_2 \frac{\partial}{\partial x_2} + \cdots + \alpha_N \frac{\partial}{\partial x_N}$$

so ein, daß

$$(8_1) \qquad \Delta t_1 = \Delta t_2 = \cdots = \Delta t_{N-1} = 0$$

und

$$(8_2) \qquad \Delta t_i \neq 0, \qquad (i \geq N)$$

ist, so folgt aus (6)

$$(9) \qquad \sum_{i=N}^{n} \mathfrak{p}_i(t_i)\, \Delta t_i = 0.$$

Wir nennen die Punkte $\mathfrak{p}_i(t_i)$ „zusammengehörig", wenn sie denselben Werten $(x_1, x_2, \cdots, x_N)$ entsprechen. Dann folgt aus (9), daß irgend $n - N + 1$ „zusammengehörige" Punkte $\mathfrak{p}_i$ linear abhängig sind.

Wir setzen im allgemeinen voraus, daß $n - N$ von den Punkten $\mathfrak{p}_1, \mathfrak{p}_2, \cdots, \mathfrak{p}_n$, z. B. $\mathfrak{p}_1, \mathfrak{p}_2, \cdots, \mathfrak{p}_{n-N}$, linear unabhängig sind. Dann liegen $\mathfrak{p}_{n-N+1}, \mathfrak{p}_{n-N+2}, \cdots, \mathfrak{p}_n$ in dem durch $\mathfrak{p}_1, \mathfrak{p}_2, \cdots, \mathfrak{p}_{n-N}$ bestimmten linearen Raume R_{n-N-1}. Somit ist gezeigt:

S_1: *Je n zusammengehörige Punkte* $\mathfrak{p}_i$ *der n Kurven* $\mathfrak{P}_i$ *liegen in einem* R_{n-N-1}.

Wir bezeichnen im folgenden den linearen Raum, welcher von einigen Punkten oder Räumen aufgespannt wird, als eine Summe davon, und zwar mit einem besonderen Summenzeichen $\dot{+}$. Nennen wir den „Summenraum" von

$$(10) \qquad \mathfrak{p}_i \dot{+} \mathfrak{p}_i' \dot{+} \cdots \dot{+} \mathfrak{p}_i^{(p)} = \mathfrak{S}_{i,p}$$

den „p-ten Schmiegraum" von $\mathfrak{P}_i$ in $\mathfrak{p}_i$ und den Summenraum zusammengehöriger

$$(11) \qquad \mathfrak{S}_{1,p} \dot{+} \mathfrak{S}_{2,p} \dot{+} \cdots \dot{+} \mathfrak{S}_{n,p} = R_{k(p)},$$

dann ist nach S_1 die erste Dimensionszahl

$$k(0) = n - N - 1. \tag{12}$$

Wir haben im allgemeinen

$$R_{n-N-1} = \mathfrak{p}_1 \dotplus \mathfrak{p}_2 \dotplus \cdots \dotplus \mathfrak{p}_{n-N}. \tag{13}$$

Wenden wir Δ auf diese Punkte an, so folgt, daß

$$R_{2n-3N} = R_{n-N-1} \dotplus \Delta\mathfrak{p}_N \dotplus \Delta\mathfrak{p}_{N+1} \dotplus \cdots \dotplus \Delta\mathfrak{p}_{n-N}$$

oder

$$R_{2n-3N} = R_{n-N-1} \dotplus \mathfrak{p}'_N \dotplus \mathfrak{p}'_{N+1} \dotplus \cdots \dotplus \mathfrak{p}'_{n-N} \tag{14}$$

ist, wenn $n - 2N + 1 \geqq 0$ ist. Da $\mathfrak{p}_{n-N+1}, \mathfrak{p}_{n-N+2}, \cdots, \mathfrak{p}_n$ lineare Kombinationen von $\mathfrak{p}_1, \mathfrak{p}_2, \cdots, \mathfrak{p}_{n-N}$ sind, liegen $\Delta\mathfrak{p}_{n-N+1}, \Delta\mathfrak{p}_{n-N+2}, \cdots, \Delta\mathfrak{p}_n$ und somit $\mathfrak{p}'_{n-N+1}, \mathfrak{p}'_{n-N+2}, \cdots, \mathfrak{p}'_n$ auch im R_{2n-3N}. Daher können wir schreiben

$$R_{2n-3N} = R_{n-N-1} \dotplus \mathfrak{p}'_N \dotplus \mathfrak{p}'_{N+1} \dotplus \cdots \dotplus \mathfrak{p}'_n. \tag{15}$$

Wir nehmen als Allgemeinfall an, daß unser R_{2n-3N} wirklich von der Dimension $2n - 3N$ ist. Dieser R_{2n-3N} ist im allgemeinen durch R_{n-N-1} und irgend $n - 2N + 1$ der Punkte $\mathfrak{p}'_N, \mathfrak{p}'_{N+1}, \cdots, \mathfrak{p}'_n$, z. B. $\mathfrak{p}'_{2N}, \mathfrak{p}'_{2N+1}, \cdots, \mathfrak{p}'_n$ bestimmt. Somit bleibt R_{2n-3N} fest, wenn wir die Marken $1, 2, \cdots, N-1$ der Reihe nach mit N vertauschen. Dies zeigt, daß die Punkte $\mathfrak{p}'_1, \mathfrak{p}'_2, \cdots, \mathfrak{p}'_{N-1}$ ebenfalls in diesem R_{2n-3N} liegen. Daher ergibt sich

$$R_{2n-3N} = R_{n-N-1} \dotplus \mathfrak{p}'_1 \dotplus \mathfrak{p}'_2 \dotplus \cdots \dotplus \mathfrak{p}'_n, \tag{16}$$

und wir können den Satz aussprechen:

S_2: *Je n zusammengehörige Tangenten $\mathfrak{p}_i \dotplus \mathfrak{p}'_i$ an die $\mathfrak{P}_i$ liegen in einem R_{2n-3N}, d. h.*

$$k(1) = 2n - 3N. \tag{17}$$

Nehmen wir allgemein an

$$R_{k(p)} = R_{k(p-1)} \dotplus \mathfrak{p}_1^{(p)} \dotplus \mathfrak{p}_2^{(p)} \dotplus \cdots \dotplus \mathfrak{p}_q^{(p)} \tag{18}$$

mit

$$q = k(p) - k(p-1). \tag{19}$$

Dann ist nach der obigen Schlußweise

$$R_{k(p+1)} = R_{k(p)} \dotplus \mathfrak{p}_N^{(p+1)} \dotplus \mathfrak{p}_{N+1}^{(p+1)} \dotplus \cdots \dotplus \mathfrak{p}_q^{(p+1)} \tag{20}$$

also

$$k(p+1) - k(p) = q - N + 1. \tag{21}$$

Nach (19) und (21) erhalten wir durch Subtraktion

$$(22) \qquad \{k(p+1)-k(p)\}-\{k(p)-k(p-1)\} = -N+1.$$

Aus (22) folgt wegen (12), (17)

$$(23) \qquad \begin{aligned} &\{k(1)-k(0)\} && = n-2N+1, \\ &\{k(2)-k(1)\}-\{k(1)-k(0)\} && = -N+1, \\ &\ldots\ldots\ldots\ldots\ldots\ldots\ldots\ldots \\ &\{k(p)-k(p-1)\}-\{k(p-1)-k(p-2)\} && = -N+1 \end{aligned}$$

und durch Addition

$$(24) \qquad k(p)-k(p-1) = n-(p+1)N+p.$$

Daraus folgt wiederum

$$\begin{aligned} k(0) &= n-N-1, \\ k(1)-k(0) &= n-2N+1, \\ &\ldots\ldots\ldots\ldots \\ k(p)-k(p-1) &= n-(p+1)N+p. \end{aligned}$$

$$(25) \qquad k(p) = (p+1)n-\frac{1}{2}(p+1)(p+2)N+\frac{1}{2}(p-1)(p+2).$$

Nach (24) wird $R_{k(p_1)}$ fest sein, wenn

$$n-(p_1+1)N+p_1 \leqq N-1$$

oder

$$(26) \qquad n-(p_1+1)N+p_1 = N-1-s,$$

wo $0 \leqq s \leqq N-2$ durch die Bedingung definiert wird, daß p_1 eine ganze Zahl sein soll. Die Bedingung wird dann und nur dann erfüllt, wenn

$$(27) \qquad \begin{aligned} s &\equiv -n+1 \mod N-1, \\ 0 &\leqq s \leqq N-2. \end{aligned}$$

So erhalten wir

$$(28) \qquad p_1 = \frac{1}{N-1}(n-2N+s+1)$$

und

$$(29) \qquad k(p_1) = \frac{1}{2(N-1)}\{(n-1)(n-N)+s(N-s-1)\}-1.$$

Somit ist der folgende Satz bewiesen:

S_8: *Die Kurven $\mathfrak{P}_i$ liegen notwendig in demselben $R_{k(p_1)}$.*

Ich behaupte: der Rang $m \leqq k(p_1)+1$. Wäre nämlich $m > k(p_1)+1$, so gäbe es eine lineare Gleichung zwischen den Koordinaten der $\mathfrak{p}_i$ mit festen Koeffizienten:

$$\sum_{k=1}^{m} c_k \frac{d f_i^{(k)}}{d t_i} = 0 \qquad (i = 1, 2, \cdots, n). \tag{30}$$

Hieraus folgt durch Integration bei geeigneter Wahl der Anfangswerte

$$\sum_{k=1}^{m} c_k f_i^{(k)} = 0 \qquad (i = 1, 2, \cdots, n). \tag{31}$$

Somit ist gezeigt:

S_4: *Der Höchstwert M für den Rang eines n-Hyperflächengewebes im linearen Raum R_N wird durch*

$$M \leqq \frac{1}{2(N-1)} \{(n-1)(n-N) + s(N-s-1)\} \tag{32}$$

gegeben.

§ 2. Die Ausnahmefälle.

Im vorhergehenden Abschnitt haben wir nur den allgemeinen Fall betrachtet, d. h. wir haben immer angenommen, daß die Räume $R_{k(p)}$ für alle p zwischen Null und p_1 wirklich von der Höchstdimension sind. Jetzt wollen wir uns überlegen, wie die Diskussion sich ändert, wenn die Dimension von $R_{k(p_0)}$, wo p_0 eine ganze Zahl zwischen Null und p_1, niedriger als die durch (25) gegebene wird.

Es sei p_0 eine ganze Zahl zwischen Null und p_1, so daß

$$k(p_0) = (p_0+1)n - \frac{1}{2}(p_0+1)(p_0+2)N + \frac{1}{2}(p_0-1)(p_0+2) - r, \qquad r \geqq 0. \tag{33}$$

Wir haben dann

$$k(p) - k(p-1) = n - (p+1)N + p - r, \qquad p \geqq p_0. \tag{34}$$

Daraus folgt

$$k(p_0) = (p_0+1)n - \frac{1}{2}(p_0+1)(p_0+2)N + \frac{1}{2}(p_0-1)(p_0+2) - r,$$

$$k(p_0+1) - k(p_0) = n - (p_0+2)N + (p_0+1) - r,$$

$$\cdots\cdots\cdots\cdots\cdots\cdots$$

$$k(p) - k(p-1) = n - (p+1)N + p - r.$$

$$k(p) = (p+1)n - \frac{1}{2}(p+1)(p+2)N + \frac{1}{2}(p-1)(p+2) - r(p-p_0+1), \qquad p \geqq p_0. \tag{35}$$

Nach (34) wird $R_{k(p_1)}$ fest sein, wenn

$$n-(p_1+1)N+p_1-r \leqq N-1.$$

Daraus

$$(36) \qquad p_1 = \frac{1}{N-1}(n-2N-r+s+1),$$

wo s durch

$$(37) \qquad \begin{aligned} &s \equiv -n+r+1, \qquad \text{mod. } N-1, \\ &0 \leqq s \leqq N-2 \end{aligned}$$

definiert wird. Durch Einsetzung des Wertes p_1 von (36) in (35) haben wir

$$(38) \qquad k(p_1) = \frac{1}{2(N-1)}\{(n-r-1)(n-r-N)+s(N-s-1)\} \\ +rp_0-1.$$

Somit ergibt sich der folgende Satz:

S_5: *Wenn die Dimensionszahl des Raumes* $R_{k(p_0)}$ *durch* (33) *gegeben wird, dann wird der Höchstwert M für den Rang des n-Hyperflächengewebes durch die Formel*

$$(39) \quad M \leqq \frac{1}{2(N-1)}\{(n-r-1)(n-r-N)+s(N-s-1)\}+rp_0$$

gegeben.

§ 3. Beispiele für Gewebe höchsten Ranges.

In diesem Abschnitt werden wir nach Castelnuovo ein Beispiel eines Gewebes geben, dessen Rang den Höchstwert in (32) wirklich erreicht.

Wir betrachten eine algebraische Kurve C von der Klasse n im linearen Raume R_N, die in keinem R_{N-1} liegt. Wir setzen voraus, daß C vom höchstmöglichen Geschlecht ist. Dann ist das Geschlecht von C gleich[2])

$$(40) \qquad \alpha = \chi\left\{n-\frac{N+1}{2}-\chi\,\frac{N-1}{2}\right\},$$

wo χ die kleinste ganze Zahl, die nicht kleiner als $\dfrac{n-N}{N-1}$ ist, bedeutet. Es ist in der Tat

$$(41) \qquad \chi = \frac{n-N}{N-1}+\frac{s}{N-1},$$

wo s durch (27) gegeben wird. Durch Einsetzung des Wertes χ in (40) erhalten wir für α gerade den Ausdruck auf der rechten Seite von (32).

[2]) G. Castelnuovo, Atti Torino **24** (1889), S. 368.

Wir beschränken uns auf das Gebiet, in dem die n Schmieghyperebenen der Kurve C durch jeden Punkt reell sind*). Die n Scharen der

*) Um zu zeigen, daß es unter diesen algebraischen Geweben reelle gibt, diene folgendes einfache Beispiel. Wir betrachten im R_N eine normale rationale Regelfläche der Ordnung $N-1$:

(A) $$x_0 = 1, \quad x_1 = \lambda, \cdots, x_m = \lambda^m, \quad x_{m+1} = \mu, \cdots, x_N = \mu\lambda^{N-m-1},$$

und eine Kurve γ der Ordnung n auf dieser Fläche, die jede Erzeugende in $\chi+1$ Punkten schneidet. Diese Kurve sei durch die folgende Gleichung gegeben:

(B) $$f(\lambda, \mu) = 0,$$

wo $f(\lambda, \mu)$ ein Polynom in λ, μ von den Graden $\eta, \chi+1$ bezüglich λ, μ ist. Der Einfachheit halber werden wir annehmen, daß das Glied $\lambda^\eta \mu^{\chi+1}$ in $f(\lambda, \mu)$ wirklich auftritt. Wenn N ungerade ist, nehmen wir $m = \frac{N-1}{2}$. Da γ von der Ordnung n sein soll, muß sie mit einer allgemeinen Hyperebene n Schnittpunkte haben, d. h. die Gleichung (B) hat mit

(C) $$c_0 + c_1\lambda + \cdots + c_m\lambda^m + \mu(c_{m+1} + \cdots + c_N\lambda^{N-m-1}) = 0,$$

wo $c_0, c_1, \cdots, c_N$ irgendwelche Konstanten sind, n Paare gemeinsamer Lösungen in λ, μ. Dann ist nach einem bekannten Theorem

$$\frac{N-1}{2}(\chi+1) + \eta = n.$$

Falls die Kurve (B) in der λ, μ-Ebene keine singulären Punkte hat, ist ihr Geschlecht

$$p = \chi(\eta - 1) = \chi\left[n - \frac{N+1}{2} - \chi\frac{N-1}{2}\right].$$

Jetzt sei N gerade. Wir setzen $m = \frac{N}{2} - 1$ und nehmen ein Polynom $f(\lambda, \mu)$ von der folgenden Form

(B') $$\begin{aligned} f(\lambda, \mu) = {} & (a_{00}\lambda^\eta + \cdots + a_{0\eta})\mu^{\chi+1} \\ & + (a_{11}\lambda^{\eta-1} + \cdots + a_{1\eta})\mu^\chi + \cdots + (a_{\chi+1,\chi+1}\lambda^{\eta-\chi-1} + \cdots + a_{\chi+1,\eta}), \end{aligned}$$

wo $a_{00}\lambda^\eta + \cdots + a_{0\eta} = 0$ und $a_{11}\lambda^{\eta-1} + \cdots + a_{1\eta} = 0$ keine gemeinsamen Wurzeln haben. Dann ist

$$\frac{N}{2}(\chi+1) + \eta = n + (\chi+1),$$

da die Gleichungen (B') und (C) jetzt $\chi+1$ feste gemeinsame Lösungen haben. Die Kurve (B') in der λ, μ-Ebene hat einen $(\chi+1)$-fachen Punkt $\lambda = \infty$, $\mu = 0$, und zwar mit lauter verschiedenen Tangenten, vorausgesetzt, daß die Gleichung

$$a_{00} + a_{11}y + \cdots + a_{\chi+1,\chi+1}y^{\chi+1} = 0$$

lauter verschiedene Wurzeln in y besitzt. Wenn (B) keine weiteren singulären Punkte hat, was wir voraussetzen können, ist ihr Geschlecht

$$p = \chi(\eta - 1) - \frac{\chi(\chi+1)}{2} = \chi\left[n - \frac{N+1}{2} - \chi\frac{N-1}{2}\right].$$

In beiden Fällen hat die Kurve γ dasselbe Geschlecht wie die Kurven (B) und (B'). Es läßt sich leicht einrichten, daß γ in keinem niederen Raum liegt und daß ihre Schnitt-

Schmieghyperebenen der Kurve C bilden ein n-Hyperflächengewebe im betrachteten Gebiet. Auf der Kurve C gibt es genau α linear unabhängige Abelsche Integrale erster Gattung $f^{(k)}$. Sind t_i die Parameterwerte der Berührungspunkte von n durch einen Punkt gehenden Schmieghyperebenen, so ist nach dem Abelschen Theorem[3]) bei geeigneter Normierung

$$(42) \qquad \sum_{i=1}^{n} f^{(k)}(t_i) = 0.$$

Daher ist unser Gewebe vom Rang α, und es gilt der folgende Satz:

S_6: *Der Höchstrang eines n-Hyperflächengewebes im linearen Raume R_N ist*

$$(43) \qquad M = \frac{1}{2(N-1)}\{(n-1)(n-N) + s(N-s-1)\}.$$

---

punkte mit einer passenden Hyperebene reell und verschieden sind. Gehen wir von γ zur dualen Figur über, so bekommen wir eine Kurve mit den gewünschten Eigenschaften.

[3]) Vgl. etwa Enzyklopädie II B 2, W. Wirtinger, Nr. 42, S. 160.

Hamburg, den 4. Dezember 1934.

Annals of Mathematics
Vol. 45, No. 4, October, 1944

A SIMPLE INTRINSIC PROOF OF THE GAUSS-BONNET FORMULA FOR CLOSED RIEMANNIAN MANIFOLDS

By Shiing-shen Chern

(Received November 26, 1943)

Introduction

C. B. Allendoerfer[1] and W. Fenchel[2] have independently given a generalization of the classical formula of Gauss-Bonnet to a closed orientable Riemannian manifold which can be imbedded in a euclidean space. Recently, Allendoerfer and André Weil[3] extended the formula to a closed Riemannian polyhedron and proved in particular its validity in the case of a general closed Riemannian manifold. In their proof use is still made of the imbedding of a Riemannian cell in a euclidean space. The object of this paper is to offer a direct intrinsic proof of the formula by making use of the theory of vector fields in differentiable manifolds.

The underlying idea of the present proof is very simple, so that a brief summary might be helpful. Let R^n be a closed orientable Riemannian manifold of an even dimension n. According to details to be given below, we define in R^n an intrinsic exterior differential form Ω of degree n, which is of course equal to a scalar invariant of R^n multiplied by the volume element. The formula of Gauss-Bonnet in question asserts that the integral of this differential form over R^n is equal to the Euler-Poincaré characteristic χ of R^n. To prove this we pass from the manifold R^n to the manifold M^{2n-1} of $2n - 1$ dimensions formed by the unit vectors of R^n.[4] In M^{2n-1} we show that Ω is equal to the exterior derivative of a differential form Π of degree $n - 1$. By defining a continuous field of unit vectors over R^n with isolated singular points, we get, as its image in M^{2n-1}, a submanifold V^n of dimension n, and the integral of Ω over R^n is equal to the same integral over V^n. The application of the theorem of Stokes shows that the latter is equal to the integral of Π over the boundary of V^n. Now, the boundary of V^n corresponds exactly to the singular points of the vector field defined in R^n, the sum of whose indices is, by a well-known theorem, equal to χ. With such an interpretation the integral of Π over the boundary of V^n can be evaluated and is easily proved to be equal to χ.

The method can of course be applied to derive other formulas of the same type and, with suitable modifications, to deduce the Gauss-Bonnet formula for a Riemannian polyhedron. We publish this proof, because it is in the present case that the main ideas of our method are most clear. Further results will be given in a forthcoming paper.

§1. Résumé of some fundamental formulas in Riemannian Geometry

Let R^n be a closed orientable differentiable manifold[5] of an even dimension $n = 2p$ and class $r \geqq 4$. In R^n suppose a Riemannian metric be defined, with

Chern, S. S., A Simple Intrinsic Proof of the Gauss–Bonnet Formula for Closed Riemannian Manifolds. Annals of Mathematics. [Vol. 45, No. 4, October 1944], pp. 747–752. © 1944 The Johns Hopkins University Press.

the fundamental tensor g_{ij}, whose components we suppose to be of class 3. Since we are to deal with multiple integrals, it seems convenient to follow Cartan's treatment of Riemannian Geometry,[6] with the theory of exterior differential forms, instead of the ordinary tensor analysis, playing the dominant rôle. The differential forms which occur below are exterior differential forms.

According to Cartan we attach to each point P of R^n a set of n mutually perpendicular unit vectors $\mathfrak{e}_1, \cdots, \mathfrak{e}_n$, with a certain orientation. Such a figure $P\mathfrak{e}_1 \cdots \mathfrak{e}_n$ is called a frame. A vector $\mathfrak{v}$ of the tangent space of R^n at P can be referred to the frame at P, thus

$$\mathfrak{v} = u_i \mathfrak{e}_i, \tag{1}$$

where the index i runs from 1 to n and repeated indices imply summation. The law of infinitesimal displacement of tangent spaces, as defined by the parallelism of Levi-Civita, is given by equations of the form

$$\begin{cases} dP = \omega_i \mathfrak{e}_i, \\ d\mathfrak{e}_i = \omega_{ij} \mathfrak{e}_j, \qquad \omega_{ij} + \omega_{ji} = 0 \end{cases} \tag{2}$$

where ω_i, ω_{ij} are Pfaffian forms. These Pfaffian forms satisfy the following "equations of structure":

$$\begin{cases} d\omega_i = \omega_j \omega_{ji}, \\ d\omega_{ij} = -\omega_{ik}\omega_{jk} + \Omega_{ij}, \qquad \Omega_{ij} + \Omega_{ji} = 0. \end{cases} \tag{3}$$

In (3) Ω_{ij} are exterior quadratic differential forms and give the curvature properties of the space.

The forms Ω_{ij} satisfy a system of equations obtained by applying to (3) the theorem that the exterior derivatives of the left-hand members are zero. The equations are

$$\begin{cases} \omega_j \Omega_{ji} = 0, \\ d\Omega_{ij} - \omega_{jk}\Omega_{ik} + \omega_{ik}\Omega_{jk} = 0, \end{cases} \tag{4}$$

and are called the Bianchi identities.

For the following it is useful to know how the Ω_{ij} behave when the frame $\mathfrak{e}_1 \cdots \mathfrak{e}_n$ undergoes a proper orthogonal transformation. In a neighborhood of P in which the same system of coordinates is valid let $\mathfrak{e}_1 \cdots \mathfrak{e}_n$ be changed to $\mathfrak{e}_1^* \cdots \mathfrak{e}_n^*$ according to the proper orthogonal transformation:

$$\mathfrak{e}_i^* = a_{ij}\mathfrak{e}_j \tag{5}$$

or

$$\mathfrak{e}_i = a_{ji}\mathfrak{e}_j^*, \tag{5'}$$

where (a_{ij}) is a proper orthogonal matrix, whose elements a_{ij} are functions of the coordinates. Suppose Ω_{ij}^* be formed from the frames $P\mathfrak{e}_1^* \cdots \mathfrak{e}_n^*$ in the same way as Ω_{ij} are formed from $P\mathfrak{e}_1 \cdots \mathfrak{e}_n$. Then we easily find

$$\Omega_{ij}^* = a_{ik}a_{jl}\Omega_{kl}. \tag{6}$$

From (6) we deduce an immediate consequence. Let $\epsilon_{i_1 \cdots i_n}$ be a symbol which is equal to $+1$ or -1 according as $i_1, \cdots, i_n$ form an even or odd permutation of $1, \cdots, n$, and is otherwise zero. Since our space R^n is of even dimension $n = 2p$, we can construct the sum

$$\Omega = (-1)^{p-1} \frac{1}{2^{2p} \pi^p p!} \epsilon_{i_1 \cdots i_{2p}} \Omega_{i_1 i_2} \Omega_{i_3 i_4} \cdots \Omega_{i_{2p-1} i_{2p}}, \tag{7}$$

where each index runs from 1 to n. Using (6), we see that Ω remains invariant under a change of frame (5) and is therefore intrinsic. This intrinsic differential form Ω is of degree n and is thus a multiple of $\omega_1 \cdots \omega_n$. As the latter product (being the volume element of the space) is also intrinsic, we can write

$$\Omega = I\omega_1 \cdots \omega_n, \tag{8}$$

where the coefficient I is a scalar invariant of the Riemannian manifold.

With all these preparations we shall write the formula of Gauss-Bonnet in the following form

$$\int_{R^n} \Omega = \chi, \tag{9}$$

χ being the Euler-Poincaré characteristic of R^n.

§2. The space of unit vectors and a formula for Ω

From the Riemannian manifold R^n we pass now to the manifold M^{2n-1} of dimension $2n - 1$ formed by its unit vectors. M^{2n-1} is a closed differentiable manifold of class $r - 1$. As its local coordinates we may of course take the local coordinates of R^n and the components u_i of the vector $\mathfrak{v}$ in (1), subjected to the condition

$$u_i u_i = 1. \tag{1'}$$

If θ_i are the components of $d\mathfrak{v}$ with respect to the frame $\mathfrak{e}_1 \cdots \mathfrak{e}_n$, we have

$$d\mathfrak{v} = \theta_i \mathfrak{e}_i, \tag{10}$$

where

$$\theta_i = du_i + u_j \omega_{ji} \tag{11}$$

and

$$u_i \theta_i = 0. \tag{12}$$

From (11) we get, by differentiation,

$$d\theta_i = \theta_j \omega_{ji} + u_j \Omega_{ji}. \tag{13}$$

As to the effect of a change of frame (5) on the components u_i, θ_i, it is evidently given by the equations

$$u_i^* = a_{ij} u_j, \qquad \theta_i^* = a_{ij} \theta_j. \tag{14}$$

We now construct the following two sets of differential forms:

$$\Phi_k = \epsilon_{i_1 \cdots i_{2p}} u_{i_1} \theta_{i_2} \cdots \theta_{i_{2p-2k}} \Omega_{i_{2p-2k+1} i_{2p-2k+2}} \cdots \Omega_{i_{2p-1} i_{2p}}, \quad k = 0, 1, \cdots, p-1, \tag{15}$$

$$\Psi_k = \epsilon_{i_1 \cdots i_{2p}} \Omega_{i_1 i_2} \theta_{i_3} \cdots \theta_{i_{2p-2k}} \Omega_{i_{2p-2k+1} i_{2p-2k+2}} \cdots \Omega_{i_{2p-1} i_{2p}}, \quad k = 0, 1, \cdots, p-1. \tag{16}$$

The forms Φ_k are of degree $2p - 1$ and Ψ_k of degree $2p$, and we remark that Ψ_{p-1} differs from Ω only by a numerical factor. Using (6) and (14), we see that Φ_k and Ψ_k are intrinsic and are therefore defined over the entire Riemannian manifold R^n.

We shall prove the following recurrent relation:

$$d\Phi_k = \Psi_{k-1} + \frac{2p - 2k - 1}{2(k + 1)} \Psi_k, \qquad k = 0, 1, \cdots, p - 1, \tag{17}$$

where we define $\Psi_{-1} = 0$. Using the property of skew-symmetry of the symbol $\epsilon_{i_1 \cdots i_{2p}}$ in its indices, we can write

$$\begin{aligned} d\Phi_k = {} & \epsilon_{(i)} du_{i_1} \theta_{i_2} \cdots \theta_{i_{2p-2k}} \Omega_{i_{2p-2k+1} i_{2p-2k+2}} \cdots \Omega_{i_{2p-1} i_{2p}} \\ & + (2p - 2k - 1) \epsilon_{(i)} u_{i_1} d\theta_{i_2} \theta_{i_3} \cdots \theta_{i_{2p-2k}} \Omega_{i_{2p-2k+1} i_{2p-2k+2}} \cdots \Omega_{i_{2p-1} i_{2p}} \\ & - k \epsilon_{(i)} u_{i_1} \theta_{i_2} \cdots \theta_{i_{2p-2k}} d\Omega_{i_{2p-2k+1} i_{2p-2k+2}} \Omega_{i_{2p-2k+3} i_{2p-2k+4}} \cdots \Omega_{i_{2p-1} i_{2p}}, \end{aligned}$$

where $\epsilon_{(i)}$ is an abbreviation of $\epsilon_{i_1 \cdots i_{2p}}$. For the derivatives du_i, $d\theta_i$, $d\Omega_{ij}$ we can substitute their expressions from (11), (13), and (4). The resulting expression for $d\Phi_k$ will then consist of terms of two kinds, those involving ω_{ij} and those not. We collect the terms not involving ω_{ij}, which are

$$\Psi_{k-1} + (2p - 2k - 1) \epsilon_{(i)} u_{i_1} u_j \Omega_{j i_2} \theta_{i_3} \cdots \theta_{i_{2p-2k}} \Omega_{i_{2p-2k+1} i_{2p-2k+2}} \cdots \Omega_{i_{2p-1} i_{2p}}. \tag{18}$$

This expression is obviously intrinsic. Its difference with $d\Phi_k$ is an expression which contains a factor ω_{ij} in each of its terms.

We shall show that this difference is zero. In fact, let P be an arbitrary but fixed point of R^n. In a neighborhood of P we can choose a family of frames $\mathfrak{e}_1 \cdots \mathfrak{e}_n$ such that at P,

$$\omega_{ij} = 0.$$

(This process is "equivalent" to the use of geodesic coordinates in tensor notation.) Hence, for this particular family of frames, the expressions (18) and $d\Phi_k$ are equal at P. It follows that they are identical, since both expressions are intrinsic and the point P is arbitrary.

To transform the expression (18) we shall introduce the abbreviations

$$\begin{cases} P_k = \epsilon_{(i)} u_{i_1}^2 \Omega_{i_1 i_2} \theta_{i_3} \cdots \theta_{i_{2p-2k}} \Omega_{i_{2p-2k+1} i_{2p-2k+2}} \cdots \Omega_{i_{2p-1} i_{2p}}, \\ \Sigma_k = \epsilon_{(i)} u_{i_1} u_{i_3} \Omega_{i_3 i_2} \theta_{i_3} \cdots \theta_{i_{2p-2k}} \Omega_{i_{2p-2k+1} i_{2p-2k+2}} \cdots \Omega_{i_{2p-1} i_{2p}}, \\ T_k = \epsilon_{(i)} u_{i_3}^2 \Omega_{i_1 i_2} \theta_{i_3} \cdots \theta_{i_{2p-2k}} \Omega_{i_{2p-2k+1} i_{2p-2k+2}} \cdots \Omega_{i_{2p-1} i_{2p}}, \end{cases} \tag{19}$$

which are forms of degree $2p$. Owing to the relations (1′) and (12) there are some simple relations between these forms and Ψ_k. In fact, we can write

$$P_k = \epsilon_{(i)}(1 - u_{i_2}^2 - u_{i_3}^2 - \cdots - u_{i_{2p}}^2)\Omega_{i_1 i_2}\theta_{i_3} \cdots \theta_{i_{2p-2k}}\Omega_{i_{2p-2k+1}i_{2p-2k+2}} \cdots \Omega_{i_{2p-1}i_{2p}}$$
$$= \Psi_k - P_k - 2(p - k - 1)T_k - 2kP_k,$$

which gives

$$\Psi_k = 2(k + 1)P_k + 2(p - k - 1)T_k. \tag{20}$$

Again, we have

$$\Sigma_k = \epsilon_{(i)} u_{i_1}\Omega_{i_3 i_2}(-u_{i_1}\theta_{i_1} - u_{i_2}\theta_{i_2} - u_{i_4}\theta_{i_4} - \cdots - u_{i_{2p}}\theta_{i_{2p}})\theta_{i_4} \cdots$$
$$\theta_{i_{2p-2k}}\Omega_{i_{2p-2k+1}i_{2p-2k+2}} \cdots \Omega_{i_{2p-1}i_{2p}}$$
$$= T_k - (2k + 1)\Sigma_k,$$

and hence

$$T_k = 2(k + 1)\Sigma_k. \tag{21}$$

The expression (18) for $d\Phi_k$ therefore becomes

$$d\Phi_k = \Psi_{k-1} + (2p - 2k - 1)\{P_k + 2(p - k - 1)\Sigma_k\}, \qquad k = 0, 1, \cdots, p - 1.$$

Using (20) and (21), we get the desired formula (17).

From (17) we can solve Ψ_k in terms of $d\Phi_0, d\Phi_1, \cdots, d\Phi_k$. The result is easily found to be

$$\psi_k = \sum_{m=0}^{k} (-1)^m \frac{2^{m+1}(k + 1)k \cdots (k - m + 1)}{(2p - 2k - 1)(2p - 2k + 1) \cdots (2p - 2k + 2m - 1)} d\Phi_{k-m}, \tag{22}$$
$$k = 0, 1, \cdots, p - 1.$$

In particular, it follows that Ω is the exterior derivative of a form Π:

$$\Omega = (-1)^{p-1}\frac{1}{2^{2p}\pi^p p!}\Psi_{p-1} = d\Pi, \tag{23}$$

where

$$\Pi = \frac{1}{\pi^p}\sum_{m=0}^{p-1}(-1)^m \frac{1}{1\cdot 3 \cdots (2p - 2m - 1)m!\,2^{p+m}}\Phi_m. \tag{24}$$

§3. Proof of the Gauss-Bonnet formula

Basing on the formula (24) we shall give a proof of the formula (9), under the assumption that R^n is a closed orientable Riemannian manifold.

We define in R^n a continuous field of unit vectors with a point 0 of R^n as the only singular point.[7] By a well-known theorem the index of the field at 0 is equal to χ, the Euler-Poincaré characteristic of R^n. This vector field defines in M^{2n-1} a submanifold V^n, which has as boundary χZ, where Z is the $(n - 1)$-

dimensional cycle formed by all the unit vectors through 0. The integral of Ω over R^n is evidently equal to the same over V^n. Applying Stokes's theorem, we get therefore

$$(25)\qquad \int_{R^n} \Omega = \int_{V^n} \Omega = \chi \int_Z \Pi = \chi \frac{1}{1\cdot 3 \cdots (2p-1)2^p \pi^p} \int_Z \Phi_0 .$$

From the definition of Φ_0 we have

$$(26)\qquad \Phi_0 = (2p-1)! \sum_{i=1}^{n} (-1)^i \theta_1 \cdots \theta_{i-1} u_i \theta_{i+1} \cdots \theta_{2p} .$$

The last sum is evidently the volume element of the $(2p - 1)$-dimensional unit sphere. Therefore

$$\int_Z \Phi_0 = (2p-1)! \frac{2\pi^p}{(p-1)!} .$$

Substituting this into (25), we get the formula (9).

INSTITUTE FOR ADVANCED STUDY, PRINCETON, N. J. AND
TSING HUA UNIVERSITY, KUNMING, CHINA.

REFERENCES

1. ALLENDOERFER, C. B., *The Euler number of a Riemann manifold*, Amer. J. Math., 62 (1940), 243–248.
2. FENCHEL, W., *On total curvatures of Riemannian manifolds I*, Jour. London Math. Soc., 15 (1940), 15–22.
3. ALLENDOERFER, C. B., AND ANDRÉ WEIL, *The Gauss-Bonnet theorem for Riemannian polyhedra*, Trans. Amer. Math. Soc., 53 (1943), 101–129.
4. For its definition and topology see, for instance, E. STIEFEL, *Richtungsfelder und Fernparallelismus in n-dimensionalen Mannigfaltigkeiten*, Comm. Math. Helv., 8 (1936), 3–51.
5. WHITNEY, H., *Differentiable manifolds*, Annals of Math., 37 (1936), 645–680.
6. CARTAN, E., *Leçons sur la géométrie des espaces de Riemann*, Paris 1928.
7. ALEXANDROFF-HOPF, Topologie I, 550.

C. R. Acad. Sci. Paris, t. 314, Série I, p. 757-761, 1992 757

Géométrie différentielle/*Differential Geometry*

On Finsler geometry

Shiing-Shen CHERN

Abstract – A new connection is introduced in a Finsler space, solving at the same time the problem of equivalence. In terms of this connection the curvature is divided into two parts, to be called the Minkowski and the Riemannian curvature respectively. A formula for the second variation of the arc length is derived, which involves only the Riemannian curvature.

Sur la géométrie finslerienne

Résumé – *On introduit une connexion nouvelle dans un espace de Finsler, qui résout en même temps le problème de l'équivalence. Dans cette connexion la courbure se partage en deux parties, dites respectivement la courbure de Minkowski et celle de Riemann. On trouve une formule pour la seconde variation de l'arc, qui ne contient que la courbure riemannienne.*

Version française abrégée – Une connexion euclidienne pour un espace de Finsler a été introduite par Elie Cartan en 1934. Pour résoudre le problème de l'équivalence on trouve avantageux de prendre comme base l'espace des éléments linéaires et de considérer une connexion dans cet espace. Ce traitement contient le cas riemannienn. En général la courbure se partage en deux parties, respectivement la courbure de Minkowski et celle de Riemann. L'espace est riemannien si et seulement si la courbure de Minkowski s'annule.

On déduit des équations de structure une formule pour la seconde variation de l'arc, qui est de même forme que dans le cas riemannien. Comme la seconde variation a joué un rôle important en géométrie riemannienne globale, il serait intéressant d'étudier les problèmes analogues dans le cas finslerien général.

---

1. The geometry based on the element of arc

(1) $$ds = F(x^1, \ldots, x^n, dx^1, \ldots, dx^n),$$

where F is positively homogeneous of degree 1 in dx^i (1), went back to Riemann himself. As a result of Finsler's thesis in 1918, it is now called Finsler geometry. From the point of view of differential geometry the most significant result was Cartan's definition of an euclidean connection in 1934. Extensive and deep work was done by H. Busemann by a purely geometric approach.

I propose to give a definition of a connection and curvature, which are natural generalizations of corresponding concepts in Riemannian geometry. Taking the place of the sectional curvature is the notion of flag curvature, to be defined below. The treatment is so simple and natural that the Riemannian restriction seems to be only a habit. In particular, we derive a formula for the second variation of the arc-length which is a replica of the Riemannian formula. This seems to open the way to Finsler manifolds of positive or negative curvature and to comparison and pinching theorems.

It is possible that the generality of the metric gives the flexibility suitable for applications. This is already the case with the Caratheodory and Kobayashi metrics in complex analysis, which are Finsler metrics. I am inclined to think that if a metric is introduced on a stratified manifold, it will be more natural to have a Finsler one.

Note présentée par Shiing-Shen CHERN.

758 C. R. Acad. Sci. Paris, t. 314, Série I, p. 757-761, 1992

2. Local Finsler geometry is basically very simple if we put ourselves from the point of view of the equivalence problem: To decide when two Finsler metrics differ by a change of coordinates. Denote by M our manifold and by TM and T*M its tangent and cotangent bundles. Let PTM be the projectivized tangent bundle (=bundle of line elements), whose "points" are the equivalence classes of non-zero tangent vectors differing from each other by a factor. The projection $p: \mathrm{PTM} \to \mathrm{M}$ induces a bundle $p^* \mathrm{TM} \to \mathrm{PTM}$. A connection in this bundle will be introduced.

We put $dx^i = y^i\, dt$. The element of arc in (1) defines the form

$$\omega = \mathrm{F}(x^i, y^i)\, dt + \lambda_i (dx^i - y^i\, dt),$$

where λ_i are arbitrary. By imposing the condition

$$d\omega \equiv 0 \mod dx^i - y^i\, dt, \tag{2}$$

we find $\lambda_i = \partial \mathrm{F}/\partial y^i$ and

$$\omega = \mathrm{F}_i\, dx^i. \tag{3}$$

Here and later, subscripts to F denote partial differentiations with respect to the y^i. The form ω is known as Hilbert's invariant integral. It is a Pfaffian form in PTM, since y^i can be considered homogeneous coordinates in PTM and F_i is homogeneous of degree zero in y^j.

Let T_x and T_x^* be respectively the tangent and cotangent spaces to M at x. The $\mathrm{Y} = y^i (\partial/\partial x^i) \in p^* \mathrm{T}_x$. Its dual space $p^* \mathrm{T}_x^*$ is spanned by the annihilator $\mathrm{Y}^\perp$ and ω. We choose a basis ω^i in $p^* \mathrm{T}_x^*$, such that $\omega^n = \omega$ and ω^α span $\mathrm{Y}^\perp$, so that ω^α is defined up to a linear transformation.

Condition (2) can be written

$$d\omega^n \equiv 0 \mod \omega^\alpha,$$

which gives

$$d\omega^n = \omega^\alpha \wedge \omega_\alpha^n.$$

On the other hand, we have

$$d\omega^\alpha \equiv \omega^n \wedge \omega_n^\alpha \mod \omega^\beta.$$

LEMMA. — *There exists* $\mathrm{A}_{\alpha\beta} = \mathrm{A}_{\beta\alpha}$ *such that*

$$\omega_\alpha^n + \mathrm{A}_{\alpha\beta}\, \omega_n^\beta \equiv 0 \mod \omega^\gamma.$$

The quadratic differential form

$$\mathrm{A}_{\alpha\beta}\, \omega^\alpha \omega^\beta + (\omega^n)^2 = \left(\frac{1}{2}\mathrm{F}^2\right)_{ij} dx^i\, dx^j \tag{4}$$

is well-defined in PTM.

In the Riemannian case we have

$$\mathrm{F}^2 = g_{ij}(x)\, y^i y^j,$$

and we recover in (4) the quadratic differential form $g_{ij}(x)\, dx^i\, dx^j$. In the general case the form in (4) depends on Y. It defines in $p^* \mathrm{T}_x$ and $p^* \mathrm{T}_x^*$ an inner product. We now make the

REGULARITY HYPOTHESIS. — *The form in* (4) *is positive definite.*

Thus it makes sense to talk about orthonormal coframes. Supposing ω^i orthonormal, we have

$$\mathrm{A}_{\alpha\beta} = \delta_{\alpha\beta}. \tag{5}$$

C. R. Acad. Sci. Paris, t. 314, Série I, p. 757-761, 1992

We shall call these the preferred coframes. Their space P is of dimension

$$(n-1)(n-2)/2+2n-1=n(n+1)/2.$$

The fundamental theorem on the existence of a connection is:

THEOREM 1. – *There exists in* P *a uniquely determined set of Pfaffian forms* ω_i^j, *which satisfy the equations*

$$(6)\qquad \begin{cases} d\omega^i=\omega^j\wedge\omega_j^i, \\ \omega_i^n+\delta_{ij}\,\omega_n^j=0, \\ \delta_{\alpha\gamma}\,\omega_\beta{}^\gamma+\delta_{\beta\gamma}\,\omega_\alpha{}^\gamma=-H_{\alpha\beta\rho}\,\omega_n{}^\rho. \end{cases}$$

The forms $\omega_i{}^j$ define a connection in P. It solves the equivalence problem in the following sense: The forms ω^i, $\omega_i{}^j$, $i<j$, are linearly independent and constitute a multiplicative basis of the algebra of exterior differential forms in P. Let $F^*(x^{*i}, dx^{*i})$ be another Finsler metric whose corresponding quantities we denote by asterisks. Then the two Finsler metrics are locally equivalent, if and only if there is a local diffeomorphism between P and P* such that

$$\omega^i=\omega^{*i},\qquad \omega_i{}^j=\omega^*{}_i{}^j.$$

Scalar invariants, to be called the curvature, are obtained from the exterior differentiation of these equations.

THEOREM 2. – *The Finsler metric is Riemannian, if and only if* $H_{\alpha\beta\rho}=0$. *The forms* $\omega_i{}^j$ *then define the Levi-Civita connection.*

We find the structure equations

$$(7)\qquad d\omega_i{}^k=\omega_i{}^j\wedge\omega_j{}^k+\Omega_i{}^k,$$

where

$$(8)\qquad \Omega_i{}^k=\frac{1}{2}R_i{}^k{}_{jl}\,\omega^j\wedge\omega^l+P_i{}^k{}_{l\alpha}\,\omega^l\wedge\omega_n^\alpha,$$

and

$$(9)\qquad R_i{}^k{}_{jl}+R_i{}^k{}_{lj}=0,\qquad P_i{}^k{}_{l\alpha}=P_l{}^k{}_{i\alpha}.$$

These equations have several distinctive features: 1. There are no quadratic terms in ω_α^n; 2. By exterior differentiation of (7), *i. e.*, by the Bianchi identities, one can show that $P_i{}^k{}_{l\alpha}$ can be expressed in terms of the covariant derivatives of $H_{\alpha\beta\rho}$; 3. $P_i{}^k{}_{l\alpha}$ is zero whenever two of the Latin indices are equal to n.

3. Let e_i be the frame in p^*T_x, which is dual to the coframe ω^k. The connection is defined by

$$(10)\qquad De_i=\omega_i{}^k\otimes e_k.$$

The analytical conditions in Theorem 1 can be interpreted geometrically by the properties.

1. The connection is without torsion;
2. The inner products (e_i, e_n) are preserved;
3. The inner products (e_α, e_β) are preserved when e_n is transported parallelly.

We put

$$(11)\qquad R_{\alpha\beta}=\delta_{\alpha\gamma}\,R_n{}^\gamma{}_{n\beta}.$$

Then $R_{\alpha\alpha}$, as a function on PTM, is the Ricci curvature. More generally, if $b=b^\alpha e_\alpha$, the quantity

$$(12)\qquad R(b)=-R_{\alpha\beta}\,b^\alpha b^\beta$$

is a function of the flag $\{x, e_n, e_n \wedge b\}$; we will call it the flag curvature. This is a natural generalization of the sectional curvature since in the Riemannian case it depends only on the plane element $\{x, b \wedge e_n\}$ and is the sectional curvature.

Thus the curvature splits into two parts, the Riemannian curvature $R_{i\ jl}^{\ k}$ and $H_{\alpha\beta\rho}$, which we will call the Minkowski curvature.

It is the flag curvature which plays an important rôle in variational problems in Finsler geometry. One finds evidence in the first and second variations of the arc length. In fact, map into P a (t, τ)-unit square, so that its projection into M gives a family of curves with the parameter t, τ being the parameter of the family. Supposing e_n be the unit tangent vectors to the curves, we can write, by pullback,

$$\omega^\alpha = b^\alpha\, d\tau, \qquad \omega^n = a^n\, dt + b^n\, d\tau, \tag{13}$$
$$\omega_i^j = a_i^j\, dt + b_i^j\, d\tau.$$

The length of the curve τ is

$$L(\tau) = \int_0^1 a^n\, dt. \tag{14}$$

The expressions (13) satisfy the structure equations (6), giving in particular

$$\frac{\partial a^n}{\partial \tau} = \frac{\partial b^n}{\partial t} + b^\alpha a_\alpha^n.$$

It follows that the first variation of the arc length is

$$\delta s = \left.\frac{\partial L(\tau)}{\partial \tau}\right|_{\tau=0} = b^n \Big|_0^1 + \int_0^1 b^\alpha a_\alpha^n\, dt. \tag{15}$$

Hence the geodesics are characterized by the condition $a_\alpha^n = 0$. In other words, the geodesics are defined by the differential system

$$\omega^\alpha = \omega_\beta^{\ n} = 0. \tag{16}$$

To derive the second variation we consider a family of geodesics, so that $a_\alpha^n = 0$. Making use of (6), we have the formula

$$\frac{\partial^2}{\partial \tau^2} L(\tau) = \frac{\partial b^n}{\partial \tau} + b^\alpha b_\alpha^n \Big|_0^1 + \int_0^1 \{(b_\alpha^n)^2 - R(b)\}\, a^n\, dt. \tag{17}$$

This is the same formula as in the Riemannian case, with the sectional curvature replaced by the flag curvature, the Minkowski curvature playing no role. A crucial property for (17) to be valid is the analytical fact that $P_{i\ l\alpha}^{\ k}$ is zero when two of its Latin indices are equal to n. This has the consequence that the P's do not appear in the second variation.

Indications are that Finsler manifolds whose flag curvature keeps a constant sign will be an interesting subject for study.

Added June 28, 1991. – Dr. Patrick Foulon kindly informed me that some of the results of this Note, including the notion of flag curvature (which he calls Jacobi endomorphism) and the formula for the second variation of the arc length, overlap with his in his work on dynamical systems. On the other hand, the main calculations were carried out in [2], a long-ignored paper published 43 years ago. The purpose of this Note is to call attention to the fact that the foundations of Finsler geometry and perhaps the behavior of its geodesics are not much different from Riemannian geometry, if properly treated. Needless to say, other questions, such as the expression of characteristic classes by curvature forms and the notion of volume, are more subtle in Finsler geometry and deserve further study. Also important are the complex Finsler spaces, which should play a rôle in several complex variables, as the Carathéodory and Kobayashi metrics of a domain are Finsler metrics. For a recent work, *cf.* J. J. Faran V, Hermitian Finsler metrics and the Kobayashi metric, *J. Diff. Geom.*, 31, 1990, pp. 601-625.

C. R. Acad. Sci. Paris, t. 314, Série I, p. 757-761, 1992 761

For Foulon's work, *cf.* Nouveaux invariants géométriques des systèmes dynamiques du second ordre, *Thèse d'État*, 1986 and Géométrie des équations différentielles du second ordre, *Ann. Inst. Henri Poincaré*, 45, 1986, pp. 1-28.

([1]) Throughout this Note we shall agree that small Latin indices run from 1 to n and small Greek indices run from 1 to $n-1$ and that repeated indices imply summation.

Note remise le 21 juin 1991, acceptée le 23 juin 1991.

REFERENCES

[1] E. CARTAN, *Les espaces de Finsler*, Actualites 79, Paris, 1934.

[2] S.-S. CHERN, Local equivalence and euclidean connections in Finsler spaces, *Sci. Rep. Nat. Tsing Hua Univ.*, Ser. A 5, 1948, pp. 95-121, or, *Selected Papers*, II, Springer 1989, pp. 194-212.

[3] H. RUND, *The Differential Geometry of Finsler Spaces*, Springer, 1959.

Mathematical Sciences Research Institute,
1000 *Centennial Drive, Berkeley, CA* 94720.

Ennio De Giorgi (1928–1996)

by J.-L. Lions and F. Murat*

Ennio De Giorgi died in Pisa on October 25, 1996.

E. De Giorgi was born in 1928 in Lecce, a city of Puglia (southern Italy) with which he always maintained deep ties. He received his *laurea*, the diploma which marks the end of the Italian undergraduate curriculum, in Rome in 1950. He began his research work in 1951 at the Istituto per le Applicazioni del Calcolo (Institute for the Applications of Calculus) in Rome. The Institute was then directed by Mauro Picone, and E. De Giorgi became one of his assistants. Professor and student were indeed an odd match: the former, a classicist, dressed with rigor and elegance; the latter, unorthodox, already wearing his strange beret. But M. Picone, a seasoned observer of the development of science, knew how to spot talent: he soon acknowledged E. De Giorgi's exceptional abilities. The assistant was freed from all constraints and worked as he pleased at his own leisurely but, in the end, frightfully efficient rhythm.

His attention was first attracted to problems of the calculus of variations. He began with the problem of minimal surfaces, for which he obtained important results as early as 1954. His views of geometric measure theory were already original when he attended a lecture by Renato Caccioppoli in Rome. Unabashed by R. Caccioppoli's great fame, E. De Giorgi did not hesitate to speak up at the end of the lecture and to suggest alternative solutions. According to Edoardo Vesentini, who was also a young researcher at the time and who was attending the lecture, R. Caccioppoli instantly recognized the exceptional character of E. De Giorgi's proposal.

The word "exceptional" comes up again and again when talking with anyone who had the opportunity of meeting E. De Giorgi. Let his work speak for itself.

Minimal Surfaces and Geometric Measure Theory

From the very beginning of his research activity, E. De Giorgi was interested in geometric measure theory. He gave a rigorous definition of the perimeter of a Borel set and applied this concept to the study of minimal surfaces, generalizing in particular to n dimensions a classical theorem of Bernstein: if $n \leq 8$, the only complete minimal graphs in R^n are hyperplanes. In 1969, in collaboration with E. Bombieri and E. Giusti, he showed that this result is false for $n \geq 9$.

In the 1980s he returned to applications of geometric measure theory to the calculus of variations. He introduced the space SBV of "special" functions of bounded

*Jacques-Louis Lions is a professor at the Collège de France. François Murat is directeur de recherches at the CNRS, Université Paris VI. This is a translation of the article in French published in *Matapli*, No. 49, January 1997, pp. 15–17, and in *Gazette des Mathématiciens*, No. 71, January 1997, pp. 31–34.

variation, i.e. functions the distributional derivatives of which are measures consisting only of a part which is absolutely continuous with respect to the Lebesgue measure and an $(n-1)$-dimensional measure concentrated on the jump set of the function. He used this space to study problems of the calculus of variations "with free discontinuities" (in the sense that the discontinuity set is not fixed *a priori*), such as problems of image segmentation. In 1989 he proved, in collaboration with M. Carriero and A. Leaci, the existence of a minimizer for the Mumford–Shah functional through the solving of a weak formulation of the problem in SBV.

More recently he developed a general theory of motion of a surface according to its mean curvature and formulated a collection of conjectures in the field. He also formulated a general Plateau problem in metric spaces of finite or infinite dimension.

Regularity of Solutions of Elliptic Equations

In 1956, E. De Giorgi proved that every solution of a scalar elliptic equation of second order in divergence form with bounded coefficients is Hölder continuous. This theorem, known as "De Giorgi's theorem", is the crucial step to solve Hilbert's 19th problem, which consists in showing that a function which minimizes a convex integral functional of the calculus of variations is analytic if the functional is analytic. This result and its proof have had a considerable influence on the study of the regularity of solutions of elliptic equations.

In 1968, E. De Giorgi gave an example that shows that this regularity does not extend to systems.

General Theory of Partial Differential Equations

In 1955, E. De Giorgi gave the first example of nonuniqueness for the Cauchy problem for linear partial differential equations of hyperbolic type with regular coefficients. In 1971 he proved, in collaboration with L. Cattabriga, the existence of solutions to linear elliptic equations with constant coefficients and analytic right-hand side, which are analytic in the whole plane.

In 1979 he proved, in collaboration with F. Colombini and S. Spagnolo, that the Cauchy problem for hyperbolic equations with coefficients which are not regular in the time variable is well-posed in Gevrey spaces.

Γ-Convergence

In 1973, in collaboration with S. Spagnolo, and later in 1975 in an article entitled "Sulla convergenza di alcune successioni di integrali del tipo dell'area" (On the convergence of some sequences of area-type integrals, *Rend. Mat.* (6) 8 (1975), 277–294). E. De Giorgi introduced a new notion of convergence for functionals. Γ-convergence, as he called it, provides an easy-to-use necessary and sufficient condition for the convergence of minimizers and values of the corresponding minimization problems. This concept has proved extremely fruitful, as witnessed by the hundreds of papers which make use of it.

Logic and the Foundations of Mathematics

Besides his work in analysis, E. De Giorgi became involved in the mid-1980s in logic and the foundations of mathematics. He developed a self-reference-oriented theory of the foundations of mathematics which tries to reconcile the hierarchical principles of the classical set theories with the wide self-referential power of natural languages.

The exceptional quality of the ideas and of the mathematical work of E. De Giorgi has been recognized through numerous Italian as well as international prizes and distinctions. He received the National Prize of the President of the Italian Republic in 1973, an Honoris Causa doctorate from the Universities of Paris in 1983, and the Wolf Prize in 1990. He was a member of the Accademia Nazionale dei Lincei, of the Pontifical Academy of Sciences, of the Accademia dei XL, of the Academy of Sciences of Turin, of the Lombard Institute of Sciences and Letters, and of the Ligurian Academy. More recently he was named a foreign member of the Academy of Sciences of Paris and of the National Academy of Sciences of the U.S.A.

In 1959, after a year as full professor in Messina, E. De Giorgi was appointed full professor at the Scuola Normale Superiore di Pisa at the age of 31. For almost 40 years, he lived there, taught there, and was a constant source of inspiration for the mathematical school that he founded. Always cheerful, always available, he enjoyed long debates with his students during which he would toss out original ideas and propose conjectures, often interrupting himself to read the newspaper before returning to the discussion, proposing new conjectures or sketching the lines of a proof. He attracted many students, young and not so young, not only from the Scuola Normale, but from all of Italy and from abroad. Well known throughout the world, his school has had a deep influence on mathematics.

Ennio De Giorgi was a mathematician of exceptional creativity. An original mind, an authentic believer in God, and a man with an innate sense of humanity, he enjoyed sharing his thinking on the connections between Mathematics and Wisdom (in the meaning of the Bible, which he often quoted). A passionate advocate of human rights, he was in particular an active member of Amnesty International. His memory will stay alive, not only because of his exceptional mathematical work, but also because of his exceptional human qualities, which have left their imprint on all those who have had the good fortune to know him.

Bibliography

1950

[1] Costruzione di un elemento di compattezza per una successione di un certo spazio metrico, *Atti Accad. Naz. Lincei Rend. Cl. Sci. Fis. Mat. Natur. (8)* **8**, 302–304.

[2] Un criterio generale di compattezza per lo spazio delle successioni, *Atti Accad. Naz. Lincei Rend. Cl. Sci. Fis. Mat. Natur. (8)* **9**, 238–242.

1952

[3] Ricerca dell'estremo di un cosiddetto funzionale quadratico, *Atti Accad. Naz. Lincei Rend. Cl. Sci. Fis. Mat. Natur. (8)* **12**, 256–260.

[4] Sulla sommabilità delle funzioni assolutamente integrabili, *Atti Accad. Naz. Lincei Rend. Cl. Sci. Fis. Mat. Natur. (8)* **12**, 507–510.

[5] Compiuta ricerca dell'estremo inferiore di un particolare funzionale, *Rend. Accad. Sci. Fis. Mat. Napoli (4)* **19**, 29–41.

1953

[6] Un nuovo teorema di esistenza relativo ad alcuni problemi variazionali, *Pubbl. IAC* **371**, CNR, Roma.

[7] Definizione ed espressione analitica del perimetro di un insieme, *Atti Accad. Naz. Lincei Rend. Cl. Sci. Fis. Mat. Natur. (8)* **14**, 390–393.

[8] Un teorema sulle serie di polinomi omogenei, *Atti Accad. Sci. Torino Cl. Sci. Fis. Mat. Natur.* **87**, 185–192.

[9] Osservazioni relative ai teoremi di unicità per le equazioni differenziali a derivate parziali di tipo ellittico, con condizioni al contorno di tipo misto, *Ricerche Mat.* **2**, 183–191.

1954

[10] Su una teoria generale della misura $(r-1)$-dimensionale in uno spazio ad r dimensioni, *Ann. Mat. Pura Appl. (4)* **36**, 191–213.

[11] (with M. Picone) *Lições sobre uma teoria das equações integrais lineares e suas aplicações segundo a orientação de Jordan–Hilbert*, Traduzido por Dante A. O. Martinelli. Escola Politécnica da Universidade de São Paulo.

1955

[12] Un teorema di unicità per il problema di Cauchy, relativo ad equazioni differenziali lineari a derivate parziali di tipo parabolico, *Ann. Mat. Pura Appl. (4)* **40**, 371–377.

[13] Un esempio di non unicità della soluzione di un problema di Cauchy, relativo ad un'equazione differenziale lineare di tipo parabolico, *Rend. Mat. e Appl. (5)* **14**, 382–387.

[14] Nuovi teoremi relativi alle misure $(r-1)$-dimensionali in uno spazio ad r dimensioni, *Ricerche Mat.* **4**, 95–113.

1956

[15] Alcune applicazioni al calcolo delle variazioni di una teoria della misura k-dimensionale, *Atti V Congresso U.M.I.* (Pavia-Torino, 1955), Cremonese, Roma.

[16] Sull'analiticità delle estremali degli integrali multipli, *Atti Accad. Naz. Lincei Rend. Cl. Sci. Fis. Mat. Natur. (8)* **20**, 438–441.

1957

[17] Sulla differenziabilità e l'analiticità delle estremali degli integrali multipli regolari, *Mem. Accad. Sci. Torino Cl. Sci. Fis. Mat. Natur. (3)* **3**, 25–43.

1958

[18] (with A. Ghizzetti) *Lezioni di Analisi Superiore: funzioni olomorfe, teoria dell'integrazione secondo Lebesgue, trasformata di Laplace*, Veschi, Roma.

[19] Sulla proprietà isoperimetrica dell'ipersfera, nella classe degli insiemi aventi frontiera orientata di misura finita, *Atti Accad. Naz. Lincei Mem. Cl. Sci. Fis. Mat. Natur. Sez. I (8)* **5**, 33–44.

1959

[20] Considerazioni sul problema di Cauchy per equazioni differenziali di tipo parabolico di ordine qualunque, *Symposium on the Numerical Treatment of Partial Differential Equations with Real Characteristic* (Roma, 1959), 136–139, Veschi, Roma.

1961

[21] Complementi alla teoria della misura $(n-1)$-dimensionale in uno spazio n-dimensionale, *Seminario di Matematica della Scuola Normale Superiore di Pisa*, 1960–61, Editrice Tecnico Scientifica, Pisa.

[22] Frontiere orientate di misura minima, *Seminario di Matematica della Scuola Normale Superiore di Pisa*, 1960–61, Editrice Tecnico Scientifica, Pisa.

1965

[23] Una estensione del teorema di Bernstein, *Ann. Scuola Norm. Sup. Pisa (3)* **19**, 79–85.

[24] (with G. Stampacchia) Sulle singolarità eliminabili delle ipersuperficie minimali, *Atti Accad. Naz. Lincei Rend. Cl. Sci. Fis. Mat. Natur. (8)* **38**, 352–357.

1966

[25] Sulle singolarità eliminabili delle ipersuperficie minimali, *Atti del Convegno sulle Equazioni alle Derivate Parziali* (Nervi, 1965), 55–58, Cremonese, Roma.

1968

[26] Hypersurfaces of minimal measure in pluridimensional Euclidean spaces, in *Proceedings of the International Congress of Mathematicians* (Moscow, 1966), 395–401, Mir, Moscow.

[27] Maggiorazioni a priori relative ad ipersuperfici minimali, *Atti del Convegno di Analisi Funzionale* (Roma, 1968), 283–285, INdAM, Symposia Mathematica II.

[28] Un esempio di estremali discontinue per un problema variazionale di tipo ellittico, *Boll. Un. Mat. Ital. (4)* **1**, 135–137.

[29] *Corso di Analisi per Chimici,* Università di Pisa, 1967–68 (reprinted ed. De Salvio, Ferrara, 1969).

1969

[30] (with E. Bombieri and E. Giusti) Minimal cones and the Bernstein problem, *Invent. Math.* **7**, 243–268.

[31] (with E. Bombieri and M. Miranda) Una maggiorazione a priori per le ipersuperfici minimali non parametriche, *Arch. Rational Mech. Anal.* **32**, 255–267.

[32] Teoremi di semicontinuità nel calcolo delle variazioni, *Lezioni tenute all'Istituto Nazionale di Alta Matematica, Roma*, 1968–69; lecture notes by U. Mosco, G. Troianiello and G. Vergara.

1971

[33] (with L. Cattabriga) Una formula di rappresentazione per funzioni analitiche in $\mathbb{R}^n$, *Boll. Un. Mat. Ital. (4)* **4**, 1010–1014.

[34] (with L. Cattabriga) Una dimostrazione diretta dell'esistenza di soluzioni analitiche nel piano reale di equazioni a derivate parziali a coefficienti costanti, *Boll. Un. Mat. Ital. (4)* **4**, 1015–1027.

1972

[35] Sur l'existence de solutions analitiques d'équations à coefficients constants, *Actes du Colloque d'Analyse Fonctionnelle* (Bordeaux, 1971), *Bull. Soc. Math. France*, **31–32**, 133–135.

[36] Solutions analitiques des équations aux derivées partielles à coefficients constants, *Séminaire Goulaouic-Schwartz 1971–72: Équations aux Dérivées Partielles et Analyse Fonctionnelle*, Exp. No. 29, École Polytechnique, Centre de Math., Paris.

[37] (with L. Cattabriga) Sull'esistenza di soluzioni analitiche di equazioni a derivate parziali a coefficienti costanti in un qualunque numero di variabili, *Boll. Un. Mat. Ital. (4)* **6**, 301–311.

[38] (with F. Colombini and L. C. Piccinini) Frontiere orientate di misura minima e questioni collegate, *Quaderni della Scuola Normale Superiore di Pisa*, Editrice Tecnico Scientifica, Pisa.

1973

[39] Problemi di superfici minime con ostacoli: forma non cartesiana, *Boll. Un. Mat. Ital. (4)* **8**, 80–88.

[40] (with S. Spagnolo) Sulla convergenza degli integrali dell'energia per operatori ellittici del secondo ordine, *Boll. Un. Mat. Ital. (4)* **8**, 391–411.

1974

[41] Introduction to minimal surface theory. *Global Analysis and its Applications*, Lectures Internat. Sem. Course (ICTP, Trieste, 1972), Vol. II, 43–45, Internat. Atomic Energy Agency, Vienna.

1975

[42] Sulle soluzioni globali di alcuni sistemi di equazioni differenziali, *Collection of Articles Dedicated to Giovanni Sansone on the Occasion of his 85th Birthday, Boll. Un. Mat. Ital. (4)* **11**, 77–79.

[43] Sulla convergenza di alcune successioni d'integrali del tipo dell'area, *Collection of Articles Dedicated to Mauro Picone on the Occasion of his 90th Birthday, Rend. Mat. (6)* **8**, 277–294.

[44] (with L. Cattabriga) Soluzioni di equazioni differenziali a coefficienti costanti appartenenti in un semispazio a certe classi di Gévrey, *Collection in Memory of Enrico Bompiani, Boll. Un. Mat. Ital. (4), Suppl.* **12**, 324–348.

[45] (with T. Franzoni) Su un tipo di convergenza variazionale, *Atti Accad. Naz. Lincei Rend. Cl. Sci. Fis. Mat. Natur. (8)* **58**, 842–850.

1976

[46] Convergenza in energia di operatori ellittici, *Lecture. Contributi del Centro Linceo Interdisciplinare di Scienze Matematiche e loro Applicazioni*, **16**.

1977

[47] Γ-convergenza e G-convergenza, *Boll. Un. Mat. Ital. (5)* **14-A**, 213–220.

[48] (with G. Letta) Une notion générale de convergence faible pour des fonctions croissantes d'ensemble, *Ann. Scuola Norm. Sup. Pisa Cl. Sci. (4)* **4**, 61–99.

1978

[49] (with F. Colombini and S. Spagnolo) Existence et unicité des solutions des équations hyperboliques du second ordre à coefficients ne dépendant que du temps, *C. R. Acad. Sci. Paris Sér. A* **286**, 1045–1048.

1979

[50] (with F. Colombini and S. Spagnolo) Sur les équations hyperboliques avec des coefficients qui ne dépendent que du temps, *Ann. Scuola Norm. Sup. Pisa Cl. Sci. (4)* **6**, 511–559.

[51] Convergence problems for functionals and operators, in *Proceedings of the International Meeting on "Recent Methods in Nonlinear Analysis"* (Roma,

1978), eds. E. De Giorgi, E. Magenes and U. Mosco, 131–188, Pitagora, Bologna.

[52] (with T. Franzoni) Su un tipo di convergenza variazionale, *Rend. Sem. Mat. Brescia* **3**, 63–101.

[53] (with L. Modica) Γ-convergenza e superfici minime, *Preprint Scuola Normale Superiore,* Pisa.

1980

[54] Quelques problèmes de Γ-convergence, *Computing Methods in Applied Sciences and Engineering,* IV Internat. Symposium (Versailles, 1979), eds. R. Glowinski and J. L. Lions, 637–643, North-Holland, Amsterdam.

[55] New problems in Γ-convergence and *G*-convergence, *Free Boundary Problems* (Pavia, 1979), Vol. II, 183–194, Istituto Nazionale di Alta Matematica, Roma.

[56] Guido Stampacchia, *Atti Accad. Naz. Lincei Rend. Cl. Sci. Fis. Mat. Natur. (8)* **68**, 619–625.

[57] (with G. Dal Maso and P. Longo) Γ-limiti di ostacoli, *Atti Accad. Naz. Lincei Rend. Cl. Sci. Fis. Mat. Natur. (8)* **68**, 481–487.

[58] (with A. Marino and M. Tosques) Problemi di evoluzione in spazi metrici e curve di massima pendenza, *Atti Accad. Naz. Lincei Rend. Cl. Sci. Fis. Mat. Natur. (8)* **68**, 180–187.

1981

[59] Γ-limiti di ostacoli, *Recent Methods in Nonlinear Analysis and Applications,* SAFA IV (Napoli, 1980), eds. A. Canfora, S. Rionero, C. Sbordone and G. Trombetti, 51–84, Liguori, Napoli.

[60] Generalized limits in calculus of variations, *Topics in Functional Analysis* 1980–81, 117–148, Quaderni della Scuola Normale Superiore, Editrice Tecnico Scientifica, Pisa.

[61] Limits of functionals and differential operators, *Analytic Solutions of Partial Differential Equations* (Trento, 1981), ed. L. Cattabriga, 153–161, *Astérisque* **89–90**, Soc. Math. France, Paris.

[62] (with G. Buttazzo) Limiti generalizzati e loro applicazioni alle equazioni differenziali, *Matematiche (Catania)* **36**, 53–64.

1982

[63] Operatori elementari di limite ed applicazioni al calcolo delle variazioni, *Atti del Convegno su "Studio di Problemi-Limite in Analisi Funzionale"* (Bressanone, 1981), 101–116, Pitagora, Bologna.

[64] Alcune osservazioni sui rapporti tra matematica pura e matematica applicata, *Atti del VI convegno AMASES* (Marina di Campo, Isola d'Elba, 1982).

[65] Nuovi risultati sulla convergenza dei funzionali, in *Proceedings of the International Conference on Partial Differential Equations, Dedicated to Luigi Amerio on his 70th Birthday* (Milano-Como, 1982), *Rend. Sem. Mat. Fis. Milano* **52**, 167–173.

[66] (with T. Franzoni) Una presentazione sintetica dei limiti generalizzati, *Portugal. Math.* **41**, 405–436.

[67] (with A. Marino and M. Tosques) Funzioni (p, q)-convesse, *Atti Accad. Naz. Lincei Rend. Cl. Sci. Fis. Mat. Natur. (8)* **73**, 6–14.

1983

[68] (with G. Dal Maso) Γ-convergence and calculus of variations, *Mathematical Theories of Optimization* (S. Margherita Ligure, 1981), ed. T. Zolezzi, 121–143, *Lecture Notes in Math. 979*, Springer-Verlag, Berlin.

[69] (with M. Degiovanni and M. Tosques) Recenti sviluppi della Γ-convergenza in problemi ellittici, parabolici ed iperbolici, *Methods of Functional Analysis and Theory of Elliptic Equations* (Napoli, 1982), 103–114, Liguori, Napoli.

[70] (with G. Buttazzo and G. Dal Maso) On the lower semicontinuity of certain integral functionals, *Atti Accad. Naz. Lincei Rend. Cl. Sci. Fis. Mat. Natur. (8)* **74**, 274–282.

[71] (with M. Degiovanni, A. Marino and M. Tosques) Evolution equations for a class of nonlinear operators, *Atti Accad. Naz. Lincei Rend. Cl. Sci. Fis. Mat. Natur. (8)* **75**, 1–8.

1984

[72] On a definition of Γ-convergence of measures, *Multifunctions and Integrands* (Catania, 1983), ed. G. Salinetti, 150–159, *Lecture Notes in Math. 1091*, Springer-Verlag, Berlin.

[73] *G*-operators and Γ-convergence, in *Proceedings of the International Congress of Mathematicians* (Warsaw, 1983), Vol. 2, 1175–1191, PWN, Varsavia and North-Holland, Amsterdam.

[74] (with M. Forti) Premessa a nuove teorie assiomatiche dei Fondamenti della Matematica, *Quaderni del Dipartimento di Matematica* **45**, Pisa.

1985

[75] Some semi-continuity and relaxation problems, *Ennio De Giorgi Colloquium* (Paris, 1983), ed. P. Krée, 1–11, *Res. Notes in Math. 125*, Pitman, London.

[76] References. *Ennio De Giorgi Colloquium* (Paris, 1983), ed. P. Krée, 191–196, *Res. Notes in Math. 125*, Pitman, London.

[77] Su alcuni nuovi risultati di minimalità, *Atti del Convegno Celebrativo del I Centenario del Circolo Matematico di Palermo* (Palermo, 1984), *Rend. Circ. Mat. Palermo (2), suppl.* **8**, 71–106.

[78] (with G. Dal Maso and L. Modica) Convergenza debole di misure su spazi di funzioni semicontinue, *Atti Accad. Naz. Lincei Rend. Cl. Sci. Fis. Mat. Natur. (8)* **79**, 98–106.

[79] Nuovi sviluppi del calcolo delle variazioni, *Metodi di Analisi Reale nelle Equazioni alle Derivate Parziali* (Cagliari, 1985), eds. O. Montaldo, A. Piro and F. Ragnedda, 83–91, Dipartimento di Matematica, Cagliari.

[80] (with M. Forti) Una teoria quadro per i Fondamenti della Matematica, *Atti Accad. Naz. Lincei Rend. Cl. Sci. Fis. Mat. Natur. (8)* **79**, 55–67.

1986

[81] Su alcuni indirizzi di ricerca nel calcolo delle variazioni, *Convegno celebrativo del centenario della nascita di Mauro Picone e di Leonida Tonelli* (Roma, 1985), 183–187, *Atti dei Convegni Lincei 77.*

[82] (with M. Forti and V. M. Tortorelli) Sul problema dell'autoriferimento, *Atti Accad. Naz. Lincei Rend. Cl. Sci. Fis. Mat. Natur. (8)* **80**, 363–372.

1987

[83] Some open problems in the theory of partial differential equations, *Hyperbolic Equations* (Padova, 1985), eds. F. Colombini and M. K. V. Murthy, 67–73, *Pitman Res. Notes in Math. 158*, Longman, Harlow.

[84] (with G. Dal Maso and L. Modica) Weak convergence of measures on spaces of lower semicontinuous functions, in *Proceedings of the International Workshop on Integral Functionals in Calculus of Variations* (Trieste, 1985), *Rend. Circ. Mat. Palermo (2), suppl.* **15**, 59–100.

[85] (with L. Ambrosio and G. Buttazzo) Integral representation and relaxation for functionals defined on measures, *Atti Accad. Naz. Lincei Rend. Cl. Sci. Fis. Mat. Natur. (8)* **81**, 7–13.

1988

[86] (with L. Ambrosio) Un nuovo tipo di funzionale del calcolo delle variazioni, *Atti Accad. Naz. Lincei Rend. Cl. Sci. Fis. Mat. Natur. (8)* **82**, 100 210.

[87] (with M. Clavelli, M. Forti and V. M. Tortorelli) A self-reference oriented theory for the foundations of mathematics, *Analyse Mathématique et Applications*, 67–115, Gauthier-Villars, Paris.

[88] (with G. Congedo and I. Tamanini) Problemi di regolarità per un nuovo tipo di funzionale del calcolo delle variazioni, *Atti Accad. Naz. Lincei Rend. Cl. Sci. Fis. Mat. Natur. (8)* **82**, 673–678.

1989

[89] New functionals in calculus of variations, in *Nonsmooth Optimization and Related Topics, Proceedings of the Fourth Course of the International School of Mathematics* (Erice, 1988), eds. F. H. Clarke, V. F. Dem'yanov and F. Giannessi, 49–59, *Ettore Majorana Internat. Sci. Ser. Phys. Sci.* **43**, Plenum Press, New York.

[90] Su alcuni problemi comuni all'analisi e alla geometria, *Giornate di Studio su Geometria Differenziale e Topologia* (Lecce, 1989), *Note Mat., suppl.* **9**, 59–71.

[91] Sviluppi dell'analisi funzionale nel Novecento, *Il Pensiero Matematico del XX Secolo e l'Opera di Renato Caccioppoli, 41–59, Istituto Italiano per gli Studi Filosofici, Seminari di Scienze*, **5**, Napoli.

[92] Introduzione ai problemi di discontinuità libera, *Symmetry in Nature. A volume in Honour of Luigi A. Radicati di Brozolo, I*, 265–285, Scuola Normale Superiore, Pisa.

[93] (with M. Carriero and A. Leaci) Existence theorem for a minimum problem with free discontinuity set, *Arch. Rational Mech. Anal.* **108**, 195–218.

1990

[94] Una conversazione su: Fondamenti della Matematica e "Teoria base". L'esempio della teoria "7 × 2", *Seminari di Didattica* 1988–89, Quaderni del Dipartimento di Matematica dell'Università di Lecce 56–62.

[95] Conversazioni di Matematica – Anni accademici 1988/90, *Quaderni del Dipartimento di Matematica dell'Università di Lecce.*

[96] Definizioni e congetture del giorno 2 maggio 1990. Typescript, Pisa.

[97] Free discontinuity problems in calculus of variations, *Proc. Israel Mathematical Union Annual Congress* 1990.

[98] Some conjectures on flow by mean curvature, *Methods of Real Analysis and Partial Differential Equations* (Capri, 1990), eds. M. L. Benevento, T. Bruno and C. Sbordone, 9–16, Liguori, Napoli (reprinted in *Quaderni dell'Accademia Pontaniana*, **14**, Napoli, 1992).

[99] (with M. Forti) "5 × 7": a basic theory for the foundations of mathematics, *Preprint Scuola Normale Superiore* **74**, Pisa.

[100] Alcune congetture riguardanti l'evoluzione delle superfici secondo la curvatura media, Typescript, 1990–91.

1991

[101] Qualche riflessione sui rapporti tra matematica ed altri rami del sapere, *Convegno Internazionale "Conoscenza e Matematica"* (Pavia, 1989), ed. L. Magnani, 241–249, Marcos y Marcos, Milano.

[102] Problemi con discontinuità libera, *Int. Symp. "Renato Caccioppoli"* (Napoli, 1989), *Ricerche Mat., suppl.* **40**, 203–214.

[103] Alcuni problemi variazionali della geometria, *Recent Developments in Mathematical Analysis and its Applications* (Bari, 1990), 113–125, *Conf. Sem. Mat. Univ. Bari,* 237–244.

[104] Some remarks on Γ-convergence and least squares method, *Composite Media and Homogeneization Theory* (ICTP, Trieste, 1990), eds. G. Dal Maso and G. F. Dell'Antonio, 135–142, Birkhäuser, Boston.

[105] Riflessioni su alcuni problemi variazionali, *Lecture Notes: Equazioni Differenziali e Calcolo delle Variazioni* (Pisa, 1991).

[106] Free discontinuity problems in calculus of variations, *Frontiers in Pure and Applied Mathemathics, a collection of papers dedicated to J. L. Lions on the occasion of his 60th birthday,* ed. R. Dautray, 55–62, North-Holland, Amsterdam.

[107] Varietà poliedrali di tipo Sobolev, *Lecture Notes: Meeting on Calculus of Variations and Nonlinear Elasticity* (Cortona, 1991).

[108] Congetture sui limiti delle soluzioni di alcune equazioni paraboliche quasi lineari, *Nonlinear Analysis. A Tribute in Honour of Giovanni Prodi*, 173–187, Quaderni della Scuola Normale Superiore, Pisa.

1992

[109] Conjectures on limits of some quasilinear parabolic equations and flow by mean curvature, *Partial Differential Equations and Related Subjects, Proceedings of a Conference dedicated to L. Nirenberg* (Trento, 1990), ed. M. Miranda, 85–95, *Pitman Res. Notes in Math. 269*, Longman, Harlow.

[110] Problemi variazionali con discontinuità libere, *Convegno Internazionale in Memoria di Vito Volterra* (Roma, 1990), 133–150, Atti dei Convegni Lincei 92.

[111] On the relaxation of functionals defined on cartesian manifolds, *Developments in Partial Differential Equations and Applications to Mathematical Physics* (Ferrara, 1991), eds. G. Buttazzo, G. P. Galdi and L. Zanghirati, 33–38, Plenum Press, New York.

[112] Movimenti minimizzanti, *Aspetti e Problemi della Matematica Oggi* (Lecce, 1992), Dipartimento di Matematica, Univ. Lecce.

[113] Un progetto di teoria unitaria delle correnti, forme differenziali, varietà ambientate in spazi metrici, funzioni a variazione limitata. Typescript, 1992–1993.

[114] Funzionali del tipo di integrali di Weierstrass – Osservazioni di Ennio De Giorgi su alcune idee di De Cecco e Palmieri. Typescript.

1993

[115] Movimenti secondo la variazione, *Problemi Attuali dell'Analisi e della Fisica Matematica, Simposio internazionale in onore di G. Fichera per il suo 70º compleanno* (Taormina, 1992), ed. P. E. Ricci, 65–70, Dipartimento di Matematica, Università "La Sapienza", Roma.

[116] New problems on minimizing movements, *Boundary Value Problems for Partial Differential Equations and Applications, in Honor of E. Magenes on the Occasion of his 70th Birthday*, eds. C. Baiocchi and J. L. Lions, 81–98, Masson, Paris.

[117] Congetture riguardanti barriere, superfici minime, movimenti secondo la curvatura. Typescript, Lecce, November 4, 1993.

[118] Congetture riguardanti barriere e sopramovimenti secondo la curvatura media, Typescript, Pisa, November 26, 1993.

[119] Recenti sviluppi nel calcolo delle variazioni, *Rend. Sem. Fac. Sci. Univ. Cagliari* **63** (1993), 47–56.

[120] Complementi ad alcune congetture riguardanti le barriere, Typescript.

1994

[121] Il valore sapienziale della matematica, *Lectio magistralis, Conferimento Laurea "Honoris Causa" in Filosofia a Ennio De Giorgi* (Lecce, 1992), 19–28, Univ. Lecce, Fac. Lettere e Filosofia, Adriatica Editrice Salentina, Lecce.

[122] New ideas in calculus of variations and geometric measure theory, *Motion by Mean Curvature and Related Topics* (Trento, 1992), eds. G. Buttazzo and A. Visintin, 63–69, de Gruyter, Berlin.

[123] Sugli assiomi fondamentali della matematica, *Lecture*, Napoli, March 3, 1994.

[124] Barriers, boundaries, motion of manifolds, *Lecture*, Dipartimento di Matematica dell'Università di Pavia, March 18, 1994.

[125] Barriere, frontiere, movimenti di varietà, Typescript, Pavia, March 19 and 26, 1994.

[126] Fundamental Principles of Mathematics, *Relazione alla Sessione Plenaria della Pontificia Accademia delle Scienze*, October 25–29, 1994.

[127] Complementi tecnici alla relazione di Ennio De Giorgi all'Accademia Pontificia delle Scienze, October 28, 1994, Typescript.

[128] Soluzioni di equazioni paraboliche convergenti verso movimenti di partizioni. *Convergence Theory* (Dijon, 1994).

[129] On the convergence of solutions of some evolution differential equations, *Set-Valued Anal.* **2**, 175–182.

[130] Un progetto di teoria delle correnti, forme differenziali e varietà non orientate in spazi metrici, in *Variational Methods, Nonlinear Analysis and Differential Equations, in Honor of J. P. Cecconi* (Genova, 1993), eds. M. Chicco, P. Oppezzi and T. Zolezzi, 67–71, ECIG, Genova.

[131] (with M. Forti and G. Lenzi) Una proposta di teorie base dei Fondamenti della Matematica, *Atti Accad. Naz Lincei Cl. Sci. Fis. Mat. Natur. Rend. Lincei (9) Mat. Appl.* **5**, 11–22.

[132] (with M. Forti and G. Lenzi) Introduzione delle variabili nel quadro delle teorie base dei Fondamenti della Matematica, *Atti Accad. Naz. Lincei Cl. Sci. Fis. Mat. Natur. Rend. Lincei (9) Mat. Appl.* **5**, 117–128.

1995

[133] New conjectures on flow by mean curvature, in *Nonlinear Variational Problems and Partial Differential Equations* (Isola d'Elba, 1990), eds. A. Marino and M. K. V. Murthy, 120–128, *Pitman Res. Notes Math. 320*, Longman, Harlow.

[134] Su alcune generalizzazioni della nozione di perimetro, *Equazioni Differenziali e Calcolo delle Variazioni* (Pisa, 1992), eds. G. Buttazzo, A. Marino and M. V. K. Murthy, 237–250, Quaderni U.M.I. **39**, Pitagora, Bologna.

[135] Problema di Plateau generale e funzionali geodetici, *Nonlinear Analysis – Calculus of Variations* (Perugia, 1993), *Atti Sem. Mat. Fis. Univ. Modena* **43**, 285–292.

[136] Congetture sulla continuità delle soluzioni di equazioni lineari ellittiche autoaggiunte a coefficienti illimitati, Typescript, Lecce, January 4, 1995.

[137] Congetture nel calcolo delle variazioni, Typescript, Pisa, January 19, 1995.

[138] Riflessioni su matematica e sapienza, *Quaderni dell'Accademia Pontaniana* **18**, Napoli.

[139] (with M. Forti, G. Lenzi and V. M. Tortorelli) Calcolo dei predicati e concetti metateorici in una teoria base dei Fondamenti della Matematica, *Atti Accad. Naz. Lincei Cl. Sci. Fis. Mat. Natur. Rend. Lincei (9) Mat. Appl.* **6**, 79–92.

1996

[140] Movimenti di partizioni, *Variational Methods for Discontinuous Structures* (Como, 1994), eds. R. Serapioni and F. Tomarelli, 1–5, Birkhäuser, Basel.

[141] (with G. Lenzi) La teoria '95. Una proposta di teoria aperta e non riduzionistica dei fondamenti della matematica, *Rend. Accad. Naz. Sci. XL Mem. Mat. Appl. (5)* **20**, 7–34.

[142] Congetture riguardanti alcuni problemi di evoluzione, *A Celebration of J. F. Nash Jr., Duke Math. J.* **81**, 255–268.

[143] (with M. Forti and G. Lenzi) Verso i sistemi assiomatici del 2000 in Matematica, Logica e Informatica, Preprint *Scuola Normale Superiore*, Pisa.

[144] Verità e giudizi in una nuova prospettiva assiomatica, *Contratto, rivista di filosofia tomista e filosofia contemporanea, volume "Il fare della Scienza: i fondamenti e le palafitte"*, eds. F. Barone, G. Basti and A. Testi, 233–252.

[145] L'analisi matematica standard e non standard rivista in una nuova prospettiva scientifica e culturale, Typescript, Lecce, June 1996.

1997

[146] Su alcuni problemi instabili legati alla teoria della visione, *Atti del convegno in onore di Carlo Ciliberto* (Napoli, 1995), eds. T. Bruno, P. Buonocore, L. Carbone and V. Esposito, 91–98, La Città del Sole, Napoli.

[147] (with M. Forti) Dalla matematica e dalla logica alla sapienza, *Pensiero Scientifico, Fondamenti di Epistemologia* (Ancona, 1996), ed. A. Repola Boatto, 17–36, *Quaderni di "Innovazione Scuola"* **29**, Ancona.

1998

[148] (with G. Buttazzo and G. Dal Maso) Calcolo delle variazioni, *Enciclopedia del Novecento, vol. XI, suppl. II*, 832–848. Istituto della Enciclopedia Italiana, Roma.

[149] Dal superamento del riduzionismo insiemistico alla ricerca di una più ampia e profonda comprensione tra matematici e studiosi di altre discipline scientifiche e umanistiche, *Atti Accad. Naz. Lincei Cl. Sci. Fis. Mat. Natur. Rend. Lincei (9) Mat. Appl.* **9**, 71–72.

De Giorgi's Summer Holidays and XIX Hilbert Problem

Mario Miranda

Ennio De Giorgi used to spend two months, every year, to relax. In August he liked to rusticate in one Alpine village, where he could enjoy the company of colleagues and their families. In September he never missed a coming back home, to the Beaches of Salento, to stay with his family and old friends.

It was August '55, when hiking in the Dolomites, Guido Stampacchia told him about the XIX Hilbert Problem. And that Summer turned out to be a no-vacation time for Ennio. In the first week of October, at the IV Congresso della Unione Matematica Italiana, he announced his solution of the XIX Problem.

A few months later he wrote a short paper edited by the Accademia dei Lincei (see Bibliography, [16]) containing some details about his proof.

A year later, ignoring the existence of De Giorgi's theorem, John F. Nash, Jr. decided to work at the regularity problem for solutions of elliptic and parabolic equations, see pp. 218–220 of "A Beatiful Mind" ed. by Sylvia Nasar (Simon and Schuster, 1998). In Spring '57, the complete proof of De Giorgi's Theorem was published by the Accademia delle Scienze di Torino (see Bibliography, [17]). In Spring '58 Nash published his results in "Continuity of Solutions of Parabolic and Elliptic Equations", *Am. J. Math.*, 80.

This is a brief account of the famous De Giorgi–Nash Theorem.

De Giorgi's method, directly applied to solve the Hilbert Problem, is presented by Cristina Mosna in "Regularity of Lipschitz Minima", Rend. Semin. Mat. Univ. Padova, 104 (2000).

Alcune applicazioni al Calcolo delle variazioni di una teoria della misura K-dimensionale*

di Ennio De Giorgi (a Roma)

La teoria cui ci riferiamo è stata esposta per la prima volta da R. Caccioppoli (v. "Misura e integrazione sugli insiemi dimensionalmente orientati" Rend. Lincei Genn. Febbr. 1952) e da me trattata in vari lavori già pubblicati (v. "Su una teoria generale della misura $(r-1)$-dimensionale in uno spazio ad r dimensioni", Ann. di Mat. pura e applicata 1954) o in via di pubblicazione. Tale teoria consente da una parte di estendere ad insiemi orientati molto generali le nozioni di misura K-dimensionale, spazio K-dimensionale tangente, sezione, proiezione, bordo, dall'altra di dare una nozione di convergenza debole di successioni di insiemi orientati assai utile nel calcolo delle variazioni. Infatti molti funzionali del calcolo delle variazioni sono semicontinui rispetto a tale convergenza, mentre da ogni successione di insiemi orientati aventi, coi loro bordi, misure equilimitate può estrarsi una successione subordinata debolmente convergente. Ne segue la possibilità di applicare a larghe classi di problemi (fra cui rientrano come casi particolari il problema di Plateau, lo studio delle proprietà isoperimetriche della sfera ecc ...) i metodi diretti del calcolo delle variazioni, per stabilire l'esistenza del massimo o del minimo.

La teoria considerata è utile anche nello studio delle proprietà differenziali delle soluzioni dei problemi variazionali; citerò in proposito un risultato relativo ad un problema propostomi dal Prof. G. Stampacchia. Sia $u(x) \equiv u(x_1, \dots, x_r)$ una funzione definita in un campo A dello spazio euclideo S_r, assolutamente continua su quasi tutti i segmenti paralleli agli assi coordinati, avente derivate parziali prime di quadrato sommabile in A. Sia $f(y_1, \dots, y_r)$ una funzione analitica in S_r, verificante per ogni punto $y \in S_r$ e per ogni vettore $\lambda \equiv (\lambda_1, \dots, \lambda_r)$ la relazione

$$\mu|\lambda|^2 \leq \sum_{h,k}^{1,r} \frac{\partial^2 f}{\partial y_h \partial y_k} \lambda_h \lambda_k \leq M|\lambda|^2$$

con μ, M costanti positive. Se $u^*(x)$ è un'estremale del funzionale

$$\int_A f\left(\frac{\partial u}{\partial x_1}, \dots, \frac{\partial u}{\partial x_r}\right) dx_1 \cdots dx_r \, ,$$

allora $u^*(x)$ è analitica in A (o almeno esiste una funzione analitica in A e quasi ovunque uguale ad $u^*(x)$).

Il contenuto della comunicazione è stato in parte pubblicato per esteso nella nota "Sull'analiticità delle estremali degli integrali multipli", Rend. Acc. Naz. dei Lincei S. VIII, Vol. XX, fasc. 4 – Aprile 1956.

*In *Atti del V congresso dell'Unione Matematica Italiana*, (Pavia–Torino, 6–9 Ottobre 1955), Edizioni Cremonese, Roma, 1956, 291–292.

Inventiones math. 7, 243 – 268 (1969)

Minimal Cones and the Bernstein Problem

E. BOMBIERI, E. DE GIORGI, and E. GIUSTI (Pisa)

I. Introduction

The main purpose of this paper is to prove that there exist complete analytic minimal graphs of sufficiently large dimension, which are not hyperplanes.

In [8], Fleming gave a new proof of Bernstein's theorem that a complete minimal graph in $\mathbb{R}^3$ must be a plane, by pointing out that the falsity of Bernstein's theorem would imply the existence of a minimal cone in $\mathbb{R}^3$, a contradiction.

The next step forward was taken by De Giorgi [6], who improved Fleming's argument by showing that the falsity of the extension of the Bernstein theorem in $\mathbb{R}^n$ would imply the existence of a minimal cone in $\mathbb{R}^{n-1}$, thus extending the Bernstein theorem to complete minimal graphs in $\mathbb{R}^4$.

In [1] Almgren proved the non-existence of minimal cones in $\mathbb{R}^4$, and in his recent paper Simons [12] extended Almgren's theorem up to $\mathbb{R}^7$. From these results it follows, among other things, an interior regularity theorem for the solutions of the codimension one Plateau problem in $\mathbb{R}^n$ for $n \leqq 7$, and the extension of Bernstein's theorem through dimension 8.

In the same paper, Simons gave examples of locally stable cones of codimension one in $\mathbb{R}^{2m}$, for each $m \geqq 4$, and raised the question whether these cones were global minima of the area function.

In this paper, we shall prove also that these cones are of least area in the large.

Simons' example is as follows.

Let $S_m(r) \subset \mathbb{R}^m$ be the sphere

$$x_1^2 + x_2^2 + \cdots + x_m^2 = r^2$$

and let $C_{2m}(r)$ be the truncated cone

$$x_1^2 + x_2^2 + \cdots + x_m^2 = x_{m+1}^2 + \cdots + x_{2m}^2 < r^2$$

of codimension one in $\mathbb{R}^{2m}$ and with boundary $S_m(r) \times S_m(r) \subset S_{2m}(\sqrt{2}r)$.

It can be proved that $C_{2m}(r)$ has mean curvature zero at every point except at the origin, which is singular.

Let $\mu_\varepsilon: \mathbb{R}^{2m} \to \mathbb{R}^{2m}$ be a Lipschitz map which reduces to the identity on $S_{2m}(\sqrt{2}r)$ and such that

$$\|\mu_\varepsilon - \mathrm{I}\|_{\mathrm{lip}} < \varepsilon$$

where I is the identity map. Then Simons proves that if $m \geqq 4$ there exists $\varepsilon_0(m) > 0$ such that for $0 < \varepsilon < \varepsilon_0(m)$:

$$\operatorname{Area}(C_{2m}(r)) < \operatorname{Area}(\mu_\varepsilon(C_{2m}(r))),$$

unless $\mu_\varepsilon = \mathrm{I}$.

We shall work in the frame of oriented boundaries of least area in the sense of De Giorgi [5] and Miranda [9]; our approach can be shown to be equivalent to that of Federer and Fleming [7] by means of integral currents.

Let E be a set in $\mathbb{R}^n$ and let φ_E be its characteristic function. The essential boundary $\mathscr{F}_e E$ of E is the set

$$\mathscr{F}_e E = \{y \in \mathbb{R}^n | 0 < \liminf_{r \to 0} r^{-n} \int_{B(y,r)} \varphi_E \, dx \leqq \limsup_{r \to 0} r^{-n} \int_{B(y,r)} \varphi_E \, dx < \omega_n\}$$

where $B(y, r)$ is the open ball of radius r, centered at y and ω_n is the measure of $B(0, 1)$.

Let $A \subset \mathbb{R}^n$ be an open set; then E has an oriented boundary of least area with respect to A, if:

(i) $\varphi_E \in BV_{\mathrm{loc}}(A)$,

(ii) for each $g \in BV_{\mathrm{loc}}(A)$ with compact support $K \subset A$ we have

$$\int_K |D\varphi_E| \leqq \int_K |D(\varphi_E + g)|$$

where, for $f \in BV_{\mathrm{loc}}(A)$, $\int_K |Df|$ is the total variation over K of the vector-valued measure Df (gradient of f in the sense of distribution theory).

If $A = \mathbb{R}^n$ we shall say that E has an oriented boundary of least area.

It is known (De Giorgi [5], Miranda [9]) that the essential boundary of a set E of oriented boundary of least area is the union of an analytical hypersurface having mean curvature zero at every point and a singular set N of Hausdorff measure $H_{n-1}(N) = 0$.

In this paper we prove the following results:

Theorem A. *Let $m \geqq 4$. The set $E \subset \mathbb{R}^{2m}$ defined by*

$$x_1^2 + x_2^2 + \cdots + x_m^2 \geqq x_{m+1}^2 + x_{m+2}^2 + \cdots + x_{2m}^2$$

has an oriented boundary of least area.

Theorem A proves that if $m \geqq 4$ the cones $C_{2m}(r)$ are of least area in the large among all hypersurfaces in $\mathbb{R}^{2m}$ having $S_m(r) \times S_m(r)$ as boundary, and thus provides a counterexample to interior regularity for parametric minimal hypersurfaces of dimension 7 or more.

Theorem B. *If $n \geqq 9$ there exists complete minimal graphs in $\mathbb{R}^n$ which are not hyperplanes.*

Theorem B and Simons' result give a complete solution to Bernstein's problem in $\mathbb{R}^n$.

Section II contains some auxiliary results needed for the proof of Theorem A and a theorem (Theorem 1) connecting the problem of minimal hypersurfaces with the problem of functions of least gradient.

In Section III, the proof of Theorem A is completed by reducing it to a question about the behaviour of solutions of an ordinary differential equation of the first order; this is solved by means of the classical methods of Bendixon and Poincaré.

In Section IV, we give our proof of Theorem B. It is self-contained and depends on usual tools in the theory of elliptic partial differential equations, including classical results on the Dirichlet problem for the minimal surface equation.

Though the motivation for our choices of subsolutions and supersolutions of the minimal surface equation will be apparent from the results obtained in our proof of Theorem A, we emphasize here that the theory of minimal cones and minimal surfaces is not needed in this section. The reader interested in the more classical aspects of non-linear elliptic partial differential equations may read Section IV of this paper independently of the other two.

II. Functions of Least Gradient and Oriented Boundaries of Least Area

Let $A \subset \mathbb{R}^n$ be an open set and let $f \in BV_{\text{loc}}(A)$. We say that f has least gradient with respect to A if for every $g \in BV_{\text{loc}}(A)$ with compact support $K \subset A$:

$$\int_K |Df| \leqq \int_K |D(f+g)|.$$

If $A = \mathbb{R}^n$ we say that f has least gradient.

The following lemma is analogous to a result of Fleming ([8], § 5, p. 83).

Lemma 1. *Let $A \subset \mathbb{R}^n$ be an open set and let $f \in H^{1,1}_{\text{loc}}(A)$. Suppose that*

(i) $$H_n(\{x \in A \mid |\nabla f| = 0\}) = 0,$$

where we have written $\nabla f = \text{grad}\, f$; also let N be a closed set in A such that

(ii) $$H_{n-1}(N)=0,$$

where H_{n-1} denotes the $(n-1)$-dimensional Hausdorff measure in $\mathbb{R}^n$;

(iii) $$\int_{A-N} |\nabla f|^{-1} \sum_{i=1}^{n} \frac{\partial f}{\partial x_i} \frac{\partial \varphi}{\partial x_i} dx=0$$

for every $\varphi \in C_0^1(A-N)$.

Then f has least gradient with respect to A.

Proof. We have to show that for every $g \in BV_{\text{loc}}(A)$ with compact support $K \subset A$ we have

$$\int_K |Df| \leqq \int_K |D(f+g)|. \tag{1}$$

Let us first suppose that $g \in C_0^1(A)$. Then in the case in which N is empty (1) follows from (iii) and the convexity of $|Df|$.

Now suppose $N \neq \emptyset$. The set $N_K = N \cap K$ is compact and $H_{n-1}(N_K)=0$; hence for each $\varepsilon>0$ there exists a finite covering of N_K with open balls $B_j = B(x_j, r_j)$, $x_j \in N_K$, $j=1, 2, \ldots, M$, $r_1 \leqq r_2 \leqq \cdots \leqq r_M$, such that

$$\sum_{j=1}^{M} r_j^{n-1} < \varepsilon.$$

Let

$$\alpha_j(x)=\begin{cases} 1 & \text{if } x \notin 2B_j \\ r_j^{-1}|x-x_j|-1 & \text{if } x \in 2B_j - B_j \\ 0 & \text{if } x \in B_j \end{cases}$$

and let

$$\alpha(x)=\min_j \alpha_j(x).$$

We have

$$\int_K |Df| \leqq \int_K |D(f+\alpha g)|$$

by our previous remark and because $\operatorname{supp}(\alpha g) \subset A-N$.

Now

$$\int_K |D(f+\alpha g)| \leqq \int_K |D(f+g)| + \int_K (1-\alpha)|Dg| + \int_K |g|\,|D\alpha|,$$

$$\int_K (1-\alpha)|Dg| \leqq (\sup_K |\operatorname{grad} g|) \int_K (1-\alpha)\,dx,$$

$$\int_K |g|\,|D\alpha| \leqq (\sup_K |g|)\Big\{\int_{2B_1} |D\alpha| + \int_{2B_2-2B_1} |D\alpha| + \cdots + \int_{2B_M - \bigcup_{j=1}^{M-1} 2B_j} |D\alpha|\Big\}$$

$$\leqq (\sup_K |g|)\, 2^n \omega_n \sum_{j=1}^{M} r_j^{n-1} < \text{cost}\ \varepsilon.$$

Hence

$$\lim_{\varepsilon\to 0}\int_K |D(f+\alpha g)| \leqq \int_K |D(f+g)|$$

and (1) is proved if $g\in C_0^1(A)$.

In order to get the full result for $g\in BV_{\text{loc}}(A)$, g with compact support, some additional argument is needed.

Let $g\in BV_{\text{loc}}(A)$ with compact support K and let $Dg=G_1+G_2$ where G_1 is completely continuous and G_2 is the singular part of Dg, with support N_g of measure zero. Then we have:

$$\int_K |D(f+g)| = \int_K |Df+G_1| + \int_K |G_2| \tag{2}$$

because $f\in H^{1,1}_{\text{loc}}(A)$.

Let $g_\varepsilon = g*\psi_\varepsilon$, where ψ_ε is a mollifier; then $g_\varepsilon\in C_0^1(A)$ and

$$\int_{K_\varepsilon} |Df| \leqq \int_{K_\varepsilon} |D(f+g_\varepsilon)| \leqq \int_{K_\varepsilon} |Df+G_1*\psi_\varepsilon| + \int_A |G_2*\psi_\varepsilon|,$$

where $K_\varepsilon=\{x\in A \mid \text{dist}(x,K)<\varepsilon\}$.

Letting $\varepsilon\to 0$ our lemma follows from (2). q.e.d.

Lemma 2. *Let $A\subset\mathbb{R}^n$ be an open set and let $S\subset A$ be a closed set in A such that*

$$H_{n-1}(S\cap K) < +\infty$$

for every compact set $K\subset A$.

Let $f\in C^2(A-S)$, let

$$|\nabla f| \neq 0 \qquad \textit{for } x\in A-S$$

and suppose that

(i) $$v_i(x) = |\nabla f|^{-1}\frac{\partial f}{\partial x_i}, \qquad i=1,2,\dots,n$$

has a continuous extension to the whole of A;

(ii) $$\sum_{i=1}^n \frac{\partial v_i}{\partial x_i} = 0 \qquad \textit{on } A-S.$$

Then we have

$$\int_A \sum_{i=1}^n v_i \frac{\partial\varphi}{\partial x_i}\,dx = 0$$

for every $\varphi\in C_0^1(A)$.

Proof. Let $\varphi \in C_0^1(A)$, $K = \operatorname{supp}(\varphi)$, $S_K = S \cap K$. The set S_K is compact and $H_{n-1}(S_K) = L < +\infty$, whence for every $r > 0$ there exists a finite covering of S_K with open balls

$$B_j = B(x_j, r_j), \qquad j = 1, 2, \ldots, M, \; x_j \in S_K, \; r \geqq r_1 \geqq r_2 \geqq \cdots \geqq r_M,$$

such that

$$\sum_{j=1}^{M} r_j^{n-1} \leqq 2L + 1 = L_1.$$

Let

$$\beta_j(x) = \begin{cases} 1 & \text{if } x \in B_j \\ 2 - r_j^{-1} |x - x_j| & \text{if } x \in 2B_j - B_j \\ 0 & \text{if } x \notin 2B_j \end{cases}$$

and let

$$\gamma_1(x) = \beta_1(x),$$

$$\gamma_j(x) = \max\{\beta_j - \max(\beta_1, \beta_2, \ldots, \beta_{j-1}); 0\}, \qquad 2 \leqq j \leqq M.$$

Then we have

$$\operatorname{supp}(\gamma_j) = 2B_j;$$

$$\gamma(x) = \sum_{j=1}^{M} \gamma_j(x) = \max_j \beta_j(x),$$

and in particular

$$\gamma(x) = 1 \qquad \text{if } x \in \bigcup_j B_j,$$

$$\gamma(x) - 0 \qquad \text{if } x \notin \bigcup_j 2B_j;$$

lastly

$$|\nabla \gamma_j| \leqq 2/r_j.$$

We have

$$\int_A \sum_{i=1}^{n} v_i \frac{\partial \varphi}{\partial x_i} dx = \int_A \sum_{i=1}^{n} v_i \frac{\partial[(1-\gamma)\varphi]}{\partial x_i} dx + \int_A \sum_{i=1}^{n} v_i \frac{\partial(\gamma \varphi)}{\partial x_i} dx;$$

the first integral in the right hand side of this equation is 0, because $\operatorname{supp}((1-\gamma)\varphi) \subset A - S$ and because of condition (iii) of Lemma 2. Hence

$$\int_A \sum_{i=1}^{n} v_i \frac{\partial \varphi}{\partial x_i} dx = \int_A \gamma \sum_{i=1}^{n} v_i \frac{\partial \varphi}{\partial x_i} dx + \sum_{i=1}^{n} \sum_{j=1}^{M} \int_A \varphi \, v_i \frac{\partial \gamma_j}{\partial x_i} dx. \tag{3}$$

We have

$$\left| \int_A \gamma \sum_{i=1}^{n} v_i \frac{\partial \varphi}{\partial x_i} dx \right| \leqq \max_K (|v| \, |\nabla \varphi|) \int_A \gamma \, dx \leqq \operatorname{cost} \sum_{j=1}^{M} r_j^n < \operatorname{cost} L_1 r. \tag{4}$$

Also

$$\left|\int_A \varphi\, v_i \frac{\partial \gamma_j}{\partial x_i}\, dx\right| = \left|\int_{2B_j} (\varphi(x)\, v_i(x) - \varphi(x_j)\, v_i(x_j)) \frac{\partial \gamma_j}{\partial x_i}\, dx\right|. \tag{5}$$

The function $\varphi\, v_i$ is uniformly continuous in K and for every $\varepsilon > 0$ there exists $r > 0$ such that if $|x-y| < 2r$ then

$$|\varphi(x)\, v_i(x) - \varphi(y)\, v_i(y)| < \varepsilon.$$

From this remark and from (5) and the inequality $|\nabla \gamma_j| \leqq 2/r_j$ we get

$$\sum_{i=1}^{n} \sum_{j=1}^{M} \left|\int_A \varphi\, v_i \frac{\partial \gamma_j}{\partial x_i}\, dx\right| \leqq \text{cost}\, \varepsilon \sum_{j=1}^{M} r_j^{n-1} < \text{cost}\, L_1\, \varepsilon. \tag{6}$$

The result of Lemma 2 now follows from inequalities (3), (4), (6) and letting $r, \varepsilon \to 0$. q.e.d.

The next result gives the connection between functions of least gradient and sets with an oriented boundary of least area.

Theorem 1. *Let f be a function of least gradient with respect to A and let*

$$E_\lambda = \{x \in A \mid f(x) \geqq \lambda\}.$$

Then the set E_λ has an oriented boundary of least area with respect to A.

Proof. Let φ_λ be the characteristic function of E_λ. By Theorem 1.6 of [9] we have $\varphi_\lambda \in BV_{\text{loc}}(A)$ for almost all λ, also

$$\int_K |Df| = \int_{-\infty}^{\infty} \left(\int_K |D\varphi_\lambda|\right) d\lambda \tag{7}$$

for each compact set $K \subset A$.

Let t be a real constant and let

$$f_1 = \max(f-t, 0),$$
$$f_2 = \min(f, t).$$

Then $f_1, f_2 \in BV_{\text{loc}}(A)$, $f = f_1 + f_2$ and by Eq. (7)

$$\int_K |Df| = \int_K |Df_1| + \int_K |Df_2|.$$

Let $g \in BV_{\text{loc}}(A)$ with compact support K; we have

$$\int_K |Df_1| + \int_K |Df_2| = \int_K |Df| \leqq \int_K |D(f+g)| \leqq \int_K |D(f_1+g)| + \int_K |Df_2|$$

which shows that f_1, f_2 have least gradient with respect to A. It follows that the functions

$$\varphi_{\lambda,\varepsilon}=\frac{1}{\varepsilon}\min\{\varepsilon;\max(f-\lambda,0)\}$$

have least gradient with respect to A for every λ, ε.

Now if $H_n(\{x\in A|f(x)=\lambda\})=0$ we have for each compact set $K\subset A$

$$\lim_{\varepsilon\to+0}\int_K|\varphi_{\lambda,\varepsilon}-\varphi_\lambda|\,dx=0$$

whence by Theorem 3 of Miranda [11] it follows that $\varphi_\lambda\in BV_{\mathrm{loc}}(A)$ and that φ_λ has least gradient with respect to A.

If $H_n(\{x\in A|f(x)=\lambda\})>0$, there exists a sequence $\{\lambda_m\}$, such that $\lambda_m<\lambda$, $\lambda_m\to\lambda$, $H_n(\{x\in A|f(x)=\lambda_m\})=0$ and for each compact set $K\subset A$

$$\lim_{m\to+\infty}\int_K|\varphi_{\lambda_m}-\varphi_\lambda|\,dx=0,$$

whence by our previous result and Theorem 3 of Miranda, Theorem 1 follows. q.e.d.

III. Proof of Theorem A

In order to prove our Theorem A it will be sufficient, in view of Theorem 1 of the previous section, to exhibit a function f of least gradient (with respect to $\mathbb{R}^{2m}$) and having the cone $x_1^2+\cdots+x_m^2=x_{m+1}^2+\cdots+x_{2m}^2$ as a level hypersurface.

This section contains the construction of such a function f.

Heuristic arguments suggest looking for a function f of least gradient which has the same orthogonal symmetries of the cone. Hence let us define new variables

$$u=(x_1^2+\cdots+x_m^2)^{\frac{1}{2}},$$

$$v=(x_{m+1}^2+\cdots+x_{2m}^2)^{\frac{1}{2}}$$

and the function $F=F(u,v)$ defined by

$$F(u,v)=f(x),$$

and let Ω be an open set of the (u,v)-plane contained in the first quadrant and let $\tilde{\Omega}$ be the corresponding open set in $\mathbb{R}^{2m}$.

If we suppose that

(i) $$f\in C^2(\tilde{\Omega}),$$

(ii) $$|\nabla f|\neq 0 \quad \text{in } \tilde{\Omega}$$

then the Euler equation for f in $\tilde{\Omega}$:

$$\sum_{i=1}^{2m} \frac{\partial}{\partial x_i} \left(\frac{\frac{\partial f}{\partial x_i}}{|\nabla f|} \right) = 0$$

is easily shown to be equivalent to

$$\frac{\partial}{\partial u} \left[\frac{(u\,v)^{m-1} \frac{\partial F}{\partial u}}{|\nabla F|} \right] + \frac{\partial}{\partial v} \left[\frac{(u\,v)^{m-1} \frac{\partial F}{\partial v}}{|\nabla F|} \right] = 0$$

which in turn is transformed in the new equation

$$F_v^2 F_{uu} - 2F_u F_v F_{uv} + F_u^2 F_{vv} + p\left(\frac{F_u}{u} + \frac{F_v}{v}\right)(F_u^2 + F_v^2) = 0 \qquad (8)$$

where

$$F_u = \frac{\partial F}{\partial u}, \ldots \quad \text{and} \quad p = m - 1.$$

Our method is based on reconstructing the function F from the knowledge of its level curves.

Suppose the generic level curve $F = \text{cost}$ has a parametric representation $u = u(t)$, $v = v(t)$ with u, v twice continuously differentiable and $u'(t) \neq 0$. On eliminating F from Eq. (8) with the help of

$$dF = 0,$$

which holds true along the generic level curve of F, we get the differential equation

$$u''v' - u'v'' + p[(u')^2 + (v')^2]\left(\frac{u'}{v} - \frac{v'}{u}\right) = 0. \qquad (9)$$

It is useful to define the angular parameters

$$\varphi = \operatorname{arctg} \frac{v}{u},$$

$$\theta = \operatorname{arctg} \frac{v'}{u'},$$

which are left invariant by the homotheties

$$u(t) \to \lambda\, u(t),$$

$$v(t) \to \lambda\, v(t),$$

$\lambda > 0$.

We then get the relation

$$2p\cos(\theta+\varphi)\,d\varphi+\sin 2\varphi\,\sin(\theta-\varphi)\,d\theta=0 \tag{10}$$

and putting

$$\sigma=\theta-3\varphi+\frac{\pi}{2},$$

$$\psi=\theta+\varphi-\frac{\pi}{2},$$

we obtain, after a suitable choice of the parameter t, $-\infty<t<+\infty$, the differential system

$$\begin{aligned}\frac{d\sigma}{dt}&=-\frac{3}{2}\sin\sigma-\left(2p+\frac{3}{2}\right)\sin\psi\\ \frac{d\psi}{dt}&=\frac{1}{2}\sin\sigma-\left(2p-\frac{1}{2}\right)\sin\psi.\end{aligned} \tag{11}$$

This system is easily studied in the phase space (σ,ψ) with the methods of Bendixon and Poincaré (see for instance [2] for the general theory of these systems); also it is convenient to consider p as a continuous parameter.

If $p>\frac{3}{2}+\sqrt{2}$ the singular points of (11) can be classified as follows:

$$(\sigma,\psi)=(h\pi,k\pi),\qquad h,k\in\mathbb{Z}, \tag{A}$$

$h+k$ odd.

The singular point is a saddle point, with principal directions having tangent

$$t_1=1,$$

$$t_2=-\frac{1}{4p+3},$$

$$(\sigma,\psi)=(h\pi,k\pi),\qquad h,k\in\mathbb{Z}, \tag{B}$$

$h+k$ even.

In this case we have a nodal point with principal direction having tangent

$$t_1=\frac{2p-2-\sqrt{\Delta}}{4p+3}$$

and exceptional direction having tangent

$$t_2=\frac{2p-2+\sqrt{\Delta}}{4p+3},$$

where $\Delta=4p^2-12p+1$. This nodal point is stable if h is even, unstable if h is odd.

We remark that this classification must be changed if $p<\frac{3}{2}+\sqrt{2}$; for example if $\frac{3}{2}-\sqrt{2}<p<\frac{3}{2}+\sqrt{2}$ then $\Delta<0$ and the singular points (B) are focal points.

Lemma 3. *Let* $p>\frac{3}{2}+\sqrt{2}$. *There exists an integral curve* γ_0, $\gamma_0(t)=(\sigma_0(t), \psi_0(t))$ *of the differential system* (11), *with the following properties*

(i)
$$\gamma_0(-\infty)=(\pi, 0),$$
$$\gamma_0(+\infty)=(0, 0);$$

(ii)
$$0<\psi_0(t)<\sigma_0(t)<\pi$$

for $-\infty<t<+\infty$;

(iii) *the straight lines* $\psi-\sigma=c$, $0<c<\pi$, *meet* γ_0 *exactly once.*

Proof. Let us first verify the existence of an integral curve γ_0 satisfying (i) and (ii) of Lemma 3 provided $p>p_0$, for some sufficiently large p_0.

Let T denote the triangle

$$T=\left\{0\leqq\psi\leqq\frac{\sigma}{2}\leqq\frac{\pi}{2}\right\}$$

with sides

$$l_1=\{0<\sigma<\pi, \psi=0\},$$

$$l_2=\left\{\sigma=\pi, 0<\psi\leqq\frac{\pi}{2}\right\},$$

$$l_3=\left\{0<\psi=\frac{\sigma}{2}<\frac{\pi}{2}\right\}$$

and let us look at the behaviour on these sides of the vector field in (σ, ψ) determined by the system (11).

On l_1 we have

$$\frac{d\sigma}{dt}=-\frac{3}{2}\sin\sigma<0,$$

$$\frac{d\psi}{dt}=\frac{1}{2}\sin\sigma>0.$$

On l_2 we have

$$\frac{d\sigma}{dt}=-\left(2p+\frac{3}{2}\right)\sin\psi<0,$$

$$\frac{d\psi}{dt}=-\left(2p-\frac{1}{2}\right)\sin\psi<0.$$

Finally on l_3 we have

$$\frac{d\sigma}{dt}=-\frac{3}{2}\sin 2\psi-\left(2p+\frac{3}{2}\right)\sin\psi=-\sin\psi\left[3\cos\psi+\left(2p+\frac{3}{2}\right)\right]<0,$$

$$\frac{d\psi}{dt}=\frac{1}{2}\sin 2\psi-\left(2p-\frac{1}{2}\right)\sin\psi=-\sin\psi\left[-\cos\psi+\left(2p-\frac{1}{2}\right)\right]<0$$

and

$$\frac{d\psi}{d\sigma}=\frac{2p-\frac{1}{2}-\cos\psi}{2p+\frac{3}{2}+3\cos\psi}\geqq\frac{2p-\frac{3}{2}}{2p+\frac{9}{2}}>\frac{1}{2}$$

for

$$p>\tfrac{15}{4}=p_0.$$

In all cases, if $p>p_0=\frac{15}{4}$, the tangent vector through a regular point of the boundary of T is directed towards the interior of T. It follows that if γ, $\gamma(t)=(\sigma(t),\psi(t))$ is an integral curve such that, for some point t_0, $\gamma(t_0)\in T$, then $\gamma(t)\in T$ for each $t\geqq t_0$.

The point $(\pi,0)$ is a saddle point thus there are two integral curves ending at, and two integral curves starting from, the point $(\pi,0)$. The two integral curves ending at $(\pi,0)$ are

$$\sigma=\psi+\pi,\qquad (0<\psi<\pi)$$

and

$$\sigma=\psi+\pi,\qquad (-\pi<\psi<0).$$

Hence the other two integral curves starting from $(\pi,0)$ have tangent there

$$t_2=-\frac{1}{4p+3}<0$$

whence one of them starts inside the triangle T. By our previous remark, this integral curve γ_0 remains trapped in T and so verifies (i) and (ii) of Lemma 3.

In order to eliminate the condition $p>p_0$ we proceed as follows.

The vector field determined by (11) depends continuously on p provided $p>\frac{3}{2}+\sqrt{2}$; it follows that there exists an unique integral curve γ_0 satisfying (i) of Lemma 3, which depends continuously on the parameter p. Let θ_p be the variation in the argument of the tangent vector to γ_0 at the point $\gamma_0(t)$, as t varies from $-\infty$ to $+\infty$. Then θ_p is finite and depends continuously on p and it is easily seen that

$$\lim_{p\to+\infty}\theta_p=0.$$

If $p>p_0=\frac{15}{4}$ the integral curve γ_0 is contained inside the triangle T and it follows that γ_0 ends at $(0,0)$ with the principal direction having tangent

$$t_1=\frac{2p-2-\sqrt{\Delta}}{4p+3};$$

by an obvious continuity argument, this remains true for $p>\frac{3}{2}+\sqrt{2}$. It follows that

$$\theta_p=\operatorname{arctg}\left(\frac{2p-2-\sqrt{\Delta}}{4p+3}\right)-\operatorname{arctg}\left(-\frac{1}{4p+3}\right)+\pi k$$

where k is an integer and arctg has its principal value. If p is sufficiently large, $k=0$ but as θ_p depends continuously on p for $p>\frac{3}{2}+\sqrt{2}$ and as

$$0<\operatorname{arctg}\left(\frac{2p-2-\sqrt{\Delta}}{4p+3}\right)-\operatorname{arctg}\left(-\frac{1}{4p+3}\right)<\pi$$

for $p>\frac{3}{2}+\sqrt{2}$, it follows that $k=0$ in every case and

$$0<\theta_p<\pi \quad \text{for} \quad p>\tfrac{3}{2}+\sqrt{2}.$$

A simple consideration of the behaviour of the vector field of the system (11) in the square Q: $-\pi<\sigma<\pi$, $-\pi<\psi<\pi$ shows that γ_0 is always contained in Q; looking at the behaviour of this vector field along the lines $\sigma=0$, $\psi=0$ one sees easily that γ_0 stays in the square Q': $0<\sigma<\pi$, $0<\psi<\pi$, for otherwise we would contradict the inequality $0<\theta_p<\pi$.

Finally, γ_0 cannot cross the line $\sigma=\psi$; for otherwise it would cross this line at least twice because γ_0 is contained in the square Q': $0<\sigma<\pi$, $0<\psi<\pi$. If this were the case, by Rolle's theorem there would be a point t_0 such that

$$\frac{d\psi_0}{d\sigma_0}(t_0)=1.$$

At such a point we have

$$\tfrac{1}{2}\sin\sigma_0-(2p-\tfrac{1}{2})\sin\psi_0=-\tfrac{3}{2}\sin\sigma_0-(2p+\tfrac{3}{2})\sin\psi_0$$

that is

$$\sin\sigma_0+\sin\psi_0=0,$$

a contradiction with the fact that $(\sigma_0,\psi_0)\in Q'$.

Hence γ_0 does not cross the line $\sigma=\psi$ if $p>\frac{3}{2}+\sqrt{2}$, which proves (ii) of Lemma 3.

The same argument proves (iii) of Lemma 3; for if γ_0 were to cross the line $\psi-\sigma=c$, $0<c<\pi$, at least twice, there would be, again by Rolle's theorem, a point t_c such that

$$\frac{d\psi_0}{d\sigma_0}(t_c)=1$$

and we would get the same contradiction as before. q.e.d.

The curve γ_0 gives rise to a family of homothetic curves Γ_λ, $\lambda>0$, in the (u,v) plane. By Lemma 3 these curves are analytic and contained in the domain $T_1=\{0\leqq v<u\}$, and there is exactly one curve of the family $\{\Gamma_\lambda\}$ passing through any given point of T_1.

Our aim is to construct a homogeneous function $F(u,v)$ of degree $2\alpha>0$ in T_1 which has the family $\{\Gamma_\lambda\}$ as its family of level curves; then by the definition of $\{\Gamma_\lambda\}$ the function F will satisfy the Eq. (8) in $\mathring{T}_1$.

By Lemma 3, (iii) the lines $\varphi=c$, where $0\leqq c<\dfrac{\pi}{4}$ intersect a given Γ_λ exactly once. Now let Γ_1 be the curve issuing from the point $(1,0)$; by our previous remark we may take φ as parameter on Γ_1 and we get a parametric representation for Γ_1

$$u=u_0(\varphi),$$
$$v-v_0(\varphi)$$

where

$$\operatorname{tg}\varphi=\frac{v_0}{u_0}.$$

If we normalize F by the requirement

$$F(u,v)=1 \qquad \text{along } \Gamma_1,$$

then if 2α is the degree of homogeneity of F we obtain if $(u,v)\in T_1$:

$$F(u,v)=(u^2+v^2)^\alpha\, G\left(\operatorname{arctg}\frac{v}{u}\right)$$

where G is determined by $F(u_0,v_0)\equiv 1$, from which it follows

$$G(t)=[u_0^2(t)+v_0^2(t)]^{-\alpha}, \qquad 0\leqq t<\frac{\pi}{4}.$$

Our results imply

$$\lim_{t\to\frac{\pi}{4}-} G(t)=0,$$

whence the function F so defined can be extended to a continuous function in the first quadrant and analytic for $u\neq v$, by means of the relation

$$F(u,v)=-F(v,u) \qquad \text{for } (v,u)\in T_1.$$

This function F is a solution of Eq. (8) in the open set

$$\Omega = \{0<u,\, 0<v,\, u \neq v\};$$

also

$$F(u,u)=0 \qquad \text{if } u>0,$$

$$|\nabla F| \neq 0 \qquad \text{if } (u,v)\in\Omega.$$

In order to apply Lemma 2 to the function $f(x)=F(u,v)$ we have to show that the vector $(\nabla F)/|\nabla F|$ can be extended to a continuous vector in the whole domain $\{(u,v)|u>0,\, v>0\}$.

We have

$$F_u=(u^2+v^2)^{\alpha-1}(2\alpha u G - v G'),$$

$$F_v=(u^2+v^2)^{\alpha-1}(2\alpha v G + u G'),$$

$$|\nabla F|=(u^2+v^2)^{\alpha-\frac{1}{2}}(4\alpha^2 G^2+G'^2)^{\frac{1}{2}},$$

whence it is enough to show that the functions

$$G(4\alpha^2 G^2+G'^2)^{-\frac{1}{2}} \quad \text{and} \quad G'(4\alpha^2 G^2+G'^2)^{-\frac{1}{2}}$$

have a limit if $t\to\pi/4$.

If $\varphi<\pi/4$ we get

$$G'(\varphi)=-2\alpha(u_0^2+v_0^2)^{-\alpha-1}(u_0 u_0'+v_0 v_0')$$

$$=-2\alpha G(\varphi)\frac{u_0 u_0'+v_0 v_0'}{u_0^2+v_0^2};$$

on the other hand

$$\varphi=\operatorname{arctg}\frac{v_0}{u_0}$$

and it follows that

$$u_0^2+v_0^2=u_0 v_0'-v_0 u_0'$$

whence

$$G'(\varphi)=-2\alpha G(\varphi)\frac{1+\dfrac{v_0}{u_0}\dfrac{v_0'}{u_0'}}{\dfrac{v_0'}{u_0'}-\dfrac{v_0}{u_0}}$$

and finally

$$G'(\varphi)=2\alpha G(\varphi)\operatorname{cotg}(\varphi-\vartheta) \tag{12}$$

where

$$\vartheta=\operatorname{arctg}\frac{v_0}{u_0'}.$$

Let us determine the constant α in such a way that

$$\lim_{\varphi\to\frac{\pi}{4}} G'(\varphi)=k\neq 0; \tag{13}$$

it will then follow that $(\nabla F)/|\nabla F|$ is a continuous non-zero vector in $\{(u,v)|u>0, v>0\}$.

The integral curve γ_0 considered in Lemma 3 had tangent at the origin

$$t_1=\frac{2p-2-\sqrt{\Delta}}{4p+3},$$

whence the asymptotic relation

$$\psi\sim t_1\sigma$$

holds as $\sigma\to 0$, and we get

$$\vartheta-\varphi\sim 2\frac{1+t_1}{1-t_1}\left(\frac{\pi}{4}-\varphi\right)$$

as $\varphi\to\pi/4$.

On the other hand we have

$$v_0=(\operatorname{tg}\varphi)\,u_0,$$
$$v_0'=(\operatorname{tg}\vartheta)\,v_0$$

whence

$$(\operatorname{tg}\vartheta)\,u_0'=(\operatorname{tg}\varphi)\,u_0'+(1+\operatorname{tg}^2\varphi)\,u_0.$$

From this equation and the previous asymptotic relation we easily find

$$\frac{u_0'}{u_0}\sim\frac{1-t_1}{2(1+t_1)}\left(\frac{\pi}{4}-\varphi\right)^{-1}$$

and finally

$$u_0\sim c\left(\frac{\pi}{4}-\varphi\right)^{-\frac{1-t_1}{2(1+t_1)}}$$

where c is a positive constant. It follows that

$$G(\varphi)\sim(2c^2)^{-\alpha}\left(\frac{\pi}{4}-\varphi\right)^{\frac{1-t_1}{1+t_1}\alpha}$$

and recalling Eq. (12) we get

$$G'(\varphi)\sim k\left(\frac{\pi}{4}-\varphi\right)^{\frac{1-t_1}{1+t_1}\alpha-1}$$

for some constant $k \neq 0$. If

$$\alpha = \frac{1+t_1}{1-t_1} = \frac{2p+1-\sqrt{\Delta}}{4}$$

we obtain Eq. (13), as required.

It is easily seen from our previous discussion that the function

$$f(x) = F(u, v)$$

has a locally bounded gradient in $\mathbb{R}^{2m}$ and satisfies the hypotheses of Lemma 2 if one takes

$$A = \{x \in \mathbb{R}^{2m} \mid u \neq 0, v \neq 0\},$$
$$S = \{x \in \mathbb{R}^{2m} \mid u = v\}.$$

Hence Lemma 2 applies and we are able to apply Lemma 1 where now

$$A = \mathbb{R}^{2m},$$
$$N = \{x \in \mathbb{R}^{2m} \mid u\,v = 0\};$$

note that

$$H_{2m-1}(N) = 0.$$

We conclude that $f(x)$ has least gradient in $\mathbb{R}^{2m}$. Now Simons' cone $x_1^2 + \cdots + x_m^2 = x_{m+1}^2 + \cdots + x_{2m}^2$ coincides with the level hypersurface $f(x) = 0$ because $F(u, v) = 0$ if and only if $u = v$, and Theorem A follows from Theorem 1. q.e.d.

IV. Proof of Theorem B

We begin with the simple remark that it is enough to prove Theorem B in the case $n = 2m \geqq 8$, because if $f(x_1, \ldots, x_n)$ is a solution in $\mathbb{R}^n$ of the equation

$$\sum_{i=1}^{n} \frac{\partial}{\partial x_i} \left(\frac{\dfrac{\partial f}{\partial x_i}}{\sqrt{1+|\nabla f|^2}} \right) = 0 \tag{14}$$

for minimal graphs over $\mathbb{R}^n$, then it is also a solution of the same equation for each $n' > n$.

Let us look for solutions f of (14) with $n = 2m$ and which are invariant with respect to the group

$$G = SO(m) \times SO(m)$$

of automorphisms of the cone

$$C_{2m} = \{x \in \mathbb{R}^{2m} \mid x_1^2 + \cdots + x_m^2 = x_{m+1}^2 + \cdots + x_{2m}^2\}.$$

We may write

$$f(x)=F(u,v)$$

where

$$u=(x_1^2+\cdots+x_m^2)^{\frac{1}{2}},$$

$$v=(x_{m+1}^2+\cdots+x_{2m}^2)^{\frac{1}{2}}$$

and now Eq. (14) can be written

$$\mathscr{E}F\equiv(1+F_v^2)F_{uu}-2F_uF_vF_{uv}+(1+F_u^2)F_{vv} \\ +p\left(\frac{F_u}{u}+\frac{F_v}{v}\right)(1+F_u^2+F_v^2)=0 \tag{15}$$

where $p=m-1$.

Let

$$T=\{(u,v)\,|\,u\geqq 0, v\geqq 0\}, \quad T_1=\{(u,v)\,|\,0\leqq v<u\}, \quad T_2=\{(u,v)\,|\,0\leqq u<v\}.$$

We shall prove in the Appendix that the function

$$F_1=(u^2-v^2)(u^2+v^2)^{\alpha-1}$$

where

$$\alpha=\frac{2p+1-\sqrt{\Delta}}{4}>1, \qquad \Delta=4p^2-12p+1$$

verifies

$$\mathscr{E}F_1>0 \quad \text{in } \mathring{T}_1, \tag{16}$$

$$\mathscr{E}F_1<0 \quad \text{in } \mathring{T}_2, \tag{17}$$

while the function

$$F_2=H\left\{(u^2-v^2)+(u^2-v^2)(u^2+v^2)^{\alpha-1}\left[1+A\left|\frac{u^2-v^2}{u^2+v^2}\right|^{\lambda-1}\right]\right\}$$

where λ is any real number subjected to

$$\frac{2p+1}{2p+2}\alpha<\lambda<\min\left(\alpha, \frac{p}{\alpha^2}\right)$$

and H is defined by

$$H(z)=\int_0^z \exp\left(B\int_{|w|}^{\infty}\frac{dt}{t^{2-\lambda}(1+t^{2\alpha\lambda-2\alpha})}\right)dw$$

and $A=A(\lambda,p)$, $B=B(\lambda,p)$ are sufficiently large positive constants, verifies for $p\geqq 3$

$$\mathscr{E}F_2<0 \quad \text{in } \mathring{T}_1, \tag{18}$$

$$\mathscr{E}F_2>0 \quad \text{in } \mathring{T}_2. \tag{19}$$

Moreover

$$0<F_1<F_2 \quad \text{in } T_1, \tag{20}$$

$$F_2<F_1<0 \quad \text{in } T_2, \tag{21}$$

$$F_1=F_2=0 \quad \text{if } u=v, \tag{22}$$

and

$$\frac{F_u}{u}, \frac{F_v}{v} \qquad (F=F_1, F_2)$$

have a continuous extension to the whole of T.

Let

$$f_1(x)=F_1(u,v),$$

$$f_2(x)=F_2(u,v),$$

$$D_1=\{x\in\mathbb{R}^{2m}\,|\,0\leqq v<u\},$$

$$D_2=\{x\in\mathbb{R}^{2m}\,|\,0\leqq u<v\};$$

from (16) and (18) we readily get

$$\int_{D_1}\sum_{i=1}^{2m}\frac{\partial f_1}{\partial x_i}\frac{\partial\varphi}{\partial x_i}(1+|\nabla f_1|^2)^{-\frac{1}{2}}\,dx\leqq 0 \tag{23}$$

for every $\varphi\in C_0^\infty(D_1)$, $\varphi\geqq 0$,

$$\int_{D_1}\sum_{i=1}^{2m}\frac{\partial f_2}{\partial x_i}\frac{\partial\varphi}{\partial x_i}(1+|\nabla f_2|^2)^{-\frac{1}{2}}\,dx\geqq 0 \tag{24}$$

for every $\varphi\in C_0^\infty(D_1)$, $\varphi\geqq 0$, and analogous inequalities in D_2.

Let us consider the Dirichlet problem

$$\begin{aligned}&\sum_{i=1}^{2m}\frac{\partial}{\partial x_i}\left(\frac{\dfrac{\partial f}{\partial x_i}}{\sqrt{1+|\nabla f|^2}}\right)=0, \qquad f\in C^2(B_R)\\ &f=f_1 \qquad \text{on } \partial B_R.\end{aligned} \tag{25}$$

The function $f_1|_{\partial B_R}$ is of class C^2 and Theorem 1.2 of [10] gives existence and uniqueness of the solution of (14). The function f_1 is invariant by the group G whence so is the solution $f^{(R)}$ of (25) and we may write

$$f^{(R)}(x)=F^{(R)}(u,v);$$

also we have $F_j(u,v)=-F_j(v,u)$ $(j=1,2)$ and it follows that

$$F^{(R)}(u,v)=-F^{(R)}(v,u)$$

and

$$f^{(R)}(x)=0 \tag{26}$$

for $x \in C_{2m} \cap B_R$.

From (22) and (26) we get

$$f_1(x) \leqq f^{(R)}(x) \leqq f_2(x)$$

for $x \in \partial(B_R \cap D_1)$. Hence we may apply the inequalities (23), (24) and the well-known maximum principle for solutions of (14) and obtain

$$f_1(x) \leqq f^{(R)}(x) \leqq f_2(x)$$

for $x \in \bar{B}_R \cap \bar{D}_1$. In the same way one finds

$$f_2(x) \leqq f^{(R)}(x) \leqq f_1(x)$$

for $x \in \bar{B}_R \cap \bar{D}_2$, and we conclude that

$$|f_1(x)| \leqq |f^{(R)}(x)| \leqq |f_2(x)| \tag{27}$$

for $x \in \bar{B}_R$.

Now consider the sequence $f^{(k)}(x)$, $k=1,2,\ldots$. By inequality (27) and Theorem A of [4] (see also [3]) we find that for every $k \geqq h$ we have

$$|\nabla f^{(k)}(x)| \leqq c_1 \exp\left(c_2 \frac{1}{2h} \sup_{B_{2h}} |f_2|\right) = c(h) \tag{28}$$

for $x \in B_h$; the essential point of this inequality is that the right hand side is independent of k. Hence there is a subsequence $\{f^{(k_\nu)}(x)\}$, $\nu = 1,2,\ldots$ of the sequence $\{f^{(k)}(x)\}$ which is uniformly convergent in the closure of the unit ball $\bar{B}_1$ to a function $f(x)$. By Remark 6 of [11] and the inequality (28) this function $f(x)$ is analytic in B_1 and verifies there the minimal surface Eq. (14).

Next, we prove that the sequence $\{f^{(k_\nu)}(x)\}$ is uniformly convergent in every compact set $K \subset \mathbb{R}^{2m}$. If this were not the case, there would exist, by the same argument as before, two subsequences uniformly convergent in a ball B_R, $R>1$, to two distinct functions f' and f'', both analytic in B_R and solution of the minimal surface Eq. (14) there. This however is impossible, because $f'=f''=f$ in B_1.

Arguing as before we conclude that the function $f(x)$, limit of the sequence $\{f^{(k_\nu)}(x)\}$, is analytic in $\mathbb{R}^{2m}$ and verifies there the Eq. (14) and the inequalities

$$|f_1(x)| \leqq |f(x)| \leqq |f_2(x)|.$$

We have

$$\limsup_{|x| \to +\infty} |f_1(x)|/|x|^{2\alpha} = 1,$$

$\alpha > 1$. Hence $f(x)$ cannot be a polynomial of the first degree, and the proof of Theorem B is complete.

Appendix

Here we prove the statements (16), ..., (22) about the two functions F_1, F_2.

It is convenient to split the differential operator $\mathscr{E}$ defined by (15) in two parts as $\mathscr{E}=\mathscr{E}_0+\mathscr{D}$, where

$$\mathscr{E}_0 F=F_v^2 F_{uu}-2F_u F_v F_{uv}+F_u^2 F_{vv}+p\left(\frac{F_u}{u}+\frac{F_v}{v}\right)(F_u^2+F_v^2)$$

is the "least gradient operator" already considered in Section III, (8), and where

$$\mathscr{D} F=F_{uu}+F_{vv}+p\left(\frac{F_u}{u}+\frac{F_v}{v}\right)$$

is related to the Laplace operator.

This decomposition suggests looking for functions F_1, F_2 having the same asymptotic behaviour as the function F constructed in Section III, for large values of $|\nabla F|$.

With conditions (16), (17) and (22) to fulfill, the simplest choice is the function

$$F_1=(u^2-v^2)(u^2+v^2)^{\alpha-1}$$

where

$$\alpha=\frac{2p+1-\sqrt{\Delta}}{4}.$$

In order to prove that F_1 is a subsolution for the differential operator $\mathscr{E}$ in the domain $T_1=\{(u,v)\,|\,0\leqq v<u\}$ it is useful to define new independent variables

$$r=u^2+v^2,$$

$$t=\frac{u^2-v^2}{u^2+v^2}, \qquad 0<t\leqq 1.$$

If we write

$$F(u,v)=G(r,t)$$

we find

$$\begin{aligned}\tfrac{1}{8}\mathscr{E}_0 F &=(1-t^2)\left\{2G_t^2 G_{rr}-4G_t G_r G_{rt}+2G_r^2 G_{tt}+\frac{2p+3}{r} G_r G_t^2-2p\frac{t}{r^2} G_t^3\right\} \quad (29)\\ &\quad +(2p+1)r G_r^3-(2p+2)t G_r^2 G_t,\end{aligned}$$

$$\tfrac{1}{8}\mathscr{D} F=2(p+1)\left(G_r-\frac{t}{r} G_t\right)+2r\left[G_{rr}+\frac{1}{r^2}(1-t^2)G_{tt}\right] \quad (30)$$

and in the special case in which

$$G(r,t)=r^{\alpha}f(t)$$

where

$$\alpha=\frac{2p+1-\sqrt{\Delta}}{4},$$

one finds

$$\begin{aligned}\tfrac{1}{8}\mathscr{E}_0 F &= r^{3\alpha-2}\{(1-t^2)[2pff'^2-2ptf'^3+2\alpha^2f^2f''] \\ &\quad +(2p+1)\alpha^3f^3-(2p+2)\alpha^2tf^2f'\} \\ &= r^{3\alpha-2}M_1(t),\end{aligned} \tag{31}$$

$$\tfrac{1}{8}\mathscr{D}F=2r^{\alpha-1}[\alpha(p+\alpha)f-(p+1)tf'+(1-t^2)f''] . \tag{32}$$

If $f(t)=t$ we have in $\mathring{T}_1$

$$\tfrac{1}{8}\mathscr{E}_0 F=2\alpha^2(\alpha^2-1)r^{3\alpha-2}t^3>0,$$

$$\tfrac{1}{8}\mathscr{D}F=2[\alpha(p+\alpha)-(p+1)]r^{\alpha-1}t>0$$

because $0<t\leqq 1$ and $2\alpha^2-(2p+1)\alpha+2p=0$.

Now take

$$G(r,t)=rt+r^{\alpha}f(t);$$

we obtain

$$\tfrac{1}{8}\mathscr{E}_0 F=r^{3\alpha-2}M_1(t)+r^{2\alpha-1}M_2(t)+r^{\alpha}M_3(t)-rt, \tag{33}$$

where $M_1(t)$ is given by (31) and

$$\begin{aligned}M_2(t)&=(1-t^2)[(4p-2)\alpha ff'-(4p-3+4\alpha)tf'^2+4\alpha tff''] \\ &\quad +(4p+1)\alpha^2tf^2-(4p+4)\alpha t^2ff',\end{aligned} \tag{34}$$

$$\begin{aligned}M_3(t)&=(1-t^2)[\alpha(2p-3+2\alpha)f-(2p-2+4\alpha)tf'+2t^2f''] \\ &\quad -(2p+2)t^3f'+(2p-1)\alpha t^2f.\end{aligned} \tag{35}$$

We choose $f(t)=t+At^{\lambda}$, supposing $0<t\leqq 1$, where λ is such that $\lambda>1$. We write $M_1(t)$ as a polynomial in A and get

$$\begin{aligned}M_1(t)&=A^3t^{3\lambda-2}\{2\lambda(\lambda-1)(\alpha^2-p\lambda)(1-t^2)+\alpha^2[(2p+1)\alpha-(2p+2)\lambda]t^2\} \\ &\quad +A^2t^{2\lambda-1}\{4\lambda(\lambda-1)(\alpha^2-p)(1-t^2) \\ &\quad +\alpha^2[(6p+3)\alpha-(2p+2)(1+2\lambda)]t^2\} \\ &\quad +At^{\lambda}\{2(\lambda-1)(\alpha^2\lambda-p)(1-t^2)+\alpha^2[(6p+3)\alpha-(2p+2)(2+\lambda)]t^2\} \\ &\quad +2\alpha^2(\alpha^2-1)t^3.\end{aligned}$$

Now if

$$\frac{2p+1}{2p+2}\alpha<\lambda<\frac{p}{\alpha^2} \tag{36}$$

it is easily seen that

$$M_1(t)\leqq -c_1(A^3t^{3\lambda-2}+A^2t^{2\lambda-1}+At^\lambda)+c_2(A^2t^{2\lambda+1}+At^{\lambda+2}+t^3)$$

for some positive constants c_1, c_2. It follows from this inequality that if in addition $\lambda\leqq 3$ then

$$M_1(t)\leqq -c_3(A^3t^{3\lambda-2}+A^2t^{2\lambda-1}+At^\lambda)$$

if only $A\geqq c_4(p,\lambda)$, for some positive c_3.

In exactly the same way one deals with $M_2(t)$ and $M_3(t)$. If λ is in the range (36) and $\lambda\leqq 3$, $A\geqq c_4$ (possibly changing the constant c_4) we still obtain

$$M_2(t)\leqq -c_5(A^2t^{2\lambda-1}+At^\lambda),$$
$$M_3(t)\leqq -c_6(At^\lambda+t),$$

provided $p\geqq 3$. This condition comes out in handling the terms of $M_2(t)$ which are independent of A, where we need now

$$(4p-6)\alpha-(4p-3)\leqq 0,$$

and this inequality is true only for $p\geqq 3$.

We conclude that with this choice of G one has in $\mathring{T}_1$:

$$\tfrac{1}{8}\mathscr{E}_0F\leqq -c_7(r^{3\alpha-2}t^\lambda+rt) \tag{37}$$

for some positive c_7, provided $A\geqq c_4$, $p\geqq 3$ and

$$\frac{2p+1}{2p+2}\alpha<\lambda<\frac{p}{\alpha^2},\qquad \lambda\leqq 3.$$

A similar but much simpler estimate gives in $\mathring{T}_1$:

$$\tfrac{1}{8}\mathscr{D}F\leqq c_8\,r^{\alpha-1}t^{\lambda-2}. \tag{38}$$

Now let $H(z)$ be a monotone increasing function for $z>0$ and of class C^2 there. Then for each positive F we have identically

$$\mathscr{E}_0(H(F))=H'(F)^3\cdot\mathscr{E}_0F,$$
$$\mathscr{D}(H(F))=H'(F)\cdot\mathscr{D}F+16r\,H''(F)\left(G_r^2+\frac{1-t^2}{r^2}G_t^2\right).$$

In the special case $G=rt+r^\alpha(t+At^\lambda)$ we easily get, if $\lambda<\alpha$ and if $0<t\leqq 1$:

$$r\left(G_r^2+\frac{1-t^2}{r^2}G_t^2\right)\geqq r+r^{2\alpha-1},$$

whence if we suppose

$$H'(z)\geqq 1, \qquad H''(z)\leqq 0 \tag{39}$$

we obtain by (37) and (38)

$$\begin{aligned}\mathscr{E}(H(F))<&-c_7(r^{3\alpha-2}t^\lambda+rt)\,H'(F)^3\\&+c_8\,r^{\alpha-1}t^{\lambda-2}H'(F)+16(r+r^{2\alpha-1})\,H''(F).\end{aligned}$$

This gives by (39):

$$\begin{aligned}\mathscr{E}(H(F))&<-r^{\alpha-\lambda}t^{\lambda-2}[c_7\,H'(F)^3\,r^{2\alpha-\lambda}t^2-c_8\,H'(F)]\\&<0\end{aligned}$$

provided $0<t\leqq 1$ and

$$r^{2\alpha-1}t^2>c_9=c_8/c_7.$$

Now suppose $0<t\leqq 1$ and

$$r^{2\alpha-1}t^2\leqq c_9. \tag{40}$$

In order to get $\mathscr{E}(H(F))<0$ in this domain too, and hence $\mathscr{E}(H(F))<0$ in the whole of $\mathring{T}_1$, it is enough to fulfill the condition

$$16(r+r^{2\alpha-1})\,H''(F)<-c_8\,r^{\alpha-1}t^{\lambda-2}H'(F),$$

which we may write

$$-\frac{H''(F)}{H'(F)}\geqq c_{10}\frac{r^{\alpha-1}t^{\lambda-2}}{r+r^{2\alpha-1}}.$$

We have

$$\frac{r+r^{2\alpha-1}}{r^{\alpha-1}t^{\lambda-2}}=r^{2-\alpha}t^{2-\lambda}+r^\alpha t^{2-\lambda}.$$

If $r\geqq 1$, then $F<c_{11}r^\alpha t$; it follows that

$$\begin{aligned}r^{2-\alpha}t^{2-\lambda}+r^\alpha t^{2-\lambda}&\geqq r^\alpha t^{2-\lambda}\\&=(r^\alpha t)^{2-\lambda+2\alpha(\lambda-1)}(t^2r^{2\alpha-1})^{-\alpha(\lambda-1)}\\&\geqq c_{12}F^{2-\lambda+2\alpha(\lambda-1)}[\min(1,F)]^{-2\alpha(\lambda-1)}\\&\geqq c_{13}(F^{2-\lambda}+F^{2-\lambda+2\alpha(\lambda-1)})\end{aligned}$$

because of the inequalities (40), $r\geqq 1$ and $r^\alpha t<F<c_{11}r^\alpha t$.

If $r<1$, we have $F<c_{14}\, r\,t<c_{14}$ and using the condition $\lambda<\alpha$ we still obtain

$$\begin{aligned} r^{2-\alpha}t^{2-\lambda}+r^{\alpha}t^{2-\lambda} &\geqq (r\,t)^{2-\lambda}\\ &\geqq c_{15}\,F^{2-\lambda}\\ &\geqq c_{16}(F^{2-\lambda}+F^{2-\lambda+2\alpha(\lambda-1)}). \end{aligned}$$

The conclusion is that if

$$-\frac{H''(z)}{H'(z)}\geqq c_{17}\,z^{\lambda-2}(1+z^{2\alpha(\lambda-1)})^{-1}$$

for some large c_{17} and if

$$H'(z)\geqq 1,$$

then we have in $\mathring{T}_1$ the inequality

$$\mathscr{E}(H(F))<0$$

where $F(u,v)=r\,t+r^{\alpha}(t+A\,t^{\lambda})$. The conditions we have put on the way are

$$\frac{2p+1}{2p+2}\alpha<\lambda<\min\left(\frac{p}{\alpha^2},\alpha\right)$$

(note that $1<\lambda\leqq 3$ is included in it), and $p\geqq 3$, $A\geqq c_4$. Note also that the interval for λ is not empty, already for $p\geqq 3$.

If one takes

$$H'(z)=\exp\left(c_{17}\int_z^{\infty}\frac{dt}{t^{2-\lambda}(1+t^{2\alpha(\lambda-1)})}\right),$$

$$H(z)=\int_0^z H'(w)\,dw$$

then if we define F_2 by $F_2=H(F)$ we get

$$\mathscr{E}F_2<0 \quad \text{in } \mathring{T}_1,$$

as required.

This choice of $H(z)$ is admissible, because

$$\int_0^{\infty}\frac{dt}{t^{2-\lambda}(1+t^{2\alpha(\lambda-1)})}<+\infty; \tag{41}$$

note that $2-\lambda<1$, $2-\lambda+2\alpha(\lambda-1)>1$. It is also clear that $0<F_1<F_2$ in $\mathring{T}_1$, and $F_1=F_2=0$ on $u=v$, this because of (41).

Thus we have proved that if $p\geqq 3$ the two functions F_1, F_2 defined in Section IV satisfy (16), (18), (20), (22). The proof of (17), (19), (21) is also immediate because of the antisymmetry of F_1, F_2 with respect to the line $u=v$. q.e.d.

References

1. Almgren, F. J. Jr.: Some interior regularity theorems for minimal surfaces and an extension of Bernstein's theorem. Ann. of Math. **85**, 277–292 (1966).
2. Birkhoff, G., and G. C. Rota: Ordinary differential equations. Boston: Ginn & Co. 1962.
3. Bombieri, E.: Nuovi risultati sulle ipersuperfici minimali non parametriche. Rend. Sem. Mat. Fis. Milano **38**, 2–12 (1968).
4. – E. De Giorgi, and M. Miranda: Una maggiorazione a priori relativa alle ipersuperfici minimali non parametriche. Archive for Rat. Mech. and Analysis (1968).
5. De Giorgi, E.: Frontiere orientate di misura minima. Sem. Mat. Sc. Norm. Sup. Pisa, A. A. 1960/61.
6. – Una estensione del teorema di Bernstein. Ann. Sc. Norm. Sup. Pisa **19**, 79–85 (1965).
7. Federer, H., and W. H. Fleming: Normal and integral currents. Ann. of Math. **72**, 458–520 (1960).
8. Fleming, W. H.: On the oriented Plateau problem. Rend. Circolo Mat. Palermo **9**, 69–89 (1962).
9. Miranda, M.: Sul minimo dell'integrale del gradiente di una funzione. Annali Sc. Norm. Sup. di Pisa **19**, 627–665 (1965).
10. – Un teorema di esistenza e unicità per il problema dell'area minima in n variabili. Annali Sc. Norm. Sup. di Pisa **19**, 233–250 (1965).
11. – Comportamento delle successioni convergenti di frontiere minimali. Rend. Sem. Mat. Padova **38**, 238–257 (1967).
12. Simons, J.: Minimal varieties in riemannian manifolds. Annals of Math. **88**, 62–105 (1968).

E. Bombieri
E. De Giorgi
E. Giusti
Istituto Matematico «Leonida Tonelli»
Università di Pisa
Via Derna 1
Pisa (Italia)

(Received April 3, 1969)

New problems in Γ-convergence and G-convergence*

Ennio De Giorgi
Scuola Normale Superiore, Pisa

Many problems arising in mathematical physics (including also free boundary problems considered during this bimester) may be stated by limiting processes of various kind: the simplest cases are perturbation methods, others concern homogenization problems, finally there are more complicated problems of which I shall try to give some examples in this lecture. They concern the theory of Γ-convergence and G-convergence: in abstract these two concepts are strictly related, in practise the G-convergence scheme was mainly used for limits of differential operators, while the Γ-convergence scheme was used for limits of functionals in the calculus of variations.

This last application is justified for the stability of minima with respect to Γ-convergence (for the proof of the following result and for definitions related to this theory we refer to the Proceedings of the Meeting in Rome on Recent Methods in Nonlinear Analysis, 1978 [21]), more precisely the following holds:

Theorem. *Let (X, τ) be a metrizable space and $f_h, f : X \to \mathbb{R}$. If f_h $\Gamma(\tau)$-converges to f on X and the f_h are equicoercive, i.e. for all $\varrho \in \mathbb{R}$ there exists a compact subset K_ϱ of X such that*

$$\bigcup_{h=1}^{\infty} \{x \in X : f_h(x) \leq \varrho\} \subseteq K_\varrho \,,$$

then $\min\limits_X f(x) = \lim\limits_h \big[\inf\limits_X f_h\big]$. Moreover, if $\{x_h\}$ is a sequence of elements of X satisfying $f_h(x_h) < \inf_X f_h + \varepsilon_h$, for $\varepsilon_h > 0$, $\lim_h \varepsilon_h = 0$, and $x_h \xrightarrow{\tau} x_0$, then $f(x_0) = \min_X f$.

Usually, as in the theorem, one studies functionals depending on a real parameter, but in the general theory of Γ-convergence, more general relations could be studied.

Examples of functionals much studied until now are functionals of energy type (in the sense of the norm in Sobolev spaces) or funtionals of area type. Likewise it is possible to deal in a similar way with other problems in mathematical physics. For instance, let us consider a problem in which the free boundary is the surface separating two fluids. Actually, an alternative physical model could be like the one outlined in this way: there is not a real surface between the two fluids, but a small layer of transition, whose thickness is $1/h$ and h is big; in the layer some physical quantities remain bounded independently from h, while their derivatives are of the same order as h. If the thickness of the layer tends to 0 (as $h \to \infty$) we get some

*In *Free Boundary Problems* (Pavia, 1979), Ist. Alta Mat. (Roma, 1980), Vol. II, 183–194. Notes written by Franco Tomarelli.

discontinuities, therefore we cannot hope to manage with the problem by expansion in uniformly convergent series of regular functions. Probably it is possible to expand the solutions in series depending upon $1/h$ separately on each side of the layer. It will be interesting to test the point of view of G- and Γ-convergence in problems of this kind; for instance, Stefan problems, fluids with cavitation, bubbles, drops and other two-phase or multi-phase problems.

I mentioned these problems in connection with the subjects developed in Albertoni's lecture, only to point out the interest of a future possible connection between the theory of Γ- and G-convergence and mathematical physics problems not studied in this way till now. On the contrary, I don't dwell upon Γ-convergence problems related to the homogenization, for they were extensively exposed in the meeting of Rome,[21] to which I refer again for the bibliography.

I'll speak about four kinds of problems, that are already schematised in a quite precise way, but, as far as I know, are not solved in general.

Remark. In the following I consider a sequence of functionals of this kind

$$F_h(u,\Omega) = \int_\Omega f_h(x,u,Du,\dots,D^\nu u)\,.$$

For our purposes it is not necessary to fix *a priori* the optimal space in which they are defined (Sobolev spaces $H^{k,p}(\Omega)$, $BV(\Omega)$, Orlicz spaces or others), but we define the functional F_h in a wider class $T(\Omega)$ (like $L^1_{\rm loc}(\Omega)$) in order that F_h takes the value $+\infty$ except in a smaller class $S(\Omega)$, like for instance $C^\infty(\Omega)$, where the functional coincides with the integral. Then we study the Γ-limit of the sequence (F_h) and finally we verify which is the subset of $T(\Omega)$ where the Γ-limit takes a finite value; in many cases this is the optimal space in the sense of the calculus of variations. It is more convenient to follow this outline for it causes no problem in standard problem (thanks to a result by De Giorgi and Franzoni,[23] in the evaluation of the Γ-limit the sequence F_h may be replaced with any sequence F'_h such that $\mathrm{sc}^- F_h \le F'_h \le F_h$, where sc^- denotes the lower semicontinuous regularization); moreover, this outline is much more convenient when the optimal space for the F_h differs from the optimal space for their Γ-limit; this may happen, as shown by Modica and Mortola in Ref. 25.

1. Obstacles

Let us take a simple example of minimization with a unilateral bound

$$\min_{u>\varphi} \int_\Omega (|\nabla u|^2 + u^2 + \psi(x,u))\,,$$

where $\Omega \subset \mathbb{R}^n$, φ belongs to $C(\Omega)$, $u : \Omega \to \overline{\mathbb{R}}$ and ψ is a smooth perturbation, e.g. $\psi \in C_0^\infty(\Omega \times \mathbb{R})$. If we want to study this problem with thin (or however not regular) obstacles, we can take sequences $\{\varphi_h\}$ of regular functions which converge in a suitable topology to φ and we pose

$$F_h(u) = \begin{cases} 0 & \text{if } u(x) \ge \varphi_h(x) \text{ for all } x \in \Omega \\ +\infty & \text{if there exists } x \in \Omega \text{ such that } u(x) < \varphi_h(x). \end{cases}$$

Now, we consider the problem

$$(P_h) \qquad \min \left| \int_\Omega (|\nabla u|^2 + u^2 + \psi(x,u)) + F_h(u) \right| .$$

We suppose that there exists a functional F such that, for every sequence $\{u_h\}$ whose elements are solutions of the corresponding problems (P_h) and for every subsequence $\{u_{h_k}\}$ converging to a function u, the function u is solution of

$$\min \left\{ \int_\Omega (|\nabla u|^2 + u^2 + \psi(x,u)) + F(u) \right\} .$$

It is then natural to ask if the functional F is always of obstacle type, and, if not, which kind of functional it might be. The answer is known only partially. Colombini and Carbone[10] gave a negative answer, with a counterexample valid if $n \geq 2$: let us consider a squared domain contained in $\mathbb{R}^2$ and divided into smaller squares; inside each one we fix a disk, we let the side of the small squares and the diameter of the disks tend to zero. We call χ_h the characteristic function of the union of the disks at the step h; the graphs of χ_h are systems of towers which become thinner. We consider the case in which the sides of the squares are of order (h^{-1}) and the diameters of order $(h^{-\sigma})$, $0 < \sigma < 1$. The result of Carbone and Colombini tells us that for some values of σ the functional is still an obstacle: say, if the towers are big enough, the obstacle F corresponds to the characteristic function χ of the whole square; if the towers are too thin in the limit we get the obstacle equal to zero. But, for intermediate values of σ, F is not an obstacle, that is, there exists u such that $F(u) \neq 0$ and $\neq +\infty$. It is clear that in this case it is not realistic to hope that solutions expressible by expansion in series of the step of the mesh exist.

It is reasonable to expect that the following conjecture is verified:

If F_h are of the obstacle type, the limit F should always be of this kind:

$$(\alpha) \qquad F(u) = \int_\Omega g(x,u)\,d\mu$$

where g is convex and monotone in u, with values in $\overline{\mathbb{R}}$, μ is a Borel measure.

Obviously, it is an obstacle if g takes only the values 0 and $+\infty$.

If, as in the previous example, the obstacles are uniformly distributed, μ is the Lebesgue measure, while it may be a surface density if the towers pile up on a surface, and so on.

This method seems to be the inverse of a penalization, in the sense that when using a penalization one approximates functions with values 0 or $+\infty$ by functions of different kind; on the contrary, here we approximated with obstacles functions which take also values different from 0 and $+\infty$. The conjecture may be stated as follows:

The type (α) of obstacle is closed with respect to Γ-convergence.

Remark that in one dimension the counterexample does not work because it is impossible to tighten the towers enough to eliminate the contribution of the integral of the gradient. For functionals like

$$\int_\Omega |\nabla u|^\beta + u^2 + \psi(x,u), \qquad u > \varphi$$

we can do analogous considerations and in particular that the conjecture is still valid if $\beta > 1$; but for the critical value $\beta = 1$ one may encounter complications.

2. Local Minima

The problem of local minima is strictly related to the solutions of partial differential equations. For instance, let

$$G(u,\Omega) = \int_\Omega \sum_{j,k=1}^{n} a_{jk} \frac{\partial u}{\partial x_j} \frac{\partial u}{\partial x_k} \,.$$

The search for solutions of the equation

$$\sum_{j,k=1}^{n} \frac{\partial}{\partial x_j} \left(a_{jk} \frac{\partial u}{\partial x_k} \right) = 0 \qquad \forall x \in A \text{ open set in } \mathbb{R}^n$$

is reduced to the problem: find the function u, defined in A, such that for every $\Omega \subset\subset A$, and for every $g \in C_0^\infty(\Omega)$

$$(\gamma) \qquad G(u,\Omega) < G(u+g,\Omega)\,,$$

where (γ) is a condition of local minimum. In general, given a functional $F(u,\Omega) = \int_\Omega f(x,u,Du)$, we say that u is a local minimum in A of the functional F if, for every $\Omega \subset\subset A$, and for every $g \in C_0^\infty(\Omega)$, one has $F(u,\Omega) < F(u+g,\Omega)$.

Now, let $F_h(u,\Omega) = \int_\Omega f_h(x,u,Du)$ be such that, for any $\Omega \subset\subset A$, $F_h(u,\Omega)$ Γ-converges to $F(u,\Omega)$. In this situation, given the open set A and a sequence of local minima u_h of F_h in A, such that u_h converges to u, we can ask if u is a local minimum for F in A. The answer is positive for uniformly elliptic quadratic forms and also for larger classes of integral functionals as Dal Maso and Modica have recently shown [20]. The following problem seems to be much more difficult, in a certain sense the inverse of the previous one: given F_h and F as before, if u is a local minimum for F in A, does there exist a sequence of local minima u_h of F_h in A which converges to u?

3. Evolution Problems

Let F_h be a sequence of functionals satisfying: $F_h(u)$ Γ-converges to $F(u)$. We denote $F' = \operatorname{grad} F$, $F'_h = \operatorname{grad} F_h$. If u_h is a solution of

$$\frac{du_h}{dt} = F'_h(u_h)$$

it is natural to ask if $\{u_h\}$ converges in any topology, at least for subsequences, to the solution of the problem

$$\frac{du}{dt} = F'(u).$$

We must observe that we are dealing with sets of solutions, as we have no hypothesis ensuring the uniqueness, either for the approximating problems or for the limit problem. For instance, we can consider functionals of this kind

$$F_h(u) = \int_a^b \left[\left(\frac{du}{dx} \right)^2 + g(hu) \right] dx$$

with $g \in C^\infty(\mathbb{R})$ periodic. In this case Buttazzo and Dal Maso proved that the Γ-limit of F_h can be expressed in this way

$$F(u) = \int_a^b \theta \left(\frac{du}{dx} \right) dx\,,$$

where θ is a convex function of which we also know the analytical expression in some simple cases, but they didn't say anything about the convergence of the solution of $\frac{du_h}{dt} = F_h'(u_h)$. We observe that in this example it is not possible to use Brézis' methods, because of the perturbation term which is not convex, and becomes more and more oscillating when h increases.

4. Hyperbolic Problems

The study of this kind of problem is more difficult

$$\frac{d^2u}{dt^2} = F_h'(u)$$

which arises in the case of the vibrating string or vibrating membrane which rebounds against an obstacle, even if F_h is convex. The problem has been studied by Amerio, Prouse and Citrini in Refs. 1–3, 12–16, but they used methods strictly related to space dimension one.

In general, let us suppose to have an obstacle like $u \geq 0$; we penalize the obstacle with a regular function

$$\psi(s) = \begin{cases} 0 & \text{if } s \geq 0 \\ > 0 & \text{if } s < 0 \end{cases}$$

and we consider the limit of the functionals as h goes to infinity

$$F_h(u) = \int_\Omega |\nabla u|^2 + h\psi(u)\,.$$

Only partial results are known and we can say only a few things about the convergence. The problem of a point bound to move along a straight line and rebounding against a wall is already complicated; this problem has been studied by Carriero and Pascali in Ref. 11.

The penalized problem has the following formulation

$$\frac{d^2u}{dt^2} + f(t) = h\psi(u)$$

with a Cauchy data given in the admissible region. When h tends to $+\infty$ we get a problem for which the existence and uniqueness of solutions have been proved only if f is piecewise constant and

$$\lim_{s\to 0+} \frac{\psi'(s)}{\psi(s)} = +\infty\,.$$

For evolution equations we recall that the parabolic problem

$$\frac{du_h}{dt} = F_h'(u_h, t)$$

with

$$F_h'(u_h,t) = \sum_{j,k}^{n} \frac{\partial}{\partial x_k}\left(a_{jk}(h^\alpha x, u, h^\beta t)\frac{\partial u}{\partial x_j}\right)$$

and a periodic in x and t, has been studied by Lions in Ref. 8. In this case the parabolic equation has different limits depending upon the values α and β.

The analogous hyperbolic problem $\frac{d^2u_h}{dt^2} = F_h'(u_h,t)$, and F_h' defined as before, has been studied by Colombini and Spagnolo in Refs. 17–19. Even in the simplest case

$$\frac{d^2u}{dt^2} = a(ht)\frac{\partial u}{\partial x^2}$$

with a periodic the solutions can diverge with h also if the initial data are C^∞. But, if $u(x,0)$, $\frac{du}{dt}(x,0)$ are analytic functions, the solutions u_h converge to the solution of the equation

$$\frac{d^2u}{dt^2} = \overline{a}\frac{\partial u}{\partial x^2}$$

where $\overline{a}$ is the average of a in the period.

An important open question is the commutativity of the Γ-limits for sequences depending upon many indices. De Giorgi and Franzoni gave a positive answer if one of the indices is a regularizing operator. For instance, let F_h be a sequence of functions Γ-converging to a functional F, and let

$$F_{h\lambda}(u) = \min w(F_h(w) + \lambda G(u-w))$$

the regularization of F. Then the following diagram commutes

$$\begin{array}{ccc}
F_h & \xrightarrow{\;h\to\infty\;} & F \\
\Big\uparrow{\scriptstyle \lambda\to\infty} & & \Big\uparrow{\scriptstyle \lambda\to\infty} \\
F_{h,\lambda} & \xrightarrow[h\to\infty]{} & F_\lambda
\end{array}$$

where the arrows denote Γ-convergence of functionals.

Finally, we come back to the initial free boundary problem formulated by means of the problems (P_h) with a thin layer of height $1/h$ between the phases. Let us study the behaviour as h tends to $+\infty$. Then we can study the following problems:

1) In which space do the problems (P_h) converge.
2) Study the speed of convergence.
3) Find possibly different but simpler mathematical models (P_h) which converge to the free boundary problem.

References

1. **L. Amerio**, *Continuous solutions of the problem of a string vibrating against an obstacle,* Rend. Sem. Mat. Univ. Padova **59** (1978), 67–96.
2. **L. Amerio**, *An unilateral problem for a nonlinear vibrating string equation,* Atti Accd. Naz. Lincei Cl. Sci. Fis. Mat. Natur. **(8) 64** (1978), 8–21.
3. **L. Amerio and G. Prouse**, *Study of the motion of a string vibrating against an obstacle,* Rend. di Mat. **(6) 8** (1975), 563–585.
4. **H. Attouch**, *Convergence des solutions d'inéquations variationnelles avec obstacles,* Proc. Int. Meeting on Recent Methods in Nonlinear Analysis (Rome, 1978), Pitagora, Bologna, 1978, 101–113.
5. **H. Attouch and Y. Konishi**, *Convergence d'opérateur maximaux monotones et inéquations variationnelles,* C. R. Acad. Sci. Paris, Sez. A **282** (1976), 467–469.
6. **H. Attouch and C. Sbordone**, *Asymptotic limits for perturbed functionals of calculus of variations,* Ric di Mat., to appear.
7. **B. Betrò and L. Gotusso**, *Calcolo numerico del moto di una corda vibrante contro un ostacolo,* Pubbl. I.A.C., Sez. III **114** (1977).
8. **A. Bensoussan, J. L. Lions and G. Papanicolau**, *Asymptotic analysis for periodic structures,* North-Holland, Amsterdam, 1978.
9. **L. Boccardo and P. Marcellini**, *Sulla convergenza delle soluzioni di disequazioni variazionali,* Ann. Mat. Pura Appl. **(4) 110** (1976), 137–159.
10. **L. Carbone and F. Colombini**, *Sur la convergence de fonctionnelles soumises à des constraintes unilatéraux,* Preprint Scuola Normale Superiore di Pisa.
11. **M. Carriero and E. Pascali**, *Il problema del rimbalzo unidimensionale e sue approssimazioni con penalizzazioni non convesse,* to appear.
12. **C. Citrini**, *Sull'urto parzialmene elastico o anelastico di una corda vibrante contro un ostacolo,* Atti Accd. Naz. Lincei Cl. Sci. Fis. Mat. Natur. **(8) 59** (1975), 368–276 and 667–676.
13. **C. Citrini**, *Energia ed impulso nell'urto parzialmente elastico od anelastico di una corda vibrante contro un ostacolo,* Istituto Lombardo Accad. Sci. Lett. Rend. All. O (1976), 271–280.
14. **C. Citrini**, *Controesempi all'unicità del moto di una corda contro una parete* Atti Accad. Naz. Lincei Cl. Sci. Fis. Mat. Natur., to appear.
15. **C. Citrini**, *Discontinuous solutions of a nonlinear hyperbolic equation with unilateral constraints,* Manuscripta Mathematica, to appear.

16. **C. Citrini and B. D'Acunto**, *Sur le choc de deux cordes,* C. R. Acad. Sci. Paris, Sez. A **295** (1979).
17. **F. Colombini and S. Spagnolo**, *Sur la convergence des solutions d'équations paraboliques avec des coéfficients qui dépendent du temps,* C. R. Acad. Sci. Paris, Sez. A **282** (1976), 735–737.
18. **F. Colombini and S. Spagnolo**, *On the convergence of solutions of hyperbolic equations,* Comm. Partial Differential Equations **3** (1978), 77–103.
19. **G. Dal Maso and L. Modica**, *Convergenza dei minimi locali,* Rend. di Mat. to appear.
20. **E. De Giorgi**, *Convergence problems for functionals and operators,* Proc. Int. Meeting on Recent Methods in Nonlinear Analysis (Rome, 1978), Pitagora, Bologna, 1978, 131–188.
21. **E. De Giorgi and T. Franzoni**, *Su un tipo di convergenza variazionale,* Atti Accd. Naz. Lincei Cl. Sci. Fis. Mat. Natur. **(8) 58** (1975), 842–850.
22. **E. De Giorgi and L. Modica**, Γ-convergenza e superfici minime, settimana di studi su superfici minime e questioni collegate, Pisa.
23. **L. Modica**, *Γ-convergence to minimal surfaces problems and global solutions of* $\Delta u = 2(u^3 - u)$, Proc. Int. Meeting on Recent Methods in Nonlinear Analysis (Rome, 1978), Pitagora, Bologna, 1978, 223–244.
24. **L. Modica and S. Mortola**, *Un esempio di Γ-convergenza,* Boll. Un. Mat. Ital. **(5) 14-A** (1977), 526–529.
25. **U. Mosco**, *Convergence of convex sets and of solutions of variational inequalities,* Ad. Math. **3** (1969), 510–585.
26. **F. Murat**, *Sur l'homogénéisation d'inéquations elliptiques du 2-ème ordre, relatives au convex* $k(\psi_1, \psi_2) = \{v \in H_0^1(\Omega) : \psi_1 < v < \psi_2$ p.p. dans $\Omega\}$, preprint Université Pierre et Marie Curie, Paris VI.
27. **M. Schatzman**, *Problèmes unilatéraux d'evolution du 2-ème ordre en temps,* Université Pierre et Marie Curie, Paris VI, thèse de Doctorat d'état, 1979.

Samuel Eilenberg (1913–1998)

Hyman Bass, Henri Cartan, Peter Freyd, Alex Heller, and Saunders Mac Lane

Samuel Eilenberg died in New York, January 30, 1998, after a two-year illness brought on by a stroke. He left no surviving family, except for his wide family of friends, students, and colleagues, and the rich legacy of his life's work, in both mathematics and as an art collector.

"Sammy", as he has long been called by all who had the good fortune to know him, was one of the great architects of twentieth-century mathematics and definitively reshaped the ways we think about topology. The ideas that accomplished this were so fundamental and supple that they took on a life of their own, giving birth first to homological algebra and in turn to category theory, structures that now permeate much of contemporary mathematics.

Born in Warsaw, Poland, Sammy studied in the Polish school of topology. At his father's urging, he fled Europe in 1939. On his arrival in Princeton, Oswald Veblen and Solomon Lefschetz helped him (as they had helped other refugees) find a position at the University of Michigan, where Ray Wilder was building up a group in topology. Wilder made Michigan a center of topology, bringing in such figures as Norman Steenrod, Raoul Bott, Hans Samelson, and others. Saunders Mac Lane's invited lecture there on group extensions precipitated the long and fruitful Eilenberg-Mac Lane collaboration.

In 1947 Sammy came to the Columbia University mathematics department, which he twice chaired and where he remained till his retirement. In 1982 he was named a University Professor, the highest faculty distinction that the university confers.

Sammy traveled and collaborated widely. For fifteen years he was a member of Bourbaki. His collaboration with Steenrod produced the book *Foundations of Algebraic Topology*, that with Henri Cartan the book *Homological Algebra*, both of them epoch-making works. The Eilenberg-Mac Lane collaboration gave birth to category theory, a field that both men nurtured and followed throughout their ensuing careers. Sammy later brought these ideas to bear in a multivolume work on automata theory. A joint work on topology with Eldon Dyer may see posthumous publication soon.

Among his many honors Sammy won the Wolf Prize (shared in 1986 with Atle Selberg), was awarded several honorary degrees (including one from the University of Pennsylvania), and was elected to membership in the National Academy of Sciences of the USA. On the occasion of the honorary degree at the University of Pennsylvania in 1985, he was cited as "our greatest mathematical stylist".

The aesthetic principles that guided Sammy's mathematical work also found expression in his passion for art collecting. Over the years Sammy gathered one of the world's most important collections of Southeast Asian art. His fame among certain art collectors overshadows his mathematical reputation. In a gesture characteristically marked by its generosity and elegance, Sammy in 1987 donated much of his collection to the Metropolitan Museum of Art in New York, which in turn was thus motivated to contribute substantially to the endowment of the Eilenberg Visiting Professorship in Mathematics at Columbia University.

—*Hyman Bass*

Henri Cartan

Samuel Eilenberg died in New York on January 30, 1998, after spending two years in a state of precarious health. I would like to write here of the mathematician and especially of the friend that I gradually discovered in the course of a close collaboration that lasted at least five years and that taught me many things.

I met Sammy for the first time at the end of December 1947: he had come to greet me at LaGuardia Airport in New York, a city buried under snow, where airplanes had been unable either to take off or to land for two days. This was my first visit to the United States; it was to last five months. Of course, Eilenberg was not unknown to me, because since the end of the war I had begun to be interested in algebraic topology. Notably I had studied the article in the 1944 *Annals of Mathematics* in which Eilenberg set forth his theory of *singular homology* (one of those theories which immediately takes on a definitive shape). I had, for my part, reflected on the "Künneth formula", which gives the Betti numbers and the torsion coefficients of the product of two simplicial complexes. In fact, that formula amounts to a calculation of the homology groups of the tensor product of two graded differential groups as a function of the homology groups of each of them. The solution involves not only the tensor product of the homology groups of the factors but also a new functor of these groups, the functor Tor. At the time of my first meeting with Sammy, I was quite happy with telling that to him.

This was the point of departure for our collaboration, by means of postal mail at first. Then Sammy came to spend the year 1950–51 in Paris. He took part in my seminar at the École Normale, devoted that year to cohomology of groups, spectral sequences, and sheaf theory. Sammy gave two lectures on spectral sequences. Armand Borel and Jean-Pierre Serre took an active part in this seminar also.

Independently of the seminar, Sammy and I had work sessions with the aim of writing an article that would develop some of the new ideas born out of the Künneth formula. We went from discovery to discovery, Sammy having an extraordinary gift for formulating at each moment the conclusions that would emerge from the discussion. And it was always he who wrote everything up as we went along in precise and concise English. After the notion of *satellites* of a functor came that of *derived functors*, with their axiomatic characterization. Gradually the theory included several existing theories (cohomology of groups, cohomology of Lie algebras, in the sense of Chevalley and Eilenberg, cohomology of associative algebras). Then came the concept of *hyperhomology*.

Photograph courtesy of Columbia University.

Samuel Eilenberg

Of course, this work together took several years. Sammy made several trips to my country houses (in Die and in Dolomieu). Outside of our work hours he participated in our family life.

Sammy knew how to put his friends to work. I think I remember that he persuaded Steenrod to contribute the preface of our book, where the evolution of the ideas is explained perfectly. He arranged also for other colleagues to collaborate in the writing of the chapter devoted to finite groups. Our initial project of a mere article for a journal was transformed; it became a book that we would propose to a publisher and for which it would be necessary to find a title that captured its content. We finally agreed on the term *Homological Algebra*. The text was given to Princeton University Press in 1953. I do not know why the book appeared only in 1956.

For fifteen years Sammy was also an active member of the Bourbaki group. It was, I think, in 1949 that André Weil, who was living in the United States, made contact with him in order to have him collaborate on a draft for use by Bourbaki, entitled "SEAW Report on Homotopy Groups and Fiber Spaces". It is therefore very natural that Eilenberg was invited to the Congress that Bourbaki held in October 1950. He was immediately appreciated and became a member of the group under the name "Sammy". It is necessary to say that he mastered the French language perfectly, which he had learned when he was living in his native Poland.

The collaboration of Sammy with Bourbaki lasted until 1966. He took part in the summer meetings, which lasted two weeks. He knew admirably how to present his point of view, and he often made us agree to it.

The above gives only a faint idea of Samuel Eilenberg's mathematical activity. The list made in 1974 of his publications comprises, besides 4 books, 111 articles; the first 37 articles are before his emigration from Poland to the United States in

Henri Cartan is professor emeritus of mathematics at Université de Paris XI. This segment is translated and adapted from the Gazette des Mathématiciens *by permission.*

1939, and almost all are written in French. He was not yet twenty years old when he began to publish. The celebrated articles written with S. Mac Lane extended from 1942 to 1954. The list of his other collaborators is long: N. E. Steenrod, J. A. Zilber, T. Nakayama, T. Ganea, J. C. Moore, G. M. Kelly, to cite only the main ones. Starting in 1966, Sammy became actively interested in the theory of automata, which led him to write a book entitled *Automata, Languages, and Machines*, published in 1974 by Academic Press.

I have not mentioned a magnificent collection of sculptures in bronze, silver, or stone, patiently collected in India, Pakistan, Indonesia, Cambodia,..., some of which dated to the third century B.C. In 1967 he gave a great part of his collection to the Metropolitan Museum in New York.

In 1982 Eilenberg retired from Columbia University, where he had taught since 1947. In 1986 his mathematical work was recognized by the award of the Wolf Prize in Mathematics, which he shared with Atle Selberg.

The last time I saw Sammy was when the Université de Louvain-la-Neuve organized a conference in his honor. Our meeting there was not without emotion. He was for me a friend whose kindness, humor, and faithfulness cannot be forgotten.

Saunders Mac Lane

Samuel Eilenberg, who made decisive contributions to topology and other areas of mathematics, died on Friday, January 30, 1998, in New York City. He had been a leading member of the department of mathematics at Columbia University since 1947. His mathematical books, ideas, and papers had a major influence.

Eilenberg was born in Poland in 1913. At the University of Warsaw he was a student of Borsuk in the active school of Polish topology. His thesis, concerned with the topology of the plane, was published in *Fundamenta Mathematica* in 1936. Its results were well received both in Poland and in the USA. In 1938 he published in the same journal another influential paper on the action of the fundamental group on the higher homotopy groups of a space. Algebra was not foreign to his topology!

Early in 1939 Sammy's father told him, "Sammy, it doesn't look good here in Poland. Get out." He did, arriving in New York on April 23, 1939, and going at once to Princeton. At that university Oswald Veblen and Solomon Lefschetz efficiently welcomed refugee mathematicians and found them suitable positions at American universities. Sammy's work in topology was well known, so a position for him was found at the University of Michigan. There Ray Wilder had an active group of topologists, including Norman Steenrod, then a recent Princeton Ph.D. Sammy immediately fitted in, did collaborative research (for example, with Wilder, O. G. Harrold, and Deane Montgomery). His 1940 paper in the *Annals of Mathematics* formulated and codified the ideas of the "obstructions" recently introduced by Hassler Whitney. He also argued with Lefschetz. Finding the Lefschetz book (1942) obscure in its treatment of singular homology, he provided an elegant and definitive treatment in the *Annals* (1944).

Sammy's idea was to dig deep and deeper till he got to the bottom of each issue. This I learned when I lectured at Ann Arbor about group extensions. I had calculated an example of the group of group extensions for an interesting factor group involving a prime number p. When I told Sammy this result, he immediately saw that it answered a question of Steenrod about the regular cycles of the p-adic solenoid (inside a solid torus, wrap another one p times around, and so on, ad infinitum). So Sammy and I stayed up all night to find out the reason for this unexpected appearance of group extensions. We found out more: it rested on a "universal coefficient theorem" which gave cohomology with any coefficient group G in terms of homology and an exact sequence involving Ext, the group of group extensions. Thus Sammy insisted on understanding this unexpected connection between algebra and topology. There was more there: the connection involved mapping topology into algebra, so we were forced to invent functors, natural transformations, and categories to describe this. All told, this led to our fifteen joint papers.

They all involved the maxim: Dig deeper and find out. For example, Hurewicz and Heinz Hopf had observed that the fundamental group of a space had effects on the higher homology and cohomology groups. Sammy, with his knowledge of his singular homology theory, had just the needed tools to understand this, which resulted in our discovery of the cohomology of groups. Sammy saw that this idea went further, so he started Gerhard Hochschild on his study of the cohomology of algebras and then went on to write, with Henri Cartan, that very influential book on homological algebra, which caught the interest of many algebraists and provided the first book presentation of the important French technique of spectral sequences.

Sammy applied his maxim in other connections. With Joe Zilber he developed the category of simplicial sets as a new type of space—using his singular simplices with face and degeneration operations. With Calvin Elgot he wrote about recursion, a topic in logic. By himself he wrote two volumes on *Automata, Languages, and Machines*. And with

Saunders Mac Lane is Max Mason Distinguished Service Professor, Emeritus, at the University of Chicago.

Eldon Dyer he prepared two volumes (not yet published) on *General and Categorical Topology*.

Algebraic topology was decisively influenced by Eilenberg's earlier 1952 work with Norman Steenrod, entitled *Foundations of Algebraic Topology*. At that time there were many different and confusing versions of homology theory, some singular, some cellular. This book used categories to show that they all could be described conceptually as presenting homology functors from the category of pairs of spaces to groups or to rings, satisfying suitable axioms such as "excision". Thanks to Sammy's insight and his enthusiasm, this text drastically changed the teaching of topology.

At Columbia University Sammy took vigorous steps to build up the department. He trained many graduate students. For example, his students and postdocs in category theory included Harry Applegate, Mike Barr, Jonathan Beck, David Buchsbaum, Peter Freyd, Alex Heller, Daniel Kan, Bill Lawvere, Fred Linton, Steve Schanuel, Myles Tierney, and others. He was an inspiring teacher.

Early in 1996 Sammy was felled by a stroke. It became hard for him to talk. In May 1997 I was able to visit him; he was lively and passed on to me a not clearly understood proposal. He was then able to spend some time in his apartment on Riverside Drive. I think his message then to me was the same maxim: Keep on pressing those mathematical ideas. This is well illustrated by his life. His ideas—singular homology, categories, simplicial sets, generic acyclicity, obstructions, automata, and the rest—will live on.

Our fifteen joint research papers have been collected in the volume *Eilenberg/Mac Lane, Collected Works*, Academic Press, Inc., New York, 1988.

Next, I comment on Eilenberg's contributions to the sources of homological algebra. The startling idea that homology theory for topological spaces could be used for algebraic objects first arose with the discovery of the cohomology groups of a group. Hurewicz had considered spaces which are aspherical (any image of a higher-dimensional sphere can be deformed into a point) and had shown that the fundamental group π_1 determines the homotopy type of the space—and hence its homology and cohomology groups. Hopf had then found explicit formulas for the homology (Betti) groups of such a space. Then Eilenberg-Mac Lane exhibited the nth cohomology group $H^n(X, A)$ of such a space with coefficients in an abelian group A as a functor of π_1 and A—the nth cohomology $H^n(\pi_1, A)$ of the group π_1 with coefficients in the π_1-module A. In particular H^1 was simply the group of "crossed homomorphisms" $f : \pi_1 \to A$ satisfying

$$f(xy) = xf(y) + f(x)$$

and taken modulo the "principal" such—those f given as $f(x) = xa - a$ for some a in A. The elements of $H^n(\pi_1, A)$ were functions $f(x_1, \ldots, x_n)$ of n elements x_i satisfying a suitable equation, modulo trivial solutions. In other words, the cohomology of π_1 was given as the cohomology of a certain chain complex, the so-called "bar resolution". In the terminology subsequently refined by Cartan-Eilenberg, $H^n(\pi_1, -)$ was the $(n-1)^{\text{st}}$ "derived" functor of $H^1(\pi_1, -)$. In other words, old functors lead to new ones.

Eilenberg very quickly saw that such cohomological methods would apply to any algebraic situation. He explained this in the 1949 paper [2]. In 1948 he wrote with Chevalley a paper on the cohomology theory of Lie algebras, and about the same time he encouraged Gerhard Hochschild, then one of Chevalley's Ph.D. students, to introduce cohomology groups for associative algebras. In each of these cases the cohomology groups in question were the derived functors of naturally occurring Hom functors. Classical questions of algebraic topology also entered by way of the Künneth formulas. These formulas originally were stated to give the Betti numbers and torsion coefficients of a product of two spaces X and Y. This really involved the tensor product of homology groups, and in the famous Eilenberg-Steenrod book it appears in the following short exact sequence:

$$\begin{aligned} 0 \to \sum_{m+q=n} & H_m(X) \otimes H_q(Y) \to H_n(X \times Y) \\ \to \sum_{m+q=n-1} & \operatorname{Tor}(H_m(X), H_q(Y)) \to 0. \end{aligned}$$

Here "exact" means that at each point the image of the incoming arrow is the kernel of the outgoing arrow. Also, $\operatorname{Tor}(A, B)$ is a functor of abelian groups, as is $\otimes$; in fact, Tor turns out to be the first derived functor of $\otimes$! The definitions of these terms do suffice for the topological task in question: elements of finite order in the groups A and B give elements in Tor. I clearly recall an occasion when I tried to explain to Professor Künneth at Erlangen University that this abstract language did indeed produce his original numerical Künneth formulas. As stated, Tor is the first derived functor of $\otimes$; it turns out for modules that there are also higher derived functors $\operatorname{Tor}_n(A, B)$ for each n. The construction of these higher torsion products and their description by generators and relations were examined by Eilenberg-Mac Lane; these products provided new examples of higher derived functors of modules. For abelian groups A and B, $\operatorname{Tor}_n(A, B) = 0$ when $n > 1$.

Now return to the functor $\operatorname{Ext}(A, B)$, the group of abelian group extensions E of B by A, so that

E appears in a short exact sequence of abelian groups:

$$0 \to B \to E \to A \to 0.$$

It turns out that the functor Ext$(A, —)$ is the first derived functor of Hom$(A, —)$ and thus that there are higher derived functors Ext$_n(A, —)$. They vanish for abelian groups A, but not generally for modules. The work of the Japanese mathematician Yoneda showed that an element of Ext$_n(A, B)$ could be represented as a long exact sequence of modules (with n intermediate terms):

$$0 \to B \to E_1 \to E_2 \to \cdots \to E_n \to A \to 0.$$

All these various examples of the construction of new functors as "derived" functors of given ones were at hand for Eilenberg. He saw how they could be used to determine a homological "dimension" for algebraic objects, and he established the connection with the Hilbert notion of a syzygy in a 1956 paper [3]. This provided the background for the influential Cartan-Eilenberg book [1] on homological algebra. This text emphasized how the derived functors for a module M could be calculated from *any* "resolution" of M by free modules, a long exact sequence ← ←

$$0 \quad M \quad X_0 \quad X_1 \quad M_2 \quad \cdots$$

with all X_j free. One simply applies the functor to the resolution with the M term dropped and then takes the homology or cohomology of the resulting complex. This effectively generalized the computation from specific "bar resolutions" used to define the cohomology of a group. The ideas of homological algebra were presented in two pioneering books by Cartan-Eilenberg [1] and Mac Lane [4]. The Cartan-Eilenberg treatise had a widespread and decisive influence in algebra. This again illustrates the genius of Eilenberg: If essentially the same idea crops up in different places, follow it out and find out where it lives.

Alex Heller

When I met Samuel Eilenberg in 1947, he was introduced as Sammy. He was always referred to as Sammy. It would be wrong to speak of him otherwise. I was then a student; I promptly became *his* student. I would like to record what drew me then to Sammy and continued over the years to do so—namely, what I perceived as his radical insistence on lucidity, order, and understanding as opposed to trophy hunting, and his idea of how that understanding was to be achieved.

Alex Heller is professor of mathematics at the Graduate School and University Center, CUNY. His e-mail address is aheller@email.gc.cuny.edu.

Perhaps I should illustrate this by a partial (in both senses) account of his mathematical career. At the end of the 1930s algebraic topology had amassed a stock of problems which its then available tools were unable to attack. Sammy was prominent among a small group of mathematicians—among them, for example, J. H. C. Whitehead, Hassler Whitney, Saunders Mac Lane, and Norman Steenrod—who dedicated themselves to building a more adequate armamentarium. Their success in doing this was attested to by the fact that by the end of the 1960s most of those problems had been solved (inordinately many of them by J. F. Adams).

Sammy's contributions appeared for the most part in a series of collaborations. With Mac Lane he developed the theory of cohomology of groups, thus providing a proper setting for the remarkable theorem of Hopf on the homology of highly connected spaces. This led them to the study of the Eilenberg-Mac Lane spaces and thus to a deeper understanding of the relations between homotopy and homology. Their most fateful invention perhaps was that of category theory, responding, no doubt, to the exigencies of algebraic topology but destined to radiate across most of mathematics.

In collaboration with Steenrod, Sammy drained the Pontine Marshes of homology theory, turning an ugly morass of variously motivated constructions into a simple and elegant system of axioms applied, for the first time, to functors. This was a radical innovation. Heretofore homology theories had been procedures for computing; henceforth they would be mathematical objects in their own right. What was especially remarkable was that in order to achieve this, Sammy and Steenrod undertook to raise the logical level of the things that might be so regarded.

The algebraic structures of the new algebraic topology were proving themselves useful in other parts of mathematics: in algebra, representation theory, algebraic geometry, and even in number theory. Together with Henri Cartan, Sammy systematized these structures under the rubric of *Homological Algebra*, once more raising the level of discourse by introducing such notions as derived functors. I am tempted to insert a parenthesis here. This latest innovation brought its authors into conflict with the "establishment" by putting in question the very notion of definition, raising a fundamental question of the relation between category theory and set theory that has yet to be put definitively to rest. Since homological algebra has proved indispensable, the honors lie, I think, with Cartan and Eilenberg. In any case, the field proliferated so rapidly that Grothendieck, only a few years later, was said to have spoken of their book as "le diplodocus", regarding it apparently as palaeontology.

The roots of homological algebra lay nevertheless in algebraic topology, and Sammy, in collaboration with John Moore, returned to these. They introduced such novelties as differential graded homological algebra and relative homological algebra to provide homes for the new techniques introduced not only by Sammy and his collaborators but also by a new generation including Serre, Grothendieck, and Adams. Notable among them are the so-called Eilenberg-Moore spectral sequences, which deal with pullbacks of fibrations and with associated fiber bundles.

Unfortunately neither Sammy nor his last collaborator, Eldon Dyer, lived to complete their ultimate project of refounding algebraic topology in the correct—which is to say, homotopical—setting. Perhaps this project was too ambitious. I learned from Eldon how much agony accompanied even such choices as that of the correct definition of a topological space. Some part of their book may yet survive, and others are already continuing their project piecemeal.

As I perceived it, then, Sammy considered that the highest value in mathematics was to be found, not in specious depth nor in the overcoming of overwhelming difficulty, but rather in providing the definitive clarity that would illuminate its underlying order. This was to be accomplished by elucidating the true structure of the objects of mathematics. Let me hasten to say that this was in no sense an ontological quest: the true structure was intrinsic to mathematics and was to be discerned only by doing more mathematics. Sammy had no patience for metaphysical argument. He was not a Platonist; equally, he was not a non-Platonist. It might be more to the point to make a different distinction: Sammy's mathematical aesthetic was classical rather than romantic.

Category theory was one of Sammy's principal tools in his search for mathematical reality. Category theory also developed into a mathematical subject with its own honorable history and practitioners, beginning with Mac Lane and including, notably, F. W. Lawvere, Sammy's most remarkable student, who saw it as a foundation for all of mathematics and justified this intuition with such innovations as categorical semantics and topos theory. Sammy did not, I think, want to be reckoned a member of this school. I believe, in fact, that he would have rejected the idea that mathematics needed a foundation. Category theory was for him only a tool—in fact, a powerful one—for expanding our understanding. It was his willingness to search for this understanding at an ever higher level that really set him apart and that made him, in my estimation, the author of a revolution in mathematics as notable as that initiated by Cantor's invention of set theory. Like Cantor, Sammy has changed the way we think about mathematics.

Peter Freyd

Thirty years ago I found myself a neighbor of Arthur Upham Pope, the master of ancient Persian art. He had retired in his nineties to an estate in the center of the city of Shiraz in southern Iran, where I lived, briefly, across the street. I found an excuse for what has to be called an audience, and I mentioned that I was a friend of Samuel Eilenberg.

"I don't know him," he said. "I know *of* him, of course. How do you know him?"

"We work in the same area of mathematics."

"You're talking about a different Eilenberg. I meant the dealer in Indian art."

"Actually, it's the same person. He's both a mathematician and a collector of Indian art."

"Don't be silly, young man. The Eilenberg I mean is not a *collector* of Indian art, he's the *dealer* in Indian art. I know him well. He established the historicity of one of the Persian kings. He certainly is not a mathematician."

End of audience.

In later years even Arthur Upham Pope would have known. In the art world, Eilenberg became universally known as "Professor". Indeed, if one walked with him in London or Zürich or even Philadelphia and one heard "Professor!", it was always Eilenberg who was being hailed, and it was always the art world hailing him.

If you heard "Sammy!", you knew it was a mathematician.

It was complicated, explaining that name. For a person who knew him first through his works, it was hard to conceive of him as "Sammy". And upon meeting him for the first time, it was even harder: He was in charge of entire fields of mathematics—indeed, he had created a number of them. Whenever he was in a room, he was in charge of the room, and it did not matter whose room it was. Sammy? The name did not fit.

But he had to have a name like Sammy. I said it was hard to explain. Here was one of the most aggressive people one might ever meet. He would challenge almost anything. If a person mentioned something about the weather, he would challenge it: once in California I heard him insist that it was not weather; it was climate. But somehow it was almost always clear: it was all right to challenge him right back. Aggressive and challenging, but not at all pompous. One cannot be pompous with a name like Sammy.

Peter Freyd is a professor of mathematics at the University of Pennsylvania. His e-mail address is `pjf@saul.cis.upenn.edu`.

Sammy kept his two worlds, mathematics and art, at something of a distance. But both worlds seemed to agree on one thing, the very one that Arthur Upham Pope had insisted upon: Sammy was the dealer.

Without question, Sammy loved playing the role of dealer. In the days when mathematicians were in demand and jobs were easy to come by, Sammy loved to tell about the math market he was going to create. The trade would be in mathematician futures: "This one's done only two lemmas and one proposition in the last year; the most recent theorem was two years ago; better sell this one at a loss." With his big cigar (expensive) and his big gold ring (in fact, a valuable Indian artifact), he could enter his dealer mode at a moment's notice. One always wondered just how many young mathematicians' careers were in his hands.

But his two worlds, mathematics and art, perceived this role of dealer quite differently. In mathematics we understood that it was a role he loved playing, but that he was only playing. His being a mathematician was what counted, and he would have been the same mathematician whether or not he played the dealer, indeed, whether or not he played —and he did—high-stakes poker. This was not so clear in his other world.

It was usually frustrating trying to explain to others how Sammy was perceived by his fellow mathematicians. Sammy had an unprintable way of saying that mathematics required both intelligence and aggression. But imagine not knowing how his mathematics—when he had finished—would totally belie that aggression. Imagine not knowing how remarkably well-behaved his mathematics always was. Imagine not knowing how his mathematics, when he had finished, always seemed preordained and how it seemed no more aggressive than, say, the sun rising at its appointed sunrise time.

Forty years ago Sammy hoped to turn the study of Indian bronzes into an equally well-behaved subject. He had already acquired a reputation for being the best detector of fakes in the business, and he believed he could axiomatize the process. He even had a provisional list of axioms, and it was truly an elegant list.

A few years later we found ourselves at a small French-style bistro in La Jolla, California. We had been out of touch: there had been an argument about mathematical ethics, but somehow we had resolved it; the dinner was something of a celebration of the resolution. I asked him about his book on bronzes.

"The axioms failed."

"What does that mean?"

"It means that I've been taken. I bought a fake."

He had suspected it only after the work had been in his bedroom for a few weeks. He had the pleasure, at least, of investigating until he found out who the master faker was and tracking him down in his studio, not to berate him, but to congratulate him.

After that, Sammy made a point of not building bridges between his two worlds. I recall just one exception. He moved from a conversation about sculpture to one about mathematics. Sculptors, he said, learn early to create from the inside out: what finally is to be seen on the surface is the result of a lot of work in conceptualizing the interior. But there are others for whom the interior is the result of a lot of work on getting the surface right. "And," Sammy asked, "isn't that the case for my mathematics?"

Style is only one part of his mathematics—as, of course, he knew—but there are, indeed, wonderful stories about Sammy, attending only to what seemed the most superficial of stylistic choices, restructuring entire subjects on the spot.

Many have witnessed this triumph of style over substance, particularly with students. But the most dramatic example had a stellar cast. D. C. Spencer gave a colloquium at Columbia in the spring of 1962, and Sammy decided it was time to demonstrate his get-rid-of-subscripts rule: "If you define it right, you won't need a subscript." Spencer, with the greatest of charm—it was for good reason that he was already affectionately known as "Uncle Don"—followed Sammy's orders and proceeded to restructure his subject while standing there at the board. One by one, the subscripts disappeared, each disappearance preceded by a Sammy-dictated redefinition. He had virtually no idea of the intended meanings of any of the symbols. He was operating entirely on the surface, looking only at the shape of the syntax.

The process went on for several minutes, until Sammy took on the one proposition on the board. "So now what does that say?"

"Sammy, I don't know. You're the one making all the definitions."

So Sammy applied his definitions, and one by one the subscripts continued to disappear, until finally the proposition itself disappeared: it became the assertion that a thing was equal—behold—to itself.

"My mother's father had the town brewery and he had one child, a daughter. He went to the head of the town yeshiva and asked for the best student," Sammy told me one day. "So my future father became a brewer instead of a rabbi."

Sammy regarded prewar Poland with some affection. He felt that he had been well nurtured by the Polish community of mathematicians, and he told me of his pleasure on being received by Stefan Banach himself, a process of being welcomed to the holy of holies, the café in which Banach spent his time during the annual Polish mathematical conferences. By the time Sammy came to the U.S. in his mid-twenties he was a well-known topologist.

When I questioned him on his attitude about prewar Poland, he answered that one must "watch the derivative": Don't judge just by how good things are, but by how fast they're becoming better.

Sammy's view of Poland since the war was more complicated. It was particularly complicated by what he viewed as its treatment of category theory as a fringe subject.

In the late 1950s Sammy began to concentrate his mathematical activities, both research and teaching, on category theory. He and Mac Lane had invented the subject, but to them it was always an applied subject, not an end in itself. Categories were defined in order to define functors, which in turn were defined in order to define natural transformations, which were defined finally in order to prove theorems that could not be proved before. In this view, category theory belonged in the mainstream of mathematics.

There was another view, the "categories-as-fringe" view. It said that categories were defined in order to *state* theorems that could not be stated before, that they were not tools but objects of nature worthy of study in their own right. Sammy believed that this counterview was a direct challenge to his role as the chief dealer for category theory. He had watched many of his inventions become standard mathematics—singular homology, obstruction theory, homological algebra—and he had no intention of leaving the future of category theory to others.

Today the language of category theory has permeated a good part of mathematics and is treated with some respect. It was not ever so. There were years before the words "category" and "functor" could be pronounced unapologetically in diverse mathematical company. One of my fonder memories comes from sitting next to Sammy in the early 1960s when Frank Adams gave one of his first lectures on how every functor on finite-dimensional vector spaces gives rise to a natural transformation on the K-functor. Frank used that construction to obtain what are now called the Adams operations, and he used those to count how many independent vector fields there could be on a sphere. It was not until then that it became permissible to say "functor" without a little snort.

In those years, Sammy was a one-man employment agency for a fresh generation of mathematicians who viewed categories not just as a language but as a potentially central mathematical subject. For the next thirty-five years he went to just about every category theory conference, and, much more important, he used his masterly expository skills to convey categorical ideas to other mathematicians. Sammy's efforts succeeded for the language of category theory, and he never abandoned his efforts for the theory itself. He was confident that the categorical view would eventually be the standard mathematical view, with or without his salesmanship. Its inevitability would be based not on Sammy's skills as a dealer but on the theorems whose proofs required category theory. That was obvious to Sammy. He wanted to make it obvious to everyone else.

Hyman Bass

Sammy visited the University of Chicago for a topology meeting while he was department chair at Columbia. I was then a graduate student, working with Irving Kaplansky on topics in homological algebra. So I was already familiar with some of Sammy's work when I first met him and we discussed mathematics. Homological algebra was insinuating itself into commutative algebra and algebraic geometry through the pioneering work of Maurice Auslander and David Buchsbaum (Sammy's student) and J.-P. Serre. Kaplansky was introducing many of my cohorts to this work.

When I graduated in 1959, in a now distant time of affluent mathematical opportunity, I contemplated a year at the Institute for Advanced Study. But Sammy, while I accompanied him to an art dealer in downtown Chicago (an errand whose significance I only later appreciated), persuaded me that it would be better first to launch my professional career as a regular faculty member, doing both research and teaching. That might now seem a difficult case to make, but it fit with my own disposition, and, in any case, Sammy had a charismatic charm and warm humor that were hard to resist.

Sammy's mentoring made me virtually his student. Columbia's was a small and intimate department, with such figures as Harish-Chandra, Serge Lang, Paul Smith, Ellis Kolchin, Dick Kadison, Edgar Lorch, Masatake Kuranishi, Lipman Bers, Joan Birman, and, briefly, Heisuke Hironaka, Steve Smale, Wilfried Schmid, and many others. The department featured some strong personalities, but Sammy, along with Lipman Bers when he arrived somewhat later, set the tone and style of the department. Research in topology, algebraic geometry, complex analysis, number theory, and the then budding category theory were quite active there. Though a faculty member, I functioned much like a student, learning about both mathematics and the intellectual culture of our discipline.

Over the years my appreciation deepened for the way Sammy worked and thought about mathematics. Though quite accomplished at computa-

Hyman Bass is professor of mathematics at Columbia University. His e-mail address is hb@math.columbia.edu.

tion and geometric reasoning, Sammy was preeminently a formalist. He fit squarely into the tradition of Hilbert, E. Artin, E. Noether, and Bourbaki; he was a champion of the axiomatic unification that so dominated the early postwar mathematics. His philosophy was that the aims of mathematics are to find and articulate with clarity and economy the underlying principles that govern mathematical phenomena. Complexity and opaqueness were, for him, signs of insufficient understanding. He sought not just theorems, but ways to make the truth transparent, natural, inevitable for the "right thinking" person. It was this "right thinking", not just facts, that Sammy tried to teach and that, in many domains, he succeeded in teaching to a whole generation of mathematicians.

Some Ph.D. Students of Samuel Eilenberg

Kuo-Tsai Chen (1950)
Alex Heller (1950)
David Buchsbaum (1954)
Ramaiyengar Sridharan (1954)
Kalathoor Varadarajan (1954)
F. William Lawvere (1963)
Harry Applegate (1965)
Estelle Goldberg (1965)
Myles Tierney (1965)
George A. Hutchinson (1967)
Jonathan M. Beck (1967)
Stephen C. Johnson (1968)
Albert Feuer (1974)
Chang-San Wu (1974)
Martin Golumbic (1975)
Alan Littleford (1979)

In some ways Sammy seemed to have a sense of the structure of mathematical thinking that almost transcended specific subject matter. I remember the uncanny sensation of this on more than one occasion when sitting next to him in department colloquia. The speaker was exposing a topic with which I knew that Sammy was not particularly familiar. Yet a half to two thirds of the way through the lecture, Sammy would accurately begin to tell me the kinds of things the speaker was going to say next.

Though his mathematical ideas may seem to have a kind of crystalline austerity, Sammy was a warm, robust, and very animated human being. For him mathematics was a social activity, whence his many collaborations. He liked to do mathematics on his feet, often prancing while he explained his thoughts. When something connected, one could read it in his impish smile and the sparkle in his eyes.

He was engaged with the world in many ways, a sophisticated and wise man who took a refined pleasure in life. His was a most satisfying and inspiring influence on my own professional life. After his stroke, it was painful to see Sammy, frail and gaunt and deprived of speech when his still active mind had so much yet to say. Yet he bravely showed the same good humor and dignity that marked his whole life. He leaves us with much to treasure, even while we miss him.

References

[1] H. Cartan and S. Eilenberg, *Homological algebra*, Princeton Univ. Press, Princeton, NJ, 1956.

[2] S. Eilenberg, *Topological methods in abstract algebra: Cohomology theory of groups*, Bull. Amer. Math. Soc. **55** (1949), 3–37.

[3] ——, *Homological dimension and syzygies*, Ann. Math. (2) **64** (1956), 328–336.

[4] S. Mac Lane, *Homology*, Springer-Verlag, Berlin, 1963.

Published Works of Samuel Eilenberg

Books

1. [with N. E. Steenrod] "Foundations of Algebraic Topology." Princeton Univ. Press, Princeton, New Jersey, 1952.
2. [with H. Cartan] "Homological Algebra." Princeton Univ. Press, Princeton, New Jersey, 1956.
3. [with C. C. Elgot] "Recursiveness." Academic Press, New York, 1970.
4. "Automata, Languages, and Machines," Vol. A. Academic Press, New York, 1974.
5. "Automata, Languages, and Machines," Vol. B. Academic Press, New York, 1976.
6. [with S. MacLane] "Eilenberg-MacLane: Collected Works" Academic Press, 1986.

Papers

1. Remarques sur les ensembles et les fonctions relativement mésurables, *C. R. Soc. Sci. Varsovie*, Chap. III, **25** (1932), 93–98.
2. Sur les transformations périodiques de la surface de sphère, *Fund. Math.* **22** (1934), 28–41.
3. Sur les transformations continues d'éspaces metriques compacts, *Fund. Math.* **22** (1934), 292–296.
4. Sur les décompositions des continus en ensembles connexes, *Fund. Math.* **22** (1934), 297–302.
5. Sur quelques propriétés des transformations localement homéomorphes, *Fund. Math.* **24** (1935), 35–42.
6. Sur le plongement des éspaces dans les continus acycliques, *Fund. Math.* **24** (1935), 65–71.
7. Deux théoremes sur l'homologie dans les éspaces compacts, *Fund. Math.* **24** (1935), 151–155.

Reprinted from Algebra, Topology, and Category Theory: A Collection of Papers in Honor of Samuel Eilenberg, eds. A. Heller and M. Tierney (Academic Press, 1976), pp. 221–225.

8. Remarque sur un théoreme de M. W. Hurewicz, *Fund. Math.* **24** (1935), 156–159.
9. Sur les transformations d'éspaces metriques en circonférence, *Fund. Math.* **24** (1935), 160–176.
10. Sur l'invariance par rapport aux petites transformations, *C. R. Acad. Sci. Paris* **200** (1935), 1003–1005.
11. [with S. Saks] Sur la dérivation des fonctions dans des ensembles denombrables, *Fund. Math.* **25** (1935), 264–266.
12. Sur quelques proprietes topologiques de la surface de sphère, *Fund. Math.* **25** (1935), 267–272.
13. Ozastosowaniach topologicznych odwzcicwan no okrag kola, *Wiadom. Mat.* **41** (1935), 1–32.
14. Transformations continues en circonférence et la topologie du plan, *Fund. Math.* **26** (1936), 61–112.
15. Sur le théoreme de décomposition de la théorie de la dimension, *Fund. Math.* **26** (1936), 146–149.
16. [with K. Borsuk] Uber stetige Abbildungen der Teilmengen euklidischer Raume auf die Kreislinie, *Fund. Math.* **26** (1936), 207–223.
17. Bemerkungen zur Pontrjagin'schen Verallgemeinerung des Alexander'schen Dualitatssatzes, *Fund. Math.* **26** (1936), 224–228.
18. Un théoreme de dualités, *Fund. Math.* **26** (1936), 280–282.
19. Sur les éspaces multicohérents I, *Fund. Math.* **27** (1936), 153–190.
20. Sur un théoreme topologique de M. L. Schnirelmann, *Rec. Math. Moscou*, **1** (1936), 557–560.
21. Uber ein Problem von H. Hopf, *Fund. Math.* **28** (1937), 58–60.
22. Sur les groupes compacts d'homéomorphies, *Fund. Math.* **28** (1937), 75–80.
23. Sur les courbes sans noeuds, *Fund. Math.* **28** (1937), 233–242.
24. Sur l'enlacement faible, *C. R. Acad. Sci. Paris* **204** (1937), 1226–1227.
25. Sur les éspaces multicohérents II, *Fund. Math.* **29** (1937), 101–122.
26. Sur les ensembles plans localement connexes, *Fund. Math.* **29** (1937), 159–160.
27. Un théoreme sur l'homotopie, *Ann. of Math.* **38** (1937), 656–661.
28. Sur la multicohérence des surfaces closes, *C. R. Soc. Sci. Varsovie*, Chap. III, **30** (1937), 109–111.
29. Sur les transformations à petites tranches, *Fund. Math.* **30** (1938), 92–95.
30. [with E. Otto] Quelques propriétés characteristiques de la dimension, *Fund. Math.* **31** (1938), 149–153.
31. Sur le prolongement des transformations en surfaces sphériques, *Fund. Math.* **31** (1938), 179–200.
32. On φ-measures, *Ann. Soc. Pol. Math.* **17** (1938), 251–252.
33. On continua of finite length 1, *Ann. Soc. Pol. Math.* **17** (1938), 253–254.
34. Cohomologies et transformations continues, *C. R. Acad. Sci. Paris* **208** (1939), 68–69.
35. Généralisation du théoreme de M. H. Hopf sur les classes des transformations en surfaces sphériques, *Compositio Math.* **6** (1939), 428–433.
36. On the relation between the fundamental group of a space and the higher homotopy groups, *Fund. Math.* **32** (1939), 167–175.
37. [with C. Kuratowski] Théoremes d'addition concernant le groupe des transformations en circonference, *Fund. Math.* **39** (1939), 193–200.
38. Cohomology and continuous mappings, *Ann. of Math.* **41** (1940), 231–251.
39. On continuous mappings of manifolds into spheres, *Ann. of Math.* **41** (1940), 662–673.
40. On a theorem of P. A. Smith concerning fixed points for periodic transformations, *Duke Math. J.* **6** (1940), 428–437.

41. On homotopy groups, *Proc. Nat. Acad. Sci.* **26** (1940), 563–565.
42. Ordered topological spaces, *Amer. J. Math.* **63** (1941), 39–45.
43. An invariance theorem for subsets of S^n, *Bull. Amer. Math. Soc.* **47** (1941), 73–75.
44. Continuous mappings of infinite polyhedra, *Ann. of Math.* **42** (1941), 459–468.
45. On spherical cycles, *Bull. Amer. Math. Soc.* **47** (1941), 432–434.
46. Extensions and classification of continuous mappings, "Lectures in Topology." Univ. of Michigan Press, Ann Arbor, (1941), 57–99.
47. [with S. Mac Lane] Infinite cycles and homologies, *Proc. Nat. Acad. Sci.* **27** (1941), 535–539.
48. [with E. W. Miller] Zero-dimensional families of sets, *Bull. Amer. Math. Soc.* **47** (1941), 921–923.
49. Banach space methods in topology, *Ann. of Math.* **43** (1942), 568–579.
50. [with R. L. Wilder] Uniform local connectedness and contractibility, *Amer. J. Math.* **64** (1942), 613–622.
51. [with S. Mac Lane] Group extensions and homology, *Ann. of Math.* **43** (1942), 757–831.
51a. [with S. Mac Lane] Appendix to S. Lefschetz's "Algebraic Topology," pp. 344–349. *Amer. Math. Soc.*, Providence, Rhode Island, 1942.
52. [with S. Mac Lane] Natural isomorphisms in group theory, *Proc. Nat. Acad. Sci.* **28** (1942), 537–543.
53. [with O. G. Harrold] Continua of finite linear measure, *Amer. J. Math.* **65** (1943), 137–146.
54. [with S. Mac Lane] Relations between homology and homotopy groups, *Proc. Nat. Acad. Sci.* **29** (1943), 155–158.
55. [with I. Niven] The "fundamental theorem of algebra" for quaternions, *Bull. Amer. Math. Soc.* **50** (1944), 246–248.
56. Continua of finite linear measure II, *Amer. J. Math.* **66** (1944), 425–427.
57. Singular homology theory, *Ann. of Math.* **45** (1944), 407–447.
58. [with N. E. Steenrod] Axiomatic approach to homology theory, *Proc. Nat. Acad. Sci.* **31** (1945), 117–120.
59. [with S. Mac Lane] General theory of natural equivalences, *Trans. Amer. Math. Soc.* **58** (1945), 231–294.
60. [with S. Mac Lane] Relations between homology and homotopy groups of spaces, *Ann. of Math.* **46** (1945), 480–509.
61. [with D. Montgomery] Fixed point theorem for multi-valued transformations, *Amer. J. Math.* **68** (1946), 480–509.
62. [with S. Mac Lane] Determination of the second homology and cohomology groups of a space by means of homotopy invariants, *Proc. Nat. Acad. Sci.* **32** (1946), 277–280.
63. [with S. Mac Lane] Cohomology theory in abstract groups I, *Ann. of Math.* **48** (1947), 51–78.
64. [with S. Mac Lane] Cohomology theory in abstract groups II, Group extensions with a non-abelian kernel, *Ann. of Math.* **48** (1947), 326–341.
65. Homology of spaces with operators I, *Trans. Amer. Math. Soc.* **61** (1947), 378–417.
66. [with S. Mac Lane] Algebraic cohomology groups and loops, *Duke Math. J.* **14** (1947), 435–463.
67. Singular homology in differentiable manifolds, *Ann. of Math.* **48** (1947), 670–681.
68. On a linkage theorem by L. Cesari, *Bull. Amer. Math. Soc.* **53** (1947), 1192–1195.
69. [with C. Chevalley] Cohomology theory of Lie groups and Lie algebras, *Trans. Amer. Math. Soc.* **63** (1948), 85–124.
70. Relations between cohomology groups in a complex, *Comment. Math. Helv.* **21** (1948), 302–320.

71. [with S. Mac Lane] Cohomology and Galois theory I, Normality of algebras and Teichmuller's cocycle, *Trans. Amer. Math. Soc.* **64** (1948), 1-20.
72. Extensions of general algebras, *Ann. Soc. Polonaise Math.* **21** (1948), 125-134.
73. [with S. Mac Lane] Homology of spaces with operators II, *Trans. Amer. Math. Soc.* **65** (1949), 49-99.
74. Topological methods in abstract algebra, Cohomology theory of groups, *Bull. Amer. Math. Soc.* **55** (1949), 3-37.
75. On the problems of topology, *Ann. of Math.* **50** (1949), 247-260.
76. [with J. A. Zilber] Semi-simplicial complexes and singular homology, *Ann. of Math.* **51** (1950), 499-513.
77. [with S. Mac Lane] Relations between homology and homotopy groups of spaces II, *Ann. of Math.* **51** (1950), 514-533.
78. [with S. Mac Lane] Cohomology theory of abelian groups and homotopy theory I, *Proc. Nat. Acad. Sci.* **36** (1950), 443-447.
79. [with S. Mac Lane] Cohomology theory of abelian groups and homotopy theory II, *Proc. Nat. Acad. Sci.* **36** (1950), 657-663.
80. [with S. Mac Lane] Cohomology theory of abelian groups and homotopy theory III, *Proc. Nat. Acad. Sci.* **37** (1951), 307-310.
81. [with S. Mac Lane] Homology theories for multiplicative systems, *Trans. Amer. Math. Soc.* **71** (1951), 294-330.
82. [with S. Mac Lane] Cohomology theory of abelian groups and homotopy theory IV, *Proc. Nat. Acad. Sci.* **38** (1952), 325-329.
82a. Homotopy groups and algebraic homology theories, *Proc. Int. Congress 1950*, 349-353.
83. [with S. Mac Lane] Acyclic models, *Amer. J. Math.* **75** (1953), 189-199.
84. [with J. A. Zilber] On products of complexes, *Amer. J. Math.* **75** (1953), 200-204.
85. [with S. Mac Lane] On the groups $H(\Pi, n)$, *Ann. of Math.* **58** (1953), 55-106.
86. [with S. Mac Lane] On the groups $H(\Pi, n)$, II, *Ann. of Math.* **60** (1954), 49-139.
87. [with S. Mac Lane] On the groups $H(\Pi, n)$, III, *Ann. of Math.* **60** (1954), 513-557.
88. [with S. Mac Lane] On the homology theory of abelian groups, *Canad. J. Math.* **7** (1955), 45-53.
89. Algebras of cohomologically finite dimension, *Comment. Math. Helv.* **28** (1954), 310-319.
90. [with H. Ikeda and T. Nakayama] On the dimension of modules and algebras I, *Nagoya Math. J.* **8** (1955), 49-57.
91. [with T. Nakayama] On the dimension of modules and algebras II, Frobenius algebras and quasi-Frobenius rings, *Nagoya Math. J.* **9** (1955), 1-16.
92. [with H. Nagao and T. Nakayama] On the dimension of modules and algebras IV, Dimension of residue rings of hereditary rings, *Nagoya Math. J.* **10** (1956), 87-95.
93. Homological dimension and syzygies, *Ann. of Math.* **64** (1956), 328-336, Errata **65** (1957), 593.
94. [with T. Nakayama] On the dimension of modules and algebras V. Dimension of residue rings, *Nagoya Math. J.* **11** (1957), 9-12.
95. [with T. Ganea] On the Lusternik-Schnirelmann category of abstract groups, *Ann. of Math.* **65** (1957), 517-518.
96. [with A. Rosenberg and D. Zelinsky] On the dimension of modules and algebras VIII, Dimension of tensor products, *Nagoya Math. J.* **12** (1957), 71-93.
97. [with H. Cartan] Foundations of fibre bundles, *Symp. Inter. Topologia Algebraica, Mexico 1958*, 16-23.
97a. Foundations of fiber bundles, *Chicago Notes.*
98. Abstract description of some basic functors, *J. Indiana Math. Soc.* **24** (1960), 231-234.
99. [with J. C. Moore] Limits and spectral sequences, *Topology* **1** (1962), 1-23.

100. [with K. Kuratowski] A remark on duality, *Fund. Math.* **50** (1962), 515–517.
101. [with J. C. Moore] Foundations of relative homological algebra, *Memoirs Amer. Math. Soc.* **55** (1965), 1–39.
102. [with J. C. Moore] Adjoint functors and triples, *Illinois J. Math.* **9** (1965), 381–398.
103. [with J. C. Moore] Homology and fibrations I, *Comment. Math. Helv.* **40** (1956), 199–236.
104. [with J. C. Moore] Homological algebra and fibrations, *Colloq. Topologie, Bruxelles 1966*, 81–90.
105. [with G. M. Kelly] A generalization of the functorial calculus, *J. Algebra* **3** (1966), 366–375.
106. [with G. M. Kelly] Closed categories, *Proc. Conf. Categorical Algebra, La Jolla 1966*, 421–562.
107. [with J. B. Wright] Automata in general algebra, *Information and Control* **II** (1967), 452–470.
108. [with C. C. Elgot] Iteration and recursion, *Proc. Nat. Acad. Sci.* **61** (1968), 378–379.
109. [with M. P. Schutzenberger] Rational sets in commutative monoids, *J. Algebra* **13** (1969), 173–191.
110. [with C. C. Elgot and J. C. Shepherdson] Sets recognized by *n*-tape automata, *J. Algebra* **13** (1969), 447–464.
111. Algebraic aspects of automata theory, *Actes Congrès Inter. Math.* **3** (1970) 265–267.
112. [with E. Dyer] An adjunction theorem for locally equiconnected spaces, *Pacific J. Math.* **41** (1972), 669–685.
113. Classes of Semigroups and Classes of Sets, *Fifth Annual ACM Symposium on Theory of Computing*, (Austin, Texas, 1973), pp. 266–267. Assoc. Computing Mach., NY, 1973.
114. Decomposition theorems, *Mathematical Systems Theory* (Proc. International Sympos., International Centre Mech. Sci., Udine, 1975), pp. 1–13. Lecture Notes in Econ. and Math. Systems, 131. Springer, Berlin, 1976.
115. [with M.-P. Schützenberger] On pseudovarieties. *Advances in Math.* **19** (1976), 413–418.
116. [with A. Heller] Limit properties of stochastic matrices. *Contributions to Algebra (collection of papers dedicated to Ellis Kolchin)*, pp. 135–149. Academic Press, NY, 1977.
117. [with E. Dyer] Globalizing Fibrations by Schedules, *Fund. Math.* **130** (1988), 125–136.
118. Karol Borsuk — personal reminiscences. *Topol. Methods Nonlinear Anal.* **1** (1993), 1–2.
119. Witold Hurewicz — personal reminiscences. *Collected Works of Witold Hurewicz*, xlv–xlvi, Amer. Math. Soc., 1995.

NATURAL ISOMORPHISMS IN GROUP THEORY

BY SAMUEL EILENBERG AND SAUNDERS MACLANE

DEPARTMENTS OF MATHEMATICS, UNIVERSITY OF MICHIGAN AND HARVARD UNIVERSITY

Communicated October 26, 1942

1. *Introduction.*—Frequently in modern mathematics there occur phenomena of "naturality": a "natural" isomorphism between two groups or between two complexes, a "natural" homeomorphism of two spaces and the like. We here propose a precise definition of the "naturality" of such correspondences, as a basis for an appropriate general theory. In this preliminary report we restrict ourselves to the natural isomorphisms of group theory; with this limitation we can present the basic concepts of our theory without developing the axiomatic approach necessary for a general treatment applicable to various branches of mathematics.

Reprinted from Proc. Nat. Acad. Sci., Vol. 28 (1942) pp. 537–543.

Properties of character groups (see the definitions in § 5 below) may serve to illustrate the ideas involved. Thus, it is often asserted that the character group of a finite group G is isomorphic to the group itself, but not in a "natural" way. Specifically, if G is cyclic of prime order p, there is for each generator of G an isomorphism of G to its character group, so that the proof furnishes $p - 1$ such isomorphisms, no one of which is in any way distinguished from its fellows. However, the proof that the character group of the character group of G is isomorphic to G itself is considered "natural," because it furnishes for each G a unique isomorphism, not dependent on any choice of generators.

To give these statements a clear mathematical meaning, we shall regard the character group $Ch(G)$ of G as a function of a variable group G, *together* with a prescription which assigns to any homomorphism γ of G into a second group G',

$$\gamma : G \to G',$$

the induced homomorphism (see (5) below)

$$Ch(\gamma) : Ch(G') \to Ch(G).$$

The functions $Ch(G)$ and $Ch(\gamma)$ jointly form what we shall call a "functor"; in this case, a "contravariant" one, because the mapping $Ch(\gamma)$ works in a direction opposite to that of γ. A natural isomorphism between two functions of groups will be an isomorphism which commutes properly with the induced mappings of the functors.

With our description of a natural isomorphism, practically all the general isomorphisms obtained in group theory and its applications (homology theory, Galois theory, etc.) can be shown to be "natural." This results in added clarity in such situations. Furthermore, there are definite proofs where the naturality of an isomorphism is needed, especially when a passage to the limit is involved. In fact, our condition ($E2$) below appears in the definition of the isomorphism of two direct or two inverse systems of groups.[1]

2. *Functors.*—The definition of a functor will be given for the typical case of a functor T which depends on two groups as arguments, and is *covariant* in the first argument and *contravariant* in the second. Such a functor is determined by two functions. The *group* function determines for each pair of topological groups G and H (contained in a given legitimate set of groups) another group $T(G, H)$. The *mapping function* determines for each pair of homomorphisms[2] $\gamma : G_1 \to G_2$ and $\eta : H_1 \to H_2$ a homomorphism $T(\gamma, \eta)$, such that

$$T(\gamma, \eta) : T(G_1, H_2) \to T(G_2, H_1). \qquad (1)$$

We require that $T(\gamma, \eta)$ be the identity isomorphism whenever γ and η are identities, and that, whenever the products $\gamma_2\gamma_1$ and $\eta_2\eta_1$ are defined,

$$T(\gamma_2\gamma_1, \eta_2\eta_1) = T(\gamma_2, \eta_1)T(\gamma_1, \eta_2). \tag{2}$$

Some functors will be defined only for special types of groups (e.g., for abelian groups) or for special types of homomorphisms (e.g., for homomorphisms "onto").

If γ and η are both isomorphisms,[3] it follows from these conditions that $T(\gamma, \eta)$ is also an isomorphism. Consequently, if the groups G_1 and G_2 and the groups H_1 and H_2 are isomorphic, the functor T gives rise to isomorphic groups $T(G_1, H_1)$ and $T(G_2, H_2)$.

3. *Examples.*—The direct product $G \times H$ of two groups may be regarded as the group function of a functor. The corresponding mapping function specifies, for each pair of homomorphisms $\gamma: G_1 \rightarrow G_2$ and $\eta: H_1 \rightarrow H_2$, an induced homomorphism $\gamma \times \eta$, defined for every element (g_1, h_1) in $G_1 \times H_1$ as

$$[\gamma \times \eta](g_1, h_1) = (\gamma g_1, \eta h_1).$$

Then

$$\gamma \times \eta : G_1 \times H_1 \rightarrow G_2 \times H_2, \tag{3}$$

and, whenever $\gamma_2\gamma_1$ and $\eta_2\eta_1$ are defined, one has

$$(\gamma_2\gamma_1) \times (\eta_2\eta_1) = (\gamma_2 \times \eta_2)(\gamma_1 \times \eta_1). \tag{4}$$

Except for the absence of contravariance, these conditions are parallel to (1) and (2), hence $G \times H, \gamma \times \eta$ define a functor, covariant in both G and H.

Whitney's tensor product[4] $G \circ H$ of two discrete groups[5] G and H is the group function of a functor. The elements of this group are all finite sums $\Sigma g_i \circ h_i$ of formal products $g_i \circ h_i$; the group operation is the obvious addition, and the relations are $g \circ (h + h') = g \circ h + g \circ h'$ $(g + g') \circ h = g \circ h + g' \circ h$. Given two homomorphisms $\gamma: G_1 \rightarrow G_2$ and $\eta: H_1 \rightarrow H_2$, there is an induced homomorphism $\gamma \circ \eta$ of $G_1 \circ H_1$ into $G_2 \circ H_2$, defined for any generator $g_1 \circ h_1$ of $G_1 \circ H_1$ as

$$[\gamma \circ \eta](g_1 \circ h_1) = (\gamma g_1) \circ (\eta h_1) \in G_2 \circ H_2.$$

Formulae (3) and (4), with the cross replaced by the circle, again hold, so that $G \circ H, \gamma \circ \eta$ determine a functor of discrete groups, covariant in both arguments.

In a similar fashion, the free product of two groups leads to a functor.

An important functor is given by the group of all homomorphisms ϕ of a fixed locally compact topological abelian group G into another topological abelian group H. The sum of two such homomorphisms ϕ_1 and ϕ_2 is defined for each $g \in G$ by setting $(\phi_1 + \phi_2)(g) = \phi_1(g) + \phi_2(g)$. Under this operation, all $\phi: G \rightarrow H$ constitute a group $Hom(G, H)$; it carries an appropriate topology, the description of which we omit. For given $\gamma: G_1 \rightarrow G_2$ and $\eta: H_1 \rightarrow H_2$ and for each $\phi \in Hom(G_2, H_1)$ we have

$$G_1 \xrightarrow{\gamma} G_2 \xrightarrow{\phi} H_1 \xrightarrow{\eta} H_2.$$

Consequently we define $Hom(\gamma, \eta)(\phi) = \eta\phi\gamma$, and verify that

$$Hom(\gamma, \eta): Hom(G_2, H_1) \rightarrow Hom(G_1, H_2),$$

$$Hom(\gamma_2\gamma_1, \eta_2\eta_1) = Hom(\gamma_1, \eta_2)\ Hom(\gamma_2, \eta_1).$$

Clearly when γ and η are identity mappings of G and H the induced mapping $Hom(\gamma, \eta)$ is the identity mapping of $Hom(G, H)$ on itself. Hence the functions $Hom(G, H)$ and $Hom(\gamma, \eta)$ determine for abelian groups a functor Hom, covariant in H and contravariant in G.

The special case when H is the group P of reals modulo 1 furnishes the character group,

$$Ch(G) = Hom(G, P), \quad Ch(\gamma) = Hom(\gamma, e)$$

where e is the identity mapping of P on itself. Therefore the character group is a contravariant functor, defined for abelian groups. Explicitly, if we express the result $\chi(g)$ of applying the character χ to the element $g \in G$ as the value (a real number modulo 1) of the bilinear form (g, χ), the definition of $Ch(\gamma)$ can be written as

$$(g, Ch(\gamma)\chi') = (\gamma g, \chi'), \quad g \in G, \quad \chi' \in Ch(G'). \tag{5}$$

4. *Equivalence of Functors.*—Let T and S be two functors which are, say, both covariant in the variable G and contravariant in H. Suppose that for each pair of groups G and H we are given a homophorism

$$\tau(G, H): T(G, H) \rightarrow S(G, H).$$

We say that τ establishes a natural *equivalence* of the functor T to the functor S and that T is naturally *equivalent* to S (in symbols, τ: $T \longleftrightarrow S$) whenever

($E1$) Each $\tau(G, H)$ is a bicontinuous isomorphism of $T(G, H)$ onto $S(G, H)$;

($E2$) For each $\gamma: G_1 \rightarrow G_2$ and $\eta: H_1 \rightarrow H_2$,
$\tau(G_2, H_1)T(\gamma, \eta) = S(\gamma, \eta)\tau(G_1, H_2)$.

The first requirement insures the term-by-term isomorphism of the two group functions $T(G, H)$ and $S(G, H)$, while the second requirement is precisely the "naturality" condition. It can be shown that the condition ($E2$) is implied by two special cases; the case when η is an identity, and the case when γ is an identity.

This relation of natural equivalence between functors is reflexive, symmetric and transitive. In many cases we dispense with condition ($E1$), and obtain a more general concept of a "transformation" of a functor T into a functor S.

5. *Examples of Natural Equivalence.*—The well-known isomorphism

$$G \cong Ch(Ch(G)) \tag{6}$$

for locally compact abelian groups, can be regarded as an equivalence of functors, and is in this sense *natural.* The right-hand side of (6) suggests the covariant functor, Ch^2, defined by iteration of the functor Ch, as

$$Ch^2(G) = Ch(Ch(G)), \quad Ch^2(\gamma) = Ch(Ch(\gamma)).$$

The left-hand side of (6) suggests the identity functor, I,

$$I(G) = G, \quad I(\gamma) = \gamma.$$

The bilinear form $(g, \chi) = \chi(g)$ determines to each character $\chi \in Ch(G)$ and each $g \in G$ a real number modulo 1; similarly the form $(\chi, h) = h(\chi)$ is defined for each $h \in Ch^2(G)$. The form (g, χ), regarded as a function of χ for fixed g, is a character h in $Ch^2(G)$ which we call $[\tau(G)]g$. Explicitly, this definition of τ reads

$$(\chi, \tau(G)g) = (g, \chi), \quad g \in G, \quad \chi \in Ch(G).$$

The validity of condition ($E1$) for $\tau(G)$ is the basic theorem of character theory. The condition ($E2$) asserts that in the diagram

$$\begin{array}{ccc} G & \xrightarrow{\tau(G)} & Ch^2(G) \\ \downarrow \gamma & & \downarrow Ch^2(\gamma) \\ G' & \xrightarrow{\tau(G')} & Ch^2(G') \end{array}$$

the two paths leading from G to $Ch^2(G')$ have the same effect, or that, for each $g \in G$, both elements $\tau(G')\gamma g$ and $Ch^2(\gamma)\tau(G)g$ are identical as elements of $Ch^2(G')$. This means that, for each $\chi \in Ch(G')$, one should have

$$(\chi', \tau(G')\gamma g) = (\chi', Ch^2(\gamma)\tau(G)g).$$

By the definition of τ, the expression on the left is simply $(\gamma g, \chi')$. By successive application to the expression on the right of the definitions of Ch, τ and Ch, we obtain

$$(\chi', Ch^2(\gamma)\tau(G)g) = (Ch(\gamma)\chi', \tau(G)g) = (g, Ch(\gamma)\chi') = (\gamma g, \chi').$$

The identity of these results shows that we do have a natural equivalence $\tau(G):G \longleftrightarrow Ch^2(G)$.

When G is finite, the isomorphism $G \rightarrow Ch\,G$ cannot be "natural" according to our definitions, for the simple reason that the functor I on the left is covariant, while the functor Ch on the right is contravariant.

As other examples of equivalences between functors, we may cite the usual isomorphisms which give the associative and commutative laws for the direct product, the tensor product and the free product. Various distributive laws, such as

$$(G_1 \times G_2) \circ H \cong (G_1 \circ H) \times (G_2 \circ H),$$

$$Hom(G_1 \times G_2, H) \cong Hom(G_1, H) \times Hom(G_2, H),$$

when established with the obvious isomorphisms, are in fact equivalences between functors.

A less obvious relation between the tensor product and the functor "*Hom*" is[6]

$$Hom(G, Hom(H, K)) \cong Hom(G \circ H, K), \tag{7}$$

where G and H are discrete abelian groups, K a topological abelian group. This isomorphism is obtained by a correspondence $\tau(G, H, K)$ which specifies for each element $\phi \in Hom(G, Hom(H, K))$ a corresponding homomorphism in $Hom(G \circ H, K)$, defined for any generator $g \circ h$ of $G \circ H$ as

$$[\tau(G, H, K)]\,(\phi)(g \circ h) = [\phi(g)](h) \text{ in } K.$$

One may show that τ does give an isomorphism, bicontinuous in the appropriate topologies. Both sides of (7) may be treated as the group functions of functors which are obtained by composition from "*Hom*" and "$\circ$." The corresponding mapping functions, for given homomorphisms

$$\gamma:G_1 \rightarrow G_2, \quad \eta:H_1 \rightarrow H_2, \quad \kappa:K_1 \rightarrow K_2,$$

are defined by a parallel composition as

$$Hom(\gamma, Hom(\eta, \kappa)), \quad Hom(\gamma \circ \eta, \kappa).$$

Both functors are contravariant in G and H, covariant in K.

The naturality condition for the isomorphism τ reads

$$\tau(G_1, H_1, K_2)\; Hom(\gamma, Hom(\eta, \kappa)) = Hom(\gamma \circ \eta, \kappa)\tau(G_2, H_2, K_1).$$

Both sides, when applied to an element $\phi \in Hom(G_2, Hom(H_2, K_1)$ yield a homomorphism in $Hom(G_1 \circ H_1, K_2)$. If each of these homomorphisms is applied to a typical generator $g_1 \circ h_1$ of the tensor product $G_1 \circ H_1$, straightforward application of the relevant definitions shows that the same element of K_2 is obtained in both cases; namely, $\kappa\{[\phi(\gamma(g_1))](\eta(h_1))\}$.

One may also see directly that this expression represents the only way of constructing an element of K_2 from the elements g_1 and h_1 and the mappings κ, ϕ, γ and η.

The natural isomorphism (7) has some interesting consequences. If K is taken to be the group P of real numbers modulo 1, $Hom(H, K)$ becomes the character group $Ch(H)$, and the formula may be written as

$$Hom(G, Ch\, H) \simeq Ch(G \circ H).$$

Applying the functor Ch to both sides and using the natural equivalence of Ch^2 and I, we obtain the equivalence

$$G \circ H \simeq Ch\, Hom(G, Ch\, H).$$

Since this is "natural," this could be used as a definition of the tensor product $G \circ H$.

6. *Generalizations.*—With the appropriate definition of a normal subfunctor S of a functor T one can construct a quotient functor T/S, whose group function has as its values quotient groups (i.e., factor groups). With this operation, all the standard constructions on groups may be represented as group functions of suitable functors.

An inspection of the concept of a functor and of a natural equivalence shows that they may be applied not only to groups with their homomorphisms, but also to topological spaces with their continuous mappings, to simplicial complexes with their simplicial transformations, and to Banach spaces with their linear transformations. These and similar applications can all be embodied in a suitable axiomatic theory. The resulting much wider concept of naturality, as an equivalence between functors, will be studied in a subsequent paper.

[1] Pontrjagin, L., "Ueber den algebraischen Inhalt der topologische Dualitätsätze," *Mathematische Ann.*, **105**, 165–205 (1931). Lefschetz, S., "Algebraic Topology," *Am. Math. Soc. Colloquium Pub.*, **27**, 55 (1942).

[2] By a homomorphism we mean a *definite* pair of groups G_1 and G_2 and a (continuous) homomorphic mapping γ_1 of the first onto a subgroup of the second. The product $\gamma_2\gamma_1$ is defined for those pairs $\gamma_1: G_1 \longrightarrow G_2$, $\gamma_2: G_2' \longrightarrow G_3$ with $G_2 = G_2'$.

[3] By an isomorphism we mean a homomorphism of G_1 *onto* G_2 which is one-one and bicontinuous.

[4] Whitney, H., "Tensor Products of Abelian Groups," *Duke Math. Jour.*, **4**, 495–528 (1938).

[5] Here and subsequently the group operation in G and in H is written as addition, whether or not the groups are abelian.

[6] This isomorphism was established by the authors; cf. *Ann. Math.*, **44** (1943).

ANNALS OF MATHEMATICS
Vol. 45, No. 3, July, 1944

SINGULAR HOMOLOGY THEORY

BY SAMUEL EILENBERG

(Received October 25, 1943)

CONTENTS

The best treatment of the singular homology theory so far has been given by Lefschetz.[6] He defines a singular simplex in a space X as a pair (s, T) where s is an oriented simplex and $T: s \rightarrow X$ is a continuous mapping. If $B: s \rightarrow s'$ is a barycentric mapping of s onto another oriented simplex of the same dimension as s, then

$$(*) \qquad (s, T) \equiv \pm (s', TB^{-1})$$

where the sign is $+$ or $-$ according as B preserves or reverses the orientation. Following a suitable definition of boundary and incidence numbers, Lefschetz arrives at what he calls the "total singular complex" $\mathfrak{S}(X)$ of the space X. In this closure finite complex homologies, cohomologies and products can be constructed.

The main difficulty with using the complex $\mathfrak{S}(X)$ is that it is not a bona fide abstract complex. Unfortunately relation (*) causes elements of order 2 to appear in the group of chains,[7] while in an abstract complex the group of chains ought to be free. There is the possibility of leaving out the elements of order 2 as degenerate, but this would make the use of the complex $\mathfrak{S}(X)$ cumbersome.

The main purpose of this paper is, by adjunction of a new idea to Lefschetz's method, to give a precise and systematic treatment of the singular homology theory. Instead of using oriented simplices as in Lefschetz's approach we shall use simplices with *ordered* vertices. In addition, the barycentric mapping $B: s \rightarrow s'$, used in defining equivalence, is required to preserve the order of the vertices. This modification leads to a much larger total singular complex $S(X)$ which, however, in contrast to $\mathfrak{S}(X)$, is a closure finite complex proper, so that the whole closure finite theory can be applied without restrictions. This program, including the product theory and the invariance theorem, is carried through in Chapters III–V.

In Chapter VI we show some connections with the homotopy groups. Two

---

[6] S. Lefschetz, Bull. Amer. Math. Soc., 39 (1933), p. 124–129; also *Algebraic Topology*, p. 311. See also P. Alexandroff and H. Hopf, *Topologie I*, Berlin, 1935, Ch. VIII, 5.

[7] See S. Lefschetz, Bull. Amer. Math. Soc., 39 (1933), p. 128.

theorems previously published by the author[8] without proofs are treated in detail.

The idea of using simplexes with a definite order of vertices turned out also to be helpful in clearing up some obscure points in the product theory in simplicial complexes. Given a simplicial polyhedron P, we choose a definite orientation for each simplex of P and arrive at an abstract complex $k(P)$. In order to introduce the products, one usually has to pick an ordering for all the vertices of P. At first the products depend upon the chosen order and only when we pass to the products of homology and cohomology classes does the order become immaterial. This results in a certain lack of naturality and uniqueness. In Chapter II we define a much larger abstract complex $K(P)$ in which each simplex appears with all the possible orderings of its vertices; we then show that the homology theories of $k(P)$ and $K(P)$ are essentially the same, while the product theory in $K(P)$ has very desirable uniqueness features (Ch. V).

Chapter I was meant as a review of the theory of abstract complexes. The technique of chain transformations and chain hemotopies is developed. Some of the concepts, like that of *chain equivalence*, are new.

CHAPTER I

ABSTRACT COMPLEXES[9]

1. Closure finite complexes

Let K be a collection of abstract elements σ^q called *cells*. With each cell there is associated an integer q called the dimension of σ^q. To any two cells σ^{q+1}, σ^q there corresponds an integer $[\sigma^{q+1}:\sigma^q]$ called the *incidence number*. K will be called a *closure finite abstract complex* provided the incidence numbers satisfy the following conditions:

(1.1) Given σ^{q+1}, $[\sigma^{q+1}:\sigma^q] \neq 0$ for only a finite number of q-cells σ^q;

(1.2) Given σ^{q+1} and σ^{q-1}, $\sum_{\sigma^q} [\sigma^{q+1}:\sigma^q][\sigma^q:\sigma^{q-1}] = 0.$

The q-cells σ^q are taken as free generators of an abelian group $C^q(K)$; the elements c^q of $C^q(K)$ will be called (finite) integral q-dimensional chains of K. The *boundary operator* ∂ is a homomorphism

(1.3) $$\partial : C^q(K) \to C^{q-1}(K)$$

defined for each generator σ^q as

(1.4) $$\partial\sigma^q = \sum_{\sigma^{q-1}} [\sigma^q:\sigma^{q-1}]\sigma^{q-1}$$

Condition (1.1) insures that the summation is finite and therefore that $\partial\sigma^q$ is a well defined $(q-1)$-chain. We verify that condition (1.2) is equivalent with

(1.5) $$\partial\partial = 0$$

[8] S. Eilenberg, Proc. Nat. Acad. Sc., U. S. A., 26 (1940), p. 563–565.

[9] General reference: S. Lefschetz, *Algebraic Topology*, Ch. III and IV.

Very often in defining a complex we define the boundary operation (1.3) and prove (1.5) before introducing the incidence numbers. The incidence numbers can then be defined by (1.4) and conditions (1.1) and (1.2) can be proved.

2. Homology groups

Let G be a discrete abelian group. The q-dimensional chains of K over G are the formal finite sums

$$c^q = \sum g_i \sigma_i^q, \qquad g_i \in G$$

They form an abelian group $C^q(K, G)$.[10] The boundary operation is a homomorphism

$$\partial : C^q(K, G) \to C^{q-1}(K, G)$$

defined as

$$\partial c^q = \sum g_i \partial \sigma_i^q .$$

Again we have $\partial\partial = 0$. The chains c^q with $\partial c^q = 0$ are called *cycles*; they form a subgroup $Z^q(K, G)$ of $C^q(K, G)$. Chains c^q such that $c^q = \partial c^{q+1}$ for some c^{q+1} are called *boundaries*; they form a subgroup $B^q(K, G)$ of $C^q(K, G)$. It follows from $\partial\partial = 0$ that every boundary is a cycle, so

$$B^q(K, G) \subset Z^q(K, G).$$

The discrete factor group

$$H^q(K, G) = Z^q(K, G)/B^q(K, G)$$

is the q-dimensional *homology group* of K over G. If G is the additive group I of all integers we will write $C^q(K)$, $Z^q(K)$, $B^q(K)$ and $H^q(K)$, omitting the symbol for the group G.

The elements of $H^q(K, G)$ are called *homology classes*. Each cycle c^q determines uniquely a homology class. Two cycles c_1^q, c_2^q in the same homology class are called *homologous* (notation: $c_1^q \sim c_2^q$). Clearly $c_1^q \sim c_2^q$ if and only if $c_1^q - c_2^q$ is a boundary.

3. Cohomology groups

Let G be a topological group.[11] The q-dimensional *cochains* of K over G are functions f^q which with each q-cell σ^q in K associate an element $f^q(\sigma^q)$ of G. The q-cochains form an abelian topological group $C_q(K, G)$.[12] Since the q-cells σ^q are

---

[10] More precisely $C^q(K, G)$ is the group generated by symbols $g\sigma^q$ with relations $(g_1 + g_2)\sigma^q = g_1\sigma^q + g_2\sigma^q$.

[11] A topological group is a group with a topology with respect to which the group operations are continuous. No separation axioms are postulated, so that the factor group by a non-closed subgroup still may be regarded as a topological group.

[12] $C_q(K, G)$ is topologized as follows: Given a q-cell σ^q and an open set V in G a neighborhood (σ^q, V) in $C_q(K, G)$ is obtained by taking all the cochains f^q such that $f^q(\sigma^q) \in V$.

the free generators of the discrete group $C^q(K)$ each cochain leads to a unique homomorphism

$$f^q: C^q(K) \to G$$

and we have

$$C_q(K, G) = \text{Hom}\{C^q(K), G\}.^{13}$$

This second point of view will prevail in the sequel. The coboundary operator δ is a homomorphism

$$\delta: C_q(K, G) \to C_{q+1}(K, G)$$

defined as

$$(\delta f^q)c^{q+1} = f^q(\partial c^{q+1})$$

for $f^q \in C_q(K, G)$, $c^{q+1} \in C^{q+1}(K)$. The cochains f^q with $\delta f^q = 0$ are called *cocycles* and form a subgroup $Z_q(K, G)$ of $C_q(K, G)$. Cochains f^q of the form $f^q = \delta g^{q-1}$ are called *coboundaries* and form a subgroup $B_q(K, G)$ of $C_q(K, G)$. Since

$$(\delta\delta f^q)c^{q+2} = (\delta f^q)(\partial c^{q+2}) = f^q(\partial\partial c^{q+2}) = f^q(0) = 0$$

we have $\delta\delta = 0$ and B_q is a subgroup of Z_q. The factor group

$$H_q(K, G) = Z_q(K, G)/B_q(K, G)$$

is the qth *cohomology group* of K over G. It carries a topology[14] induced by that of $C_q(K, G)$. The elements of $H_q(K, G)$ are called *cohomology classes*. Each cocycle f^q determines uniquely a cohomology class. Two cocycles f_1^q, f_2^q in the same cohomology class are called *cohomologous* (notation $f_1^q \backsim f_2^q$). Clearly $f_1^q \backsim f_2^q$ if and only if $f_1^q - f_2^q$ is a coboundary.

4. Chain transformations

Let K_1 and K_2 be two abstract closure finite complexes and let τ be a collection of homomorphisms, one for each dimension q

$$\tau: C^q(K_1) \to C^q(K_2). \tag{4.1}$$

We say that τ is a *chain transformation*

$$\tau: K_1 \to K_2$$

if

$$\tau\partial = \partial\tau. \tag{4.2}$$

[13] By Hom $\{H, G\}$ we denote the additive group of all homomorphisms $\varphi: H \to G$ with G abelian. A topology in Hom $\{H, G\}$ is obtained by taking a compact subset X of H and a neighborhood V of 0 in G and considering the set of all φ such that $\varphi(X) \subset V$ as a neighborhood of 0 in Hom $\{H, G\}$.

[14] Since B_q need not be a closed subgroup of Z_q, the group H_q may not satisfy any separation axioms. Cf. footnote 11.

More specifically, consider the diagram

$$\begin{array}{ccc} C^q(K_1) & \xrightarrow{\tau} & C^q(K_2) \\ \partial\downarrow & & \downarrow\partial \\ C^{q-1}(K_1) & \xrightarrow{\tau} & C^{q-1}(K_2) \end{array}$$

Condition (4.2) means that the two mappings of $C^q(K_1)$ into $C^{q-1}(K_2)$ that can be derived from the diagram are equal.

The homomorphisms (4.1) can be extended to homomorphisms

$$\tau : C^q(K_1, G) \to C^q(K_2, G)$$

by putting

$$\tau(\sum g_i \sigma_i^q) = \sum g_i \tau(\sigma_i^q).$$

Condition (4.2) being still valid, we obtain

$$\tau : Z^q(K_1, G) \to Z^q(K_2, G), \qquad \tau : B^q(K_1, G) \to B^q(K_2, G);$$

consequently τ induces homomorphisms

$$\tau : H^q(K_1, G) \to H^q(K_2, G)$$

of the homology groups of K_1 into those of K_2.

For the cochains we have induced homomorphisms

$$\tau^* : C_q(K_2, G) \to C_q(K_1, G)$$

defined by

$$(\tau^* f^q)(c^q) = f^q(\tau c^q)$$

for $f^q \in C_q(K_2, G)$, $c^q \in C^q(K_1)$. We verify that

$$\tau^* \delta = \delta \tau^*$$

and therefore

$$\tau^* : Z_q(K_2, G) \to Z_q(K_1, G) \qquad \tau^* : B_q(K_2, G) \to B_q(K_1, G)$$

and consequently τ induces homomorphisms

$$\tau^* : H_q(K_2, G) \to H_q(K_1, G)$$

of the cohomology groups of K_2 into those of K_1. These homomorphisms will be called the dual homomorphisms induced by τ.

Given two chain transformations

$$\tau_1 : K_1 \to K_2, \qquad \tau_2 : K_2 \to K_3$$

the composite chain transformations

$$\tau_2 \tau_1 : K_1 \to K_3$$

is defined by $\tau_2\tau_1(c^q) = \tau_2[\tau_1(c^q)]$. We notice that for the induced dual homomorphisms we have $(\tau_2\tau_1)^* = \overset{*}{\tau_1}\overset{*}{\tau_2}$.

Let K be a complex and K_1 a collection of cells of K. We shall say that K_1 is a closed subcomplex whenever $\sigma^q \in K$ and $[\sigma^q : \sigma^{q-1}] \neq 0$ implies $\sigma^{q-1} \in K$. This condition is equivalent with the condition that for every chain in K_1 the boundary also is in K_1. If K_1 is a closed subcomplex we define a chain transformation

$$\epsilon : K_1 \to K$$

by setting $\epsilon(c^q) = c^q$ for every chain c^q in K_1.

The dual homomorphisms of the cochains

$$\epsilon^* : C_q(K, G) \to C_q(K_1, G)$$

can be described as follows. Given a q-cochain f^q in K i.e. given a homomorphism

$$f^q : C^q(K) \to G$$

the homomorphism

$$\epsilon^* f^q : C^q(K_1) \to G$$

is obtained by considering f^q on the subgroup $C^q(K_1)$ of $C^q(K)$.

5. Chain homotopy

Let two chain transformations

$$\tau_1 : K_1 \to K_2 \qquad \tau_2 : K_1 \to K_2$$

be given. A collection D of homomorphisms

$$D : C^q(K_1) \to C^{q+1}(K_2)$$

(one for each q) will be called a chain homotopy between τ_1 and τ_2 provided

$$\partial D c^q = \tau_2 c^q - \tau_1 c^q - D\partial c^q \tag{5.1}$$

for all $c^q \in C^q(K_1)$.

If such a chain homotopy D exists we say that τ_1 and τ_2 are *chain homotopic* (notation: $\tau_1 \simeq \tau_2$). It is clear that the relation $\simeq$ is symmetric, reflexive and transitive.

It follows from (5.1) that if c^q is a cycle then $\tau_2 c^q$ and $\tau_1 c^q$ are homologous; consequently we have

(5.2) *Chain homotopic transformations induce identical homomorphisms of the homology groups.*

The homomorphisms D induce dual homomorphisms

$$D^* : C_q(K_2, G) \to C_{q-1}(K_1, G)$$

defined as

$$(D^* f^q) c^{q-1} = f^q(Dc^{q-1})$$

for $f^q \in C_q(K_2, G)$, $c^{q-1} \in C^{q-1}(K_1)$. We verify that

$$\delta D^* f^q = \overset{*}{\tau_2} f^q - \overset{*}{\tau_1} f^q - D^* \delta f^q.$$

Consequently if f^q is a cocycle then $\overset{*}{\tau_2} f^q$ and $\overset{*}{\tau_1} f^q$ are cohomologous. Hence we get

(5.3) *Chain homotopic transformations induce identical dual homorphisms of the cohomology groups.*

We further remark that if only one of the chain transformations, say τ_1, and the homotopy operator D are given, then the other chain transformation could be defined using (5.1) as

$$\tau_2 c^q = \tau_1 c^q + \partial D c^q + D \partial c^q.$$

τ_2 will automatically be a chain transformation since

$$\begin{aligned} \partial \tau_2 c^q &= \partial \tau_1 c^q + \partial \partial D c^q + \partial D \partial c^q \\ &= \tau_1 \partial c^q + [\tau_2 \partial c^q - \tau_1 \partial c^q - D \partial \partial c^q] \\ &= \tau_2 \partial c^q. \end{aligned}$$

Clearly τ_1 and τ_2 will be chain homotopic with D as the homotopy operator.

6. Chain equivalences

Let K_1 and K_2 be two closure finite abstract complexes. We will denote by the symbol 1 the identity chain transformation

$$1 c^q = c^q$$

of K_1 into itself or of K_2 into itself.

Two chain transformations

$$\tau : K_1 \to K_2 \qquad \rho : K_2 \to K_1$$

will be said to form an *equivalence pair* if and only if

$$\rho\tau \simeq 1 \qquad \text{and} \qquad \tau\rho \simeq 1.$$

Each of the transformations τ, ρ will be then called a *chain equivalence.*

The homomorphisms τ and ρ induced on the homology groups are then inverses of one another; hence they are isomorphisms. The same holds for the dual homomorphisms τ^* and ρ^* of the cohomology groups. It follows that

(6.1) *A chain equivalence induces isomorphisms of the homology groups and of the cohomology groups of the two complexes involved.*

(6.2) *Let $\tau : K_1 \to K_2$ be a chain equivalence. The chain transformation $\rho : K_2 \to K_1$ such that τ and ρ form an equivalence pair is determined uniquely up to chain homotopy.*

In fact, if $\tau_1\rho$ amd $\tau_1\rho'$ are both equivalence pairs then

$$\rho \simeq \rho'\tau\rho \simeq \rho'$$

because $\rho'\tau \simeq 1$ and $\tau\rho \simeq 1$.

(6.3) *If* τ, ρ *is an equivalence pair and*

$$\tau \simeq \tau' \quad \text{and} \quad \rho \simeq \rho'$$

then τ', ρ' *is also an equivalence pair.*

The proof is obvious. The previous two propositions show that the concept of an equivalence pair applies to chain homotopy classes of chain transformations, and that then the elements of the pair determine each other uniquely, and could justly be called inverses of one another.

Chapter II

Simplicial complexes

8. The complexes $k(P)$ and $K(P)$

Let P be a locally finite polyhedron with a fixed simplicial decomposition.[16] After choosing a definite orientation for each simplex of P we define the complex $k(P)$ as follows. The cells σ^q of $k(P)$ are the oriented q-simplices of P (the 0-simplices are not oriented). The incidence numbers $[\sigma^q:\sigma^{q-1}]$ are defined in the usual way and are $+1$, -1 or 0 according to whether σ^{q-1} is a positively oriented face, a negatively oriented face or is not a face at all of σ^q. The resulting complex $k(P)$ is closure finite.

We shall now define another complex $K(P)$ as follows. A q-cell of $K(P)$ is an ordered array $v_0 \cdots v_q$ of vertices of P with possible repetitions and with the condition that all the vertices in question have to lie on one geometric simplex of P. A q-cell without repetition will be called *proper*, with repetitions, it will be called *degenerate*.

Next we define the boundary operation as[17]

$$\partial(v_0 \cdots v_q) = \sum_{i=0}^{q} (-1)^i v_0 \cdots \hat{v}_i \cdots v_q \tag{8.1}$$

for the cells. We extend this operation to a homomorphism

$$\partial : C^q(K(P)) \to C^{q-1}(K(P)).$$

We verify by a straightforward computation from (8.1) that

$$\partial\partial = 0.$$

---

[16] For more details see Lefschetz, *Algebraic Topology*, pp. 93–98.

[17] The dash over v_i indicates that v_i is omitted.

Using the operation ∂ we can define the incidence numbers and get an abstract closure finite complex $K(P)$.

Both complexes $k(P)$ and $K(P)$ will be taken "unaugmented." That means that there are no cells of dimension -1. Consequently every 0-chain is a cycle and for a connected P we have $H^0(k(P), G) \approx G$ and $H^0(K(P), G) \approx G$.

Given a cell σ^q in either of the two complexes $k(P)$ or $K(P)$ we denote by $|\sigma^q|$ the smallest geometric simplex of the decomposition of P, containing the cell σ^q.

9. The chain transformation α

We now define a chain transformation

$$\alpha: K(P) \to k(P) \tag{9.1}$$

as follows. Given a q-cell $v_0 \cdots v_q$ in $K(P)$ we distinguish two cases.

(1) If $v_0 \cdots v_q$ is improper we set

$$\alpha(v_0 \cdots v_q) = 0$$

(2) If $v_0 \cdots v_q$ is proper then it determines uniquely a q-cell σ^q in $k(P)$. The vertices $v_0 \cdots v_q$ determine a chain $[v_0 \cdots v_q]$ of $k(P)$

$$[v_0 \cdots v_q] = \pm \sigma^q$$

according as the orientation of σ^q agrees with the order $v_0 \cdots v_q$ or not. We define

$$\alpha(v_0 \cdots v_q) = [v_0 \cdots v_q].$$

This defines α for each cell in $K(P)$. We extend α by linearity to get homomorphisms

$$\alpha: C^q(K(P)) \to C^q(k(P)).$$

To show that these homomorphisms determine a chain transformation we have to show that

$$\partial\alpha = \alpha\partial. \tag{9.2}$$

It is sufficient to verify (9.2) for each cell of $K(P)$. If $v_0 \cdots v^q$ is a proper q-cell of $K(P)$ then

$$\begin{aligned}\partial\alpha(v_0 \cdots v_q) &= \partial[v_0 \cdots v_q] = \textstyle\sum(-1)^i[v_0 \cdots \hat{v}_i \cdots v_q]\\ &= \textstyle\sum(-1)^i\alpha(v_0 \cdots v_i \cdots v_q) = \alpha(\sum(-1)^i v_0 \cdots \hat{v}_i \cdots v_q)\\ &= \alpha\partial(v_0 \cdots v_q)\end{aligned}$$

If $v_0 \cdots v_q$ is improper with $v_k = v_l$ for $k < l$ then by definition $\partial\alpha(v_0 \cdots v_q) = 0$. We compute $\alpha\partial$

$$\alpha\partial(v_0 \cdots v_q) = \sum (-1)^i\alpha(v_0 \cdots \hat{v}_i \cdots v_q)$$

in this summation all the cells will be improper except perhaps for $i = k$ and $i = l$. Hence

$$\alpha\partial(v_0 \cdots v_q) = (-1)^k[v_0 \cdots \hat{v}_k \cdots v_q] + (-1)^l[v_0 \cdots \hat{v}_l \cdots v_q]$$

since $v_k = v_l$ we have

$$[v_0 \cdots \hat{v}_k \cdots v_q] = (-1)^{l-k+1}[v_0 \cdots \hat{v}_l \cdots v_q]$$

and therefore $\alpha\partial(v_0 \cdots v_q) = 0$ as desired.

THEOREM 9.1. *The chain transformation*

$$\alpha: K(P) \to k(P)$$

is a chain equivalence.

COROLLARY 9.2. *The chain transformation α induces isomorphisms of the homology groups of $K(P)$ onto those of $k(P)$.*

COROLLARY 9.3. *The chain transformation α induces dual isomorphisms α^* of the cohomology groups of $k(P)$ onto those of $K(P)$.*

Theorem 9.1 will be proved in §11.

10. The join

Let $\hat{P}$ be the polyhedron obtained from P by taking the join of P with a vertex v_0 outside of P. We may regard $\hat{P}$ as a cone with P as base and v_0 as vertex. Given any q-cell σ^q of $K(\hat{P})$ we denote by $v_0\sigma^q$ the $(q+1)$-cell obtained by writing out the array of vertices defining σ^q and placing v_0 in front of them all. For each q-chain c^q of $K(\hat{P})$ the $(q+1)$-chain v_0c^q of $K(\hat{P})$ is then well defined. A direct computation shows that

$$\partial(v_0c^q) = c^q - v_0(\partial c^q) \qquad \text{for } q > 0 \tag{10.1}$$

In particular if c^q is a cycle we have $c^q = \partial(v_0c^q)$ and hence $c^q \sim 0$. This proves

LEMMA 10.1. *The homology groups of the complex $K(\hat{P})$ vanish for dimensions $q > 0$.*

In particular a geometric simplex s may be regarded as a join of any of its faces with the opposite vertex. Consequently the symbol vc^q is well defined for any vertex v of s and any q-chain c^q of $K(s)$.

COROLLARY 10.2. *If s is a geometric simplex, the homology groups of $K(s)$ vanish for all dimensions $q > 0$.*

11. Proof of Theorem 9.1

In order to construct a chain transformation

$$\bar{\alpha}: k(P) \to K(P) \tag{11.1}$$

such that α, $\bar{\alpha}$ form an equivalence pair, we choose a definite ordering of the vertices of the polyhedron P. Each q-cell σ^q of $k(P)$ can then be written uniquely as

$$\sigma^q = \pm [v_0 \cdots v_q] \text{ with } v_0 < \cdots < v_q .$$

We define

$$\bar{\alpha}\sigma^q = \pm v_0 \cdots v_q .$$

It is obvious that $\bar{\alpha}\partial = \partial\bar{\alpha}$ and therefore $\bar{\alpha}$ is a chain transformation. Clearly

$$\alpha\bar{\alpha}\sigma^q = \sigma^q$$

which means that

(11.2) $$\alpha\bar{\alpha} = 1.$$

In order to prove that

(11.3) $$\bar{\alpha}\alpha \simeq 1$$

we will define a homotopy operator

$$D: C^q(K(P)) \to C^{q+1}(K(P))$$

subject to the following conditions

(11.4) $$\partial D c^q = c^q - \bar{\alpha}\alpha c^q - D\partial c^q$$

(11.5) $$D\sigma^q \subset |\sigma^q| .$$

We proceed by induction. Define $D\sigma^0 = 0$. Since $\bar{\alpha}\alpha\sigma^0 = \sigma^0$ conditions (11.4) and (11.5) are satisfied. Suppose that D has been defined for all dimensions less than q so that (11.4) and (11.5) hold. Consider

$$c^q = \sigma^q - \bar{\alpha}\alpha\sigma^q - D\partial\sigma^q.$$

Since $\bar{\alpha}\alpha\sigma^q \subset |\sigma^q|$ and $D\partial\sigma^q \subset |\sigma^q|$ we have

$$c^q \subset |\sigma^q| .$$

Further

$$\partial c^q = \partial\sigma^q - \partial\bar{\alpha}\alpha\sigma^q - \partial D\partial\sigma^q = \partial\sigma^q - \bar{\alpha}\alpha\partial\sigma^q - (\partial\sigma^q - \bar{\alpha}\alpha\partial\sigma^q - D\partial\partial\sigma^q) = 0.$$

Hence c^q is a cycle in $K(|\sigma^q|)$. Since by Corollary 10.2 every cycle in $K(|\sigma^q|)$ bounds in $K(|\sigma^q|)$ there is a $(q+1)$-chain $D\sigma^q$ in $K(P)$ such that

$$D\sigma^q \subset |\sigma^q| \qquad \partial D\sigma^q = c^q.$$

This completes the definition of D and proves Theorem 9.1.

12. Proper and degenerate chains

We recall that a cell σ^q of $K(P)$ is called proper if its vertices do not contain repetitions, and degenerate otherwise. Accordingly a chain of $K(P)$ will be called *proper* or *degenerate* if it is composed of proper or degenerate cells only.

(12.1) *Each cycle in $K(P)$ is homologous to a proper cycle.*

(12.2) *Each degenerate cycle in $K(P)$ is bounding.*

Given a cycle c^q in $K(P)$ consider the cycle $\bar{\alpha}\alpha c^q$. Since $\bar{\alpha}\alpha \simeq 1$ we have $c^q \sim \bar{\alpha}\alpha c^q$. This proves (12.1) since $\bar{\alpha}\sigma^q$ is proper for any σ^q in $k(P)$. If c^q is degenerate then $\alpha c^q = 0$ and $\bar{\alpha}\alpha c^q = 0$. Hence $c^q \sim 0$.

The proper cells of $K(P)$ form a closed subcomplex $K_p(P)$. The inclusion relation induces homomorphic mappings of the homology groups of K_p into those of K. It follows from (12.1) that these homomorphisms are all mappings *onto*. We will show that generally they are *not* isomorphisms i.e. a cycle in K_p may bound in K without bounding in K_p. Let P be an interval with endpoints v_0, v_1. The chain

$$c^1 = v_0v_1 + v_1v_0$$

is then a proper cycle and

$$c^1 = \partial(v_0v_1v_0 - v_0v_0v_0).$$

However c^1 cannot be the boundary of proper 2-chain since $K(P)$ contains no proper 2-cells.

Chapter III

Singular homologies in a space

13. The singular complex $S(X)$

Let s be a non-degenerate q-dimensional geometric simplex in some Euclidean space. If the vertices of s are given in a definite order $p_0 < p_1 < \cdots < p_q$ we shall say that s is an *ordered simplex* and write $s = \langle p_0 \cdots p_q \rangle$. We shall denote by $s^{(i)}$ the face of s opposite the ith vertex $s^{(i)} = \langle p_0, \cdots, \hat{p}_i, \cdots p_q \rangle$. Taken with the same order of vertices as in s, the faces $s^{(i)}$ are ordered simplices.

Given two ordered q-simplices s_1 and s_2 we denote by B_{s_1,s_2} the barycentric mapping

$$B_{s_1,s_2}: s_1 \rightarrow s_2$$

preserving the order of the vertices. Clearly B_{s_1,s_2} is unique.

Let X be a topological space.[18] By a *singular* q-simplex in X we understand a continuous mapping

$$T: s \rightarrow X$$

of an ordered q-dimensional geometric simplex s.

Two singular q-simplices

$$T_1: s_1 \rightarrow X, \qquad T_2: s_2 \rightarrow X$$

are called *equivalent* (notation: $T_1 \equiv T_2$) provided

$$T_2B_{s_1,s_2} = T_1.$$

We verify that this relation is reflexive, symmetric, and transitive. Consequently the totality of all singular q-simplices in X is split into disjoint equivalence classes.

[18] The topological spaces considered here are of the most general type, with no separation axioms postulated.

We remark here that if $T:s \to X$ is a singular q-simplex, then given any ordered q-simplex s' there is a *unique* $T':s' \to X$ such that $T \equiv T'$.

Let $C^q(X)$ be the free abelian group generated by these equivalence classes. Alternatively, $C^q(X)$ may be defined as the group with the singular q-simplices in X as generators and $T_1 \equiv T_2$ as relations. The elements of the group $C^q(X)$ will be called the *integral singular q-chains in X*.

We now proceed with the definition of the boundary operator for singular chains. Given a singular q-simplex

$$T:s \to X, \qquad s = < p_0 \cdots p_q >,$$

consider the singular $(q - 1)$-simplices

$$T^{(i)}:s^{(i)} \to X$$

defined by[19] $T^{(i)} = T \mid s^{(i)}$. We define the boundary of T to be

$$\partial T = \sum_{i=0}^{q} (-1)^i T^{(i)}.$$

It is clear that if $T_1 \equiv T_2$ then $T_1^{(i)} \equiv T_2^{(i)}$ and therefore $\partial T_1 = \partial T_2$ in $C^{q-1}(X)$. Therefore we get a homomorphism

$$\partial:C^q(X) \to C^{q-1}(X).$$

We further verify that

$$\partial\partial = 0.$$

Consequently the boundary operation ∂ can be used to define incidence numbers and leads to a closure finite abstract complex that we will denote by $S(X)$ and call the *singular complex* of the space X. By definition we have

$$C^q(S(X)) = C^q(X).$$

The groups of the complex $S(X)$ will be called the singular groups of X:

$$H^q(X, G) = H^q(S(X), G) \qquad G \text{ discrete}$$

$$H_q(X, G) = H_q(S(X), G) \qquad G \text{ topological}$$

We notice that a cochain

$$f \in C_q(S(X), G)$$

can be considered either as a homomorphism

$$f:C^q(X) \to G$$

or as a function associating with each singular q-simplex T an element $f(T)$ of G, so that $f(T_1) = f(T_2)$ for $T_1 \equiv T_2$.

---

19 Given a mapping $\varphi: X \to Y$ and a subset A of X we denote by $\varphi \mid A$ the mapping φ restricted to the subset A.

14. The invariance theorem

Let P be a simplicial polyhedron as described in §8. For P we have constructed three abstract closure finite complexes: $k(P)$, $K(P)$, and $S(P)$. We have already compared the complexes $k(P)$ and $K(P)$ using the chain equivalence $\alpha: K(P) \to k(P)$. We now proceed to compare the complexes $K(P)$ and $S(P)$.

We define a chain transformation

$$\beta: K(P) \to S(P) \tag{14.1}$$

as follows. Given a q-cell $v_0 \cdots v_q$ in $K(P)$, we consider an ordered q-dimensional simplex $s = < p_0 \cdots p_q >$ and a barycentric mapping $T: s \to P$ such that $T(p_i) = v_i$. Clearly T exists since $v_0, \cdots, v_q$ are all in the same simplex of P and T is unique. We set

$$\beta(v_0 \cdots v_q) = T$$

to get (14.1). It is clear that $\beta\partial = \partial\beta$ so that β is a chain transformation.

The reason the simplex s had to be constructed outside of P is because of the possibility of repetitions among the vertices $v_0, \cdots v_q$. If $v_0 \cdots v_q$ were nondegenerate we could have taken s to be contained in P and T to be the identity mapping.

One of the basic results of this paper is the following

THEOREM 14.1. *The chain transformation*

$$\beta: K(P) \to S(P)$$

is a chain equivalence.

The proof of this theorem requires preparatory considerations concerning barycentric subdivisions, and will be postponed until the next chapter.

This theorem jointly with Theorem 9.1 implies that the three complexes $k(P)$, $K(P)$ and $S(P)$ have isomorphic homology and cohomology groups. In particular since $S(P)$ is a topological invariant of P as a *space* it follows that the homology theories of $k(P)$ and $K(P)$ depend only upon P as a *space* and not upon the particular representation of P as a simplicial polyhedron.

15. Continuous mappings

Let

$$\varphi: X \to Y$$

be a continuous mapping of a space X into a space Y. Given a singular q-simplex in X

$$T: s \to X$$

we have the singular q-simplex in Y

$$\varphi T: s \to Y.$$

This leads to a chain transformation

(15.1) $$\varphi : S(X) \to S(Y).$$

If X is a simplicial polyhedron P and

$$\varphi : P \to Y$$

is a continuous mapping, then combining β with φ we obtain a chain transformation

$$\varphi\beta : K(P) \to S(Y).$$

This observation leads to the following convention. Let P be a simplicial polyhedron, c^q a q-chain in the complex $K(P)$ and $\varphi : P \to Y$ a continuous mapping. Then $\varphi\beta(c^q)$ is a q-chain in $S(Y)$, we shall write

$$(P, c^q, \varphi) = \varphi\beta(c^q).$$

This symbolism will prove to be advantageous in many proofs and applications.

16. Prisms

The singular homology theory is particularly useful in connection with problems dealing with homotopy. In order to prepare the ground for such applications we introduce a few definitions.

A *prism* of dimension $q + 1$ is the cartesian product $s \times I$ of an ordered q-simplex s and the closed interval $0 \leqq t \leqq 1$. The *bases* $s \times 0$ and $s \times 1$ of the prism $s \times I$ are q-simplices, in which the vertices are ordered just as in s. The *sides* $s^{(i)} \times I$ are prisms of dimension q.

We turn the prism into a simplicial polyhedron as follows. Let $s \times 0 = < p_0 \cdots p_q >$ and let $s \times 1 = < \bar{p}_0 \cdots \bar{p}_q >$ be the bases of $s \times I$. We subdivide $s \times I$ by considering the $(q + 1)$-simplexes

$$p_0 \cdots p_i \bar{p}_i \cdots \bar{p}_q$$

and all their faces. The simplicial subdivision of $s \times I$ so obtained will be called the *standard division*. Its main property is consistency; i.e., the standard divisions of the sides $s^{(i)} \times I$ agree with the division of $s \times I$. We also notice that the bases $s \times 0$ and $s \times 1$ are not subdivided.

In the complex $K(s \times I)$, with $s \times I$ taken in the standard division, we consider the $(q + 1)$-chain

$$d(s \times I) = \sum_{i=0}^{q} (-1)^i p_0 \cdots p_i \bar{p}_i \cdots \bar{p}_q$$

which we will call the *basic* $(q + 1)$-*chain on the prism* $s \times I$. We verify that

(16.1) $$\partial d(s \times I) = s \times 1 - s \times 0 - \sum_{i=0}^{q} (-1)^i d(s^{(i)} \times I)$$

A *singular* $(q + 1)$-*dimensional prism* in a space X is a continuous mapping

$$R : s \times I \to X$$

of a $(q+1)$-prism $s \times I$ into X. The *bases*

$$R(0) = R \mid s \times 0 \qquad R(1) = R \mid s \times 1$$

are singular q-simplexes in X. The sides

$$R^{(i)} = R \mid s^{(i)} \times I$$

are singular q-prisms in X.

Each singular prism $R: s \times I \to X$ generates a singular $(q+1)$-chain in X:

$$c(R) = (s \times I,\ d(s \times I),\ R)$$

where $s \times I$ is in its standard division and $d(s \times I)$ is the basic $(q+1)$-chain of $K(s \times I)$. We verify at once that (16.1) implies

$$\partial c(R) = R(1) - R(0) - \sum_{i=0}^{q} (-1)^i c(R^{(i)}). \tag{16.2}$$

Given two ordered q-simplices s_1 and s_2 we have denoted by B_{s_1,s_2} the unique barycentric mapping $s_1 \to s_2$ preserving the order of the vertices. The mapping B_{s_1,s_2} obviously induces a mapping

$$\tilde{B}_{s_1,s_2}: s_1 \times I \to s_2 \times I.$$

Two singular prisms

$$R_1: s_1 \times I \to X, \qquad R_2: s_2 \times I \to X$$

will be called *equivalent* (notation $R_1 \equiv R_2$) provided

$$R_2 \tilde{B}_{s_1,s_2} = R_1 .$$

It is clear that if $R_1 \equiv R_2$ then $c(R_1) = c(R_2)$.

17. Homotopy

The constructions of the previous section will be used in the proof of the following

THEOREM 17.1 *If the continuous mappings $\varphi: X \to Y$ and $\psi: X \to Y$ are homotopic then the induced chain transformations $\varphi: S(X) \to S(Y)$ and $\psi: S(X) \to S(Y)$ are chain homotopic*

PROOF. Since φ and ψ are homotopic there is a mapping

$$\eta: X \times I \to Y$$

such that

$$\eta(x, 0) = \varphi(x) \qquad \eta(x, 1) = \psi(x).$$

Given a singular q-simplex of X

$$T: s \to X$$

we define a singular prism

$$R_T: s \times I \to Y$$

as follows

$$R_T(x, t) = \eta(T(x), t) \qquad x \in s, \qquad t \in I.$$

We notice that

$$R_T(0) = \varphi(T), \qquad R_T(1) = \psi(T)$$

hence putting $D(T) = c(R_T)$ we get by (16.2)

$$\partial D(T) = \psi(T) - \varphi(T) - D(\partial T).$$

This proves the theorem.

According to Hurewicz, two spaces X and Y are said to have the same *homotopy type* if there are two continuous mappings

$$\varphi: X \to Y \qquad \psi: Y \to X$$

such that $\psi\varphi$ is homotopic to the identity mapping of X into itself, and similarly $\varphi\psi$ is homotopic to the identity mapping of Y into itself. If we consider the induced chain transformations

$$\varphi: S(X) \to S(Y) \qquad \psi: S(Y) \to S(X)$$

we get from Theorem 17.1 that

$$\psi\varphi \simeq 1 \quad \text{and} \quad \varphi\psi \simeq 1$$

so that φ and ψ form a chain equivalence pair and the complexes $S(X)$ and $S(Y)$ are chain equivalent.

18. Relative homologies

We briefly outline the procedure leading to the homology and cohomology groups of a space X modulo a subset A. Every singular simplex in A may also be regarded as a singular simplex in X and this way the singular complex $S(A)$ may be considered a subcomplex of the larger complex $S(X)$. Clearly $S(A)$ is a closed subcomplex of $S(X)$ and therefore homology and cohomology groups of $S(X)$ mod $S(A)$ may be considered. These groups will be, by definition, the groups of the space X modulo the subset A.

Accordingly, a singular chain c^q in X will be called a cycle mod A if ∂c^q is a singular chain in A. We shall say that c^q bounds mod A if there is a singular chain c^{q+1} such that $c^q - \partial c^{q+1}$ is a chain in A. This way we get the homology groups of X mod A.

A cochain f^q in X will be called a cochain in X mod A if $f^q(T) = 0$ for every singular simplex T in A. Using cochains mod A throughout we arrive at the cohomology groups of X mod A.

With these definitions the properties of the Kronecker index and the duality between homology and cohomology are preserved.

If P is a simplicial polyhedron and Q a closed subpolyhedron then similar considerations can be applied to the complexes $k(Q)$ and $k(P)$ as well as the complexes $K(Q)$ and $K(P)$. Since the chain transformation $\alpha: K(P) \to k(P)$ agrees on $K(Q)$ with the chain transformation $\alpha: K(Q) \to k(Q)$ and similarly for the chain transformation β, we conclude as before that the chain transformations α and β induce isomorphisms of the relative homology and cohomology groups.

CHAPTER V

PRODUCTS[20]

24. Augmentable complexes

The purpose of introducing the definition of an augmentable complex here is to secure that the products that will subsequently be studied have a unit.

[20] General reference: S. Lefschetz, *Algebraic Topology*, Chapters III and IV; H. Whitney, *On products in a complex*, Ann. of Math., 39 (1938), pp. 397–432.

A closure finite abstract complex K will be called *augmentable* provided the following two properties hold.

(24.1) *All the cells of K have dimension $\geqq 0$ and there are cells of dimension* 0.

(24.2) *The integral 0-cochain f_0 defined by $f_0(\sigma^0) = 1$ for all σ^0 in K is a cocycle.*

An augmentable complex can be *augmented* by the addition of a single (-1)-dimensional cell σ^{-1} with incidence numbers $[\sigma^0, \sigma^{-1}] = 1$ for all 0-cells σ^0. Then f_0 becomes a coboundary.

Let K be augmentable. Given a 0-chain

$$c^0 = \sum g_i \sigma_i^0 \in C^0(K, G)$$

we define the *index* $I(c^0)$ of c^0 as

$$I(c^0) = \sum g_i \in G$$

and verify at once that

(24.3) $$I(c^0) = KI(f_0, c^0).$$

We verify by inspection that

(24.4) *The complexes $k(P)$, $K(P)$ and $S(X)$ are augmentable.*

As we have remarked earlier these complexes will be left unaugmented.

25. The cup product

The complexes that are studied in this paper do not all fall in the class that Whitney calls "complexes admitting a product theory." Particularly our complex $S(X)$ is not such a complex. As a result Whitney's theory establishing in a general fashion the existence and the uniqueness of the products cannot be used here. The products will have to be defined individually for each class of complexes $k(P)$, $K(P)$, $S(X)$, and others that will appear later. Nevertheless all these products have enough common features to permit a uniform axiomatic treatment that will be developed in the sequel.

An augmentable closure finite abstract complex K will be called a *complex with products* if a rule is given which with any three groups G_1, G_2, G such that G_1 and G_2 are paired to G (see §7), and with any two cochains

$$f_1^p \in C_p(K, G_1) \qquad f_2^q \in C_q(K, G_2)$$

associates a third cochain, called the *cup product of f_1^p and f_2^q*

$$f_1^p \cup f_2^q \in C_{p+q}(K, G)$$

subject to the following five axioms:

(∪ 1) $f_1^p \cup f_2^q$ is additive and continuous in each variable.

This axiom merely states that under the cup product as multiplication the groups $C_p(K, G_1)$ and $C_q(K, G_2)$ are paired to the group $C_{p+q}(K, G)$.

(∪ 2) Associativity: $f_1^p \cup (f_2^q \cup f_3^r) = (f_1^p \cup f_2^q) \cup f_3^r$.

More precisely if G_1, G_2, G_3, G_{12}, G_{23} and G are six groups with pairings

$$g_1g_2 \in G_{12} \qquad g_2g_3 \in G_{23}$$

$$g_1g_{23} \in G \qquad g_{12}g_3 \in G$$

such that

$$g_1(g_2g_3) = (g_1g_2)g_3$$

then (∪ 2) holds for

$$f_1^p \in C_p(K, G_1), \qquad f_2^q \in C_q(K, G_2), \qquad f_3^r \in C^r(K, G_3).$$

(∪ 3) $$f \cup f_0 = f$$

(∪ 4) $$f_0 \cup f = f.$$

In these two axioms f_0 is the integral 0-cocycle described in the previous section. The group I of integers is considered paired with any group G in the obvious fashion.

(∪ 5) $$\delta(f_1^p \cup f_2^q) = \delta f_1^p \cup f_2^q + (-1)^p f_1^p \cup \delta f_2^q$$

It follows from (∪ 5) that

$$\begin{aligned} \text{cocycle} \cup \text{cocycle} &= \text{cocycle} \\ \text{cocycle} \cup \text{coboundary} &= \text{coboundary} \\ \text{coboundary} \cup \text{cocycle} &= \text{coboundary}; \end{aligned}$$

consequently the cup product is defined for cohomology classes and gives a pairing of the cohomology groups $H_p(K, G_1)$ and $H_q(K, G_2)$ to the group $H_{p+q}(K, G)$.

We can now see why we do not consider the complexes K augmented. Let f be any cocycle. If K were augmented than f_0 would be a coboundary and consequently $f = f \cup f_0$ would also be a coboundary.

26. The cap product

The cap product $f^q \cap c^{p+q}$ defined for a cochain f^q and a chain c^{p+q} can be derived from the cup product using characters and the Kronecker index.

27. Chain transformations preserving products

Let K_1 and K_2 be two complexes with products, and let

$$\tau: K_1 \to K_2$$

be a chain transformation. We shall say that τ *preserves the products* provided

$$\tau^*(f_1^p \cup f_2^q) = \tau^* f_1^p \cup \tau^* f_2^q \tag{27.1}$$

for $f_1^p \in C^p(K_2, G_1)$ and $f_2^p \in C^p(K_2, G_2)$, with G_1 and G_2 paired to some group G.

If τ preserves the products then for the cap product we have the formula

$$\tau(\tau^* f^q \cap c^{p+q}) = f^q \cap \tau c^{p+q} \tag{27.2}$$

for $f^q \in C_q(K_2, G_1)$, $c^{p+q} \in C^{p+q}(K_1, G_2)$ with G_1 and G_2 paired to G.

If formula (27.1) is not necessarily true for all cochains f_1^p, f_2^q but is true for any two cohomology classes, then we say that τ *weakly* preserves the products. Formula (27.2) is then still valid for cohomology and homology classes.

If τ weakly preserves the product, so does every chain transformation chain homotopic with τ. If τ and ρ are an equivalence pair and τ weakly preserves the products, then the same is true for ρ.

28. Products in $S(X)$, $K(P)$ and $k(P)$

As before we assume that the groups G_1 and G_2 are paired to G.

We first define the cup product in the complex $S(X)$. Let then two cochains

$$f_1^p \in C_p(X, G_1) \qquad \text{and} \qquad f_2^q \in C_q(X, G_2)$$

be given. Consider a singular $(p + q)$-simplex

$$T: s \to X, \qquad s = \langle p_0 \cdots p_{p+q} \rangle$$

and define

$$_pT = T \mid s_1 \to X \quad \text{where} \quad s_1 = \langle p_0 \cdots p_p \rangle$$

$$T_q = T \mid s_2 \to X \quad \text{where} \quad s_2 = \langle p_p \cdots p_{p+q} \rangle.$$

Clearly T_1 is a singular p-simplex and T_2 is a singular q-simplex in X. We define

$$(f_1^p \cup f_2^q)(T) = f_1^p({}_pT) f_2^q(T_q). \tag{28.1}$$

Clearly $f_1^p \cup f_2^q$ is a well defined $(p+q)$-cochain in X with coefficients in G. Axioms ($\cup$ 1)–($\cup$ 5) can be verified by straightforward computation, and thus $S(X)$ is a complex with products.

As shown in §26 the definition of the cup product automatically implies a definition of the cap product. In this case the cap product can easily be seen to be

(28.2) $$f^q \cap (g_2 T) = [f^q(T_q)g_2]_p T$$

with T, ${}_pT$ and T_q related as above.

Next we turn to the complex $K(P)$. In this case the definition of $f_1^p \cup f_2^q$ is even simpler. It reads

(28.3) $$(f_1^p \cup f_2^q)(v_0 \cdots v_{p+q}) = f_1^p(v_0 \cdots v_p)f_2^q(v_p \cdots v_{p+q}).$$

The verification of axioms ($\cup$ 1) – ($\cup$ 5) is immediate. The related definition of the cap product is

(28.4) $$f^q \cap g_2 v_0 \cdots v_{p+q} = [f^q(v_p \cdots v_{p+q})g_2]v_0 \cdots v_p\,.$$

If we now return to the definition of the basic chain transformation (§14)

$$\beta{:}K(P) \to S(P)$$

it becomes evident that

(28.5) *β preserves the products.*

Since β was proved to be a chain equivalence it follows that the complexes $K(P)$ and $S(P)$ not only have isomorphic homology and cohomology groups but also isomorphic product theories. This proves the topological invariance of the products in $K(P)$.

In order to introduce the products in the complex $k(P)$ we choose a definite order for all the vertices of P. Each $(p+q)$-cell σ^{p+q} of $k(P)$ can then uniquely be written as (cf. §9 and 11)

$$\sigma^{p+q} = \pm [v_0 \cdots v_{p+q}], \qquad v_0 < \cdots < v_{p+q}$$

We define

(28.5) $$(f_1^p \cup f_2^q)([v_0 \cdots v_{p+q}]) = f_1^p([v_0 \cdots v_p])f_2^q([v_p \cdots v_{p+q}])$$

Axioms ($\cup$ 1)–($\cup$ 5) are easy to verify. The related definition of the cap product reads

(28.6) $$f^q \cap g_2[v_0 \cdots v_{p+q}] = f^q([v_p \cdots v_q])g_2[v_0 \cdots v_p]$$

We recall now that in order to prove that the chain transformation

$$\alpha{:}K(P) \to k(P)$$

defined in §9 was a chain equivalence we have constructed (§11) a chain transformation

$$\bar{\alpha}{:}k(P) \to K(P)$$

such that α and $\bar{\alpha}$ were an equivalence pair. In order to define $\bar{\alpha}$ we have assumed that a definite order of the vertices of P was chosen. If we now agree to use the same order of vertices of P for the definition of $\bar{\alpha}$ as for the definition of the product in $k(P)$ it becomes clear by inspection that

(28.7) *$\bar{\alpha}$ preserves the products.*

By §27 this implies that

(28.8) *α weakly preserves the products.*

This shows that as far as cohomology and cohomology classes are concerned the products in $k(P)$ do not depend upon the choice of the order of the vertices of P, initially used in the definition of $f_1^p \cup f_2^q$. It also proves the topological invariance of the products in $k(P)$.

If X and Y are topological spaces and

$$\varphi: X \to Y$$

a continuous mapping, then it is easy to see that the chain transformation

$$\varphi: S(X) \to S(Y)$$

preserves the products.

29. Commutativity of the products

Let the (topologized) groups G_1 and G_2 be paired to G. For two cohomology classes

$$f_1^p \in H_p(X, G_1) \quad \text{and} \quad f_2^q \in H_q(X, G_2)$$

the cohomology class

$$f_1^p \cup f_2^q \in H_{p+q}(X, G)$$

is defined.

We also define a pairing of G_2 and G_1 to G by setting

$$g_2 g_1 = g_1 g_2$$

and consequently get the cohomology class

$$f_2^q \cup f_1^p \in H^p{}_{+q}(X, G)$$

THEOREM 29.1.

$$f_1^p \cup f_2^q = (-1)^{pq} f_2^q \cup f_1^p.$$

We emphasize that the theorem is valid only for cohomology classes and not for individual cocycles.

Reprinted from the Proceedings of the NATIONAL ACADEMY OF SCIENCES,
Vol. 31, No. 4, pp. 117–120. April, 1945

AXIOMATIC APPROACH TO HOMOLOGY THEORY

BY SAMUEL EILENBERG AND NORMAN E. STEENROD

DEPARTMENT OF MATHEMATICS, UNIVERSITY OF MICHIGAN

Communicated February 21, 1945

1. *Introduction.*—The present paper provides a brief outline of an axiomatic approach to the concept: homology group. It is intended that a full development should appear in book form.

The usual approach to homology theory is by way of the somewhat complicated idea of a complex. In order to arrive at a purely topological concept, the student of the subject is required to wade patiently through a large amount of analytic geometry. Many of the ideas used in the constructions, such as orientation, chain and algebraic boundary, seem artificial. The motivation for their use appears only in retrospect.

Since, in the case of homology groups, the definition by construction is so unwieldy, it is to be expected that an axiomatic approach or definition by properties should result in greater logical simplicity and in a broadened point of view. Naturally enough, the definition by construction is not eliminated by the axiomatic approach. It constitutes an existence proof or proof of consistency.

2. *Preliminaries.*—The concepts of a topological space and of a group are assumed to be known. The symbol (X, A) stands for a pair consisting of a topological space X and a closed subset A. A map $f:(X, A) \to (Y, B)$ of one such pair into another is a continuous map of X into Y which maps A into B. In case A is the vacuous set (X, A) is written as (X). If f_0, f_1 are two maps of (X, A) into (Y, B), they are homotopic if there exists a homotopy $f(x, t)$ connecting the two maps of X into Y such that $f(x, t) \in B$ for any $x \in A$ and all t.

3. *Basic Concepts.*—The fundamental concept to be axiomatized is a function $H_q(X, A)$ (called the *q-dimensional, relative homology group of X mod A*) defined for all triples consisting of an integer $q \geqq 0$ and a pair (X, A). The value of the function is an abelian group.

The first subsidiary concept is that of boundary. For each $q \geqq 1$ and each (X, A), there is a homomorphism

$$\partial : H_q(X, A) \to H_{q-1}(A)$$

called the *boundary operator.*

The second subsidiary concept is that of the induced homomorphism. If f is a map of (X, A) into (Y, B) and $q \geqq 0$, there is an attached homomorphism

$$f_*: H_q(X, A) \to H_q(Y, B)$$

called the *homomorphism induced by f*.

4. *Axioms.*—These three concepts have the following properties.

AXIOM 1. *If f = identity, then f_* = identity.*

That is to say, if f is the identity map of (X, A) on itself, then f_* is the identity map of $H_q(X, A)$ on itself.

AXIOM 2. $(gf)_* = g_*f_*$.

Explicitly, if $f:(X, A) \to (Y, B)$ and $g:(Y, B) \to (Z, C)$, then the combination of the induced homomorphisms $f_*: H_q(X, A) \to H_q(Y, B)$ and $g_*: H_q(Y, B) \to H_q(Z, C)$ is the induced homomorphism $(gf)_*: H_q(X, A) \to H_q(Z, C)$.

An immediate consequence of Axioms 1 and 2 is that homeomorphic pairs (X, A) and (Y, B) have isomorphic homology groups.

AXIOM 3. $\partial f_* = f_* \partial$.

Explicitly, if $f:(X, A) \to (Y, B)$ and $q \geqq 1$, the axiom demands that two homomorphisms of $H_q(X, A)$ into $H_{q-1}(B)$ shall coincide. The first is the combination of $\partial: H_q(X, A) \to H_{q-1}(A)$ followed by $(f|A)_*: H_{q-1}(A) \to H_{q-1}(B)$. The second is the combination of $f_*: H_q(X, A) \to H_q(Y, B)$ followed by $\partial: H_q(Y, B) \to H_{q-1}(B)$.

AXIOM 4. *If f is homotopic to g, then $f_* = g_*$.*

Definition: The *natural system* of the pair (X, A) is the sequence of groups and homomorphisms

$$\ldots \to H_q(X) \to H_q(X, A) \to H_{q-1}(A) \to H_{q-1}(X) \to \ldots \to H_0(X, A)$$

where $H_q(X) \to H_q(X, A)$ is induced by the identity map $(X) \to (X, A)$. $H_q(X, A) \to H_{q-1}(A)$ is the boundary operation, and $H_{q-1}(A) \to H_{q-1}(X)$ is induced by the identity map $(A) \to (X)$.

AXIOM 5. *In the natural system of (X, A) the last group, $H_0(X, A)$, is the image of $H_0(X)$. In any other group of the sequence, the image of the preceding group coincides with the kernel of the succeeding homomorphism.*

At first glance, this axiom may seem strange even to one familiar with homology theory. It is equivalent to three propositions usually stated as follows: (1) the boundary of a cycle of X mod A bounds in A if and only if the cycle is homologous mod A to a cycle of X; (2) a cycle of A is homologous to zero in X if and only if it is the boundary of a cycle of X mod A; (3) a cycle of X is homologous to a cycle of A if and only if it is homologous to zero mod A.

Definition: An open set U of X is strongly contained in A, written $U \Subset A$, if the closure $\overline{U}$ is contained in an open set $V \Subset A$.

AXIOM 6. *If $U \Subset A$, then the identity map: $(X - U, A - U) \to (X, A)$ induces isomorphisms $H_q(X - U, A - U) \cong H_q(X, A)$ for each $q \geqq 0$.*

This axiom expresses the intuitive idea that $H_q(X, A)$ is pretty much independent of the internal structure of A.

Axiom 7. *If P is a point, then $H_q(P) = 0$ for $q \geqq 1$.*

A particular reference point P_0 is selected, and $H_0(P_0)$ is called the *coefficient group* of the homology theory.

5. *Uniqueness.*—On the basis of these seven axioms, one can deduce the entire homology theory of a complex in the usual sense. Some highlights of the procedure are the following.

If σ is an n-simplex, and $\dot{\sigma}$ its point-set boundary, then $H_n(\sigma, \dot{\sigma})$ is isomorphic to the coefficient group. Further, $H_q(\sigma, \dot{\sigma}) = 0$ for $q \neq n$, and the boundary operator $\partial: H_n(\sigma, \dot{\sigma}) \rightarrow H_{n-1}(\dot{\sigma})$ is an isomorphism onto for $n > 1$, and into for $n = 1$.

Let f be the simplicial map of σ on itself which interchanges two vertices and leaves all others fixed. Then, for any $z \in H_n(\sigma, \dot{\sigma})$, we have $f_*(z) = -z$. This permits the usual division of permutations into the classes of even and odd, and leads naturally to a definition of orientation—a concept which is quite troublesome in the usual approach.

Definition: Let H, H' be two homology theories satisfying Axioms 1 through 7. A *homomorphism*

$$h: H \rightarrow H'$$

is defined to be a system of homomorphisms

$$h(q, X, A): H_q(X, A) \rightarrow H_q'(X, A)$$

defined for all q, (X, A), which commute properly with the boundary operator and induced homomorphisms:

$$h(q-1, A)\partial = \partial' h(q, X, A), \qquad h(q, Y, B)f_* = f_*' h(q, X, A). \qquad \text{(I)}$$

If h gives an isomorphism of the coefficient groups $h(0, P_0): H_0(P_0) \cong H_0'(P_0)$, then h is called a *strong homomorphism*. If each $h(q, X, A)$ is an isomorphism, then h is called an *equivalence* and H and H' are called *equivalent*.

Since the usual homology theory of complexes is deducible from the axioms, there follows the

Uniqueness Theorem: *Any two homology theories having the same coefficient group coincide on complexes.*

Explicitly, if $i: H_0(P_0) \cong H_0'(P_0)$ is an isomorphism between the coefficient groups of H and H', then isomorphisms

$$h(q, X, A): H_q(X, A) \cong H_q'(X, A)$$

can be defined for X a complex, A a subcomplex such that $h(0, P_0)$ coincides with i, and the relations (I) hold in so far as they are defined (f need not be simplicial). Indeed, there is just one way of constructing $h(q, X, A)$. The uniqueness theorem implies that any strong homomorphism $h: H \rightarrow$

H' is an equivalence as far as complexes are concerned. In view of Axiom 4, the uniqueness theorem holds for spaces having the same homotopy type as complexes. These include the absolute neighborhood retracts.

6. *Existence.*—As is to be expected, homology theories exist which satisfy the axioms. Both the Čech homology theory H^1 and the singular homology theory H^0 satisfy the axioms. This is fairly well known, although the proofs of some of the axioms are only implicitly contained in the literature. It is well known that the two homology theories differ for some pairs (X, A). Thus, the axioms do not provide uniqueness for all spaces.

The surprising feature of H^0 and H^1 that appears in this development is that they play extreme roles in the family of all homology theories, and have parallel definitions. They can be defined as follows: The homology groups of the simplicial structure of a complex (using chains, etc.) are defined as usual. (As a first step of an existence proof, this is quite natural since the definition has been deduced from the axioms.) Using maps $K \to X$ of complexes into the space X, the singular homology groups $H^0{}_q(X, A)$ can be defined using a suitable limiting process. Similarly using maps $X \to K$ of the space into complexes, the Čech homology groups $H^1{}_q(X, A)$ are obtained. It is then established that H^0 and H^1 are minimal and maximal in the family of all homology theories with a prescribed coefficient group in the sense that, if H is any homology theory, there exist strong homomorphisms $H^0 \to H \to H^1$. This is an indication of how it is possible to characterize H^0 or H^1 by the addition of a suitable Axiom 8.

7 *Generalizations.*—A suitable refinement of the axioms will permit the introduction of topologized homology groups.

Cohomology can be axiomatized in the same way as homology. It is only necessary to reverse the directions of the operators ∂ and f_* in the above axioms and make such modifications in the statements as these reversals entail. The analogous uniqueness theorems can be proved.

The products of elements of two cohomology groups with values in a third (in the usual sense) may also be axiomatized and characterized uniquely.

The Life and Work of Paul Erdős*

Béla Bollobás

January 4, 1999

"Every human activity, good or bad, *except mathematics*, must come to an end"—this was a favorite saying of Paul Erdős, mathematician extraordinaire, who passed away on September 20th 1996, aged 83, after a life devoted to mathematics. He was one of the giants of 20th century mathematics: he proved fundamental results in number theory, probability theory, approximation theory, geometry, interpolation theory, real and complex analysis, set theory and combinatorics. He practically created probabilistic number theory, partition calculus for large cardinals, extremal graph theory, and the theory of random graphs, and no one did more to develop and advocate the use of probabilistic methods throughout mathematics. He had a wonderful talent for interacting with other mathematicians, and so many of his best results were obtained in collaboration. Altogether, he wrote close to 1500 papers, about five times as many as other prolific mathematicians, and he had about 500 collaborators. The rate of his output peaked in later years: like Leonhard Euler (1707–1783), who produced more than half of his works after 1765 in spite of being blind, Erdős wrote more than half of his papers in the last 20 years of his life.

He was a personal friend of more mathematicians than anybody ever; he was eager to help whomever he came in touch with, and a large number of successful mathematicians today owe their careers to him. His love of mathematics, permeating his existence, was infectious, and fired the imagination of many a young colleague. He was eager to share his ideas and to pass on his love of mathematics: there is no doubt that he attracted me to combinatorics when, as a young schoolboy, I first met him. Having known him for almost forty years, it is difficult to imagine that there is mathematics without him.

Although he was interested in medicine, history and politics, and would put incisive questions to casual acquaintances on these subjects, he had a total devotion to mathematics. He never had a 'proper' teaching job, but traveled around the world, collaborating with mathematicians of all fields. His travels started long before traveling became commonplace for mathematicians, and he had many collaborators, years before that became as common a practice as it is today. It was claimed that on a longer

*This is an abbreviated version of my artice *To Prove and Conjecture: Paul Erdős and his Mathematics*, in Amer. Math. Monthly **105** (1998), 209–237.

train journey he would write a joint paper with the conductor, and this was not much of an exaggeration. He never had a check book or credit card, never learned to drive, and was happy to travel for years on end with two half-empty suitcases. "La propriété, c'est le vol", wrote Pierre Joseph Proudhon—"Property is a nuisance", echoed Paul Erdős, and his life was a testimony to this statement. For a while he did travel with a small transistor radio, but later he abandoned even that.

With his motto, 'another roof, another proof', he would arrive on the doorstep of a mathematical friend, bringing news of discoveries and problems: "declaring 'his brain open"' he would plunge into discussions about the work of his hosts, and after a few days of furious work on their problems, he would take off for another place, often leaving his exhausted hosts to work out the details and write up the papers.

In addition to producing an immense body of results, Erdős contributed to mathematics in three important ways: he championed *elementary methods*, he introduced *probabilistic methods* and turned them into powerful tools in many branches of mathematics, and, perhaps above all, he gave mathematics hundreds of exciting *problems*.

Let us start with *problems*, as the name of Erdős is practically synonymous with problems. If one classifies mathematicians into "problem solvers" and "theory builders" then Erdős was the purest of problem solvers. Not that he had anything against building big machines, but, from his early youth on, his taste ran to problems; and this tendency was reinforced by his mentor, Louis Mordell, and close friends Harold Davenport, Paul Turán, Richard Rado, Alfréd Rényi, and others. Unlike Einstein, who chose physics instead of mathematics for fear of working on the "wrong" questions, Erdős was happy to be seduced by any beautiful problem that came his way. However, he not only solved problems, but also created many: as a *problem poser*, the world has not seen anybody remotely like him. He had a wonderful talent for keeping an open mind, for asking searching questions, the answers to which uncovered hidden connections. As Ernst Straus put it, Erdős was the prince of problem solvers and the undisputed monarch of problem posers. Through his problems Erdős has greatly influenced many branches of mathematics, especially combinatorics, and will continue to do so for many years to come. To indicate his estimate of the difficulty of his problems, from the 1950s onwards he attached to them monetary rewards. These rewards tended to increase in time, as it emerged that the problems were rather hard, after all. Ernst Specker was the first to claim a reward, and Endre Szemerédi collected the largest sum, \$1000. It is hardly worth saying that nobody ever tackled an Erdős problem in order to earn money.

When, 65 years ago, Erdős started his career, *elementary methods* were out of fashion. Today this is even more so: the mathematical world is ruled by big theories straddling several branches of mathematics, and these big theories have had amazing successes, most notably the proof of Fermat's Last Theorem by Andrew Wiles. But Erdős believed that no matter how important sophisticated theories are, they cannot constitute all of mathematics. There are remarkably many natural questions in mathematics which one is unable to attack with sophisticated machines: rather than declaring that these problems are thereby uninteresting, we should be happy to answer them by whatever means. Erdős proved over and over again that elementary methods are frequently effective in attacking these 'untractable' problems, and that they often provide enlightening proofs of main-line results. Needless to say, an elementary proof need not be simple: often the opposite is the case.

The third major contribution of Erdős to mathematics was that he was the first to fully appreciate and exploit the power of *random methods* in order to attack a host of problems which have nothing to do with randomness and probability theory. In its simplest form, the probabilistic method is nothing but double counting, but at a higher level it is a delicate, complicated, and powerful tool. Just as every analyst automatically interchanges the order of integration, and every combinatorialist worth his salt double counts at the drop of a hat, Erdős was constantly on the lookout for opportunities to apply the probabilistic method. By now we all sense that frequently the best way of showing the existence of an object with seemingly contradictory properties is to work in an appropriate probability space, but this instinct has been instilled in us largely by Erdős, the apostle of random methods, who was the first to recognize their power and apply them repeatedly, many years before they became accepted. For many years, the probabilistic method was known as the *Erdős method.*

Paul Erdős was born into an intellectual Hungarian-Jewish family on March 26th 1913, in Budapest, amidst tragic circumstances: when his mother returned home from the hospital with the little Paul, she found that her two daughters had died of scarlet fever. Both his parents were teachers of mathematics and physics: his father was born Louis (Lajos) Engländer, but changed his name to the Hungarian Erdős ("of the forest", a fairly common name in Hungary), and his mother was born Anna Wilhelm. The following year, the first great Austro-Hungarian offensive of the First World War quickly turned into a disaster, and when the South flank was driven back through Lemberg (26–30 August), the present day Lvov in the Ukraine, many Hungarians were taken prisoner by the Russians. Among them was Lajos Erdős, who returned home from Siberia only six years later. In the absence of his father, the young Erdős was brought up by his mother and a German Fräulein. As a result of her terrible loss, Mrs. Erdős felt excessively protective towards her son throughout her life, and there was always an exceptionally strong bond between the two.

In 1919, at the end of the war, the Hungarian government could not accept the harsh demands of the victorious Entente, and in the ensuing turmoil a *Dictatorship of the Proletariat* was proclaimed. Practically the entire economy and cultural life were placed under state supervision, and everything was run by *Revolutionary Soviets.* Although many changes were introduced within a few weeks, the results did not come close to what had been planned. Béla Kun's Red Army could not withstand the relentless attack of the Romanians: the Dictatorship collapsed after four and a half months, and a counterrevolutionary terror followed. Mrs. Erdős, who had been a member of the soviet running her school, was dismissed from her job, and could never teach again. It is not surprising that this traumatic experience shaped Erdős's political outlook: throughout his life he remained sympathetic towards the left in every shape or form.

Erdős was a child prodigy: at the age of three he could multiply three digit numbers, and at the age of four he discovered negative numbers on his own. At the fashionable spa his mother took him to, he would ask the guests how old they were and tell them how many seconds they had lived.

His mother was so worried that he would catch diseases in a school that, for much of his school years, he was educated at home, mostly by his father, who taught him not only mathematics and physics, but English as well. As his father never really spoke English, having learned it from books, the young Erdős acquired a somewhat idiosyncratic pronunciation. Besides German and English, he learned French, Latin and Ancient Greek; later in life he picked up a smattering of Hebrew. Erdős spent two brief periods in school: first in the Tavaszmező Gymnasium and then in the St. Stephen Gymnasium. Many years later, when he met my father, they were astonished to discover that they had common classmates: my father joined the class just after Erdős had left it.

In addition to his parents, an important role was played by the *Journal of Mathematics and Physics for High Schools* (*Középiskolai Matematikai és Fizikai Lapok*) in nurturing and developing his interest in mathematics. The journal, founded by the young teacher Dániel Arany in 1893 as a mathematics journal (*Journal of Mathematics for High Schools*), appears ten times a year and specializes in publishing problems of various levels of difficulty, somewhat like the *American Mathematical Monthly.* In a subsequent issue model solutions are published and the photographs of the best problem solvers are printed in the final issue of the year. The readers are encouraged to generalize and strengthen the results, so the journal provides an exciting introduction to mathematical research. Later a physics section was added, and this was also acknowledged in the title in 1925. Erdős and many of his later friends were avid readers of the Journal and cut their mathematical teeth on its problems. A photograph of Erdős was published in each of his high school years, and a model solution printed under the joint names of Paul Erdős and Paul Turán was his "first joint paper" with Turán, whom he met only some years later and who became one of Erdős's closest collaborators and best friends.

In 1930, at the age of 17, Paul Erdős entered the *Pázmány Péter Tudományegyetem*, the science university of Budapest, founded by Peter Pázmány, Primate of Hungary, in 1635, and soon became the focal point of a small group of extremely talented Jewish mathematicians, all studying mathematics and physics. The group included Paul Turán, Dezső Lázár, George Szekeres, Esther Klein, László Alpár, Martha Svéd, and others: they discussed mathematics not only at the university, but also in the afternoons and evenings. One of their favorite meeting places was the Statue of Anonymus, the chronicler of Béla III (1173–1196).

It was during this period that Erdős started to develop his own special language: he called a child an *epsilon*, a woman a *boss*, a man a *slave*; *Sam* (or better still, *sam*) was the U.S. and *Joe* (or *joe*) was the Soviet Union, and so on. In his words, a slave could be *captured* and later *liberated*, one could drink a little *poison* and listen to *noise*, and a mathematician could *preach*, usually to the converted.

At the university, Erdős learned most from Lipót (Leopold) Fejér, the great analyst, and Dénes (Dennis) König, the author of the first book on graph theory. Erdős immersed himself in number theory, which remained his love all his life, and as a second year undergraduate obtained a new proof of Chebyshev's theorem, which later prompted Nathan Fine to write the rhyme:

"Chebyshev said, and I say it again,
There's always a prime between n and $2n$."

In 1934, Erdős finished university and, on the basis of his thesis on primes in arithmetic progressions, was awarded a doctorate as well. Much of the work for his thesis was done in his second year. Like most well-off Hungarians of talent, he intended to continue his studies in Germany but, as "Hitler got there first", he decided to go to England, to join Louis Mordell's exceptional group of number theorists in Manchester. On his way to Manchester, he stopped in Cambridge, where he met Harold Davenport and Richard Rado, who later became two of his closest friends.

In Manchester Erdős held various fellowships, which did not require him to teach, and he was free to do research under Mordell's guidance. In 1937, Davenport left Cambridge to join Mordell's group, and soon Erdős and Davenport became the best of friends.

I have a special reason to be grateful for the Erdős-Davenport friendship: many years later, Erdős helped me to get to Trinity College, Cambridge, precisely because Davenport was a Fellow of that college.

In 1938 Erdős left Manchester for the Institute for Advanced Study in Princeton. In the stimulating atmosphere of the place, his talent blossomed as never before; even almost 60 years later he thought that 1938/39 was his *annus mirabilis*. He wrote outstanding papers with Mark Kac and Aurel Wintner, which practically created *probabilistic number theory*, he wrote a major paper with Paul Turán on *approximation theory*, and he solved an important problem of Witold Hurewicz in *dimension theory*. Despite the hints and verbal assurances which followed this tremendous output, his Fellowship at the Institute was not continued. The memory of this pained him even to the end, although he always added that the Institute did not have much money.

Without a Fellowship to support him, he was forced to embark on his travels: he visited Philadelphia, Purdue, Notre Dame, Stanford, Syracuse, Johns Hopkins, Ann Arbor, and elsewhere, and like a Wandering Scholar of the Middle Ages, he never stopped again. The war years were very hard on him, as it was difficult to hear from his parents and the news was distressing, but he kept producing mathematics at a prodigious rate. In addition to the many important papers he produced by himself, his genius for collaboration blossomed: he wrote outstanding papers with Mark Kac, Kai Lai Chung, Ivan Niven, Arye Dvoretzky, Shizuo Kakutani, Arthur Stone, Leon Alaoglu, Alfred Tarski, Irving Kaplansky, Gábor Szegő, William Feller, Fritz Herzog, George Piranian, and others.

In the fall of 1948 Erdős returned to Europe for the first time in a decade: after two months in Holland, where he worked with Nicolaas de Bruijn and Jurgen Koksma, he arrived in Budapest on 2nd December, to be reunited with his beloved mother. The great joy of seeing his mother was tinged with sadness, as his father had died of a stroke in 1942, and most of his relatives had been murdered in the Holocaust. In the Yalta Agreement, Hungary was given to Stalin, and the changes were already felt in the country: following a brief period of freedom and democracy, Hungary was sinking once more into a dictatorship. After a stay of about two months, Erdős left Hungary for England, before returning to the U.S. two months later.

For Erdős, 1949 was a momentous and bittersweet year. Atle Selberg found an ingenious elementary proof for his asymptotic formula concerning the distribution of primes, and after a substantial contribution by Erdős, Selberg completed an elementary proof of the Prime Number Theorem (PNT). Later both Erdős and Selberg found simpler elementary proofs. This elementary proof of the PNT was the great mathematical event of 1949, since the search for an elementary proof had been on for over 50 years, ever since 1896, when Hadamard and de la Vallée Poussin proved the PNT. There was much hope that the ideas involved in the proof would penetrate and revolutionize number theory. Subsequent events showed that this hope was unfounded. It was a great pity for Erdős that the planned back-to-back publications did not materialize, and Selberg [46] and Erdős [49.02] published their contributions in different journals. No doubt Erdős could have handled the delicate situation with more tact, but unquestionably he had no base motives. At the International Congress of Mathematicians at Harvard in 1950, Atle Selberg was awarded a Fields Medal, and much of the citation was about the elementary proof of the PNT.

Following the Congress, Erdős left the U.S. for Europe. After a year in Aberdeen, he spent the academic year 1951/52 at University College, London; while there, he renewed his friendship and collaboration with Harold Davenport and Richard Rado. He then returned to the U.S. and for the next two years he was mostly at Notre Dame.

The year 1954 brought a great trauma for Erdős: he left the U.S. for the International Congress of Mathematicians in Amsterdam without having obtained a reentry permit. In later life, he frequently claimed that "sam tried to starve him to death" by not allowing him to return to the American universities where he was supported. However, his claim was probably strongly colored by his desire to show that America was almost as bad as the Soviet Union. He could have gone to Amsterdam and returned to the U.S. without any difficulty, had he not insisted that he would do so with a *Hungarian passport.* The relations between the U.S. and the Hungary of Rákosi (the

Hungarian Stalin) were very bad at the time and Erdős was absolutely inflexible. He left, saying that neither sam nor joe could restrict his right to travel.

In his distress, having been left with neither a country nor any means of supporting himself, he turned to Israel for support. He was received with open arms: the Hebrew University in Jerusalem offered him a job, and the state of Israel offered him a passport. He accepted the employment, but when the officials asked him whether he wanted to become an Israeli citizen, he politely refused, saying that he did not believe in citizenship. Nevertheless, from then on he visited Israel almost every year.

In the communist Hungary of the 1950s, ordinary Hungarians were not allowed to visit a Western country, not even for a short period. Although Westerners were permitted to visit Hungary, they were viewed with hostility. Thus it was a tremendous achievement when in 1955 George Alexits, a good friend of Erdős, managed to persuade the officials to permit Erdős to enter the country *and* to leave again. From then on, Erdős returned to Hungary once or twice a year, partly in order to spend more and more time with his mother, and also to collaborate with Hungarian mathematicians, especially Paul Turán and Alfréd Rényi. For many a Hungarian mathematician, Erdős was the window to the West.

As a young pupil, I first heard him lecture during one of his earlier brief visits when, in 1957, he addressed the Junior Mathematical Society in Budapest. But it was during his next visit to Hungary, in 1958, that I first came to meet him: he summoned me to the elegant hotel where he was staying with his mother, and told me numerous beautiful results and exciting problems in combinatorics, geometry and number theory. It was impossible not to fall under his spell. His mother, whom by then most people called *Annus Néni* (Aunt Anna), was constantly by his side: they looked after each other with great affection, and she was a hostess to all his mathematical visitors. In 1961 they got to know my parents, and from then on they were frequent visitors to our house, especially for Sunday lunches. My father, who was a physician, looked after them both.

His ground-breaking work on random graphs with Alfréd Rényi started in the late fifties: in a series of brilliant papers they laid the foundation of the theory of random graphs. Their main discovery was that, for many a monotone increasing property, there is a *sharp threshold*: graphs of order n with slightly fewer edges than a certain function $f(n)$ are very unlikely to have the property, while graphs with slightly more than $f(n)$ edges are almost certain to have the property. They also discovered the *phase transition* in the component structure: for $c < 1$, in most graphs with n vertices and $\lfloor cn \rfloor$ edges every component has $O(\log n)$ vertices, while for $c > 1$ most graphs with n vertices and $\lfloor cn \rfloor$ edges have a *giant component* with at least $\gamma(c)n$ vertices, where $\gamma(c) > 0$.

The late 1950s also saw the start of much research in *transfinite combinatorics*: in 1958 he published the first of over fifty joint papers with András Hajnal, and in 1965, with Rado and Hajnal, in a difficult paper of over 100 pages they named the *giant triple paper*, he founded *partition calculus*, a detailed study of the relative sizes of large cardinals and ordinals. Erdős was rather disappointed when later 'independence reared its ugly head', and many of the natural questions turned out to be undecidable.

In addition to his work with Rényi on random graphs, the highlight of the 1960s was his collaboration with András Sárközi and Endre Szemerédi on divisibility properties of sequences. This continued the work he started with Davenport in the 1930s. Another major topic of the day was *statistical group theory*: in a series of substantial papers, Erdős and Turán laid the foundations of this subject.

Erdős's brief stays in Hungary drove home the painful truth that for three decades he had seen rather little of his mother, so they started to travel together. Their first trip was to Israel in November 1964 and the second to England in 1965. They traveled all over Europe, the United States and Canada; in 1968 they even went to Australia. Erdős was truly happy then: he was with his beloved mother, his mathematics was in full bloom, and he was respected by all. If only time could have stood still and these halcyon days could have lasted longer! It was a terrible tragedy for Erdős when, in January 1971, Mrs. Erdős died during a trip to Calgary. He never recovered from the blow, and from then on was perhaps never really happy. It did not help that he lost many of his closest friends: Harold Davenport in 1969, Alfréd Rényi in 1970, and, above all, Paul Turán in 1976. One of his favorite sayings, "Pusztulunk, veszünk!" (We are perishing, vanishing!—in poor translation) seemed less and less of a joke.

Since the fifties Erdős never had anything resembling a permanent home, and even a stay of a few months was a rarity. In the mid 70s he made an exception when he came to visit me for a semester in Cambridge, where I had moved in 1969. Nevertheless, he considered Budapest his permanent base, even though he spent only a small fraction of the year there. There was only one substantial break in his visits to Hungary: when his friends from Israel were not allowed to enter Hungary for the international conference for his 60th birthday, he got so upset that for three years he did not return to Hungary. While in Budapest, he liked to stay in the Guest House of the Academy, two doors away from my mother's house, and always set up his headquarters in the office of the Director of the Mathematical Institute. Most days he had a steady flow of collaborators, so that he was likely to be working at the same time on several problems, with several sets of collaborators. He was especially happy working with Vera Sós, András Hajnal, András Sárközy, Endre Szemerédi, Miklós Simonovits, and András Gyárfás.

For the last twenty years of Erdős's life, Ron Graham of AT&T Bell Laboratories provided a fixed point in his life by looking after his finances, forwarding his mail, taking care of manuscripts, and so on. When he was in the New York area, he frequently stayed with Ron Graham and Fan Chung, who arranged one of their rooms for him, but he was also keen to work with János Pach, Joel Spencer, Melvyn Nathanson, and Peter Winkler as well, to mention only some of his collaborators. He had a permanent position in Memphis, where he did much work with Ralph Faudree, Cecil Rousseau, and Dick Schelp. There were numerous other permanent ports of call: Calgary with

Eric Milner, Richard Guy, and Norbert Sauer, Kalamazoo with Yousef Alavi, Atlanta with Dick Duke and Vojtěch Rödl, Lyon with Jean-Louis Nicolas, Bielefeld with Walter Deuber, Poznań with Michal Karoński, Tomasz Łuczak, and Andrzej Ruciński, Prague with Jarik Nešetřil, Athens, Georgia, with Carl Pomerance and Andrew Granville, and so on. He frequently came to see me in Cambridge and Baton Rouge; when he was in England, he always looked up the Rado family in Reading, Yael Dowker in London, and Mrs. Davenport in Cambridge.

Given his lifestyle, it is not surprising that he was a *conference-junkie*: the only reason why he would miss a conference he was interested in was that he went to another meeting. Most of us are totally exhausted if we have to attend back-to-back conferences, whereas he thrived on three or more meetings without any break. He was happy that every fifth year from his 65th birthday, I organized a meeting in his honor in Cambridge, England, making sure that his birthday took place during the conference. As he liked to say, over the years he progressed from *prodigy* to *dotigy*; despite his good humor it was sad to witness this transformation. Although from his late teens he had been saying that he was old, he felt that at 80 the joke was wearing thin. Throughout 1993, the year of his 80th birthday, there was a slew of international conferences dedicated to him, with the largest by far in Hungary, where most of his friends were in attendance. He did not care for honors, but was happy to receive honorary degrees from a good many places, including the ancient universities of Cambridge and Prague, and he was delighted to be awarded the Wolf Prize.

Occasionally Erdős talked of *The Book*: a transfinite book whose pages contain all the theorems and their best possible proofs. Unfortunately, went on Erdős, The Book is held by God who, being malicious towards us, only very rarely allows us to catch a glimpse of a page. But when that happens, then we see mathematics in all its beauty.

In writing and talking about Erdős, The Book is frequently overemphasized: he himself always insisted that this is only a joke, which should not be taken seriously, lest it damage the mathematics we like. By its very nature, a book proof tends to be short and snappy: it is as if lightning allowed us to see some detail clearly. When it comes to substantial results, the best we can feel is that the *global idea* of the proof is from The Book.

Nevertheless, we shall present four proofs Erdős himself considered to be straight from The Book: all have close connections with him.

First, we shall prepare the ground for the *fourth* example. Anning and Erdős [45.01] proved the following simple but surprising result: for every n one can find n points on a circle such that their distances are all integral, but it is impossible to find infinitely many points in the plane not all on a line such that all distances are integers.

To prove the first assertion, let $p_1, p_2, \dots$ be the primes of the form $4k+1$, so that there are $1 \le a_i < b_i \le p$ such that $p_i^2 = a_i^2 + b_i^2$. Let (x_i, y_i) be a point on the circle $x^2+y^2 = 1/4$, which is at distance a_i/p_i from $(1/2, 0)$ and so b_i/p_i from $(-1/2, 0)$. Then the sequence of points $(-1/2, 0)$, $(1/2, 0)$, (x_1, y_1), (x_2, y_2), ... is such that the distance between any two is rational. Indeed, this is an immediate consequence of Ptolemy's theorem since for any four concyclic points $(-1/2, 0)$, $(1/2, 0)$, (x_i, y_i), (x_j, y_j), five of the distances are rational by our choice of the points, so the sixth distance, that between (x_i, y_i) and (x_j, y_j), is also rational. By enlarging the radius of the circle we can obtain n points on a circle with integral distances.

Although this argument is certainly pretty, it is not nearly as striking as the book-proof Erdős gave of the second part a little later [45.02]. As some readers may not have seen this proof, to give them a chance to think about the problem a little and thereby better appreciate the proof Erdős gave, we postpone it to last.

Our first example of a book-proof also needs some preparation. Starting with one of his earliest papers [35.01], written with George Szekeres, Erdős was a great devotee of Ramsey's theorem, which he generalized, applied, and popularized on numerous occasions throughout his life; it also led him to partition calculus. A special case of Ramsey's theorem claims that, for every natural number s, if n is large enough then every red-blue coloring of the edges of K_n, a complete graph on n vertices, contains a monochromatic K_s, that is a complete graph K_s, with all edges red or all edges blue. The smallest value of n that will do is denoted by $R(s)$, so that $R(s) - 1$ is the maximal value of n for which some red-blue coloring of K_n has no monochromatic K_s. In particular, $R(1) = 1$, $R(2) = 2$ and $R(3) = 6$. The upper bound Ramsey had given for $R(s)$ was greatly improved by Erdős and Szekeres [35.01], when they showed that

$$R(s) \le \binom{2s-2}{s-1} \le 2^{2s-2}.$$

Although over sixty years have passed since then, this very simple bound has hardly been improved: the best result to date is due to Thomason [51], who proved that there is an absolute constant A such that

$$R(s) \le e^{A\sqrt{\log s}} \frac{1}{s} 2^{2s}.$$

But what about a lower bound? Polynomial lower bounds are easy to come by, but as one of the first unexpected applications of his probabilistic method, Erdős [47.09] obtained an exponential lower bound. In [59.06, 61.06] he proved more sophisticated results in the same vein, about off-diagonal Ramsey numbers, and about graphs of large chromatic number and large girth. Instead of *constructing* an appropriate graph, he simply considered a random red-blue coloring of (the edges of) K_n. The expected number of monochromatic K_s subgraphs is $\binom{n}{s}2^{-\binom{s}{2}+1}$, since we have $\binom{n}{s}$ choices for the vertex set and the probability that all $\binom{s}{2}$ edges of a K_s subgraph get the same color is $2^{-\binom{s}{2}+1}$: having colored one edge, the remaining $\binom{s}{2} - 1$ edges must get the same color. Now if

$$\binom{n}{s} 2^{-\binom{s}{2}+1} < 1$$

then *some* coloring of K_n has *no* monochromatic K_s, so $R(s) > n$. In particular, $R(s) > cs2^{s/2}$ for some $c > 0$.

The next result, proved by Erdős, Ko, and Rado [61.07] in 1961, is of fundamental importance in extremal combinatorics, and was the starting point of much research. Let $2 \leq r < n/2$, and let $\mathcal{A}$ be a collection of r-subsets of $[n] = \{1, 2, \ldots, n\}$. Suppose that $\mathcal{A}$ is an *intersecting* family, that is $A \cap B \neq \emptyset$ for all $A, B \in \mathcal{A}$. Then

$$|\mathcal{A}| \leq \binom{n-1}{r-1}.$$

The family $\mathcal{A} = \{A \subset [n] : 1 \in A,\ |A| = r\}$ shows that the inequality cannot be improved; in fact, this is the unique extremal family.

The Erdős-Ko-Rado theorem has many beautiful proofs: here we shall present one from The Book, given by Katona [27]. Arrange the elements of $[n]$ in a *cyclic order*. How many of the sets $A \in \mathcal{A}$ can form *intervals*, i.e. sets consisting of consecutive elements? At most r, since if $(a_1, a_2, \ldots, a_r)$ is one of these intervals then for every i, $1 \leq i \leq r-1$, at most one interval from $\mathcal{A}$ contains precisely one of a_i and a_{i+1}.

Now the probability that a set $A \in \mathcal{A}$ is an interval in a *random* cyclic order is clearly $r!(n-r)!/(n-1)!$, so

$$|\mathcal{A}| \leq r\frac{(n-1)!}{r!(n-r)!} = \binom{n-1}{r-1},$$

as claimed.

In 1946, Erdős [46.03] conjectured that the number of unit distances among n points in the plane is at most $n^{1+o(1)}$ and proved that this number is at most $cn^{3/2}$. After various improvements by Józsa and Szemerédi [25], and Beck and Spencer [5], the best bound to date, $cn^{4/3}$, was proved by Spencer, Szemerédi, and Trotter [47]. Recently, Székely [49] found a beautiful proof: it was probably the last proof Erdős judged to have come from The Book.

The proof is based on a result of Leighton [32] and Ajtai, Chvátal, Newborn, and Szemerédi [1], stating that every planar graph with n vertices and $m \geq 4n$ edges has crossing number at least $m^3/100n^2$, i.e., that every drawing of the graph in the plane contains at least this many crossing pairs of edges. Here the constant 100 is unimportant and is easily improved, but m^3/n^2 cannot be improved.

Given n points in the plane with m unit distances, let us draw a multigraph in the plane with this set of vertices as follows. Draw a unit circle about every point that has at least three points at unit distance from it. Join the consecutive points on the unit circles by the circular arcs. In this way some pairs of points are joined by two circular arcs: discarding one of them we obtain a graph drawn in the plane with at least $m - n$

edges. The number of crossings of this graph is at most $n(n-1)$ since any two circles intersect in at most two points. Therefore either $m - n \leq 4n$ and so $m \leq 5n$, or else $(m-n)^3/100n^2 \leq n^2$, and so $m \leq 6n^{4/3}$.

Now, as our fourth and final example, we present the book proof from [45.02] we have postponed. This proof is so short that in his review of the paper, Irving Kaplansky [26] simply *reproduced the entire paper!* We can do no better either. Here is Kaplansky's review:

"In the note under the same title [Erdős is referring to his paper with Anning [45.01]] it was shown that there does not exist in the plane an infinite set of noncollinear points with all mutual distances integral.

"It is possible to give a shorter proof of the following generalization: if A, B, C are three points not in line and $k = [\max(AB, BC)]$, then there are at most $4(k+1)^2$ points P such that $PA - PB$ and $PB - PC$ are integral. For $|PA - PB|$ is at most AB and therefore assumes one of the values $0, 1, \ldots, k$, that is, P lies on one of $k+1$ hyperbolas. Similarly P lies on one of the $k+1$ hyperbolas determined by B and C. These (distinct) hyperbolas intersect in at most $4(k+1)^2$ points. An analogous theorem clearly holds for higher dimensions."

It is rather ironic that in the collection of papers of Erdős [73.27] the original paper of Anning and Erdős [45.01] is reproduced, but not the brief paper just quoted.

No matter how beautiful these proofs are, the reputation of Erdős rests on his more substantial results. It would be wonderful to concentrate on a dozen or so papers, but that would be most unfair to a mathematician with about 1500 papers. He himself always found it impossible to select his "Top Ten": whenever he tried to, he ended up with at least forty.

His first paper [32.01], written as a first-year undergraduate, was on Bertrand's postulate: *for every* $n \geq 1$, *there is a prime* p *satisfying* $n < p \leq 2n$. This was first proved by Chebyshev, and Ramanujan [38] gave a considerably simpler proof of it in 1919. Later the great German mathematician Edmund Landau [31, pp. 66–68] gave a particularly simple proof of the assertion that for some $q > 1$ there is always a prime between n and qn. Landau did not give an estimate for q, but it was clear that his proof did not permit q to be taken to be 2. By sharpening Landau's argument, and concentrating on the prime divisors of $\binom{2n}{n}$, Erdős gave a simple and elementary proof of Bertrand's postulate.

Erdős further developed these ideas in [35.10] in order to apply them to primes in arithmetic progressions. In this paper the nineteen year old Erdős gave elementary proofs of extensions of some very recent results of Breusch [9]. Independently of Erdős, similar results were obtained by Ricci [41], [42].

Writing $\sigma(n)$ for the sum of the positive divisors of a natural number n, we call n *perfect* if $\sigma(n) = 2n$, *abundant* if $\sigma(n) \geq 2n$, and *deficient* if $\sigma(n) < 2n$. Several of the early papers of Erdős concerned abundant numbers, especially their density $\lim_{x\to\infty} \tilde{A}(x)/x$, where $\tilde{A}(x)$ is the number of abundant numbers not exceeding x. Schur conjectured that the abundant numbers have positive density. Using Fourier analysis, this was proved by Behrend, Chowla, and Davenport; Erdős [34.04] gave an elegant elementary proof.

A number is *primitive abundant* if it is abundant but every divisor of it is deficient. Denoting by $A(x)$ the number of primitive abundant numbers not exceeding x, Erdős [35.05] proved that for some positive constants one has

$$xe^{-c_1(\log x \log\log x)^{1/2}} < A(x) < xe^{-c_2(\log x\log\log x)^{1/2}}. \tag{1}$$

From (1) it follows that the sum of the reciprocals of primitive abundant numbers is convergent, and that implies that $\lim_{x\to\infty} \tilde{A}(x)/x$ exists. Fifty years later, Ivič [23] simplified the proof of (1) and obtained better constants; recently Avidon [3] proved that here any $c_1 > \sqrt{2}$ and $c_2 < 1$ will do, provided x is large enough. However, it is still open, whether

$$A(x) = xe^{-(c+o(1))(\log x \log\log x)^{1/2}}$$

for some c, $1 \le c \le \sqrt{2}$. Erdős [96.17] also wondered whether one could prove that $\tilde{A}(2x)/\tilde{A}(x) \to 2$ as $x \to \infty$.

The study of abundant numbers led Erdős to a variety of problems concerning real-valued additive arithmetical functions, that is, to functions $f : \mathbb{N} \to \mathbb{R}$ such that $f(ab) = f(a) + f(b)$ whenever a and b are relatively prime. Hardy and Ramanujan [20] proved that if $g(n) \to \infty$ then

$$|\nu(n) - \log\log n| < g(n)(\log\log n)^{1/2} \tag{2}$$

holds for almost every n, where $\nu(n)$ denotes the number of distinct prime factors of n. In other words, the density of n satisfying (2) is 1. Erdős [37.02] extended this result by showing that the median is about $\log\log n$: the number of integers $m \le n$ for which $\nu(m) > \log\log n$ is $\frac{1}{2}n + o(n)$.

In 1934, Turán [52] gave a brilliant elementary proof of this theorem. To be precise, Turán proved considerably more, namely that:

$$\sum_{m=1}^{n}\{\nu(m) - \log\log n\}^2 = n\log\log n + o(n\log\log n). \tag{3}$$

What (3) says is that if we turn $[n] = \{1, 2, \ldots, n\}$ into a probability space by giving each $m \in [n]$ probability $1/n$, then ν becomes a *random variable* with mean and variance $(1 + o(1))\log\log n$. In particular, (2) is an immediate consequence of (3).

Surprisingly, G.H. Hardy, one of the greatest mathematicians of the day, did not appreciate the proof, which he acknowledged as a curiosity. The young Erdős, on the other hand, was very receptive, and was ready to develop the ideas further. In particular, by making use of Turán's ideas, he proved [35.06], [37.05] that if $f(m)$ is any non-negative arithmetical function then, for every constant c,

$$\frac{1}{n}|\{m \leq n : f(m) \leq c\}|$$

tends to a limit as $n \to \infty$. As a slight extension of his theorem, Turán also proved [35.06] that if f is a non-negative additive arithmetical function satisfying

$$\sum_{p \leq n} \frac{f(p)}{p} = \psi(n) \to \infty,$$

where the summation is over primes p not exceeding n, then, for every $\varepsilon > 0$, there are only $o(n)$ integers $m \leq n$ for which

$$|f(m) - \psi(n)| > \varepsilon\psi(n).$$

The case when $f(p^\alpha) = 1$ for every prime power is precisely the Hardy-Ramanujan theorem.

In 1939 Erdős joined forces with Kac to write a ground-breaking paper (see [39.09] for an announcement and [40.12] for a detailed exposition) on additive arithmetical functions, strongly extending the results above. With this paper, Erdős, the number theorist, and Kac, the probabilist, founded probabilistic number theory, although the subject did not really take off until several years later. Among others, they proved that if a bounded real-valued additive arithmetical function $f(m)$ satisfies $\sum_p f(p)^2/p = \infty$ then, for every fixed $x \in \mathbb{R}$,

$$\lim_{m \to \infty} A_x(m)/m = \Phi(x),$$

where $A_x(n)$ is the number of positive integers $m \leq n$ satisfying

$$f(m) < \sum_{p \leq n} f(p)/p + x\left(\sum_{p \leq n} f(p)^2/p\right)^{1/2}.$$

Here, as usual,

$$\Phi(x) = \frac{1}{\sqrt{2\pi}} \int_{-\infty}^{x} e^{-t^2/2}\, dt$$

is the *standard normal distribution.* In other words, under very mild conditions, the arithmetical function $f(m)$ satisfies the Gaussian law of error!

Given a real-valued function $f(m)$ and a real number c, denote by $N(f; x, n)$ the number of integers $m \leq n$ at which $f(m) < x$. If there is a monotone increasing function $\sigma : \mathbb{R} \to [0,1]$ with $\lim_{x\to-\infty} \sigma(x) = 0$ and $\lim_{x\to\infty} \sigma(x) = 1$, such that at every point of continuity x of σ we have

$$\lim_{n\to\infty} N(f;x,n)/n = \sigma(x),$$

then f is said to have an *asymptotic distribution function*, σ. Culminating in [38.09], Erdős proved that an additive arithmetical function $f(m)$ has an asymptotic distribution function, provided

$$\sum_p \frac{f'(p)}{p} \quad \text{and} \quad \sum_p \frac{(f'(p))^2}{p} \tag{4}$$

both converge, where $f'(p) = f(p)$ for $|f(p)| \le 1$ and $f'(p) = 1$ otherwise. Also, if $\sum_{f(p)\ne 0} 1/p$ diverges then the distribution function is continuous. Following his work with Kac, Erdős proved with Wintner [39.01] that the convergence of the series in (4) is also necessary for the existence of a distribution function.

Some years later, Erdős [46.06] returned to the problem, and considered the case when the second series in (4) is divergent. He proved the stunning result that if $f(p) \to 0$ as $p \to \infty$ and $\sum_p (f'(p))^2/p = \infty$ then the distribution function of $F(m) = f(m) - \lfloor f(m) \rfloor$ is x. (To be precise, the distribution function $\sigma(x)$ is 0 for $x \le 0$, x for $0 \le x \le 1$, and 1 for $x \ge 1$.) For many related results, see the two-volume treatise on probabilistic number theory by Elliot [12, 13].

The last paper Erdős wrote in Budapest before leaving for Manchester, concerned another favorite topic of his, the difference between consecutive primes. Writing p_n for the nth prime, he proved [35.07] that for some $c > 0$ and infinitely many n one has

$$p_{n+1} - p_n > \frac{c \log n \log\log n}{(\log\log\log n)^2}.$$

As Erdős used to say, a little later Rankin [39] smuggled in a factor $\log\log\log\log n$, so that

$$p_{n+1} - p_n > \frac{c \log n \log\log n \log\log\log\log n}{(\log\log\log n)^2} \tag{5}$$

for some $c > 0$ and infinitely many n. The original value was improved by Schönhage [45] and later by Rankin [40]. Recently, Maier and Pomerance [35] proved (5) with a slightly larger constant: this is the best result to date. Although this improvement given by Maier and Pomerance seems to be slight, they introduced new tools, which may lead to substantial improvements.

Again as he used to say, "somewhat rashly" Erdős [90.29] offered $10,000 for a proof that (5) holds for every c. Most unusually for him, in [97.18] he *reduced* the offer to $5000, and instead offered $10,000 for a proof that

$$p_{n+1} - p_n > (\log n)^{1+\varepsilon} \tag{6}$$

holds for some $\varepsilon > 0$ and infinitely many values of n. As he remarked, a proof of (6) would have actually cost him \$15,000! How sad that by the time [97.18] appeared he was not with us, so these offers became void!

Some time in the early 1800s, it was conjectured that the product of two or more consecutive integers is never a square, cube or any higher perfect number. To be precise, the equation

$$(n+1)(n+2)\cdots(n+k) = x^\ell \tag{7}$$

has no solution in integers with $k, \ell \geq 2$ and $n \geq 0$. In 1939, Erdős [39.03, 39.04] and Rigge [43] proved the conjecture for $\ell = 2$, and they showed also that for any $\ell \geq 2$ there are at most finitely many solutions to (7). Later Erdős [55.08] found a different proof of this result. By making use of these ideas in [55.08], Erdős and Selfridge [75.46] proved the full conjecture. Although the work was done in the mid-sixties, the paper was published only in 1975.

In 1940 Turán [53] proved a beautiful result concerning graphs, vaguely related to Ramsey's theorem: every red-blue coloring of a complete graph with many red edges contains a large red complete subgraph. To be precise, *for* $3 \leq r \leq n$, *every graph of order* n *that has more edges than any* $(r-1)$*-partite graph of order* n *contains a complete graph of order* r. With this result as the starting point, Erdős and his collaborators founded a large and lively branch of combinatorics, *extremal graph theory*. In order to formulate the basic problem of extremal graph theory, we write $|G|$ for the *order* (number of vertices) and $e(G)$ for the *size* (number of edges) of a graph G. Given graphs G and H, $H \subset G$ means that H is a *subgraph* of G. Let F be a fixed graph, usually called the *forbidden graph*. Set

$$ex(n; F) = \max\{e(G) : |G| = n \text{ and } F \not\subset G\}$$

and

$$EX(n; F) = \{G : |G| = n,\ e(G) = ex(n; F), \text{ and } F \not\subset G\}.$$

We call $ex(n; F)$ the *extremal function*, and $EX(n; F)$ the *set of extremal graphs* for the *forbidden graph* F. Then the first problem of extremal graph theory is to determine, or at least estimate, $ex(n; F)$ for a given graph F and, if possible, to determine $EX(n; F)$. A slight extension of this is problem is to exclude several forbidden graphs, i.e., the problem of determine the functions $ex(n; F_1, \dots, F_k)$ and $EX(n; F_1, \dots, F_k)$ for a finite family $F_1, \dots, F_k$ of forbidden graphs.

Writing $T_k(n)$ for the unique k-partite graph of order n and maximal size (so that $T_k(n)$ is the *k-partite Turán graph* of order n), Turán proved, in fact, that $EX(n; K_r) = \{T_{r-1}(n)\}$, i.e., $T_{r-1}(n)$ is the *unique* extremal graph, and so $ex(n; K^r) = t_{r-1}(n)$, where $t_{r-1}(n) = e(T_{r-1}(n))$ is the size of $T_{r-1}(n)$.

As Erdős frequently said, he came very close to founding extremal graph theory *before* Turán proved his theorem: in 1938, in connection with sequences of integers no one of which divided the product of two others, he proved that for a *quadrilateral* C^4 we have $ex(n; C^4) = O(n^{3/2})$. However, at the time he failed to see the significance of problems of this type: he was "blind".

The importance of Turán's theorem is greatly enhanced by the fact that it is not very far from the *fundamental theorem of extremal graph theory*, proved by Erdős and Stone [46.08]. By Turán's theorem, the maximal size of a K_r-free graph of order n is about $(r-2)\binom{n}{2}/(r-1)$; in fact, trivially,

$$\frac{r-2}{r-1}\binom{n}{2} \leq t_r(n) \leq \frac{r-2}{r-1}\frac{n^2}{2}.$$

Writing $K_r(t)$ for the *complete r-partite graph with t vertices in each class*, so that $K_r(t) = T_r(rt)$, Erdős and Stone proved in 1946 that if $r \geq 2$, $t \geq 1$ and $\epsilon > 0$ are fixed and n is sufficiently large then *every graph of order n and size at least* $((r-2)/(r-1)+\epsilon)\binom{n}{2}$ *contains a $K_r(t)$*. In other words, *even ϵn^2 more edges* than can be found in a Turán graph *guarantee not only a K_r but a "thick" K_r*, one in which every vertex has been replaced by a group of t vertices.

The significance of the paper is that not only does it give much information about the *size* of extremal graphs, but it is also the starting point for the study of the *structure* of extremal graphs. If F is a non-empty r-chromatic graph, i.e., $\chi(F) = r \geq 2$, then, precisely by the definition of the chromatic number, F is not a subgraph of $T_{r-1}(n)$, so $ex(n; F) \geq t_{r-1}(n) \geq (r-2)\binom{n}{2}/(r-1)$. On the other hand, $F \subset K_r(t)$ if t is large enough (say, $t \geq |F|$), so if $\epsilon > 0$ and n is large enough then

$$ex(n; F) < \left(\frac{r-2}{r-1}+\epsilon\right)\binom{n}{2}.$$

In particular, if $\chi(F) = r \geq 2$ then

$$\lim_{n\to\infty} ex(n; F)/\binom{n}{2} = \frac{r-2}{r-1},$$

that is, the asymptotic density of the extremal graphs with forbidden subgraph F is $(r-2)/(r-1)$. Needless to say, the same argument can be applied to the problem of forbidding any finite family of graphs: given graphs $F_1, F_2, \ldots, F_k$, with $\min \chi(F_i) = r \geq 2$, we have

$$\lim_{n\to\infty} ex(n; F_1, \ldots, F_k)/\binom{n}{2} = \frac{r-2}{r-1}.$$

In an important series of papers, starting in 1966, Erdős and Simonovits went considerably further than noticing this instant consequence of the Erdős-Stone theorem.

Among other results, Erdős and Simonovits proved [66.01] that if $G \in EX(n;F)$, with $\chi(F) = r \geq 2$, then G can be obtained from $T_{r-1}(n)$ by adding and deleting $o(n^2)$ edges. Later this was refined to several results concerning the *structure* of extremal graphs.

Returning to the Erdős-Stone theorem itself, let us remark that Erdős and Stone also gave a bound for the *speed* of growth of t for which $K_r(t)$ is guaranteed to be a subgraph of every graph with n vertices and at least $((r-2)/(r-1)+\epsilon)\binom{n}{2}$ edges. Thirty years later, Erdős and I [73.18] gave a substantially better bound for the speed, which was essentially best possible. Further refinements were given in [75.28], [11] and [8].

Starting in the 1940s, Erdős frequently applied random graphs to extremal problems, but it was only in the late 1950s that he embarked, with Rényi, on a *systematic* study of random graphs. Here let us mention only one of their results, the one concerning the emergence of the 'giant component.' Let us write $G_{n,M}$ for a *random graph* on n distinguishable vertices, with M edges. Thus every graph with M edges on these n vertices has the same probability, $1/\binom{N}{M}$, where $N = \binom{n}{2}$. Erdős and Rényi [60.10] proved that if $M(n) = \lfloor cn \rfloor$ for some constant $c > 0$ then, with probability tending to 1, the largest component of $G_{n,M}$ is of order $\log n$ if $c < \frac{1}{2}$, it jumps to order $n^{2/3}$ if $c = \frac{1}{2}$, and it jumps again, this time right up to order n if $c > \frac{1}{2}$. Quite understandably, Erdős and Rényi considered this phenomenon to be one of the most striking features of random graphs.

By now, all this is well known, especially since we know of similar phenomena in percolation theory, but in 1960 this was a striking discovery indeed. In fact, for over two decades not much was added to our knowledge of this *phase transition*. The investigations were reopened in 1984 in [6] with the main aim of deciding what happens around $M = \lfloor n/2 \rfloor$; in particular, what *scaling*, what *magnification* we should use to see the giant component growing continuously. Among other properties of the phase transition, it was shown in [6] that if $M = n/2 + s$ and $s = o(n)$ but slightly larger than $n^{2/3}$ then, with probability tending to 1, there is a unique largest component, with about $4s$ vertices, and the second largest component has no more than $(\log n)n^2/s^2$ vertices. Thus, in a rather large range, on average every new edge adds four new vertices to the giant component!

With this result, the floodgates opened, and many more precise studies of the behavior of the components near the point of phase transition were published, notably by Kolchin [28], Stepanov [48], Flajolet, Knuth and Pittel [15], Łuczak [33], and others. The crowning achievement is a monumental paper by Janson, Knuth, Łuczak and Pittel [24], which gives very detailed information about the random graph $G_{n,M}$ near its phase transition. Łuczak, Pittel, and Wierman [34] studied the structure of the random graph precisely at the critical point.

Rather than giving more results in number theory, combinatorics, probability theory, or any of the other areas closely associated with Erdős, let us present some results in fields tackled by Erdős only occasionally.

In the 1930s much research was done on the dimension of product spaces [37], [21], [22]. In those days mostly one kind of dimension was studied, that introduced by Menger and Urysohn; now it is called the *small inductive dimension*, and is denoted by *ind*. It is defined as follows: $ind\, X = -1$ if and only if $X = \emptyset$. For $n = 0, 1, \ldots$, we define $ind\, X \leq n$ inductively: if for every $x \in X$ and every neighborhood U of x, there is a neighborhood V of x that is contained in U and whose boundary has dimension at most $n-1$ then $ind\, X \leq n$. Also, $ind\, X = n$ if $ind\, X \leq n$ and $ind\, X \leq n-1$ does not hold. Finally, $ind\, X = \infty$ if $ind\, X \leq n$ fails for every natural number n; see [14] for an introduction to dimension theory. It is rather trivial that

$$ind(X \times Y) \leq ind\, X + ind\, Y, \tag{8}$$

so the natural question arises whether we have equality in (8). By constructing two compact spaces of dimension 2 whose product has dimension 3, Pontrjagin [37] proved that we may have strict inequality in (8). Nevertheless, the problem remained open whether in (8) we have to have equality for spaces of dimension 1. In 1939, while writing his fundamental book on dimension theory with Wallman [22], Hurewicz encouraged Erdős to work on the problem, who promptly solved it, proving that there is a separable complete metric space X with $ind(X \times X) = ind\, X = 1$. In fact, Erdős proved that for X we can take a very simple subset of the Hilbert space ℓ_2, namely the closure of

$$\{x = (x_i)_1^\infty \in \ell_2 : x_i = 1/n_i,\ n_i \in \mathbb{Z},\ i = 1, 2, \ldots\}.$$

Similarly, the set of rational points of ℓ_2 also has dimension 1. Clearly, not only do we have $ind(X \times X) = ind\, X$, but even $X \times X \cong X$, so it is rather surprising that $ind\, X = 1$.

It is easily seen that an algebraically closed field of characteristic 0 is determined (up to isomorphism) by its cardinality. For real-closed fields this is not the case: an additional invariant is the order type of the field. In 1955, Erdős, Gillman, and Henriksen [55.01] gave a characterization theorem for real-closed fields. As a consequence of this result they proved, among others, that, assuming the continuum hypothesis, all hyper-real fields of cardinality of the continuum are isomorphic. (A *hyper-real* field is a real-closed field of the form $C(X)/M$, where $C(X)$ is the ring of all continuous functions on a completely regular space X, and M is a maximal ideal.) The authors made use of the following beautiful result on almost disjoint sets Erdős that proved in 1934, but published only in [55.01]: For every infinite set X is of cardinality m, there exists a set of more than m subsets of X, each of cardinality m, such that the intersection of any two has cardinality less than m.

Throughout his career, Erdős was ready to apply Ramsey's theorem and its extensions. In 1976, he found an exciting application in the theory of Banach spaces. An infinite matrix of reals $A = (a_{ij})_{i,j\in\mathbb{N}}$ is a *regular method of summability* if for every sequence $(e_i)_{i\in\mathbb{N}}$ of elements of a Banach space X, converging in norm to an element e, the sequence $e_i' = \sum_{j=1}^{\infty} a_{ij}e_j$ also converges to e. Call a sequence $(x_i)_{i\in\mathbb{N}}$ *A-summable* if $x_i' = \sum_{j=1}^{\infty} a_{ij}x_j$ converges in norm. Erdős and Magidor [76.01] proved that, given a regular method of summability A, and a bounded sequence $\mathbf{x} = (x_i)_{i\in\mathbb{N}}$ in a Banach space, there is an infinite subsequence $\mathbf{x}' = (x_{i_k})_{k\in\mathbb{N}}$ such that either every infinite subsequence of $\mathbf{x}'$ is A-summable or no infinite subsequence of $\mathbf{x}'$ is A-summable. The proof is based on the Galvin-Prikry [17] partition theorem from set theory. The Erdős-Magidor theorem was the starting point of much research; see, e.g., [36].

Every mathematical paper about Erdős should present some of his unsolved problems. We mentioned one or two; now let us give four more. The first is relatively unknown, but the second two are among his best known problems; the last carried the largest award Erdős ever offered. Erdős himself wrote many papers about his unsolved problems, including [90.29], [97.18], [97.21]. Recently Chung [10] published an excellent collection of unsolved problems of Erdős in graph theory.

At the 'Quo Vadis, Graph Theory?' conference in Fairbanks, Alaska, in 1990, while admiring the playful bears, we came up with the following problem. Let us run a graph process 'in reverse' as follows. Start with a complete graph K_n on n vertices, select one of the $\binom{n}{3}$ triangles at random, and delete its edges. Next, select one of the remaining $\binom{n}{3} - 3n + 8$ triangles at random, and delete its edges. Continue in this way: if at time t we have a graph G_t, and G_t has a triangle then select one of these at random, and delete its edges to obtain G_{t+1}. If G_t has no triangles, stop the process with G_t, and write S_n for the number of edges in G_t. (Note that $S_n = \binom{n}{3} - 3t$.) Thus the random variable S_n is the number of edges in a random triangle-free graph obtained in this peculiar way. What can we say about the random variable S_n? With Erdős we conjectured that the expectation of S_n has order $n^{3/2}$, and for some $c > 0$ the probability that $S_n > cn^{3/2}$ tends to 0 as $n \to \infty$. Grable [18] has proved the beautiful result that for every $\varepsilon > 0$ one has $\lim_{n\to\infty} \mathbb{P}(S_n > n^{7/4+\varepsilon}) = 0$, but it does not seem to be easy to improve the exponent 7/4 to 3/2.

Needless to say, the problem has many variants. Instead of triangles, we may delete a random K_4, K_5, or any graph H. Also, we need not start with a complete graph but just a graph containing many subgraphs isomorphic to H. In yet another variation, we do not delete all edges of a random copy of H but only a random edge.

In their study of intersecting set systems, Erdős and Rado [60.04, 69.02] called a family of sets $\mathcal{A} = \{A_1, A_2, \dots, A_k\}$ a *strong Δ-system* if all the intersections $A_i \cap A_j$, $1 \le i < j \le k$, are identical. Denoting by $f(n,k)$ the smallest integer m for which every family of n-sets $\{A_1, A_2, \dots, A_m\}$ contains k sets forming a strong Δ-system, Erdős and Rado proved that

$$2^n < f(n,3) < 2^n n!. \qquad (9)$$

Furthermore, they conjectured that

$$f(n,3) < c_3^n \tag{10}$$

for some constant c_3. As, almost certainly, a similar assertion holds for $f(n,k)$, inequality (10) does not seem to be too much to ask. Nevertheless, the conjecture has turned out to be amazingly resistant, so eventually Erdős offered \$1000 for a proof or disproof of (10).

For many years, there was virtually no progress with (10), but shortly before Erdős died, Kostochka [30], [29] won a consolation prize of \$100 for the first substantial improvement of (9), and recently Axenovich, Fon-Der-Flass, and Kostochka [2] proved the further improvement that

$$f(n,3) < (n!)^{1/2+\varepsilon}$$

holds for every $\varepsilon > 0$, provided n is sufficiently large.

The problems Erdős probably liked best concern arithmetic progressions. In 1936, Erdős and Turán [36.05] conjectured the far-reaching extension of van der Waerden's theorem on arithmetic progressions that if $a_1 < a_2 < \cdots$ is a sequence of natural numbers such that, for some $c > 0$, we have $a_n < cn$ for infinitely many values of n, then the sequence contains arbitrarily long arithmetic progressions. In 1952, Roth [44] proved the existence of three-term progressions, and in 1974 Szemerédi [50] proved the full conjecture, amply deserving the \$1000 reward from Erdős, the largest he ever paid. In 1977, Fürstenberg [16] gave another proof of Szemerédi's theorem, based on ergodic theory; the proof and its ramifications revolutionized ergodic theory.

Erdős conjectured that the sequence of primes contains arbitrarily long arithmetic progressions and, even more, that every sequence $1 \le a_1 < a_2 < \cdots$ with $\sum_{n=1}^{\infty} 1/a_n = \infty$ contains arbitrarily long arithmetic progressions. He offered \$3000 for this conjecture.

The final problem is again about the differences between consecutive primes. Writing $d_n = p_{n+1} - p_n$, Erdős and Turán [48.05] conjectured that $d_n < d_{n+1} < d_{n+2}$ holds for infinitely many values of n, but could not come close to proving it. Forty years later, Erdős [90.29] offered \$100 for a proof and \$25,000 for a *disproof.* Needless to say, he had no doubt that the conjecture was true, and it would be hard to find anybody who would disagree.

At this point we bring to an end this brief and ruefully inadequate sketch of the life and the achievements of Paul Erdős. We have not come close to doing justice to his enormous contribution to the mathematics of this century nor, in so few pages, could we.

In spite of his tremendous achievements, Paul Erdős remained outside the mathematical establishment. His eagerness to help everybody, especially those whom he thought had been badly treated, worked, to some extent, against him, as he gave the impression of not being discriminating enough. However, in private he was a remarkably shrewd and perceptive judge of mathematicians and their abilities. He was passionately, almost pathologically, keen to be free, to do as he liked, when he liked.

Did he die too early? I certainly think so, although at the very the end there were signs of deterioration: occasionally even his once remarkable memory skipped a beat. It is a minor consolation that he died as he wished to, with his mathematical boots on. He always lived up to his own high moral principles, and expected others to do the same. He had a passionate desire to be free in every way and so he strongly disliked the oppressive political systems; nevertheless, he was remarkably free of personal hate. Sadly, for much of his life, he knew loneliness and sorrow, and he needed the constant stimulus of new mathematical companions and ideas to keep his unhappiness at bay. But for us, mathematicians, this companionship was a treasured gift. He was a never-failing font of magical inspiration; by his incisive mind and generous heart he enriched us all. His message to posterity, paraphrasing lines of the Hungarian poet Endre Ady, clearly showed his love for mathematics: "*Let him be blessed, who takes my place!*" Needless to say, his place will never be taken: we shall never see his like again.

Acknowledgements

In the preparation of this paper I made use of numerous interviews conducted with Paul Erdős in Cambridge, Budapest, and Memphis, together with the articles [4], [7], [97.18], [96.17].

Bibliography*

[1] M. Ajtai, V. Chvátal, M. Newborn and E. Szemerédi, Crossing-free subgraphs, *Annals of Discrete Mathematics* **12** (1982), 9–12.

[2] M. Aksenovich, D. Fon-Der-Flaass and A. Kostochka, On set systems without weak 3-Delta-subsystems, *Discrete Math.*, **138** (1995), 57–62.

[3] M. Avidon, On the distribution of primitive abundant numbers, *Acta Arithmetica* **77** (2) (1996), 195–205.

[4] L. Babai, In and out of Hungary: Paul Erdős, his friends and times, in "Combinatorics, Paul Erdős is Eighty", vol. II, D. Miklós, V.T. Sós and T. Szőnyi, eds, János Bolyai Mathematical Society, Budapest, 1996, pp. 7–95.

[5] J. Beck and J. Spencer, Unit distances, *Journal of Combinatorial Theory A* **37** (1984), 231–238.

[6] B. Bollobás, The evolution of random graphs, *Transactions of the American Mathematical Society* **286** (1984), 257–274.

*In the bibliography we list only the papers referred to above that do not have Paul Erdös as the author or a coauthor; the papers involving Paul Erdös are listed in the *Selected Publications of Paul Erdös* following this article.

[7] B. Bollobás, Paul Erdős—Life and Work, in "The Mathematics of Paul Erdős", vol. I, R.L. Graham and J. Nešetřil, eds, Springer-Verlag, Berlin, 1996, pp. 1–41.

[8] B. Bollobás and Y. Kohayakawa, An extension of the Erdős-Stone theorem, *Combinatorica* **14** (1994), 279–286.

[9] R. Breusch, Zur Verallgemeinung des Bertrandschen Postulates, dass zwischen x and $2x$ stets Primzahlen liegen, *Mathematische Zeitschrift* **34** (1932), 505–526.

[10] F.R.K. Chung, Open problems of Paul Erdős in graph theory, *J. Graph Theory* **25** (1997), 3-36.

[11] V. Chvátal and E. Szemerédi, On the Erdős-Stone theorem, *Jounral of the London Mathematical Society* **23** (1981), 297–319.

[12] P.D.T.A. Elliot, *Probabilistic Number Theory I, Mean-Value Theorems*, Springer-Verlag, New York, 1979.

[13] P.D.T.A. Elliot, *Probabilistic Number Theory II, Central Limit Theorems*, Springer-Verlag, New York, 1980.

[14] R. Engelking, *Dimension Theory*, North-Holland, Amsterdam, 1978, x+314pp.

[15] Ph. Flajolet, D.E. Knuth and B. Pittel, The first cycle in an evolving graph, *Discrete Mathematics* **75** (1989), 167–215.

[16] H. Fürstenberg, *Recurrence in Ergodic Theory and Combinatorial Number Theory*, Princeton University Press, 1981.

[17] F. Galvin and K. Prikry, Borel sets and Ramsey's theorem, *Journal of Symbolic Logic* **38** (1973), 193–198.

[18] D.A. Grable, On random greedy triangle packing, *Electronic Journal of Combinatorics* **4** (1997), #R11, 1–19.

[19] H. Halberstam and H.-E. Richert, *Sieve Methods*, Academic Press, London, 1974.

[20] G.H. Hardy and S. Ramanujan, *Quarterly Journal of Mathematics* (77)**48** (1920), 76–92.

[21] P.W. Hurewicz, Sur la dimension des produits Cartésiens, *Annals of Mathematics* (1) **36** (1935), 194–197.

[22] P.W. Hurewicz and H. Wallman, *Dimension Theory*, Princeton, 1941.

[23] A. Ivič, The distribution of primitive abundant numbers, *Studia Sci. Math. Hungar.* **20** (1985), 183–187.

[24] S. Janson, D.E. Knuth, T. Łuczak and B. Pittel, The birth of the giant component, *Random Structures and Algorithms* **4** (1993), 233–358.

[25] S. Józsa and E. Szemerédi, The number of unit distances on the plane, in: *Infinite and Finite Sets* Coll. Math. Soc. J. Bolyai **10** (1973), 939–950.

[26] I. Kaplansky, *Mathematical Reviews* **7** (1946), 164.

[27] G.O.H. Katona, A simple proof of the Erdős-Ko-Rado theorem, *Journal of Combinatorial Theory (B)* **13** (1972), 183–184.

[28] V.F. Kolchin, On the limit behaviour of a random graph near the critical point, *Theory Probability Appl.* **31** (1986), 439–451.

[29] A. Kostochka, An intersection theorem for systems of sets, *Random Structures and Algorithms*, **9** (1996), 213–221.

[30] A. Kostochka, A bound of the cardinality of families not containing Δ-systems, in "The Mathematics of Paul Erdős", vol. II, R.L. Graham and J. Nešetřil, eds, Springer-Verlag, Berlin, 1997, pp. 229–235.

[31] E. Landau, Vorlesungen über Zahlentheorie, Leipzig, 1927.

[32] F.T. Leighton, *Complexity Issues in VLSI*, M.I.T. Press, Cambridge, MA, 1983.

[33] T. Łuczak, Component behavior near the critical point of the random graph process, *Random Structures and Algorithms* **1** (1990), 287–310.

[34] T. Łuczak, B. Pittel and J.C. Wierman, The strucure of a random graph at the critical point of the phase transition, *Trans. Amer. Math. Soc.*, **341** (1994), 721-748.

[35] H. Maier and C. Pomerance, Unusually large gaps between consecutive primes, *Transactions of the American Mathematical Society* **322** (1990), 201–237.

[36] J.R. Partington, Almost sure summability of subsequences in Banach spaces, *Studia Mathematica* **71** (1982), 27–35.

[37] L.S. Pontrjagin, Sur une hypothèse fondamentale de la théorie de la dimension, *Comptes R. Acad. Paris* **190** (1930), 1105–1107.

[38] S. Ramanujan, A proof of Bertrand's postulate, *Journal of the Indian Mathematical Society* **11** (1919), 181–182.

[39] R.A. Rankin, The difference between consecutive prime numbers, *Journal of the London Mathematical Society* **13** (1938), 242–247.

[40] R.A. Rankin, The difference between consecutive prime numbers, V, *Proceedings of the Edinburgh Mathematical Society* (2) **13** (1962/63), 331–332.

[41] G. Ricci, Sul teorema di Dirichlet relativo alla progressione aritmetica, *Bolletino della Unione Matematica Italiana* **12** (1933), 304–309.

[42] G. Ricci, Sui teoremi di Dirichlet e di Bertrand-Tchebychef relativi alla progressione aritmetica, *Bolletino della Unione Matematica Italiana* **13** (1934), 1–11.

[43] O. Rigge, Über ein diophantisches Problem, 9th *Congress Math. Scand.* (1939), 155–160.

[44] K.F. Roth, Sur quelques ensembles d'entiers, *Comptes R. Acad. Sci. Paris* **234** (1952), 388–390.

[45] A. Schönhage, Eine Bemerkung zur Konstruktion grosser Primzahllücken, *Arch. Math.* **14** (1963), 29–30.

[46] A. Selberg, An elementary proof of the prime-number theorem, *Annals of Mathematics* (2)50 (1949), 305–313.

[47] J. Spencer, E. Szemerédi and W.T. Trotter, Unite distances in the Euclidean plane, in: *Graph Theory and Combinatorics* (B. Bollobás, ed), Academic Press, London, 1984, 293–308.

[48] V.E. Stepanov, Some features of the structure of random graphs in the neighborhood of the critical point, *Theory Probab. Appl.* **32** (1987), 573–594.

[49] L. Székely, Crossing numbers and hard Erdős problems in discrete geometry, *Combinatorics, Probability and Computing* **6** (1997).

[50] E. Szemerédi, On sets of integers containing no k elements in arithmetic progression, *Acta Arithmetica* **27** (1975), 199–245.

[51] A.G. Thomason, An upper bound for some Ramsey numbers, *Journal of Graph Theory* **12** (1988), 509–517.

[52] P. Turán, On a theorem of Hardy and Ramanujan, *London Mathematical Society* **9** (1934), 274–276.

[53] P. Turán, On an extremal problem in graph theory (in Hungarian), *Mat. Fiz. Lapok* **48** (1941), 436–452.

University of Memphis, Memphis TN 38152
and Trinity College, Cambridge CB2 1TQ, England

SELECTED PUBLICATIONS OF PAUL ERDÖS[1]

[32.01] Beweis eines Satzes von Tschebyschef (in German), *Acta Litt. Sci. Szeged* **5** (1932), 194–198.

[32.02] Egy Kürschák-féle elemi számelméleti tétel általánosítása (Generalization of an elementary number-theoretic theorem of Kürschák, in Hungarian), *Mat. Fiz. Lapok* **39** (1932), 17–24.

[34.01] A theorem of Sylvester and Schur, *J. London Math. Soc.* **9** (1934), 282–288.

[34.02] Bizonyos számtani sorok törzsszámairól (On primes in some arithmetic progressions, in Hungarian), *Bölcsészdoktori értekezés*, Sárospatak, 1934, 1–20.

[34.03] On a problem in the elementary theory of numbers, *Amer. Math. Monthly* **41** (1934), 608–611 (with P. Turán).

[34.04] On the density of the abundant numbers, *J. London Math. Soc.* **9** (1934), 278–282.

[34.05] Über die Anzahl der Abelschen Gruppen gegebener Ordnung und über ein verwandtes zahlentheoretisches Problem (in German), *Acta Litt. Sci. Szeged* **7** (1934), 95–102 (with G. Szekeres).

[35.01] A combinatorial problem in geometry, *Compositio Math.* **2** (1935), 463–470 (with G. Szekeres).

[35.02] Ein zahlentheoretischer Satz (in German), *Mitt. Forsch.-Inst. Math. Mech. Univ. Tomsk* **1** (1935), 101–103 (with P. Turán).

[35.03] Note on consecutive abundant numbers, *J. London Math. Soc.* **10** (1935), 128–131.

[35.04] Note on sequences of integers no one of which is divisible by any other, *J. London Math. Soc.* **10** (1935), 126–128.

[35.05] On primitive abundant numbers, *J. London Math. Soc.* **10** (1935), 49–58.

[35.06] On the density of some sequences of numbers, *J. London Math. Soc.* **10** (1935), 120–125.

[35.07] On the difference of consecutive primes, *Quart. J. Math., Oxford Ser.* **6** (1935), 124–128.

[35.08] On the normal number of prime factors of $p-1$ and some related problems concerning Euler's φ-function, *Quart. J. Math., Oxford Ser.* **6** (1935), 205–213.

[35.09] The representation of an integer as the sum of the square of a prime and of a square-free integer, *J. London Math. Soc.* **10** (1935), 243–245.

[35.10] Über die Primzahlen gewisser arithmetischer Reihen (in German), *Math. Z.* **39** (1935), 473–491.

[1]As Paul Erdős had about 1500 papers, here we give only a selection of his publications, concentrating on his earlier papers. The Erdős Number Project maintained by Jerry Grossman on the World Wide Web, `http://www.oakland.edu/~grossman/erdoshp.html`, has the best list available: the reference numbers below are in agreement with that list.

[35.11] Über die Vereinfachung eines Landauschen Satzes (in German), *Mitt. Forsch.-Inst. Math. Mech. Univ. Tomsk* **1** (1935), 144–147 (with P. Turán).

[36.01] A generalization of a theorem of Besicovitch, *J. London. Math. Soc.* **11** (1936), 92–98.

[36.02] Note on some additive properties of integers, *Publ. de Congrès International des Math.*, Oslo, 1936, 1–2.

[36.03] On a problem of Chowla and some related problems, *Proc. Cambridge Philos. Soc.* **32** (1936), 530–540.

[36.04] On sequences of positive integers, *Acta Arith.* **2** (1936), 147–151 (with H. Davenport).

[36.05] On some sequences of integers, *J. London Math. Soc.* **11** (1936), 261–264 (with P. Turán).

[36.06] On the arithmetical density of the sum of two sequences one of which forms a basis for the integers, *Acta Arith.* **1** (1936), 197–200.

[36.07] On the integers which are the totient of a product of three primes, *Quart. J. Math., Oxford Ser.* **7** (1936), 16–19.

[36.08] On the integers which are the totient of a product of two primes, *Quart. J. Math., Oxford Ser.* **7** (1936), 227–229.

[36.09] On the representation of an integer as the sum of k kth powers, *J. London Math. Soc.* **11** (1936), 133–136.

[36.10] Sur le mode de convergence pour l'interpolation de Lagrange (in French), *C. R. Acad. Sci. Paris* **203** (1936), 913–915 (with E. Feldheim).

[36.11] Végtelen gráfok Euler vonalairól (On Euler lines of infinite graphs, in Hungarian), *Mat. Fiz. Lapok* **43** (1936), 129–140 (with T. Grünwald [= T. Gallai] and E. Weiszfeld [= E. Vázsonyi]).

[37.01] Eine Bemerkung über lineare Kongruenzen (in German), *Acta Arith.* **2** (1937), 214–220 (with V. Jarník).

[37.02] Note on the number of prime divisors of integers, *J. London Math. Soc.* **12** (1937), 308–314.

[37.03] Note on the transfinite diameter, *J. London Math. Soc.* **12** (1937), 185–192 (with J. Gillis).

[37.04] On interpolation, I. Quadrature and mean convergence in the Lagrange interpolation, *Ann. of Math. (2)* **38** (1937), 142–155 (with P. Turán).

[37.05] On the density of some sequences of numbers, II., *J. London Math. Soc.* **12** (1937), 7–11.

[37.06] On the easier Waring problem for powers of primes, I., *Proc. Cambridge Philos. Soc.* **33** (1937), 6–12.

[37.07] On the sum and difference of squares of primes, *J. London Math. Soc.* **12** (1937), 133–136.

[37.08] On the sum and difference of squares of primes, II., *J. London Math. Soc.* **12** (1937), 168–171.

[37.09] Über diophantische Gleichungen der Form $n! = x^p \pm y^p$ und $n! \pm m! = x^p$ (in German), *Acta Litt. Sci. Szeged* **8** (1937), 241–255 (with R. Obláth).

[38.01] Note on an elementary problem of interpolation, *Bull. Amer. Math. Soc.* **44** (1938), 515–518 (with G. Grünwald).

[38.02] Note on the Euclidean algorithm, *J. London Math. Soc.* **13** (1938), 3–8 (with Chao Ko [=Zhao Ke]).

[38.03] On additive properties of squares of primes, I., *Nederl. Akad. Wetensch., Proc.* **41** (1938), 37–41.

[38.04] On definite quadratic forms which are not the sum of two definite or semi-definite forms, *Acta Arith.* **3** (1938), 102–122 (with Chao Ko [=Zhao Ke]).

[38.05] On fundamental functions of Lagrangean interpolation, *Bull. Amer. Math. Soc.* **44** (1938), 828–834 (with B. A. Lengyel).

[38.06] On interpolation, II. On the distribution of the fundamental points of Lagrange and Hermite interpolation, *Ann. of Math. (2)* **39** (1938), 703–724 (with P. Turan).

[38.07] On sequences of integers no one of which divides the product of two others and on some related problems, *Mitt. Forsch.-Inst. Math. Mech. Univ. Tomsk* **2** (1938), 74–82.

[38.08] On the asymptotic density of the sum of two sequences one of which forms a basis for the integers, II., *Trav. Inst. Math. Tbilissi* **3** (1938), 217–223.

[38.09] On the density of some sequences of numbers, III., *J. London Math. Soc.* **13** (1938), 119–127.

[38.10] On the number of integers which can be represented by a binary form, *J. London Math. Soc.* **13** (1938), 134–139 (with K. Mahler).

[38.11] Some results on definite quadratic forms, *J. London Math. Soc.* **13** (1938), 217–224 (with Chao Ko [=Zhao Ke]).

[38.12] Über die arithmetischen Mittelwerte der Lagrangeschen Interpolationspolynome (in German), *Studia Math.* **7** (1938), 82–95 (with G. Grünwald).

[38.13] Über die Reihe $\sum \frac{1}{p}$ (in German), *Mathematica, Zutphen B* **7** (1938), 1–2.

[38.14] Über einen Faber'schen Satz (in German), *Ann. of Math. (2)* **39** (1938), 257–261 (with G. Grünwald).

[38.15] Über Euler-Linien unendlicher Graphen (in German), *J. Math. Phys. Mass. Inst. Tech.* **17** (1938), 59–75 (with T. Grünwald [=T. Gallai] and E. Vázsonyi).

[39.01] Additive arithmetical functions and statistical independence, *Amer. J. Math.* **61** (1939), 713–721 (with A. Wintner).

[39.02] An extremum-problem concerning trigonometric polynomials, *Acta Litt. Sci. Szeged* **9** (1939), 113–115.

[39.03] Note on products of consecutive integers, *J. London Math. Soc.* **14** (1939), 194–198.

[39.04] Note on the product of consecutive integers, II., *J. London Math. Soc.* **14** (1939), 245–249.

[39.05] On a family of symmetric Bernoulli convolutions, *Amer. J. Math.* **61** (1939), 974–976.

[39.06] On polynomials with only real roots, *Ann. of Math. (2)* **40** (1939), 537–548 (T. Grünwald =[T. Gallai]).

[39.07] On sums of positive integral kth powers, *Ann. of Math. (2)* **40** (1939), 533–536 (with H. Davenport).

[39.08] On the easier Waring problem for powers of primes, II., *Proc. Cambridge Philos. Soc.* **35** (1939), 149–165.

[39.09] On the Gaussian law of errors in the theory of additive functions, *Proc. Nat. Acad. Sci. U. S. A.* **25** (1939), 206–207 (with M. Kac).

[39.10] On the integers of the form $x^k + y^k$, *J. London Math. Soc.* **14** (1939), 250–254.

[39.11] On the smoothness of the asymptotic distribution of additive arithmetical functions, *Amer. J. Math.* **61** (1939), 722–725.

[39.12] Some arithmetical properties of the convergents of a continued fraction, *J. London Math. Soc.* **14** (1939), 12–18 (with K. Mahler).

[40.01] Additive functions and almost periodicity (B^2), *Amer. J. Math.* **62** (1940), 635–645 (with A. Wintner).

[40.02] Note on some elementary properties of polynomials, *Bull. Amer. Math. Soc.* **46** (1940), 954–958.

[40.04] On extremal properties of the derivatives of polynomials, *Ann. of Math. (2)* **41** (1940), 310–313.

[40.05] On interpolation, III. Interpolatory theory of polynomials, *Ann. of Math. (2)* **41** (1940), 510–553 (with P. Turán).

[40.06] On the distribution of normal point groups, *Proc. Nat. Acad. Sci. U. S. A.* **26** (1940), 294–297.

[40.07] On the smoothness properties of a family of Bernoulli convolutions, *Amer. J. Math.* **62** (1940), 180–186.

[40.08] On the uniformly-dense distribution of certain sequences of points, *Ann. of Math. (2)* **41** (1940), 162–173 (with P. Turán).

[40.09] Ramanujan sums and almost periodic functions, *Studia Math.* **9** (1940), 43–53 (with M. Kac, E. R. van Kampen and A. Wintner).

[40.10] The difference of consecutive primes, *Duke Math. J.* **6** (1940), 438–441.

[40.11] The dimension of the rational points in Hilbert space, *Ann. of Math. (2)* **41** (1940), 734–736.

[40.12] The Gaussian law of errors in the theory of additive number theoretic functions, *Amer. J. Math.* **62** (1940), 738–742 (with M. Kac).

[41.01] On a problem of Sidon in additive number theory and on some related problems, *J. London Math. Soc.* **16** (1941), 212–215 (with P. Turán).

[41.02] On divergence properties of the Lagrange interpolation parabolas, *Ann. of Math. (2)* **42** (1941), 309–315.

[41.03] On some asymptotic formulas in the theory of the "factorisatio numerorum", *Ann. of Math. (2)* **42** (1941), 989–993.

[41.04] The distribution of the number of summands in the partitions of a positive integer, *Duke Math. J.* **8** (1941), 335–345 (with J. Lehner).

[42.01] On a problem of I. Schur, *Ann. of Math. (2)* **43** (1942), 451–470 (with G. Szegő).

[42.02] On an elementary proof of some asymptotic formulas in the theory of partitions, *Ann. of Math. (2)* **43** (1942), 437–450.

[42.03] On the asymptotic density of the sum of two sequences, *Ann. of Math. (2)* **43** (1942), 65–68.

[42.04] On the law of the iterated logarithm, *Ann. of Math. (2)* **43** (1942), 419–436.

[42.05] On the uniform distribution of the roots of certain polynomials, *Ann. of Math. (2)* **43** (1942), 59–64.

[42.06] Some set-theoretical properties of graphs, *Univ. Nac. Tucumán. Revista A.* **3** (1942), 363–367.

[43.01] A note on Farey series, *Quart. J. Math., Oxford Ser.* **14** (1943), 82–85.

[43.02] Approximation by polynomials, *Duke Math. J.* **10** (1943), 5–11 (with J. A. Clarkson).

[43.03] Corrections to two of my papers, *Ann. of Math. (2)* **44** (1943), 647–651.

[43.04] On families of mutually exclusive sets, *Ann. of Math. (2)* **44** (1943), 315–329 (with A. Tarski).

[43.05] On non-denumerable graphs, *Bull. Amer. Math. Soc.* **49** (1943), 457–461 (with S. Kakutani).

[43.06] On some convergence properties in the interpolation polynomials, *Ann. of Math. (2)* **44** (1943), 330–337.

[43.07] On the convergence of trigonometric series, *J. Math. Phys. Mass. Inst. Tech.* **22** (1943), 37–39.

[43.08] Some remarks on set theory, *Ann. of Math. (2)* **44** (1943), 643–646.

[44.01] A conjecture in elementary number theory, *Bull. Amer. Math. Soc.* **50** (1944), 881–882 (with L. Alaoglu).

[44.02] Addendum. On a problem of Sidon in additive number theory and on some related problems [*J. London Math. Soc.* **16** (1941), 212–215], *J. London Math. Soc.* **19** (1944), 208.

[44.03] On highly composite and similar numbers, *Trans. Amer. Math. Soc.* **56** (1944), 448–469 (with L. Alaoglu).

[44.04] On highly composite numbers, *J. London Math. Soc.* **19** (1944), 130–133.

[44.05] On the maximum of the fundamental functions of the ultraspherical polynomials, *Ann. of Math. (2)* **45** (1944), 335–339.

[44.06] Some remarks on connected sets, *Bull. Amer. Math. Soc.* **50** (1944), 442–446.

[45.01] Integral distances, *Bull. Amer. Math. Soc.* **51** (1945), 598–600 (with N. H. Anning).

[45.02] Integral distances, *Bull. Amer. Math. Soc.* **51** (1945), 996.

[45.03] Note on the converse of Fabry's gap theorem, *Trans. Amer. Math. Soc.* **57** (1945), 102–104.

[45.04] On a lemma of Littlewood and Offord, *Bull. Amer. Math. Soc.* **51** (1945), 898–902.

[45.05] On certain variations of the harmonic series, *Bull. Amer. Math. Soc.* **51** (1945), 433–436 (with I. Niven).

[45.06] On the least primitive root of a prime p, *Bull. Amer. Math. Soc.* **51** (1945), 131–132.

[45.07] Some remarks on almost periodic transformations, *Bull. Amer. Math. Soc.* **51** (1945), 126–130 (with A. H. Stone).

[45.08] Some remarks on Euler's φ-function and some related problems, *Bull. Amer. Math. Soc.* **51** (1945), 540–544.

[45.09] Some remarks on the measurability of certain sets, *Bull. Amer. Math. Soc.* **51** (1945), 728–731.

[46.01] Note on normal numbers, *Bull. Amer. Math. Soc.* **52** (1946), 857–860 (with A. H. Copeland).

[46.02] On certain limit theorems of the theory of probability, *Bull. Amer. Math. Soc.* **52** (1946), 292–302 (with M. Kac).

[46.03] On sets of distances of n points, *Amer. Math. Monthly* **53** (1946), 248–250.

[46.04] On some asymptotic formulas in the theory of partitions, *Bull. Amer. Math. Soc.* **52** (1946), 185–188.

[46.05] On the coefficients of the cyclotomic polynomial, *Bull. Amer. Math. Soc.* **52** (1946), 179–184.

[46.06] On the distribution function of additive functions, *Ann. of Math. (2)* **47** (1946), 1–20.

[46.07] On the Hausdorff dimension of some sets in Euclidean space, *Bull. Amer. Math. Soc.* **52** (1946), 107–109.

[46.08] On the structure of linear graphs, *Bull. Amer. Math. Soc.* **52** (1946), 1087–1091 (with A. H. Stone).

[46.09] Sequences of plus and minus, *Scripta Math.* **12** (1946), 73–75 (with I. Kaplansky).

[46.10] Some properties of partial sums of the harmonic series, *Bull. Amer. Math. Soc.* **52** (1946), 248–251 (with I. Niven).

[46.11] Some remarks about additive and multiplicative functions, *Bull. Amer. Math. Soc.* **52** (1946), 527–537.

[46.12] The asymptotic number of Latin rectangles, *Amer. J. Math.* **68** (1946), 230–236 (with I. Kaplansky).

[46.13] The $\alpha + \beta$ hypothesis and related problems, *Amer. Math. Monthly* **53** (1946), 314–317 (with I. Niven).

[46.14] Toeplitz methods which sum a given sequence, *Bull. Amer. Math. Soc.* **52** (1946), 463–464 (with P. C. Rosenbloom).

[47.01] A note on transforms of unbounded sequences, *Bull. Amer. Math. Soc.* **53** (1947), 787–790 (with G. Piranian).

[47.02] On the connection between gaps in power series and the roots of their partial sums, *Trans. Amer. Math. Soc.* **62** (1947), 53–61 (with H. Fried).

[47.03] On the lower limit of sums of independent random variables, *Ann. of Math. (2)* **48** (1947), 1003–1013 (with K.-L. Chung).

[47.04] On the number of positive sums of independent random variables, *Bull. Amer. Math. Soc.* **53** (1947), 1011–1020 (with M. Kac).

[47.05] Over-convergence on the circle of convergence, *Duke Math. J.* **14** (1947), 647–658 (with G. Piranian).

[47.06] Some asymptotic formulas for multiplicative functions, *Bull. Amer. Math. Soc.* **53** (1947), 536–544.

[47.08] Some remarks on polynomials, *Bull. Amer. Math. Soc.* **53** (1947), 1169–1176.

[47.09] Some remarks on the theory of graphs, *Bull. Amer. Math. Soc.* **53** (1947), 292–294.

[48.01] On a combinatorial problem, *Nederl. Akad. Wetensch., Proc.* **51** (1948), 1277–1279 = *Indag. Math.* **10** (1948), 421–423 (with N. G. de Bruijn).

[48.02] On a problem in the theory of uniform distribution, I., *Nederl. Akad. Wetensch., Proc.* **51** (1948), 1146–1154 = *Indag. Math.* **10** (1948), 370–478 (with P. Turán).

[48.03] On a problem in the theory of uniform distribution, II., *Nederl. Akad. Wetensch., Proc.* **51** (1948), 1262–1269 = *Indag. Math.* **10** (1948), 406–413 (with P. Turán).

[48.05] On some new questions on the distribution on prime numbers, *Bull. Amer. Math. Soc.* **54** (1948), 371–378 (with P. Turán).

[48.06] On the density of some sequences of integers, *Bull. Amer. Math. Soc.* **54** (1948), 685–692.

[48.07] On the difference of consecutive primes, *Bull. Amer. Math. Soc.* **54** (1948), 885–889.

[48.08] On the integers having exactly k prime factors, *Ann. of Math. (2)* **49** (1948), 53–66.

[48.09] On the representation of $1, 2, \ldots, N$ by differences, *Nederl. Akad. Wetensch., Proc.* **51** (1948), 1155–1158 = *Indag. Math.* **10** (1949), 379–382 (with I. S. Gál).

[48.10] On the roots of a polynomial and its derivative, *Bull. Amer. Math. Soc.* **54** (1948), 184–190 (with I. Niven).

[48.11] Some asymptotic formulas in number theory, *J. Indian Math. Soc. (N.S.)* **12** (1948), 75–78.

[48.12] Some remarks on Diophantine approximations, *J. Indian Math. Soc. (N.S.)* **12** (1948), 67–74.

[48.13] The set on which an entire function is small, *Amer. J. Math.* **70** (1948), 400–402 (with R. P. Boas, Jr. and R. C.Buck).

[49.01] A property of power series with positive coefficients, *Bull. Amer. Math. Soc.* **55** (1949), 201–204 (with W. Feller and H. Pollard).

[49.02] On a new method in elementary number theory which leads to an elementary proof of the prime number theorem, *Proc. Nat. Acad. Sci. U. S. A.* **35** (1949), 374–384.

[49.03] On a Tauberian theorem connected with the new proof of the prime number theorem, *J. Indian Math. Soc. (N.S.)* **13** (1949), 131–144.

[49.04] On a theorem of Hsu and Robbins, *Ann. Math. Statistics* **20** (1949), 286–291.

[49.05] On some applications of Brun's method, *Acta Univ. Szeged. Sect. Sci. Math.* **13** (1949), 57–63.

[49.08] On the number of terms of the square of a polynomial, *Nieuw Arch. Wiskunde (2)* **23** (1949), 63–65.

[49.09] On the strong law of large numbers, *Trans. Amer. Math. Soc.* **67** (1949), 51–56.

[49.10] On the uniform distribution modulo 1 of lacunary sequences, *Nederl. Akad. Wetensch., Proc.* **52** (1949), 264–273 = *Indag. Math.* **11** (1949), 79–88 (with J. F. Koksma).

[49.11] On the uniform distribution modulo 1 of sequences $(f(n,\theta))$, *Nederl. Akad. Wetensch., Proc.* **52** (1949), 851–854 = *Indag. Math.* **11** (1949), 299–302 (with J. F. Koksma).

[49.12] Problems and results on the differences of consecutive primes, *Publ. Math. Debrecen* **1** (1949), 33–37.

[49.13] Sequences of points on a circle, *Nederl. Akad. Wetensch., Proc.* **52** (1949), 14–17 = *Indag. Math.* **11** (1949), 46–49 (with N. G. de Bruijn).

[50.01] A combinatorial theorem, *J. London Math. Soc.* **25** (1950), 249–255 (with R. Rado).

[50.03] Convergence fields of row-finite and row-infinite Toeplitz transformations, *Proc. Amer. Math. Soc.* **1** (1950), 397–401 (with G. Piranian).

[50.04] Double points of paths of Brownian motion in n-space, *Acta Sci. Math. Szeged* **12** (1950), Leopoldo Fejér et Frederico Riesz LXX annos natis dedicatus, Pars B, 75–81 (with A. Dvoretzky and S. Kakutani).

[50.08] On the distribution of roots of polynomials, *Ann. of Math. (2)* **51** (1950), 105–119 (with P. Turán).

[50.10] Remarks on the size of $L(1,\chi)$, *Publ. Math. Debrecen* **1** (1950), 165–182 (with P. T. Bateman abd S. Chowla).

[50.13] Some remarks on set theory, *Proc. Amer. Math. Soc.* **1** (1950), 127–141.

[51.02] A theorem on the distribution of the values of L-functions, *J. Indian Math. Soc. (N.S.)* **15** (1951), 11–18 (with S. Chowla).

[51.10] Probability limit theorems assuming only the first moment, I., *Mem. Amer. Math. Soc.* (1951) no. 6, 19 pp. (with K.-L. Chung).

[51.14] Some problems on random walk in space, *Proceedings of the Second Berkeley Symposium on Mathematical Statistics and Probability, 1950*, pp. 353–367, University of California Press, Berkeley and Los Angeles, 1951 (with A. Dvoretzky).

[52.07] On the greatest prime factor of $\prod_{k=1}^{x} f(k)$, *J. London Math. Soc.* **27** (1952), 379–384.

[52.11] The distribution of quadratic and higher residues, *Publ. Math. Debrecen* **2** (1952), 252–265 (with H. Davenport).

[52.12] The distribution of values of the divisor function $d(n)$, *Proc. London Math. Soc. (3)* **2** (1952), 257–271 (with L. Mirsky).

[53.01] A problem on ordered sets, *J. London Math. Soc.* **28** (1953), 426–438 (with R. Rado).

[53.02] Arithmetical properties of polynomials, *J. London Math. Soc.* **28** (1953), 416–425.

[53.03] Changes of sign of sums of random variables, *Pacific. J. Math.* **3** (1953), 673–687 (with G. A. Hunt).

[53.06] On linear independence of sequences in a Banach space, *Pacific J. Math.* **3** (1953), 689–694 (with E. G. Straus).

[53.08] The covering of n-dimensional space by spheres, *J. London Math. Soc.* **28** (1953), 287–293 (with C. A.Rogers).

[54.03] Multiple points of paths of Brownian motion in the plane, *Bull. Res. Council Israel* **3** (1954), 364–371 (with A. Dvoretzky and S. Kakutani).

[54.05] On Taylor series of functions regular in Gaier regions, *Arch. Math.* **5** (1954), 39–52 (with F. Herzog and G. Piranian).

[54.06] Rearrangements of C_1-summable series, *Acta Math.* **92** (1954), 35–53 (with F. Bagemihl).

[54.07] Sets of divergence of Taylor series and of trigonometric series, *Math. Scand.* **2** (1954), 262–266 (with F. Herzog and G. Piranian).

[54.10] The insolubility of classes of diophantine equations, *Amer. J. Math.* **76** (1954), 488–496 (with N. C. Ankeny).

[55.01] An isomorphism theorem for real-closed fields, *Ann. of Math. (2)* **61** (1955), 542–554 (with L. Gillman and M. Henriksen).

[55.05] On power series diverging everywhere on the circle of convergence, *Michigan Math. J.* **3** (1955), 31–35 (with A. Dvoretzky).

[55.06] On the law of the iterated logarithm, I., *Nederl. Akad. Wetensch. Proc. Ser. A.* **58** = *Indag. Math.* **17** (1955), 65–76 (with I. S. Gál).

[55.07] On the law of the iterated logarithm, II., *Nederl. Akad. Wetensch. Proc. Ser. A.* **58** = *Indag. Math.* **17** (1955), 77–84 (with I. S. Gál).

[55.08] On the product of consecutive integers, III., *Nederl. Akad. Wetensch. Proc. Ser. A.* **58** = *Indag. Math.* **17** (1955), 85–90.

[55.09] On the role of the Lebesgue functions in the theory of the Lagrange interpolation, *Acta Math. Acad. Sci. Hungar.* **6** (1955), 47–66 (with P. Turán).

[55.10] Partitions of the plane into sets having positive measure in every non-null measurable product set, *Trans. Amer. Math. Soc.* **79** (1955), 91–102 (with J. C. Oxtoby).

[55.11] Polynomials whose zeros lie on the unit circle, *Duke Math. J.* **22** (1955), 347–351 (with F. Herzog and G. Piranian).

[55.16] The existence of a distribution function for an error term related to the Euler function, *Canad. J. Math.* **7** (1955), 63–76 (with H. N. Shapiro).

[56.01] A limit theorem for the maximum of normalized sums of independent random variables, *Duke Math. J.* **23** (1956), 143–156 (with D. A. Darling).

[56.02] A partition calculus in set theory, *Bull. Amer. Math. Soc.* **62** (1956), 427–489 (with R. Rado).

[56.05] Monotonicity of partition functions, *Mathematika* **3** (1956), 1–14 (with P. T. Bateman).

[56.06] On a high-indices theorem in Borel summability, *Acta Math. Acad. Sci. Hungar.* **7** (1956), 265–281.

[56.07] On a problem of additive number theory, *J. London Math. Soc.* **31** (1956), 67–73 (with W. H. J. Fuchs).

[56.13] On the number of real roots of a random algebraic equation, *Proc. London Math. Soc. (3)* **6** (1956), 139–160 (with A. C. Offord).

[56.19] Sur la majorabilité C des suites de nombres réels (in French), *Acad. Serbe. Sci. Publ. Inst. Math.* **10** (1956), 37–52 (with J. Karamata).

[56.20] Über arithmetische Eigenschaften der Substitutionswerte eines Polynoms für ganzzahlige Werte des Arguments (in German), *Revue Math. Pures et Appl.* **1** (1956), 189–194.

[57.01] A probabilistic approach to problems of diophantine approximation, *Illinois J. Math.* **1** (1957), 303–315 (with A. Rényi).

[57.10] On the set of points of convergence of a lacunary trigonometric series and the equidistribution properties of related sequences, *Proc. London Math. Soc.* **7** (1957), 598–615 (with S. J. Taylor).

[57.15] Triple points of Brownian paths in 3-space, *Proc. Cambridge Philos. Soc.* **53** (1957), 856–862 (with A. Dvoretzky, S. Kakutani and S. J. Taylor).

[54] Über eine Fragestellung von Gaier und Meyer-König (in German), *Jber. Deutsch. Math. Verein.* **60** (1957), Abt. 1, 89–92.

[58.12] On the structure of set-mappings, *Acta Math. Acad. Sci. Hungar.* **9** (1958), 111–131 (with A. Hajnal).

[58.13] Points of multiplicity c of plane Brownian paths, *Bull. Res. Council Israel, Sect. F* **7F** (1958), 175–180 (with A. Dvoretzky and S. Kakutani).

[58.18] Some remarks on Euler's φ function, *Acta Arith.* **4** (1958), 10–19.

[58.20] The topologization of a sequence space by Toeplitz matrices, *Michigan Math. J.* **5**, 139–148 (with G. Piranian).

[59.04] Divergence of random power series, *Michigan Math. J.* **6** (1959), 343–347 (with A. Dvoretzky).

[59.06] Graph theory and probability, *Canad. J. Math.* **11** (1959), 34–38.

[59.10] On maximal paths and circuits of graphs, *Acta Math. Acad. Sci. Hungar.* **10** (1959), 337–356 (with T. Gallai).

[59.11] On random graphs, I., *Publ. Math. Debrecen* **6** (1959), 290–297 (with A. Rényi).

[59.14] On the distribution of primitive lattice points in the plane, *Canad. Math. Bull.* **2** (1959), 91–96 (with J. H. Chalk).

[59.15] On the Lipschitz condition for Brownian motion, *J. Math. Soc. Japan* **11** (1959), 263–274 (with K.-L. Chung and T. Sirao).

[59.23] Some examples in ergodic theory, *Proc. London Math. Soc. (3)* **9** (1959), 227–241 (with Y. N. Dowker).

[60.02] Additive properties of random sequences of positive integers, *Acta Arith.* **6** (1960), 83–110 (with A. Rényi).

[60.04] Intersection theorems for systems of sets, *J. London Math. Soc.* **35** (1960), 85–90 (with R. Rado).

[60.10] On the evolution of random graphs, *Magyar Tud. Akad. Mat. Kutató Int. Közl.* **5** (1960), 17–61 (with A. Rényi).

[60.16] Some intersection properties of random walk paths, *Acta Math. Acad. Sci. Hungar.* **11** (1960), 231–248 (with S. J. Taylor).

[60.17] Some problems concerning the structure of random walk paths, *Acta Math. Acad. Sci. Hungar.* **11** (1960), 137–162 (with S. J. Taylor).

[60.19] Some remarks on set theory, VIII., *Michigan Math. J.* **7** (1960), 187–191 (with A. Hajnal).

[60.21] Válogatott Fejezetek a Számelméletből (Selected Chapters from Number Theory, in Hungarian), *Tankönyvkiadó Vállalat*, Budapest, 1960, 250 pp. (with J. Surányi).

[61.06] Graph theory and probability, II., *Canad. J. Math.* **13** (1961), 346–352.

[61.07] Intersection theorems for systems of finite sets, *Quart. J. Math. Oxford Ser. (2)* **12** (1961), 313–320 (with Chao Ko [=Zhao Ke] and R. Rado).

[61.08] Nonincrease everywhere of the Brownian motion process, *Proc. 4th Berkeley Sympos. Math. Statist. and Prob., Vol. II*, pp. 103–116, Univ. California Press, Berkeley, 1961 (with A. Dvoretzky and S. Kakutani).

[61.15] On the evolution of random graphs, *Bull. Inst. Internat. Statist.* **38** (1961) no. 4, 343–347 (with A. Rényi).

[61.19] On the strength of connectedness of a random graph, *Acta Math. Acad. Sci. Hungar.* **12** (1961), 261–267 (with A. Rényi).

[55] Über einige Probleme der additiven Zahlentheorie, *J. Reine Angew. Math.* **206** (1961), 61–66.

[62.04] Gráfelméleti szélsőértékekre vonatkozó problémákról (Extremal problems in graph theory, in Hungarian), *Mat. Lapok* **13** (1962), 143–152 (with B. Bollobás).

[62.09] On a theorem of Rademacher-Turán, *Illinois J. Math.* **6** (1962), 122–127.

[62.13] On the maximal number of disjoint circuits of a graph, *Publ. Math. Debrecen* **9** (1962), 3–12 (with L. Pósa).

[62.23] Számelméleti megjegyzések, IV. Extremális problémák a számelméletben, I. (Remarks on number theory, IV. Extremal problems in number theory, I., in Hungarian), *Mat. Lapok* **13** (1962), 228–255.

[63.10] On the maximal number of independent circuits in a graph, *Acta Math. Acad. Sci. Hungar.* **14** (1963), 79–94 (with G. A. Dirac).

[63.19] The Hausdorff measure of the intersection of sets of positive Lebesgue measure, *Mathematika* **10** (1963), 1–9 (with S. J. Taylor).

[64.05] Arithmetical Tauberian theorems, *Acta Arith.* **9** (1964), 341–356 (with A. E.Ingham).

[64.18] On the addition of residue classes mod p, *Acta Arith.* **9** (1964), 149–159 (with H. Heilbronn).

[65.08] On some problems of a statistical group-theory, I., *Z. Wahrscheinlichkeitstheorie und Verw. Gebiete* **4** (1965), 175–186 (with P. Turán).

[65.12] On the mean value of nonnegative multiplicative number-theoretical functions, *Michigan Math. J.* **12** (1965), 321–338 (with A. Rényi).

[66.01] A limit theorem in graph theory, *Studia Sci. Math. Hungar.* **1** (1966), 51–57 (with M. Simonovits).

[66.06] On a problem of graph theory, *Studia Sci. Math. Hungar.* **1** (1966), 215–235 (with A. Rényi and V. T.Sós).

[56] On the divisibility properties of sequences of integers, I., *Acta Arith.* **11** (1966), 411–418 (with A. Sárközy and E. Szemerédi).

[66.16] On the existence of a factor of degree one of a connected random graph, *Acta Math. Acad. Sci. Hungar.* **17** (1966), 359–368 (with A. Rényi).

[57] On the divisibility properties of sequences of integers, II., *Acta Arith.* **14** (1967/1968), 1–12 (with A. Sárközy and E. Szemerédi).

[58] Some problems on the prime factors of consecutive integers, *Illinois J. Math.* **11** (1967), 428–430 (with J. L. Selfridge).

[68.07] On random matrices, II., *Studia Sci. Math. Hungar.* **3** (1968), 459–464 (with A. Rényi)

[68.11] On some problems of a statistical group-theory, IV., *Acta Math. Acad. Sci. Hungar.* **19** (1968), 413–435 (with P. Turán).

[69.02] Intersection theorems for systems of sets, II., *J. London Math. Soc.* **44** (1969), 467–479 (with R. Rado).

[69.17] Über Folgen ganzer Zahlen (in German), *Number Theory and Analysis (Papers in Honor of Edmund Landau)*, pp. 77–86, Plenum, New York, 1969 (with A. Sárközy and E. Szemerédi).

[70.07] On a new law of large numbers, *J. Analyse Math.* **23** (1970), 103–111 (with A. Rényi).

[70.19] Set mappings and polarized partition relations, *Combinatorial theory and its applications, I (Proc. Colloq., Balatonfüred, 1969)*, pp. 327–363, North-Holland, Amsterdam, 1970 (with A. Hajnal and E. C. Milner).

[71.10] On some general problems in the theory of partitions, I., *Acta Arith.* **18** (1971), 53–62 (with P. Turán).

[71.28] Unsolved problems in set theory, *Axiomatic Set Theory (Proc. Sympos. Pure Math., Vol. XIII, Part I, Univ. California, Los Angeles, Calif,. 1967)*, pp. 17–48, Amer. Math. Soc., Providence, R.I., 1971 (with A. Hajnal).

[72.11] On some applications of graph theory, I., *Discrete Math.* **2** (1972) no. 3, 207–228 (with A. Meir, V. T. Sós and P. Turán).

[72.11] On some applications of graph theory, III., *Canad. Math. Bull.* **15** (1972), 27–32 (with A. Meir, V. T. Sós and P. Turán).

[72.21] Separability properties of almost-disjoint families of sets, *Israel J. Math.* **12** (1972), 207–214 (with S. Shelah).

[73.16] On the number of solutions of $f(n) = a$ for additive functions, Collection of articles dedicated to Carl Ludwig Siegel on the occasion of his seventy-fifth birthday, I., *Acta Arith.* **24** (1973), 1–9 (with I. Z. Ruzsa and A. Sárközy).

[73.18] On the structure of edge graphs, *Bull. London Math. Soc.* **5** (1973), 317–321 (with B. Bollobás).

[73.27] *The Art of Counting: Selected writings*, edited by Joel Spencer and with a dedication by Richard Rado, Mathematicians of Our Time, Vol. 5, xxiii+742 pp., MIT Press, Cambridge, Mass.-London, 1973.

[74.25] *Probabilistic Methods in Combinatorics*, Probability and Mathematical Statistics, **17**, 106 pp., Academic Press, New York–London, 1974 (with J. Spencer).

[74.32] Some problems on random intervals and annihilating particles, *Ann. Probability* **2** (1974), 828–839 (with P. Ney).

[75.18] Méthodes probabilistes en théorie des nombres (in French), *Séminaire Delange-Pisot-Poitou (15e année: 1973/74), Théorie des nombres, Fasc. 1, Exp. No. 1*, 4 pp., Secrétariat Mathematique, Paris, 1975.

[75.28] On the structure of edge graphs, II., *J. London Math. Soc. (2)* **12** (1975/76) 219–224 (with B. Bollobás and M. Simonovits).

[75.34] Problems and results on 3-chromatic hypergraphs and some related questions, *Infinite and finite sets (Colloq., Keszthely, 1973; dedicated to P. Erdős on his 60th birthday), Vol. II; Colloq. Math. Soc. János Bolyai, Vol. 10*, pp. 609–627, North-Holland, Amsterdam, 1975 (with L. Lovász).

[75.39] Some additive and multiplicative problems in number theory, Collection of articles in memory of Juriĭ Vladimirovič Linnik, *Acta Arith.* **27** (1975), 37–50 (with S. L. G. Choi and E. Szemerédi).

[75.46] The product of consecutive integers is never a power, *Illinois J. Math.* **19** (1975), 292–301 (with J. L. Selfridge).

[76.01] P. Erdős and M. Magidor, A note on regular methods of summability and the Banach-Saks property, *Proceedings of the American Mathematical Society* (2) **59** (1976), 232–234.

[76.41] Proof of a conjecture about the distribution of divisors of integers in residue classes, *Math. Proc. Cambridge Philos. Soc.* **79** (1976), 281–287 (with R. R. Hall).

[77.05] Bases for sets of integers, *J. Number Theory* **9** (1977) no. 4, 420–425 (with D. J. Newman).

[77.13] Nonbases of density zero not contained in maximal nonbases, *J. London Math. Soc. (2)* **15** (1977) 403–405 (with M. B. Nathanson).

[78.06] Biased positional games, *Ann. Discrete Math.* **2** (1978), 221–229 (with V. Chvátal).

[78.12] Intersection properties of systems of finite sets, *Proc. London Math. Soc. (3)* **36** (1978) no. 2, 369–384 (with M. Deza and P. Frankl).

[78.48] The size Ramsey number, *Period. Math. Hungar.* **9** (1978), 145–161 (with R. J. Faudree, C. C. Rousseau and R. H. Schelp).

[79.02] Bases and nonbases of squarefree integers, *J. Number Theory* **11** (1979) no. 2, 197–208 (with M. B. Nathanson).

[79.05] Evolution of the n-cube, *Comput. Math. Appl.* **5** (1979) no. 1, 33–39 (with J. H. Spencer).

[79.07] Old and new problems and results in combinatorial number theory: van der Waerden's theorem and related topics, *Enseign. Math. (2)* **25** (1979) 325–344 (with R. L. Graham).

[80.34] Random graph isomorphism, *SIAM J. Comput.* **9** (1980) 628–635 (with L. Babai andS. M. Selkow).

[81.10] Minimal decompositions of graphs into mutually isomorphic subgraphs, *Combinatorica* **1** (1981), 13–24 (with F. R. K. Chung and R. L. Graham).

[81.34] Sur la fonction: nombre de facteurs premiers de N (in French), *Enseign. Math (2)* **27** (1981) 3–27 (with J.-L. Nicolas).

[82.15] On Ramsey-Turán type theorems for hypergraphs, *Combinatorica* **2** (1982) 289–295 (with V. T. Sós).

[83.28] Supersaturated graphs and hypergraphs, *Combinatorica* **3** (1983) no. 2, 181–192 (with M. Simonovits).

[84.15] On the statistical theory of partitions, *Topics in classical number theory, Vol. I, II (Budapest, 1981), Colloq. Math. Soc. János Bolyai, 34*, pp. 397–450, North-Holland, Amsterdam-New York, 1984 (with M. Szalay).

[85.19] On the length of the longest excursion, *Z. Wahrsch. Verw. Gebiete* **68** (1985) 365–382 (with E. Csáki and P. Révész).

[86.11] On the number of false witnesses for a composite number, *Math. Comp.* **46** (1986) 259–279 (with C. Pomerance).

[86.17] The asymptotic number of graphs not containing a fixed subgraph and a problem for hypergraphs having no exponent, *Graphs Combin.* **2** (1986) 113–121 (with P. Frankl and V. Rödl).

[87.22] On the residues of products of prime numbers, *Period. Math. Hungar.* **18** (1987) 229–239 (with A. M. Odlyzko and A. Sárközy).

[87.23] On unavoidable hypergraphs, *J. Graph Theory* **11** (1987) 251–263 (with F. R. K. Chung).

[88.29] Random walks on $\mathbf{Z}_2^n$, *J. Multivariate Anal.* **25** (1988) 111–118 (with R. W. Chen).

[89.24] On the number of distinct induced subgraphs of a graph, Graph theory and combinatorics (Cambridge, 1988), *Discrete Math.* **75** (1989) 145–154 (with A. Hajnal).

[90.29] Some of my favourite unsolved problems, in "A Tribute to Paul Erdős", A. Baker, B. Bollobás and A. Hajnal, eds, Cambridge University Press, 1990, pp. 467–478.

[91.25] On sums of a Sidon-sequence, *J. Number Theory* **38** (1991) 196–205 (with R. Freud).

[92.02] Arithmetic progressions in subset sums, *Discrete Math.* **102** (1992) 249–264 (with A. Sárközy).

[93.16] Ramsey problems in additive number theory, *Acta Arith.* **64** (1993) 341–355 (with B. Bollobás and G. P. Jin).

[94.19] On the densities of sets of multiples, *J. Reine Angew. Math.* **454** (1994), 119–141 (with R. R. Hall and G. Tenenbaum).

[95.07] Extremal problems in combinatorial geometry, *Handbook of Combinatorics, Vol. 1, 2*, pp. 809–874, Elsevier, Amsterdam, 1995 (with G. B. Purdy).

[95.08] Independence of solution sets and minimal asymptotic bases, *Acta Arith.* **69** (1995) no. 3, 243–258 (with M. B. Nathanson and P. Tetali).

[96.17] On some of my favourite theorems, in *Combinatorics, Paul Erdős is Eighty*, vol. II, D. Miklós, V.T. Sós and T. Szőnyi, eds, János Bolyai Mathematical Society, Budapest, 1996, pp. 97–132.

[97.18] Some of my favorite problems and results, in "The Mathematics of Paul Erdős", vol. I, R.L. Graham and J. Nešetřil, eds, Springer-Verlag, Berlin, 1996, pp. 47–67.

[97.21] Some unsolved problems, in: *Combinatorics, Geometry and Probability: in tribute to Paul Erdős*, B. Bollobás and A. Thomason, eds, pp. 1–10, Cambridge University Press, 1997.

THE GAUSSIAN LAW OF ERRORS IN THE THEORY OF ADDITIVE NUMBER THEORETIC FUNCTIONS.* [1]

By P. Erdös and M. Kac.

The present paper concerns itself with the applications of statistical methods to some number-theoretic problems. Recent investigations of Erdös and Wintner [2] have shown the importance of the notion of statistical independence in number theory; the purpose of this paper is to emphasize this fact once again.

It may be mentioned here that we get as a particular case of our main theorem the following result:

If $\nu(m)$ denotes the number of prime divisors of m, and K_n the number of those integers from 1 up to n for which $\nu(m) < \lg\lg n + \omega\sqrt{2\lg\lg n}$ (ω an arbitrary real number), then

$$\lim_{n\to\infty} \frac{K_n}{n} = \pi^{-\frac{1}{2}} \int_{-\infty}^{\omega} \exp(-u^2)\,du.$$

This theorem refines some known results of Hardy, Ramanujan [3] and Erdös.[4]

1. In what follows p will denote a prime and ω will denote a real number.

Let $f(m)$ be an additive number-theoretic function, so that $f(mn) = f(m) + f(n)$ if $(m, n) = 1$. Suppose that $f(p^\alpha) = f(p)$ and $|f(p)| \leqq 1$. Obviously

$$f(m) = \sum_{p|m} f(p).$$

Furthermore put $\sum_{p<n} p^{-1}f(p) = A_n$ and $(\sum_{p<n} p^{-1}f^2(p))^{1/2} = B_n$. Then our main theorem may be stated as follows:

* Received December 7, 1939.

[1] A preliminary account appeared in the *Proceedings of the National Academy*, vol. 25 (1939), pp. 206-207.

[2] P. Erdös and A. Wintner, "Additive arithmetic functions and statistical independence," *American Journal of Mathematics*, vol. 61 (1939), pp. 713-722.

[3] Srinivasa Ramanujan, *Collected Papers* (1927), pp. 262-275.

[4] P. Erdös, "Note on the number of prime divisors of integers," *Journal of the London Mathematical Society*, vol. 12 (1937), pp. 308-314.

THEOREM. *If $B_n \to \infty$ as $n \to \infty$, and K_n denotes the number of integers m from 1 up to n for which*

$$f(m) < A_n + \omega\sqrt{2}\, B_n$$

then

$$\lim_{n\to\infty} \frac{K_n}{n} = \pi^{-\frac{1}{2}} \int_{-\infty}^{\omega} \exp(-u^2)\,du = D(\omega).$$

2. We first prove the following

LEMMA 1. *Let*

$$f_l(m) = \sum_{\substack{p|m \\ p<l}} f(p).$$

Then denoting by δ_l the density of the set of integers m for which $f_l(m) < A_l + \omega\sqrt{2}\, B_l$ one has

$$\lim_{l\to\infty} \delta_l = D(\omega).$$

Let $\rho_p(n)$ be 0 or $f(p)$ according as p does not or does divide n. Then

$$f_l(m) = \sum_{p<l} \rho_p(m).$$

Since the $\rho_p(n)$ are statistically independent, $f_l(m)$ behaves like a sum of independent random variables and consequently the distribution function of $f_l(m) - A_l/\sqrt{2}\, B_l$ is a convolution (Faltung) of the distribution functions of $\rho_p(m) - p^{-1}f(p)/\sqrt{2}\, B_l$ [5] $(p < l)$. It is easy to see that the "central limit theorem of the calculus of probability" can be applied to the present case,[6] and this proves our lemma.

3. Lemma 1 is the only "statistical" lemma in the proof. Using this lemma, the main result will be established by purely number-theoretical methods.

LEMMA 2. *If m_n tends to ∞* (as $n \to \infty$) *more rapidly than any fixed*

---

[5] *Loc. cit.* 2, where statistical independence of arithmetical functions is defined and discussed. See also P. Hartman, E. R. van Kampen and A. Wintner, *American Journal of Mathematics*, vol. 61 (1939), pp. 477-486.

[6] Cf. for instance the first chapter of S. Bernstein's paper, "Sur l'extension du théorème limite du calcul des probabilités aux sommes de quantités dépendantes," *Mathematische Annalen*, vol. 97, pp. 1-59. See also M. Kac and H. Steinhaus, "Sur les fonctions indépendantes II," *Studia Math.*, vol. 6 (1936), pp. 59-66.

power of s_n, then the number of integers from 1 up to m_n which are not divisible by any prime less than s_n is equal to

$$\frac{m_n e^{-C}}{\lg s_n} + o\left(\frac{m_n}{\lg s_n}\right),$$

where C denotes Euler's constant.

The proof of this statement is implicitly contained in the reasoning of V. Brun on page 21 of his famous memoir "Le crible d'Erasosthène et le théorème de Goldbach"[7] and may therefore be omitted.

Let $\phi(n)$ represent a function which tends, as $n \to \infty$, to 0 in such a way that $n^{\phi(n)} \to \infty$. The function $n^{\phi(n)}$ will be denoted by α_n and $n^{\sqrt{\phi(n)}}$ by β_n. Let $a_1(n), a_2(n), \cdots$ be the integers whose prime factors are all less than α_n, and let $\psi(m; n)$ be the greatest a_i which divides m. We then have the following

LEMMA 3. *The number of integers $m \leqq n$ for which $\psi(m; n) = a_i(n)$, where $a_i(n) \leq \beta_n$ is equal to*

$$\frac{e^{-C} n}{a_i(n)\phi(n) \lg n} + o\left(\frac{n}{a_i(n)\phi(n) \lg n}\right).$$

This is a direct consequence of Lemma 2. For consider all those integers $\leqq n$ which are of the form $r \cdot a_i(n)$ and such that r is not divisible by any prime $< \alpha_n$. Evidently, the integers thus defined are all the integers $\leqq n$ for which $\psi(m; n) = a_i(n)$. Their number is equal to the number of integers r which are $\leqq n/a_i(n)$ and not divisible by any prime $< \alpha_n$. The restriction $a_i(n) < \beta_n$ makes $n/a_i(n)$ tend to ∞ more rapidly than any power of α_n and therefore Lemma 2 can be applied (put $m_n = n/a_i(n)$ and $s_n = \alpha_n$). This completes the proof.

LEMMA 4. *The number y of integers $\leq M$ divisible by an $a_i(n) > \beta_n$ is less than $bM\sqrt{\phi(n)}$, where b is an absolute constant. (It follows from this that the density of the integers which are divisible by an $a_i(n) > \beta_n$ is less than $b\sqrt{\phi(n)}$.)*

We have

$$\prod_{m=1}^{M} \psi(m; n) = \prod_{p<\alpha_n} p^{\sum_{r=1}^{\infty}[M/p^r]} < \prod_{p<\alpha_n} p^{2M/p};$$

and since

$$\lg \prod_{p<\alpha_n} p^{2M/p} = 2M \sum_{p<\alpha_n} p^{-1} \lg p \sim 2M\phi(n) \lg n$$

[7] *Skrifter Videns*, Kristiania, 1920.

one has

$$\prod_{m=1}^{M} \psi(m;n) < n^{bM\phi(n)}.$$

Hence, finally

$$(\beta_n)^y = (n^{\sqrt{\phi(n)}})^y < n^{bM\phi(n)}, \quad \text{i. e.,} \quad y < bM\sqrt{\phi(n)}.$$

4. LEMMA 5. *Denote by l_n the number of integers from 1 up to n for which*

(i) $$f_{a_n}(m) < A_{a_n} + \omega\sqrt{2}\,B_{a_n}.$$

Then

$$\lim_{n\to\infty} \frac{l_n}{n} = D(\omega).$$

Divide the integers from 1 up to n which satisfy (i) into classes $E_1, E_2, \cdots$ so that m belongs to E_i if and only if $\psi(m;n) = a_i(n)$; and denote by $|E_i|$ the number of integers in E_i. One obviously has

$$l_n = \sum_i |E_i| = \sum_{a_i \leq \beta_n} |E_i| + \sum_{a_i > \beta_n} |E_i|.$$

By Lemma 4 $\sum_{a_i>\beta_n} |E_i| < bn\sqrt{\phi(n)}$ and therefore it is sufficient to prove that $n^{-1} \sum_{a_i \leq \beta_n} |E_i| \to D(\omega)$ as $n \to \infty$. On the other hand by Lemma 3

(ii) $$\sum_{a_i\leq\beta_n} |E_i| = \left(\frac{e^{-C}\cdot n}{\phi(n)\lg n} + o\left(\frac{n}{\phi(n)\lg n}\right)\right) \sum_{a_i\leq\beta_n}{}' \frac{1}{a_i(n)},$$

where the dash in the summation indicates that it is extended over the a_i's satisfying $f_{a_n}(a_i) < A_{a_n} + \omega\sqrt{2}\,B_{a_n}$. In order to evaluate Σ', divide all the integers into classes $F_1, F_2, \cdots$ having the property that m belongs to F_i if and only if $\psi(m;n) = a_i(n)$ and let $\{F_i\}$ denote the density of F_i. Consider now the set $\Sigma' F_i$, where the dash in summation has the same meaning as above. By putting $l = \alpha_n$ and using Lemma 1 we have that $\{\Sigma' F_i\} \to D(\omega)$ as $n \to \infty$ or $\{\Sigma' F_i\} = D(\omega) + o(1)$. Now

(iii) $$\Sigma' F_i = \sum_{a_i\leq\beta_n}{}' F_i + \sum_{a_i>\beta_n}{}' F_i$$

and by Lemma 4

(iv) $$\{\sum_{a_i>\beta_n}{}' F_i\} < b\sqrt{\phi}(n).$$

Furthermore there is only a finite number of a_i's which are less than β_n and therefore $\{\sum_{a_i<\beta_n}{}' F_i\} = \sum_{a_i<\beta_n}{}' \{F_i\}$. But

$$\{F_i\} = \frac{1}{a_i(n)} \prod_{p<a_n} \left(1 - \frac{1}{p}\right) = \frac{1}{a_i(n)} \left(\frac{e^{-C}}{\phi(n)\lg n} + o\left(\frac{1}{\phi(n)\lg n}\right)\right)$$

and this implies that

(v) $$\{\sum_{a_i<\beta_n}' F_i\} = \left(\frac{e^{-C}}{\phi(n)\lg n} + o\left(\frac{1}{\phi(n)\lg n}\right)\right) \sum_{a_i<\beta_n}' \frac{1}{a_i(n)}.$$

Finally (iii), (iv) and (v) give

$$D(\omega) - b\sqrt{\phi(n)} < \left(\frac{e^{-C}}{\phi(n)\lg n}\right) + o\left(\frac{1}{\phi(n)\lg n}\right)\Big) \sum_{a_i<\beta_n}' \frac{1}{a_i(n)} < D(\omega) + 0(1).$$

The combination of this formula with (ii) completes the proof of our Lemma.

5. We now come to the proof of the main theorem. Notice first that for $m \leq n$, $|f(m) - f_{a_n}(m)| < 1/\phi(n)$. In fact, $|f(p)| \leq 1$ implies that $|f(m) - f_{a_n}(m)|$ is less than the number of those prime divisors of m which are $\geq \alpha_n$. This number is obviously $< 1/\phi(n)$, since $(\alpha_n)^{1/\phi(n)} = n$. Notice furthermore that $|f(p)| \leq 1$ and the well known results concerning the sum $\sum_{p<n} p^{-1}$ imply that $|A_n - A_{a_n}| < -C_1 \lg \phi(n)$ and $|B_n - B_{a_n}| < -C_2 \lg \phi(n)$, where C_1 and C_2 are absolute constants.

Now choose $\phi(n)$ so that $1/\phi(n) = o(B_n)$. Evidently every $m \leq n$ satisfying the inequality $f(m) < A_n + \omega\sqrt{2}\, B_n$ also satisfies, for sufficiently large n, the inequality $f_{a_n}(m) < A_{a_n} + (\omega + \epsilon)\sqrt{2}\, B_{a_n}$. In addition every $m \leq n$ satisfying $f_{a_n}(m) < A_{a_n} + (\omega - \epsilon)\sqrt{2}\, B_{a_n}$ satisfies, for sufficiently large n, the inequality $f(m) < A_n + \omega\sqrt{2}\, B_n$. Hence, by Lemma 5,

$$D(\omega - \epsilon) \leq \liminf \frac{K_n}{n} \leq \limsup \frac{K_n}{n} \leq D(\omega + \epsilon);$$

and this proves the theorem, since $\epsilon > 0$ is arbitrary.

6. The theorem mentioned in the introduction is obviously a particular case of our main theorem. It corresponds to the case $f(p) = 1$. Because of the large number of applications of $\nu(m)$ it is of special interest. It should be mentioned that the assumption $f(p^a) = f(p)$ can be removed; also $|f(p)| \leq 1$ may be replaced by a much weaker condition. This however, would complicate the statement of the main theorem

We may perhaps point out that Lemma 2 (Brun) is the "deepest" part of the proof and that the "statistical" part is relatively superficial. However, the statistical considerations seemed to be suggestive and fruitful in leading to new and perhaps striking results.

INSTITUTE FOR ADVANCED STUDY AND CORNELL UNIVERSITY.

THE PRODUCT OF CONSECUTIVE INTEGERS IS NEVER A POWER

BY

P. ERDÖS AND J. L. SELFRIDGE

We dedicate this paper to the memory of our friends H. Davenport, Ju.V. Linnik, L. J. Mordell, L. Moser, A. Rényi and W. Sierpiński, all of whom were alive when we started our work in 1966 at the University of Illinois at Urbana.

0. Introduction

It was conjectured about 150 years ago that the product of consecutive integers is never a power. That is, the equation

$$(n + 1)\cdots(n + k) = x^l \tag{1}$$

has no solution in integers with $k \geq 2$, $l \geq 2$ and $n \geq 0$. (These restrictions on k, l and n will be implicit throughout this paper.) The early literature on this subject can be found in Dickson's history and the somewhat later literature in the paper of Obláth [5].

Rigge [6], and a few months later Erdös [1], proved the conjecture for $l = 2$. Later these two authors [1] proved that for fixed l there are at most finitely many solutions to (1). In 1940, Erdös and Siegel jointly proved that there is an absolute constant c such that (1) has no solutions with $k > c$, but this proof was never published. Later Erdös [2] found a different proof; by improving the method used, we can now completely establish the old conjecture. Thus we shall prove:

THEOREM 1. *The product of two or more consecutive positive integers is never a power.*

In fact we shall prove a stronger result:

THEOREM 2. *Let k, l, n be integers such that $k \geq 3$, $l \geq 2$ and $n + k \geq p^{(k)}$, where $p^{(k)}$ is the least prime satisfying $p^{(k)} \geq k$. Then there is a prime $p \geq k$ for which $\alpha_p \not\equiv 0 \pmod{l}$, where α_p is the power of p dividing $(n + 1)\cdots(n + k)$.*

Theorem 2 implies Theorem 1, since it is easy to see that $(n + 1)(n + 2)$ is never an lth power and if $n \leq k$ then by Bertrand's postulate the largest prime factor of $(n + 1)\cdots(n + k)$ divides this product to exactly the first power. Moreover, this shows that in proving Theorem 2 it will suffice to assume $n > k$.

One could conjecture the following strengthening of Theorem 2: if $k \geq 4$ and $n + k \geq p^{(k)}$, then there is at least one prime greater than k which divides

Received May 24, 1974.

Reprinted from Illinois J. Math. (June '75), pp. 292–301. Reprinted with permission of the Illinois Journal of Mathematics, University of Illinois at Urbana/Champaign.

$(n+1)\cdots(n+k)$ to the first power. This conjecture, if true, seems very deep. The requirement of $k \geq 4$ is motivated by

$$\binom{50}{3} = 140^2.$$

Now we start the proof of Theorem 2. We suppose that Theorem 2 is false for some particular k, l and n, and show that in every case this leads to a contradiction. As noted above, we assume $n > k$.

1. Basic lemmas

First observe that by the well-known theorem of Sylvester and Schur [3] there is always a prime greater than k which divides $(n+1)\cdots(n+k)$, since $n > k$. Such a prime divides only one of the k factors, so $n + k \geq (k+1)^l$, whence

$$n > k^l. \tag{2}$$

Furthermore since we suppose Theorem 2 is false, for $1 \leq i \leq k$ we have

$$n + i = a_i x_i^l, \tag{3}$$

where a_i is lth-power free and all its prime factors are less than k.

In the proof [1] for the case $l = 2$, it was shown that $a_i \neq a_j$ if $i \neq j$. In fact for $l > 2$ it is also known that the products $a_i a_j$ are all distinct. In this paper we need the stronger result:

LEMMA 1. *For any $l' < l$, the products $a_{i_1} \cdots a_{i_{l'}}$ $(i_1 \leq \cdots \leq i_{l'})$ are all distinct.*

In fact we prove that the ratio of two such products cannot be an lth power. First we show that (2) ensures

$$(n + i_1)\cdots(n + i_{l-1}) \neq (n + j_1)\cdots(n + j_{l-1}), \tag{4}$$

provided the two products are not identical.

Cancel any equal factors. Since $(n + i, n + j) < k$ and $n > k^l$, it follows that no factor of one member of (4) divides the product of the factors remaining in the other member, so the nonequality in (4) is proved.

Now we prove the lemma. For some rational t, suppose that

$$a_{i_1} \cdots a_{i_{l-1}} = a_{j_1} \cdots a_{j_{l-1}} t^l. \tag{5}$$

We shall show that (5) implies the subscripts must all match. Assume without loss of generality that $(n + i_1)\cdots(n + i_{l-1}) > (n + j_1)\cdots(n + j_{l-1})$, and put $t = u/v$, with $(u, v) = 1$. Then

$$(n + i_1)\cdots(n + i_{l-1}) = a_{i_1} \cdots a_{i_{l-1}} \frac{x^l}{u^l}$$

and

$$(n + j_1)\cdots(n + j_{l-1}) = a_{j_1} \cdots a_{j_{l-1}} \frac{y^l}{v^l},$$

where $x = ux_{i_1} \cdots x_{i_{l-1}}$ and $y = vx_{j_1} \cdots x_{j_{l-1}}$ in the notation of (3). By (5), we may put

$$A = \frac{a_{i_1} \cdots a_{i_{l-1}}}{u^l} = \frac{a_{j_1} \cdots a_{j_{l-1}}}{v^l},$$

so $Ax^l > Ay^l$ and therefore $x \geq y + 1$. Thus

$$(n + i_1) \cdots (n + i_{l-1}) - (n + j_1) \cdots (n + j_{l-1}) \geq A\{(y + 1)^l - y^l\} > Aly^{l-1}. \tag{6}$$

Note that (5) implies A is a positive integer. Also $Ay^l = (n + j_1) \cdots (n + j_{l-1}) > n^{l-1}$, so with (6) we have

$$(n + i_1) \cdots (n + i_{l-1}) - (n + j_1) \cdots (n + j_{l-1}) > Al\left(\frac{n^{l-1}}{A}\right)^{(l-1)/l} \geq ln^{(l-1)^2/l}. \tag{7}$$

On the other hand,

$$(n + i_1) \cdots (n + i_{l-1}) - (n + j_1) \cdots (n + j_{l-1}) < (n + k)^{l-1} - n^{l-1} < kln^{l-2}, \tag{8}$$

where the last inequality is obvious if $l = 2$ and for $l \geq 3$ it may be seen as follows. Clearly it suffices to show that

$$kn^{l-2} > \sum_{i=2}^{l-1} \binom{l-1}{i} n^{l-1-i}k^i,$$

that is

$$1 > \sum_{i=2}^{l-1} \binom{l-1}{i} \left(\frac{k}{n}\right)^{i-1}.$$

Now

$$\binom{l-1}{i} < \frac{l^i}{2^{i-1}};$$

also $n > k^l$, $k \geq 3$ and $l \geq 3$, so $n > kl$ and moreover $n > kl^2$. Therefore

$$\sum_{i=2}^{l-1} \binom{l-1}{i} \left(\frac{k}{n}\right)^{i-1} < l \sum_{i=2}^{\infty} \left(\frac{kl}{2n}\right)^{i-1} = \frac{kl^2}{2n - kl} < \frac{kl^2}{n} < 1.$$

The lemma now follows, since (7) and (8) require $k > n^{1/l}$, contrary to (2). Now we prove:

LEMMA 2. *By deleting a suitably chosen subset of* $\pi(k - 1)$ *of the numbers* $a_i(1 \leq i \leq k)$, *we have*

$$a_{i_1} \cdots a_{i_{k'}} \mid (k - 1)! \tag{9}$$

where $k' = k - \pi(k - 1)$.

For each prime $p \leq k - 1$ we omit an a_m for which $n + m$ is divisible by p to the highest power. If $1 \leq i \leq k$ and $i \neq m$, the power of p dividing $n + i$

is the same as the power of p dividing $i - m$. Thus $p^\alpha \| a_{i_1} \cdots a_{i_{k'}}$ implies $p^\alpha \mid (k - m)!(m - 1)!$, so $p^\alpha \mid (k - 1)!$ and our lemma is proved.

Change of notation. In the remainder of this paper it will be convenient to have the a's renumbered so that $a_1 < a_2 < \cdots < a_k$. We shall employ this new notation in Sections 2 and 3.

Note also that to prove Theorem 2 for any particular l it is enough to prove it for some divisor of l, so it suffices to consider only prime l.

2. The case $l > 2$

2.1. *The case $k \geq 30000$.* Now we show that (9) leads to a contradiction for $k \geq 30000$, using only the distinctness of the products $a_i a_j$. It is known [4] that the number of positive integers $b_1 < \cdots < b_r \leq x$ for which the products $b_i b_j$ are all distinct satisfies

$$r < \pi(x) + \frac{c_1 x^{3/4}}{(\log x)^{3/2}}, \tag{10}$$

and this is best possible apart from the value of c_1. However, when r is small this result is not adequate for our needs, so we shall now establish a bound which is sharper for small r.

First we need a graph theoretic lemma. A subgraph of a graph is called a *rectangle* if it comprises two pairs of vertices, with each member of one pair joined to each member of the other. We prove:

LEMMA 3. *Let G be a bipartite graph of s white and t black vertices which contains no rectangles. Then the number of edges of G is at most*

$$s + \binom{t}{2}.$$

Call a subgraph of G comprising one vertex joined to each of two others a *V-subgraph.* Since G contains no rectangle, there can be at most one V-subgraph with any given pair of black vertices as its endpoints. Let s_i be the number of white vertices of valence i, so $\sum_{i \geq 1} s_i = s$. Counting the number of V-subgraphs with black endpoints gives

$$\sum_{i \geq 2} s_i \binom{i}{2} \leq \binom{t}{2}. \tag{11}$$

If E is the number of edges of G, then by (11)

$$E = \sum_{i \geq 1} i s_i = s + \sum_{i \geq 2} (i - 1) s_i \leq s + \sum_{i \geq 2} s_i \binom{i}{2} \leq s + \binom{t}{2}$$

which proves Lemma 3.

Now let $u_1 < \cdots < u_s \leq x$ and $v_1 < \cdots < v_t \leq x$ be two sequences of positive integers such that every positive integer up to x can be written in the form $u_i v_j$. If $b_1 < \cdots < b_r \leq x$ are positive integers such that all the products $b_i b_j$ are distinct, form the bipartite graph G with s white vertices labelled $u_1, \ldots, u_s$ and t black vertices labelled $v_1, \ldots, v_t$ and an edge between u_i and v_j if $u_i v_j = b_m$ for some m. Distinctness of the products $b_i b_j$ ensures that G has no rectangles so by Lemma 3,

$$r \leq s + \binom{t}{2}. \tag{12}$$

Lemma 1 shows in particular that the bound (12) applies to the sequence $a_1 < \cdots < a_k$. Using (12) we next prove that the product of any $k - \pi(k)$ of the a's exceeds $k!$ provided $k \geq 30000$. Because of Lemma 2 this implies Theorem 2 for $k \geq 30000$ and $l > 2$. Evidently it suffices to prove

$$\prod_{i=1}^{k-\pi(k)} a_i > k! \quad \text{if} \quad k \geq 30000. \tag{13}$$

We shall now obtain lower bounds on a_i $(1 \leq i \leq k)$. We clearly have

$$a_i \geq i, \tag{14}$$

and using (12) we shall show two further inequalities:

$$a_i \geq 3.5694(i - 304), \tag{15}$$

$$a_i \geq 4.3402(i - 1492). \tag{16}$$

Of these, (14) is sharpest for $i \leq 422$, (15) is sharpest for $422 < i \leq 6993$, and (16) is sharpest for $i > 6993$. With these inequalities, a routine calculation using Stirling's formula suffices to verify (13) when $k = 30000$, and (16) ensures that (13) holds when $k > 30000$.

To prove (15), we take $v_1 < \cdots < v_t$ to be the $t = 25$ positive integers up to 36 which have no prime factor greater than 7 (so $v_1 = 1$ and $v_{25} = 36$). Next we obtain a suitable set of positive integers $u_1 < \cdots < u_s \leq x$ so that every positive integer $m \leq x$ is expressible in the form $u_i v_j$. For convenience, let V denote the set of v's. Clearly any positive integer $m \leq x$ with all prime factors greater than 7 must be included in the u's: let U_1 denote the set of such numbers. Next, suppose $m \leq x$ is a positive multiple of 7 and $m = dd'$, where d is the largest divisor of m with no prime factor greater than 7. If $d \notin V$ then $d \geq 42$, since $7 \mid d$. Thus $x \geq m = dd' > 42d'$, so $7d' \leq x/6$. Hence we include in the u's all positive integers of the form $7d' \leq x/6$ with least prime factor 7: let U_2 denote this set of numbers. Similarly, if $m \leq x$ is a positive multiple of 5 and $m = dd'$, where d is the largest divisor of m with no prime factor greater than 5, then $d \notin V$ requires $d \geq 40$ and $5d' \leq x/8$. Hence we include in the u's all positive integers of the form $5d' \leq x/8$ with least prime factor 5, and let U_3 denote this set. Likewise we include in the u's all positive integers of the form $3d' \leq x/14$ with least prime factor 3, and all positive integers of the form $2d' \leq x/20$, denoting these sets by U_4 and U_5 respectively.

Now every positive integer $m \le x$ is expressible in the form $m = u_i v_j$ for some $u_i \in U$ and $v_j \in V$, where U denotes the union of $U_1, \ldots, U_5$.

The number of u's in each U_i can readily be calculated. For example

$$|U_2| = \frac{\phi(30)}{30} \cdot \frac{x}{42} + \varepsilon_2(x) = \frac{2x}{315} + \varepsilon_2(x),$$

where the error term has the bound $\varepsilon_2(x) \le 14/15$. Likewise $\varepsilon_1(x) \le 53/35$, $\varepsilon_3(x) \le 2/3$, $\varepsilon_4(x) \le 1/2$ and $\varepsilon_5(x) \le 0$. Thus the total number of u's is

$$s = |U| = \sum_{i=1}^{5} |U_i| = \left(\frac{8}{35} + \frac{2}{315} + \frac{1}{120} + \frac{1}{84} + \frac{1}{40}\right) x$$
$$+ \sum_{i=1}^{5} \varepsilon_i(x) < \frac{353}{1260} x + 4.$$

Now (12) implies that the number of a's up to x is less than $353x/1260 + 304$, whence (15).

To prove (16), we take the v's to be the $t = 55$ positive integers up to 100 with no prime factor greater than 11, and the u's to be all positive integers up to x with all prime factors greater than 11, together with all those up to $x/10$ with least prime factor 11, all those up to $x/15$ with least prime factor 7, all those up to $x/21$ with least prime factor 5, all those up to $x/35$ with least prime factor 3, and finally all even integers up to $x/54$. The first of these subsets of u's contains $16x/77 + \varepsilon_0(x)$ numbers, where $\varepsilon_0(x) \le 194/77$. The error terms in counting the other subsets of u's are the same as before, so the total error is less than 7. With (12), this leads to (16). Now we shall work upwards from small k to resolve the cases with $k < 30000$.

2.2. *The case* $k = 3$. It is easy to see that (1) has no solution when $k = 3$, for $(n + 1)(n + 2)(n + 3) = m(m^2 - 1)$, where $m = n + 2$, shows that the product could only be an lth power if m and $m^2 - 1$ are lth powers, but $m^2 - 1$ and m^2 cannot both be lth powers. But for Theorem 2 we need to show $\alpha_p \not\equiv 0 \pmod{l}$ for some prime $p \ge 3$, where α_p is the power of p in $(n + 1)(n + 2)(n + 3)$. Suppose there is no such p. If n is even, $(n + 1, n + 3) = 1$ ensures $a_1 = a_2 = 1$, contradicting Lemma 1. If n is odd, $(n + 1, n + 3) = 2$ ensures $a_1 = 1$, $a_2 = 2$ and $a_3 = 2^\alpha$, with $1 < \alpha < l$, and Lemma 1 is contradicted by $a_1^{\alpha-1} a_3 = a_2^\alpha$.

2.3. *The case* $4 \le k \le 1000$, $l = 3$. Here we restrict attention to those a's with no prime factor greater than the mth prime, say $f(k, m)$ in number. If u and v are positive integers with prime factors similarly restricted, there are 3^m rationals u/v no two of which differ by a factor which is the cube of a rational. The number of formally distinct expressions a_i/a_j is $f(k, m)\{f(k, m) - 1\}$, so there are two whose quotient yields a solution to (5), thus contradicting Lemma 1, if

$$f(k, m)\{f(k, m) - 1\} > 3^m. \tag{17}$$

Since the a's arise as divisors of k consecutive integers, and have all prime factors less than k, it is straightforward to calculate a lower bound for $f(k, m)$. Thus we verify (17) for $4 \le k \le 10$ with $m = 2$, for $10 < k \le 28$ with $m = 3$, for $28 < k \le 77$ with $m = 4$, for $77 < k \le 143$ with $m = 5$, for $143 < k \le 340$ with $m = 6$, for $340 < k \le 646$ with $m = 7$, and for $646 < k \le 1000$ with $m = 8$.

This method could be continued beyond $k = 1000$, but certainly fails before reaching $k = 10000$. Fortunately we have an improvement available, and we now proceed with it.

2.4. *The case* $1000 < k < 30000$, $l = 3$. Let $q_1 < \cdots < q_r$ be the r largest primes satisfying $q_i \le k^{1/2}$, where r is to be suitably chosen. We now restrict attention to those a's, say $F(k, r)$ in number, which have no prime factor greater than $k^{1/2}$, and at most one prime factor (counting multiplicity) among the q's. If u and v are positive integers with prime factors similarly restricted, there are $3^{\pi(q_1)-1} R$ rationals u/v no two of which differ by a factor which is the cube of a rational. In this count the factor $R = r^2 + r + 1$ arises from the fact that u and v each contain at most one of the q's as a divisor. As in (17), the number of formally distinct expressions a_i/a_j is enough to ensure that there are two whose quotient yields a solution to (5), and therefore contradicts Lemma 1, if

$$F(k, r)\{F(k, r) - 1\} > 3^{\pi(q_1)-1}(r^2 + r + 1). \tag{18}$$

To obtain a lower bound for $F(k, r)$, note that for each prime p in $(k^{1/2}, k)$ we omit at most $[k/p] + 1$ of the a's; similarly for the products q_i^2 and $q_i q_j$, so

$$F(k, r) \ge k - \sum_{k^{1/2} < p < k} \left(\left[\frac{k}{p} \right] + 1 \right) - \sum_{1 \le i \le j \le r} \left(\left[\frac{k}{q_i q_j} \right] + 1 \right)$$

$$\ge k - \sum_{k^{1/2} < p < k} \left(\left[\frac{k}{p} \right] + 1 \right) - \left[\frac{k}{2} \left\{ \sum_{i=1}^{r} \frac{1}{q_i^2} + \left(\sum_{i=1}^{r} \frac{1}{q_i} \right)^2 \right\} \right] - \binom{r+1}{2}.$$

For example, with $k = 175^2 = 30625$ and $r = 31$ (so $q_1 = 29$) this bound is adequate to verify (18). Indeed, for $1000 < k < 30000$ we can readily verify (18), in each case taking q_1 around $k^{0.3}$.

2.5. *The case* $4 \le k \le 30000$, $l > 3$. Here it is inconvenient to work with ratios of products of a's, so we work directly with the products themselves, since we do not need the extra sharpness.

With the a's selected as in Section 2.3, the inequality corresponding to (17) is

$$\binom{f(k, m) + l - 2}{l - 1} > l^m. \tag{19}$$

The left member of (19), derived by counting the number of nondecreasing sequences of $l - 1$ a's, is the number of formally distinct products of $l - 1$ a's, and the right member is the number of lth-power free positive integers with all prime factors among the first m primes. When (19) holds, (5) has a solution, contradicting Lemma 1. It is easy to verify by direct computation that (17) implies (19) for $4 \le k \le 1000$ and m chosen as in Section 2.3.

Similarly, with the a's selected as in Section 2.4, the inequality corresponding to (18) is

$$\binom{F(k, r) + l - 2}{l - 1} > l^{\pi(q_1) - 1}\binom{l + r - 1}{l - 1}. \tag{20}$$

The left member of (20) is the number of formally distinct products of $l - 1$ a's, and the right member is the number of lth-power free positive integers with no prime factor greater than $k^{1/2}$ and at most $l - 1$ prime factors among the q's (counted by multiplicity). When (20) holds, (5) has a solution, contradicting Lemma 1. For $1000 < k < 30000$ and the values of r chosen as in Section 2.4, (20) easily holds when (18) holds.

This completes the proof of Theorem 2 for $l > 2$. It seems certain that one could get a more general inequality than (19) and (20), leading to a more elegant method valid for all k.

3. The case $l = 2$

It remains to prove that $(n + 1) \cdots (n + k)$ always contains a prime $p \ge k$ to an odd exponent. (We already know that the product is not a square, by the results of Rigge and Erdös cited earlier.)

The a's are now square-free and, by Lemma 1, all distinct. So, by Lemma 2,

$$\prod_{i=1}^{k} a_i \,\Big|\, (k - 1)! \prod_{p<k} p \tag{21}$$

We shall now show that for $k \ge 71$ this leads to a contradiction.

3.1. *The case* $k \ge 71$. Since 12 of every 36 consecutive integers are divisible by 4 or 9, at most 24 of any 36 consecutive integers are square-free. Thus for $k \ge 64$ we have

$$\prod_{i=1}^{k} a_i > k!\left(\frac{3}{2}\right)^k. \tag{22}$$

For any positive integer m and prime p, the power to which p divides $p^m!$ is $(p^m - 1)/(p - 1)$. From this we can deduce that if the powers to which 2 and 3 divide $(k - 1)!$ are α and β respectively, then

$$\alpha \ge k - 1 - \log_2 k \quad \text{and} \quad \beta \ge \tfrac{1}{2}(k - 1) - \log_3 k.$$

On the other hand, since the a's arise from k consecutive integers and are square-free, we calculate that if the powers to which 2 and 3 divide $a_1 \cdots a_k$ are γ and δ respectively, then

$$\gamma \le \tfrac{1}{3}\{k + \log_2 (3k + 1)\} \quad \text{and} \quad \delta \le \tfrac{1}{4}\{k + 1 + 2 \log_3 (2k + 1)\}.$$

Since (21) implies that $a_1 \cdots a_k \leq (k-1)!2^{\gamma-\alpha}3^{\delta-\beta}\prod_{p<k} p$, we now deduce from (22) that for $k \geq 64$,

$$\left(\frac{3}{2}\right)^k 2^{2k/3}3^{k/4} < \frac{14}{3} k^2 \prod_{p<k} p. \tag{23}$$

However $\prod_{p<k} p < 3^k$, so (23) fails for $k \geq 297$. Indeed, $\prod_{p<k} p < e^k$ for $k \leq 10^8$ by Theorem 18 of [7], so (23) fails for $k \geq 71$.

3.2. *The case* $k < 71$. For $k = 3$, it is impossible for the a's to be distinct. For $k = 4$ the only possibility is $a_1 = 1$, $a_2 = 2$, $a_3 = 3$ and $a_4 = 6$. Then $a_1a_2a_3a_4 = 6^2$, so

$$(n+1)(n+2)(n+3)(n+4)$$

must be a square; but this product equals $(n^2 + 5n + 5)^2 - 1$ and we have a contradiction.

For $5 \leq k \leq 20$, we will count the number of a's with no prime factor greater than 3; if this is at least 5, it is impossible for the a's to be distinct, and we have a contradiction. This works unless $k = 6$ and $5 \mid n + 1$, or $k = 8$, $7 \mid n + 1$ and $5 \mid n + 2$. But in either of these cases we have four consecutive integers whose product is a square, and this was shown above to be impossible.

Similarly we obtain a contradiction for $20 < k \leq 56$ by noting that there are at least 9 a's with no prime factor greater than 5, and for $56 < k < 176$, where there are at least 21 a's with no prime factor greater than 7. (This method could be extended. For example, with $176 < k \leq 416$ there are at least 42 a's with no prime factor greater than 11, and with $416 < k < 823$ there are at least 65 a's with no prime factor greater than 13.)

This completes the proof of Theorem 2.

4. Remarks and further problems

No doubt our method would suffice to show that the product of consecutive odd integers is never a power, in the sense of (1). In fact, the proof would probably be simpler. More generally, for any positive integer d there must be an integer t_d such that $(n + d)(n + 2d) \cdots (n + td)$ is never a perfect power if $t > t_d$. Without t_d this result fails since $x(x + d)(x + 2d) = y^2$ has infinitely many solutions.

By our methods we can prove that for fixed t,

$$(n + d_1) \cdots (n + d_k) = x^l, \quad 1 = d_1 < \cdots < d_k \leq k + t \tag{24}$$

has only a finite number of solutions. Our theorem shows that there is no solution with $t = 0$. With $t = 1$ we have the solutions $4!/3$, $6!/5$ and $10!/7$; perhaps there are no others. Suppose that t is a function of k, or of k and l. How fast must t grow to give an infinite number of solutions to (24)? The Thue-Siegel theorem implies that (24) has only a finite number of solutions when d_k and l are fixed, with $l > 2$. For fixed k it seems probable that $\lim_{l\to\infty} d_k = \infty$.

Another question which arises naturally from our method is the following. Let $a_i^{(l)}$ be the largest divisor of $n + i$ which is lth-power free and has all prime factors less than k. Our proof for $l = 2$ implies that for $1 \leq i \leq k$, the $a_i^{(2)}$ are not all distinct when $k \neq 4, 6, 8$. An easy argument also shows that the $a_i^{(2)}$ cannot all be distinct when $k = 8$. To what extent do these results extend to $l > 2$? For how many consecutive values of i can the $a_i^{(l)}$ be distinct?

We mention one final problem. Let a_i be the largest divisor of $n + i$ which has all prime factors less than k. Our proof of Theorem 2 shows that for any $n \geq 0$ and $k \geq 30000$, the products $a_i a_j$ cannot all be distinct. Very likely this holds for much smaller values of k, perhaps as small as $k \geq 16$. To see that it does not hold for $3 \leq k < 16$, it suffices to check for $k = 3, 5, 7, 11, 13, 15$. We conclude with a table of examples for these cases.

k	a_1	a_2	a_3	a_4	a_5	a_6	a_7	a_8	a_9	a_{10}	a_{11}	a_{12}	a_{13}	a_{14}	a_{15}
3	2	1	2^{α}												
5	12	1	2	3^{α_1}	2^{α_2}										
7	60	1	2	3	2^{α_1}	5^{α_2}	2.3^{α_3}								
11	90	7^{α_1}	2^{α_2}	3	2	5^{α_3}	12	1	14	3^{α_4}	40				
13	90	11^{α_1}	2^{α_2}	3	14	5^{α_3}	12	1	2	3^{α_4}	40	7^{α_5}	66		
15	104	11^{α_1}	18	7^{α_2}	20	3	2	1	3.2^{α_3}	5	14	3^{α_4}	44	13^{α_5}	6.5^{α_6}

Acknowledgements. We wish to thank the referee for his comments and suggestions, and R. B. Eggleton for reorganizing and writing the paper in its final form.

References

1. P. Erdös, *Notes on the product of consecutive integers: I* and *II*, J. London Math. Soc., vol. 14 (1939), pp. 194–198 and 245–249.
2. ———, *On the product of consecutive integers III*, Indagationes Math., vol. 17 (1955), pp. 85–90.
3. ———, *A theorem of Sylvester and Schur*, J. London Math. Soc., vol. 9 (1934), pp. 282–288
4. ———, *On some applications of graph theory to number theoretic problems*, Publ. Ramanujan Inst., vol. 1 (1968–69), pp. 131–136.
5. R. Obláth, *Über Produkte aufeinanderfolgender Zahlen*, Tohoku Math. J., vol. 38 (1933), pp. 73–92.
6. O. Rigge, *Über ein diophantisches Problem*, 9th Congress Math. Scand., pp. 155–160. (Rigge, in fact, proves a sharper theorem, but it is a special case of our Theorem 2.)
7. J. B. Rosser and L. Schoenfeld, *Approximate formulas for some functions of prime numbers*, Illinois J. Math., vol. 6 (1962), pp. 64–94.

Northern Illinois University
DeKalb, Illinois

F. Hirzebruch

Preliminary Remarks

In many ways I am lazy. For example, I shall never write an autobiography. Also for this volume I did not want to write on myself and on my work. I prefered to use existing reports of others. I included the "Curriculum vitae mathematicae" written by Grauert, Harder, and Remmert for the volume of Mathematische Annalen dedicated to me on the occasion of my sixtieth birthday. I included also "Friedrich Hirzebruch - An appreciation" by Michael Atiyah written for the Proceedings of the Conference which Mina Teicher organized at Bar-Ilan University in Israel on the occasion of my sixty-fifth birthday.

I am aware that such laudations by good friends for birthdays contain a lot of statements which are approximations of the truth from above. They can be simultaneously enjoyable and embarrassing for the addressee. But they offer information relevant for the purpose of this volume.

I included three original papers which are representative of my early work. The third one is the first of nine joint papers with Michael Atiyah. It stands as a symbol for how much I owe to him, mathematically and in many other ways. Without him the *Arbeitstagungen* would not have developed as successfully as they did.

The papers "Learning Complex Analysis in Münster-Paris, Zürich and Princeton from 1945 to 1953" and "Kunihiko Kodaira: Mathematician, Friend and Teacher" concern mainly my early mathematical life from the beginning of my studies in Münster directly after the war until the end of my two years at the Institute for Advanced Study 1952 to 1954. They will show how much my work depended on the interaction with other mathematicians.

My bibliography until 1987 was published in my "Gesammelte Werke" (Collected Papers), Springer-Verlag, Berlin-Heidelberg-New York (1987). Also a Curriculum Vitae is included there. Here I only give the references not contained in the Collected Papers.

76. Elliptic genera of level N for complex manifolds, In: Differential Geometrical Methods in Theoretical Physics, Kluwer Academic Publishers, Dordrecht/Boston/London 1988, p. 37-63.

77. Codierungstheorie und ihre Beziehungen zu Geometrie und Zahlentheorie (Ausarbeitung von N.-P. Skoruppa), Vorträge N 370 der Rheinisch-Westfälischen Akademie der Wissenschaften, Westdeutscher Verlag GmbH Opladen 1989.

78. (with Th. Höfer) On the Euler number of an orbifold. Math. Ann. 286 (1990), 255-260.

79. (with P. Slodowy) Elliptic genera, involutions and homogeneous spin-manifolds, Geometriae Dedicata 35 (1990), 309-343.

80. Axiome, Definitionen, Begriffe in der Mathematik. Nova acta Leopoldina 63, Nr. 272, 61-77 (1990).

81. Centennial of the German Mathematical Society (Bremen, September 16th-22nd 1990) In: Miscellanea mathematica, Eds. P. Hilton, F. Hirzebruch, R. Remmert, Springer-Verlag, Heidelberg, 1991, p. 177-194.

82. Mannigfaltigkeiten und Modulformen. Jahresberichte der Deutschen Mathematiker-Vereinigung, 1992, B. G. Teubner, Stuttgart, Jubiläumstagung 1990, S. 20-38.

83. Kombinatorik in der Geometrie, Jahrbuch der Heidelberger Akademie der Wissenschaften für 1991, Heidelberg 1992, p. 96-98.

84. (with P. Beazley Cohen) Book Review on Pierre Deligne and G. Daniel Mostow, Commensurabilities among lattices in PU(1,n). Bulletin Amer. Math. Soc. 32 (1995), 88-105.

85. Regular polyhedra and the football, Sugaku Tsushin, Februar 1997, p. 18-40

86. German-Russian Cooperation in Mathematics, Mitteilungen der DMV, 4-1997, p. 54-58.

87. Learning Complex Analysis in Münster-Paris, Zürich and Princeton from 1945 to 1953; Gazette des Mathématiciens 74 (1997), 27-39.

88. Opening address as honorary president, International Congress of Mathematicians, Berlin 1998, Proceedings Vol. 1, p. 31-33, Documenta Mathematica 1998.

89. Kunihiko Kodaira: Mathematician, Friend, and Teacher, Notices of the American Mathematical Society (1998), 1456-1462.

90. Emmy Noether and Topology, Israel Math. Conf. Proceedings Vol. 12, 1999, p. 57-65.

M11. (mit Th. Berger und R. Jung) Manifolds and modular forms.
Aspects of Mathematics, E20, Braunschweig: Vieweg 1992.
Second Edition 1994.

Mathematische
Annalen, Vol. 278

Curriculum vitae mathematicae

Friedrich Ernst Peter Hirzebruch first came into contact with mathematical research immediately after the Second World War, when he was at the Institute of Mathematics of Münster University in Westphalia from 1945 to 1949. During this period, as an advanced student of Heinrich Behnke he became familiar with the function theory of several complex variables: this became the basis of his life as a scientist, which subsequently took him far beyond this theory.

The first widening of his horizons was in Zürich in 1949–1950, with his meeting with Heinz Hopf, the great topologist and geometer. Hirzebruch saw that the results of algebraic topology and algebraic geometry would contribute to complex analysis. He found that some things carried over from these areas, he resolved all two-dimensional singularities, and he discovered the so-called Hirzebruch surfaces.

Later he was invited to spend two years carrying out research at the Institute for Advanced Study in Princeton, New Jersey, starting in the winter term of 1952. He became acquainted with A. Borel, K. Kodaira, and D.C. Spencer and, through them, with mathematical methods that were unknown in Germany. He became familiar with coherent analytic sheaves, vector bundles, characteristic classes and Thom's cobordism. In this stimulating environment he was able to prove the Riemann-Roch theorem in arbitrary dimension; this brought him fame overnight, for this was a problem that many of the top mathematicians of the day had been working on.

The Riemann-Roch theorem is concerned with calculating the Euler characteristic of a holomorphic bundle. It was already clear that this number was dependent only on topological data; the problem was to express it in terms of these data – namely, of Chern classes. Hirzebruch developed the formalism of multiplicative sequences (A-polynomial, Â-genus), combined this with Thom's results, and arrived at the correct formulae. Since 1954 the Riemann-Roch-Hirzebruch theorem has played a key role in algebraic geometry. The theorem now represents the decisive step in the proof of innumerable further published results.

Hirzebruch's proof was based on topological methods. Grothendieck realized, once Hirzebruch's results were available, that it is possible to prove a far more general theorem using purely algebraic methods. He introduced the so-called Grothendieck ring of vector bundles, which proved to be very useful in topology and led to topological and algebraic K-theory.

The Riemann-Roch theorem is also of relevance for the solution spaces of certain linear partial differential equations: a certain difference of dimensions of solution spaces depends solely on topological data and can be calculated explicitly on the basis of these data. This type of assertion was generalized by Atiyah and Singer, a development that culminated in the general Atiyah-Singer theorem. A connection between quite distinct mathematical areas becomes apparent, the beginnings of which were noted in Hirzebruch's work.

During his stay in Princeton, Hirzebruch not only learned new mathematical methods, but also came to realize that, in order to develop further, mathematics requires personal contact, exchange of ideas, and a stimulating environment. Consequently, he consistently devoted a great deal of energy and political influence to creating such an environment and, in this way, did much to promote mathematics.

In 1957, Hirzebruch founded his Arbeitstagung which takes place in Bonn almost every summer. In so doing, he became an organizer of international research. The number of participants was small at first but has increased every year; now mathematicians congregate from East and West to discuss new results and to exchange ideas. The Arbeitstagung was the starting point for the Sonderforschungsbereich "Theoretische Mathematik" that was set up at the Mathematics Institute in Bonn in 1969. Under Hirzebruch's leadership it soon gained international recognition and attracted mathematicians with a wide range of interests to Germany.

Finally using his powers of persuasion and backed by the success of his Sonderforschungsbereich, he managed to convince the powers that be of the need for a Max Planck Institute. The Institute was founded in 1980 and Hirzebruch is its director.

Hirzebruch has also continued to devote himself unwaveringly to mathematical research. He returned to his "first love": singularities. In 1966 he discovered that exotic spheres can occur in the boundary of neighbourhoods of singularities. From about 1970 onwards he began to take an interest in Hilbert modular forms. He supplied explicit resolutions of cusp singularities of surfaces, making it possible to study these surfaces systematically. Wonderful connections became apparent between the geometry of these surfaces and the number theory of the underlying number fields. This has developed further; at present one of the focal topics of study at the Max Planck Institute concerns algebraic surfaces from the point of view of differential geometry, of moduli theory and of algebraic geometry.

Friedrich Hirzebruch has been an editor of Mathematische Annalen since 1961. During this period he has devoted a great deal of time and effort to promoting the journal both as an editor and as an author.

H. Grauert G. Harder R. Remmert

ISRAEL MATHEMATICAL CONFERENCE PROCEEDINGS Vol. 9, 1996

FRIEDRICH HIRZEBRUCH - AN APPRECIATION

by MICHAEL ATIYAH
Trinity College, Cambridge CB2 ITQ, U.K.

§1. Introduction.

Although, to my regret, I was not present at the 65th birthday conference for Fritz Hirzebruch that this volume commemorates, I am delighted to have the opportunity of contributing here a personal appreciation. I have been a friend and close collaborator of Fritz for 40 years and I am happy to pay tribute to his remarkable influence over Mathematics during that time.

I first met Fritz briefly in 1954 when he visited Cambridge to talk about the exciting work he had been doing in Princeton. I was a research student at the time, just beginning to get to grips with the revolution in algebraic geometry brought about by the introduction of sheaf theory. But it was at Princeton in 1955-56 that I really got to know Fritz well and began to learn from him the skills of dealing with characteristic classes. That year in Princeton, in the company of Serre, Bott and Singer, attending the lectures of Kodaira and encouraged by the enthusiasm of Spencer, appears now like a golden age. The galaxy of talent and the wealth of ideas created an environment ideally suited to spurring on the aspiring young mathematicians at the Institute for Advanced Study.

When Fritz returned to Germany to the chair in Bonn, much of the excitement of the Princeton era followed him there. For many of us Bonn became the focal point of our interests, and the annual Arbeitstagung the major event of the year. The free and easy atmosphere at Bonn generated many active collaborations both formal and informal, and my own mathematical development, like that of many others, was heavily dependent on these contacts. Over the years Fritz and I have written nine joint papers and have worked together in Bonn, Oxford and Princeton. Chronologically I was the junior partner since Fritz was a full Professor while I was just beginning my academic career. I learnt much from him in mathematical content as well as in style and approach. In fact I have absorbed so much of his point of view that it is hard now for me to think back to a pre-Hirzebruch era. On this occasion I would like to survey briefly the many contributions Fritz has made to the development of mathematics. By his ideas, his example, his influence and his leadership he has played a distinctive and unique role in the mathematics of our time.

§2. The Riemann-Roch Theorem.

There is no doubt that the generalized Rieman-Roch Theorem was Fritz's first major achievement. The development of sheaf theory for complex manifolds in the hands of Cartan and Serre, together with its exploitation by Kodaira and Spencer in the early 1950s, had begun a revolution in algebraic geometry. It was, however, left to Hirzebruch to put all this together in the Riemann-Roch Theorem for holomorphic vector bundles. This Theorem assists in brief the equality

$$\chi(X,E) = T(X,E)$$

where

$$\chi(X,E) = \sum_q (-1)^q \dim H^q(X,E)$$

is the Euler-characteristic of the sheaf of holomorphic sections of a complex vector bundle E over a compact complex manifold X, and $T(X,E)$ is a cohomological invariant constructed from the Chern classes of X and E.

In addition to the analytical foundations, there were severe calculational difficulties related to the complicated nature of the formula for $T(X,E)$. It was Fritz's great contribution to tame this complication. By going back to the papers of J. A. Todd on the precursors of the Clern classes and developing his formal ideas on multiplicative invariants, Fritz produced a powerful machine for controlling and manipulating the forbidding algebraic formulae.

The Riemann-Roch Theorem marks the turning point from the old style algebraic geometry to the new. Important in its own right, it also opened up new doors for the future. The first of these was the generalization of Grothendieck, involving the introduction of his K-groups and the construction of a functorial homomorphism

$$f_! : K(X) \to K(Y)$$

for a proper holomorphic map $f : X \to Y$ of algebraic varieties. Grothendieck's theorem gave a formula for the Chern character of $f_!$ and it reduced to the Hirzebruch-Riemann Roch Theorem when Y was a point.

Emerging from the works of Grothendieck and Hirzebruch came first the development of topological K-theory, and then the index theorem for elliptic differential operators. In various guises these have gone on developing into the field of C*-algebras. The Hirzebruch-Riemann-Roch Theorem is, through these descendants, still alive and kicking.

3. Characteristic Classes.

Both in the Riemann-Roch Theorem and elsewhere, characteristic classes of vector bundles have been a dominating theme of Fritz's work. Beginning with his collaboration with Borel, characteristic classes were thoroughly mastered and integrated with the theory of Lie groups. The proof of the signature theorem, hinging on Thom's cobordism theory, was one of the motivations for a proper development of the formalism of characteristic classes. Together with the earlier work of Todd, this led to Fritz's deep understanding of the role of characteristic classes in the theory of manifolds.

A surprising by-product of these investigations was an unexpected relation between the Steenrod power operations and the Todd polynomials. This work spanning the gap from mod p cohomology operations to complex algebraic geometry is typical of Fritz at his most magical. He revelled in such surprising and spectacular results, especially if they appeared in the form of elaborate but beautiful formulae.

When, in the aftermath of Grothendieck's work in algebraic geometry and Bott's periodicity theorem, topological K-theory was being developed, Fritz's experience and insight into characteristic classes were quite essential. Many of the early applications depended on a skillful use of representations and their associated Chern classes and this was an area where Fritz excelled.

Modern mathematics has sometimes been criticized (notably by Siegel) for being too abstract and not as concretely computational as the great mathematics of the past. There is some justice in this criticism, but the best mathematicians of our time have been able to combine abstract ideas with a complete mastery and control of complicated formulae that compares favourably with the best of our predecessors. One such example is Fritz Hirzebruch. My former teacher J.A. Todd was no mean calculator of the old school, as I discovered when we worked together on a problem in number theory. However, in his own field of the famous 'Todd polynomials' he was totally outgunned by Fritz. Incidentally, another mathematician of our time, who could have held his own with Euler or Jacobi is an old friend and colleague of mine, Ian Macdonald. It is reassuring to know that the algebraic virtuosity of our ancestors is still present in our genes.

§4. Number Theory.

Although Fritz would certainly describe himself as a geometer, his facility with intricate algebraic formulae has on many occasions led him into number theory. The most notable example of this has been his long involvement with Hilbert modular surfaces. This is an area which brings together all of Fritz's main interests: topology, algebraic geometry and number theory. The story begins with his explicit description of the resolution of the cusp singularities, in which continued fractions get beautifully realized by configurations of algebraic curves. Subsequently, together with Zagier, he made a detailed study of Hilbert modular surfaces associated with particular primes,. relating them to the traditional classification of surfaces.

Other aspects of number theory that have emerged from Fritz's work include properties of the Bernoulli numbers (related to the Todd polynomials) and of Dedekind sums, which turn up in quotient singularities.

§5. Style.

In both his writing and in his lectures Fritz has an elegant and lucid style. Listening to his seminars I always admired the skill of his presentation where, bit by bit, the pieces of the jig-saw were put into place just in time for the finale. Studying his technique I realized that it was all carefully stage-managed, with meticulous planning in the choice of material and the order in which it was presented. Rabbits do not come out of hats unless they are carefully hidden there in advance.

Fritz's papers are always models of clear exposition. Well-organized, with every equation and theorem properly numbered and referenced, one gets the impression of a well-oiled machine (like the Mercedes that he likes to drive) cruising along smoothly and effortlessly. Those of us who are not fluent in German always find papers in that language particularly severe and demanding. It is perhaps the ultimate tribute to Fritz's style that his papers in German present me with few problems, and delude me into believing that I have finally mastered the language!

An apprentice picks up the skills of his trade by watching his master at work, and I very soon found myself modelling my style of writing on that of Fritz. In fact, when we wrote our first joint papers, I was carefully instructed in a few elementary points of presentation and taught to avoid phrases or punctuation that might confuse the reader.

The motto of New College Oxford (where I was once a Fellow and where Fritz visited me) is "Manners Makyth Man" which, in a mathematical context, one can freely interpret as asserting that style is the outward reflection of the inner mathematician. Certainly I feel that Fritz searches for elegance in mathematics at a fundamental level and then ensures that his results emerge and are presented in a correspondingly elegant fashion.

§6. The Arbeitstagung.

In 1957, after returning from Princeton, I went to Bonn for the tentative first version of what later developed into the famous Arbeitstagung. At this first meeting we were still a small group, and most of the time was spent listening to Grothendieck expounding his new ideas on K-theory and the Riemann-Roch Theorem. From this auspicious if small beginning the Arbeitstagung developed into the most famous and well-attended series of mathematical meetings. The secret of its success lay in the informal and impromptu character of the meeting, the high quality of the participants and the open-ended nature of the programme. There was no fixed subject, except as defined by the interests of those attending, and over the years the focus shifted in line with the latest developments.

Fritz, of course, was the maestro. His interests, in the broad sense, tended to define the participants and the programme. His organizational skill turned potential chaos into good planning, and the record of topics lectured on at the Arbeitstagung is a faithful reflection of the excitements of the time.

As one of the original members, I tended to be asked to give the first lecture. This enabled Fritz to play for time so that miraculously, at the end of my lecture, the initial programme for the Arbeitstagung was handed out at the door. The lectures themselves were selected by a pseudo-democratic choice in which Fritz carefully orchestrated the discussion so that a sensible and balanced programme finally emerged. This process needed all of Fritz's skills, a command of the mathematics, a sense of occasion, a good rapport with the audience and the tact and finesse to put it all together without causing offence.

The boat trips on the Rhine were an invariable part of the Arbeitstagung, and they provided a marvellous crucible in which ideas were forged. I have clear memories of long discussions on the latest mathematics, accompanied by sizeable portions of strawberry tarts and generous amounts of liquid refreshment. In the early days we took modest boats but, as the Arbeitstagung grew and its fame spread, Fritz commandeered ever more imposing vessels. I assume that, by the end, we must have had an admiral on the bridge!

§7. The Max Planck Institute.

As the Arbeitstagung developed, there was a simultaneous build-up in the total research capacity in Bonn. The establishment of the SFB meant that a constant stream of longer-term visitors came to Bonn and maintained a high background level of activity leading to the Arbeitstagung as the annual climax.

In due course the SFB got converted into the more permanent form of the Max Planck Institute, a worthy competitor of the Princeton Institute where Fritz had spend so much time.

Research Institutes of this kind are not easy to create unless there is a wealthy benefactor around. Great persistence and persuasive powers are called for, and it is quite remarkable that Fritz was able to set up the Max Planck Institute while at the same time carrying on with his full load of teaching and research at the University. In fact, Fritz's contribution to German mathematics rests as much on the many students he inspired and trained as on his administrative skills. The role of a great mathematical centre is not only to generate ideas but also to train the mathematicians of the future. In this respect Bonn has, since the war, played the leading role that Göttingen once played.

Fritz's administrative skills and mathematical influence have in recent years spread far beyond Bonn. He has been involved with the IMU and International Mathematical Congresses and has just finished serving as the crucial first President of the European Mathematical Society. Finally, he is playing a key part in the re-integration of East German mathematics in the new unified Germany.

With all these many burdens, Fritz remains remarkably cheerful and unflustered. He always seems to enjoy what he is doing and his enthusiasm is infectious.

Reprinted from the Proceedings of the NATIONAL ACADEMY OF SCIENCES,
Vol. 39, No. 9, pp. 951–956. September, 1953

ON STEENROD'S REDUCED POWERS, THE INDEX OF INERTIA, AND THE TODD GENUS

BY FRIEDRICH HIRZEBRUCH

THE INSTITUTE FOR ADVANCED STUDY

Communicated by S. Lefschetz, July 3, 1953

Introduction.—This note is a preliminary report on some results concerning the Steenrod reduced powers[1] in *oriented* manifolds M^m and the index τ of such a manifold.

Definition of τ: If the dimension m of M is not divisible by 4, then $\tau(M) = 0$. If $m = 4k$, then $\tau(M)$ is equal to the number of positive squares minus the number of negative squares of the normalized quadratic form defined by the cup-product $x \smile x$ where $x \in H^{2k}(M^{4k}, R)$.

We apply the results to almost complex manifolds, in particular we give a definition of the Todd[2] genus of an almost complex manifold and state properties of this genus. Since a special algebraic formalism will be used throughout the note, we outline this formalism in the first section. All manifolds occurring in this note are compact, differentiable, and *oriented* unless stated to the contrary. Full details with further applications will appear elsewhere.

1. Multiplicative Γ*-Sequences.*—Let $\sum\limits_{i=0}^{\infty} a_i x^i$, $a_0 = 1$, be the power series with the indeterminates a_i as coefficients, and let Γ be a field. Let $\{K_j\}$ be a sequence of polynomials, K_j being of weight j in the a_i and having coefficients in Γ $(K_0 = 1)$. If $C = \sum_{i=0}^{\infty} c_i x^i$, $c_0 = 1$, is an arbitrary power series, we denote by $K(C)$ the series $\sum_{j=0}^{\infty} K_j(c_1, c_2, \ldots, c_j)x^j$. We call $\{K_j\}$ a multiplicative Γ-sequence provided that K is a homomorphism, i.e., $K(AB) = K(A)K(B)$, where A and B are power series with indeterminates a_i, b_i as coefficients $(a_0 = b_0 = 1)$. We construct for every given power series $Q(x) = \sum_{i=0}^{\infty} \gamma_i x^i$, $(\gamma_i \epsilon \Gamma, \gamma_0 = 1)$, a multiplicative Γ-sequence. Writing formally

$$1 + a_1x + \ldots + a_mx^m = (1 + \alpha_1 x)(1 + \alpha_2 x) \ldots (1 + \alpha_m x),$$

we express $Q(\alpha_1 x)Q(\alpha_2 x) \ldots Q(\alpha_m x)$ as a power series with coefficients which are polynomials in the a_i:

$$Q(\alpha_1 x)Q(\alpha_2 x) \ldots Q(\alpha_m x) = \sum_{j=0}^{\infty} K_{j,\, m}(a_1, \ldots, a_j)x^j.$$

One verifies easily that $K_{j,\, m}$ does not depend on m for $j \leq m$. We write $K_{j,\, j} = K_j$ and obtain the unique multiplicative Γ-sequence with $K(1 + x) = Q(x)$.

Now let Γ be the rationals. We denote by $\{T_j\}$ the multiplicative sequence belonging to $Q(x) = -x(e^{-x} - 1)^{-1}$ and call it the Todd sequence.[2]

$$2T_1 = a_1,\ 12T_2 = a_1^2 + a_2,\ 24T_3 = a_1a_2,\ 720T_4 = -a_4 + a_3a_1 + 3a_2^2 + 4a_2a_1^2 - a_1^4.$$

For a prime $q \geq 2$ the coefficients of $q^r T_{(q-1)r}$ are integers mod q (i.e., do not contain q in the denominators).

We also consider the multiplicative sequence belonging to $Q(x) = \sqrt{x}(\text{tgh}\sqrt{x})^{-1}$ which we denote by $\{L_j\}$. We have

$$3L_1 = a_1,\ 45L_2 = 7a_2 - a_1^2,\ 945L_3 = 62a_3 - 13a_2a_1 + 2a_1^3,\ \ldots$$

For a prime $q \geq 3$ the coefficients of $q^r L_{1/2(q-1)r}$ are integers mod q.

2. *Reduced Powers.*[1]—Let M^m be a compact *oriented* manifold. The reduced powers are defined for every odd prime q:

$$\mathcal{P}_q^r: H^k(M^m, Z_q) \rightarrow H^{k+2(q-1)r}(M^m, Z_q).$$

In case $k + 2(q - 1)r = m$, there exists by Poincaré duality an element $s_q^r \epsilon H^{2(q-1)r}(M^m, Z_q)$ such that

$$\mathcal{P}_q^r u = s_q^r u \qquad \text{for all} \qquad u \epsilon H^k(M^m, Z_q).$$

-Let $p_1, p_2, \ldots$ be the Pontrjagin classes[3] of M^m where $p_i \epsilon H^{4i}(M^m, Z)$ In the notation of Wu[3] $p_i = P_0^{4i}$.

THEOREM 2.1. *The class s_q^r can be expressed as a polynomial in the Pontrjagin classes:*

$$s_q^r = q^r L_{1/2(q-1)r}(p_1, p_2, \ldots, p_{1/2(q-1)r}) \bmod q.$$

THEOREM 2.2. *If M^m is an almost complex manifold of n complex dimensions ($m = 2n$), then we can express s_q^r as a polynomial in the Chern classes c_i where $c_i \in H^{2i}(M^m, Z)$:*

$$s_q^r = q^r T_{(q-1)r}(c_1, c_2, \ldots, c_{(q-1)r}) \bmod q.$$

We have assumed that the prime q is odd, but in the case $q = 2$ we may consider the Steenrod squares Sq^i. For a manifold M^m (not necessarily oriented) Wu[4] defined the class $U^i \in H^i(M^m, Z_2)$ by $\mathrm{Sq}^i v = U^i v$ for all $v \in H^{m-i}(M^m, Z_2)$.

THEOREM 2.3. *Let $w_i \in H^i(M^m, Z_2)$ be the Stiefel-Whitney classes of M^m. We have*

$$U^i = 2^i T_i(w_1, w_2, \ldots, w_i) \bmod 2.$$

Remark:[4] T_{2i+1} is divisible by a_1. Hence $w_1 = 0$ implies $U_{2i+1} = 0$.

THEOREM 2.4. *If M^m is almost complex, we have*

$$U^{2i+1} = 0, \qquad U^{2i} = 2^i T_i(c_1, c_2, \ldots, c_i) \bmod 2.$$

The proofs of Theorems 2.1–2.4 are based on the "diagonal" method of Thom[5] and Wu,[4, 6] and on a topological interpretation of the multiplicative sequences which uses the Borel-Serre[7] method of regarding the classes of Stiefel-Whitney, Chern, Pontrjagin as elementary symmetric functions, and which also uses the Whitney duality theorem. The new point in the Theorems 2.1–2.4 lies in the explicit construction of the polynomials and in the fact that for all primes q these polynomials are obtainable from one and the same (rational) multiplicative sequence by reduction mod q.

3. *The Index.*—For an oriented M^{4k} we can regard the class $L_k(p_1, \ldots, p^k) \in H^{4k}(M^{4k}, R)$ as a rational number. By using a strong theorem of Thom[8] (p. 1735, Theorem 7) we obtain:

THEOREM 3.1. *We have $\tau(M^{4k}) = L_k(p_1, \ldots, p_k)$. Hence L_k for an M^{4k} is always an integer. For example,*

$$3\,\tau(M^4) = p_1,\ 45\,\tau(M^8) = 7p_2 - p_1^2,\ 945\,\tau(M^{12}) = 62p_3 - 13p_2p_1 + 2p_1^3.$$

Remarks: It was known to Thom that τ can be expressed as a polynomial in the Pontrjagin classes.[9] Theorem 3.1 implies for an M^4: $p_1 \equiv 0(3)$ (see Wu[6]) and for an M^8: $7p_2 - p_1^2 \equiv 0(45)$. From Theorem 2.1 we obtain $p_1 \equiv 0(3)$ and $7p_2 - p_1^2 \equiv 0(15)$. Analogously for all dimensions. In case M^{4k} is almost complex, we can express the Pontrjagin classes by the Chern classes[3]

$$\sum_{i=0}^{\infty} (-1)^i p_i x^{2i} = \left(\sum_{i=0}^{\infty} c_i x^i \right) \left(\sum_{i=0}^{\infty} (-1)^i c_i x^i \right)$$

and obtain polynomials for τ in the Chern classes.

We now state two theorems about the index which follow from formal properties of the polynomials L_j.

THEOREM 3.2. *If the manifold M^m is fibred in complex projective spaces P_n of n complex dimensions with the group of all projective transformations as structure group and the manifold B^{m-2n} as base ($m \geq 2n$), then*

$$\tau(M^m) = \tau(B^{m-2n})\tau(P_n).$$

Remarks: $\tau(P_n) = 1$, if n is even; $\tau(P_n) = 0$, if n is odd (see the Introduction). For the direct product of two manifolds M, M' we have $\tau(M \times M') = \tau(M) \cdot \tau(M')$.

Consider a manifold M^{4k+2}. Every element $x \in H^2(M^{4k+2}, Z)$ can be represented by a subvariety V^{4k} of M^{4k+2} (Thom[8], p. 573, Theorem 2). The index $\tau(V^{4k})$ only depends on x and may be denoted by $\tau(x)$. If $x_1, \ldots, x_r \in H^2(M^{4k+2}, Z)$, then x_1 can be represented by a subvariety V^{4k} of M^{4k+2}, the restriction of x_2 to V^{4k} can be represented by a subvariety V^{4k-2} of V^{4k}, etc. Finally, the restriction of x_r to $V^{4k-2r+4}$ can be represented by a subvariety $V^{4k-2r+2}$ of $V^{4k-2r+4}$. The index of $V^{4k-2r+2}$ only depends on the non-ordered r-tpl $(x_1, \ldots, x_r)$ and may be denoted by $\tau(x_1, \ldots, x_r)$.

THEOREM 3.3. *We have for $x_1, x_2 \in H^2(M^{4k+2}, Z)$*

$$\tau(x_1 + x_2) = \tau(x_1) + \tau(x_2) - \tau(x_1, x_2, x_1 + x_2).$$

4. *The Todd Genus.*—Let M_n be an almost complex manifold of n complex dimensions and $c_i \in H^{2i}(M_n, Z)$ its Chern classes. We can regard $T_n(c_1, \ldots, c_n) \in H^{2n}(M_n, R)$ as a rational number which we denote by $T(M_n)$. We call $T(M_n)$ the Todd genus of M_n. Kodaira[10] proved for all algebraic varieties M_n which are a complete non-singular intersection of hypersurfaces in some projective space that

$$T(M_n) = 1 - g_1 + g_2 - \ldots + (-1)^n g_n$$

where g_i is the number of linearly independent i-pl differentials of the first kind attached to M_n. From the results of Todd,[2] Hodge,[11] and Kodaira-Spencer[12] it seems very likely that the last formula is true for all non-singular algebraic varieties. But in the present moment it is not known, even for algebraic varieties, whether $T(M_n)$ is always an integer. Therefore it seems to be interesting that one can prove by the Theorems 2.4 and 3.1:

THEOREM 4.1. *The number $2^{n-1}T(M_n)$ is an integer for every almost complex manifold. This means for $n = 1, 2, 3, 4$:*

$$M_1\colon\ c_1 \equiv 0(2),\quad M_2\colon\ c_1^2 + c_2 \equiv 0(6),\quad M_3\colon\ c_1\cdot c_2 \equiv 0(6),$$

$$M_4\colon\ -c_4 + c_3c_1 + 3c_2^2 + 4c_2c_1^2 - c_1^4 \equiv 0\ (90).$$

We state two theorems about the Todd genus which follow from formal properties of the polynomials T_j and which are analogous to the Theorems 3.2 and 3.3.

THEOREM 4.2. *If the almost complex manifold M_n is fibred in complex projective spaces P_k of k complex dimensions with the group of all projective transformations of P_k as structure group and the almost complex manifold B_{n-k} as base, the fibering being compatible with the almost complex structures of M_n, P_k and B_{n-k} then*

$$T(M_n) = T(B_{n-k})T(P_k) = T(B_{n-k}).$$

Remarks: $T(P_k) = 1$ for all k. Since $\{T_j\}$ is a multiplicative sequence we have for the direct product of two almost complex manifolds M, M'

$$T(M \times M') = T(M)\cdot T(M').$$

This (in the algebraic case) was checked by Todd[2] for dimensions not exceeding 6.

For a class $x \in H^2(M_n, Z)$ we can define a virtual Todd genus $T(x)$, which is a polynomial of weight n in x and the Chern classes of M_n, such that $T(x)$ is the Todd genus of every admissible almost complex subvariety V_{n-1} of M_n representing x. Moreover, if $x_1, x_2 \in H^2(M_n, Z)$ we can define a virtual Todd genus $T(x_1, x_2)$ with $T(x_1, x_2) = T(x_2, x_1)$ such that for every admissible almost complex subvariety V_{n-1} representing x_1, the number $T(x_1, x_2)$ is the virtual genus with respect to V_{n-1} of the restriction of x_2 to V_{n-1}. The virtual genus $T(x_1, x_2)$ is a polynomial of weight n in x_1, x_2 and the Chern classes of M_n. The following theorem is well known in algebraic geometry:

THEOREM 4.3. *For $x_1, x_2 \in H^2(M_n, Z)$, we have*

$$T(x_1 + x_2) = T(x_1) + T(x_2) - T(x_1, x_2).$$

[1] Steenrod, N. E., these PROCEEDINGS, **39**, 213–223 (1953).

[2] Todd, J. A., *Proc. London Math. Soc.*, (2), **43**, 190–225 (1937).

[3] Wu, Wen-Tsun, and Reeb, G., "Sur les espaces fibrés et les variétés feuilletées," *Actual. sci. industr.*, 1183, (1952).

[4] Wu, Wen-Tsun, *Compt. rend. acad. sci., Paris*, **230**, 508–511 (1950).

[5] Thom, R., *Ann. sci. Écol. norm. sup.* (3), **69**, 109–182 (1952).

[6] Wu, Wen-Tsun, "Sur les puissances de Steenrod," Colloque de Topologie de Strasbourg, 1951; (mimeographed notes).

[7] Borel, A., and Serre, J. P., *Compt. rend. acad. sci., Paris*, **233**, 680–682 (1951); Borel, A., *Ann. Math.*, **57**, 115–207 (1953); Borel, A., and Serre, J. P., *Am. J. Math.*, **75**, 409–448 (1953).

[8] Thom, R., *Compt. rend. acad. sci., Paris,* **236**, 453, 573, 1733 (1953).

[9] Thom, R., "Quelques propriétés globales des variétés différentiables"; (to appear in *Comm. Math. Helv.*).

[10] Kodaira, K., "The Theory of Harmonic Integrals and Their Application to Algebraic Geometry," Notes, Princeton University, 1953.

[11] Hodge, W. V. D., *Proc. London Math. Soc.*, (3), **1**, 138–151 (1951).

[12] Kodaira, K., and Spencer, D. C., "On Arithmetic Genera of Algebraic Varieties," these PROCEEDINGS, **39**, 641–649 (1953).

Reprinted from the Proceedings of the NATIONAL ACADEMY OF SCIENCES,
Vol. 40, No. 2, pp. 110–114. February, 1954

ARITHMETIC GENERA AND THE THEOREM OF RIEMANN-ROCH FOR ALGEBRAIC VARIETIES

BY FRIEDRICH HIRZEBRUCH

THE INSTITUTE FOR ADVANCED STUDY

Communicated by S. Lefschetz, December 21, 1953

Introduction.—In a preceding note[1] we posed the problem whether the arithmetic genus $\Pi(V_n)$ of a (non-singular) algebraic variety[2] coincides with the Todd genus $T(V_n)$. The purpose of the present note is to prove that this is actually the case. Moreover we prove a main theorem (M) which gives a formula for the Euler-Poincaré characteristic of V_n with respect to the cohomology[3] of V_n with coefficients in the sheaf (faisceau) of local holomorphic cross-sections of any complex analytic bundle W over V_n which has the complex vector space C_q as fibre and the linear group $GL(q, C)$ as structure group. The main theorem expresses this Euler-Poincaré characteristic as a polynomial in the Chern classes of the tangential bundle of V_n and in the Chern classes of the bundle W. As special cases one gets the Todd formula $\Pi(V_n) = T(V_n)$ and the "Theorem of Riemann-Roch for arbitrary dimensions." The author wishes to extend his hearty thanks to Professors A. Borel, K. Kodaira, and D. C. Spencer, with all of whom he had many valuable discussions. The notes of Kodaira and Spencer[4] are essentially used. The author also wants to point out that the main theorem (in a slightly different formulation) was conjectured by J. P. Serre in a letter to Kodaira and Spencer. The proof of the main theorem uses the index theorem of the author[1] which involves essentially the theory of "cobordisme" due to R. Thom. Full details of the proof of the main theorem will appear elsewhere.

1. *The Main Theorem and Some of Its Consequences.*—Let V_n be an algebraic variety (non-singularly imbedded in some complex projective space). Let W be a complex analytic bundle over V_n with the complex vector space C_q as fibre and the linear group $GL(q, C)$ as structure group. Denote by c_i the Chern classes of the

tangential bundle of V_n and by d_j the Chern classes of the bundle W. ($c_i \in H^{2i}(V_n, Z)$, $d_j \in H^{2j}(V_n, Z)$, $0 \leqq i \leqq n$, $0 \leqq j \leqq q$, $c_0 = d_0 = 1$.) We introduce the formal roots γ_i, δ_j:

$$\sum_{i=0}^{n} c_i x^i = \prod_{i=1}^{n} (1 + \gamma_i x), \qquad \sum_{=0}^{q} d_j x^j = \prod_{j=1}^{q} (1 + \delta_j x).$$

Every formal power series which is symmetric in the γ_i as well as in the δ_j will be considered as a power series in the Chern classes c_i and d_j and hence as an element of the cohomology ring of V_n. Denoting by Q the rationals we define the operator κ_n on the cohomology ring $H^*(V_n, Q)$ as follows: For every $u \in H^*(V_n, Q)$ we take the component of topological dimension $2n$ which we consider in the unique fashion as a rational number. This number is denoted by $\kappa_n[u]$.

MAIN THEOREM (M). *Using the above notations we put*

$$\chi(V_n, W) = \sum_{i=0}^{n} (-1)^i \dim H^i(V_n, W),$$

where $H^i(V_n, W)$ is the i-dimensional cohomology group of V_n with coefficients in the sheaf (faisceau) of local holomorphic cross-sections of W. We have

$$\chi(V_n, W) = \kappa_n \left[(e^{\delta_1} + e^{\delta_2} + \ldots + e^{\delta_q}) \prod_{i=1}^{n} \frac{-\gamma_i}{e^{-\gamma_i} - 1} \right] =$$
$$\kappa_n \left[e^{c_1/2} (e^{\delta_1} + e^{\delta_2} + \ldots + e^{\delta_q}) \prod_{i=1}^{n} \frac{\gamma_i/2}{\sinh \gamma_i/2} \right].$$

Remark: $x/\sinh x$ is a power series in x^2. Since the γ_i^2 can be regarded as the formal roots of the Pontrjagin polynomial[1] of V_n, we see that $\chi(V_n, W)$ is a polynomial in c_1, the Pontrjagin classes of V_n, and the Chern classes of the bundle W.

Now let W be the trivial bundle over V_n with C_1 as fibre. The main theorem implies[5]

THEOREM 1. *The Todd genus of an algebraic variety is equal to the arithmetic genus:*

$$\Pi(V_n) = \sum_{i=0}^{n} (-1)^i g_i = T(V_n).$$

More generally, let F be an arbitrary complex line bundle (fibre C_1) over V_n. Denote by $f = c(F)$ its characteristic class ($f \in H^2(V_n, Z)$). The main theorem implies

THEOREM 2. *We have*

$$\chi(V_n, F) = \kappa_n \left[e^{f + c_1/2} \prod_{i=1}^{n} \frac{\gamma_i/2}{\sinh \gamma_i/2} \right].$$

Hence $\chi(V_n, F)$ is a polynomial in $f + c_1/2$ and the Pontrjagin classes of V_n.

Theorem 2 can be considered as a generalization of the theorem of Riemann-Roch to arbitrary dimensions. It contains the known Riemann-Roch theorems for $n = 1, 2, 3$. In the case $n = 2$ the term $\dim H^1(V_n, F)$ has to be identified with the superabundance of F. A similar remark applies to $\dim H^1(V_n, F)$ and $\dim H^2(V_n, F)$ in the case $n = 3$. In general, the term $\dim H^n(V_n, F)$ is equal to dim $H^0(V_n, K - F)$, where K is the canonical bundle of V_n. The characteristic class of K is equal to $-c_1$. Kodaira[6] proved that $H^i(V_n, F) = 0$ for $i > 0$, if the characteristic cohomology class of the bundle $F - K$ "contains" a closed positive definite Hermitian form. In this case $F - K$ is called ample. An arbitrary divisor D of V_n represents a complex line bundle,[7] which we also denote by D. Obviously dim $|D| + 1 =$ dim $H^0(V_n, D)$. Denoting the cohomology class of D by d we get

THEOREM 2*. *If $D - K$ is ample (in the sense of Kodaira), then we have*

$$\dim |D| + 1 = \kappa_n \left[e^{d+c_1/2} \prod_{i=1}^{n} \frac{\gamma_i/2}{\sinh \gamma_i/2} \right].$$

(For $n = 1$, this is the classical formula: dim $|D| = d - p$, where $c_1 = 2 - 2p$.)

Applying (M) to the case where W is the bundle of complex covariant p-vectors, yields[5]

THEOREM 3. *Let $h^{p,q}$ be the number of harmonic forms of type (p, q) on V_n. We have*

$$\chi^p(V_n) = \chi(V_n, W) = \sum_{q=0}^{n} (-1)^q h^{p,q} = \kappa_n \left[\left(\sum e^{-(\gamma_{i_1} + \ldots + \gamma_{i_p})} \right) \cdot \prod_{i=1}^{n} \frac{-\gamma_i}{e^{-\gamma_i} - 1} \right].$$

The sum on the right side has to be extended over all the $\binom{n}{p}$ possible combinations.

We attach to every algebraic variety the polynomial

$$\chi_y(V_n) = \sum_{p=0}^{n} \chi^p(V_n) y^p$$

and get by easy calculations

THEOREM 3*. *We have*

$$\chi_y(V_n) = \kappa_n \left[\prod_{i=1}^{n} Q_y(\gamma_i) \right],$$

where

$$Q_y(x) = \frac{-x(y+1)}{e^{-x(y+1)} - 1} - yx.$$

Theorem 3* is for $y = 0$ a restatement of Theorem 1. For $y = -1$ it gives that

$$\chi_{-1}(V_n) = \sum_{p,q=0}^{n} (-1)^{p+q} h^{p,q}$$

is equal to the Euler-Poincaré characteristic c_n of V_n. For $y = 1$ we obtain

THEOREM 4. *We have*

$$\chi_1(V_n) = \sum_{p,q=0}^{n} (-1)^q h^{p,q} = \kappa_n \left[\prod_{i=1}^{n} \frac{\gamma_i}{\operatorname{tgh} \gamma_i} \right].$$

Theorem 4 is known even for Kähler manifolds, because the left side of the equation is equal to the index $\tau(V_n)$.[8] The right side is also equal to $\tau(V_n)$ by the index theorem of the author.[1] The main theorem admits many further applications which we do not mention in this note.

2. A Sketch of the Proof of the Main Theorem.—In section 1 we derived the Theorems 1–4 from the main theorem (M). Actually the proof of (M) goes almost in the opposite direction and involves the following steps.

2.1. Proof of Theorem 1 by Means of the Known Theorem 4.—A complex manifold M_m is called a split manifold, if the structure group $GL(m, C)$ of its tangential bundle can be reduced complex analytically to the group $\Delta(m, C)$ of all triangular matrices (all entries above the diagonal are 0). Let V_n be an arbitrary (non-singular) algebraic variety, the tangential bundle has $GL(n, C)$ as structure group. Hence we can construct the associated bundle with $GL(n, C)/\Delta(n, C)$ as fibre. We obtain in this way an algebraic variety V_m^* of dimension $m = n + {}^1/_2 n(n-1)$. The fact that V_m^* is algebraic can be proved, for example, by using the general theorem of Kodaira[9] that every complex projective bundle over an algebraic variety is an algebraic variety. It can be proved that V_m^* is a split manifold. It is almost obvious that $\Pi(V_m^*) = \Pi(V_n)$. On the other hand, it follows from Theorem 4.2 in the preceding note[1] that $T(V_m^*) = T(V_n)$. Hence it suffices to prove Theorem 1 for all *algebraic* split varieties. Now let V_m^* be an arbitrary algebraic split variety. Then we have over V_m^* an increasing sequence of bundles W_i (in the sense of section 1), where W_i has C_i as fibre and where W_m is the tangential bundle of V_m^* $(0 \leqq i \leqq m)$. We denote the complex line bundle W_i/W_{i-1} by A_i $(1 \leqq i \leqq m)$ and its cohomology class by a_i. We use the notation of the preceding note and obtain readily the formula[10]

$$2^m T(V_m^*) = \sum_{j=0}^{m} \sum \tau(a_{r_1}, a_{r_2}, \ldots, a_{r_j}). \tag{1}$$

The inner sum has to be extended always over all the possible $\binom{m}{j}$ combinations. By the four-term formula of Kodaira-Spencer[11] we define in the faisceau theory virtual indices χ_1. This can be done for all algebraic varieties. (It can be done not only for χ_1, but even for χ_y.) We know that $\chi_1 = \tau$ for all non-singular algebraic varieties, (Theorem 4). Since χ_1 and τ both fulfill the "functional equation" 3.3 of the preceding note,[1] we get by an induction argument on the dimension that χ_1 and τ also coincide in the virtual case.[12]

By some rather complicated calculations based on exact sequences of sheaves we can prove the following formula which corresponds to (1)

$$2^m \Pi(V_m^*) = \sum_{j=0}^{m} \sum \chi_1(A_{r_1} \circ A_{r_2} \circ \ldots \circ A_{r_j}) \tag{2}$$

Here the terms χ_1 denote virtual indices.[13] Formulae (1) and (2) imply Theorem 1 for split varieties, and hence Theorem 1 is proved in general.

2.2. Proof of Theorem 2.—The virtual Π and the virtual T both fulfill the "functional equation" 4.3 of the preceding note. Hence the usual induction argument already used for χ_1 in 2.1 proves that Π and T also coincide in the virtual case. This proves Theorem 2 (i.e., the main theorem for $q = 1$). Moreover, the main theorem follows immediately for all cases in which the bundle W admits the triangular group $\Delta(q, C)$ as structure group.

2.3. Proof of the Main Theorem.—We use the notation of the first section. We construct over V_n the bundle V_m^* with fibre $GL(q, C)/\Delta(q, C)$ which is associated with W. The manifold V_m^* is, according to Kodaira's theorem,[9] algebraic ($m = n + {}^1/_2q(q-1)$). We lift the bundle W up to V_m^* and call the lifted bundle W^*. The bundle W^* splits, i.e., admits $\Delta(q, C)$ as structure group. Hence Theorem 2 implies that $\chi(V_m^*, W^*)$ is given correctly by the formula of the main theorem. A spectral sequence argument shows that $\chi(V_m^*, W^*) = \chi(V_n, W)$. On the other hand, we can prove by a slight generalization of the Theorem 4.2 of the preceding note[1] that the formula of the main theorem gives for $\chi(V_m^*, W^*)$ and $\chi(V_n, W)$ the same values. This concludes the proof of the main theorem.

Remark.—The formulae (1) and (2) are special cases of more general formulae valid for χ_y with y regarded as indeterminate. The "functional equations" for $\Pi = \chi_0 = T$ and for $\chi_1 = \tau$ are special cases of a "functional equation" valid for χ_y.

[1] These PROCEEDINGS, **39**, 951–956 (1953). We use the notations of the preceding note.

[2] Complex dimensions are indicated by a subscript.

[3] All the cohomology groups occurring in this note have finite dimensions. See Cartan, H., and Serre, J. P., *Compt. rend. acad. sci., Paris*, **237**, 128–130 (1953), and Kodaira, K., these PROCEEDINGS, **39**, 865–868 (1953).

[4] Kodaira, K., and Spencer, D. C., these PROCEEDINGS, **39**, 641–649, 868–877 (1953). K. Kodaira, *loc. cit.* in ref. 3.

[5] Here we use the theorem of Dolbeault (*Compt. rend. acad. sci., Paris*, **236**, 175–177 (1953)) which states that $H^q(V_n, W^{(p)}) \cong H^{p,q}$, where $W^{(p)}$ is the bundle of covariant p-vectors and where $H^{p,q}$ is the linear space of all harmonic forms of type (p, q) on V_n.

[6] Kodaira, K., "On a Differential-Geometric Method in the Theory of Analytic Stacks," these PROCEEDINGS, **39**, 1268–1273 (1953).

[7] Conversely every complex line bundle can be represented by a divisor. See Kodaira, K., and Spencer, D. C., *loc. cit.* in ref. 4, p. 874.

[8] Hodge, W. V. D., *Proc. Int. Congr. Math.*, **I**, 182–192 (1952), and *Proc. Lond. Math. Soc.* (*3*), **1**, 104–117 (1951).

[9] Kodaira, K., not yet published.

[10] The sum on the right side starts for $j = 0$ with $\tau(V_m^*)$. The formula (1) was used by the author to prove Theorem 4.1 of the preceding note (*loc. cit.* in ref. 1). See also Hirzebruch, F., "The Index of an Oriented Manifold and the Todd Genus of an Almost Complex Manifold," Notes, Princeton University, 1953 (mimeographed).

[11] Kodaira, K., and Spencer, D. C., "On a Theorem of Lefschetz and the Lemma of Enriques-Severi-Zariski," these PROCEEDINGS, **39**, 1273–1278 (1953), formula (14).

[12] The fact that χ_1 fulfills the functional equation can be obtained from the four-term formula by some calculations.

[13] The small circle $\circ$ denotes "virtual intersection."

Reprinted from the
BULLETIN OF THE AMERICAN MATHEMATICAL SOCIETY
July 1959, Vol. 65, No. 4
Pp. 276-281

RIEMANN-ROCH THEOREMS FOR DIFFERENTIABLE MANIFOLDS

BY M. F. ATIYAH AND F. HIRZEBRUCH

Communicated by Hans Samelson, May 11, 1959

1. **Introduction.** The Riemann-Roch Theorem for an algebraic variety Y (see [7]) led to certain divisibility conditions for the Chern classes of Y. It was natural to ask whether these conditions held more generally for any compact almost complex manifold. This question, and various generalizations of it, were raised in [8] and most of these have since been answered in the affirmative in [2] and [11].

More recently Grothendieck [3] has obtained a more general Riemann-Roch Theorem for a map $f: Y \rightarrow X$ of algebraic varieties. This reduces to the previous Riemann-Roch Theorem on taking X to be a point. Grothendieck's Theorem implies many conditions on characteristic classes, and again it is natural to ask if these conditions hold more generally for almost complex or even differentiable manifolds. The purpose of this note is to enunciate certain differentiable analogues of Grothendieck's Theorem. These "differentiable Riemann-Roch Theorems" yield, as special cases, the divisibility conditions mentioned above and also certain new homotopy invariance properties of Pontrjagin classes. As an application of the latter we get a new proof (and slight improvement) of the result of Kervaire-Milnor [10] on the stable J-homomorphism.

Another differentiable Riemann-Roch Theorem, with applications to embeddability problems of differentiable manifolds, will be found in [1].

The proofs of our theorems rely heavily on the Bott periodicity of the classical groups [4; 5; 6], and are altogether different from the earlier methods of [2] and [11], which were based on Thom's cobordisme theory and Adams' spectral sequence.

2. **Definitions.** The spaces X, Y considered will be countable finite-dimensional CW-complexes, and for simplicity we will suppose them connected.

Following Grothendieck [3] we define an abelian group $K(X)$ as follows. Let $F(X)$ be the free abelian group generated by the set of all isomorphism classes of complex vector bundles over X. To every triple $t=(\xi, \xi', \xi'')$ of vector bundles with $\xi \cong \xi' \oplus \xi''$ we assign the element $[t]=[\xi]-[\xi']-[\xi'']$ of $F(X)$, where $[\xi]$ denotes the isomorphism class of ξ. The group $K(X)$ is defined as the quotient of $F(X)$ by the subgroup generated by all the elements of the form $[t]$; it has an augmentation $\epsilon: K(X)\to \boldsymbol{Z}$ obtained by assigning to each vector bundle the dimension of its fibre. It is then not difficult to show that the kernel of ϵ may be identified with the group of "stable" vector bundles, i.e. with the homotopy classes of maps $X\to B_U$, where B_U is the classifying space of the "infinite" unitary group U (see [**4**]).

The tensor product of vector bundles, being distributive with respect to direct sums, defines a commutative ring structure in $K(X)$; the unit 1 is given by the trivial bundle of dimension one.

If Y is a (nonempty) subcomplex of X we define the relative group $K(X, Y)$ to be the subgroup of $K(X/Y)$ of augmentation zero, where X/Y is the space obtained from X by identifying Y to a point. Then, as for cohomology, one can show that $K(X, Y)$ is a $K(X)$-module.

In a similar way, using real or quaternionic vector bundles we define groups $KO(X)$ or $KSP(X)$ respectively; $KO(X)$ is also a ring (but not $KSP(X)$). The inclusions $i: O\to U$, $j: SP\to U$ induce homomorphisms $i_*: KO(X)\to K(X)$ and $j_*: KSP(X)\to K(X)$.

The Chern character (see [**2**]) defines a ring homomorphism $\mathrm{ch}: K(X)\to H^*(X, \boldsymbol{Q})$ (rational cohomology) which commutes with augmentation; its image will be denoted by $\mathrm{ch}(X)$. The images of $\mathrm{ch}\circ i_*$ and $\mathrm{ch}\circ j_*$ will be denoted by $\mathrm{ch}O(X)$ and $\mathrm{ch}SP(X)$ respectively.

Let S^2 denote the 2-sphere. Then the Bott periodicity for B_U (see [**4**; **5**; **6**]) may be reinterpreted to give[1]

PROPOSITION 1. *We have a commutative diagram*:

$$\begin{array}{ccc} K(X)\otimes K(S^2) & \xrightarrow{\beta} & K(X\times S^2) \\ {\scriptstyle \mathrm{ch}\otimes\mathrm{ch}}\downarrow & & \downarrow{\scriptstyle \mathrm{ch}} \\ H^*(X,\boldsymbol{Q})\otimes H^*(S^2,\boldsymbol{Q}) & \xrightarrow{\alpha} & H^*(X\times S^2,\boldsymbol{Q}), \end{array}$$

where β is induced by the tensor product of bundles, α by the cup product, and both are isomorphisms.

By iteration the same result holds with S^2 replaced by S^{2n}.

From now on X, Y will denote compact connected oriented differentiable manifolds.

According to [**2**, §26.5], an $SO(n)$-bundle is the image of a $\mathrm{Spin}(n)$-bundle if and only if the second Stiefel-Whitney class w_2 vanishes. This motivates the following definition: a continuous map $f: Y\to X$

[1] There are similar results for O and SP [see **6**].

is called a *Spin-map* if $w_2(Y)=f^*w_2(X)$. Y will be called a *Spin-manifold* if the map $Y\to$point is a Spin-map, i.e. if $w_2(Y)=0$. More generally $f: Y\to X$ will be called a *c_1-map* if we are given an element $c_1\in H^2(Y, \mathbf{Z})$ such that $c_1\equiv w_2(Y)-f^*w_2(X)$ mod 2. Y will be called a *c_1-manifold* if $Y\to$point is a c_1-map. Thus a Spin-map (or manifold) is a c_1-map (or manifold) with $c_1=0$.

As in [2], if ξ is a real vector bundle we define

$$\hat{\mathfrak{A}}(\xi) = \prod_i \frac{x_i/2}{\sinh x_i/2},$$

where the Pontrjagin classes of ξ are the elementary symmetric functions in the x_i^2. If ξ is the tangent bundle of Y we write $\hat{\mathfrak{A}}(Y)$ instead of $\hat{\mathfrak{A}}(\xi)$. For any $d\in H^2(Y, \mathbf{Q})$ and any $y\in \mathrm{ch}(Y)$ we define $\hat{A}(Y, d, y)$ to be the rational number obtained by evaluating the top-dimensional component of $y\cdot e^d\cdot\hat{\mathfrak{A}}(Y)$ on the fundamental cycle of Y. If $d=0$, $y=1$, we put $\hat{A}(Y, 0, 1)=\hat{A}(Y)$.

3. **Statement of results.**

THEOREM 1. *Let* dim $Y\equiv$dim X mod 2. *Then a c_1-map $f: Y\to X$ induces a (group) homomorphism $f_!$*: $\mathrm{ch}(Y)\to\mathrm{ch}(X)$ *given by*

$$f_!(y)\cdot\hat{\mathfrak{A}}(X) = f_*(y\cdot e^{c_1/2}\cdot\hat{\mathfrak{A}}(Y)), \qquad y\in \mathrm{ch}(Y),$$

where f_* is the Gysin homomorphism (the Poincaré dual of the homology homomorphism).

REMARKS. (1) It is probable that $f_!$ is actually induced by a functorial homomorphism $K(Y)\to K(X)$.

(2) If Y and X are complex manifolds and f is a holomorphic map, then f is a c_1-map in a natural way: $c_1=c_1(Y)-f^*c_1(X)$. Theorem 1 can then be rewritten as

$$f_!(y)\cdot\mathfrak{T}(X) = f_*(y\cdot\mathfrak{T}(Y)),$$

where $\mathfrak{T}$ is the (total) Todd class (see [2]). This is the Grothendieck formula (see [3]). However, in the Grothendieck theory (for algebraic varieties) $f_!$ has a direct definition in terms of coherent algebraic sheaves. This has no counterpart in the differentiable theory.

THEOREM 2. (i) *Let* dim $Y\equiv$dim X mod 2. *Then a Spin-map $f: Y\to X$ induces a (group) homomorphism $f_!$*: $\mathrm{ch}(Y)\to\mathrm{ch}(X)$ *given by*

$$f_!(y)\cdot\hat{\mathfrak{A}}(X) = f_*(y\cdot\hat{\mathfrak{A}}(Y)).$$

(ii) *If moreover*

(a) $$\dim Y - \dim X \equiv 0 \bmod 8,$$

then $f_!\mathrm{ch}O(Y) \subset \mathrm{ch}O(X)$, *while if*

(b) $$\dim Y - \dim X \equiv 4 \bmod 8,$$

then $f_!\mathrm{ch}O(Y) \subset \mathrm{ch}SP(X)$.

REMARK. Theorem 2(i) is a special case of Theorem 1, but (ii) is a further refinement.

COROLLARY 1. *Let Y be a c_1-manifold with* $\dim Y \equiv 0 \bmod 2$, *and let* $y \in \mathrm{ch}(Y)$. *Then* $\hat{A}(Y, c_1/2, y)$ *is an integer.*

This follows at once from Theorem 1 by taking X to be a point.

COROLLARY 2. (i) *Let Y be a Spin-manifold with* $\dim Y \equiv 0 \bmod 2$, *and let* $y \in \mathrm{ch}(Y)$. *Then* $\hat{A}(Y, 0, y)$ *is an integer; in particular* $\hat{A}(Y)$ *is an integer.*

(ii) *If moreover* $\dim Y \equiv 4 \bmod 8$ *and* $y \in \mathrm{ch}O(Y)$, *then* $\hat{A}(Y, 0, y)$ *is an even integer; in particular* $\hat{A}(Y)$ *is an even integer.*

(i) is a special case of Corollary 1. For (ii) we use Theorem 2(ii)b, with X = point; we need only observe that $\mathrm{ch}SP(\text{point}) = 2\mathbf{Z}$ (since the inclusion $Sp(n) \to U(2n)$ doubles the fibre dimension).

REMARKS. Corollary 1 was proved by Borel-Hirzebruch [2, Part II], except for the prime 2. A complete proof, including the prime 2, follows by the methods of [2] from results of Milnor [11] (see [2, Part III]). Corollary 2(ii) is new. It was conjectured in [2, Part II], where it was pointed out that it provides a satisfactory generalization of Rohlin's Theorem [12]: *if* $\dim Y = 4$ *and* $w_2(Y) = 0$, *then* $p_1(Y) \equiv 0 \bmod 48$. Kervaire and Milnor have used Corollary 2(i) for their result on the J-homomorphism [10]; their result can now be sharpened by using Corollary 2(ii) (see also Corollary 5 below).

COROLLARY 3. *Let* $f: Y \to X$ *be a homotopy equivalence, and identify* $H^*(Y, \mathbf{Q})$ *with* $H^*(X, \mathbf{Q})$ *under* f^* (*or* f_*). *Then* $\hat{\mathfrak{A}}(Y)/\hat{\mathfrak{A}}(X) \in \mathrm{ch}O(X)$.

This follows from Theorem 2(ii)a. We need only remark that, w_2 being a homotopy type invariant [13], f is a Spin-map. Corollary 3 contains new information on the Pontrjagin classes, for example:

COROLLARY 4. (i) *The first Pontrjagin class p_1 of a compact oriented differentiable manifold X (regarded as an element of $H^4(X, \mathbf{Z})$ modulo its torsion subgroup) is a homotopy type invariant* mod 24.

(ii) *If moreover* $H^2(X, \mathbf{Z}_2) = 0$, *then p_1 is invariant* mod 48.

PROOF. Equating 4-dimensional components in Corollary 3 we get $p_1(Y)-p_1(X)=24p_1(\xi)$ for some orthogonal bundle ξ. This proves (i). Also $p_1(\xi)\equiv w_2^2(\xi)$ mod 2, see [2, §30]; thus $p_1(\xi)$ is even if $H^2(X, \mathbf{Z}_2)=0$. This proves (ii).

REMARK. Wu [14] has proved that the p_i are homotopy invariants mod 12. Thus, for p_1, Corollary 4 is an improvement on Wu's results. Moreover the example of S^4 bundles over S^4 shows that Corollary 4(ii) is best possible.

COROLLARY 5 (KERVAIRE-MILNOR [10]). *The order of the image of the stable homomorphism* $J\colon \pi_{4k-1}(SO(q+1))\rightarrow\pi_{4k+q}(S^{q+1})$ $(q\geqq 4k)$ *is divisible by the denominator of* $B_k/4k$, *where* B_k *is the kth Bernoulli number.*

This follows easily from Corollary 3 and the results of James and Whitehead [9] on stable sphere bundles over spheres.

4. **Outline of proof.** We sketch the proof of Theorem 2(i). The other proofs are similar. Since only the homotopy class of f is relevant to the theorem we may suppose that f is differentiable. Let $g\colon Y\rightarrow S^{2n}$ be a differentiable embedding of Y. Then $f\colon Y\rightarrow X$ can be factored into the differentiable embedding $F=f\times g\colon Y\rightarrow X\times S^{2n}$ followed by the projection $\pi\colon X\times S^{2n}\rightarrow X$. Since Theorem 2(i) is transitive it is sufficient to prove it for the case of an embedding and for the projection π separately. For π the theorem is an elementary consequence of Proposition 1 (with S^{2n} replacing S^2), i.e. it follows from the Bott theory.

For an embedding $f\colon Y\rightarrow X$ Theorem 2(i) states that

$$f_*(\mathrm{ch}(Y)\cdot\hat{\mathfrak{A}}(\xi)^{-1})\subset\mathrm{ch}(X)$$

where ξ is the normal bundle of Y in X (and $w_2(\xi)=0$ since f is a Spin-map). Let A be a closed tubular neighborhood of Y in X whose boundary E is the sphere bundle associated to ξ. Because of the natural homomorphism $\mathrm{ch}(A, E)\rightarrow\mathrm{ch}(X)$ it is sufficient to show that

$$\phi_*(\mathrm{ch}(Y)\cdot\mathfrak{A}(\xi)^{-1})\subset\mathrm{ch}(A, E),$$

where $\phi_*\colon H^*(Y, \mathbf{Q})\rightarrow H^*(A, E; \mathbf{Q})$ is the Thom-Gysin isomorphism [13].

Since $K(A, E)$ is a $K(A)$-module, and so a $K(Y)$-module (since Y is a deformation retract of A) it is sufficient finally to construct an element $\eta\in K(A, E)$ such that

$$\mathrm{ch}\eta = \phi_*\hat{\mathfrak{A}}(\xi)^{-1}.$$

This is done by a universal construction on the classifying space of Spin($2n$), using the difference between the two spinor representations Δ^+ and Δ^- of Spin($2n$). The formula for chη is a consequence of the character formula:

$$\text{ch}\Delta^+ - \text{ch}\Delta^- = \prod_{i=1}^{n} (e^{x_i/2} - e^{-x_i/2}),$$

where we adopt the notation of [2].

References

1. M. F. Atiyah and F. Hirzebruch, *Some non-embeddability theorems for differentiable manifolds*, Colloque de Topologie, Lille, 1959.

2. A. Borel and F. Hirzebruch, *Characteristic classes and homogeneous spaces* I, Amer. J. Math. vol. 80 (1958) pp. 458–538; II and III to appear.

3. A. Borel and J-P. Serre, *Le théorème de Riemann-Roch (d'après Grothendieck)*. Bull. Soc. Math. France vol. 86 (1958), pp. 97–136.

4. R. Bott, *The stable homotopy of the classical groups*, Proc. Nat. Acad. Sci. U.S.A. vol. 43 (1957) pp. 933–935.

5. ———, *The space of loops on a Lie group*, Mich. Math. J. vol. 5 (1958) pp. 35–61.

6. ———, *Some remarks on the periodicity theorems*, Colloque de Topologie, Lille, 1959.

7. F. Hirzebruch, *Neue topologische Methoden in der algebraischen Geometrie*, Springer, 1956.

8. ———, *Some problems on differentiable and complex manifolds*, Ann. of Math. vol. 60 (1954) pp. 213–236.

9. I. M. James and J. H. C. Whitehead, *The homotopy theory of sphere bundles over spheres*, Proc. London Math. Soc. vol. 4 (1954) pp. 196–218.

10. M. A. Kervaire and J. Milnor, *Bernoulli numbers, homotopy groups and a theorem of Rohlin*. Proceedings of the International Congress of Mathematicians, 1958.

11. J. Milnor, *On the cobordisme ring* Ω^*, *and a complex analogue*, (in preparation).

12. V. A. Rohlin, *New results in the theory of 4-dimensional manifolds*, Dokl. Acad. Nauk. S.S.S.R. vol. 84 (1952) pp. 221–224 (in Russian).

13. R. Thom, *Espacés fibrès en sphères et carrés de Steenrod*, Ann. Sci. Ecole Norm. Sup. vol. 69 (1952) pp. 109–182.

14. W. T. Wu, *On the Pontrjagin classes* III, Acta Math. Sinica vol. 4 (1954) (in Chinese).

Pembroke College, Cambridge and
University of Bonn

Reprinted from Gazette des Mathematiciens (Oct. '97), No. 74, pp. 27–39.

—— LEARNING COMPLEX ANALYSIS IN MÜNSTER-PARIS, —— ZÜRICH AND PRINCETON FROM 1945 TO 1953

Friedrich HIRZEBRUCH

Max-Planck-Institut für Mathematik

Liebe Frau Cartan, lieber Herr Cartan!

Dear friends and colleagues!

It is a great pleasure and honour for me to lecture on this day dedicated to Henri Cartan. Many of us here are students of Cartan directly or indirectly. Henri Cartan is very happy about his students. This can be seen from the following picture, taken during his retirement meeting and party in Orsay and Bures in June 1975.

Photo W. Scharlau

Here are four students of Henri Cartan.

Photo W. Scharlau

They are from left to right : Karl Stein, Reinhold Remmert, Hans Grauert, Friedrich Hirzebruch. This picture was also taken at the retirement party in 1975. Actually these are four of Heinrich Behnke's 57 doctoral students. Because of the close cooperation and friendship between Heinrich Behnke and Henri Cartan, these four mathematicians are also students of Cartan. Because of this cooperation I wrote Münster – Paris with a hyphen. It lasted from 1931 to October 10, 1979, when Behnke died. On the occasion of Behnke's 80th birthday on October 9, 1978, Henri Cartan wrote a beautiful article "Quelques souvenirs" wich I shall use for my report. He adressed these souvenirs to Behnke during the dinner in Münster for Behnke's 80th birthday. Sadly enough, Heinrich Behnke unexpectedly could not attend because of illness.

Henri Cartan visited Münster for the first time in May 1931. He was 26 years old. Behnke had been appointed "ordentlicher Professor" in Münster already in 1927. Cartan gave four lectures in German and one in French. There was a long walk in the forest. One of the talks concerned circle domains. A domain in $\mathbb{C}^2$ (complex coordinates x, y) is called a circle domain if it contains the origin and admits the elements of the circle group

$$x' = xe^{i\alpha}$$

$$y' = ye^{i\alpha}$$

as automorphisms. Cartan had proved in a CRAS note of 1930 that every biholomorphic map of a bounded circle domain to a circle domain fixing the origin is linear. Behnke had proved a related result. There was great interest in "Abbildungstheorie" in Münster at the time. Cartan met the young assistant of Behnke, Peter Thullen, who had received his doctoral degree in 1930. The famous book "Theorie der Funktionen mehrerer komplexer Veränderlichen" by Behnke and Thullen (Ergebnisse der Mathematik, Springer–Verlag) appeared in 1934. On page 35, the "Cartansche Körper" (generalisations of circle domains) are introduced. The bibliography of the book contains 11 papers of H. Cartan. Chapter 7 (Abbildungstheorie) is full of references to Cartan. In the introduction, the authors thank Professors H. Cartan and H. Kneser and his excellency Professor Severi for extensive criticisms which led to many improvements. The book of Behnke – Thullen is a symbol of the cooperation Behnke – Cartan.

In May 1938, Cartan visited Münster again. Times had changed. Peter Thullen had left in 1933. He had a professorship in Equador. Behnke had applied to the "Herr Reichs-und Preussische Minister für Wissenschaft, Erziehung und Volksbildung" for permission to invite H. Cartan. This was necessary in these days, though Behnke did not need any money. The financial support came from the "Stiftung Still". The "Firma Still" was founded in 1898 by Carl Still. The Still family had close relations with many scientists and supported the work of mathematicians and physicists from 1920 until to-day, for example Max Planck, Richard Courant, Max Born, Heinrich Behnke. On October 11, 1938 Behnke reported to the Kurator of the University of Münster about how he used the money of the Stiftung Still. For the summer term of 1938, H. Cartan (Strassburg) is recorded with RM 135,37, Marston Morse (Princeton) with RM 100, – and Karl Stein with RM 30,40. Karl Stein had received his doctoral degree with Behnke in 1936. In July 1938, he had no salary and obtained RM 30,40 for his transportation from Münster to his home town Hamm (Westfalen), thirty kilometers away. Let me mention that Hamm is also my home town. My father was Karl Stein's teacher in mathematics in a secondary school in Hamm for six years. It is important to note that Cartan met Karl Stein in Münster in 1938 for the first time. Fifteen years later the *Colloque sur les fonctions de plusieurs variables* was held in Bruxelles where Behnke, Cartan, Serre and Stein attended. Stein manifolds played a fundamental role in the Bruxelles lectures of Cartan and Serre. But we shall come to this later.

The second world war began a little more than one year after Cartan's second visit to Münster. But the friendship and cooperation between Behnke and Cartan was not interrupted. Cartan writes in "Quelques Souvenirs" : « *[...], en février 1941, je reçois une lettre de mon ami Behnke. Il me fait part d'une lettre de Oka (datée de décembre 1940) qui annonce qu'il a résolu le problème de la pseudo-convexité globale (problème de Levi) [...] Heinrich Behnke prend la peine de recopier de sa main le lettre de Oka, écrite en français [...]* ».

Let us recall that Eugenio Elia Levi (1883 – 1917) published in 1911 a paper with the title : *Sulle ipersuperficie dello spazio a quattro dimensioni che possono essere frontiera del campo di esistenza di una funzione analitica di due variabili complesse.*

A domain in the space of two complex variables has to satisfy certain local conditions (pseudoconvexity) in its boundary points if it is the domain of existence of a holomorphic function (domain of holomorphy). The problem of Levi asks wether these local boundary conditions are also sufficient to ensure that the domain is a domain of holomorphy. The same problem can be formulated in higher dimensions. Oka published his paper under the title *Domaines pseudoconvexes* in the Tôhoku Mathematical Journal in 1942. Papers by Cartan, Cartan-Thullen, Behnke-Stein and the book by Behnke-Thullen are used. The Collected Papers of Kiyoshi Oka (1901 – 1978) were published by Springer–Verlag in 1984 and edited by Reinhold Remmert with commentaries by Henri Cartan. R. Narasimhan translated the papers from French to English. The commentary of H. Cartan to *Pseudoconvex domains* begins as follows.

«*Soit D un domaine de $\mathbb{C}^n$. Hartogs et Levi ont donné des conditions nécessaires pour que D soit domaine d'holomorphie; ces conditions ont un caractère local au voisinage de chaque point-frontière de D. Le problème était resté ouvert de savoir si réciproquement ces conditions entraînent que D est un domaine d'holomorphie.* OKA *se propose de résoudre ce problème; pour simplifier il se borne au cas $n = 2$. Le cas général a été ensuite résolu par* BREMERMANN *et par* NORGUET *(indépendamment l'un de l'autre).* OKA *prouve d'abord le théorème suivant : si un domaine $D \subset \mathbb{C}^2$ est "pseudo-convexe au sens de* CARTAN*" (i.e. : si tout point-frontière de D possède un voisinage ouvert V tel que $V \cap D$ soit un domaine d'holomorphie), alors D est un domaine d'holomorphie...*»

Hans Bremermann (1926 – 1996) received his doctoral degree with Behnke in 1951.

As we saw scientific contacts between Behnke and Cartan continued during the war. But Behnke was also a friend who tried to help in other ways. In 1943 Cartan's brother Louis was deported to Germany. About this tragedy Cartan says in "Quelques Souvenirs" adressed to Heinrich Behnke the following : «*Je ne puis pas non plus oublier toutes les démarches que vous avez faites durant les années 1943 et 1944 (en vain, hélas) pour tenter de retrouver la trace de mon frère Louis, déporté en Allemagne au mois de février 1943, et qui ne devait jamais revenir* ».

After the war Cartan came to Germany already in the fall of 1946. He travelled to Wolfach and started to walk from there to Oberwolfach, but fortunately got a lift for part of the way from a French officer in charge of the Oberwolfach Institute. In Oberwolfach, Behnke and Cartan met again. The Oberwolfach guest book records on November 1, 1946 that Cartan participated actively in a concert (Bach, Wohltemperiertes Klavier I, b-moll, II, fis moll, and Haydn, Die Sinfonie mit dem Paukenschlag, with H. Boerner) and that he lectured on November 4 on *Théorie de Galois pour les corps non commutatifs* . Two further visits with lectures are recorded in the guest book of the fifties.

November 26,1950 : *Sur la théorie des foncteurs*. The abstract begins as follows. «*Il s'agit d'exposer les éléments d'une théorie actuellement développée par S. Eilenberg et H. Cartan*». The famous book *Homological Algebra* by Cartan and Eilenberg appeared in 1956. The following picture shows the authors presenting the book, with Frank Adams looking on.

Photo anonymous

On March 3, 1957 Cartan lectured on *Sous-ensembles analytiques réels* (Résultats obtenus en collaboration avec F. Bruhat ou par Bruhat seul).

My first visit of Oberwolfach was in 1951. I lectured on *Eine Verallgemeinerung der Plückerschen Formel für das Geschlecht einer algebraischen Kurve*. As far as I remember, I met Cartan for the first time at this occasion. I had begun to understand a few facts on characteristic classes under the influence of H. Hopf and how to apply them to well-known questions in Algebraic Geometry. At this time I was Scientific Assistant in Erlangen.

In December 1949, Cartan came to Münster for the first time after the war. His visit was inspiring for many young mathematicians like Grauert and Remmert. I was not there. In the summer term of 1949 I had begun my studies in Zürich.

Cartan's early visits to Germany gave many German mathematicians help, encouragement and inspiration. It was the beginning of the reconciliation. Cartan contributed essentially to the reintegration of German mathematicians into the international mathematical community.

I began my studies in Münster in November 1945. The city had been destroyed. Very few buildings of the University had survived. Thanks to Behnke's energy the students of mathematics could use a lecture hall every three weeks. We got many problems, went home and returned three weeks later when the solutions were discussed and new problems given out. The situation improved very fast. We lived soon in an active mathematical atmosphere. I learned much about Complex Analysis in the areas I mentioned earlier in this lecture. I profited a lot from the cooperation Münster – Paris and realised from the very beginning that mathematics progresses through international cooperation and exchange of ideas.

Soon Karl Stein joined the faculty as Dozent. We frequently travelled jointly by train from Münster to our home town Hamm and discussed mathematics. He told me that complex spaces in dimension 2 are topologically non-trivial. In dimension 1, the function $w = \sqrt[2]{z}$ determines a Riemann surface branched over $\mathbb{C}$ with the origin as point of ramification. Then $z = t^2, w = t$ defines a "uniformising parameter" t. However, in two complex variables $w = \sqrt[2]{z_1 z_2}$ gives a ramification over $\mathbb{C}^2$ branched at $z_1 = 0$ and $z_2 = 0$. Away from the origin the ramification is not worse than in dimension 1. In the origin $z_1 = t_1^2, z_2 = t_2^2$ $w = t_1 t_2$ is a a parametrisation of the singularity. But in the t_1, t_2 space the points t_1, t_2 and $-t_1, -t_2$ have to be identified because z_1, z_2, w are invariant under this involution. The singularity has a neighborhood whose 3-dimensional boundary is the 3-dimensional sphere with antipodal points identified. It is therefore the 3-dimensional real projective space.

Of course, the singularity can also be written as $w^2 = z_1^2 - z_2^2 = (z_1 - z_2)(z_1 + z_2)$, and $z_1 = u^2 + v^2, z_2 = u^2 - v^2, w = 2uv$ is a parametrisation for which $c = z_1$, $a = z_2$, $b = w$ gives $c^2 = a^2 + b^2$ and the well-known formulas for Pythagorean triples. Fermat's problem is much in the air with A. Wiles receiving the Wolfskehl prize in Göttingen on June 26. In my lecture I wanted to make the following remark which had to be omitted because of lack of time :

Felix Klein in 1884 gave a parametric solution of the singularity $T^2 + H^3 = 1728 f^5$ (using the icosahedron) by homogeneous polynomials T, H, f in u, v of degrees 30, 20, 12 with integral coefficients. (See my Bourbaki lecture of 1962/63). Don Zagier, my immediate successor as managing director of the Max-Planck-Institut für Mathematik, told me recently that Felix Klein's parametrisation can be modified to give several parametric solutions by polynomials with integral coefficients of $a^2 + b^3 = c^5$ to obtain analogues of Pythagorean triples in this case (see Frits Beukers, *The diophantine equation* $Ax^p + By^q = Cz^r$, to appear in Duke Math. Journal).

Felix Klein's singularity has higher dimensional liftings :

$$z_1^5 + z_2^3 + z_3^2 + z_4^2 + z_5^2 = 0$$

$$\sum z_i \bar{z}_i = 1$$

This is the Brieskorn description of one of Milnor's exotic 7-dimensional spheres(see my Bourbaki lecture 1966/67).

From the summer of 1949 to the summer of 1950 I studied in Zürich at the Eidgenössische Technische Hochschule. I learned topology from Heinz Hopf and Beno Eckmann. Hopf studied examples of complex manifolds and the question wether certain differentiable manifolds admit complex structures. To increase his repertoire of examples of complex manifolds he introduced the method of "blowing up a point" rediscovering well-known processes in algebraic geometry (for example the Cremona birational transformation). He analysed the purely local character and the topological features of "blowing-up a point" (Hopf's σ-process).

If P is a point in a complex surface M we consider the tangent space $T_P(M)$. This is a 2-dimensional complex vector space. The lines in $T_P(M)$ through the origin constitute a projective line σ_P. It is possible to remove P and replace it by σ_P. For the new complex surface M' we have a natural holomorphic map

$$\pi: M' \to M \quad \text{with } \pi: M' - \sigma_P \to M - \{P\} \text{biholomorphic and } \pi(\sigma_P) = \{P\}$$

In M' the smooth rational curve σ_P has self-intersection number -1. In the old days I proved this in the terminology of complex analysis. Without loss of generality we assume $M = \mathbb{C}^2$ and $P = (0,0)$. If we lift the coordinate z_1 of $\mathbb{C}^2$ to M' we get a holomorphic function which defines a zero divisor (Cousin II distribution) in M' consisting of a non-compact divisor E_1 and of the compact divisor σ_P with intersection number $E_1 \cdot \sigma_P = 1$. Then $(E_1 + \sigma_P) \cdot \sigma_P = 0$ which I took from Karl Stein's *Topologische Bedingungen für die Existenz analytischer Funktionen zu vorgegebenen Nullstellenflächen* (Math. Ann. 117 (1941)). We come back to the Cousin problems later.

My dissertation was written in Zürich under Hopf with close contact with Münster. I resolved the singularities of complex spaces in dimension 2 without knowing the corresponding results in Algebraic Geometry! A complex space was given for me locally by an algeroid function defining a ramified cover over $\mathbb{C}^2$. The ramification was considered to generalise Riemann surfaces and correspondingly the title was *Über vierdimensionale Riemannsche Flächen mehrdeutiger analytischer Funktionen von zwei komplexen Veränderlichen.* I received my doctoral degree in 1950 in Münster. The paper appeared much later (Math. Ann 126 (1953)). In this thesis I blow up points of the ramification locus until it has only transversal intersections and then reduce everything (following H. W. E. Jung 1908) to algebroid functions

$$w = \sqrt[n]{z_1^q / z_2}$$

with $0 < q < n$ and $(n, q) = 1$. The function w defines an n-fold covering of $\mathbb{C}^2$ ramified in the coordinate axes with the orgin as the unique singular point. The resolution is described by the continued fraction

$$\frac{n}{q} = b_1 - \cfrac{1}{b_2 - \cfrac{}{\ddots - \cfrac{1}{b_s}}} \quad \text{with} \quad 1 < b_i$$

The isolated singular point over the origin is blown up into a chain of smooth rational curves with selfintersection numbers $-b_i$. Each curve intersects the next one transversally in one point

The dotted lines are the transforms of the coordinate axes. The blown up surface X is covered by $s+1$ coordinate systems (u_k, v_k) with $0 \leqq k \leqq s$ centered in the marked intersection points related to each other by

$$u_k = u_{k-1}^{b_k} v_{k-1} \qquad (1 \leqq k \leqq s)$$

$$v_k = u_{k-1}^{-1}$$

In X the function

$$w = \sqrt[n]{z_1^q / z_2}$$

is meromorphic and univalued.

Recently I studied the following problem. The $N - th$ root of w (assume for simplicity $(N, q) = 1$) defines an N-fold ramified cover of X with possisble singularities in the above marked centers of the coordinate system. The resolution of all these singularities should relate to the direct resolution of $\sqrt[Nn]{z_1^q/z_2}$ and also relate the continued fractions of $\frac{n}{q}$ and $\frac{Nn}{q}$ (see Michel Mendès France and Henri Cohen in *Acta Arithmetica 1973 and 1974)*. Here is an example as an exercise for resolving and blowing down. I also wish to demonstrate once more (as I did before with $a^2 + b^3 = c^5$) how old things of 50 years ago sometimes become active again in one's life. We have $\frac{4}{3} = 2 - \cfrac{1}{2 - \cfrac{1}{2}}$ and $\frac{20}{3} = 7 - \frac{1}{3}$ and consider

$$\sqrt[4]{z_1^3/z_2} \quad \text{and} \quad \sqrt[20]{z_1^3/z_2}.$$

The resolution of the first function is

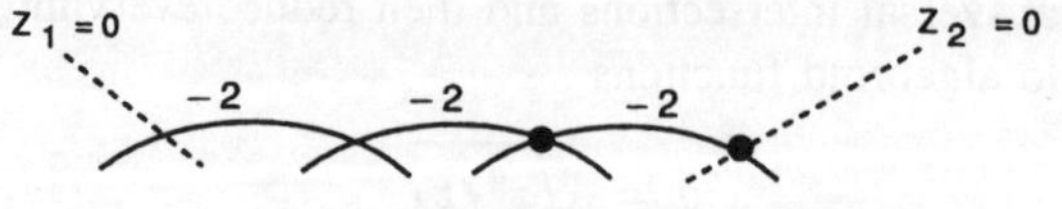

It turns out that the five fold ramified cover has singularities over the two marked points of type $\frac{5}{3} = 2 - \frac{1}{3}$ and $\frac{5}{1} = 5$ respectively. In the resolved five fold cover the first (-2)-curve has selfintersection number -10 whereas the two others become exceptional with selfintersection number -1. Thus the resolution of the two singular points leads to

Blowing down the two exceptional curves and then the curves which become exceptional after the blow down etc. gives

Henri Cartan reported about my thesis in the Bourbaki seminar of December 1953. The title was *Functions et variétés algébroides* [d'après F. Hirzebruch]. He introduced the notion of *Espace analytique général de dimension n* and made it clear in which category I was working. Hans Grauert and Reinhold Remmert published a long paper *Komplexe Räume* in Math. Annalen 136 (1958) *(Heinrich Behnke, in Dankbarkeit und Verehrung zum 60. Geburtstag gewidmet)*. In the Introduction, they report about the concepts of complex space in the sense of Behnke and Stein on the one hand and of Henri Cartan and later J.-P. Serre on the other hand. In the case of Behnke and Stein, complex spaces have as local models finite analytically ramified coverings over domains in $\mathbb{C}^n$. For Cartan and Serre, complex spaces are locally given by analytic subsets of $\mathbb{C}^N$. Grauert and Remmert show that the analytically ramified coverings of Behnke and Stein can be given by algebroid functions which implies that they are complex spaces in the sense of Cartan and Serre (where I neglected here questions of normality). The fundamental paper of Grauert and Remmert is the first paper in *Grauert's Selected Papers* (Springer-Verlag 1994).

Hopf published a paper *Schlichte Abbildungen und lokale Modifikationen 4-dimensionaler Mannigfaltigkeiten* in Comm. Math. Helvetici 29 (1955). His main result (in complex dimension 2) states that for a point in a complex surface all local modifications (the point is replacaed by a compact 1-dimensional analytic set) are iterations of σ-processes (blowing-up of points). Hopf relates this to earlier work of O. Zariski in Algebraic Geometry. In his paper Hopf quotes a paper of H. Cartan *(Sur une extension d'un théorème de Radó.* Math. Ann.125 (1952)) with the following theorem :

If X is a complex manifold and $g : X \to \mathbb{C}$ a continuous map such that g is holomorphic in $X - g^{-1}(0)$, then g is holomorphic.

Cartan's paper is part of a letter to Behnke and Stein. See Nr. 40 in *Cartan's Collected Works Volume II* (Springer-Verlag 1979).

On August 17, 1952 I arrived in Princeton. Immediately I had intensive contact with K. Kodaira and D.C. Spencer who taught me sheaves and their cohomology theory and applications in Algebraic Geometry. I spoke about this period in Nice in January 1996, at the colloque *Matériaux pour l'Histoire au XXe siècle (in honour of Jean-Alexandre Dieudonné)*. H. Cartan was present. The lecture was not published. A video is available at CNRS. I reported about the correspondence between Kodaira-Spencer and Serre, between A. Borel and Serre, between René Thom and myself and how the proof of the Riemann-Roch theorem was finalised in November 1953.

J.-P. Serre lectured on parts of my results in the Séminaire Bourbaki in December 1953 *(Travaux de Hirzebruch sur la topologie des variétés)*. My book *Neue topologische Methoden in der Algebraischen Geometrie* appeared in 1956. It became my Habilitationsschrift. I can be glad that Dissertation and Habilitationsschrift appeared in the same Bourbaki meeting with Cartan and Serre reporting.

My "Antrittsvorlesung" for the Habilitation took place in Münster in February 1955. Henri Cartan was present. When preparing the lecture I asked Behnke for advice. He said « *This is very simple. The dean who is a professor of Pharmacy, should understand everything and Henri Cartan should find it interesting.* »

The following Cartan Seminars at the École Normale Supérieure occur in the bibliography of my book

1949-50 *Espaces fibrés et homotopie*

1950-51 *Cohomologie des groupes, suites spectrales, faisceaux*

1951-52 *Fonctions analytiques de plusieurs variables complexe*

1953-54 *Fonctions automorphes et espaces analytiques*

Many things Kodaira and Spencer taught me had their origin in these seminars. In the introduction of my book I speak of the cohomology groups of a compact complex manifold with coefficients in the sheaf of local holomorphic sections of a holomorphic complex line bundle and point out that these are finite-dimensional vector spaces. I refer to Cartan-Serre, *Un théorème de finitude concernant les variétés analytiques compactes*, CRAS 237 (1953), and to K. Kodaira, *On cohomology groups of compact analytic varieties with coefficients in some analytic faisceaux*, Proc. Nat. Acad. 39 (1953). Cartan-Serre use coverings of the manifold by Stein manifolds and apply Theorem B (see below). Kodaira uses finiteness theorems for linear elliptic operators in the spirit of Hodge theory. The cohomology groups studied here vanish in dimensions greater than the dimension of the manifold. The alternating sum of their dimensions (holomorphic Euler number) is well-defined. Since one is basically interested in the dimension of the space of global holomorphic sections (Riemann-Roch problem), the vanishing of the higher dimensional cohomology groups is of importance. For this I refer in the Introduction of my book to Kodaira, *On a differential geometric method in the theory of analytic stacks*, Proc. Nat. Acad. 39 (1953) and to *Exposé XVIII* (J.-P. Serre) in Cartan's 1953-54 seminar (« *Voici les théorèmes qui, dans le cas de l'espace projectif, jouent le même rôle que les théorèmes A et B de la théorie des variétés de Stein)*». Karl Stein lectured twice in the 1953/54 seminar. These few remarks show how my Princeton time was influenced by the Cartan school in Paris.

The lectures of H. Cartan *(Variétés analytiques complex et cohomologie)* and J.-P. Serre *(Quelques problèmes globaux relatifs aux variétés de Stein)* in the *Colloque sur les fonctions des plusieurs variables (Bruxelles, March 1953)* show even more clearly how much "we in Princeton" could learn from the Paris school. In Cartan's paper the Stein manifolds (Karl Stein 1951) are introduced which generalise the domains of holomorphy. All smooth analytic subvarieties of $\mathbb{C}^n$ are Stein manifolds (and vice versa, Remmert 1957). The famous Theorems A and B are formulated *(Proofs in the Cartan seminar 1951/52 Exposé XIX)*. I restrict to the Theorem B and quote from Cartan.

Théorème B. - *Soit X une variété de Stein, et F un faisceaux analytique cohérent sur X. Alors, pour tout entier $q > 0$, les groupes de cohomologie $H^q(X, F)$ sont nuls.*

Theorem A and Theorem B for $q = 1$ occur essentially in a paper of Cartan of 1950. «*La formulation cohomologique du Théorème B, et l'idée d'étudier non seulement le cas $q = 1$, mais le cas $q > 0$ quelconque sont dues à J.-P. Serre*».

Let me recall briefly the Cousin problems (P. Cousin 1895) which play such an important role in the work of Cartan, Oka and the Münster school. Consider an open covering $\{U_i\}$ of the complex manifold X. A Cousin I problem associates to each U_i a meromorphic function g_i such that $g_i - g_j$ is holomorphic in $U_i \cap U_j$. The g_i define a distribution of principal parts of meromorphic functions and $g_{ij} = g_i - g_j$ an element of $H^1(X, \mathcal{O})$ where $\mathcal{O}$ is the sheaf of local holomorphic functions. The problem is to find a global meromorphic function g with the given principal parts, i.e. the difference $g_i - g$ should be a holomorphic function h_i in U_i. The h_i with $g_{ij} = h_i - h_j$ exist if $H^1(X, \mathcal{O}) = 0$. Therefore Cousin I is always solvable for Stein manifolds, but also for projective algebraic manifolds with vanishing first Betti number.

The Cousin II problem associates to each U_i of an open covering a meromorphic function g_i (nowhere identically 0) such that $g_{ij} = g_i/g_j$ is holomorphic and different from 0 in $U_i \cap U_j$. The g_i (or rather their locus of zeros and poles) define a divisor and g_{ij} an element of $H^1(X, \mathcal{O}^*)$ where $\mathcal{O}^*$ is the sheaf of local holomorphic non vanishing functions. The problem is to find a global meromorphic function g such that $h_i = g_i/g$ is holomorphic and not zero in U_i. In other words the divisor of g is the given divisor of the Cousin II problem. The Cousin II problem is solvable if and only if the corresponding element in $H^1(X, \mathcal{O}^*)$ vanishes ($g_{ij} = h_i/h_j = g_i/g_j$ and $g = g_i/h_i = g_j/h_j$). In his Bruxelles lecture Serre explains the exact sequence of sheaves

$$0 \longrightarrow \mathbb{Z} \longrightarrow \mathcal{O} \xrightarrow{\exp} \mathcal{O}^* \longrightarrow 1$$

(with $exp(f) = e^{2\pi i f}$) and the exact cohomology sequence

$$\ldots \to H^1(X,\mathbb{Z}) \to H^1(X,\mathcal{O}) \to H^1(X,\mathcal{O}^*) \xrightarrow{\delta} H^2(X,\mathbb{Z}) \to H^2(X,\mathcal{O}) \to \ldots$$

The cohomology group $H^1(X,\mathcal{O}^*)$ is also the group of isomorphism classes of holomorphic complex line bundles. The homomorphism δ associates to each complex line bundle its first Chern class. For a divisor D (considered as element of $H^1(X,\mathcal{O}^*)$) the cohomology class δD is the image of the (non-compact) homology class of D under the Poincaré isomorphism. The above exact cohomology sequence implies for a Stein manifold by Theorem B that $H^1(X,\mathcal{O}^*)$ and $H^2(X,\mathbb{Z})$ are isomorphic under δ. A Cousin II problem (divisor D) is solvable if and only if $\delta D = 0$. A necessary condition is that the intersection number of D with any compact 2-dimensional homology class vanishes (Stein 1941, loc. cit.). The cohomology class δD vanishes if and only if the complex line bundle associated to D is topologically trivial. In fact, $H^2(X,\mathbb{Z})$ is the group of isomorphism classes of toplogical complex line bundles. (Replace in the above cohomology sequence the sheaves $\mathcal{O}$ and $\mathcal{O}^*$ by the correponding sheaves of continuous functions.) Therefore $\delta D = 0$ means that the Cousin II problem has a continuous solution, namely there exists continuous non vanishing complex valued functions $h_i : U_i \to \mathbb{C}$ with $h_i/h_j = g_{ij}$. Already in 1939 Oka *(Collected papers III. The Cousin II problem)* had shown that for a domain of holomorphy a Cousin II problem has a solution if and only if it has a continuous solution. It is worth reading the commentaries of H. Cartan on this paper and on the preceding paper (concerning Cousin I) in the Collected Papers of Oka.

In his Habilitationsschrift, Hans Grauert proved that for a Stein manifold X and a complex Lie group L the classification of topological principal fibre bundles over X with structural group L coincides with the classification of analytic principal fibre bundles over X with structural group L. This generalises the isomorphism $H^1(X,\mathcal{O}^*) \simeq H^2(H,\mathbb{Z})$, because $H^2(X,\mathbb{Z})$ is the group of isomorphism classes of topological complex line bundles ($L = \mathbb{C}^*$). Grauert published his Habilitationsschrift in three parts in Mathematische Annalen 133 (1957) and 135 (1958). See his Selected Papers (Nr. 11, 12, 16 in Vol. I). In his Habilitationsschrift Grauert thanks Cartan for advice.

Henri Cartan lectured on Grauert's results in the *Symposium Internacional de Topologia Algebraica, Mexico 1956.* I also participated in this Symposium. In my lecture *Automorphe Formen und der Satz von Riemann-Roch* I presented the *Proportionality theorem for Chern numbers* using the bounded homogeneous symmetric domains classified by Elie Cartan.

It was my goal to show the cooperation between Münster and Paris, between the Behnke school and the Cartan school. An official recognition of this cooperation is the fact that Cartan received his first honorary doctoral degree from the University of Münster in 1952. It was the fiftieth anniversary of Münster as a University being called before «*Königlich theologisch-philosophische Akademie*» and the twenty-fifth anniversary of Behnke as «*ordentlicher Professor*» in

Münster. The short laudatio of the Doctor honoris causa document reads : « *Sie [die Fakultät] ehrt durch diese Verleihung einen bahnbrechenden Forscher, der durch seine vielseitigen und tiefsinnigen Untersuchungen in den Bereichen der Analysis, Topologie und Algebra die Mathematik um eine grosse Fülle neuer Ergebnisse bereichert hat. Sie ehrt den Gelehrten, der sein reiches Wissen und sein vielgestaltiges Können im Geiste der Völkerverständigung dienstbar gemacht hat.* »

At the end of this talk a few words about Cartan as a European. Emil Artin, Heinrich Behnke and I were the three German mathematicans nominated by Cartan for a European Committee. Cartan's letter to me dated June 23, 1960 begins as follow :

« *Cher Collègue et Ami,*

Depuis plusieurs années les gouvernements des pays d'Europe sont préoccupés du problème de l'equivalence des périodes d'études universitaires et de l'equivalence des diplômes universitaires. Ils ont, au Conseil de l'Europe, signé plusieurs conventions à ce sujet. Mais il faut reconnaître que ces conventions n'ont pas été suivies de beaucoup d'effet; et il est probable que toutes les tentatives gouvernementales seront condamnées à l'impuissance tant que les professeurs d'Université n'auront pas pris conscience de l'importance du problème, dont la solution dépend avant tout d'eux-mêmes, me semble-t-il. »

Our committee agreed on minimal requirements for basic courses and published a description of them in a European Study Book *(Livret Européen de l'étudiant)* which students could use when changing from a university in one country to a university in some other country. The professor in one university would mark in the booklet contents of courses attended. The professor in the next university would be able to advise the student in which courses to enroll. The study book was not used very much, unfortunately. For me it was often useful when reforms of courses were discussed in Germany. In this way it certainly had an influence also in other countries. In any case, this early activity of Cartan is a good example of his European endeavours. It was very appropriate that we could celebrate Cartan's 88th birthday during the first European Congress of Mathematics held in Paris five years ago. He had put much effort also into this enterprise. Part of the birthday celebrations took place in the Palais Beauharnais, the residence of the German ambassador.

Kunihiko Kodaira: Mathematician, Friend, and Teacher

F. Hirzebruch

Kunihiko Kodaira died July 26, 1997, and a memorial article appeared in the March 1998 *Notices*, pp. 388–389.

Kunihiko Kodaira

Kunihiko Kodaira was friend and teacher for me. My wife and I remember our last visit to the Kodairas' house in Tokyo. He was working at the kitchen table on textbooks for secondary schools. Seiko Kodaira had to push the papers away when preparing the meal. In 1995 I congratulated him on his eightieth birthday. He answered in his charming way. But when we came to Tokyo in 1996, he was already in the hospital. We could not talk to him anymore.

I have read the obituaries by D. C. Spencer in the *Notices of the AMS* (March 1998) and by M. F. Atiyah for the London Mathematical Society. Both say much about our mutual friend; I do not have to repeat it.

I want to report about the influence which Kodaira had on my mathematical work. I shall emphasize the period from 1952 to 1954 when I was a member of the Institute for Advanced Study in Princeton. On Monday, August 18, 1952, I arrived in Hoboken, New Jersey, on the *Ryndam* of the Holland America Line. D. C. Spencer and Newton Hawley picked me up. On Saturday, August 23, I wrote to my parents that I had worked every day in the Institute with Kodaira, Spencer, and Hawley. When I read this letter again after forty-six years, I was surprised to see that my mathematical training in Princeton under Kodaira and Spencer started immediately after my arrival in spite of the Princeton summer.

When I arrived, I knew nothing about sheaves and very little about algebraic geometry and characteristic classes. This improved fast. Our heavy work was made easier by a fine picnic given by Kunihiko and Seiko Kodaira.

In 1975 Kodaira's *Collected Works* appeared in three volumes (Iwanami Shoten, Publishers, and Princeton University Press) with a preface by his student Walter L. Baily Jr. giving a survey and appreciation of Kodaira's work until then.

At the end of this paper I shall reproduce twenty-six entries from the table of contents of the *Collected Works* using the numbering there. These are mostly the papers quoted in my book *Topological Methods in Algebraic Geometry* (Translation and Appendix One by R. L. E. Schwarzenberger, Appendix Two by A. Borel), which was published by Springer-Verlag in 1966 as the English version of *Neue topologische Methoden in der algebraischen Geometrie* (Springer-Verlag, 1956). I added reference [28] ("Work done at Princeton University, 1952"). These are the lecture notes of his course at Princeton University which I attended, at least partially, in the winter 1952–53. I do not know how much of [28] he covered in his course, but this

Friedrich Hirzebruch is professor of mathematics and retired founding director of the Max Planck Institute for Mathematics in Bonn.

rich material certainly occurred in the many conversations and private seminars of Kodaira, Spencer, and me. In September 1952 Spencer picked me up by car quite regularly at 9 a.m. and drove me to the Institute, where we worked until 5 p.m., mostly with Kodaira, whose course began at the end of September. I also added [37], which is the announcement of his great result that the Hodge manifolds are all projective algebraic, which is fully presented in [38]. I added [63, 66, 68], which together with [60] are the four papers of the famous series "On the structure of compact complex analytic surfaces", which I quoted in my paper *Hilbert modular surfaces* (Enseign. Math. **19** (1973), 183-281) and which I used so much in teaching and research. This paper on Hilbert modular surfaces grew out of my International Mathematical Union lectures, Tokyo, February-March 1972. I remember vividly that Kodaira and Kawada picked us up at the airport. This was the first journey to Japan by my wife and me. Kodaira, having returned to Japan in 1967, was in full action as dean at the University of Tokyo. He introduced me to many of his brilliant students, who later became research visitors in Bonn. For the first time we enjoyed Kunihiko's and Seiko's hospitality in Japan. Finally, I added [64] to the list because it gave rise to my paper *The signature of ramified coverings*, published in the volume *Global Analysis* (University of Tokyo Press, Princeton University Press, 1969), dedicated to Kodaira at the time when he left the United States for Japan.

I have explained how my selection of papers from Kodaira's *Collected Works* was motivated. Among them are eight joint papers by Kodaira and Spencer. Atiyah characterized the collaboration between the two, from which I profited so much: "The Kodaira-Spencer collaboration was more than just a working relationship. The two had very different personalities which were complementary. Kodaira's shyness and reticence were balanced by Spencer's dynamism. In the world of university politics Spencer was able to exercise his talents on Kodaira's behalf, providing a protective environment in which Kodaira's mathematical talents could flourish."

I now begin to go into more detail concerning some of the selected papers. In [28] results of earlier papers are incorporated. Sheaves do not occur yet. The theory of harmonic integrals is used to study the vector space of all meromorphic differentials W of degree n on an n-dimensional Kähler manifold V_n of dimension n which satisfy $(W)+S \geq 0$, where S is a given divisor of V_n. The dimension of this space equals $\dim \mid K+S \mid +1$ if K is a canonical divisor, where $\mid K+S \mid$ is the complete linear system of divisors linearly equivalent to $K+S$ and is called the *adjoint system* of S. Several Riemann-Roch type formulas for $\dim \mid K+S \mid$ are proved. Following Severi, Kodaira introduces the numerical characteristic $a(V_n)$ by the formula

$$(1)\quad a(V_n)=g_n(V_n)-g_{n-1}(V_n)+\ldots+(-1)^{n-1}g_1(V_n),$$

where $g_i(V_n)$ is the dimension of the space of holomorphic differentials of degree i. He formulates a Riemann-Roch theorem for adjoint systems (assuming now that V_n is projective algebraic and E is a smooth hyperplane section for some embedding). He proves (Theorem 2.3.1 in [28])

$$(2)\qquad \dim \mid K+E \mid = a(V_n)+a(E)-1.$$

Not much later we would say that it is better to consider the holomorphic Euler number

$$(3)\qquad \chi(V_n)=\sum_{i=0}^{n}(-1)^i g_i(V_n)$$

($g_0=1$ if V_n is connected) and for a divisor D the number

$$(4)\qquad \chi(V_n,D)=\sum_{i=0}^{n}(-1)^i \dim H^i(V_n,\Omega(D)),$$

where $\Omega(D)$ is the sheaf of local meromorphic functions f with $(f)+D\geq 0$. Then $\dim H^0(V_n,\Omega(D))=\dim \mid D \mid +1$. By the Kodaira vanishing theorem [35] the spaces $H^i(V_n,\Omega(K+E))$ are zero for $i>0$. Hence $\dim \mid K+E \mid +1 = \chi(V_n,K+E)$, but by Serre duality $\chi(V_n,K+E) = (-1)^n\chi(V_n,-E)$ (true for any divisor E and for the individual terms in the alternating sum). Serre duality is mentioned in [34] and [35]. Therefore (2) becomes

$$(5)\qquad \chi(V_n,-E)=\chi(V_n)-\chi(E),$$

which follows from an easy exact sequence of sheaves.

But let us go back to [28]. Kodaira proves that $\dim \mid D+hE \mid$ is a polynomial $\nu(h,D)$ in h for large h (often called a Hilbert polynomial) and that $\nu(0,D)$ depends only on D. It is called the virtual dimension of $\mid D \mid$. Now we follow [31]. There are two distinct ways in which arithmetic genera may be defined. In the first place we may define the *arithmetic genus* $P_a(V_n)$ to be the virtual dimension $\nu(0,K)$ of $\mid K \mid$ increased by $1-(-1)^n$ and alternatively the *arithmetic genus* $p_a(V_n)$ by $(-1)^n\nu(0,0)$. In [31] the authors point out that $p_a(V_n)=P_a(V_n)$ has not been established before for $n\geq 5$. They prove it using sheaves. In [28] Kodaira showed $P_a(V_n)=a(V_n)$ in general and $P_a(V_n)=p_a(V_n)$ for a special class of varieties, including complete intersections in projective spaces. A little later we would say

$$\dim \mid D+hE \mid +1=\chi(V_n,D+hE)$$

for h large by the Kodaira vanishing theorem. But $\chi(V_n,D+hE)$ is a polynomial for all h. Therefore $\nu(0,D)+1=\chi(V_n,D)$ and

$$\nu(0,0)+1=(-1)^n(\nu(0,K)+1)$$

At the Institute for Advanced Study, August 1952. Left to right: F. Hirzebruch, Mrs. Kodaira, K. Kodaira.

by Serre duality. But this is the equality $p_a = P_a$. I pointed out "A little later we would say...." Here I must mention the papers [31]-[36] from which I could learn so much (of course, before the papers were written). These papers were all communicated to the *Proceedings of the National Academy of Sciences* by S. Lefschetz. Let us look briefly at [32] and [35].

In [32] Kodaira works with a compact complex analytic variety V of complex dimension n and a holomorphic line bundle F over V and studies the sheaf (faisceau, stack) $\Omega^p(F)$ over V of germs of holomorphic p forms with coefficients in F. I quote from [32]: "The faisceau $\Omega^p(F)$ introduced recently by D. C. Spencer and, independently, by J.-P. Serre turned out to be of importance to applications of faisceaux to the theory of compact analytic varieties. However, for these applications, we need a basic theorem to the effect that the cohomology groups $H^q(V;\Omega^p(F))$ of V with coefficients in $\Omega^p(F)$ have finite dimension. The purpose of the present short note is to give an outline of a proof of this basic theorem." Kodaira uses a Hermitian metric and has to generalize Hodge theory on Kähler manifolds to this more general case using the complex Laplace-Beltrami operator studied earlier by Garabedian and Spencer in the case where F is trivial. The Laplace-Beltrami operator is elliptic. Solution spaces are finite dimensional. With respect to this operator $H^q(V;\Omega^p(F))$ can be identified with the vector space $H^{p,q}(F)$ of all harmonic forms of type (p,q) on V with coefficients in F. In [35] Kodaira writes, "In the present note we shall prove by a differential-geometric method due to Bochner some sufficient conditions for the vanishing of $H^q(V;\Omega^p(F))$ in terms of the characteristic class of the bundle F." In particular he proves: If the characteristic class of F is positive in the sense of Kodaira (representable by a Kähler form), then $H^q(V,\Omega^n(F))$ and $H^{n-q}(V,\Omega^0(-F))$ (Serre duality) both vanish for $1 \le q \le n$. Bochner's papers (*Curvature and Betti numbers, I and II*) appeared in the *Annals of Mathematics* in 1948 and 1949. In [36] Kodaira and Spencer study the holomorphic Euler numbers

$$\text{(6)} \qquad \chi_V^p(F) = \sum_q (-1)^q \dim H^q(V;\Omega^p(F))$$

and prove the "form term formula"

$$\text{(7)} \quad \chi_V^p(F) = \chi_V^p(F-\{S\}) + \chi_S^p(F_S) + \chi_S^{p-1}(F_S - \{S\}_S)$$

(for a line bundle or divisor F and a smooth hypersurface S), which was very important for me when I studied the polynomial

$$\text{(8)} \qquad \chi_y(V,F) = \sum_{p=0}^{n} \chi_V^p(F) y^p$$

(also for a holomorphic vector bundle F), where the χ_y-genus $\chi_y(V)$ is the polynomial obtained if F is the trivial line bundle. These polynomials occur in the proof of my Riemann-Roch theorem.

The χ_y-genus is a generalization of the holomorphic Euler number $\chi(V)$ (for $y = 0$) and $\chi(V)$ equals $(-1)^n a(V) + 1$ if $n = \dim V$ (see (3)).

At this point let me emphasize the Paris-Princeton relations of the early 1950s. I recommend reading the letter of Serre to Borel of April 16, 1953 (published in Serre's *Collected Papers*, Vol. 1, No. 20, Springer-Verlag, 1986), and Serre's comments (Vol. 1, p. 588).

In my recent lecture "Learning Complex Analysis in Münster-Paris, Zürich and Princeton from 1945 to 1953" (Journée en l'Honneur d'Henri Cartan, June 14, 1997; *Gazette des Mathématiciens* **74** (1997), 27-39) I talk about Paris-Princeton on pp. 35-36.

In the introduction of my book I speak of four definitions of the arithmetic genus

$$p_a(V), P_a(V), a(V) = g_n - g_{n-1} + \ldots + (-1)^{n-1} g_1$$

and the Todd genus. The basic reference is J. A. Todd, *The arithmetical invariants of algebraic loci*, Proc. London Math. Soc. **43** (1937), 190-225, where Todd uses the characteristic classes K_i of Eger and Todd, which are $(2n-2i)$-dimensional cycles ($K_1 = K$), to express $(-1)^n P_a(V) + 1$ as a polynomial in the K_i. The proof relies on an unproved lemma of Severi from which Todd concludes that such polynomials must exist. He characterizes them by requesting that they give the correct values on the complete intersection of smooth hypersurfaces of degrees $n_1, n_2, \ldots, n_d$ in the projective space of dimension $2d$. Todd's formalism of his polynomials is very difficult to read. Kodaira ([28], (6.1.1)) gave a formula for the Todd polynomial which is close to my multiplicative sequences and stems from his careful analysis of Todd's paper. However, I do not remember whether I realized this in the old days. Clearly, the power series $(e^x - 1)/x$ is recognizable in his formula, as it is in Todd's formula (22), which can be interpreted as a formula for the arithmetic genus of the complete intersection V of smooth hypersurfaces of degrees $n_1, \ldots, n_d$ in $P_{2d}(\mathbb{C})$ involving the total Todd class of the normal bundle of V in $P_{2d}(\mathbb{C})$ and the total Todd class of $P_{2d}(\mathbb{C})$. Staying close to Todd's formalism, Kodaira proves that the Todd polynomial gives the arithmetic genus $P_a(V)$ for a class of varieties including complete intersections in projective spaces. At one point, relating the characteristic classes of the tangent bundle of a hypersurface to those of the ambient variety and of the normal bundle, he needs the help of S. S. Chern, who had just proved his "duality theorem" for Chern classes. Kodaira was aware of the fact that the K_i of Eger and Todd coincide up to the factor $(-1)^i$ with the Chern classes $c_i \in H^{2i}(V, \mathbb{Z})$. With the use of multiplicative sequences, the inductive proof for Kodaira's result that the Todd polynomials give the arithmetic genus on complete intersections became very simple. At a time when I had formulated the Riemann-Roch theorem but could not yet prove it, I also conjectured as a special case of the Riemann-Roch theorem that the polynomial $\chi_y(V)$ is the genus belonging to the power series

$$\frac{x(y+1)}{1-e^{-x(y+1)}} - xy = \frac{x}{f_y(x)}, \tag{9}$$

and I proved it for complete intersections. This leads me to the following story. (Compare my *Collected Papers*, Vol. 1, Commentaries, Springer-Verlag, 1987, p. 785.)

A. Weil wrote to Kodaira on October 22, 1953, asking in particular for the Hodge numbers of the complete intersection of two quadrics. Kodaira answered on November 4, 1953, explaining my result on the χ_y-genus of complete intersections by which all Hodge numbers of complete intersections are known. At the end of this letter Kodaira writes:

"Recently I could prove that every Hodge variety (i.e. a Kähler variety whose fundamental form $i\Sigma g_{\alpha\bar{\beta}} dz^\alpha \wedge dz^{\bar{\beta}}$ is homologous to an integral cocycle) is an algebraic variety imbedded in a projective space. I believe that my proof is correct; however, I am afraid that my result is too good. I would appreciate very much your comment on this result."

Here Kodaira announces one of his most famous and deep results ([37], communicated by S. Bochner on February 23, 1954, and [38]). I do not know Weil's answer. He must have reacted, because Kodaira wrote him on November 18, 1953, thanking him for a letter and explaining to him in all detail my formulas, which make the calculation of the $h^{p,q}$ of complete intersections more explicit and which are published in the *Proceedings of the International Congress of Mathematicians, Amsterdam, 1954* (my *Collected Papers*, Vol. 1, No. 12, formula (1)).

In the introduction to [38] Kodaira recalls that Hodge had introduced in 1951 the Kähler manifolds of special type and A. Weil had called them Hodge manifolds (A. Weil, *On Picard varieties*, Amer. J. Math. **74** (1952), 865-894). A. Weil proves theorems on Hodge manifolds and recalls Hodge's result that the Picard variety of a Hodge manifold is projective algebraic.

For the proof of his fundamental result Kodaira has to use results of earlier papers, for example, his vanishing theorem. He proves that for a Hodge variety V there exists a real (1, 1)-form β such that, for any complex line bundle F whose characteristic class contains a closed real (1, 1)-form $\gamma > \beta$, the holomorphic sections of F define a biregular mapping of V into a projective space (Theorem 3 in [38]).

Kodaira on Princeton campus, 1952.

Kodaira's fundamental theorem generalizes classical results characterizing those complex tori which are projective algebraic. He gives several applications. One was especially important for me. Section 18 of my book *Topological Methods in Algebraic Geometry* carries the title "Some fundamental theorems of Kodaira". I quote Theorem 18.3.1 (Kodaira):

"A complex analytic fiber bundle L over the projective algebraic manifold V with the projective space $P_r(\mathbb{C})$ as fiber and $\mathrm{PGL}(r+1, \mathbb{C})$ as structure group is itself a projective algebraic manifold."

This is used for the proof of my Riemann-Roch theorem, which was completed on December 10, 1953, and announced in the *Proceedings of the National Academy of Sciences* (communicated by S. Lefschetz on December 21, 1953). I had to reduce everything to complex split manifolds where the structural group is the triangular group contained in the general linear group. Then the arithmetic genus can be expressed by virtual signatures which (by the signature theorem as a consequence of Thom's cobordism theory) can be expressed by characteristic classes. But for certain inductive processes I had to stay in the category of projective algebraic manifolds. For a projective algebraic manifold the total space of the flag manifold bundle associated to the tangent bundle is a split manifold. It is projective algebraic by repeated applications of Kodaira's theorem 18.3.1. In my announcement I refer to Kodaira in footnote 9 ("Kodaira, K., not yet published"). I also needed results on the behavior of genera in fiber bundles. The best result is in Appendix Two (by A. Borel) of my book :

"Let $\xi = (E, B, F, \pi)$ be a complex analytic fiber bundle with connected structure group, where E, B, F are compact connected, and F is Kählerian. Let W be a complex analytic fiber bundle over B. Then $\chi_y(E, \pi^* W) = \chi_y(B, W)\chi_y(F)$."

The cooperation with A. Borel in Princeton was of great importance for Kodaira, Spencer, and me in learning characteristic classes and in many other ways, as can be seen, for example, by remarks of Kodaira in [38].

Of course, I am very proud to have one joint paper with Kodaira [41], which was published only in 1957, though I had announced the result already in my talk in Amsterdam in 1954 (loc. cit.). One of my main discoveries (standard joke) is the formula

$$\frac{x}{1-e^{-x}} = e^{x/2} \cdot \frac{x/2}{\sinh x/2},$$

which showed that the Todd genus is expressible by the first Chern class c_1 and the Pontryagin classes. The latter ones do not depend on the complex analytic structure. For a divisor D the number $\chi(V, D)$ depends on the cohomology class $d + c_1/2$ where d is the cohomology class of D and otherwise only on the oriented differentiable manifold V. This we used in [41]. This remark led to the introduction of the $\hat{A}$-genus which is defined for oriented differentiable manifolds. It equals $\chi(V, D)$ if $2d + c_1 = 0$. From here a new development started whose beginning for me was Atiyah's lecture at the Bonn *Arbeitstagung* in 1962, where it was conjectured that for a spin-manifold the $\hat{A}$-genus is the index of the Dirac operator. This was proved a little later by Atiyah and Singer as a special case of their general index theorem for linear elliptic operators. The index theorem also included my Riemann-Roch theorem as a special case even for complex manifolds (used by Kodaira in [60]). The paper [41] was for me a sign of the importance of the $\hat{A}$-genus.

One more word about the χ_y-genus. If S and T are smooth hypersurfaces in the projective algebraic manifold V and if the divisor $S + T$ is also represented by a smooth hypersurface such that the intersections $S \cdot T$ and $S \cdot T \cdot (S + T)$ are transversal and hence smooth, then

$$\text{(10)} \quad \chi_y(S+T) = \chi_y(S) + \chi_y(T) + (y-1)\chi_y(S \cdot T) - y\chi_y(S \cdot T \cdot (S+T)),$$

which I deduced from the four-term formula (7). The functional equation (10) is also true for the χ_y-genus in terms of characteristic classes using the power series (9). It follows from a corresponding elementary functional equation of $f_y(x)$. Kodaira often proved and used (10) for $y = 0$. (See his concept of A-functional in [28], Sections 2.7 and 6.3). It is clear that (10) is useful for the study of complete intersections (inductive proofs).

Kodaira's and Spencer's joint work on deformations of complex analytic structures ([43], [48], and several other papers) is perhaps the greatest achievement of their cooperation. It is an enlightment to read in the introduction of [43], "... we *define* a differentiable family of compact complex structures (manifolds) as a fiber space $\mathcal{V}$ over a connected differentiable manifold M whose structure is a mixture of differentiable and complex structures." Kodaira and Spencer introduce the sheaf Θ on $\mathcal{V}$, the corresponding sheaf of cohomology $\mathcal{H}^1(\Theta)$ on M, and a homomorphism (the Kodaira-Spencer map) $\rho : T_M \to \mathcal{H}^1(\Theta)$ where T_M is the sheaf of germs of differentiable vector fields of M. The complex structure V_t, $t \in M$, is independent of t if and only if ρ vanishes. By restricting Θ to V_t (fixed fiber over $t \in M$) they obtain the homomorphism $\rho_t : (T_M)_t \to H^1(V_t, \Theta_t)$ of Frölicher and Nijenhuis, where $(T_M)_t$ is the tangent space of M at t and Θ_t is the sheaf of germs of holomorphic vector fields on V_t. The vanishing of ρ_t for all t does not imply the vanishing of ρ, as "jumps" show, for example, from the smooth quadric surface to the singular quadric with a node and the node blown up (Atiyah, Brieskorn). Now I quote again from the introduction of [43]: "Next we extend Riemann's concept of number of moduli to higher dimensional complex manifolds (Section 11). The main point here is to avoid the use of the concept of the space of moduli of complex manifolds which cannot be defined in general for higher dimensional manifolds (Section 14, (γ)). Moreover, a necessary condition for the existence of a number $m(V_0)$ of moduli of a complex manifold V_0 is that $H^1(V_0, \Theta_0)$ contain only one deformation space; hence $m(V_0)$ is not defined for all compact complex manifolds...." Kodaira and Spencer find it surprising that $m(V_0) = \dim H^1(V_0, \Theta_0)$ for so many examples and consider a better understanding of this fact as the main problem in deformation theory. I do not want to say more about their deformation theory. Surveys are in Baily's preface to Kodaira's *Collected Works* and in the introduction by K. Ueno and T. Shioda to the volume *Complex Analysis and Algebraic Geometry* (Iwanami Shoten, Publishers, and Cambridge University Press, 1977), dedicated to Kodaira on the occasion of his sixtieth birthday. Anyhow, this report is personal and concerns those aspects of Kodaira's work related to my own. Hence, for lovers of Riemann-Roch, I write what this theorem gives for Θ_0 in dimension $n = 1$ (Riemann) and $n = 2$ (Max Noether).

$n = 1$:

$$\dim H^0(V_0, \Theta_0) - \dim H^1(V_0, \Theta_0) = 3 - 3g.$$

The number of moduli equals $3g - 3+$ dimension of the group of automorphisms of V_0.

$n = 2$:

$$\dim H^0(V_0, \Theta_0) - \dim H^1(V_0, \Theta_0) + \dim H^2(V_0, \Theta_0) = -10\chi(V_0) + 2c_1^2,$$

where χ is the holomorphic Euler number.

Let us remark that by Serre duality

$$H^i(V_0, \Theta_0) \simeq H^{n-i}(V_0, \Omega^1(K)),$$

which, for $n = 1$ and $i = 1$, is the isomorphism to the space of holomorphic quadratic differentials (see the obituary by Spencer). We have

$$\chi(V_0, \Theta_0) = (-1)^n \chi^1_{V_0}(K).$$

These numbers can be calculated by the Riemann-Roch theorem as linear combinations of Chern numbers. For a Kähler manifold with trivial canonical bundle, $\dim H^i(V_0, \Theta_0)$ equals the Hodge number $h^{1,n-i}$. For a $K3$-surface we have $h^{1,1} = 20$. Kodaira and Spencer discuss many more examples. For the complex projective space $P_n(\mathbb{C})$ we have $\dim H^1(P_n(\mathbb{C}), \Theta_0) = 0$ in agreement with the result in [41].

With the exception of three papers, the whole Volume III of Kodaira's *Collected Works* is concerned with complex analytic surfaces. His work in this area is overwhelming. He can use his earlier papers on complex manifolds and on deformations. I have used the papers in Volume III very often. Looking, for example, at my joint paper with A. Van de Ven, *Hilbert modular surfaces and the classification of algebraic surfaces* (Invent. Math. 23 (1974), 1–29), I find that we used the following:

1. *Rough classification of surfaces, Kodaira dimension* ([68], Theorem 55). Kodaira proves that the compact complex surfaces free from exceptional curves can be divided into seven classes. Class 5 comprises the minimal algebraic surfaces of general type. Class 7 surfaces are mysterious surfaces with first Betti number equal to 1. Van de Ven and I specialize Kodaira's classification to algebraic surfaces, where this classification in broad outline was known to the Italian school, but many of the proofs are due to Kodaira.
2. *Kodaira's proof of Castelnuovo's criterion for the rationality of algebraic surfaces.*
3. *Study of elliptic surfaces, their multiple fibers, and a formula for the canonical divisor.*
4. *Classification of the exceptional fibers in elliptic surfaces.*
5. *The fact that all K3-surfaces are homeomorphic and hence simply connected.* Kodaira proves more ([60], Theorem 13): Every $K3$-surface is a deformation of a nonsingular quartic surface in a projective 3-space.

The surfaces in Class 7 are also called VII_0-surfaces ([60], Theorem 21). Masahisa Inoue (*New surfaces with no meromorphic functions II*, in the volume dedicated to Kodaira's sixtieth birthday) has constructed such surfaces using my resolution of the cusp singularities of Hilbert modular surfaces. Such a surface has only finitely many curves on it. They are rational and arranged in two disjoint cycles.

Now a last case where a paper of Kodaira was especially close to my interest. In [64] he constructed algebraic surfaces with positive signature whose total spaces are differentiable fiber bundles with compact Riemann surfaces as base and fiber. In the early 1950s we did not know a single surface with signature greater than 1 and often talked about it at Princeton. How the situation developed over the years can be seen, for example, in the book *Geradenkonfigurationen und Algebraische Flächen* (by Gottfried Barthel, Thomas Höfer, and me, Vieweg, 1987). Also, Kodaira's surfaces give examples in which the signature of the total space of a fibration is not equal to the product of the signatures of base and fiber. The multiplicativity of the signature in fiber bundles (of oriented manifolds) was proved by S. S. Chern, J.-P. Serre, and me under the assumption that the fundamental group of the base operates trivially on the real cohomology of the fiber (*On the index of a fibered manifold*, Proc. Amer. Math. Soc. 8 (1957), 587–596).

Grauert, Hirzebruch, Remmert return to Germany after the Conference on Analytic Functions at the Institute for Advanced Study, September 1957. From left to right: F. Hirzebruch, Joan Frankel (Mrs. Theodore Frankel), D. C. Spencer, K. Kodaira, A. Borel, W.-L. Chow, H. Grauert. Foreground: R. Remmert.

Far from attempting to give a thorough appreciation of Kodaira's great mathematical work, I wanted to show how much I am indebted to him and where our mathematical lives crossed.

Selected Papers from Kodaira's Collected Works

[22] *The theorem of Riemann-Roch on compact analytic surfaces*, Amer. J. Math. **73** (1951), 813–875.

[24] *The theorem of Riemann-Roch for adjoint systems on 3-dimensional algebraic varieties*, Ann. of Math. **56** (1952), 298–342.

[28] *The theory of harmonic integrals and their applications to algebraic geometry*, Work done at Princeton University, 1952.

[31] *On arithmetic genera of algebraic varieties* (collaborated with D. C. Spencer), Proc. Nat. Acad. Sci. U.S.A. **39** (1953), 641–649.

[32] *On cohomology groups of compact analytic varieties with coefficients in some analytic faisceaux*, Proc. Nat. Acad. Sci. U.S.A. **39** (1953), 865–868.

[33] *Groups of complex line bundles over compact Kähler varieties* (collaborated with D. C. Spencer), Proc. Nat. Acad. Sci. U.S.A. **39** (1953), 868–872.
[34] *Divisor class groups on algebraic varieties* (collaborated with D. C. Spencer), Proc. Nat. Acad. Sci. U.S.A. **39** (1953), 872–877.
[35] *On a differential-geometric method in the theory of analytic stacks*, Proc. Nat. Acad. Sci. U.S.A. **39** (1953), 1268–1273.
[36] *On a theorem of Lefschetz and the Lemma of Enriques-Severi-Zariski* (collaborated with D. C. Spencer), Proc. Nat. Acad. Sci. U.S.A. **39** (1953), 1273–1278.
[37] *On Kähler varieties of restricted type*, Proc. Nat. Acad. Sci. U.S.A. **40** (1954), 313–316.
[38] *On Kähler varieties of restricted type (an intrinsic characterization of algebraic varieties)*, Ann. of Math. **60** (1954), 28–48.
[39] *Some results in the transcendental theory of algebraic varieties*, Proc. Internat. Congr. Math., Vol. III, 1954, pp. 474–480.
[40] *Characteristic linear systems of complete continuous systems*, Amer. J. Math. **78** (1956), 716–744.
[41] *On the complex projective spaces* (collaborated with F. Hirzebruch), J. Math. Pures Appl. **36** (1957), 201–216.
[42] *On the variation of almost-complex structure* (collaborated with D. C. Spencer), Algebraic Geometry and Topology, Princeton Univ. Press, 1957, pp. 139–150.
[43] *On deformations of complex analytic structures*, I–II (collaborated with D. C. Spencer), Ann. of Math. **67** (1958), 328–466.
[45] *A theorem of completeness for complex analytic fibre spaces* (collaborated with D. C. Spencer), Acta Math. **100** (1958), 281–294.
[48] *On deformation of complex analytic structures*, III, *Stability theorems for complex structures* (collaborated with D. C. Spencer), Ann. of Math. **71** (1960), 43–76.
[52] *On compact complex analytic surfaces*, I, Ann. of Math. **71** (1960), 111–152.
[55] *On stability of compact submanifolds of complex manifolds*, Amer. J. Math. **85** (1963), 79–94.
[56] *On compact analytic surfaces*, II–III, Ann of Math. **77** (1963), 563–626; **78** (1963), 1–40.
[60] *On the structure of compact complex analytic surfaces*, I, Amer. J. Math. **86** (1964), 751–798.
[63] *On the structure of compact complex analytic surfaces*, II, Amer. J. Math. **88** (1966), 682–721.
[64] *A certain type of irregular algebraic surfaces*, J. Analyse Math. **19** (1967), 207–215.
[66] *On the structure of compact complex analytic surfaces*, III, Amer. J. Math. **90** (1968), 55–83.
[68] *On the structure of complex analytic surfaces*, IV, Amer. J. Math. **90** (1968), 1048–1066.

WOLF PRIZE CEREMONY
THE KNESSET
JERUSALEM, MAY 12, 1988
RESPONSE
FOR LARS HÖRMANDER AND FRIEDRICH HIRZEBRUCH
BY

FRIEDRICH HIRZEBRUCH

Mr. President, Ladies and Gentlemen:

For Lars Hörmander and myself, it was a great surprise when we learned at the end of January that the International Committee of the Wolf Foundation had chosen us as recipients of the 1988 Prize. We are delighted and deeply honoured for this award. We think with gratitude of Ricardo Wolf and thank the members of his Foundation for all their work and hospitality. We are happy now to belong to an impressive series of prize-winners but, at the same time, also a little ashamed because there are many excellent mathematicians who would have deserved the prize just as much.

The prize is a recognition for past work, which extended over a period of several decades; but we hope still to have years ahead of us, during which it will spur us on to further work in the service of mathematics and mathematicians.

Today we are seeing an impressive ceremony in the Knesset; the 12th of May, 1988, and the week we are spending in Israel will certainly remain in our memories as one of the outstanding events of our lives.

We have also gained an impression of the great achievements which the people of Israel have accomplished. We hope with all our hearts that this country and its neighbours may be granted a peaceful future.

Finally, let me add something which concerns only myself. As a professor at the University of Bonn, I am one of the successors of the famous mathematicians Felix Hausdorff and Otto Toeplitz. Hausdorff committed suicide in 1942, together with his wife, when deportation to a concentration camp was imminent. Toeplitz emigrated to Israel with his family in 1939 and died there the following year. The memory of these mathematicians is with me always on this day.

Friedrich Hirzebruch

Honorary President of the ICM'98

Many thanks for the honour just bestowed on me. At the closing session in Zürich, I invited the congress to Berlin on behalf of the German Mathematical Society (DMV). The Organizing Committee in Berlin under Professor Martin Grötschel has worked hard and very efficiently using the most modern developments of electronic communication. As honorary president of this committee I had to do very little, but I had ample chance to admire their work. I wish to thank Professor Grötschel and all members of his committee very much, especially for making the honorary presidency so easy for me. In 1904 the Congress was in Heidelberg, supported by Kaiser Wilhelm and the Grand Duke of Baden. This time our support comes from the Federal Republic of Germany and the Land Berlin. We are grateful for the generous support. I welcome Staatssekretär Wilhelm Staudacher, who will read a message of the President of Germany, who agreed to be the protector of this Congress. The Federal support comes through the Minister of Education, Science, Research, and Technology. I welcome the Minister Dr. Jürgen Rüttgers. The Land Berlin is represented by its Governing Mayor Eberhard Diepgen. We thank the Technical University and its president Professor Hans-Jürgen Ewers for letting us use the University as venue of the Congress. In 1990 the German Mathematical Society (DMV) celebrated its 100th anniversary. Our application to issue a special postage stamp on this event was turned down. We are all the happier that for this congress a special stamp will be issued and Staatssekretär Hansgeorg Hauser will present it to us.

I mentioned the 100th anniversary of the DMV. Its first president was Georg Cantor, the founder of set theory. He was an ardent fighter for the establishment of the International Mathematical Congress. From the founding years of the DMV up to Nazi times, mathematics in Germany was leading internationally. Among the presidents of the Society in this period were Felix Klein, Alexander Wilhelm von Brill, Max Noether, David Hilbert, Alfred Pringsheim, Friedrich Engel, Kurt Hensel, Edmund Landau, Erich Hecke, Otto Blumenthal, and Hermann Weyl.

Alfred Pringsheim died in Zürich in 1941 at the age of 90 after having escaped from Germany. Edmund Landau lost his chair in Göttingen in 1934. Otto Blumenthal was deported to the concentration camp Theresienstadt, where he died in 1944. Hermann Weyl, president of our society in 1932, emigrated to the United States in 1933. He worked at the Institute for Advanced Study in Princeton together with Albert Einstein, Kurt Gödel, John von Neumann, who were all members of our society.

David Hilbert died in Göttingen in 1943. Hermann Weyl wrote an obituary published in the middle of the war in Great Britain and the United States. I quote: "Not until many years after the first world war, after Felix Klein had gone and Richard Courant had succeeded him, towards the end of the sadly brief period of the German Republic, did Klein's dream of the Mathematical Institute at Göttingen come true. But soon the Nazi storm broke and those who had laid the plans and who taught there besides Hilbert where scattered over the earth, and the years after 1933 became for Hilbert years of ever deepening tragic loneliness."

To those "scattered over the earth" belongs Emmy Noether, the famous Göttingen mathematician, daughter of Max Noether, president of the German Mathematical Society in 1899.

It is not possible for me here to analyse the behaviour of the DMV and its members during the Nazi time, or its reaction to the Nazi time after the war. When we began to prepare the present congress, it was clear for us that we "must not forget." My generation should be unable to forget. Many of my age have good friends all over the world where parents or other family members were killed in Auschwitz. We must teach the next generation "not to forget." The German Mathematical Society has announced a special activity during this congress to honour the memory of the victims of the Nazi terror. I read from this announcement and ask you to participate:

In 1998, the ICM returns to Germany after an intermission of 94 years. This long interval covers the darkest period in German history. Therefore, the DMV wants to honour the memory of all those who suffered under the Nazi terror. We shall do this in the form of an exhibition presenting the biographies of 53 mathematicians from Berlin who were victims of the Nazi regime between 1933 and 1945. The fate of this small group illustrates painfully well the personal sufferings and the destruction of scientific and cultural life; it also sheds some light on the instruments of suppression and the mechanism of collaboration.

In addition, there will be a special session entitled "Mathematics in the Third Reich and Racial and Political Persecution" with two talks given by Joel Lebowitz (Rutgers University), "Victims, Oppressors, Activists, and Bystanders: Scientists' Response to Racial and Political Persecution," and Herbert Mehrtens (Technische Hochschule Braunschweig), "Mathematics and Mathematicians in Nazi Germany. History and Memory."

Of the 53 mathematicians from Berlin honoured in the exhibition, three are here with us as guests of the Senate of Berlin and the German Mathematical Society. I greet them with pleasure and thanks. They are

Michael Golomb, United States,

Walter Ledermann, Great Britain,

Bernhard Neumann, Australia.

The last student of the famous Berlin mathematician Issai Schur is Feodor Theilheimer who lives in the United States. It is a pleasure to welcome his daughter Rachel Theilheimer. Schur and Theilheimer both belong to the 53 mathematicians honoured in the exhibition.

In addition, I welcome

Franz Alt,

driven away from Vienna, who emigrated to the United States and is with us today as a guest of the DMV.

In 1961 I became president of the DMV as successor of Ott-Heinrich Keller from Halle in the German Democratic Republic (DDR). The wall had just been built. The Mathematical Society of the DDR was founded. In 1990 I was president again and had to work for the reintegration of the DDR society into the DMV.

We look hopefully into the future and are happy as the reunited DMV to host the congress.

Progress and future of mathematics are represented by the laureates of the Fields medal and the Nevanlinna prize. It will be a great honour and pleasure for me to hand over the Fields medals to the winners.

Lars Hörmander
Curriculum Vitae

Born: January 24, 1931, Mjällby, Blekinge län, Sweden

University Education

University of Lund, Sweden, 1948–1955.
Fil.kand. (B.Sc.) 1949, Fil.mag. (M.Sc.) 1950, Fil.lic. 1951, Fil.dr. (Ph.D.) 1956 (1955).

Employment

Docent at the University of Lund, 1956;
Research Associate at University of Chicago and New York University, 1956;
Professor at the University of Stockholm, 1957–1964;
Member of Institute for Advanced Study, Princeton, 1960–1961, Spring 1971 and 1977–1978;
Professor at Stanford University, 1963–1964;
Professor at Institute for Advanced Study, Princeton, 1964–1968;
Professor at the University of Lund, 1968–1995;
Professor Emeritus at the University of Lund since 1996;
Visiting Professor at New York University, Fall 1970;
Visiting Professor at Stanford University, Summer Quarters 1967, 1971, 1977, 1982;
Acting Director of Institute Mittag-Leffler in Stockholm, Fall 1974;
Director of Institute Mittag-Leffler, 1984–1986;
Visiting Professor at University of California, San Diego, Winter 1990.

Professional Honors and Positions

Awarded the Fields Medal at the International Congress of Mathematicians in Stockholm, 1962;
Awarded the Celsius Medal in Uppsala, 1970;
Awarded the Wolf Prize in Jerusalem, 1988;
Awarded Fysiografiska Sällskapets Minnesmedalj in Lund, 1990;
Member of the American Academy of Arts and Sciences, 1967;
Member of Kungliga Svenska Vetenskapsakademien, 1968;
Foreign Member of Danske Videnskabernes Selskab, 1969;
Member of Kungliga Fysiografiska Sällskapet i Lund, 1969;
Foreign Member of Accad. di Szienze di Palermo, 1972;
Foreign Associate of National Academy of Sciences, Washington, 1977;
Foreign Member of Suomalainen Tiedeakatemia, 1979;
Member of Academia Europaea, 1990;
Foreign Member of Accademia Nazionale dei Lincei, 1990;
Doctor h.c. at Université de Paris-Sud, 1982;

Chairman of the program committee of the International Mathematical Union 1971–74;
Vice President of the International Mathematical Union, 1987–1990.

Bibliography

1953

[1] *Uniqueness theorems and estimates for normally hyperbolic partial differential equations of the second order*, C. R. 12e Congr. Math. Scand. Lund, 105–115.

1954

[2] *A new proof and a generalization of an inequality of Bohr*, Math. Scand. **2**, 33–45.

[3] *On a theorem of Grace*, Math. Scand. **2**, 55–64.

[4] *Sur la fonction d'appui des ensembles convexes dans un espace localement convexe*, Ark. för Mat. **3**, 181–186.

1955

[5] *La transformation de Legendre et le théorème de Paley-Wiener*, C. R. Acad. Sci. Paris **240**, 392–395.

[6] *Some inequalities for functions of exponential type*, Math. Scand. **3**, 21–27.

[7] *On the theory of general partial differential operators*, Acta Math. **94**, 161–248.

1956

[8] (with J. L. Lions) *Sur la complétion par rapport à une intégrale de Dirichlet*, Math. Scand. **4**, 259–270.

1957

[9] *Local and global properties of fundamental solutions*, Math. Scand. **5**, 27–39.

1958

[10] *On interior regularity of the solutions of partial differential equations*, Comm. Pure Appl. Math. **11**, 197–218.

[11] *On the regularity of the solutions of boundary problems*, Acta Math. **99**, 225–264.

[12] *Definitions of maximal differential operators*, Ark. för Mat. **3**, 501–504.

[13] *Differentiability properties of solutions of systems of differential equations*, Ark. för Mat. **3**, 527–535.

[14] *On the division of distributions by polynomials*, Ark. för Mat. **3**, 555–568.

[15] *On the uniqueness of the Cauchy problem*, Math. Scand. **6**, 213–225.

1959

[16] *On the uniqueness of the Cauchy problem II*, Math. Scand. **7**, 177–190.

1960

[17] *Differential operators of principal type*, Math. Ann. **140**, 124–146.

[18] *Differential equations without solutions*, Math. Ann. **140**, 169–173.

[19] *Null solutions of partial differential equations*, Arch. Rat. Mech. Anal. **4**, 255–261.

[20] *Estimates for translation invariant operators in L^p spaces*, Acta Math. **104**, 93–140.

1961

[21] *On existence of solutions of partial differential equations*, in Partial Differential Equations and Continuum Mechanics (Madison), pp. 233–240.

[22] *Hypoelliptic convolution equations*, Math. Scand. **9**, 178–184.

[23] *Hypoelliptic differential operators*, Ann. Inst. Fourier **11**, 477–492.

[24] *Weak and strong extensions of differential operators*, Comm. Pure Appl. Math. **14**, 371–379.

1962

[25] *On the range of convolution operators*, Ann. Math. **76**, 148–170.

[26] *Differential operators with nonsingular characteristics*, Bull. Amer. Math. Soc. **68**, 354–359.

[27] *Existence, uniqueness and regularity of solutions of linear differential equations*, in Proceedings of the International Congress of Mathematicians, Stockholm, pp. 339–346.

1963

[28] *Supports and singular supports of convolutions*, Acta Math. **110**, 279–302.

[29] *Linear partial differential operators, Grundlehren d. Math. Wiss. 116* (Springer-Verlag), 1963, 1964, 1969, Russian translation (MIR) 1965.

1964

[30] *L^2 estimates and existence theorems for the $\bar{\partial}$ operator*, International Colloquium on Differential Analysis, Bombay, 65–79.

[31] *The Frobenius–Nirenberg theorem*, Ark. för Mat. **5**, 425–432.

[32] (with L. Gårding) *Strongly subharmonic functions*, Math. Scand. **15**, 93–96.

1965

[33] *L^2 estimates and existence theorems for the $\bar{\partial}$ operator*, Acta Math. **113**, 89–152.

[34] *Pseudo-differential operators*, Comm. Pure Appl. Math. **18**, 501–517.

1966

[35] *Pseudo-differential operators and non-elliptic boundary problems*, Ann. Math. **83**, 129–209.

[36] *An introduction to complex analysis in several variables* (D. van Nostrand Publ. Co.), Russian translation (MIR) 1968, Japanese translation 1973. Second extended edition 1973 and third extended edition (North-Holland) 1990.

[37] *Pseudo-differential operators and hypoelliptic equations*, Amer. Math. Soc. Symp. Sing. Int. Op. **10**, 138–183.

[38] *On the Riesz means of spectral functions and eigenfunction expansions for elliptic differential operators*, The Belfer Graduate School Science Conference Nov. 1966, 1969, pp. 155–202.

1967

[39] *L^p estimates for (pluri-)subharmonic functions*, Math. Scand. **20**, 65–78.

[40] *Generators for some rings of analytic functions*, Bull. Amer. Math. Soc. **73**, 943–949.

[41] *Hypoelliptic second order differential equations*, Acta Math. **119**, 147–171.

1968

[42] *Convolution equations in convex domains*, Inv. Math. **4**, 306–317.

[43] (with J. Wermer) *Uniform approximation on compact sets in* $\mathbf{C}^n$, Math. Scand. **23**, 5–21.

[44] *On the characteristic Cauchy problem*, Ann. Math. **88**, 341–370.

[45] *The spectral function of an elliptic operator*, Acta Math. **121**, 193–218.

1969

[46] *The Cauchy problem for differential equations with constant coefficients*, Springer Lecture Notes in Math. **103**, 60–71.

[47] *On the singularities of solutions of partial differential equations*, in International Conference on Functional Analysis and Related Topics, Tokyo, pp. 31–40.

[48] *On the index of pseudodifferential operators*, Elliptische Differentialgleichungen II, Schriftenreihe der Inst. für Math., Deutsche Akad. d. Wiss. zu Berlin, pp. 127–146.

1970

[49] *On the singularities of solutions of partial differential equations*, Comm. Pure Appl. Math. **23**, 329–358.

[50] *Linear differential operators*, Actes Congrès Int. Math., 1:121–133.

1971

[51] *The calculus of Fourier integral operators*, Prospects in Math., Ann. Math. Studies **70**, 33–57.

[52] *Fourier integral operators I*, Acta Math. **127**, 79–183.

[53] *On the L^2 continuity of pseudo-differential operators*, Comm. Pure Appl. Math. **24**, 529–535.

[54] *Uniqueness theorems and wave front sets for solutions of linear differential equations with analytic coefficients*, Comm. Pure Appl. Math. **24**, 671–704.

[55] *A remark on Holmgren's uniqueness theorem*, J. Diff. Geom. **6**, 129–134.

[56] *On the existence and the regularity of solutions of linear pseudo-differential equations*, L'Ens. Math. **17**, 99–163.

1972

[57] (with J. J. Duistermaat) *Fourier integral operators II*, Acta Math. **128**, 183–269.

[58] *On the singularities of solutions of partial differential equations with constant coefficients*, Israel J. Math. **13**, 82–105.

1973

[59] *Oscillatory integrals and multipliers on FL^p*, Ark. för Mat. **11**, 1–11.

[60] *On the existence of real analytic solutions of partial differential equations with constant coefficients*, Inv. Math. **21**, 151–182.

[61] *Lower bounds at infinity for solutions of differential equations with constant coefficients*, Israel J. Math. **16**, 103–116.

1974

[62] *The spectral analysis of singularities*, Proc. Fifth National Math. Conf., Shiraz, Iran, 129–139.

[63] *Non-uniqueness for the Cauchy problem*, Springer Lecture Notes in Math. **459**, 36–72.

1975

[64] *A class of hypoelliptic pseudodifferential operators with double characteristics*, Math. Ann. **217**, 165–188.

1976

[65] (with S. Agmon) *Asymptotic properties of solutions of differential equations with simple characteristics*, J. d'Analyse Math. **30**, 1–38.

[66] *The existence of wave operators in scattering theory*, Math. Z. **146**, 69–91.

[67] *The boundary problems of physical geodesy*, Arch. Rat. Mech. Anal. **62**, 1–52.

1977

[68] *The Cauchy problem for differential equations with double characteristics*, J. d'Analyse Math. **32**, 118–196.

1978

[69] (with A. Melin) *Free systems of vector fields*, Ark. för Mat. **16**, 83–88.

[70] *Propagation of singularities and semiglobal existence theorems for (pseudo-) differential operators of principal type*, Ann. Math. **108**, 569–609.

1979

[71] *Spectral analysis of singularities*, Seminar on the Sing. of Sol. of Diff. Eq., Ann. Math. Studies **91**, 3–49.

[72] *Subelliptic operators*, Seminar on the Sing. of Sol. of Diff. Eq., Ann. Math. Studies **91**, 127–208.

[73] *The Weyl calculus of pseudo-differential operators*, Comm. Pure Appl. Math. **32**, 359–443.

[74] *A remark on singular supports of convolutions*, Math. Scand. **45**, 50–54.

[75] *On the asymptotic distribution of eigenvalues of pseudodifferential operators in* $\mathbb{R}^n$, Ark. för Mat. **17**, 297–313.

1980

[76] (with D. Gilbarg) *Intermediate Schauder estimates*, Arch. Rat. Mech. Anal. **74**, 297–318.

[77] *On the subelliptic test estimates*, Comm. Pure Appl. Math. **33**, 339–363.

[78] *On the asymptotic distribution of eigenvalues*, in A Tribute to Åke Pleijel, Uppsala, pp. 146–154.

1981

[79] *Théorie de la diffusion à courte portée pour des opérateurs à caractéristiques simples*, Sem. Goulaouic, Meyer, Schwartz, 1980–1981 Exp. XIV, pp. 1–18.

[80] *Pseudo-differential operators of principal type*, Nato Adv. Study Inst. on Sing. in Boundary Value Problems (Reidel), pp. 69–96.

[81] *Symbolic calculus and differential equations*, Proc. 18th Scand. Congress of Mathematicians Århus 1980 (Birkhäuser), pp. 56–81.

1983

[82] *L^2 estimates for Fourier integral operators with complex phase*, Ark. för Mat. **21**, 283–307.

[83] *Uniqueness theorems for second order elliptic differential equations*, Comm. Partial Diff. Eq. **8**, 21–64.

[84] *The analysis of linear partial differential operators. I. Distribution theory. Grundlehren d. Math. Wiss. 256* (Springer-Verlag), Russian translation (MIR) 1986. Second expanded edition and Springer Study edition 1990.

[85] *The analysis of linear partial differential operators. II. Differential operators with constant coefficients. Grundlehren d. Math. Wiss. 257* (Springer-Verlag), Russian translation (MIR) 1986. Second revised printing 1990.

1985

[86] *Between distributions and hyperfunctions*, Astérisque **131**, 89–106.

[87] *On the Nash–Moser implicit function theorem*, Ann. Acad. Sci. Fenn. **10**, 355–359.

[88] *The analysis of linear partial differential operators. III. Pseudo-differential operators. Grundlehren d. Math. Wiss. 274* (Springer-Verlag), Russian translation (MIR) 1987.

[89] *The analysis of linear partial differential operators. IV. Fourier integral operators. Grundlehren d. Math. Wiss. 275* (Springer-Verlag), Russian translation (MIR) 1988.

[90] *The propagation of singularities for solutions of the Dirichlet problem*, Proc. Amer. Math. Soc. Symp. in Pure Math. **43**, 157–165.

1986

[91] *Differential operators of principal type and scattering theory*, in Proceedings of the 1982 Changchun Symposium on Differential Geometry and Differential Equations (Science Press), pp. 113–184.

[92] *On Sobolev spaces associated with some Lie algebras*, in Current Topics in Partial Differential Equations (Kinokuniya), pp. 261–287.

1987

[93] *The lifespan of classical solutions of nonlinear hyperbolic equations*, Springer Lecture Notes in Math. **1256**, 214–280.

[94] *Remarks on the Klein–Gordon equation*, Journées Equations aux dérivées partielles Saint-Jean-de-Monts, pp. I-1–I-9.

[95] *Nonlinear hyperbolic differential equations. Lectures 1986–1987*, (275 pages).

1988

[96] *L^1, L^∞ estimates for the wave operator*, Analyse Mathématique et Applications (Gauthier-Villars), pp. 211–234.
[97] *Pseudo-differential operators of type 1,1*, Comm. Partial Diff. Eq. **13**, 1085–1111.

1989

[98] (with R. Sigurdsson) *Limit sets of plurisubharmonic functions*, Math. Scand. **65**, 308–320.
[99] *On the fully nonlinear Cauchy problem with small data*, Bol. Soc. Brasil. Mat. **20**, 1–27.
[100] *Continuity of pseudo-differential operators of type 1,1*, Comm. Partial Diff. Eq. **14**, 231–243.

1990

[101] *The Nash–Moser theorem and paradifferential operators*, Analysis et cetera (Academic Press), pp. 429–449.
[102] (with G. Grubb) *The transmission property*, Math. Scand. **67**, 273–289.
[103] *A remark on the characteristic Cauchy problem*, J. Funct. Anal. **93**, 270–277.

1991

[104] *On the fully nonlinear Cauchy problem with small data. II*, Microlocal analysis and nonlinear waves (Springer-Verlag), pp. 51–81, IMA Volumes in Mathematics and its Applications, Vol. 30.
[105] *Quadratic hyperbolic operators*, Springer Lecture Notes in Mathematics **1495**, 118–160, CIME conference July 1989.
[106] *A uniqueness theorem of Beurling for Fourier transform pairs*, Ark. för Mat. **29**, 237–240.

1992

[107] *The wave front set of the fundamental solution of a hyperbolic operator with double characteristics*, J. d'Analyse Math. **59**, 1–36.
[108] *A uniqueness theorem for second order hyperbolic differential equations*, Comm. Partial Diff. Eq. **17**, 699–714.

1993

[109] *Hyperbolic systems with double characteristics*, Comm. Pure Appl. Math. **46**, 261–301.

[110] (with Bo Bernhardsson) *An extension of Bohr's inequality*, in Boundary Value Problems for Partial Differential Equations and Applications (J.-L. Lions and C. Baiocchi, eds.) (Masson), pp. 179–194.

[111] *Remarks on Holmgren's uniqueness theorem*, Ann. Inst. Fourier Grenoble **43**, 1223–1251.

1994

[112] (with A. Melin) *A remark on perturbations of compact operators*, Math. Scand. **75**, 255–262.

[113] *Notions of convexity* (Birkhäuser).

1995

[114] *Symplectic classification of quadratic forms, and general Mehler formulas*, Math. Z. **219**, 413–449.

1996

[115] *On the solvability of pseudodifferential equations*, Structure of Solutions of Differential Equations (M. Morimoto and T. Kawai, eds.) (World Scientific), pp. 183–213.

1997

[116] *Nonlinear hyperbolic differential equations* (Springer-Verlag), Mathématiques & Applications 26.

[117] *On the uniqueness of the Cauchy problem under partial analyticity assumptions*, Geometrical Optics and Related Topics (F. Colombini and N. Lerner, eds.) (Birkhäuser), pp. 179–219.

[118] *Remarks on the Klein–Gordon and Dirac equations*, Contemporary Math. **205**, 101–125.

[119] *On the Legendre and Laplace transformations*, Ann. Scuola Norm. Sup. Pisa **25**, 517–568.

1998

[120] *Looking forward from ICM 1962*, Fields Medalists Lectures (World Scientific), pp. 86–103.

[121] (with Ragnar Sigurdsson) *Growth properties of plurisubharmonic functions related to Fourier–Laplace transforms*, J. Geometric Analysis **8**, 251–311.

1999

[122] *On local integrability of fundamental solutions*, Ark. för Mat. **37**, 121–140.

My Mathematical Education

When I entered the University of Lund in the Fall of 1948, I had already passed the first semester mathematics courses with the encouragement and help of my high school mathematics teacher. This meant that I was out of phase with the courses given at the university and had to continue to study on my own apart from some problem solving sessions. The result was that after a year I received a bachelor's degree in mathematics and theoretical physics. The second year at the university was mainly devoted to experimental physics which led to a master's degree. I could then begin as a graduate student in mathematics with a licentiate degree as the next goal. This required writing a short thesis in addition to not very precisely defined reading courses. (Further courses were not required for the doctorate.) No systematic courses were given, there were very few graduate students and very little formal teaching beyond the undergraduate level. Marcel Riesz was the dominating spirit of the Mathematics Department. He usually gave a three-hour course every week on some subject. In addition, there was a seminar every other week on various topics. It was scheduled for the evening hours 18:15–20:00 which made it convenient for everybody to go afterwards under the leadership of Riesz to a restaurant where we stayed usually until closing time. This was an essential part of the graduate teaching. I encountered many topics there the first time and could study them in the library the next day. (I received a modest salary as "amanuens". The main duties consist in taking care of the library, which was a very useful occupation for a fresh graduate student.)

During my two undergraduate years, I had already followed the Riesz lectures which were then devoted to various topics in harmonic analysis and the theory of analytic functions of one complex variable. In the Fall of 1950 the subject of his lectures was geometrical optics. Of course I followed them conscientiously but little did I realize that the topic would become very important to me almost two decades later. In 1950 I was more interested in the themes from classical analysis in the preceding two years of the Riesz lectures, such as approximation theory. Among the graduate students there was also much interest in convexity theory, no doubt stimulated by Werner Fenchel in Copenhagen who had spent some years during the war as refugee in Lund and continued to visit occasionally to give lectures which I liked very much. The combination of these interests made me realise that the Tchebycheff approximation theorem could be obtained as a consequence of Helly's theorem on intersection of convex sets. I wrote a paper on this which Riesz accepted as thesis for the licentiate degree. That it was approved was surprising to me, for A. Zygmund visited Lund during the Spring of 1951 and he had a vague idea that something like that had been done not so long ago by Rademacher. True enough, I found a paper by H. Rademacher and I. J. Schoenberg in Can. J. Math. **2** (1950) written in a somewhat less general framework but making the same main point. Riesz must have been convinced of my independence, but of course I could not think of publishing my paper. When I look at the manuscript now the contents seem very meager, but it may have seemed more substantial in 1951. Perhaps I was just awarded for being enterprising.

I continued of course to broaden my mathematical knowledge. In the Fall of 1951 Riesz decided to lecture on the Lebesgue integral. This was a topic which I had studied carefully so I was well prepared to act as Riesz' secretary when he decided to write a didactic paper on the subject. The Mathematics Department did not yet have a secretary, and since I was a reasonably competent typist, I had acted as his secretary since the Fall of 1950, typing letters and short manuscripts for him. However, this time I was present at the creation. It was educational for me to see the care with which he weighed every word and detail of the presentation. I hope that it has left a lasting impression on my own work.

My secretarial work sometimes required me to bring a portable typewriter to Riesz' home in the evening so that I could type for him there. He loved to teach, and always taught me some mathematics at the same time. I was therefore not very surprised when he asked me to go there one evening in October 1951. However, it turned out that the agenda that time was different. He started to examine my knowledge of various mathematical topics, filled in as usual with explanations of his own where I was ignorant, and to my surprise he declared finally that I had now finished the examinations for the licentiate degree! It must have been rather premature but it was undoubtedly a correct judgement of my appetite for mathematics.

The next goal should then be to write a Ph.D. thesis. At first I continued to pursue my interests in classical analysis and wrote a couple of short papers but did not get started on something really worthwhile. Riesz probably saw this, anyway he told me some time in the early Spring term in 1952 that he thought that I would have fulfilled my duties as "amanuens" by the middle of the term and that he could provide me with a travel grant to allow me to study abroad during the rest of the term. This was of course very exciting for me but the question was where to go. As usual I consulted my high school mathematics teacher who suggested that I should go to Zürich where he had spent some time a couple of years earlier. I decided to go there, but I knew so little of what to expect that I arrived between terms so that there was no activity at all for a while. Fortunately I soon met Rolf Nevanlinna who was a professor at the University of Zürich. He arranged permission for me to borrow books in the ETH library and allowed me to take part in a small informal seminar on Hodge theory which he ran. Unfortunately my background in the calculus of differential forms was weak and I did not follow very well. I also attended a seminar on convexity theory organised by Heinz Hopf, who lectured himself on a proof of the two-dimensional isoperimetric inequality due to Erhard Schmidt. This appealed very much to me, and I wrote down a couple of pages where I extended the idea to general Minkowski metrics. I gave a copy to Hopf who received it kindly, but I realised that the paper was not worth much.

At the university I attended the lectures of B. L. van der Waerden for a while, until I realised that he was just following his book *Einführung in die Algebraische*

Geometrie. It was probably not the first time he did so for I found a good copy cheaply in a second-hand bookstore close to the university and dropped out of the lectures after reading the book. However, I do remember something which was not in his book — he was very sarcastic in mentioning the recent, and in his view, excessively pedantic and formal presentation of algebraic geometry by André Weil. A dozen years later or so I heard André Weil during our walks in the woods at the Institute for Advanced Study voice similar views about Grothendieck!

In the early morning hours at ETH there were also introductory lectures on topology by B. Eckman. They were extremely polished, absolutely perfect, and a bit dull which made it hard to stay awake so early in the morning. Altogether, Zürich was rather disappointing to me but in a bookstore I came across the first volume of *Théorie des Distributions* by Laurent Schwartz which I read enthusiastically. I had hesitated, back in Lund, whether to go to Zürich or to Nancy, which Lars Gårding had spoken well of, and now I thought that I had made a mistake and decided to leave Zürich and go to Nancy instead. However, the semester was over when I arrived there. The planning of this first trip abroad could not have been worse, but the visit to Nancy was not a complete failure. It was then that I started to read the Bourbaki books. For many years I continued to read the new volumes as they appeared, and they were an essential part of my mathematical education as for many mathematicians of my generation.

In the Fall of 1952 R. Godement gave a beautiful series of lectures in Lund, starting with purely algebraic representation theory for algebras and ending with the theory of spherical functions on a Lie group with a given compact subgroup. The lectures covered a lot of material in less than two months — my notes contain 130 dense pages. As a result the audience diminished steadily, and I think only Riesz, Gårding and I were left at the end. The Godement lectures were a unique exception during my years as a graduate student. At the time I was looking eagerly for some topic which could lead to a decent thesis. Although I enjoyed the Godement lectures very much, I realised that the subject was in a very active stage and could not be pursued in isolation in Lund. When I asked Werner Fenchel about trends in convexity theory, he said that the most interesting current directions were connected with the theory of functions of several complex variables, and he lent me some preprints (of Bremermann?). However, I lacked the motivation and/or strength to try my luck in this area — almost a decade would pass before I got seriously involved in it. I decided then to leave my old interests in classical analysis and devote myself to partial differential equations. Since Lars Gårding and Åke Pleijel had just been appointed to the two professorships of mathematics in Lund, and both were working in this field, it was clearly the area where the department in Lund was best equipped to keep in touch with the international development.

As a student of Riesz I was already familiar with his work on hyperbolic differential equations, but I had no broad knowledge of the subject. One of the first things I started to look at closely was energy estimates for second-order hyperbolic equations. I wrote a paper [1]* on the subject, presented in August 1953 at the

*The number within the brackets corresponds to that in the bibliography.

Scandinavian Congress of Mathematicians in Lund. The main point was an invariant form for the energy identities and its application to global energy estimates for some mixed problems. It contained little news in substance, but it became an important step toward my thesis.

I finished [1] during some weekends when I was in military service, and managed to get a week off to go to the congress in Lund. After basic military training during the Summer of 1953, I was stationed in Stockholm as some sort of assistant to a professor of theoretical physics at KTH (the Royal Institute of Technology in Stockholm), who worked as a consultant for FOA (the Defence Research Institute). The official purpose of the program was to study nuclear weapons to give a basis for protection against such attacks. I did not contribute anything valuable to that program but the military service gave me an opportunity to learn something about nonlinear differential equations. Actually I could devote most of my time to reading pure mathematics, and it was during that year that I started to get a broader understanding of the theory of partial differential equations.

Following advice from Gårding I tried in vain to get a grip on Petrowsky's L^2 estimates for hyperbolic operators with variable coefficients. These were not derived by energy integral arguments but by some kind of Fourier analysis. I never managed to understand the 18 pages of estimates in the proof, and probably nobody else did at the time. However, in the light of the later development of pseudodifferential operator techniques, it is no longer difficult to follow Petrowsky's arguments. In fact, they can be seen as a forerunner of pseudodifferential operator methods.

During the weekends I sometimes took the train to Djursholm to read in the library of the Mittag-Leffler Institute. At that time there was no activity at all besides very rare guest lectures, but the library was well maintained and on the whole much better than that at the university (Stockholm Högskola), which in turn was far better than that at KTH. Two things which I read there became particularly important for my thesis: The thesis of Mark Višik, *On general boundary problems for elliptic differential equations, Trudy Moskov. Mat. Obsč.* **1** (1952), 187–246, and the notes by Bernard Malgrange, *Équations aux dérivées partielles à coefficients constants, C. R. Acad. Sci. Paris* **237** (1953), 1620–1622 and **238** (1954), 196–198. Višik's paper concerned second-order elliptic differential equations, but the operator theoretical arguments were quite general and were presented in an introductory Chap. I of my thesis [7] as a motivation for the later work. Chapters II and III there concerned differential operators $P(D)$, $D = -i\partial/\partial x$, with constant coefficients. Malgrange had proved in the notes quoted above that $P(D)$ has a continuous inverse when acting on the Schwartz space $\mathcal{S}$ modified to an exponential decrease at infinity. His method of proof could give estimates of the form

$$\|u\|_{L^2} \le C_K \|P(D)u\|_{L^2}, \qquad u \in C_0^\infty(K),$$

for every compact set $K \subset \mathbb{R}^n$. However, for elliptic operators it was well known that one could also estimate $Q(D)u$ in the same way for every $Q(D)$ of order at most equal to that of $P(D)$, and for strictly hyperbolic operators it was known that this was possible when the order of $Q(D)$ is strictly smaller. This was the essence of the Petrowsky estimates mentioned above. A proof using an extension of the energy integral method had just been given by Leray, and it had been simplified further by Gårding, *L'inégalité de Friedrichs et Lewy pour les équations hyperboliques linéaires d'ordre supérieur, C. R. Acad. Sci. Paris* **239** (1954), 849–850. By a systematic analysis of what can be achieved by a partial integration in a sesquilinear form involving the derivatives of a function I found that the method could be generalised to proving that

$$\|Q(D)u\|_{L^2} \le C_{K,Q}\|P(D)u\|_{L^2}\,, \qquad u \in C_0^\infty(K)\,,$$

$$\text{if and only if} \qquad |Q(\xi)| \le C\widetilde{P}(\xi) = \Big(\sum |P^{(\alpha)}(\xi)|^2\Big)^{1/2}, \qquad \xi \in \mathbb{R}^n.$$

Q was then said to be weaker than P. One source of inspiration for this work was that in [1] I had observed that the standard energy estimates gave such bounds for ultra-hyperbolic operators acting on functions with vanishing Cauchy data on the whole boundary of a bounded domain. This was only mentioned briefly since it could not lead to an existence theorem for the Cauchy problem. However, such an estimate for an operator and for its adjoint is exactly what is required for the existence of correctly posed abstract boundary problems, so the viewpoints of Višik's work were precisely what were made this fact interesting.

In particular, a derivative $P^{(\alpha)}(D)$ is always weaker than $P(D)$, which means that the minimal domain of $P(D)$ obtained as the L^2 closure of $P(D)$ acting on $C_0^\infty(\Omega)$ for some bounded domain Ω is always a C^∞ module. In Chap. III, I took up the question whether the maximal domain of $P(D)$ consisting of all $u \in L^2(\Omega)$ with $P(D)u \in L^2(\Omega)$ in the weak sense (of distribution theory) is a C^∞ module. This is true for ordinary differential operators but I found that it is essentially only true then. A far more interesting question solved in my thesis was to decide when $P(D)$ is of *local type* in the sense that the maximal domain of the operator $P(D)$ in $L^2(\Omega)$ is a $C_0^\infty(\Omega)$ module, i.e. φu is in the minimal domain if u is in the maximal domain and $\varphi \in C_0^\infty(\Omega)$. If one introduces suitable coordinates, one can assume that $P(D)$ is *complete*, i.e. operates effectively in all variables in the sense that there is no *real* $\eta \ne 0$ such that $P(\xi + t\eta) \equiv P(\xi)$. Then I proved that $P(D)$ is of local type if and only if $\operatorname{Im}\zeta \to \infty$ when $\zeta \to \infty$ while $P(\zeta) = 0$, or equivalently, $P^{(\alpha)}(\xi)/P(\xi) \to 0$ as $\mathbb{R}^n \ni \xi \to \infty$, if $\alpha \ne 0$. I also proved that all (distribution) solutions of the homogeneous equation $P(D)u = 0$ are in C^∞ if and only if $P(D)$ is complete and of local type, so I had characterised these operators, later called hypoelliptic. This answered a question raised by Laurent Schwartz, which was undoubtedly more interesting than the question when an operator is of local type. Incidentally, Riesz reproached me for using the dubious notion of distribution in the proof and urged me to work some more to prove the existence of a locally integrable fundamental solution. This was quite typical of the attitudes prevailing at the time; I had in fact avoided appealing explicitly to distribution theory when I could do so, although I thought in such terms. Long after these prejudices have disappeared I proved in [122] that distributions are in fact essential in this context.

Chapter IV was very short, only five pages, but it is the basis for much later work on solvability of differential equations, and it is included below. I knew from

[1] that the energy integral method worked well for ultra-hyperbolic second-order differential equations with variable coefficients, and it was natural to try the energy integral techniques of Chap. II also for differential operators $\mathcal{P}$ of order $m > 2$ with variable coefficients. In doing so I assumed that the operator was of principal type, i.e. that it was stronger than all operators of order $< m$ if the coefficients were frozen at a point, for this should make lower order terms harmless. However, to my great surprise I found that there was another problem, the commutator $[\mathcal{P}, \mathcal{P}^*] = \mathcal{P}\mathcal{P}^* - \mathcal{P}^*\mathcal{P}$ is in general of order $2m-1$ and then I could not control it. This does not happen if the principal symbol is real, so I added that assumption and proved that for functions u of small support it was then possible to estimate $m-1$ derivatives of u in L^2 by the L^2 norm of $\mathcal{P}u$. By duality this gave a local existence theorem for the adjoint: The equation $\mathcal{P}^*u = f$ has locally a solution in L^2 for every $f \in L^2$ (in fact, for every f in the Sobolev space $H_{(1-m)}$, but as already mentioned, I hesitated to provoke the prejudices against distribution theory by saying so). In the introduction I mentioned that it was sufficient to assume that the order of $[\mathcal{P}, \mathcal{P}^*]$ is at most $2m-2$, but I took it for granted that it was only a flaw of my methods which made that assumption necessary. A look at the simplest example where the hypothesis is not fulfilled should have given me the Lewy example of an unsolvable equation $(D_1 + iD_2 - 2i(x_1 + ix_2)D_3)u = f$! However, a couple of years later, the obstacle encountered in my thesis was converted to the first general necessary conditions for solvability, in the second paper [18] included below. (The third paper included here is an example of how nontrivial commutators can give rise to hypoellipticity of operators which are far from those with constant coefficients identified in Chap. III of my thesis.)

I defended my thesis in October 1955, which marks the end of my formal mathematical education. However, it is reasonable to include in my education also the year 1956 which I spent in the United States. The first half year at the University of Chicago did not contribute to my education in the theory of partial differential equations, but it gave me an opportunity to get familiar with the work of the Zygmund school on singular integrals. During the Summer spent at the University of Kansas and the University of Minnesota I picked up much more in my special field, but particularly the Fall spent in New York, at what later became the Courant Institute, was very fruitful. I learned much particularly from Louis Nirenberg, Peter Lax and Fritz John, and continued to profit from contact with them for many years. When I went back to Sweden in January 1957 my student days were definitely over, for I returned as full professor at the University of Stockholm.

ON THE THEORY OF GENERAL PARTIAL DIFFERENTIAL OPERATORS

BY

LARS HÖRMANDER

FIL. LIC., BL.

BY DUE PERMISSION OF THE PHILOSOPHICAL FACULTY OF THE UNIVERSITY OF LUND TO BE PUBLICLY DISCUSSED IN FRENCH AT THE INSTITUTE OF MATHEMATICS ON SATURDAY OCTOBER 22ND, 1955, AT 10 A.M., FOR THE DEGREE OF DOCTOR OF PHILOSOPHY

UPPSALA 1955

ALMQVIST & WIKSELLS BOKTRYCKERI AB

Reprinted from Acta Math., Vol. 94 (1955), pp. 161–248.

In submitting my thesis, I wish firstly to thank my teacher Professor Marcel Riesz to whom I owe my mathematical education, and whose stimulating guidance and encouragement have meant so much to me. I would also like to thank the Professors Lars Gårding and Åke Pleijel for their kind and very valuable interest in my work.

Lund, September 1955.

Lars Hörmander

ON THE THEORY OF GENERAL PARTIAL DIFFERENTIAL OPERATORS

BY

LARS HÖRMANDER

in Lund

CONTENTS

11 – 553810. *Acta Mathematica*. 94. Imprimé le 26 septembre 1955.

PREFACE

0.1. The main interest in the theory of partial differential equations has always been concentrated on elliptic and normally hyperbolic equations. During the last few years the theory of these equations has attained a very satisfactory form, at least where Dirichlet's and Cauchy's problems are concerned. There is also a vivid interest in other differential equations of physical importance, particularly in the mixed elliptic-hyperbolic equations of the second order. Very little, however, has been written concerning differential equations of a general type. Petrowsky ([25], p. 7, pp. 38–39) stated in 1946 that "it is unknown, even for most of the very simplest non-analytical equations, whether even one solution exists", and "there is, in addition, a sizable class of equations for which we do not know any correctly posed boundary problems. The so-called ultra-hyperbolic equation

$$\frac{\partial^2 u}{\partial x_1^2} + \cdots + \frac{\partial^2 u}{\partial x_p^2} = \frac{\partial^2 u}{\partial y_1^2} + \cdots + \frac{\partial^2 u}{\partial y_p^2},$$

with $p \geqq 2$ appears, for example, to be one of these." Some important papers have appeared since then. In particular, we wish to mention the proof by Malgrange [19] that any differential equation with constant coefficients has a fundamental solution. (Explicit constructions of distinguished fundamental solutions have been performed for the ultra-hyperbolic equations by de Rham [27] and others.) Apart from this result, however, no efforts to explore the properties of general differential operators seem to have been made. The principal aim of this paper is to make an approach to such a study. The general point of view may perhaps illuminate the theory of elliptic and hyperbolic equations also.

0.2. A pervading characteristic of the modern theory of differential equations is the use of the abstract theory of operators in Hilbert space. Our point of view here is also purely operator theoretical. To facilitate the reading of this paper we have included an exposition of the necessary abstract theory in the first chapter, where we introduce our main problems.[1] Using the abstract methods we prove that the answer to our questions depends on the existence of so-called a priori inequalities. The later chapters are to a great extent devoted to the proof of such inequalities. In Chapters II and IV the proofs are based on the energy integral method in a general form, i.e. on the study of the integrals of certain quadratic forms in the derivatives of a function. For the wave equation, where it has a physical interpretation as the conservation of energy, this method was introduced by Friedrichs and Lewy [6]. Recently Leray [19] has found a generalization which applies to normally hyperbolic equations of higher order. In Chapter II we study systematically the algebraic aspects of the energy integral method. This chapter deals only with equations with constant coefficients. The extension to a rather wide class of equations with variable coefficients is discussed in Chapter IV.

In Chapter III we chiefly study a class of differential operators with constant coefficients, which in several respects appears to be the natural class for the study of problems usually treated only for elliptic operators. For example, Weyl's lemma holds true in this class, i.e. all (weak) solutions are infinitely differentiable. Our main arguments use a fundamental solution which is constructed there. The results do not seem to be accessible by energy integral arguments in the general case, although many important examples can be treated by a method due to Friedrichs [5].

0.3. A detailed exposition of the results would not be possible without the use of the concepts introduced in Chapter I. However, this chapter, combined with the introductions of each of the following ones, gives a summary of the contents of the whole paper.

0.4. It is a pleasure for me to acknowledge the invaluable help which professor B. L. van der Waerden has given me in connection with the problems of section 3.1. I also want to thank professor A. Seidenberg, who called my attention to one of his papers, which is very useful in section 3.4.

CHAPTER I

Differential Operators from an Abstract Point of View

1.0. Introduction

In the preface we have pointed out that the present chapter has the character of an introduction to the whole paper. Accordingly we do not sum up the contents here, but

[1] Chapter I, particularly section 1.3, overlaps on several points with a part of an important paper by Višik ([34]) on general boundary problems for elliptic equations of the second order.

merely present the general plan. First, in section 1.1, we recall some well-known theorems and definitions from functional analysis. Then in section 1.2 we define differential operators in Hilbert space and specialize the theorems of section 1.1 to the case of differential operators. A discussion of the meaning of boundary data and boundary problems is given in section 1.3. This study has many ideas in common with Višik [34]. It is not logically indispensable for the rest of the paper but it serves as a general background.

1.1. Definitions and results from the abstract theory of operators

Let B_0 and B_1 be two complex Banach spaces, i.e. two normed and complete complex vector spaces. A linear transformation (operator) T from B_0 to B_1 is a function defined in a linear set $\mathcal{D}_T$ in B_0 with values in B_1 such that

$$T(\alpha x + \beta y) = \alpha Tx + \beta Ty \tag{1.1.1}$$

for $x, y \in \mathcal{D}_T$ and complex α, β. It follows from (1.1.1) that the range of values $\mathcal{R}_T$ is a linear set in B_1.

The set $B_0 \times B_1$ of all pairs $x = [x_0, x_1]$ with $x_i \in B_i$ $(i = 0, 1)$, where we introduce the natural vector operations and the norm[1]

$$|x| = (|x_0|^2 + |x_1|^2)^{\frac{1}{2}}, \tag{1.1.2}$$

is also a Banach space, called the direct sum of B_0 and B_1. If T is a linear transformation from B_0 to B_1, the set in $B_0 \times B_1$ defined by

$$G_T = \{[x_0, Tx_0], x_0 \in \mathcal{D}_T\} \tag{1.1.3}$$

is linear and contains no element of the form $[0, x_1]$ with $x_1 \neq 0$. The set G_T is called the graph of T. A linear set G in $B_0 \times B_1$, containing no element of the form $[0, x_1]$ with $x_1 \neq 0$, is the graph of one and only one linear transformation T.

A linear transformation T is said to be *closed*, if the graph G_T is closed. We shall also say that a linear transformation T is *pre-closed*, if the closure $\bar{G}_T$ of the graph G_T is a graph, i.e. does not contain any element of the form $[0, x_1]$ with $x_1 \neq 0$. The transformation with the graph $\bar{G}_T$ is then called the closure of T. Thus T is pre-closed if and only if, whenever $x_n \to 0$ in B_0 and $Tx_n \to y$ in B_1, we have $y = 0$. We also note that any linear restriction of a linear pre-closed operator is pre-closed.

The following theorem gives a useful form of the theorem on the closed graph, which states that a closed transformation from B_0 to B_1 must be continuous, if $\mathcal{D}_T = B_0$. (Cf. Bourbaki, *Espaces vectoriels topologiques*, Chap. I, § 3 (Paris 1953).)

[1] Any equivalent norm in $B_0 \times B_1$ can be used, but this choice has the advantage of giving a Hilbert norm, if B_0 and B_1 have Hilbert norms.

THEOREM 1.1. *Let B_i $(i = 0, 1, 2)$ be Banach spaces and T_i $(i = 1, 2)$ be linear transformations from B_0 to B_i. Then, if T_1 is closed, T_2 pre-closed and $\mathcal{D}_{T_1} \subset \mathcal{D}_{T_2}$, there exists a constant C such that*

$$|T_2 u|^2 \leq C(|T_1 u|^2 + |u|^2), \quad u \in \mathcal{D}_{T_1}. \tag{1.1.4}$$

PROOF. The graph G_{T_1} of T_1 is by assumption closed. Hence the mapping

$$G_{T_1} \ni [u, T_1 u] \to T_2 u \in B_2 \tag{1.1.5}$$

is defined in a Banach space. We shall prove that the mapping is closed. Thus suppose that $[u_n, T_1 u_n]$ converges in G_{T_1} and that $T_2 u_n$ converges in B_2. Since T_1 is closed, there is an element $u \in \mathcal{D}_{T_1}$ such that $u_n \to u$ and $T_1 u_n \to T_1 u$. In virtue of the assumptions, u is in $\mathcal{D}_{T_2}$ and, since T_2 is pre-closed, the existing limit of $T_2 u_n$ can only be $T_2 u$. Hence the mapping (1.1.5) is closed and defined in the whole of a Banach space, so that it is continuous in virtue of the theorem on the closed graph. This proves the theorem.

Theorem 1.1 is the only result we need for other spaces than Hilbert spaces; it will also be used when some of the spaces B_i are spaces of continuous functions with uniform norm. In the rest of this section we shall only consider transformations from a Hilbert space H to itself. In that case the graph is situated in $H \times H$, which is also a Hilbert space, the inner product of $x = [x_0, x_1]$ and $y = [y_0, y_1]$ being given by

$$(x, y) = (x_0, y_0) + (x_1, y_1).$$

For the definition of adjoints, products of operators and so on, we refer the reader to Nagy ([23], p. 27 ff.).

LEMMA 1.1. *The range $\mathcal{R}_T$ of a closed densely defined linear operator T is equal to H if and only if T^{*-1} exists and is continuous, and consequently is defined in a closed subspace.*

PROOF. We first establish the necessity of the condition. Thus suppose that $\mathcal{R}_T = H$. Since $T^* u = 0$ implies that $(Tv, u) = (v, T^* u) = 0$ for every $v \in \mathcal{D}_T$, it follows that $T^* u = 0$ only if $u = 0$. Hence T^{*-1} is defined. Now for any element v in H we can find an element w such that $Tw = v$. Hence we have, if $u \in \mathcal{D}_{T^*}$,

$$(u, v) = (u, Tw) = (T^* u, w),$$

so that for fixed v

$$|(u, v)| \leq C \| T^* u \|, \quad u \in \mathcal{D}_{T^*}.$$

Let u_n be a sequence of elements in $\mathcal{D}_{T^*}$ such that $\| T^* u_n \|$ is bounded. Since $|(u_n, v)|$ is then bounded for every fixed v, it follows from Banach-Steinhaus' theorem (cf. Nagy [23],

p. 9) that $\|u_n\|$ must be bounded. Hence T^{*-1} is continuous, and since it is obviously closed, we conclude that T^{*-1} is defined in a closed subspace.

The sufficiency of the condition is easily proved directly but follows also as a corollary of the next lemma.

LEMMA 1.2. *The densely defined closed operator T has a bounded right inverse S if and only if T^{*-1} exists and is continuous.*[1]

PROOF. Since $TS = I$ implies that $\mathcal{R}_T = H$, it follows from the part of Lemma 1.1, which we have proved, that a bounded right inverse can only exist if T^{*-1} is continuous. The remaining part of Lemma 1.1 will also follow when we have constructed the right inverse in Lemma 1.2.

In virtue of a well-known theorem of von Neumann [24], the operator TT^* is self-adjoint and positive. Under the conditions of the lemma we have

$$(TT^*u, u) = (T^*u, T^*u) \geqq C^2(u, u), \quad u \in \mathcal{D}_{TT^*},$$

where C is a positive constant. Hence $TT^* \geqq C^2 I$. Let A be the positive square root of TT^*. Since $A^2 \geqq C^2 I$, it follows from the spectral theorem that $0 < A^{-1} \leqq C^{-1} I$. The operator A^{-1} is bounded and self-adjoint, $\|A^{-1}\| \leqq C^{-1}$. Furthermore, the operator $T^* A^{-1}$ is isometric according to von Neumann's theorem. Now we define

$$S = T^*(TT^*)^{-1} = T^* A^{-1} A^{-1}. \tag{1.1.6}$$

Since S is the product of an isometric operator and A^{-1}, it must be bounded, and we have $\|S\| \leqq C^{-1}$. Finally, it is obvious that $TS = I$.

LEMMA 1.3. *The densely defined closed operator T has a completely continuous right inverse S if and only if T^{*-1} exists and is completely continuous.*

PROOF. We first note that the operator S given by (1.1.6) is completely continuous if T^{*-1} and consequently A^{-1} is completely continuous. This proves one half of the lemma. Now suppose that there exists a completely continuous right inverse S. If $u \in \mathcal{D}_{T^*}$, we have for any $v \in H$

$$(u, v) = (u, TSv) = (S^* T^* u, v),$$

and therefore $u = S^* T^* u$. Hence, if $v \in \mathcal{R}_{T^*}$, we have $T^{*-1} v = S^* v$, which proves that T^{*-1} is completely continuous, since it is a restriction of a completely continuous operator.

[1] This means that S is continuous and defined in the whole of H, and satisfies the equality $TS = I$, where I is the identity operator.

1.2. The definition of differential operators

Let Ω be a ν-dimensional infinitely differentiable manifold. We shall denote by $C^\infty(\Omega)$ the set of infinitely differentiable functions defined in Ω, and by $C_0^\infty(\Omega)$ the set of those functions in $C^\infty(\Omega)$ which vanish outside a compact set in Ω. When no confusion seems to be possible, we also write simply C^∞ and C_0^∞.

A transformation P from $C^\infty(\Omega)$ to itself is called a differential operator, if, in local coordinate systems $(x^1, \ldots, x^\nu)$, it has the form

$$P\,u = \sum a^{\alpha_1 \cdots \alpha_k}(x) \frac{1}{i}\frac{\partial}{\partial x^{\alpha_1}} \cdots \frac{1}{i}\frac{\partial}{\partial x^{\alpha_k}}\, u, \tag{1.2.1}$$

where the sum contains only a finite number of terms $\neq 0$, and the coefficients $a^{\alpha_1 \cdots \alpha_k}$ are infinitely differentiable functions of x which do not change if we permute the indices α_j.[1] We shall denote the sequence $(\alpha_1, \ldots, \alpha_k)$ of indices between 1 and ν by α and its length k by $|\alpha|$. Furthermore, we set

$$D_\iota = \frac{1}{i}\frac{\partial}{\partial x^\iota}, \quad D_\alpha = D_{\alpha_1} \cdots D_{\alpha_k}.$$

Formula (1.2.1) then takes a simplified form, which will be used throughout:

$$P\,u = \sum a^\alpha(x)\, D_\alpha\, u. \tag{1.2.2}$$

Here the summation shall be performed over all sequences α.

We shall say that we have a differential operator with constant coefficients, if Ω is a domain in the ν-dimensional real vector space R^ν, and the coefficients in (1.2.2) are constant, when the coordinates are linear.

Let ϱ be a fixed density in Ω, i.e. $\varrho(x)$ is a positive function, defined in every local coordinate system, such that $\varrho(x)\,dx^1 \ldots dx^\nu$ is an invariant measure, which will be denoted dx. We require that $\varrho(x)$ shall be infinitely differentiable, and, in cases where P has constant coefficients, we always take $\varrho(x) = \text{constant}$.

The differential operators shall be studied in the Hilbert space L^2 of all (equivalence classes of) square integrable functions with respect to the measure dx, the scalar product in this space being

$$(u, v) = \int u(x)\,\overline{v(x)}\,dx. \tag{1.2.3}$$

With respect to this scalar product we define the *algebraic adjoint* $\overline{P}$ of P as follows.

[1] We restrict ourselves to the infinitely differentiable case for simplicity in statements; most arguments and results are, however, more general and will later, in Chapter IV, be used under the weaker condition of a sufficient degree of differentiability.

Let $v \in C^\infty$ and let u be any function in C_0^∞. Integrating $(\mathcal{P} u, v)$ by parts, we find that there is a unique differential operator $\overline{\mathcal{P}}$ such that

$$(\mathcal{P}\, u, v) = (u, \overline{\mathcal{P}}\, v). \tag{1.2.4}$$

In fact, we obtain

$$\overline{\mathcal{P}}\, v = \varrho^{-1} \sum D_\alpha (\varrho\, \bar{a}^\alpha v).$$

When the coefficients are constant we thus obtain $\overline{\mathcal{P}}$ by conjugating the coefficients, which motivates our notation.

LEMMA 1.4. *The operator $\mathcal{P}$, defined for those functions u in C^∞ for which u and $\mathcal{P} u$ are square integrable, is pre-closed in L^2.*

PROOF. Let u_n be a sequence of functions in this domain such that $u_n \to 0$ and $\mathcal{P}\, u_n \to v$ (with L^2-convergence). Then we have for any $f \in C_0^\infty$

$$(v, f) = \lim\, (\mathcal{P}\, u_n, f) = \lim\, (u_n, \overline{\mathcal{P}} f) = 0.$$

Hence $v = 0$, which proves the lemma.

REMARK. It follows from the trivial proof that Lemma 1.4 would also hold if, for example, we consider $\mathcal{P}$ as an operator from L^2 to C, the space of continuous functions with the uniform norm.

Lemma 1.4 justifies the following important definition.

DEFINITION 1.1. *The closure P_0 of the operator in L^2 with domain C_0^∞, defined by $\mathcal{P}$, is called the minimal operator defined by $\mathcal{P}$. The adjoint P of the minimal operator $\overline{P}_0$, defined by $\overline{\mathcal{P}}$, is called the maximal operator defined by $\mathcal{P}$.*

The definition of the maximal operator means that u is in $\mathcal{D}_P$ and $Pu = f$ if and only if u and f are in L^2, and for any $v \in C_0^\infty$ we have

$$(f, v) = (u, \overline{\mathcal{P}} v).$$

Operators defined in this way are often called weak extensions. In terms of the more general concept of distributions (see Schwartz [28]), we might also say that the domain consists of those functions u in L^2 for which $\mathcal{P} u$ in the sense of the theory of distributions is a function in L^2.

If $u \in C^\infty$ and u and $\mathcal{P} u$ are square integrable, it follows from (1.2.4) that Pu exists and equals $\mathcal{P} u$. This is of course the idea underlying the definition. Since P is an adjoint operator, it is closed and therefore an extension of P_0.

It is unknown to the author whether in general P is the closure of its restriction to $\mathcal{D}_P \cap C^\infty$. For elliptic second order equations in domains with a smooth boundary this follows from the results of Birman [1]. If P is a homogeneous operator with constant coefficients and Ω is starshaped with respect to every point in an open set, it is also easily proved by regularization. In section 3.9 we shall prove an affirmative result for a class of differential operators with constant coefficients, when Ω is any domain.

We now illustrate Definition 1.1 by an elementary example. Let Ω be the finite interval (a, b) of the real axis, and let P be the differential operator d^n/dx^n. It is immediately verified that the domain of P consists of those n–1 times continuously differentiable functions for which $u^{(n-1)}$ is absolutely continuous and has a square integrable derivative. The domain of P_0 consists of those functions in the domain of P for which

$$u(a) = \cdots = u^{(n-1)}(a) = 0, \quad u(b) = \cdots = u^{(n-1)}(b) = 0,$$

that is, those which have vanishing Cauchy data in the classical sense at a and b with respect to the differential operator P.

The same result is true under suitable regularity conditions for any ordinary differential operator of order n. Hence, in general, the maximal (minimal) domain of an ordinary differential operator is contained in the maximal (minimal) domain of any ordinary differential operator of lower or equal order. For partial differential operators, this result is no longer valid, but we shall find a satisfactory substitute. Our results are most conveniently described in terms of the following definition.

DEFINITION 1.2. *If $\mathcal{D}_{P_0} \subset \mathcal{D}_{Q_0}$, we shall say that the operator P is stronger than the operator Q and that Q is weaker than P. If P is both weaker and stronger than Q, i.e., if $\mathcal{D}_{P_0} = \mathcal{D}_{Q_0}$, we shall say that P and Q are equally strong.*[1]

We now pose the problem to determine the set of those operators Q which are weaker than a given operator P. It is clear that the answer is closely connected with the regularity properties and the boundary properties of the functions in $\mathcal{D}_{P_0}$. The question is reduced to a concrete problem by the following lemma.

LEMMA 1.5. *The operator Q is weaker than the operator P if and only if there is a constant C such that*

$$\| Qu \|^2 \leq C(\| Pu \|^2 + \| u \|^2), \quad u \in C_0^\infty. \tag{1.2.5}$$

PROOF. If Q is weaker than P, it follows from Theorem 1.1 that

$$\| Q_0 u \|^2 \leq C(\| P_0 u \|^2 + \| u \|^2), \quad u \in \mathcal{D}_{P_0},$$

[1] Note that these notions depend on the basic manifold Ω.

which implies (1.2.5). On the other hand, suppose that (1.2.5) is valid. If $u \in \mathcal{D}_{P_0}$, we can find a sequence u_n of functions in C_0^∞ such that

$$u_n \to u, \quad \mathcal{P} u_n \to P_0 u.$$

Applying (1.2.5) to the functions $u_n - u_m$ we find that $Q u_n$ is a Cauchy sequence. Since Q_0 is closed, it follows that $u \in \mathcal{D}_{Q_0}$.

We shall repeatedly use the criterion given by Lemma 1.5 in the following chapters. In Chapter II we shall find a simple and complete description of the operators Q with constant coefficients which are weaker than a given operator $\mathcal{P}$ with constant coefficients, when Ω is a bounded domain in R^ν. (The answer is then independent of Ω.) In Chapter IV analogous results will be proved for a class of operators with variable coefficients.

Remark. If $\mathcal{D}_{P_0} \subset \mathcal{D}_Q$, it follows from Theorem 1.1 in the same way as in the proof of Lemma 1.5 that (1.2.5) is valid. Hence $\mathcal{D}_{P_0} \subset \mathcal{D}_{Q_0}$, so that Q is weaker than $\mathcal{P}$. This shows that in Definition 1.2 we might replace the condition $\mathcal{D}_{P_0} \subset \mathcal{D}_{Q_0}$ by the apparently weaker condition $\mathcal{D}_{P_0} \subset \mathcal{D}_Q$. It should also be observed that, in Definition 1.2 and in most of our arguments here, we use the minimal and not the maximal differential operators in view of the fact that the relation $\mathcal{D}_P \subset \mathcal{D}_Q$ is very exceptional for partial differential operators, as will be proved in Chapter III.

We shall next deduce the conditions in order that Qu should be continuous after correction on a null set for every $u \in \mathcal{D}_{P_0}$, the operator Q being interpreted in the distribution sense. Such results form a stepping-stone from the weak concept of a solution of a differential equation to the classical one. Sobolev has studied similar questions (see [30]), but our results overlap very little with his.

Lemma 1.6. *In order that Qu should equal a bounded function in the distribution sense for every $u \in \mathcal{D}_{P_0}$, it is necessary and sufficient that there is a constant C such that*

$$\sup_\Omega |Qu|^2 \leq C(\|\mathcal{P}u\|^2 + \|u\|^2), \quad u \in C_0^\infty. \tag{1.2.6}$$

If (1.2.6) is satisfied, Qu is a uniformly continuous function in Ω after correction on a null set, if $u \in \mathcal{D}_{P_0}$, and Qu tends to zero at the boundary in the sense that to every $\varepsilon > 0$ there is a compact set K in Ω, so that $|Qu(x)| < \varepsilon$ in $\Omega - K$.

Proof. That (1.2.6) is a necessary condition follows, if we consider Q as an operator from L^2 to L^∞ and apply Theorem 1.1, which is possible in virtue of the remark following Lemma 1.4. Conversely, let (1.2.6) be satisfied. If u_n is a sequence of functions in C_0^∞ such that $u_n \to u$ and $\mathcal{P} u_n \to P_0 u$, where u is an arbitrary function in $\mathcal{D}_{P_0}$, it follows that Qu_n

is uniformly convergent. Since the limit must equal Qu a.e., the last statement of the lemma follows.

The last assertion of the lemma may also be formulated as follows: Qu is continuous and vanishes at infinity in the Alexandrov compactification of Ω.

We now turn to another matter, the existence of solutions of differential equations. Lemma 1.1 and the definition of P as the adjoint of $\bar{P}_0$ prove the following result.

LEMMA 1.7. *The equation $Pu = f$ has, for any $f \in L^2$, at least one solution $u \in \mathcal{D}_P$, if, and only if, $\bar{P}_0$ has a continuous inverse, i.e., if*

$$(u, u) \leq C^2(\bar{P}u, \bar{P}u), \quad u \in C_0^\infty, \tag{1.2.7}$$

where C is a constant.

In Chapters II and IV it will be proved that (1.2.7) is valid under very mild assumptions about P.

1.3. Cauchy data and boundary problems

The example on page 169 makes it justifiable to say that the functions in $\mathcal{D}_{P_0}$ are those which have vanishing Cauchy data with respect to the operator P, and we are thus led to the following definition.

DEFINITION 1.3. *The quotient space*

$$C = G_P / G_{P_0} \tag{1.3.1}$$

with the quotient norm is called the Cauchy space of P. If $u \in \mathcal{D}_P$, the residue class of the pair $[u, Pu]$ is an element of C, which is called the Cauchy datum of u and is denoted by Γu.

It follows from the definition that two functions in $\mathcal{D}_P$, which only differ by a function in $C_0^\infty(\Omega)$, have the same Cauchy data. When the coefficients are constant it is easy to prove (Lemma 2.11) that every function in $\mathcal{D}_P$, which vanishes outside a compact set in Ω, is also in $\mathcal{D}_{P_0}$. It then follows that two functions in $\mathcal{D}_P$, which are identical outside a compact set in Ω, have the same Cauchy data. It is of course natural to expect that this is valid for very general operators though we have not obtained any proof.

The example on page 169 also suggests the following definition.

DEFINITION 1.4. *Let B be a linear manifold in the Cauchy space C of P. The problem to find a solution f of*

$$Pf = g, \quad \Gamma f \in B, \tag{1.3.2}$$

for arbitrarily given $g \in L^2$ *is called a linear homogeneous boundary problem.* $\Gamma f \in B$ *is the boundary condition.*

Let $\hat{P}$ be the restriction of P to those f for which $\Gamma f \in B$. Then $\hat{P}$ is linear and

$$P_0 \subset \hat{P} \subset P. \tag{1.3.3}$$

Conversely, any linear operator $\hat{P}$ with this property corresponds to exactly one linear manifold B in C.

DEFINITION 1.5. *The boundary problem* (1.3.2) *is said to be* (*completely*) *correctly posed, if* $\hat{P}$ *has a* (*completely*) *continuous inverse, defined in the whole of* L^2.

This definition and the following result are essentially due to Višik [34], who also considers less restrictive definitions.

THEOREM 1.2. *There exist* (*completely*) *correctly posed boundary problems for the operator* P *if and only if* P_0 *and* $\overline{P}_0$ *have* (*completely*) *continuous inverses.*

PROOF. Suppose that there exists a (completely) correctly posed boundary problem, and let $\hat{P}$ be the corresponding operator. Since $\hat{P}^{-1}$ is (completely) continuous and $\hat{P} \supset P_0$, it follows that P_0^{-1} must be (completely) continuous, and since $\hat{P}^{-1}$ is a right inverse of P, it follows from Lemma 1.2 (Lemma 1.3) that $\overline{P}_0^{-1}$ is (completely) continuous.

Now assume that P_0^{-1} and $\overline{P}_0^{-1}$ are (completely) continuous. In virtue of the continuity of P_0^{-1}, the range $\mathcal{R}_{P_0}$ of P_0 is closed. Let π be the orthogonal projection on $\mathcal{R}_{P_0}$. If S is the right inverse of P constructed in Lemma 1.2 (Lemma 1.3), the operator T defined by

$$Tf = P_0^{-1}(\pi f) + S((I-\pi)f), \quad f \in L^2,$$

is (completely) continuous. Since

$$PTf = \pi f + (I-\pi)f = f,$$

the operator T has an inverse $\hat{P}$, and $\hat{P} \subset P$. Furthermore, $T \supset P_0^{-1}$ and hence $\hat{P} \supset P_0$, so that $P_0 \subset \hat{P} \subset P$. Since $\hat{P}^{-1}$ is (completely) continuous and defined in the whole of L^2, the proof is completed.

We shall next derive a description of the correctly posed boundary conditions, which differs from Višik's. Let U be the set of all solutions u of the homogeneous equation $Pu = 0$. This is a closed subspace of L^2, since P is a closed operator.

LEMMA 1.8. *Suppose that* P_0^{-1} *is continuous. Then the restriction* γ *of the boundary operator* Γ *to* U *maps* U *topologically onto a closed subspace* ΓU *of* C.

PROOF. Let A be a constant such that

$$\| P_0 f \| \geqq A \| f \|, \quad f \in \mathcal{D}_{P_0}.$$

Then we have, if $u \in U$,

$$\| u \|^2 \geqq \| \Gamma u \|^2 = \inf_{f \in \mathcal{D}_{P_0}} (\| u - f \|^2 + \| P_0 f \|^2) \geqq \inf_{f \in L^2} (\| u - f \|^2 + A^2 \| f \|^2)$$

$$= \inf \left\{ (1 + A^2) \left\| f - \frac{u}{1 + A^2} \right\|^2 + \| u \|^2 A^2/(1 + A^2) \right\} = \| u \|^2 A^2/(1 + A^2).$$

This proves the lemma.

THEOREM 1.3. *Suppose that P_0^{-1} and $\overline{P}_0^{-1}$ are continuous. Let B be a linear manifold in C, and let $\hat{P}$ be the corresponding operator. Then $\hat{P}$ is closed if and only if B is closed. $\hat{P}^{-1}$ exists if and only if B and ΓU have only the origin in common. $\hat{P}^{-1}$ is continuous and defined in the whole of L^2 if and only if C is the topological sum of B and ΓU.*

PROOF. The first assertion follows at once from the definition of the topology in quotient spaces. In fact, a set in a quotient space is by definition open (closed) if and only if its inverse image is open (closed).

$\hat{P}^{-1}$ has a sense if and only if $\hat{P} f \neq 0$ when $0 \neq f \in \mathcal{D}_{\hat{P}}$, that is, if no solution $u \neq 0$ of $P u = 0$ satisfies the boundary condition. But this means that 0 is the only common element of ΓU and B.

Now suppose that C is the topological sum of ΓU and B. From the preceding remark it follows that $\hat{P}^{-1}$ exists, and we have to prove that it is bounded. The assumption means that there exists a bounded (oblique) projection π of C on ΓU along B. Let S be the bounded right inverse of P, which was constructed in Lemma 1.2, and let γ be the restriction of Γ to U, which was studied in Lemma 1.7. Then the operator

$$T g = S^{-1} g - \gamma^{-1} \pi \Gamma S^{-1} g$$

is defined in the whole of L^2 and is a continuous operator. Obviously, $T g \in \mathcal{D}_P$ and $P T g = g - 0$. Furthermore,

$$\Gamma T g = \Gamma S^{-1} g - \pi \Gamma S^{-1} g \in B,$$

so that $T g \in \mathcal{D}_{\hat{P}}$ and $\hat{P} T g = P T g = g$. Hence $\hat{P}^{-1} = T$, which proves the assertion.

On the other hand, suppose that $\hat{P}^{-1}$ is continuous and defined everywhere. Then the mapping

$$G_P \ni [f, P f] \to f - \hat{P}^{-1} P f \in U$$

is continuous. We have $f - \hat{P}^{-1} P f = 0$ if and only if $f \in \mathcal{D}_{\hat{P}}$. The mapping

$$G_P \ni [f, P f] \to \Gamma (f - \hat{P}^{-1} P f) \in \Gamma U$$

is also continuous and, since it vanishes in G_{P_0}, it defines a continuous mapping π from $G_P/G_{P_0}=C$ to ΓU. We have $\pi\varphi=0$ if and only if $\varphi\in B$. Now π leaves the elements of ΓU invariant. Hence π is a projection on ΓU along B, and from the continuity of π our assertion follows.

We finally sketch a similar study of the completely correctly posed boundary problems, when P_0 *and* $\bar{P}_0$ *have completely continuous inverses*, by introducing a new mode of convergence in C. We shall say that a sequence $[u_n, v_n]$ of elements in $L^2\times L^2$ is w-convergent, if u_n converges strongly and v_n converges weakly. In C we define the quotient w-convergence: a sequence φ_n of elements in C is w-convergent, if there exists a w-convergent sequence $[f_n, Pf_n]\in G_P$ such that $\Gamma f_n=\varphi_n$.

We shall prove that the operator γ from U to ΓU transforms the L^2- convergent sequences in U and the w-convergent sequences in ΓU into each other. In fact, Γu_n is obviously w-convergent if u_n is convergent. Conversely, if Γu_n is w-convergent, there exist elements $f_n\in\mathcal{D}_{P_0}$, so that u_n-f_n converges strongly and $P_0 f_n$ converges weakly. Since we have $f_n=P_0^{-1}(P_0 f_n)$, it follows from the weak convergence of $P_0 f_n$ and the complete continuity of P_0^{-1} that f_n is strongly convergent. Hence u_n is strongly convergent, which proves our assertion. Using Lemma 1.8 we now see that in ΓU the w-convergence is equivalent to strong convergence.

A slight modification of the proof of Theorem 1.3 shows that the operator $\hat{P}$, corresponding to a linear manifold B in C, has a completely continuous inverse, defined in the whole of L^2, if and only if C is the direct sum of B and ΓU, and the projection π of C on ΓU along B is w-continuous in the sense that it transforms w-convergent sequences into w-convergent (and hence strongly convergent) sequences.

Chapter II

Minimal Differential Operators with Constant Coefficients

2.0. Introduction

Let P be a differential operator with constant coefficients and let Ω be a domain in R^ν. In Chapter I we introduced the minimal differential operator P_0 in $L^2(\Omega)$, defined by P. The object of this chapter is to study P_0 more closely, We first restrict ourselves to the case where Ω is bounded, and can then obtain fairly complete results. Some remarks on the case of non-bounded domains are given at the end of the chapter.

We first establish the boundedness of the inverse of a minimal differential operator with constant coefficients for bounded Ω by means of the Laplace transformation, using a lemma by Malgrange [20]. This result shows that Lemma 1.7 is always applicable, i.e. that the equation $Pu = f$ has a square integrable solution u for any $f \in L^2(\Omega)$.

We then turn to the exact determination of the differential operators which are weaker than $\mathcal{P}$. With $D = (D_1, \dots, D_\nu)$, where $D_\iota = i^{-1}\, \partial / \partial x^\iota$, we may write $\mathcal{P} = P(D)$, where $P(\xi)$ is a polynomial in the vector $\xi = (\xi_1, \dots, \xi_\nu)$. Now set

$$\tilde{P}(\xi) = (\Sigma\, |P^{(\alpha)}(\xi)|^2)^{\frac{1}{2}},$$

where $P^{(\alpha)}$ are derivatives of P, and the summation extends over all α. Then Q is weaker than $\mathcal{P}$ if and only if

$$\frac{\tilde{Q}(\xi)}{\tilde{P}(\xi)} < C. \tag{2.0.1}$$

To prove this result we use a generalization of the energy integral method. For equations of higher order than two, this method was first used by Leray [19]. In the general case considered here, where the lower order terms of the operators have great importance, it has been necessary to develop an algebra of energy integrals in a systematic manner. It may be remarked that, for some special second order equations, similar questions have been posed and solved by Ladyzenskaja [18], even under less restrictive boundary conditions.

As a consequence of our result we find that the product of a function $u \in \mathcal{D}_{P_0}$ and a function ψ, which is C^∞ in a neighbourhood of $\overline{\Omega}$, is in $\mathcal{D}_{P_0}$. Hence we find that the relation $u \in \mathcal{D}_{P_0}$ has a local character. We then study this relation in the interior and at the boundary of $\overline{\Omega}$.

The inequalities derived by the energy integral method also make it possible to determine those operators Q for which Qu is continuous after correction on a null set for every $u \in \mathcal{D}_{P_0}$. In fact, this is the case if and only if

$$\int \frac{\tilde{Q}(\xi)^2}{\tilde{P}(\xi)^2}\, d\xi < \infty. \tag{2.0.2}$$

The inequalities (2.0.1) and (2.0.2) only involve the quotient $\tilde{Q}(\xi)/\tilde{P}(\xi)$. In section 2.8 we also give conditions in terms of this quotient in order that $Qu \in L^q$ for every $u \in \mathcal{D}_{P_0}$ and in order that Qu should exist in manifolds of dimension less than ν.

We can also prove that the inverse of P_0 is completely continuous, if $P(\xi)$ really depends on all variables. More generally, we prove that the operator $Q_0 P_0^{-1}$ is completely continuous if and only if

$$\frac{\tilde{Q}(\xi)}{\tilde{P}(\xi)} \to 0 \quad \text{when } \xi \to \infty. \tag{2.0.3}$$

2.1. Notations and formal properties of differential operators with constant coefficients

Let R^ν be the real ν-dimensional space with elements $x = (x^1, \dots, x^\nu)$ and let C_ν be the complex ν-dimensional space with elements $\zeta = (\zeta_1, \dots, \zeta_\nu)$. In precisely one way we can write $\zeta = \xi + i\eta$, where ξ and η, as in the whole paper, denote real vectors. The variables x and ζ will be considered as dual with respect to the bilinear form

$$\langle x, \zeta \rangle = \sum_1^\nu x^k \zeta_k.$$

A polynomial $P(\zeta)$ can be written as a finite sum

(2.1.1) $$P(\zeta) = \sum a^\alpha \zeta_\alpha,$$

where $\alpha = (\alpha_1, \dots, \alpha_k)$ is a sequence of indices between 1 and ν, the a^α are complex constants which do not change, if the indices in α are permuted, and $\zeta_\alpha = \zeta_{\alpha_1} \cdots \zeta_{\alpha_k}$. The length k of the sequence α is denoted by $|\alpha|$. The polynomial (2.1.1) defines a differential operator $\mathcal{D} = P(D)$ operating on the functions in R^ν,

(2.1.2) $$P(D) = \sum a^\alpha D_\alpha$$

(see section 1.2). The polynomials in C_ν and the differential operators in R^ν are thus in a one-to-one correspondence, and this correspondence is in fact independent of the choice of coordinates since

$$P(D)\, e^{i\langle x, \zeta \rangle} = P(\zeta)\, e^{i\langle x, \zeta \rangle}.$$

By $\mathcal{S}$ we denote the space of infinitely differentiable rapidly decreasing functions introduced by L. Schwartz [28]. Denoting the Fourier transform of a function u in $\mathcal{S}$ by $\hat{u}$,

(2.1.3) $$\hat{u}(\xi) = (2\pi)^{-\nu/2} \int u(x)\, e^{-i\langle x, \xi \rangle}\, dx,$$

the Fourier transform of $P(D)u$ is $P(\xi)\hat{u}(\xi)$, and it follows from Parseval's formula that

(2.1.4) $$\int |P(D)u|^2\, dx = \int |P(\xi)\hat{u}(\xi)|^2\, d\xi.$$

We shall repeatedly need the analogue of Leibniz' formula for general differential polynomials

(2.1.5) $$P(D)(uv) = P(D_u + D_v)uv.$$

The interpretation of this formula is that, after $P(D_u + D_v)$ has been expanded in powers of D_u and D_v, we shall let D_u operate only on u and D_v operate only on v. Formula (2.1.5)

is, of course, an immediate consequence of the rule for differentiating a product. Now we have by Taylor's formula

$$P(\xi+\eta)=\sum \frac{\eta_\alpha}{|\alpha|!} P^{(\alpha)}(\xi),$$

where

$$P^{(\alpha)}=\frac{\partial^{|\alpha|} P}{\partial \xi_\alpha}=\frac{\partial^k P}{\partial \xi_{\alpha_1}\cdots\partial \xi_{\alpha_k}}.$$

For $|\alpha|=k$ the k indices in α shall run independently form 1 to ν. Leibniz' formula (2.1.5) now takes the more explicit form

$$P(D)(uv)=\sum \frac{D_\alpha v}{|\alpha|!} P^{(\alpha)}(D)u. \tag{2.1.6}$$

2.2. Estimates by Laplace transforms

Let Ω be a *bounded* domain in R^ν, and let $P(D)$ be a differential operator with constant coefficients. We shall prove the continuity of the inverse of the minimal operator P_0.

THEOREM 2.1. *The operator P_0 has a continuous inverse, i.e. there exists a constant C such that*

$$\|u\| \leq C\|P(D)u\|, \quad u \in C_0^\infty(\Omega). \tag{2.2.1}$$

PROOF. We form the Laplace transform of u, defined by

$$\hat{u}(\zeta)=\hat{u}(\xi+i\eta)=(2\pi)^{-\nu/2}\int e^{-i\langle x,\, \xi+i\eta\rangle} u(x)\,dx.$$

This is an entire analytic function since u has compact support. The Laplace transform of $P(D)u$ is $P(\zeta)\hat{u}(\zeta)$. Now the proof of (2.2.1) follows easily from the following lemma on analytic functions of one variable, analogous to one used previously by Malgrange [20].

LEMMA 2.1. *If $g(z)$ is an analytic function of a complex variable z for $|z|\leq 1$, and $r(z)$ is a polynomial with highest coefficient A, then*

$$|Ag(0)|^2 \leq (2\pi)^{-1}\int_0^{2\pi} |g(e^{i\theta})\,r(e^{i\theta})|^2\,d\theta. \tag{2.2.2}$$

PROOF OF LEMMA 2.1. Let z_j be the zeros of $r(z)$ in the unit circle and set

$$r(z)=q(z)\prod_j \frac{z-z_j}{\bar{z}_j z-1}.$$

On the unit circle we have $|r(z)|=|q(z)|$, and $q(z)$ is analytic in the circle. Hence we have

$$(2\pi)^{-1}\int |g(e^{i\theta})\,r(e^{i\theta})|^2\,d\theta=(2\pi)^{-1}\int |g(e^{i\theta})\,q(e^{i\theta})|^2\,d\theta \geq |g(0)\,q(0)|^2.$$

Now $q(0)/A$ is, apart from a factor ± 1, the product of the zeros of $r(z)$ outside the unit circle. Hence $|q(0)| \geqq |A|$, which proves the lemma.

We now complete the proof of the theorem. Choose a real vector ξ_0 such that $p(\xi_0) \neq 0$, where p is the principal part of P, that is, the homogeneous part of highest degree of P. Applying the lemma to the analytic function $\hat{u}(\zeta + t\xi_0)$ and the polynomial $P(\zeta + t\xi_0)$, considered as functions of the complex variable t, we get

$$|\hat{u}(\zeta)\, p(\xi_0)|^2 \leqq (2\pi)^{-1} \int |\hat{u}(\zeta + e^{i\theta}\xi_0)\, P(\zeta + e^{i\theta}\xi_0)|^2\, d\theta.$$

Letting $\zeta = \xi$ be real and integrating with respect to ξ we obtain

$$|p(\xi_0)|^2 \int |\hat{u}(\xi)|^2\, d\xi \leqq (2\pi)^{-1} \iint |\hat{u}(\xi + e^{i\theta}\xi_0)\, P(\xi + e^{i\theta}\xi_0)|^2\, d\xi\, d\theta$$
$$= (2\pi)^{-1} \int d\theta \int |\hat{u}(\xi + i\xi_0 \sin\theta)\, P(\xi + i\xi_0 \sin\theta)|^2\, d\xi.$$

By Parseval's formula we can calculate the integrals with respect to ξ, which gives

$$|p(\xi_0)|^2 \int |u(x)|^2\, dx \leqq (2\pi)^{-1} \int d\theta \int |P(D)\, u(x)|^2\, e^{2\langle x, \xi_0\rangle \sin\theta}\, dx.$$

Let C be the supremum of $e^{|\langle x, \xi_0\rangle|}/|p(\xi_0)|$ when $x \in \Omega$. Then we have

$$\int |u(x)|^2\, dx \leqq C^2 \int |P(D)\, u(x)|^2\, dx,$$

which proves (2.2.1).

By choosing ξ_0 in a suitable fashion we could get a good estimate of the magnitude of the constant C. We shall not do so, since still better results can be obtained by a different method later in this chapter.

2.3. The differential operators weaker than a given one

Let $P(D)$ be a differential operator with constant coefficients and let Ω be a *bounded* domain. We shall determine those operators $Q(D)$ with constant coefficients which are weaker than $P(D)$ in the sense of Definition 1.2, i.e. such that with some constant C

(2.3.1) $$\|Q(D)u\|^2 \leqq C(\|P(D)u\|^2 + \|u\|^2), \quad u \in C_0^\infty(\Omega).$$

In virtue of Theorem 2.1 this is equivalent to

(2.3.1)′ $$\|Q(D)u\|^2 \leqq C' \|P(D)u\|^2, \quad u \in C_0^\infty(\Omega).$$

In formulating the result it is convenient to use the function

(2.3.2) $$\tilde{P}(\xi) = (\sum |P^{(\alpha)}(\xi)|^2)^{\frac{1}{2}}.$$

This notation will be retained in the whole chapter, also with P replaced by other letters.

THEOREM 2.2. *A necessary and sufficient condition in order that $Q(D)$ should be weaker than $P(D)$ in a bounded domain Ω is that*

$$\frac{\tilde{Q}(\xi)}{\tilde{P}(\xi)} < C, \tag{2.3.3}$$

for every real ξ, where C is a constant.

REMARK. We shall even prove that $Q(D)$ is weaker than $P(D)$ if $|Q(\xi)|/\tilde{P}(\xi) < C$. Hence this condition is equivalent to (2.3.3), with a different constant C.

Theorem 2.2 has a central role in this chapter. The full proof is long and will fill the next sections. That (2.3.3) follows from (2.3.1) is proved in this section. In 2.4 we develop some algebraic aspects of energy integrals, and the analytical consequences are given in section 2.5. Using these results we complete the proof of Theorem 2.2 in section 2.6. At the same time we get a new proof of Theorem 2.1, that does not use Laplace transforms.

We now prove that (2.3.3) follows, if we suppose that (2.3.1) holds true. To make use of this inequality, take a function $\psi \in C_0^\infty(\Omega)$, $\psi \neq 0$, and set with real constant ξ

$$u(x) = \psi(x)\, e^{i\langle x, \xi\rangle}. \tag{2.3.4}$$

This function is in $C_0^\infty(\Omega)$, and from Leibniz' formula (2.1.6) it follows that

$$P(D)\,u(x) = e^{i\langle x,\xi\rangle} \sum P^{(\alpha)}(\xi)\frac{D_\alpha \psi(x)}{|\alpha|!}, \tag{2.3.5}$$

and similarly with P replaced by Q. If we introduce the notation

$$\psi_{\alpha\beta} = \frac{1}{|\alpha|!\,|\beta|!}\int D_\alpha \psi\, \overline{D_\beta \psi}\, dx, \tag{2.3.6}$$

the inequality (2.3.1) gives

$$\sum Q^{(\alpha)}(\xi)\,\overline{Q^{(\beta)}(\xi)}\,\psi_{\alpha\beta} \leqq C\left(\sum P^{(\alpha)}(\xi)\,\overline{P^{(\beta)}(\xi)}\,\psi_{\alpha\beta} + \psi_{00}\right). \tag{2.3.7}$$

If m is the highest of the orders of P and Q, the sums in (2.3.7) only contain terms with $|\alpha| \leqq m$ and $|\beta| \leqq m$. Now let $t = (t_\alpha)$ be an "array" of complex numbers, $0 \leqq |\alpha| \leqq m$, such that $t_\alpha = t_{\alpha'}$, when α' is a permutation of α. The quadratic form in t defined by

$$\sum_{|\alpha|\leqq m}\sum_{|\beta|\leqq m} t_\alpha \bar{t}_\beta \psi_{\alpha\beta} = \int \left|\sum_{|\alpha|\leqq m} \frac{t_\alpha D_\alpha \psi}{|\alpha|!}\right|^2 dx = \int \left|\sum_{|\alpha|\leqq m}\frac{t_\alpha \xi_\alpha}{|\alpha|!}\right|^2 |\hat{\psi}(\xi)|^2\, d\xi \tag{2.3.8}$$

is positive, unless the polynomial $\sum t_\alpha \xi_\alpha / |\alpha|!$ vanishes identically, i.e. every $t_\alpha = 0$. Hence it follows that there is a constant C' such that

(2.3.9)
$$\sum_{|\alpha|\leq m} |t_\alpha|^2 \leqq C' \sum_{|\alpha|\leq m} \sum_{|\beta|\leq m} t_\alpha \bar{t}_\beta \psi_{\alpha\beta}.$$

With $t_\alpha = Q^{(\alpha)}(\xi)$ we now get from (2.3.9) and (2.3.7) that

$$\sum |Q^{(\alpha)}(\xi)|^2 \leqq C' \sum Q^{(\alpha)}(\xi) \overline{Q^{(\beta)}(\xi)} \psi_{\alpha\beta} \leqq C C' (\sum P^{(\alpha)}(\xi) \overline{P^{(\beta)}(\xi)} \psi_{\alpha\beta} + \psi_{00}),$$

so that with a third constant C''

$$\tilde{Q}(\xi) \leqq C'' \tilde{P}(\xi).$$

2.4. The algebra of energy integrals

In this section we shall study some algebraic aspects of quadratic forms with constant coefficients in the derivatives of a function u. Such a form can be written

(2.4.1)
$$\sum_{\alpha,\beta} a^{\alpha\beta} D_\alpha u \overline{D_\beta u},$$

where D_α and D_β are defined in section 1.2, and $a^{\alpha\beta}$ is invariant for permutations within α or β. With this quadratic differential form we associate the polynomial

(2.4.2)
$$F(\zeta, \bar{\zeta}) = \sum a^{\alpha\beta} \zeta_\alpha \bar{\zeta}_\beta,$$

where $\zeta = \xi + i\eta$ and $\bar{\zeta} = \xi - i\eta$. Since the value of the form (2.4.1) for $u(x) = e^{i\langle x, \zeta\rangle}$ is $e^{-2\langle x, \eta\rangle} F(\zeta, \bar{\zeta})$, the correspondence between the form (2.4.1) and the polynomial (2.4.2) is one to one and invariant for coordinate transformations. This justifies the following shorter notation

(2.4.3)
$$F(D, \bar{D}) u \bar{u} = \sum_{\alpha,\beta} a^{\alpha\beta} D_\alpha u \overline{D_\beta u}.$$

In section 2.1 we introduced a correspondence between the differential operators in R^ν and the complex-valued polynomials in C_ν, considered as a ν-dimensional vector space with complex structure. We have now seen that the quadratic differential forms in R^ν can be associated with the complex-valued polynomials in C_ν, considered as a 2ν-dimensional vector space with real structure.

If $\bar{F}(\zeta, \bar{\zeta})$ is the polynomial whose coefficients are the complex conjugates of those of $F(\zeta, \bar{\zeta})$, it is readily verified that

(2.4.4)
$$\overline{F(D, \bar{D}) u \bar{u}} = \bar{F}(\bar{D}, D) u \bar{u}.$$

Hence $F(D, \bar{D}) u \bar{u}$ is real for every u if and only if

$$F(\zeta, \bar{\zeta}) = \bar{F}(\bar{\zeta}, \zeta) = \overline{F(\zeta, \bar{\zeta})},$$

i.e. if $F(\zeta, \bar{\zeta})$ is always real.

We shall need a formula for the differentiation of a quadratic differential form $F(D, \bar{D})\, u\bar{u}$. Elementary product differentiation gives

$$\frac{\partial}{\partial x^k}(F(D, \bar{D})\, u\bar{u}) = i(D_k - \bar{D}_k)\, F(D, \bar{D})\, u\bar{u}.$$

Hence, if $G = (G^k)$ is a vector whose components are quadratic differential forms, we have

$$\text{div}\,(G(D, \bar{D})\, u\bar{u}) = \sum_1^{\nu} \frac{\partial}{\partial x^k}(G^k(D, \bar{D})\, u\bar{u}) = F(D, \bar{D})\, u\bar{u}, \tag{2.4.5}$$

where

$$F(\zeta, \bar{\zeta}) = i\sum_1^{\nu}(\zeta_k - \bar{\zeta}_k)\, G^k(\zeta, \bar{\zeta}) = -2\sum_1^{\nu}\eta_k\, G^k(\zeta, \bar{\zeta}). \tag{2.4.6}$$

LEMMA 2.2. *A polynomial $F(\zeta, \bar{\zeta})$ in $\zeta = \xi + i\eta$ and $\bar{\zeta} = \xi - i\eta$ can be written in the form*

$$F(\zeta, \bar{\zeta}) = -2\sum_1^{\nu}\eta_k\, G^k(\zeta, \bar{\zeta}),$$

where G^k are polynomials, if and only if $F(\xi, \xi) = 0$ when ξ is real.

PROOF. That $F(\xi, \xi) = 0$ is a necessary condition is obvious. To prove its sufficiency we observe that if $F(\xi + i\eta, \xi - i\eta) = 0$ when $\eta = 0$, there are no terms free from η in the expansion of $F(\xi + i\eta, \xi - i\eta)$ in powers of ξ and η. Hence we can write

$$F(\xi + i\eta, \xi - i\eta) = -2\sum_1^{\nu}\eta_k\, g^k(\xi, \eta),$$

where g^k are polynomials. Returning to the variables ζ and $\bar{\zeta}$ in g^k, the lemma is proved.

From the proof it follows that the vector $(G^1(\zeta, \bar{\zeta}), \ldots, G^{\nu}(\zeta, \bar{\zeta}))$ is not uniquely determined in general. We shall now determine the degree of indeterminacy, that is, we shall find all vector differential forms with divergence zero.

LEMMA 2.3. *If the polynomials $G^i(\zeta, \bar{\zeta})$ satisfy the identity*

$$\sum_1^{\nu}\eta_i\, G^i(\zeta, \bar{\xi}) \equiv 0,$$

then there exist polynomials $G^{ik}(\zeta, \bar{\zeta})$ such that $G^{ik}(\zeta, \bar{\zeta}) = -G^{ki}(\zeta, \bar{\zeta})$ and

$$G^i(\zeta, \bar{\zeta}) = -2\sum_1^{\nu}\eta_k\, G^{ik}(\zeta, \bar{\zeta}).$$

PROOF. If we write $G^i(\xi+i\eta, \xi-i\eta)=g^i(\xi,\eta)$, the assumption means that

$$\sum_1^\nu \eta_i g^i(\xi,\eta)=0. \tag{2.4.7}$$

Since the identity (2.4.7) must also be satisfied by the parts of $g^i(\xi,\eta)$, which are homogeneous of the same degree with respect to η, we may suppose in the proof that the g^i are all homogeneous of degree m with respect to η. Then Euler's identity gives

$$m\, g^k=\sum_1^\nu \eta_i \frac{\partial g^k}{\partial \eta_i}.$$

Differentiation of (2.4.7) with respect to η_k gives

$$g^k+\sum_1^\nu \eta_i \frac{\partial g^i}{\partial \eta_k}=0.$$

Now addition of these two relations shows that

$$(m+1)\, g^k=\sum_\nu^1 \eta_i \left(\frac{\partial g^k}{\partial \eta_i}-\frac{\partial g^i}{\partial \eta_k}\right),$$

and therefore

$$g^{ik}=\frac{1}{2(m+1)}\left(\frac{\partial g^k}{\partial \eta_i}-\frac{\partial g^i}{\partial \eta_k}\right)$$

has the desired properties when ζ and $\bar\zeta$ are introduced as variables again.

From Lemma 2.3 it follows, in particular, that, although the polynomials $G^i(\zeta,\bar\zeta)$ figuring in Lemma 2.2 are not uniquely determined, the values $G^i(\xi,\xi)$ for real arguments are. This is also easily proved directly. For differentiating (2.4.6) and putting $\eta=0$ afterwards gives

$$G^k(\xi,\xi)=-\tfrac{1}{2}\frac{\partial F(\xi+i\eta, \xi-i\eta)}{\partial \eta_k}\Bigg|_{(\eta=0)}. \tag{2.4.8}$$

This formula is most important in the application below.

2.5. Analytical properties of energy integrals

Let u be a function in S and let $\hat u$ be its Fourier transform. Using the definition (2.4.3) and Parseval's formula, we obtain

$$\int F(D,\bar D)\, u\bar u\, dx=\int F(\xi,\xi)\, |\hat u(\xi)|^2\, d\xi. \tag{2.5.1}$$

As a first application of this formula we prove

LEMMA 2.4. *If for every $u \in C_0^\infty(\Omega)$, where Ω is a fixed domain, we have*

$$\int F(D, \bar{D})\, u \bar{u}\, dx = 0, \tag{2.5.2}$$

then it follows that

$$F(\xi, \xi) = 0, \text{ for real } \xi. \tag{2.5.3}$$

Conversely, (2.5.3) *implies* (2.5.2) *for any* Ω.

PROOF. The last statement follows at once from (2.5.1). On the other hand, let (2.5.2) be valid. Let $u \neq 0$ be a fixed function in $C_0^\infty(\Omega)$. For fixed η, the function $u(x)e^{i\langle x, \eta\rangle}$ is in $C_0^\infty(\Omega)$ and has the Fourier transform $\hat{u}(\xi - \eta)$, so that it follows from (2.5.1) that

$$\int F(\xi + \eta, \xi + \eta)\, |\hat{u}(\xi)|^2\, d\xi = 0. \tag{2.5.4}$$

Denote the polynomial $F(\xi, \xi)$ by $\pi(\xi)$, and the principal part of $\pi(\xi)$ by $\pi_m(\xi)$. It follows from (2.5.4), which is valid for every η, that

$$\pi_m(\eta) \int |\hat{u}|^2\, d\xi = 0$$

for every η. Hence π_m and consequently π is identically zero.

Combining Lemmas 2.4 and 2.2 we obtain the following lemma.

LEMMA 2.5. *A quadratic differential form $F(D, \bar{D})\, u \bar{u}$ is the divergence of a quadratic differential vector form if and only if*

$$\int F(D, \bar{D})\, u \bar{u}\, dx = 0,$$

when $u \in C_0^\infty(\Omega)$ for some domain Ω.

We could also deduce from Lemma 2.3:

LEMMA 2.6. *A quadratic differential vector with the components $G^k(D, \bar{D})\, u \bar{u}$ is the divergence of a quadratic differential skew symmetric tensor form if and only if for any $u \in C^\infty$ and any closed surface S we have*

$$\int_S (G^k(D, \bar{D})\, u \bar{u})\, dS_k = 0.$$

The analogy between these two lemmas and the theory of exterior differential forms is obvious. In order to show this connection we have in fact proved more results on the energy integrals than we really need to prove Theorem 2.2.

2.6. Estimates by energy integrals

Let $P(D)$ and $Q(D)$ be two differential operators with constant coefficients and form

$$F(D, \bar{D})\, u \bar{u} = (P(D)Q(\bar{D}) - Q(D)P(\bar{D}))\, u \bar{u}. \tag{2.6.1}$$

We have $F(\xi, \xi) = P(\xi)Q(\xi) - Q(\xi)P(\xi) = 0$, so that in virtue of Lemma 2.2 we can write

$$F(D, \bar{D})\, u\bar{u} = \sum_{k=1}^{\nu} \frac{\partial}{\partial x^k} (G^k (D, \bar{D})\, u\bar{u}).$$

Formula (2.4.8) gives that

$$G^k (\xi, \xi) = -i\, (P^{(k)} (\xi)\, Q(\xi) - Q^{(k)} (\xi)\, P(\xi)), \tag{2.6.2}$$

where, in accordance with our general notations, $P^{(k)}$ and $Q^{(k)}$ are the partial derivatives of P and Q with respect to ξ_k.

Let Ω be a fixed *bounded* domain, and let u be a function in $C_0^\infty(\Omega)$. We shall integrate the identity

$$-i x^k F(D, \bar{D})\, u\bar{u} = -i x^k \sum_j \frac{\partial}{\partial x^j} (G^j (D, \bar{D})\, u\bar{u})$$

over Ω. In doing so, we can integrate the right-hand side by parts, so that the integral equals $i \int G^k (D, \bar{D})\, u\bar{u}\, dx$. Now it follows from (2.6.2) and Lemma 2.4 that

$$\begin{aligned}\int G^k (D, \bar{D})\, u\bar{u}\, dx &= -i \int (P^{(k)} (D)\, Q(\bar{D}) - P(D)\, Q^{(k)} (\bar{D}))\, u\bar{u}\, dx \\ &= -i \{(P^{(k)} (D)\, u, \bar{Q}(D)\, u) - (P(D)\, u, \bar{Q}^{(k)} (D)\, u)\},\end{aligned}$$

where (,) denotes scalar product in $L^2(\Omega)$. Hence we get the formula

$$\begin{aligned}&(P^{(k)} (D)\, u, \bar{Q}(D)\, u) - (P(D)\, u, \bar{Q}^{(k)} (D)\, u) \\ &\qquad = \int -i x^k (P(D)\, u\, \overline{\bar{Q}(D)\, u} - Q(D)\, u\, \overline{\bar{P}(D)\, u})\, dx.\end{aligned} \tag{2.6.3}$$

By estimating the right-hand side of the equality (2.6.3) we can obtain a useful inequality. In fact, noting that it follows from (2.1.4) that

$$\| P(D)\, u \| = \| \bar{P}(D)\, u \|, \quad \| Q(D)\, u \| = \| \bar{Q}(D)\, u \|,$$

and denoting by δ an upper bound of $|x^k|$ in Ω, we obtain

$$|(P^{(k)} (D)\, u, \bar{Q}(D)\, u)| \leq \| P(D)\, u \| (\| \bar{Q}^{(k)} (D)\, u \| + 2\delta \| \bar{Q}(D)\, u \|) \tag{2.6.4}$$

by using Schwarz' inequality. When $Q = \bar{P}^{(k)}$ this inequality reduces to

$$\| P^{(k)} (D)\, u \|^2 \leq \| P(D)\, u \| (\| P^{(kk)} (D)\, u \| + 2\delta \| P^{(k)} (D)\, u \|), \tag{2.6.5}$$

where $P^{(kk)}(\xi)$ is the second derivative of $P(\xi)$ with respect to ξ_k. The inequality (2.6.5) gives a proof of the following lemma.

LEMMA 2.7. *Let B^k be the breadth of Ω in the direction x^k, i.e.*

$$B^k = \sup_{x, y \in \Omega} |x^k - y^k|.$$

Then, if $P(\xi)$ is of degree m with respect to ξ_k, we have

$$\| P^{(k)} (D)\, u \| \leq m B^k \| P(D)\, u \|, \quad u \in C_0^\infty(\Omega). \tag{2.6.6}$$

PROOF. After a convenient choice of the origin we may suppose that $|x^k| \leq B^k/2$ in Ω, so that we may put $2\delta = B^k$ in inequality (2.6.5). If $m = 1$, the second derivative $P^{(kk)}$ is zero, and (2.6.5) reduces to (2.6.6), if we delete a factor $\| P^{(k)}(D)u \|$. Now suppose that the inequality (2.6.6) has already been proved for all polynomials of smaller degree than m in ξ_k. Then we have, in particular,

$$\| P^{(kk)}(D)u \| \leq (m-1) B^k \| P^{(k)}(D)u \|.$$

If we use this estimate in the right-hand side of (2.6.5), it follows that

$$\| P^{(k)}(D)u \| \leq m B^k \| P(D)u \|,$$

which completes the proof.

It follows from the proof that (2.6.6) remains valid for non-bounded domains Ω if only $B^k < \infty$. We shall later come back to the case of infinite domains (section 2.11), but for the moment we confine ourselves to the simpler case of a bounded domain Ω.

LEMMA 2.8. *For any derivative $P^{(\alpha)}$ of P there is a constant C such that*

$$\|P^{(\alpha)}(D)u\| \leq C \|P(D)u\|, \quad u \in C_0^\infty(\Omega). \tag{2.6.7}$$

PROOF. Iteration of the result of Lemma 2.7 proves Lemma 2.8 immediately, and also gives an estimate of the constant C, which we do not care to write out explicitly.

Since a suitable derivative of P is a constant, Lemma 2.8 contains Theorem 2.1, which has thus been proved without the use of the Laplace transform.

We can now complete the proof of Theorem 2.2 and the remark following it. Thus suppose that

$$|Q(\xi)|^2 \leq C^2 \sum |P^{(\alpha)}(\xi)|^2. \tag{2.6.8}$$

If $\hat{u}$ is the Fourier transform of a function $u \in C_0^\infty(\Omega)$, we have in virtue of (2.1.4) and (2.6.8)

$$\int |Q(D)u|^2 dx = \int |Q(\xi)|^2 |\hat{u}|^2 d\xi \leq C^2 \sum \int |P^{(\alpha)}(\xi)|^2 |\hat{u}|^2 d\xi = C^2 \sum \int |P^{(\alpha)}(D)u|^2 dx.$$

It now follows from (2.6.7) that with a suitable constant C'

$$\| Q(D)u \|^2 \leq C' \| P(D)u \|^2, \quad u \in C_0^\infty(\Omega),$$

so that (2.3.1)′ is proved.

2.7. Some special cases of Theorem 2.2

The problem of finding all differential operators $Q(D)$, which are weaker than a given differential operator $P(D)$, has been reduced by Theorem 2.2 to the purely algebraic study of inequality (2.3.3). In studying this inequality, it is convenient to say that the polynomial P is stronger than the polynomial Q, if this inequality is valid. We shall first give two explicit examples.

EXAMPLE 1. The Schrödinger equation for a free particle corresponds to the polynomial $P(\xi)=\xi_1^2+\cdots+\xi_{\nu-1}^2-\xi_\nu$. This polynomial is stronger than those polynomials $Q(\xi)$ for which

$$|Q(\xi)|^2<C((\xi_1^2+\cdots+\xi_{\nu-1}^2-\xi_\nu)^2+\xi_1^2+\cdots+\xi_{\nu-1}^2+1). \tag{2.7.1}$$

Evidently (2.7.1) requires that $Q(\xi)$ is of degree two at most and not of higher degree than one in ξ_ν, so we may write

$$Q(\xi)=a_0+\sum_{i=1}^{\nu} a_i\,\xi_i+\sum_{i,k=1}^{\nu} a_{ik}\,\xi_i\,\xi_k, \tag{2.7.2}$$

where $a_{ik}=a_{ki}$ and $a_{\nu\nu}=0$. If we set $\xi_\nu=\xi_1^2+\cdots+\xi_{\nu-1}^2$, it follows from (2.7.1) that (2.7.2) must become a polynomial of degree one at most in the remaining variables $\xi_1,\ldots,\xi_{\nu-1}$. Hence $Q(\xi)$ must have the form

$$Q(\xi)=a_0+\sum_{k=1}^{\nu-1} a_k\,\xi_k+a_\nu(\xi_\nu-\xi_1^2-\cdots-\xi_{\nu-1}^2). \tag{2.7.3}$$

Conversely, it is obvious that every polynomial of the form (2.7.3) satisfies the inequality (2.7.1).

EXAMPLE 2. The equation of heat corresponds to the polynomial $P(\xi)=\xi_1^2+\cdots+\xi_{\nu-1}^2+i\xi_\nu$. This polynomial is stronger than those polynomials $Q(\xi)$ for which

$$|Q(\xi)|^2<C((\xi_1^2+\cdots+\xi_{\nu-1}^2)^2+\xi_1^2+\cdots+\xi_{\nu-1}^2+\xi_\nu^2+1). \tag{2.7.4}$$

This inequality is evidently fulfilled if and only if $Q(\xi)$ has the form

$$Q(\xi)=a_0+\sum_{k=1}^{\nu} a_k\,\xi_k+\sum_{i,k=1}^{\nu-1} a_{ik}\,\xi_i\,\xi_k. \tag{2.7.5}$$

The two examples show clearly that the lower order terms may have a decisive influence on the strength of an operator.[1] It is this fact that compelled us to develop such a strong generalization of the usual technique of energy integrals, which essentially works with the principal part of the operator, i.e. the homogeneous part of highest degree. The usual technique would, however, be successful within the class of operators satisfying the following definition.

DEFINITION 2.1. *The differential operator $P(D)$ (and the polynomial $P(\xi)$) is said to be of principal type, if it is equally strong as any other operator with the same principal part.*

[1] A similar fact has been observed by GÅRDING [8], who has shown that the correctness of Cauchy's problem can be affected by lower order terms.

The definition only involves restrictions on the principal part. This fact is explicitly expressed by the following theorem.

THEOREM 2.3. *A necessary and sufficient condition in order that $P(\xi)$ should be of principal type is that the partial derivatives $\partial p(\xi)/\partial \xi_i$ of the principal part $p(\xi)$ do not vanish simultaneously for any real $\xi \neq 0$.*

PROOF. Let $P(\xi)$ be of principal type. Then the same is true of $p(\xi)$, so that $p(\xi)$ is stronger than $p(\xi) + \xi_\alpha$ and consequently stronger than ξ_α, if $|\alpha| = m - 1$, where m is the degree of $p(\xi)$. Hence it follows from Theorem 2.2 that

$$\frac{(\xi_1^2 + \cdots + \xi_\nu^2)^{m-1}}{\Sigma\,|p^{(\alpha)}(\xi)|^2} < C'. \tag{2.7.6}$$

Suppose that all the derivatives $\partial p(\eta)/\partial \eta$ vanish for some real $\eta \neq 0$. Then we have also $p(\eta) = 0$ in virtue of Euler's formula for homogeneous polynomials. Hence, if we set $\xi = t\eta$ in (2.7.6), the denominator is of degree less than $2(m-1)$ in t, which gives a contradiction when $t \to \infty$. This proves one half of the theorem.

Now suppose that $P(\xi)$ satisfies the condition in Theorem 2.3 and let $Q(\xi)$ have the same principal part as $P(\xi)$. Dropping positive terms in the definition of $\tilde{P}(\xi)^2$, we obtain

$$\tilde{P}(\xi)^2 > \sum_1^\nu \left| \frac{\partial P(\xi)}{\partial \xi_k} \right|^2 = \pi(\xi) + r(\xi),$$

where $r(\xi)$ is of degree less than $2(m-1)$ and

$$\pi(\xi) = \sum_1^\nu \left| \frac{\partial p(\xi)}{\partial \xi_k} \right|^2 .$$

In virtue of the assumptions we have $\pi(\xi) \neq 0$, if $\xi \neq 0$, and therefore $r(\xi)/\pi(\xi) \to 0$ when $\xi \to \infty$, so that $|r(\xi)|/\pi(\xi) < \frac{3}{4}$ for large ξ. We note that

$$\frac{|Q(\xi)|}{\tilde{P}(\xi)} \leqq \frac{|Q(\xi) - P(\xi)|}{\tilde{P}(\xi)} + \frac{|P(\xi)|}{\tilde{P}(\xi)}.$$

Since the last term is always less than 1 and

$$\overline{\lim_{\xi\to\infty}} \frac{|Q(\xi) - P(\xi)|}{\tilde{P}(\xi)} \leqq \overline{\lim_{\xi\to\infty}} \frac{2\,|Q(\xi) - P(\xi)|}{\sqrt{\pi(\xi)}} < \infty,$$

it follows that $Q(\xi)$ is weaker than $P(\xi)$. Changing P for Q we conclude that P and Q are equally strong.

Our interest in differential operators of principal type is due to the fact that they have simple properties even when the coefficients are variable. We postpone the study of this case to Chapter IV, and pass to another class of differential operators with constant coefficients. As is well known, a differential operator $P(D)$ is called *elliptic*, if the principal part $p(\xi)$ does not vanish for any real $\xi \neq 0$. We give an equivalent property:

THEOREM 2.4. *The differential operator $P(D)$ is elliptic if and only if it is stronger than any operator of order not exceeding that of P.*

This is an almost obvious consequence of Theorem 2.2, so that we may omit the proof. In particular, Theorem 2.4 shows that all elliptic operators of the same order are equally strong.

We shall now study an operator with separable variables,

$$P(\xi)=P(\xi_1,\ldots,\xi_\nu)=P_1(\xi_1,\ldots,\xi_\mu)\,P_2(\xi_{\mu+1},\ldots,\xi_\nu) \quad (\mu<\nu).$$

The vector ξ is the sum of the two components

$$\xi'=(\xi_1,\ldots,\xi_\mu,0,\ldots,0),\ \xi''=(0,\ldots,0,\xi_{\mu+1},\ldots,\xi_\nu).$$

Let W' be the set of polynomials $Q(\xi')$, which are weaker than $P_1(\xi')$, and let W'' be the set of polynomials $Q(\xi'')$ which are weaker than $P_2(\xi'')$.

THEOREM 2.5. *The set W of polynomials $Q(\xi)$ weaker than $P(\xi)$ is the linear hull of the set $W'W''$ of products of polynomials in W' and W''.*

PROOF. Since $\tilde{P}(\xi)^2=\sum|P^{(\alpha)}(\xi)|^2$ differs from

$$\tilde{P}_1(\xi')^2\,\tilde{P}_2(\xi'')^2=\sum|P_1^{(\beta)}(\xi')|^2\,|P_2^{(\gamma)}(\xi'')|^2$$

only in the magnitude of the coefficients, we have

$$0<A\leq\frac{\tilde{P}(\xi)}{\tilde{P}_1(\xi')\,\tilde{P}_2(\xi'')}\leq B<\infty.$$

Hence $Q(\xi)$ is weaker than $P(\xi)$ if and only if

$$\frac{|Q(\xi',\xi'')|}{\tilde{P}_1(\xi')\,\tilde{P}_2(\xi'')}<C. \tag{2.7.7}$$

It now follows that the linear hull of $W'W''$ is in W. Inequality (2.7.7) also shows, if $Q\in W$, that $Q(\xi',\xi'')$ is in W' as a function of ξ', for fixed ξ'', and in W'' as a function of ξ'', for fixed ξ'. Let $p_1(\xi'),\ldots,p_n(\xi')$ be a basis in the finite dimensional vector space W' and set

$$Q(\xi',\xi'')=\sum_{k=1}^{n}a_k(\xi'')\,p_k(\xi').$$

It remains to prove that the coefficients $a_k(\xi'')$ are in W''. Since $p_k(\xi')$ are linearly independent functions, there exist values $\xi_1', \ldots, \xi_n'$ such that the matrix $(p_k(\xi_l'))$ is not singular. Then the system of equations

$$Q(\xi_l', \xi'') = \sum_{k=1}^{n} a_k(\xi'')\, p_k(\xi_l'), \quad l=1, \ldots, n,$$

can be solved for $a_k(\xi'')$. Hence $a_k(\xi'')$ is a linear combination of the functions $Q(\xi_l', \xi'')$ and consequently is in W''.

It is obvious how the theorem can be generalized, if a polynomial decomposes in this way into several factors.

2.8. The structure of the minimal domain

The first topic in this section concerns the continuity of the functions in $\mathcal{D}_{P_0}$ and their derivatives. From an abstract point of view this was already studied in Chapter I. We shall assume in the whole section that Ω is a *bounded* domain.

THEOREM 2.6. *If $Q(D)u$ is a continuous function after correction on a null set, for any $u \in \mathcal{D}_{P_0}$, then*

$$\int \frac{|Q(\xi)|^2}{\tilde{P}(\xi)^2}\, d\xi < \infty. \tag{2.8.1}$$

Conversely, if (2.8.1) *is valid, then $Q(D)u$ is uniformly continuous after correction on a null set and tends to zero at the boundary of Ω, for any $u \in \mathcal{D}_{P_0}$, in the sense that to every $\varepsilon > 0$ there exists a compact set K in Ω such that $|Q(D)u(x)| < \varepsilon$ in $\Omega - K$.*

PROOF. First suppose that $Q(D)u$ is always continuous when $u \in \mathcal{D}_{P_0}$. There is then only one obstacle to using Lemma 1.6: although the functions are continuous they need not a priori be bounded. Therefore we take a function $\psi(x) \in C_0^\infty(\Omega)$ and apply Lemma 1.6 to the differential operators $P(D)$ and

$$Q = \psi(x) Q(D).$$

It follows that there is a constant C such that

$$\sup_{x \in \Omega} |\psi(x) Q(D) u(x)|^2 \leq C(\|P(D)u\|^2 + \|u\|^2), \quad u \in C_0^\infty(\Omega).$$

We may suppose without restriction that $0 \in \Omega$ and that $\psi(0) = 1$. Then it follows that

$$|Q(D)u(0)|^2 \leq C(\|P(D)u\|^2 + \|u\|^2), \quad u \in C_0^\infty(\Omega). \tag{2.8.2}$$

Now take a function $\varphi(\xi) \in S$ and form

$$v(x) = (2\pi)^{-\nu/2} \int \frac{\varphi(\xi)}{\tilde{P}(\xi)} e^{i\langle x, \xi \rangle}\, d\xi.$$

Parseval's formula gives

$$\|P^{(\alpha)}(D)v\|^2 = \int \frac{|\varphi(\xi)|^2}{\tilde{P}(\xi)^2}|P^{(\alpha)}(\xi)|^2 d\xi \leqq \int |\varphi(\xi)|^2 d\xi. \tag{2.8.3}$$

Furthermore, v is also in $\mathcal{S}$. Now take a fixed function $\chi(x) \in C_0^\infty(\Omega)$, which equals 1 in a neighbourhood of the origin, and set

$$u(x) = \chi(x)v(x).$$

We then have $u \in C_0^\infty(\Omega)$ and, in virtue of Leibniz' formula and (2.8.3),

$$\|P(D)u\| \leqq C\|\varphi\|. \tag{2.8.4}$$

Noting that $Q(D)u(0) = Q(D)v(0)$, we deduce from (2.8.2) and (2.8.4) that

$$|(2\pi)^{-\nu/2}\int \frac{Q(\xi)}{\tilde{P}(\xi)}\varphi(\xi)\, d\xi|^2 \leqq C'^2 \int |\varphi(\xi)|^2 d\xi.$$

But this inequality implies that $Q(\xi)/\tilde{P}(\xi)$ is square integrable, which proves (2.8.1).

Now assume that (2.8.1) is valid. Estimating by Schwarz' inequality we get for $u \in C_0^\infty(\Omega)$

$$|Q(D)u(x)|^2 = |(2\pi)^{-\nu/2}\int Q(\xi)\hat{u}(\xi)e^{i\langle x,\xi\rangle}\, d\xi|^2$$

$$\leqq (2\pi)^{-\nu}\int \frac{|Q(\xi)|^2}{\tilde{P}(\xi)^2}d\xi \int \tilde{P}(\xi)^2|\hat{u}(\xi)|^2 d\xi = C'^2 \sum \|P^{(\alpha)}(D)u\|^2.$$

Lemma 2.8 now shows that

$$|Q(D)u(x)|^2 \leqq C'^2\|P(D)u\|^2, \quad u \in C_0^\infty(\Omega), \tag{2.8.5}$$

for any x. Hence the second half of the theorem follows from Lemma 1.6.

The formulations of Theorems 2.2 and 2.6 are closely related. This leads us to the following theorem.

THEOREM 2.7. *$Q(D)u$ is a function in L^p $(2 \leqq p \leqq \infty)$ for every $u \in \mathcal{D}_{P,\circ}$, if $Q(\xi)/\tilde{P}(\xi) \in L^{2p/(p-2)}$ in R_ν.*

PROOF. In virtue of the theorem of Titchmarsh and M. Riesz on Fourier transforms of functions in L^p (cf. Zygmund [35], p. 316), we have for $u \in C_0^\infty(\Omega)$

$$\|Q(D)u\|_p \leqq C\|Q(\xi)\hat{u}(\xi)\|_{p'},$$

where p' is defined by $p^{-1} + p'^{-1} = 1$. We may suppose that $2 < p < \infty$, since the extreme cases have already been treated. Then we have $p' < 2$, and Hölder's inequality proves that

$$\int |Q(\xi)\hat{u}(\xi)|^{p'} d\xi \leqq \left(\int |\tilde{P}(\xi)\hat{u}(\xi)|^2 d\xi\right)^{p'/2} \left(\int \left|\frac{Q(\xi)}{\tilde{P}(\xi)}\right|^{2p'/(2-p')} d\xi\right)^{1-\frac{p'}{2}}.$$

Since $2p'/(2-p') = 2p/(p-2)$, we obtain, if $u \in C_0^\infty(\Omega)$,

$$\|Q(D)u\|_p \leqq C' \|\tilde{P}(\xi)\hat{u}(\xi)\| \leqq C'' \|P(D)u\|, \tag{2.8.6}$$

where the last estimate follows from the proof of Theorem 2.6. It is clear that (2.8.6) gives the asserted result.

The theorem cannot be reversed since Sobolev's results (cf. [30], p. 64) are stronger for *elliptic* operators. We give two examples of non-elliptic operators.

EXAMPLE 1. If $P(\xi) = \xi_1^2 + \cdots + \xi_{\nu-1}^2 - \xi_\nu$, we have $1/\tilde{P}(\xi) \in L^q$ if and only if $q > \nu$. In particular, when $\nu = 2$, it follows that the functions in $\mathcal{D}_{P_0}$ are in L^q for every $q < \infty$ but are not all continuous.

EXAMPLE 2. If $P(\xi) = \xi_1^2 + \cdots + \xi_{\nu-1}^2 + i\xi_\nu$, we have $1/\tilde{P}(\xi) \in L^q$ if and only if $q > \frac{1}{2}(\nu+1)$. In particular, every function in the domain of P_0 is continuous when $\nu = 2$.

In the proof of Theorem 2.6 we found that $Q(D)u$ is continuous for any $u \in \mathcal{D}_{P_0}$ if and only if (2.8.2) is fulfilled, i.e., if the value of $Q(D)u$ at a fixed point is a continuous function of $[u, P(D)u] \in G_{P_0}$ $(u \in C_0^\infty(\Omega))$. When we now pass to studying $Q(D)u$ on varieties of dimensions between 1 and $\nu - 1$, we examine a condition similar to (2.8.2) from the outset. Thus let Σ be a variety in Ω and let $d\sigma$ be the element of area of Σ.[1] If the inequality

$$\int_\Sigma |Q(D)u|^2 \, d\sigma \leqq C(\|P(D)u\|^2 + \|u\|^2), \quad u \in C_0^\infty(\Omega), \tag{2.8.7}$$

holds good, the restriction of $Q(D)u$ to Σ may be defined when $u \in \mathcal{D}_{P_0}$ in the ususal way: We take a sequence $u_n \in C_0^\infty$ such that $u_n \to u$ and $P(D)u_n \to P_0 u$. In virtue of (2.8.7) the sequence $Q(D)u_n$ is convergent in $L^2(\Sigma)$. The limit in $L^2(\Sigma)$, which does not depend on the sequence u_n, which we have have chosen, is the desired restriction of $Q(D)u$ to Σ. Somewhat roughly we may say that *$Q(D)u$ exists in Σ for $u \in \mathcal{D}_{P_0}$, when the inequality (2.8.7) is valid.*

Our methods only permit us to study the case when Σ is a linear variety of dimension μ, $1 \leqq \mu \leqq \nu - 1$. We may of course assume that Σ has points in common with Ω. By Σ' we denote any one of the varieties in R_ν, orthogonal to Σ. The surface element in Σ' is denoted $d\sigma'$.

THEOREM 2.8. *A necessary and sufficient condition in order that $Q(D)u$ should exist in Σ for $u \in \mathcal{D}_{P_0}$ is that $Q(\xi)/\tilde{P}(\xi)$ is uniformly square integrable in the varieties Σ', i.e.*

[1] For simplicity in statements we may suppose that R^ν and R_ν have (dual) euclidean geometries. Then surface elements and norms of vectors are defined.

(2.8.8) $$\int_{\Sigma'} \frac{|Q(\xi)|^2}{\tilde{P}(\xi)^2} d\sigma' < C,$$

where the constant does not depend on the choice of the variety Σ', orthogonal to Σ.

The statement is still true, if Σ has dimension 0 or ν. It then reduces to Theorem 2.6 and Theorem 2.2, respectively.

PROOF. Passing, if necessary, to another system of coordinates, we may assume that Σ is defined by the equations

$$x^{\mu+1} = 0, \ldots, x^{\nu} = 0.$$

First suppose that (2.8.7) is valid. In virtue of Theorem 2.1 we then also have (with a different constant C)

(2.8.9) $$\int_{\Sigma} |Q(D)u|^2 dx^1 \cdots dx^{\mu} \leqq C \int_{\Omega} |P(D)u|^2 dx^1 \cdots dx^{\nu} \quad (u \in C_0^{\infty}(\Omega)).$$

By using a combination of the arguments in the proofs of Theorems 2.2 and 2.6, we shall prove that (2.8.8) follows.

Take a function $\varphi(\xi)$ in $\mathcal{S}$ and set for fixed $\xi_1, \ldots, \xi_{\mu}$

(2.8.10) $$v(x) = \int \frac{\varphi(\xi)}{\tilde{P}(\xi)} e^{i\langle x, \xi\rangle} d\sigma' = e^{i(x^1\xi_1 + \cdots + x^{\mu}\xi_{\mu})} \int \frac{\varphi(\xi)}{\tilde{P}(\xi)} e^{i(x^{\mu+1}\xi_{\mu+1} + \cdots + x^{\nu}\xi_{\nu})} d\sigma',$$

where $d\sigma' = d\xi_{\mu+1} \cdots d\xi_{\nu}$. Thus $v(x)$ is a function with spectrum in a variety Σ', orthogonal to Σ.

Differentiation under the integral sign gives

$$P^{(\alpha)}(D)v(x) = \int \frac{\varphi(\xi) P^{(\alpha)}(\xi)}{\tilde{P}(\xi)} e^{i\langle x, \xi\rangle} d\sigma',$$

and since $|P^{(\alpha)}(\xi)| \leqq \tilde{P}(\xi)$, it follows from Parseval's formula that

(2.8.11) $$\int |P^{(\alpha)}(D)v(x)|^2 dx^{\mu+1} \cdots dx^{\nu} \leqq (2\pi)^{\nu-\mu} \int |\varphi(\xi)|^2 d\sigma'.$$

Let ψ be a function in $C_0^{\infty}(\Omega)$ and set

(2.8.12) $$u(x) = v(x)\psi(x).$$

It is clear that $u \in C_0^{\infty}(\Omega)$, and by virtue of (2.8.11) and Leibniz' formula we have

(2.8.13) $$\int |P(D)u|^2\,dx \leq C\int |\varphi(\xi)|^2\,d\sigma'.$$

(Thus far, the argument is parallel to the proof of Theorem 2.6.)

In the plane Σ we have

(2.8.14) $$Q^{(\alpha)}(D)\,v(x^1, \ldots, x^\mu, 0, \ldots, 0) = e^{i(x^1\xi_1+\cdots+x^\mu\xi_\mu)}\int \frac{Q^{(\alpha)}(\xi)}{\tilde{P}(\xi)}\varphi(\xi)\,d\sigma'.$$

Assuming, as we may, that the function $\psi \in C_0^\infty(\Omega)$ does not vanish identically in Σ and that ψ is a function of $x^1, \ldots, x^\mu$ only in a neighbourhood of Σ, we can argue as in section 2.3. For Leibniz' formula shows that, when $x \in \Sigma$,

$$Q(D)\,u(x) = \sum{}^{*} \frac{D_\alpha \psi}{|\alpha|!} Q^{(\alpha)}(D)\,v(x),$$

where $\sum^*$ means a sum only over sequences of the indices $1, \ldots, \mu$. Setting

(2.8.15) $$t_\alpha = \int \frac{Q^{(\alpha)}(\xi)}{\tilde{P}(\xi)}\varphi(\xi)\,d\sigma',$$

we deduce from (2.8.14) that

(2.8.16) $$\int_\Sigma |Q(D)\,u(x)|^2\,dx^1\cdots dx^\mu = \sum_\alpha{}^{*}\sum_\beta{}^{*} \psi_{\alpha\beta}\,t_\alpha\,\bar{t}_\beta,$$

where

$$\psi_{\alpha\beta} = \frac{1}{|\alpha|!\,|\beta|!}\int_\Sigma D_\alpha\psi\,\overline{D_\beta\psi}\,dx^1\cdots dx^\mu.$$

Now we proved in section 2.3 that the quadratic form $\sum^* \psi_{\alpha\beta}\,t_\alpha\,\bar{t}_\beta$ is a positive definite form in the array $t = (t_\alpha)$, where α only contains the indices $1, \ldots, \mu$. In particular,

$$|t_0|^2 \leq C\sum_\alpha{}^{*}\sum_\beta{}^{*}\psi_{\alpha\beta}\,t_\alpha\,\bar{t}_\beta,$$

and this inequality combined with (2.8.16), (2.8.9), (2.8.13) and the definition (2.8.15) of t gives that

(2.8.17) $$\left|\int \frac{Q(\xi)}{\tilde{P}(\xi)}\varphi(\xi)\,d\sigma'\right|^2 \leq C\int |\varphi(\xi)|^2\,d\sigma'$$

for any choice of the function $\varphi(\xi) \in S$ and for any $\xi_1, \ldots, \xi_\mu$. (We denote by C different constants, different times.) Hence (2.8.8) follows.

Now suppose that (2.8.8) is fulfilled. For $u \in C_0^\infty(\Omega)$ we have

$$Q(D)\,u(x^1, \ldots, x^\mu, 0, \ldots, 0) = (2\pi)^{-\nu/2}\int Q(\xi)\,\hat{u}(\xi)\,e^{i(x^1\xi_1+\cdots+x^\mu\xi_\mu)}\,d\xi,$$

so that the Fourier transform of the function $Q(D)u(x^1, \ldots, x^\mu, 0, \ldots, 0)$ as a function of $x^1, \ldots, x^\mu$ is

$$(2\pi)^{-(\nu-\mu)/2} \int Q(\xi)\,\hat{u}(\xi)\,d\xi_{\mu+1} \cdots d\xi_\nu .$$

Schwarz' inequality and (2.8.8) show that the square of this function of $\xi_1, \ldots, \xi_\mu$ is less than

$$(2\pi)^{-(\nu-\mu)} \int \frac{|Q(\xi)|^2}{\tilde{P}(\xi)^2}\,d\sigma' \int \tilde{P}(\xi)^2\,|\hat{u}(\xi)|^2\,d\xi_{\mu+1} \cdots d\xi_\nu$$
$$\leqq C \int \tilde{P}(\xi)^2\,|\hat{u}(\xi)|^2\,d\xi_{\mu+1} \cdots d\xi_\nu .$$

It now follows from Parseval's formula that

$$\int_\Sigma |Q(D)u|^2\,d\sigma \leqq C \int d\xi_1 \cdots d\xi_\mu \int \tilde{P}(\xi)^2\,|\hat{u}(\xi)|^2\,d\xi_{\mu+1} \cdots d\xi_\nu = C \int \tilde{P}(\xi)^2\,|\hat{u}(\xi)|^2\,d\xi,$$

and using Lemma 2.8 as in the proof of Theorem 2.6 we obtain

$$\int_\Sigma |Q(D)u|^2\,d\sigma \leqq C \int |P(D)u|^2\,dx, \quad u \in C_0^\infty(\Omega),$$

which completes the proof.

The special case of Theorem 2.8, where Σ is a hyperplane, is most important. In that case $Q(D)u$ exists in Σ for every $u \in \mathcal{D}_{P_0}$ if and only if

$$\int_{-\infty}^{+\infty} \frac{|Q(\xi + t\mathrm{N})|^2}{\tilde{P}(\xi + t\mathrm{N})^2}\,dt \leqq C \tag{2.8.18}$$

for every real ξ, where N is the normal of Σ.

LEMMA 2.9. *If $p(t)$ is a polynomial of degree n in a real variable t, we have*

$$\int_{-\infty}^{+\infty} \frac{|p'(t)|^2}{|p(t)|^2 + |p'(t)|^2}\,dt \leqq 4n^2 . \tag{2.8.19}$$

PROOF. Logarithmic differentiation gives that

$$\frac{p'(t)}{p(t)} = \sum_1^n \frac{1}{t - t_k},$$

where t_k are the zeros of $p(t)$. The integral (2.8.19) can be divided into two parts I_1 and I_2, where I_1 is the integral over the intervals where $|\mathrm{Re}(t - t_k)| \leqq 1$ for some k, and I_2 is the integral over the rest of the real axis. Since the integrand is $\leqq 1$ everywhere, and the total length of the intervals, over which the integral I_1 is extended, is at most $2n$, we have $I_1 \leqq 2n$. In the integral I_2 we have the estimate

$$\frac{|p'(t)|^2}{|p(t)|^2+|p'(t)|^2} \leqq \left|\frac{p'(t)}{p(t)}\right|^2 \leqq n \sum \frac{1}{|t-t_k|^2}$$

in virtue of Cauchy's inequality. Since $|t-t_k|^2 \geqq |t - \operatorname{Re} t_k|^2$, this gives

$$I_2 \leqq n^2 \int\limits_{|t|\geqq 1} \frac{dt}{t^2} = 2\,n^2,$$

so that $I_1 + I_2 \leqq 4\,n^2$.

Using this lemma and Theorem 2.8 in the form (2.8.18) we obtain

THEOREM 2.9. *If Σ is a hyperplane with normal* N, *we have* $\int |P_{\mathrm{N}}^{(\alpha)}(D)\,u|^2\,d\sigma \leqq$ $\leqq C \int |P(D)\,u|^2\,dx,\ u \in C_0^\infty(\Omega)$, *where* $P_{\mathrm{N}}^{(\alpha)}(\xi) = \sum \mathrm{N}_k\,\partial P^{(\alpha)}(\xi)/\partial \xi_k$. *Thus the restrictions to* Σ *of all* $P_{\mathrm{N}}^{(\alpha)}(D)\,u$ *can be defined when* $u \in \mathcal{D}_{P_0}$.

In the case where $P(\xi) = (\xi, \xi)$ is a regular quadratic form, the covariant vector N has contravariant components also, and $P_{\mathrm{N}}(D)$ is the operator of differentiating along the contravariant normal. For operators of higher order than two, the operator P_{N} appears to be an appropriate substitute for the normal derivative.

An element u in $\mathcal{D}_{P_0}$ is, strictly speaking, an equivalence class of square integrable functions, and by the function $u(x)$ we have always meant any representative of this equivalence class. Thus $u(x)$ has only been defined for almost all x. It is then obvious that the function $u(x)$, $x \in \Sigma$, does not in general define the restriction of u to Σ, if the restriction exists in the above sense.

A representative $u(x)$ of an element u in $\mathcal{D}_{P_0}$ will be called *distinguished*, if the restriction of u to any variety Σ is defined by the function $u(x)$, $x \in \Sigma$, whenever it exists in the above sense. We shall prove that every element u in $\mathcal{D}_{P_0}$ has a distinguished representative. In fact, we can find a sequence of functions $u_n \in C_0^\infty(\Omega)$ such that

$$\|u_n - u\| \to 0, \quad \|P(D)\,u_n - P_0\,u\| \to 0,$$

and

$$\sum 2^n \|u_n - u_{n+1}\| < \infty, \quad \sum 2^n \|P(D)\,u_n - P(D)\,u_{n+1}\| < \infty.$$

If the restriction of u to Σ exists, the inequality (2.8.7) is valid with $Q=1$, and it follows that

$$\sum 2^n \|u_n - u_{n+1}\|_\Sigma < \infty.$$

Denoting the open set in Σ where $|u_n(x)-u_{n+1}(x)|>2^{-n}$ by e_n, we have the estimate $2^{-n}\sigma(e_n)\leqq\|u_n-u_{n+1}\|_\Sigma$, where $\sigma(e_n)$ is the Lebesgue measure in Σ of e_n. Writing $\omega_n=\bigcup_{k\geqq n} e_k$, we obtain

$$\sigma(\omega_n)\leqq\sum_n^\infty\sigma(e_k)\leqq\sum_n^\infty 2^k\|u_k-u_{k+1}\|_\Sigma,$$

which tends to 0 with n^{-1}. Hence the set $\omega=\bigcap\omega_n$ has measure zero in Σ, and $\lim u_n(x)$ obviously exists if $x\in\Sigma-\omega$. Now set $u(x)=\lim u_n(x)$ for any $x\in\Omega$ such that the limit exists, and define $u(x)$ arbitrarily elsewhere. We have proved that the limit exists almost everywhere in any variety Σ such that (2.8.7) is valid. Hence it follows that the strong limit of u_n in $L^2(\Sigma)$, which by definition is the restriction of u to Σ, is defined by the function $u(x)$, $x\in\Sigma$.

The same arguments apply to the definition of $Q_0 u$ when $u\in\mathcal{D}_{P_0}$ and $Q(D)$ is weaker than $P(D)$. Thus the equivalence class $Q_0 u$ always contains a distinguished function $Q_0 u(x)$; the restriction of $Q_0 u$ to a variety Σ is then defined by the function $Q_0 u(x)$, $x\in\Sigma$, whenever it exists. Note that, in particular, the distinguished function $Q_0 u(x)$ is continuous, if (2.8.1) is valid.

More precise results have been obtained by Deny and Lions [4] for the Beppo Levi functions. The results proved here could probably be improved in the same direction by means of a generalized notion of capacity, but the results already proved are sufficient for us.

We now prove a result which in particular contains a localization principle for $\mathcal{D}_{P_0}$.

THEOREM 2.10. *The product of a function $u\in\mathcal{D}_{P_0}$ and a function $\psi\in C^\infty(\bar\Omega)$*[1] *is in $\mathcal{D}_{P_0}$, and there is a constant C depending on ψ such that*

$$\|P_0(\psi u)\|\leqq C\|P_0 u\|,\quad u\in\mathcal{D}_{P_0}. \tag{2.8.20}$$

PROOF. Using Leibniz' formula and Lemma 2.8 we obtain the inequality (2.8.20) if $u\in C_0^\infty(\Omega)$. This evidently gives the desired result.

Theorem 2.10 may seem evident at first sight, but to display its significance we give two examples showing that, if a function u is in $\mathcal{D}_P$, where P is the maximal operator defined by $P(D)$, and $\psi\in C_0^\infty(\bar\Omega)$, it need not be true that $\psi u\in\mathcal{D}_P$, even for the simplest operators.

EXAMPLE 1. Let $P(D)$ be the Laplace operator in two variables, and let $u(x^1,x^2)$ be a harmonic function in the circle $r=(x^{1 2}+x^{2 2})^{\frac12}<1$ such that $u\in L^2$ but $\partial u/\partial r\notin L^2$.

[1] This means that ψ is C^∞ in a neighbourhood of $\bar\Omega$.

A well-known example of a function with these properties is due to Hadamard. Now let ψ be a function in C^∞ such that $\psi = r$ outside a neighbourhood of the origin. We have

$$\Delta(\psi u) = u\Delta\psi + 2(\operatorname{grad} u, \operatorname{grad} \psi).$$

The first term is square integrable but the second is not, since it equals $2\partial u/\partial r$ outside a neighbourhood of the origin. Hence $u \in \mathcal{D}_P$ but $\psi u \notin \mathcal{D}_P$.

EXAMPLE 2. Let $P(D)$ be the wave operator $\partial^2/\partial x^1 \partial x^2$ in two variables, and let $u = u(x^1)$ be an absolutely continuous function of x^1, whose derivative is not square integrable in the neighbourhood of any point. Since we have

$$P(D)(\psi u) = \frac{\partial u}{\partial x^1}\frac{\partial \psi}{\partial x^2} + u\frac{\partial^2 \psi}{\partial x^1 \partial x^2},$$

it follows that $\psi u \notin \mathcal{D}_P$ unless ψ is a function of x^1, although we have $u \in \mathcal{D}_P$. In particular, $\psi u \notin \mathcal{D}_P$, if $0 \neq \psi \in C_0^\infty(\Omega)$.

After these two examples we leave the maximal operators, which will be discussed in the next chapter. However, to clarify the contents of Theorem 2.10, we shall also prove that Lemma 2.8, which was the essential tool in the proof of Theorem 2.2, is a consequence of Theorem 2.10. In fact, if we take $\psi(x) = e^{i\langle x, \eta\rangle}$, Theorem 2.10 shows that the polynomial $P(\xi)$ is stronger than the polynomial $P(\xi + \eta)$. Hence $P(\xi)$ is also stronger than any linear combination of the translated polynomials $P(\xi + \eta)$, with fixed η, and our assertion follows from the following lemma.

LEMMA 2.10. *A linear set I of polynomials is invariant for differentiation if and only if it is invariant for translation.*

PROOF. That invariance for differentiation implies invariance for translation follows at once from Taylor's formula. On the other hand, if I is invariant for translation and $P \in I$ is of degree μ, the set I contains all functions of the form

$$\sum_{i=1}^m t_i P(\xi + \eta^i) = \sum_{|\alpha| \leq \mu} P^{(\alpha)}(\xi) \frac{1}{|\alpha|!} \sum_{i=1}^m t_i \eta_\alpha^i,$$

where η^i are arbitrary vectors, and t_i are arbitrary complex numbers. Now the coefficients $\sum_{i=1}^m t_i \eta_\alpha^i$, $|\alpha| \leq \mu$, can be given arbitrary values, which are symmetric in α, by a convenient choice of m, t_i and η^i. For otherwise there would exist constants c_α, $|\alpha| \leq \mu$, symmetric in α and not all equal to zero, such that

$$\sum_{|\alpha| \leq \mu} c_\alpha \eta_\alpha = 0 \text{ for every } \eta.$$

But this is impossible. Hence I contains all $P^{(\alpha)}(\xi)$, which was to be proved.

THEOREM 2.11. *The conditions for a function u to be in $\mathcal{D}_{P_0}$ have a local character in $\overline{\Omega}$. More precisely, if u is a function such that to every point in $\overline{\Omega}$ there exists a neighbourhood U, and a function $v_U \in \mathcal{D}_{P_0}$, so that $u(x) = v_U(x)$ a. e. in $U \cap \Omega$, then $u \in \mathcal{D}_{P_0}$.*

PROOF. We can cover $\overline{\Omega}$ by a finite number of neighbourhoods U_i, $i = 1, \ldots, m$, of the type given by the theorem. Now take functions $\alpha_i(x) \in C_0^\infty(U_i)$ such that

$$\sum_1^m \alpha_i(x) = 1, \quad x \in \overline{\Omega}.$$

Since $u(x) = \sum u(x)\alpha_i(x)$, and $u(x)\alpha_i(x) = v_{U_i}(x)\alpha_i(x)$ is in $\mathcal{D}_{P_0}$ in virtue of Theorem 2.10, it follows that u is in $\mathcal{D}_{P_0}$.

The properties of the functions in $\mathcal{D}_{P_0}$ in the neighbourhood of a point in Ω are described by the following theorem.

THEOREM 2.12. *A function u in $L^2(\Omega)$ is equal to a function in $\mathcal{D}_{P_0}$ in a neighbourhood of a point $x \in \Omega$ if and only if all $P^{(\alpha)}(D)u$ are square integrable functions in a neighbourhood of x.*

PROOF. First suppose that u equals a function v in $\mathcal{D}_{P_0}$ in a neighbourhood of the point x. Then we have in this neighbourhood

$$P^{(\alpha)}(D)u = P^{(\alpha)}(D)v$$

and, since $P^{(\alpha)}(D)v$ is square integrable over Ω in virtue of Theorem 2.2, the assertion of the theorem follows.

Conversely, suppose that $P^{(\alpha)}(D)u$ is square integrable for every α in a neighbourhood U of x. Let $\psi \in C_0^\infty(U)$ equal 1 in a neighbourhood V of x. Then $v(x) = u(x)\psi(x)$ equals $u(x)$ in V, and in virtue of Leibniz' formula we have in the weak sense

$$P(D)v = \sum \frac{D_\alpha \psi}{|\alpha|!} P^{(\alpha)}(D)u \in L^2.$$

Hence the proof reduces to the proof of the following lemma, already referred to in Chapter I.

LEMMA 2.11. *A function $u \in \mathcal{D}_P$, which has compact support in Ω, is in $\mathcal{D}_{P_0}$.*

PROOF. Let $\psi \in C_0^\infty(R^\nu)$ and $\int \psi(x)dx = 1$. We form the convolutions $u_\varepsilon = u * \psi_\varepsilon$, where $\psi_\varepsilon(x) = \varepsilon^{-\nu}\psi(x/\varepsilon)$. When ε is sufficiently small, we have $u_\varepsilon \in C_0^\infty(\Omega)$, and it is well known that $u_\varepsilon \to u$ in L^2. Furthermore, when $\varepsilon \to 0$, we have

$$P(D)u_\varepsilon = (P(D)\,u) * \psi_\varepsilon \to P(D)\,u$$

in L^2. Hence by definition $u \in \mathcal{D}_{P_0}$.

We shall now deduce a corresponding result for a point x on the boundary. In doing so we restrict ourselves to a point on a plane portion of the boundary, where we can use our

Theorem 2.9. It would no doubt be possible to treat a much more general case by generalizing that theorem, but we shall refrain from studying that question here.

Let Σ be a plane surface with compact closure in Ω. It follows from Theorem 2.12 and Theorem 2.9 that $P_N^{(\alpha)}(D)u$ exists in Σ and is square integrable there, if u is such that $P^{(\alpha)}(D)u$ is square integrable in a neighbourhood of Σ for every α. We can now announce our result.

THEOREM 2.13. *Let x_0 be a point on a plane portion Σ of the boundary of Ω, the distance from x_0 to the rest of the boundary being positive. Then a function u in $L^2(\Omega)$ equals a function in $\mathcal{D}_{P_0}$ in a neighbourhood of x_0 if and only if all $P^{(\alpha)}(D)u$ are square integrable functions in a neighbourhood of x_0 in $\overline{\Omega}$, and the restrictions of $P_N^{(\alpha)}(D)u$ to parallel surfaces to Σ tend to zero strongly in a neighbourhood of x_0 when the surfaces approach Σ.*

The last statement needs perhaps some explanation. Let y be a fixed transversal direction to Σ, i.e. $\langle y, N\rangle \neq 0$. We may suppose that y points from Σ to Ω. If x is in a suitable neighbourhood U of x_0 in Σ and δ is a sufficiently small positive number, the function $P_N^{(\alpha)}(D)u(x+\delta y)$ is square integrable in U. The second half of the condition in the theorem is that this function tends strongly to zero in $L^2(U)$ when $\delta\to 0$. — Note that Sobolev [30] has given similar results in connection with elliptic operators.

PROOF OF THE THEOREM. First, let v be a function in $\mathcal{D}_{P_0}$. For given ε we can find a function $v_\varepsilon \in C_0^\infty(\Omega)$ such that $\| P(D)(v-v_\varepsilon)\| < \varepsilon$. In virtue of Theorem 2.9 there is a constant C such that on all planes Σ_1 with normal N we have

$$\| P_N^{(\alpha)}(D)(v-v_\varepsilon)\|_{\Sigma_1} < C\varepsilon$$

for every ε. Since v_ε vanishes in a neighbourhood of the boundary, we have with the notation introduced above

$$\| P_N^{(\alpha)}(D)v(x+\delta y)\|_U < C\varepsilon,$$

if δ is small enough. This proves the necessity of the conditions given in the theorem.

Conversely, let the conditions of the theorem be fulfilled. Since they are still valid for the function $u\varphi$, where $\varphi\in C^\infty$ and vanishes outside a neighbourhood of x, we may suppose that u vanishes outside the neighbourhoods mentioned in the theorem. Let $\psi\in C_0^\infty$ with respect to the half space $\langle x, N\rangle\langle y, N\rangle > 0$, where y is the vector mentioned above, and $\int\psi\,dx=1$. It is then easily proved that the convolution $u_\varepsilon = u * \psi_\varepsilon$, where $\psi_\varepsilon(x) = \varepsilon^{-\nu}\psi(x/\varepsilon)$, is in $C_0^\infty(\Omega)$ for small ε and that $u_\varepsilon\to u$ and $P(D)u_\varepsilon = P(D)u * \psi_\varepsilon \to P(D)u$ when $\varepsilon\to 0$. This completes the proof. The details may be left to the reader.

In particular we may note that a function u which is sufficiently differentiable in $\overline{\Omega}$ equals a function in $\mathcal{D}_{P_0}$ in a neighbourhood of a point on a plane portion of the boundary of $\bar{\Omega}$ if and only if it vanishes there together with $m-1$ transversal derivatives, where m is the degree of $P(\xi + tN)$ in t when ξ is an indeterminate.

One of the most important results in the theory of partial differential operators is the lemma of H. Weyl to the effect that any solution of the equation $Pu = 0$, where P is a maximal elliptic operator, must be infinitely differentiable (after correction on a null set). This is only true for a certain class of differential operators P, which will be determined in the next chapter. For that class it will turn out that, more generally, any function, which is in the domain of P^n for any n, is infinitely differentiable. This result has an analogue for general minimal differential operators, which we shall now discuss. We start with two definitions and a lemma, showing how the strength of the powers of an operator increases.

DEFINITION 2.2. *The linear manifold*

(2.8.21) $$\Lambda(P) = \{\eta;\ \eta \text{ is real and } P(\xi + t\eta) = P(\xi) \text{ for any } \xi \text{ and } t\}$$

is called the lineality space of the polynomial P.

DEFINITION 2.3. *A polynomial P is called complete, if the lineality space consists of the origin only.*

Thus P is complete, if it really depends on all variables. The two definitions are essentially borrowed from Gårding [8].

LEMMA 2.12. *The operator $P(D)^n$ is stronger than any product $Q_1(D)\ldots Q_k(D)$, $k \leqq n$, of operators which are weaker than P.*

PROOF. First note that for $0 \leqq k \leqq n$ we have

$$\|P(D)^k u\|^2 \leqq \frac{k}{n}\|P(D)^n u\|^2 + \left(1 - \frac{k}{n}\right)\|u\|^2, \quad u \in C_0^\infty(\Omega).$$

In fact, this inequality is equivalent to

$$\int |P(\xi)|^{2k}|\hat{u}(\xi)|^2 d\xi \leqq \frac{k}{n}\int |P(\xi)|^{2n}|\hat{u}(\xi)|^2 d\xi + \left(1 - \frac{k}{n}\right)\int |\hat{u}(\xi)|^2 d\xi,$$

which follows from the inequality between geometric and arithmetic means. Hence to prove the lemma it is sufficient to show that for any k

(2.8.22) $$\|Q_1(D)\cdots Q_k(D)u\|^2 \leqq C(\|P(D)^k u\|^2 + \cdots + \|u\|^2), \quad u \in C_0^\infty(\Omega).$$

For $k = 1$, this is only the definition of a weaker operator. Assuming as we may that (2.8.22) has already been proved when k is replaced by $k-1$, we find by substituting $P(D)u$ for u that

$$\|Q_1(D)\cdots Q_{k-1}(D)P(D)u\|^2 \leqq C(\|P(D)^k u\|^2 + \cdots + \|P(D)u\|^2), \quad u \in C_0^\infty(\Omega).$$

Using the fact that the operators all commute, and having recourse to the definition of a weaker operator again, we obtain (2.8.22).

THEOREM 2.14. *A function u which is in the minimal domain of $P(D)^n$ for every n, where P is a complete polynomial, is infinitely differentiable in Ω, and every derivative tends to zero at the boundary.*

PROOF. Let R be the algebra generated by the polynomials which are weaker than P. It follows from Lemma 2.12 that the function u of the theorem is in the minimal domain of $Q(D)$, if $Q \in R$. Now the assumption that P is complete implies that the algebra R is the whole polynomial ring. We shall prove this assertion in section 2.10. Since to any polynomial Q we can find another Q_1 such that $|Q(\xi)|/\tilde{Q}_1(\xi)$ is square integrable, it follows from Theorem 2.6 that $Q(D)u$ is continuous after correction on a null set and tends to zero at the boundary of Ω. It now easily follows (see also Schwartz [28], Tome I, p. 62) that u is infinitely differentiable in the classical sense.

REMARK. We can also prove that $u \in C^\infty$, if we suppose that u is in the domain of P_0^n for every n. For if Ω' is a bounded domain which contains $\bar{\Omega}$, and we extend u to a function u' in Ω' by setting $u' = u$ in Ω and $u' = 0$ elsewhere, the assumptions of Theorem 2.14 are satisfied by u' in $L^2(\Omega')$. (After the above was written, the question whether the domain of P_0^n always coincides with the minimal domain of $P(D)^n$ was answered in the negative by J. L. Lions.)

2.9. Some theorems on complete continuity

Theorem 2.2 gave the necessary and sufficient conditions for the continuity of the mapping

$$\mathcal{R}_{P_0} \ni P_0 u \to Q_0 u \in \mathcal{R}_{Q_0}. \tag{2.9.1}$$

We shall now derive the conditions for complete continuity. Such results are important in proving that vibration problems have a discrete spectrum. We remark that some results, similar to the theorems which we are going to prove, have been given previously by Kondrachov (see Sobolev [30]) with different proofs, based on potential theory.

THEOREM 2.15. *The mapping* (2.9.1) *is completely continuous if and only if*

$$\frac{\tilde{Q}(\xi)}{\tilde{P}(\xi)} \to 0 \quad when \quad \xi \to \infty. \tag{2.9.2}$$

PROOF. We first prove that the complete continuity of the mapping (2.9.1) follows from (2.9.2). The proof is a combination of Theorem 2.2 with the proof by Gårding ([9], p. 59) of a special case.

Suppose that (2.9.2) is fulfilled. Then we have also $|Q(\xi)|/\tilde{P}(\xi) < C$. Take any sequence $u_n \in C_0^\infty(\Omega)$ such that

$$\|P(D)u_n\| \leqq 1. \tag{2.9.3}$$

We shall prove that $Q(D)u_{n'}$ converges if n' is a suitable subsequence of the sequence n. In virtue of Theorem 2.2 we have

$$\|Q(D)u_n\| \leqq C. \tag{2.9.4}$$

Since all $Q(D)u_n$ vanish outside the bounded set Ω, it follows from (2.9.4), if we denote the Fourier transform of u_n by $\hat{u}_n$, that the functions $Q(\xi)\hat{u}_n(\xi)$ are uniformly bounded and uniformly continuous. Hence we can find a subsequence n' such that $Q(\xi)\hat{u}_{n'}(\xi)$ is uniformly convergent on every compact set. Now we have

$$\|Q(D)u_{n'} - Q(D)u_{m'}\|^2 = \int |Q(\xi)\hat{u}_{n'}(\xi) - Q(\xi)\hat{u}_{m'}(\xi)|^2 d\xi.$$

Let K be a compact set such that $|Q(\xi)|/\tilde{P}(\xi) < \varepsilon$ in the complement K' of K. Then it follows from (2.9.3) and Lemma 2.8 (see proof at the end of section 2.6) that

$$\int_{K'} |Q(\xi)|^2 |\hat{u}_{n'}(\xi) - \hat{u}_{m'}(\xi)|^2 d\xi \leqq \varepsilon^2 \int \tilde{P}(\xi)^2 |\hat{u}_{n'}(\xi) - \hat{u}_{m'}(\xi)|^2 d\xi \leqq A\varepsilon^2,$$

where A is a constant. Furthermore,

$$\int_K |Q(\xi)\hat{u}_{n'}(\xi) - Q(\xi)\hat{u}_{m'}(\xi)|^2 d\xi \to 0, \text{ when } n' \text{ and } m' \to \infty,$$

in virtue of the uniform convergence. Hence

$$\overline{\lim_{n', m' \to \infty}} \|Q(D)u_{n'} - Q(D)u_{m'}\|^2 \leqq A\varepsilon^2$$

for every ε, which proves that $Q(D)u_{n'}$ is convergent. This proves the complete continuity of the mapping (2.9.1), since the functions $P(D)u$, $u \in C_0^\infty(\Omega)$, are dense in $\mathcal{R}_{P_0}$.

Now suppose that the mapping (2.9.1) is completely continuous. We have to prove that (2.9.2) must be valid. This can be achieved by modifying the technique of section 2.3. It is obviously sufficient to prove that, if ξ_n is a sequence tending to infinity, such that $\tilde{Q}(\xi_n)/\tilde{P}(\xi_n)$ tends to a limit, then the limit must be zero. Since we may, if necessary, pass to a subsequence, it is also permitted to suppose that

$$\xi_n - \xi_m \to \infty \text{ when } n, m \to \infty \text{ and } n \neq m. \tag{2.9.5}$$

This assumption is essential in the proof.

Let ψ be a fixed function in $C_0^\infty(\Omega)$, and form the sequence of functions

$$u_n(x) = \psi(x)\frac{e^{i\langle x, \xi_n\rangle}}{\tilde{P}(\xi_n)}. \tag{2.9.6}$$

In virtue of Leibniz' formula there is a constant C such that

$$\|P(D)u_n\| \leqq C. \tag{2.9.7}$$

Using (2.3.5), we can write

$$\|Q(D)u_n - Q(D)u_m\|^2 = \|Q(D)u_n\|^2 + \|Q(D)u_m\|^2 - \delta_{nm}, \tag{2.9.8}$$

where

$$\delta_{nm} = 2\,\mathrm{Re}\left\{\sum_{\alpha,\beta} \frac{Q^{(\alpha)}(\xi_n)}{\tilde{P}(\xi_n)} \frac{\overline{Q^{(\beta)}(\xi_m)}}{\tilde{P}(\xi_m)} \frac{1}{|\alpha|!\,|\beta|!} \int D_\alpha \psi \, \overline{D_\beta \psi}\, e^{i\langle x, \xi_n - \xi_m\rangle}\, dx\right\}. \tag{2.9.9}$$

Now, since the mapping (2.9.1) is completely continuous, it is also continuous, so that $Q^{(\alpha)}(\xi)/\tilde{P}(\xi)$ is bounded in virtue of Theorem 2.2. Hence it follows from (2.9.5) and Riemann-Lebesgue's lemma in its very simplest form that $\delta_{nm} \to 0$ when n, $m \to \infty$ and $n \neq m$.

By the assumption, there is a sequence n' such that $Q(D)u_{n'}$ is convergent. Then it follows from (2.9.8) that

$$\|Q(D)u_{n'}\|^2 = \sum_{\alpha,\beta} \frac{Q^{(\alpha)}(\xi_{n'})\,\overline{Q^{(\beta)}(\xi_{n'})}}{\tilde{P}(\xi_{n'})^2} \psi_{\alpha\beta} \to 0 \quad \text{when } n' \to \infty, \tag{2.9.10}$$

where $\psi_{\alpha\beta}$ is defined by (2.3.6). It now follows from (2.3.9) that also

$$\sum_\alpha \frac{|Q^{(\alpha)}(\xi_{n'})|^2}{\tilde{P}(\xi_{n'})^2} = \frac{\tilde{Q}(\xi_{n'})^2}{\tilde{P}(\xi_{n'})^2} \to 0 \quad \text{when } n' \to \infty.$$

The proof is complete.

THEOREM 2.16. *Let Σ be a plane of dimension less than ν. Then the mapping*

$$\mathcal{R}_{P_0} \ni P(D)u \to \{Q(D)u\}_\Sigma \in L^2(\Sigma), \tag{2.9.11}$$

where $\{Q(D)u\}_\Sigma$ is the restriction of $Q(D)u$ to Σ, is completely continuous if and only if

$$\int_\Sigma \frac{|Q(\xi)|^2}{\tilde{P}(\xi)^2}\, d\sigma' \to 0 \tag{2.9.12}$$

when the normal variety $\Sigma' \to \infty$.

The reader will have no difficulty in carrying out the proof, which does not require any ideas beyond the proofs of Theorem 2.8 and Theorem 2.15.

THEOREM 2.17. *The inverse P_0^{-1} of a minimal differential operator is completely continuous if and only if P is a complete polynomial.*

PROOF. If P is not a complete polynomial, there exists a real vector $\eta \neq 0$ such that $P(\xi + t\eta) \equiv P(\xi)$. Differentiating repeatedly with respect to ξ, we obtain that $\tilde{P}(\xi + t\eta) =$

$=\tilde{P}(\xi)$, so that $\tilde{P}(\xi+t\eta)$ is bounded when $t\to\infty$. Hence Theorem 2.15 with $Q(\xi)\equiv 1$ shows that P_0^{-1} is not completely continuous.

Now suppose that P is a complete polynomial, i.e. that $\Lambda(P)=\{0\}$, where $\Lambda(P)$ is defined by (2.8.21). We shall prove that $\tilde{P}(\xi)\to\infty$ with ξ or, equivalently, that

$$M_C=\{\xi;\ \tilde{P}(\xi)\leqq C\}$$

is a bounded set for every C. The polynomial $P(\xi)$ can be written as the sum of its homogeneous parts,

$$P(\xi)=\sum_0^m P_k(\xi),$$

where $P_k(\xi)$ is homogeneous of degree k and $P_m(\xi)\not\equiv 0$. It is easy to prove that, for every polynomial P,

(2.9.13) $$\Lambda(P)=\bigcap_{k=1}^m \Lambda(P_k).$$

For, if $\eta\in\bigcap_{k=1}^m\Lambda(P_k)$, we have $P_k(\xi+t\eta)\equiv P_k(\xi)$ and consequently $P(\xi+t\eta)\equiv P(\xi)$, so that $\eta\in\Lambda(P)$. Hence $\bigcap_{k=1}^m\Lambda(P_k)\subset\Lambda(P)$. Now let $\eta\in\Lambda(P)$. Then we have

$$\sum_0^m P_k(\xi+t\eta)\equiv\sum_0^m P_k(\xi).$$

Replacing ξ by $\tau\xi$ and t by τt and identifying the powers of τ, we obtain $P_k(\xi+t\eta)\equiv$ $\equiv P_k(\xi)$, so that $\eta\in\Lambda(P_k)$. Hence $\Lambda(P)\subset\bigcap_{k=1}^m\Lambda(P_k)$, which proves (2.9.13).

From our assumption that P is complete, it thus follows that $\bigcap_{k=1}^m\Lambda(P_k)=\{0\}$. Hence it will follow that the set M_C is bounded, if we only prove that M_C is bounded modulo $\Lambda(P_k)$ for every k. We need a simple lemma on homogeneous polynomials in the proof.

LEMMA 2.13. *Let Q be a homogeneous polynomial of degree m. Then a real vector η, such that $D_\alpha Q(\eta)=0$ for every α of length $m-1$, is in $\Lambda(Q)$.*

PROOF. The lemma is obvious, if $m=1$, and we shall prove it in general by induction over m. Suppose that the assumptions of the lemma are satisfied and $m>1$. Then the assumptions hold good also for the polynomials $\partial Q/\partial\xi_i$. Assuming, as we may, that the lemma is already proved for polynomials of degree less than m, we obtain

$$\partial Q(\xi+t\eta)/\partial\xi_i\equiv\partial Q(\xi)/\partial\xi_i.$$

Hence $Q(\xi+t\eta)-Q(\xi)$ is independent of ξ, so that we have

$$Q(\xi+t\eta)-Q(\xi)\equiv Q(t\eta).$$

Setting $t = 1$ and $\xi = \eta$ we obtain $2Q(\eta) = 2^m Q(\eta)$. Hence $Q(\eta) = 0$ and thus $Q(\xi + t\eta) = Q(\xi)$, which means that $\eta \in \Lambda(Q)$.

We can now prove that M_C is bounded modulo $\Lambda(P_k)$ for every k, if P is any polynomial. Since this is obvious if the degree m of P is 1, it is sufficient to prove that it is true for P, if it is true for all polynomials of degree less than m.

That $\tilde{P}(\xi) \leq C$ in M_C, implies that $|P^{(\alpha)}(\xi)| \leq C$ there. In particular, this is true when $|\alpha| = m - 1$, and then $P^{(\alpha)}(\xi)$ differs from $P_m^{(\alpha)}(\xi)$ only by a constant term. Hence we have $|P_m^{(\alpha)}(\xi)| < C'$ in M_C if $|\alpha| = m - 1$. In virtue of Lemma 2.13, the linear forms $P_m^{(\alpha)}(\xi)$ vanish simultaneously only in $\Lambda(P_m)$. Hence M_C is bounded modulo $\Lambda(P_m)$. But since $P_m^{(\alpha)}(\xi)$ is constant modulo $\Lambda(P_m)$ for every α, we conclude that $\tilde{P}_m(\xi)$ is bounded in M_C. Now form the polynomial

$$R(\xi) = P(\xi) - P_m(\xi) = P_{m-1}(\xi) + \cdots + P_0.$$

$R(\xi)$ is of degree $m - 1$, and since

$$\tilde{R}(\xi) \leq \tilde{P}(\xi) + \tilde{P}_m(\xi),$$

we have $\tilde{R}(\xi) < C''$ in M_C. Using the assumption that our assertion is proved for polynomials of degree less than m, it follows that M_C is bounded modulo $\Lambda(P_k)$, $k \leq m - 1$. This completes the proof of Theorem 2.17.

The proof also shows that, if P is complete, there exists a constant $c > 0$ such that $\tilde{P}(\xi) > (\xi_1^2 + \cdots + \xi_\nu^2)^c$. Hence $1/\tilde{P}(\xi)$ is in L^q for large q, which permits the use of Theorem 2.7. A fairly precise result is given by the following lemma, which includes Theorem 2.17 but has a much more difficult proof.

LEMMA 2.14. *If $P^1(\xi), \ldots, P^n(\xi)$ is a set of polynomials such that*

$$\bigcap_{k=1}^{n} \Lambda(P^k) = \{0\},$$

then $(\tilde{P^1}^2 + \cdots + \tilde{P^n}^2)^{-\frac{1}{2}}$ is in L^q if $q > \nu$.

Note that Example 1 on page 191 shows that the constant ν of the lemma cannot be replaced by any smaller one. We also remark, that we shall only use Lemma 2.14 when $n = 1$, but the more general statement is necessary for our proof. Using this special case of Lemma 2.14 and Theorem 2.7, we obtain

THEOREM 2.18. *We have $u \in L^q$, if u is in the minimal domain of a complete differential operator and $q < 2\nu/(\nu - 2)$, if $\nu > 2$, $q < \infty$. if $\nu = 2$.*

As a preparation for the proof of Lemma 2.14 we introduce a new notation, which supplements the definition of $\Lambda(P)$,

$$\underline{\Lambda}(P) = \bigcap_{k=2}^{m} \Lambda(P_k) = \Lambda(P - P_1). \tag{2.9.14}$$

The last equality follows from (2.9.13). We shall prove that

$$\underline{\Lambda}(P) = \bigcap_{i=1}^{\nu} \Lambda(\partial P/\partial \xi_i). \tag{2.9.15}$$

First suppose that P is homogeneous of degree $m > 1$. Since it is obvious that $\Lambda(\partial P/\partial \xi_i) \supset \Lambda(P)$, we obtain by using Lemma 2.13

$$\Lambda(P) \subset \bigcap_{i=1}^{\nu} \Lambda(\partial P/\partial \xi_i) \subset \bigcap_{|\alpha|=m-1} \Lambda(D_\alpha P) \subset \Lambda(\mathrm{P}).$$

Hence (2.9.15) is valid in this case. Using this result and (2.9.13), we obtain for a general $P = \Sigma P_k$

$$\bigcap_{i=1}^{\nu} \Lambda(\partial P/\partial \xi_i) = \bigcap_{i=1}^{\nu} \bigcap_{k=1}^{m} \Lambda(\partial P_k/\partial \xi_i) = \bigcap_{k=1}^{m} \bigcap_{i=1}^{\nu} \Lambda(\partial P_k/\partial \xi_i) = \bigcap_{k=2}^{m} \Lambda(P_k) = \underline{\Lambda}(P),$$

since all $\partial P_1/\partial \xi_i$ are constants. This proves (2.9.15).

Before the proof we also extend our terminology slightly. We shall say that a system $P^1, \ldots, P^n$ of polynomials is *complete*, if $\bigcap_{i=1}^{n} \Lambda(P^i) = \{0\}$. The system $Q^1, \ldots, Q^l$ will be said to be *weaker* than the system $P^1, \ldots, P^n$, if we have

$$\tilde{P^1}^2 + \cdots + \tilde{P^n}^2 \geqq \tilde{Q^1}^2 + \cdots + \tilde{Q^l}^2.$$

PROOF OF LEMMA 2.14. By repeated application of the following two operations, we shall construct a system, which is weaker than the given one and for which the assertion of the lemma is valid:

A) If $\bigcap_{k \neq l} \Lambda(P_k) = \{0\}$, we obtain a weaker complete system by omitting P_l.

B) If $\underline{\Lambda}(P_l) \cap (\bigcap_{k \neq l} \Lambda(P_k)) = \{0\}$, we obtain a weaker complete system, if we replace P_l by all the polynomials $\partial P_l/\partial \xi_i$ $(i = 1, \ldots, \nu)$. This follows from formula (2.9.15).

To the system of polynomials, given in the formulation of Lemma 2.14, we first apply operation A until this is no longer possible. Then we apply operation B—if possible—to one of the remaining polynomials of highest degree, and then apply the operation A again as many times as it is possible. The new system is still complete, and either the highest degree occurring among the polynomials in the system, or else the number of polynomials of highest degree, has diminished. Hence we must after a finite number of steps come to a system $Q^1, \ldots, Q^l$, which is complete and weaker than the original system, such that A cannot be applied any more and B is not applicable to some of the polynomials—in fact to none of those of highest degree. Let one of these be Q^1. Then we have

$$\Lambda = \underline{\Lambda}(Q^1) \cap \Lambda(Q^2) \cap \cdots \cap \Lambda(Q^l) \neq \{0\}, \tag{2.9.16}$$

whereas

$$\Lambda \cap \Lambda(Q^1{}_1) = \{0\}, \tag{2.9.17}$$

where $Q^1{}_1$ is the linear part of Q^1. Since $\Lambda(Q^1{}_1)$ cannot have a co-dimension greater than one, it follows from (2.9.16) and (2.9.17) that Λ is one-dimensional. Let us suppose that the coordinates are chosen such that Λ is the ξ_1-axis. In virtue of (2.9.16), the polynomials $Q^1 - Q^1{}_1$, $Q^2, \ldots, Q^l$ are then independent of ξ_1, that is, they are polynomials in $\xi_2, \ldots, \xi_\nu$. It follows from (2.9.17) that $Q^1{}_1$ is not independent of ξ_1, so we can set

$$Q^1(\xi) = c\xi_1 + R(\xi),$$

where $c \neq 0$ and R is independent of ξ_1. Now we can write

$$\tilde{Q}^2 = \tilde{Q}^{1\,2} + \cdots + \tilde{Q}^{l\,2} = |c\xi_1 + R|^2 + \sum_{i=1}^{\nu} (\widetilde{\partial Q^1/\partial \xi_i})^2 + \tilde{Q}^{2\,2} + \cdots + \tilde{Q}^{l\,2} = |c\xi_1 + R|^2 + \tilde{Q}'^2.$$

This gives, if we perform the integration over ξ_1 explicitly,

$$\int \frac{1}{\tilde{Q}^q}\, d\xi_1 \cdots d\xi_\nu \leqq \frac{1}{|c|} \int_{-\infty}^{+\infty} \frac{d\tau}{(\tau^2+1)^{q/2}} \int \frac{1}{\tilde{Q}'^{(q-1)}}\, d\xi_2 \cdots d\xi_\nu .$$

Since $q > \nu \geqq 1$, the first integral is convergent. Furthermore, since it follows from (2.9.16) that the polynomials $\partial Q^1/\partial\xi_1, \ldots, \partial Q^1/\partial \xi_\nu, Q^2, \ldots, Q^l$ form a complete system in the variables $\xi_2, \ldots, \xi_\nu$, the convergence of the last integral follows from the validity of Lemma 2.14 in a space of dimension $\nu - 1$. Hence the lemma is true for any number of variables, since it is true when $\nu = 1$.

2.10. On some sets of polynomials

Let P be a fixed polynomial. We have studied the set of polynomials Q such that $Q(D)u$ exists for $u \in \mathcal{D}_{P_0}$ in one sense or another (Theorems 2.2, 2.6, 2.7, 2.8, 2.15, 2.16). In all cases, the set I of polynomials Q, which we have obtained, has the following two properties:

a) *I is linear and invariant for translation.*

b) *If Q is a polynomial such that*

$$|Q(\xi)| \leqq \sum_{i=1}^{n} |Q_i(\xi)|$$

for every real ξ, where $Q_1(\xi), \ldots, Q_n(\xi) \in I$, then it follows that $Q(\xi) \in I$.

In virtue of Lemma 2.10, the property (a) is equivalent to

a′) *I is linear and invariant for differentiation.*

That (b) is fulfilled is evident in all the cases, so that the only thing we need to prove is the invariance for translation. Let us verify this for the set of polynomials Q such that $Q(\xi)/\tilde{P}(\xi)$ is in L^q. Let $Q(\xi)/\tilde{P}(\xi) \in L^q$. Then, for fixed η, the function

$$\frac{Q(\xi+\eta)}{\tilde{P}(\xi)} = \frac{Q(\xi+\eta)}{\tilde{P}(\xi+\eta)} \frac{\tilde{P}(\xi+\eta)}{\tilde{P}(\xi)}$$

is also in L^q, for it follows at once from Taylor's formula that $\tilde{P}(\xi+\eta)/\tilde{P}(\xi)$ is bounded for fixed η.

We also remark, without performing the comparatively easy proof, that, if $Q(\xi)/\tilde{P}(\xi)$ is in L^q, it follows that $Q(\xi)/\tilde{P}(\xi) \to 0$ when $\xi \to \infty$, and hence that $Q(\xi)/\tilde{P}(\xi)$ is in L^r for $r \geqq q$. This can partly be deduced also from our theorems above.

The invariance for translation and differentiation proves the fact, already noticed in a remark following Theorem 2.2, that, for instance, the assumption that $Q(\xi)/\tilde{P}(\xi) \to 0$ is equivalent to $\tilde{Q}(\xi)/\tilde{P}(\xi) \to 0$ when $\xi \to \infty$. The same remark applies to the other theorems.

We now prove a result which was already used in section 2.8.

LEMMA 2.15. *The algebra R, generated by the polynomials weaker than a polynomial P, consists of all polynomials with the lineality manifold $\Lambda(P)$.*

PROOF. The statement is obvious, if P is of degree 1. To prove it for a polynomial $P = \sum_{k=0}^{m} P_k$, we may assume that it has already been proved for polynomials of degree less than m. Now the polynomials which are weaker than $\partial P/\partial \xi_i$ are also weaker than P, and hence R contains all polynomials with the lineality manifold $\Lambda(\partial P/\partial \xi_i)$. Thus R contains all polynomials with the lineality manifold

$$\bigcap_{i=1}^{\nu} \Lambda(\partial P/\partial \xi_i) = \underline{\Lambda}(P) = \Lambda(P - P_1).$$

Since R contains P and $P - P_1$, the polynomial P_1 is also in R, which proves that R also contains all polynomials with the lineality manifold $\Lambda(P_1)$. This completes the proof.

2.11. Remarks on the case of non-bounded domains

We shall here study the minimal operator P_0, defined by a differential operator $P(D)$, when Ω is not bounded, a case which has been excluded in all the previous theorems of this chapter. It seems difficult to give a perfect generalization of Theorem 2.2, but we can prove two theorems which replace Theorem 2.2 in some important cases.

THEOREM 2.19. *Let Ω be a domain, which contains the direct sum of an open set Ω' in the plane $x^{\mu+1} = \cdots = x^\nu = 0$ $(\mu < \nu)$ and the space $G = \{(0,\ldots,0,x^{\mu+1},\ldots,x^\nu)\}$. Then, if*

$$\|Q(D)u\|^2 \leq C(\|P(D)u\|^2 + \|u\|^2), \quad u \in C_0^\infty(\Omega), \tag{2.11.1}$$

it follows that

$$|Q(\xi)|^2 < C'(\sum_\alpha{}^* |P^{(\alpha)}(\xi)|^2 + 1), \tag{2.11.2}$$

where Σ^* means a sum only over sequences of the indices $1,\ldots,\mu$.

PROOF. Let ψ be a function in $C_0^\infty(\Omega' \times G)$ and consequently in $C_0^\infty(\Omega)$. Then the formula (2.3.7) must be valid. Now replace ψ by ψ^ε,

$$\psi^\varepsilon(x^1, \ldots, x^\nu) = \varepsilon^{(\nu-\mu)/2}\,\psi(x^1, \ldots, x^\mu, \varepsilon x^{\mu+1}, \ldots, \varepsilon x^\nu).$$

An easy calculation shows that $\psi^\varepsilon{}_{\alpha\beta} = \varepsilon^k \psi_{\alpha\beta}$, where k is the total number of indices occurring in α and β, which are not between 1 and μ. Hence in the limit when $\varepsilon \to 0$, it follows from (2.3.7) that

$$\sum_{\alpha,\beta}{}^* Q^{(\alpha)}(\xi)\overline{Q^{(\beta)}(\xi)}\,\psi_{\alpha\beta} \leq C(\sum_{\alpha,\beta}{}^* P^{(\alpha)}(\xi)\overline{P^{(\beta)}(\xi)}\,\psi_{\alpha\beta} + \psi_{00}). \tag{2.11.3}$$

Now our result follows at once from (2.3.9).

REMARK. It is easy to see that the same result remains valid, if we replace G by an open set in G, which contains arbitrarily large spheres.

The same argument also gives that, *if Ω satisfies the assumptions of Theorem 2.19 and the operator P_0 has a continuous inverse, we must have*

$$1 \leq C' \sum_\alpha{}^* |P^{(\alpha)}(\xi)|^2. \tag{2.11.4}$$

THEOREM 2.20. *If $x^1,\ldots,x^\mu$ are bounded in Ω, it follows from* (2.11.2) *that* (2.11.1) *is valid. It also follows from* (2.11.4) *that the inverse of P_0 is continuous.*

PROOF. It was remarked on page 185 that Lemma 2.7 is also true for infinite domains. This gives at once a proof of Theorem 2.20, if we repeat the arguments at the end of the proof of Theorem 2.2.

If Ω satisfies both the condition of Theorem 2.19 and that of Theorem 2.20, we may of course conclude that (2.11.2) is a necessary and sufficient condition for the validity of (2.11.1), and that (2.11.4) is a necessary and sufficient condition for the continuity of the inverse of P_0. The result concerning the continuity of P_0^{-1} could partly be obtained from the proof of Theorem 2.1, but it is easy to give examples where that method does not work.

CHAPTER III

Maximal Differential Operators with Constant Coefficients

3.0. Introduction

Let P and Q be two maximal differential operators with constant coefficients. Our first question is: When is it true that $\mathcal{D}_P \subset \mathcal{D}_Q$? The corresponding problem for minimal differential operators was solved by Theorem 2.2. For the maximal operators we obtain the negative result that $\mathcal{D}_P \subset \mathcal{D}_Q$ implies either that $Q = aP + b$, with constant a and b, or else that P and Q are ordinary differential operators, such that the degree of Q is not greater than the degree of P. This is proved in section 3.1.

Although there exist no operators Q (except for the trivial ones), such that $Qu \in L^2(\Omega)$ for every $u \in \mathcal{D}_P$, there may be operators Q, such that Qu is locally square integrable in Ω for every $u \in \mathcal{D}_P$. There is in fact a class of operators P—the operators of *local type*—for which this is the case for every Q weaker than P in the sense of Chapter I. In that case the functions in $\mathcal{D}_P$ have the same regularity properties as the functions in $\mathcal{D}_{P_0}$. The class of operators of local type is determined in sections 3.3, 3.4 and 3.5. The main point is the construction of a fundamental solution in section 3.4. Elliptic operators are of local type. The complete operators of local type also turn out to possess all essential properties of elliptic operators. For instance, all solutions of the equation $Pu = 0$ are infinitely differentiable if and only if P is complete and of local type. (Operators with this property are called elliptic by some authors, cf. Malgrange [21]. Thus our results give simple necessary and sufficient conditions for an operator to be elliptic in this sense.) We also estimate the magnitude of high derivatives of solutions, thus generalizing Holmgren's results for the equation of heat. As an application this gives us a result on the growth of null solutions. (The existence of null solutions is completely discussed for general operators in section 3.2.) Finally, in section 3.7, we establish a spectral theory of self-adjoint operators of local type. Examples of operators of local type are given in section 3.8.

A study of the asymptotic properties of the eigenfunctions (or rather spectral functions) of self-adjoint boundary problems, parallel to that given by Gårding [13] for elliptic operators, was originally planned. However, our results were not complete, since the Tauberian theorem of Ganelius [7], which was used by Gårding, is not sufficient in our general case. The author has therefore postponed the publication to another occasion.

3.1. Comparison of the domains of maximal differential operators

Let P and Q be two maximal differential operators with constant coefficients in $L^2(\Omega)$, where Ω is a *bounded* domain. Theorem 1.1 shows that, if $\mathcal{D}_P \subset \mathcal{D}_Q$, we must have

$$\|Qu\|^2 \leq C(\|Pu\|^2 + \|u\|^2), \quad u \in \mathcal{D}_P, \tag{3.1.1}$$

where, as always, the norm is L^2-norm in Ω. The condition (3.1.1) leads to the following theorem.

THEOREM 3.1. *If the domain of P is part of the domain of Q, we have either $Q = aP + b$ with constant a and b, or else $P(\xi) = p(\langle x_0, \xi\rangle)$ and $Q(\xi) = q(\langle x_0, \xi\rangle)$, where x_0 is a fixed real vector and the degree of the polynomial p is not less than the degree of the polynomial q.*

In the first case it is obvious that $\mathcal{D}_Q \supset \mathcal{D}_P$, with equality unless $a = 0$. In the second case the same result follows from well-known facts concerning ordinary differential operators (see the example on page 169), if, for example, Ω is a cylinder with axis in the x_0-direction.

To prove the theorem, we first note that (3.1.1) must hold for any infinitely differentiable function u. Hence we may set $u = e^{i\langle x, \zeta\rangle}$ with arbitrary complex ζ, and then obtain

$$|Q(\zeta)|^2 \leq C(|P(\zeta)|^2 + 1). \tag{3.1.2}$$

Another necessary condition is obtained, if we set $u(x) = x^k e^{i\langle x, \zeta\rangle}$ in (3.1.1):

$$\int_\Omega |x^k Q(\zeta) + i^{-1} Q^{(k)}(\zeta)|^2 e^{-2\langle x, \eta\rangle} dx \leq C \int_\Omega (1 + |x^k P(\zeta) + i^{-1} P^{(k)}(\zeta)|^2) e^{-2\langle x, \eta\rangle} dx. \tag{3.1.3}$$

Using (3.1.2) and the boundedness of x^k in Ω, we now obtain

$$|Q^{(k)}(\zeta)|^2 \leq C'(|P(\zeta)|^2 + |P^{(k)}(\zeta)|^2 + 1). \tag{3.1.4}$$

The inequalities (3.1.2) and (3.1.4) are independent of each other. We first examine the consequences of (3.1.2) by algebraic methods.

LEMMA 3.1. *Let $P(\zeta)$ and $Q(\zeta)$ be two polynomials in $\zeta = (\zeta_1, \ldots, \zeta_\nu)$ such that (3.1.2) is fulfilled for every complex ζ. Then the polynomials must be algebraically dependent, that is, there exists a polynomial $R(s, t)$ in two complex variables s and t such that $R \neq 0$ and*

$$R(P, Q) = 0. \tag{3.1.5}$$

PROOF. We may suppose without restriction that the polynomials are not constant, and choose the coordinate system such that in the developments

$$P(\zeta)=\sum_{0}^{n} a_k(\zeta_2, \ldots, \zeta_\nu)\zeta_1^k, \quad Q(\zeta)=\sum_{0}^{m} b_k(\zeta_2, \ldots, \zeta_\nu)\zeta_1^k \tag{3.1.6}$$

the highest coefficients a_n and b_m are constants $\neq 0$. Denote the resultant with respect to ζ_1 of the two polynomials $P-\alpha$ and $Q-\beta$ by $R(\alpha, \beta, \zeta_2, \ldots, \zeta_\nu)$. The resultant is a polynomial in $\alpha, \beta, \zeta_2, \ldots, \zeta_\nu$, and does not vanish identically. If the zeros of $P(\zeta)-\alpha$ for fixed $\zeta_2, \ldots, \zeta_\nu$ are $\zeta_1=t_1, \ldots, \zeta_1=t_n$, we have

$$R=a_n^m \prod_{1}^{n}(Q(t_k, \zeta_2, \ldots, \zeta_\nu)-\beta).$$

Since $P(t_k, \zeta_2, \ldots, \zeta_\nu)-\alpha=0$, it follows from (3.1.2) that $|Q(t_k, \zeta_2, \ldots, \zeta_\nu)|^2 \leqq C(1+|\alpha|^2)$. Hence R is bounded for fixed α and β, which proves that R is independent of $\zeta_2, \ldots, \zeta_\nu$, so that we may write $R=R(\alpha, \beta)$. By definition, we have $R(\alpha, \beta)=0$ if $P-\alpha$ and $Q-\beta$ have a common zero ζ_0, that is, if $\alpha=P(\zeta_0)$, $\beta=Q(\zeta_0)$. Thus we obtain

$$R(P(\zeta_0), Q(\zeta_0))=0,$$

which completes the proof.

To proceed further we need a lemma, which is essentially a special case of Lüroth's theorem (cf. van der Waerden [33], § 63).

LEMMA 3.2. *Let R be a ring over a field K such that $K \subset R \subset K[x]$, where $K[x]$ is the ring of polynomials in an indeterminate x with coefficients in K. Then there is a polynomial $\vartheta \in R$ such that $R=K[\vartheta]$.*

PROOF. Let ϑ be a not constant polynomial in R of minimal degree. Then the polynomial $\vartheta(z)-\vartheta(x)$, considered as a polynomial in a second indeterminate z, has coefficients in R and is irreducible in $R[z]$. For suppose that it decomposes in $R[z]$. The factors are then polynomials in z with coefficients in R, so that a factor which is not independent of x must be of at least the same degree in x as ϑ is. Hence all factors except one must have coefficients in K, and since it is obvious that there are no such factors, the irreducibility follows. Hence, if $\eta(x)$ is any polynomial in R, the polynomial $\eta(z)-\eta(x)$ must be divisible by $\vartheta(z)-\vartheta(x)$ in $R[z]$, since both have the zero $z=x$. Denoting the term in the quotient, which is independent of z, by $\eta_1(x)$, we have $\eta_1 \in R$ and

$$\eta(x)-\eta(0)=(\vartheta(x)-\vartheta(0))\eta_1(x).$$

Assuming as we may that $\vartheta(0)=0$, we obtain

$$\eta(x)=\eta(0)+\vartheta(x)\,\eta_1(x), \quad \eta_1(x)\in R.$$

Now we can apply this result to the polynomial η_1 and write

$$\eta_1(x)=\eta_1(0)+\vartheta(x)\,\eta_2(x), \quad \eta_2(x)\in R,$$

and so on. Since the degrees of the polynomials $\eta_1, \eta_2, \ldots$ decrease, we must after a finite number of steps come to a constant polynomial, which proves that $\eta\in K[\vartheta]$.

LEMMA 3.3.[1] *If two polynomials $P(\zeta)$ and $Q(\zeta)$ of $\zeta=(\zeta_1, \ldots, \zeta_\nu)$ are algebraically dependent, there exists a polynomial $W(\zeta)$ and two polynomials $p(t)$, $q(t)$ in one variable, so that*

$$P(\zeta)=p(W(\zeta)),\ Q(\zeta)=q(W(\zeta)). \tag{3.1.7}$$

PROOF. By assumption we have

$$F(P(\zeta), Q(\zeta))=0,$$

where $F(x, y)$ is a polynomial which may be supposed to be irreducible. Assuming as we may that P and Q are of the form (3.1.6) and setting $\zeta_1=t$, $\zeta_2=\cdots=\zeta_\nu=0$, we find that the irreducible curve $F(x, y)=0$ has a parametric representation $x=x(t)$, $y=y(t)$, where $x(t)$ and $y(t)$ are polynomials in t. Now we apply Lemma 3.2 to the ring of polynomials generated by $x(t)$ and $y(t)$. It follows that there is a polynomial $\vartheta(t)$ in this ring, that is, $\vartheta(t)=f(x(t), y(t))$, where f is a polynomial, so that $x(t)=p(\vartheta(t))$, $y(t)=q(\vartheta(t))$. Hence we have for any point on the curve

$$x=p(f(x, y)), \quad y=q(f(x, y)),$$

since this is true for a generic point. Setting $x=P(\zeta)$, $y=Q(\zeta)$ and denoting $f(P(\zeta), Q(\zeta))$ by $W(\zeta)$, we obtain the desired result.

Combining Lemma 3.1 and Lemma 3.3, we conclude that the inequality (3.1.2) is valid if and only if there exists a polynomial $W(\zeta)$ and two polynomials $p(t)$ and $q(t)$, such that the degree of q is not greater than the degree of p and

$$P(\zeta)=p(W(\zeta)), \quad Q(\zeta)=q(W(\zeta)). \tag{3.1.7}$$

[1] This lemma and another much deeper one, needed in an earlier version of this paper, were proved by Professor B. L. VAN DER WAERDEN in reply to a question from the author. His proof, which is based on Lüroth's theorem, includes in fact both Lemma 3.2 and Lemma 3.3, and differs only formally from the one given here.

Polynomials of this form satisfy the inequality (3.1.4) if

$$\left|\frac{\partial W}{\partial \zeta_k}\right|^2 (|q'(W)|^2 - C'|p'(W)|^2) \leqq C'(|p(W)|^2 + 1). \tag{3.1.8}$$

In studying this inequality we have to distinguish between two different cases.

I) If $|q'(t)|^2 - C'|p'(t)|^2 \leqq 0$ for every complex t, it follows that any zero of p' is a zero of q' with at least the same multiplicity. Now q' has not higher degree than p', so it follows that $q'(t) = a\,p'(t)$ with some constant a. Hence $q(t) = a\,p(t) + b$ and $Q(\zeta) = a\,P(\zeta) + b$, so that we have one of the cases mentioned in Theorem 3.1.

II) Now suppose that the open set U of all t such that $|q'(t)|^2 - C'|p'(t)|^2 > 0$ is not empty. Then it follows from (3.1.8), if α and β are fixed complex numbers such that $\alpha \in U$, that $|\partial W/\partial \zeta_k - \beta| < C''$ when $W - \alpha = 0$. Since the arguments of the proof of Lemma 3.1 apply under this weaker assumption, it follows that W and $\partial W/\partial \zeta_k$ are algebraically dependent for any k. Hence $\partial W/\partial \zeta_k$ is constant for any k on a piece of surface where $W(\zeta) = \text{constant}$. Thus the surface is a portion of a plane, and W must be constant in the whole plane. Since two planes, where W has different constant values, cannot meet, it follows that W is constant in a set of parallel planes $\langle z_0, \zeta\rangle = \text{constant}$. Hence W is a polynomial in $\langle z_0, \zeta\rangle$, and using (3.1.7) we obtain

$$P(\zeta) = p(\langle z_0, \zeta\rangle), \quad Q(\zeta) = q(\langle z_0, \zeta\rangle), \tag{3.1.9}$$

where p and q may not be the same polynomials as in (3.1.7). Polynomials of the form (3.1.9) satisfy both (3.1.2) and (3.1.4). To prove the remaining part of the theorem, namely that z_0 must be proportional to a real vector unless $q = a\,p + b$, we must therefore go back to the original condition (3.1.1).

Thus suppose that the polynomials P and Q are of the form (3.1.9) and that z_0 is not proportional to any real vector. We shall prove that $q'(t) = a\,p'(t)$, or, equivalently, that a zero τ of p' with multiplicity k is a zero of q' with the same multiplicity. It is sufficient to suppose that $\tau = 0$. With a suitable complex vector ζ and a real vector η we shall set

$$u(x) = \langle x, \eta\rangle^k e^{i\langle x, \zeta\rangle}. \tag{3.1.10}$$

It easily follows from Leibniz' formula that

$$P(D)\,u = p(\langle z_0, D\rangle)\,u = \sum_{j=0}^{k} \binom{k}{j} \langle z_0, \eta\rangle^j \langle x, \eta\rangle^{k-j} p^{(j)}(\langle z_0, \zeta\rangle)\, e^{i\langle x, \zeta\rangle}, \tag{3.1.11}$$

where $p^{(j)}$ is the jth derivative of p.

Since z_0 is not proportional to a real vector, there exists a vector $\zeta_0 = \xi_0 + i\,\eta_0$ such that $\langle z_0, \zeta_0\rangle = 0$ but $\langle z_0, \eta_0\rangle \neq 0$. Denote by u_t the function obtained by setting $\eta = \eta_0$ and $\zeta = t\,\zeta_0$ with real fixed t in (3.1.10). Since we have assumed that

$$p'(0) = \cdots = p^{(k)}(0) = 0,$$

it follows from (3.1.11) that

$$P(D)\,u_t = p(0)\,\langle x, \eta_0\rangle^k\, e^{i\langle x, t\zeta_0\rangle}.$$

With the notation

$$f(u) = \sum_{j=0}^{k} \binom{k}{j} \langle z_0, \eta_0\rangle^j\, u^{k-j}\, q^{(j)}(0)$$

we also obtain

$$Q(D)\,u_t = f(\langle x, \eta_0\rangle)\, e^{i\langle x, t\zeta_0\rangle},$$

and to show that $q'(0) = \cdots = q^{(k)}(0) = 0$ we have to prove that $f(u)$ cannot contain any term of lower order than u^k.

The inequality (3.1.1) gives when applied to the functions u_t

$$\text{(3.1.12)} \quad \int_\Omega |f(\langle x, \eta_0\rangle)|^2\, e^{-2t\langle x, \eta_0\rangle}\, dx \leqq C\,(1 + |p(0)|^2) \int_\Omega |\langle x, \eta_0\rangle|^{2k}\, e^{-2t\langle x, \eta_0\rangle}\, dx.$$

Translating Ω, if necessary, we may suppose that

$$\text{(3.1.13)} \quad \inf_{x\in\Omega} \langle x, \eta_0\rangle = 0.$$

Let $\alpha(u)$ be the measure of the set

$$\{x;\ x \in \Omega,\ \langle x, \eta_0\rangle \leqq u\}.$$

In virtue of (3.1.13) we have $\alpha(u) = 0$, if $u \leqq 0$, and $\alpha(u) > 0$, if $u > 0$. Furthermore, $\alpha(u)$ is constant for large values of u. The inequality (3.1.12) now takes the form

$$\int_0^\infty |f(u)|^2\, e^{-2tu}\, d\alpha(u) \leqq C' \int_0^\infty u^{2k}\, e^{-2tu}\, d\alpha(u), \quad 0 < t < \infty.$$

Suppose that u^k is not a factor of $f(u)$. Then we can find $\varepsilon > 0$ such that $|f(u)|^2 > 2\,C'\,u^{2k}$ for $0 < u < \varepsilon$. Hence

$$2\,C' \int_0^\varepsilon u^{2k}\, e^{-2tu}\, d\alpha(u) \leqq \int_0^\infty |f(u)|^2\, e^{-2tu}\, d\alpha(u) \leqq C' \int_0^\infty u^{2k}\, e^{-2tu}\, d\alpha(u),$$

and consequently

$$\int_0^{\varepsilon} u^{2k} e^{-2tu}\, d\alpha(u) \leq \int_{\varepsilon}^{\infty} u^{2k} e^{-2tu}\, d\alpha(u).$$

Estimating the two sides of this inequality in an obvious fashion, we obtain

$$e^{-t\varepsilon} \int_0^{\frac{\varepsilon}{2}} u^{2k}\, d\alpha(u) \leq e^{-2t\varepsilon} \int_{\varepsilon}^{\infty} u^{2k}\, d\alpha(u),$$

which gives a contradiction when $t \to \infty$, since the integral on the left-hand side does not vanish. Hence u^k is a factor of $f(u)$, so that $q'(t)$ has a zero of multiplicity k for $t=0$. This completes the proof.

Remark. It also follows from the proof that there exists a uniformly continuous function u, so that $P(D)u$ is uniformly continuous but $Q(D)u$ is not uniformly continuous, unless we have one of the two exceptional cases of Theorem 3.1. In fact, if we substitute for $L^2(\Omega)$ the space C of uniformly continuous functions in Ω, we still get the conditions (3.1.2) and (3.1.4), and we can also give a modification of the discussion at the end of the proof.

Somewhat roughly, we might formulate the result of this section as follows: Maximal partial differential operators with constant coefficients are characterized by their domains, apart from a linear combination with the identity operator.

3.2. The existence of null solutions

We shall call a function $u \neq 0$ a null solution of P, if it is infinitely differentiable, satisfies the equation $Pu=0$, and vanishes in a half-space $\langle x, \xi \rangle \geq 0$, where ξ is a given fixed vector $\neq 0$. It follows from Holmgren's uniqueness theorem (cf. John [16]) that a null solution cannot exist, unless the plane $\langle x, \xi \rangle = 0$ is characteristic, that is, $p(\xi)=0$, where p is the principal part of P. If P is homogeneous, it is obvious that any function $f(\langle x, \xi \rangle)$, where $0 \neq f \in C^\infty$ and $f(t)=0$ for $t>0$, is then a null solution. For equations with lower order terms the existence of null solutions seems to have been proved only for special equations, in particular, the heat equation (Tychonov [32], Täcklind [31], see also Hille [14]). Following the proof of Hille [14] and using some series developments from Petrowsky [26], we can prove the following general existence theorem.

Theorem 3.2. *There exist null solutions of P for every characteristic ξ.*[1]

[1] For equations with variable coefficients it may happen, as has been proved by Mychkis [22], that a solution can only be continued in one way across the whole of some real characteristic, even if it can be locally continued in different ways.

Proof. Let us consider the equation $P(s\xi + t\eta) = 0$, where η is a fixed non-characteristic vector and s and t are complex numbers. Since $p(\xi) = 0$, it easily follows as in Petrowsky [26] that there is a root $t = t(s)$, such that $t/s \to 0$ when $s \to \infty$, and we can develop $t(s)$ in a Puiseux series

$$t = s^{k/p} \sum_{j=0}^{\infty} c_j s^{-j/p},$$

where k and p are positive integers and $k < p$. Hence $t(s)$ is analytic outside a circle $|s| = M$, and when $|s| \to \infty$ we have

$$|t(s)| = O(|s|^{\varrho}),$$

where $\varrho < 1$. Let ϱ' be a number such that $\varrho < \varrho' < 1$, and set with $\tau > M$

$$u(x) = \int_{i\tau-\infty}^{i\tau+\infty} e^{i\langle x,\, s\xi + t(s)\eta\rangle} e^{-(s/i)^{\varrho'}} ds \quad (s = \sigma + i\tau).$$

Here we define $(s/i)^{\varrho'}$ so that it is real and positive when s is on the positive imaginary axis, and use a fixed branch of $t(s)$. The integral is obviously convergent and independent of τ, for when x is in a fixed bounded set we have

$$\operatorname{Re}\left(i t(s) \langle x, \eta\rangle - \left(\frac{s}{i}\right)^{\varrho'}\right) \leq C|s|^{\varrho} - |s|^{\varrho'} \sin \frac{\pi \varrho'}{2} < -c|s|^{\varrho'} \quad (\operatorname{Im} s > M),$$

for large $|s|$, c being a positive constant. This estimate also proves that the integral is uniformly convergent after an arbitrary number of differentiations with respect to x, so that $u(x)$ is infinitely differentiable and solves the equation $P(D)u = 0$. It is also obvious that $u \neq 0$. Now we have for sufficiently large τ

$$|u(x)| \leq e^{-\tau\langle x,\, \xi\rangle} \int_{-\infty}^{+\infty} e^{-c|\sigma|^{\varrho'}} d\sigma.$$

Hence, letting $\tau \to +\infty$, we conclude that $u(x) = 0$ if $\langle x, \xi\rangle > 0$.

The following corollary is a theorem by Petrowsky [26], who also considered systems of differential equations.

Corollary 3.1. *If y is a direction which cuts some characteristic plane of the operator P, then there is a solution u of $P(D)u = 0$ such that $u(x + ty)$ is not an analytic function of t.*

In fact, a null solution u, which vanishes on one side of the characteristic plane, will possess the required property, since we could otherwise prove by analytic continuation that $u = 0$.

3.3. Differential operators of local type

From Theorem 3.1 it follows that, if the operator $P(D)$ depends on more than one variable and Ω is a bounded domain, we can find a function $u \in \mathcal{D}_P$ and a function $\psi \in C^\infty(\bar{\Omega})$ such that $\psi u \notin \mathcal{D}_P$. For suppose that this were not possible, so that whenever $u \in \mathcal{D}_P$ and $\psi \in C^\infty(\bar{\Omega})$ we have $\psi u \in \mathcal{D}_P$. With $\psi = e^{i\langle x, \xi\rangle}$, $\xi \neq 0$, it would follow that any function $u \in \mathcal{D}_P$ is also in the maximal domain of the operator $P(D+\xi)$, which would contradict Theorem 3.1. This negative result, which contrasts with Theorem 2.10, was also proved in section 2.8 by means of explicit examples, when P is the Laplace operator or the wave operator in two variables. For the wave operator we saw that $P(D)(\psi u)$ does not even need to be locally square integrable in Ω, but for the Laplace operator we only proved that $P(D)(\psi u)$ may not be square integrable over the whole of Ω. We now raise the problem to determine those operators for which only this situation can appear. More precisely, we seek those operators P which satisfy the following definition.

Definition 3.1. *A differential operator $P(D)$ is said to be of local type, if the product of any function in $\mathcal{D}_P$ by any function in $C_0^\infty(\Omega)$ is in $\mathcal{D}_P$, and consequently, in virtue of Lemma 2.11, in $\mathcal{D}_{P_0}$.*[1]

An equivalent definition is that *P is of local type, if the functions in $\mathcal{D}_P$ and the functions in $\mathcal{D}_{P_0}$ have the same local regularity properties*, that is, if any function in $\mathcal{D}_P$ equals some function in $\mathcal{D}_{P_0}$ in an arbitrary compact subset of Ω. That this property follows from Definition 3.1 is obvious, for we can choose $\psi \in C_0^\infty$ such that $\psi = 1$ on any given compact set. Conversely, if P has this property, it follows from Theorem 2.10 that Definition 3.1 is fulfilled. Thus Theorem 2.12 proves that a necessary and sufficient condition for an operator to be of local type is that $P^{(\alpha)}(D)u$ is a locally square integrable function for any α and any $u \in \mathcal{D}_P$. If Ω' is a domain with compact closure in Ω, we can hence apply Theorem 1.1 to the mapping

$$\mathcal{D}_P \ni u \to P^{(\alpha)}(D)u \in L^2(\Omega'),$$

and then obtain the following lemma.

Lemma 3.4. *If $P(D)$ is of local type and the domain Ω' has compact closure in Ω, there exists a constant C such that*

[1] Observe that we require this property of the operator P for any domain Ω. It will however follow from our results that it is sufficient to assume that the definition is fulfilled for one bounded domain Ω, it then follows for any domain, bounded or not bounded.

$$\int_{\Omega'} |P^{(\alpha)}(D)u|^2\,dx \leq C\Big(\int_{\Omega} |P(D)u|^2\,dx + \int_{\Omega} |u|^2\,dx\Big), \quad u \in \mathcal{D}_P. \tag{3.3.1}$$

Let Ω be a bounded domain. Setting $u(x) = e^{i\langle x,\zeta\rangle}$ in (3.3.1), $\zeta = \xi + i\eta$, we obtain

$$|P^{(\alpha)}(\zeta)|^2 \int_{\Omega'} e^{-2\langle x,\eta\rangle}\,dx \leq C(1+|P(\zeta)|^2)\int_{\Omega} e^{-2\langle x,\eta\rangle}\,dx. \tag{3.3.2}$$

If δ is the supremum of $2|x|$ when $x \in \Omega$, we have the two estimates

$$\int_{\Omega'} e^{-2\langle x,\eta\rangle}\,dx \geq e^{-\delta|\eta|}\int_{\Omega'} dx = e^{-\delta|\eta|}\,m(\Omega'), \quad \int_{\Omega} e^{-2\langle x,\eta\rangle}\,dx \leq e^{\delta|\eta|}\,m(\Omega).$$

Hence it follows from (3.3.2) that

$$|P^{(\alpha)}(\zeta)|^2 \leq C\,\frac{m(\Omega)}{m(\Omega')}\,e^{2\delta|\eta|}(1+|P(\zeta)|^2). \tag{3.3.3}$$

Adding the inequalities (3.3.3) for all α, and using the notation $\tilde{P}(\zeta) = (\sum |P^{(\alpha)}(\zeta)|^2)^{\frac{1}{2}}$ again, we obtain the following lemma.

LEMMA 3.5. *Let P be of local type. Then for any A there is a constant C such that*

$$\tilde{P}(\zeta)^2 \leq C(1+|P(\zeta)|^2), \tag{3.3.4}$$

when $|\mathrm{Im}\,\zeta| < A$.

The necessary condition for an operator to be of local type, which we have now derived, is in fact also sufficient. Before proving this, we shall deduce other equivalent conditions, which seem to be more natural and useful.

LEMMA 3.6. *If a polynomial P satisfies (3.3.4), we have*

$$|P(\xi+i\eta)| \to \infty \text{ when } \xi \to \infty \text{ modulo } \Lambda(P), \tag{3.3.5}$$

and the convergence is uniform in η, if $|\eta| < A$, where A is an arbitrary fixed positive number.

PROOF. Examination of the proof of Theorem 2.17 shows that

$$\tilde{P}(\zeta) \to \infty \text{ when } \zeta \to \infty \text{ modulo } \Lambda^*(P),$$

where $\Lambda^*(P)$ is the complex lineality space of P, defined by (2.8.21), if we omit the word "real". Since $\Lambda(P)$ is the set of real vectors in $\Lambda^*(P)$, the assertion now follows from (3.3.4).

We shall next prove two lemmas, which give a converse of Lemma 3.6 in a sharp form, which will be used later. For convenience we only formulate them for complete polynomials.

LEMMA 3.7. *Suppose that for any positive number A there exists a number B such that*

$$P(\xi + i\eta) \neq 0, \text{ when } |\eta| < A \text{ and } |\xi| > B. \tag{3.3.6}$$

Then the polynomial P is complete, and for any fixed real vector ϑ we have

$$\frac{P(\xi+\vartheta)}{P(\xi)} \to 1 \text{ when } \xi \to \infty. \tag{3.3.7}$$

PROOF. That P must be complete is obvious. In proving (3.3.7) we may assume that the coordinates are so chosen that $\vartheta = (1, 0, \ldots, 0)$. Now let ε be a fixed small positive number. In virtue of the assumptions we can find a number B such that

$$P(\xi + i\eta) \neq 0 \text{ when } |\eta| < \varepsilon^{-1} \text{ and } |\xi| > B.$$

Then the inequality $|\xi - \zeta'| \geqq \varepsilon^{-1}$ is valid, if $|\xi| > B + \varepsilon^{-1}$ and $P(\zeta') = 0$. For setting $\zeta' = \xi' + i\eta'$ we have either $|\eta'| \geqq \varepsilon^{-1}$, or else $|\xi'| \leqq B$ so that $|\xi - \xi'| \geqq \varepsilon^{-1}$. Giving constant values to $\xi_2, \ldots, \xi_\nu$ we can write

$$P(\xi) = E\prod_1^m (\xi_1 - t_k),$$

where $(t_k, \xi_2, \ldots, \xi_\nu)$ is a zero of P. Hence we have $|t_k - \xi_1| \geqq \varepsilon^{-1}$ if $|\xi| > B + \varepsilon^{-1}$. Using this estimate in the formula

$$\frac{P(\xi+\vartheta)}{P(\xi)} = \prod_1^m \frac{\xi_1 + 1 - t_k}{\xi_1 - t_k} = \prod_1^m \left(1 + \frac{1}{\xi_1 - t_k}\right)$$

we obtain

$$\left|\frac{P(\xi+\vartheta)}{P(\xi)} - 1\right| \leqq m\,\varepsilon\,(1+\varepsilon)^{m-1}, \quad |\xi| > B + \varepsilon^{-1},$$

which proves the assertion.

LEMMA 3.8. *If for every constant real vector ϑ*

$$\frac{P(\xi+\vartheta)}{P(\xi)} \to 1 \text{ when } \xi \to \infty, \tag{3.3.7}$$

then (3.3.7) is valid for every complex ϑ, and the convergence is uniform in ϑ, if $|\vartheta| < A$ for some fixed A. Furthermore, we have, if $|\alpha| \neq 0$,

$$\frac{P^{(\alpha)}(\xi+\vartheta)}{P(\xi+\vartheta)} \to 0 \textit{ when } \xi \to \infty, \tag{3.3.8}$$

uniformly in ϑ, *if* $|\vartheta| < A$.

PROOF. In virtue of Lemma 2.10 we can write

$$P^{(\alpha)}(\xi) = \sum_1^m t_i P(\xi+\vartheta_i), \tag{3.3.9}$$

where ϑ_i are real vectors. Since the principal parts on the right-hand side must cancel out, if $|\alpha| \neq 0$, we have $\sum t_i = 0$. Hence we obtain in virtue of (3.3.7)

$$\frac{P^{(\alpha)}(\xi)}{P(\xi)} \to \sum_1^m t_i = 0 \text{ when } \xi \to \infty. \tag{3.3.10}$$

From Taylor's formula it follows that

$$\frac{P(\xi+\vartheta)}{P(\xi)} = 1 + \sum_{|\alpha| \neq 0} \frac{P^{(\alpha)}(\xi)}{P(\xi)} \frac{\vartheta_\alpha}{|\alpha|!},$$

which proves that (3.3.7) is valid for arbitrary complex ϑ, and also exhibits the asserted uniform convergence. Using this result and (3.3.9), we obtain

$$\frac{P^{(\alpha)}(\xi+\vartheta)}{P(\xi+\vartheta)} = \sum_1^m t_i \frac{P(\xi+\vartheta+\vartheta_i)}{P(\xi)} \frac{P(\xi)}{P(\xi+\vartheta)} \to \sum_1^m t_i = 0 \text{ when } \xi \to \infty,$$

uniformly in ϑ.

THEOREM 3.3. *The following five conditions on a polynomial P are all equivalent:*

I) *For an arbitrary given A, the polynomial* $P(\xi + i\eta)$ *does not vanish, if* $|\eta| < A$ *and the distance from* ξ *to* $\Lambda(P)$ *is sufficiently large.*

II) *For every real vector* ϑ *we have*

$$\frac{P(\xi+\vartheta)}{P(\xi)} \to 1$$

when ξ *is real and* $\to \infty$ *modulo* $\Lambda(P)$. *The convergence is uniform in* ϑ, *if* $|\vartheta|$ *is bounded.*

III) *For every* α *with* $|\alpha| \neq 0$ *we have*

$$\frac{P^{(\alpha)}(\xi)}{P(\xi)} \to 0,$$

when ξ *is real and* $\to \infty$ *modulo* $\Lambda(P)$.

IV) *For any A there is a constant C such that when $|\eta| \leqq A$*

$$\tilde{P}(\xi + i\eta)^2 \leqq C(1 + |P(\xi + i\eta)|^2).$$

V) *When $\xi \to \infty$ modulo $\Lambda(P)$ we have $|P(\xi + i\eta)| \to \infty$, and the convergence is uniform in η when $|\eta| \leqq A$.*

Each of these conditions is a necessary and sufficient condition for the operator P to be of local type.

PROOF. We first prove the equivalence of the five conditions. Lemmas 3.7 and 3.8 show that I implies II and that II implies III and IV. Furthermore, Lemma 3.6 proves that IV implies V, and I is obviously a consequence of V. Hence the conditions I, II, IV, V are all equivalent. Since III follows from II, and the proof of Lemma 3.8 shows that II follows from III, the equivalence of the conditions is established. In virtue of Lemma 3.5 the condition IV is a *necessary* condition for P to be of local type.

We note that, if P is complete, we may omit "modulo $\Lambda(P)$" from the statement, and that the theorem states that a polynomial is of local type, if the complete polynomial which it induces in $R^\nu/\Lambda(P)$ is of local type. The easy but space-consuming verification of this fact may be left to the reader. Thus in proving the sufficiency of the conditions I – V, we may restrict ourselves to the case of complete polynomials. In that case we shall carry out the proof in section 3.5, by means of a fundamental solution, which will be constructed in the next section.

3.4. Construction of a fundamental solution of a complete operator of local type

In this section we shall consistently use the theory of distributions, without explicit reference at every point. The definitions and results, which we use, can of course be found in Schwartz [28]. Our purpose is to construct a fundamental solution, that is, a distribution E such that

$$E * (P(D)u) = u, \quad u \in C_0^\infty(R^\nu), \tag{3.4.1}$$

and to prove certain regularity properties of E. The results are stated in the following theorem.

THEOREM 3.4. *Let P be complete and satisfy the conditions* I–V *of Theorem* 3.3. *Then $P(D)$ has a fundamental solution E with the properties:*

I) *In the domain $x \neq 0$ the distribution E is an infinitely differentiable function $E(x)$.*

II) *If u is square integrable and has compact support, the convolution*

$$P^{(\alpha)}(D)\, E * u$$

is a locally square integrable function.

REMARKS. 1. Every fundamental solution has the properties I and II. In fact, we shall see later that the difference between two fundamental solutions is infinitely differentiable.

2. Schwartz [28] has called a function $E(x)$, which is infinitely differentiable for $x \neq 0$ and integrable over a neighbourhood of the origin, a "noyau élémentaire", if the distribution E defined by

$$E(u) = \int E(x)\, u(x)\, dx$$

is a fundamental solution. He proved that all solutions of the equation $Pu = 0$ are infinitely differentiable if P possesses a "noyau élémentaire". We shall not prove here that the fundamental solution of a complete operator of local type is a "noyau élémentaire", but we shall nevertheless prove that all solutions are infinitely differentiable.

If $P(\xi)$ did not vanish for any real ξ, we could obtain a fundamental solution by writing

$$E * u(x) = (2\pi)^{-\nu/2} \int e^{i\langle x, \xi\rangle} \frac{\hat{u}(\xi)}{P(\xi)}\, d\xi, \quad u \in C_0^\infty, \tag{3.4.2}$$

or equivalently

$$E(\check{u}) = (2\pi)^{-\nu/2} \int \frac{\hat{u}(\xi)}{P(\xi)}\, d\xi, \quad u \in C_0^\infty, \tag{3.4.3}$$

where $\check{u}$ is defined by $\check{u}(x) = u(-x)$. Now the polynomial $P(\xi)$ has in general real zeros, and we must then give (3.4.3) a generalized sense.

We shall define (3.4.3) as a repeated integral, first an integral in the complex domain with respect to ξ_1, and then an integral with respect to the other real variables. We may then assume that the coordinates are chosen such that the highest power of ξ_1 in $P(\xi)$ has a constant coefficient.

In virtue of the condition V in Theorem 3.3 we have $|P(\xi)| \geqq 1$, if ξ is real and $|\xi| \geqq C$, where C is a suitable constant. Thus $|P(\xi)| \geqq 1$, if $\xi_2^2 + \cdots + \xi_\nu^2 \geqq C^2$. Since the zeros of a polynomial vary continuously with parameters which do not occur in the highest order term, we can find a second constant C' such that $|P(\zeta_1, \xi_2, \cdots, \xi_\nu)| \geqq 1$, if $\xi_2, \cdots, \xi_\nu$ are real, $\xi_2^2 + \cdots + \xi_\nu^2 \leqq C^2$, and $|\zeta_1| \geqq C'$.

Now we set for $u \in C_0^\infty(R^\nu)$

$$(3.4.4) \qquad E(\check{u}) = (2\pi)^{-\nu/2} \int d\xi_2 \cdots d\xi_\nu \oint \frac{\hat{u}(\xi)}{P(\xi)} d\xi_1 = (2\pi)^{-\nu/2} \oint \frac{\hat{u}(\xi)}{P(\xi)} d\xi.$$

The integral with respect to ξ_1 *shall be extended over the real axis, if* $\xi_2^2 + \cdots + \xi_\nu^2 \geq C^2$, *and over the real axis with the interval* $(-C', C')$ *replaced by a semi-circle in the lower half-plane, if* $\xi_2^2 + \cdots + \xi_\nu^2 < C^2$.[1] Thus we have $|P(\xi)| \geq 1$ everywhere in the integral.

Since $u \in C_0^\infty$, it follows that $\hat{u}$ is an entire analytic function, which decreases rapidly in the real domain. Hence the integral (3.4.4) is convergent. It is plain that the formula (3.4.2) is valid, if we interpret the integral in the way just defined. Thus, if $u \in C_0^\infty(R^\nu)$, we have

$$E * (P(D)u) = (2\pi)^{-\nu/2} \oint e^{i\langle x, \xi\rangle} P(\xi)\hat{u}(\xi)/P(\xi)\, d\xi = (2\pi)^{-\nu/2} \oint e^{i\langle x, \xi\rangle} \hat{u}(\xi)\, d\xi.$$

Since the integrand is an analytic function of ξ_1, we may shift the integration path back to the real axis. Hence we obtain

$$E * (P(D)u)(x) = (2\pi)^{-\nu/2} \int e^{i\langle x, \xi\rangle} \hat{u}(\xi)\, d\xi = u(x),$$

which proves that E is a fundamental solution.

We now divide the integral (3.4.4) into two parts in the following manner. If $R = \sqrt{C^2 + C'^2}$, we have $|\xi| < R$ in the part of the integral (3.4.4), where ξ is not real. Thus if we write

$$(3.4.5) \qquad E = E_1 + E_2,$$

$$(3.4.6) \qquad E_1(\check{u}) = (2\pi)^{-\nu/2} \int_{|\xi| \geq R} \frac{\hat{u}(\xi)}{P(\xi)} d\xi, \quad E_2(\check{u}) = (2\pi)^{-\nu/2} \oint_{|\xi| \leq R} \frac{\hat{u}(\xi)}{P(\xi)} d\xi,$$

the variable ξ only assumes real values in the integral defining E_1. The distribution E_2 is an entire analytic function, for when $u \in C_0^\infty$ we obtain in virtue of the definition of $\hat{u}$

$$E_2(\check{u}) = (2\pi)^{-\nu} \oint_{|\xi| \leq R} \frac{d\xi}{P(\xi)} \int u(x) e^{-i\langle x, \xi\rangle} dx = (2\pi)^{-\nu} \int \check{u}(x)\, dx \oint_{|\xi| \leq R} \frac{e^{i\langle x, \xi\rangle}}{P(\xi)} d\xi.$$

The change of the order of integrations is justified by the fact that both integrals are only extended over compact sets. Hence E_2 equals the function

[1] There is a very large freedom in the choice of integration paths, and different choices give different fundamental solutions. Note that ξ is here a complex variable, whereas ξ always denotes a real vector elsewhere in this paper.

(3.4.7)
$$E_2(x) = (2\pi)^{-\nu} \oint_{|\xi| \leq R} \frac{e^{i\langle x, \xi\rangle}}{P(\xi)}\, d\xi,$$

which is an entire analytic function, since the integral is uniformly convergent when $|x|$ is bounded.

Let $u \in L^2$ have a compact support. The convolution $P^{(\alpha)}(D)\, E_2 * u$ is then an analytic function. Thus the assertion II of Theorem 3.4 will follow, if we prove that $P^{(\alpha)}(D)\, E_1 * u$ is square integrable. Let $\varphi \in C_0^\infty$. Then the function $u * \varphi$ is also in C_0^∞, and in virtue of (3.4.6) we obtain

(3.4.8)
$$(P^{(\alpha)}(D)\, E_1 * u)(\check{\varphi}) = P^{(\alpha)}(D)\, E_1 * u * \varphi(0) = \int_{|\xi| \geq R} \frac{P^{(\alpha)}(\xi)}{P(\xi)}\, \hat{u}(\xi)\, \hat{\varphi}(\xi)\, d\xi,$$

so that the Fourier transform of $P^{(\alpha)}(D)\, E_1 * u$ is a function which vanishes when $|\xi| < R$ and equals $\hat{u}(\xi)\, P^{(\alpha)}(\xi)/P(\xi)$ when $|\xi| \geqq R$. Noting that $P^{(\alpha)}(\xi)/P(\xi)$ is bounded when $|\xi| \geqq R$ in virtue of condition III of Theorem 3.3, and that $\hat{u}(\xi)$ is square integrable, we conclude that the Fourier transform of $P^{(\alpha)}(D)\, E_1 * u$ is a square integrable function. Hence $P^{(\alpha)}(D)\, E_1 * u$ is also square integrable, which completes the proof of the assertion II of the theorem.

We now turn to the proof of assertion I. Since we have already proved that E_2 is an entire analytic function, it remains to prove that E_1 is an infinitely differentiable function for $x \neq 0$. We need the following algebraic lemma, which gives a precise form of the condition I of Theorem 3.3.

LEMMA 3.9. *Let $y \neq 0$ be a fixed vector in R^ν, and set*

(3.4.9)
$$M(\tau) = \inf_{\zeta, \xi} |\zeta - \xi|,$$

where ζ is a vector in C_ν such that $P(\zeta) = 0$, and ξ is a vector in R_ν such that $|\langle y, \xi\rangle| = \tau$. Then there exist positive numbers a and b such that

$$M(\tau)\, \tau^{-b} \to a \text{ when } \tau \to \infty.$$

PROOF. It follows from condition I of Theorem 3.3 that the infimum in (3.4.9) is attained, and that $M(\tau)$ is a continuous function of τ. The system of equations

(3.4.10)
$$P(\zeta) = 0, \; \langle y, \xi\rangle^2 = \tau^2, \; |\zeta - \xi|^2 = \mu^2$$

has a solution $\zeta \in C_\nu$, $\xi \in R_\nu$ if and only if $\mu \geqq M(\tau)$. Considering C_ν as a 2ν-dimensional real vector space and the equation $P(\zeta) = 0$ as two real equations, we can eliminate the variables ζ and ξ from (3.4.10) by means of Theorem 3 of Seidenberg [29].[1] We then obtain a finite number of finite sets $G_1, \dots, G_s$ of polynomial equalities and inequalities in μ and τ such that there exist vectors ζ and ξ satisfying (3.4.10) if and only if all equalities and inequalities of G_i are satisfied by μ and τ, for at least one $i=1, \dots, s$. Since the existence of solutions ζ, ξ of (3.4.10) is also equivalent to the inequality $\mu \geqq M(\tau)$, we may assume that G_i is of the form

$$G_{ik}(\mu, \tau) \geqq 0, \quad k=1, \dots, k_i.$$

$\mu = M(\tau)$ must make some of these inequalities to an equality. Let $G(\mu, \tau)$ be the product of all the polynomials $G_{ik}(\mu, \tau)$, which do not vanish identically, and let $H(\mu, \tau)$ be the polynomial with the same irreducible factors as $G(\mu, \tau)$ but all with multiplicity 1. Then we have $H(M(\tau), \tau)=0$ for every τ. For sufficiently large τ, the degree in μ of $H(\mu, \tau)$ is independent of τ, and the zeros $\mu_k(\tau)$ are different continuous functions of τ, since H has no multiple factors. Thus the index k, such that $M(\tau) = \mu_k(\tau)$, is independent of τ, since $M(\tau)$ is continuous. Hence $M(\tau)$ is an algebraic function of τ for large τ, and can be developed in a Puiseux series. In virtue of condition I of Theorem 3.3, we have $M(\tau) \to \infty$ with τ. Hence the highest power of τ in the Puiseux series must be positive, which proves the assertion.[2]

Lemma 3.10. *There exist positive constants c and d such that for sufficiently large $|\xi|$ we have*

$$|\zeta - \xi| \geqq c|\xi|^d$$

for any real ξ and any ζ with $P(\zeta)=0$.

Proof. Choosing the vector y of Lemma 3.9 as $(0, \dots, 0, 1, 0, \dots, 0)$, we obtain for large $|\xi_i|$

$$|\zeta - \xi| \geqq a_i |\xi_i|^{b_i}$$

where a_i and b_i are positive numbers. Hence, if $c' = \min a_i$ and $d = \min b_i$, we have

$$|\zeta - \xi| \geqq c' (\max |\xi_i|)^d \geqq c|\xi|^d.$$

Lemma 3.11. *Let $y \in R^\nu$ and $\eta \in R_\nu$ be two fixed vectors. Then there is a constant C such that*

$$\left| \langle D, \eta \rangle^{k+j} \left(\frac{1}{P(\xi)} \right) \right| < \frac{(k+j)!\, C^{k+j}}{|\xi|^{kd} (1 + |\langle y, \xi \rangle|)^{bj}}, \quad |\xi| \geqq R, \ j, k = 1, 2, \dots, \tag{3.4.11}$$

where b and d are the constants of the two preceding lemmas.

[1] The restriction in this theorem that the coefficients must be rational is removed on page 372.

[2] This result bears some analogy to a lemma in Gårding [8].

PROOF. The quantity, which we shall estimate, is

$$\langle D, \eta\rangle^{k+j}\left(\frac{1}{P(\xi)}\right) = i^{-(k+j)}\frac{d^{k+j}}{d\,t^{k+j}}\left(\frac{1}{P(\xi+t\,\eta)}\right)_{t=0}.$$

We can write $P(\xi+t\,\eta)=A\prod_1^m (t-t_r)$. Since $\zeta=\xi+t_r\,\eta$ is a zero of P and $\zeta-\xi=t_r\,\eta$, the numbers t_r can be estimated by either of Lemma 3.9 and Lemma 3.10:

$$|t_r|\geqq a'(1+|\langle y, \xi\rangle|)^b, \quad |t_r|\geqq c'|\xi|^d, \quad (|\xi|\geqq R). \tag{3.4.12}$$

Now the $(k+j)^{\text{th}}$ derivative of $1/P(\xi+t\,\eta)$ for $t=0$ is a sum of terms which are each of the form A^{-1} divided by a product of $k+j+m$ of the zeros t_r. The number of terms is

$$m(m+1)\cdots(m+k+j-1)=(k+j)!\binom{m+k+j-1}{k+j}<(k+j)!\;2^{m+k+j-1}.$$

Furthermore, A is independent of ξ as will be proved in section 3.8. Hence the lemma follows, if we estimate j of the zeros by the first inequality in (3.4.12), k of them by the second inequality, and the remaining m by a constant.

Let y and η be two fixed vectors, and let b and d be the same numbers as in the previous lemmas. We shall prove that the distribution

$$F=\langle x, \eta\rangle^l\langle y, D\rangle^k E_1 \tag{3.4.13}$$

is a continuous function, if

$$l\geqq\frac{k}{b}+r, \tag{3.4.14}$$

where r is the least integer $>\nu/d$. This will complete the proof of Theorem 3.4, and estimating the absolute value of F we shall get an interesting refinement of this theorem.

The definition of F means that

$$F(\check{u})=(2\,\pi)^{-\nu/2}\int\limits_{|\xi|\geqq R}\frac{\langle y, \xi\rangle^k}{P(\xi)}(\langle D, \eta\rangle^l\,\hat{u}(\xi))\,d\,\xi,$$

where D now denotes differentiation with respect to ξ. Integrating by parts, we obtain $F(\check{u})=G(\check{u})+I(\check{u})$, where

$$G(\check{u})=(2\,\pi)^{-\nu/2}\int\limits_{|\xi|\geqq R}\hat{u}(\xi)\left(\langle -D, \eta\rangle^l\left(\frac{\langle y, \xi\rangle^k}{P(\xi)}\right)\right)d\,\xi \tag{3.4.15}$$

and, dS being the vectorial element of area on the sphere $|\xi|=R$,

$$(3.4.16)\quad I(\check{u}) = i^{-1}(2\pi)^{-\nu/2}\sum_{j=0}^{l-1}\int_{|\xi|=R}\left(\langle -D,\eta\rangle^{j}\left\{\frac{\langle y,\xi\rangle^{k}}{P(\xi)}\right\}\right)(\langle D,\eta\rangle^{l-1-j}\,\hat{u}(\xi))\,\langle dS,\eta\rangle.$$

In virtue of Lemma 3.11 we have the following estimate of the integrand in (3.4.15), which we denote by $g(\xi)$,

$$|g(\xi)| = \left|\sum_{j=0}^{k}\binom{l}{j}\frac{k!}{(k-j)!}\langle y, i\eta\rangle^{j}\langle y,\xi\rangle^{k-j}\left(\langle -D,\eta\rangle^{l-j}\frac{1}{P(\xi)}\right)\right|$$
$$\leqq \sum_{j=0}^{k}\binom{l}{j}\frac{k!}{(k-j)!}|\langle y,\eta\rangle|^{j}\,|\langle y,\xi\rangle|^{k-j}\frac{(l-j)!\,C^{l-j}}{|\xi|^{rd}\,(1+|\langle y,\xi\rangle|)^{b(l-j-r)}}.$$

Now it follows from (3.4.14) that

$$b(l-j-r)-(k-j) = b\left(l-\frac{k}{b}-r\right)+j(1-b)>0,$$

so that we obtain

$$|g(\xi)| \leqq |\xi|^{-rd}\,l!\sum_{j=0}^{k}\binom{k}{j}|\langle y,\eta\rangle|^{j}C^{l-j} = |\xi|^{-rd}\,l!\,C^{l}(|\langle y,\eta\rangle|/C+1)^{k}.$$

The function $|\xi|^{-rd}$ is integrable over the domain $|\xi|>R$, since $rd>\nu$. Thus the distribution G must equal the continuous function

$$G(x) = (2\pi)^{-\nu}\int_{|\xi|\geqq R} g(\xi)\,e^{i\langle x,\xi\rangle}\,d\xi.$$

With a new constant C we have the estimate

$$(3.4.17)\qquad |G(x)| \leqq C^{l}\,l!.$$

Using in (3.4.16) the definition of $\hat{u}$, we find that the distribution I is defined by the analytic function

$$(3.4.18)\quad I(x) = i^{-1}(2\pi)^{-\nu}\sum_{j=0}^{l-1}\int_{|\xi|=R}\langle x,\eta\rangle^{l-1-j}e^{i\langle x,\xi\rangle}\left(\langle -D,\eta\rangle^{j}\left\{\frac{\langle y,\xi\rangle^{k}}{P(\xi)}\right\}\right)\langle dS,\eta\rangle.$$

The proof is quite parallel to the previous study of the distribution E_2. Since $1/P(\xi)$ is analytic in a complex neighbourhood of $|\xi|=R$, we have

$$\left|\langle -D,\eta\rangle^{s}\frac{1}{P(\xi)}\right| \leqq s!\,A^{s},\quad |\xi|=R,$$

with some constant A. Hence we obtain, when x is in a compact domain K, that, with suitable constants B and C

$$|I(x)| \leq \sum_{j=0}^{l-1} C^{l-1-j} \sum_{s=0}^{\min(k,j)} \binom{j}{s} (j-s)!\, A^{j-s} B^k\, k!/(k-s)!$$

$$\leq \sum_{j=0}^{l-1} C^{l-1-j} j! \sum_{s=0}^{k} \binom{k}{s} A^{j-s} B^k = \sum_{j=0}^{l-1} C^{l-1-j} j!\, A^j (B+A^{-1}B)^k.$$

Since $k<l$ and $\sum_0^{l-1} j! < (l-1)!\,l = l!$, we have with a new constant C

(3.4.19) $$|I(x)| \leq C^l\, l!.$$

Now $F = G + I$, so that we have proved that the distribution F, defined by (3.4.13), is a continuous function, if (3.4.14) is valid. We have also proved that the absolute value of F has an estimate of the form (3.4.17), (3.4.19), when x is in a compact set K. If we now choose l as the smallest integer such that (3.4.14) is valid, and recall that $E(x) = E_1(x) + E_2(x)$, where $E_2(x)$ is an entire analytic function, the following theorem is proved.

THEOREM 3.5. *Let y be a vector in R^ν and b the number introduced in Lemma 3.9. Then, for any compact set K, which does not contain the origin, there exists a constant C such that*

(3.4.20) $$|\langle y, D\rangle^k E(x)| \leq C^k\, \Gamma\left(\frac{k}{b}\right), \quad x \in K,$$

where $E(x)$ is the function which defines the fundamental solution of Theorem 3.4.

In constructing the fundamental solution we have used several ideas from the literature. The idea of estimating an expression of the form (3.4.13) has been taken over from a study of elliptic operators by Gårding [10]. For references to the very rich older literature on this subject, the reader should consult Schwartz [28].

3.5. Proof of Theorem 3.3

Let P be complete and satisfy the conditions I–V of Theorem 3.3, and let Ω and Ω' be any domains such that Ω' has compact closure in Ω. The domain Ω may be bounded or not be bounded. Then there exists a positive number ε such that a sphere with radius ε and centre at any point in Ω' is contained in Ω. Let $\varrho(x)$ be a function in C_0^∞, which vanishes for $|x| \geq \varepsilon$ and equals 1 in a neighbourhood of the origin. Instead of the fundamental solution constructed in Theorem 3.4, we shall use the "parametrix"

(3.5.1) $$F = \varrho E.$$

The support of F is contained in the sphere $|x| \leq \varepsilon$, and

$$P(D)F = \delta_0 + \omega(x), \tag{3.5.2}$$

where δ_0 is the Dirac measure at the origin and $\omega(x)$ is an infinitely differentiable function, which vanishes for $|x| \geqq \varepsilon$ and also in a neighbourhood of the origin. In fact, in a neighbourhood of the origin, where $\varrho = 1$, we have $F = E$ and thus $P(D)F = P(D)E = \delta_0$. Since $P(D)E$ is infinitely differentiable for $x \neq 0$, the formula (3.5.2) follows.

Now let $u \in \mathcal{D}_P$, which means that u and $P(D)u$ are square integrable functions in the sense of the theory of distributions. In Ω' we have

$$u = u * \delta_0 = u * (P(D)F - \omega) = F * (P(D)u) - \omega * u \tag{3.5.3}$$

and consequently

$$P^{(\alpha)}(D)u = (P^{(\alpha)}(D)F) * (Pu) - (P^{(\alpha)}(D)\omega) * u. \tag{3.5.4}$$

Since $P^{(\alpha)}(D)\omega$ is continuous, the last term is bounded and hence square integrable in Ω'. To study the other term in (3.5.4), we denote by φ the function which equals Pu in points with distance $< \varepsilon$ to Ω' and equals 0 elsewhere. φ is square integrable and has compact support. In Ω' we have

$$(P^{(\alpha)}(D)F) * (Pu) = (P^{(\alpha)}(D)F) * \varphi = (P^{(\alpha)}(D)E) * \varphi + (P^{(\alpha)}(D)\{(\varrho - 1)E\}) * \varphi.$$

Now $P^{(\alpha)}(D)E * \varphi$ is a square integrable function in Ω' in virtue of the assertion II of Theorem 3.4. Since $(\varrho - 1)E$ is an infinitely differentiable function and φ has compact support, it follows in particular that also $(P^{(\alpha)}(D)\{(\varrho - 1)E\}) * \varphi$ is square integrable in Ω'. Hence $P^{(\alpha)}(D)u$ is locally square integrable in Ω, for any $u \in \mathcal{D}_P$, and thus the remarks following Definition 3.1 show that the operator P is of local type.

We may also note that (3.5.3) shows that all distributions u, such that $P(D)u = 0$, are infinitely differentiable functions. We shall refine this result in the next section.

3.6. The differentiability of the solutions of a complete operator of local type

We observed at the end of the previous section that all solutions of the equation $Pu = 0$, where P is complete and of local type, are infinitely differentiable. More generally we can prove:

THEOREM 3.6. *If u belongs to the domain of the operator P^k for every k, where P is a complete differential operator of local type, it follows that u is an infinitely differentiable function after correction on a null set.*

PROOF. It follows from Theorem 3.3 (or else directly from Definition 3.1) that P^k is also complete and of local type. Hence, if $\psi \in C_0^\infty(\Omega)$, the function ψu is in the minimal domain of $P(D)^k$ in any bounded domain Ω', containing the support of ψ. Thus ψu equals an infinitely differentiable function in virtue of Theorem 2.14. Since ψ is an arbitrary function in C_0^∞, we obtain the desired result.

The proof of Theorem 2.14 also gives the following more precise result: For any differential operator $Q(D)$ there exists an integer k such that, if Ω is a bounded domain, we have with some constant C

$$\sup_{x\in\Omega} |Q(D)\,u(x)|^2 \leq C(\|(P^k)_0\,u\|^2 + \|u\|^2),$$

when u is in the minimal domain of $P(D)^k$. Using this result and the proof of Theorem 3.6, we obtain the following useful estimate.

LEMMA 3.12. *Let P be complete and of local type, and let Q be any differential operator with constant coefficients. Then there exists an integer k with the following property: If $u \in \mathcal{D}_{P^k}$, the function $Q(D)\,u$ is continuous in Ω, and for any domain Ω' with compact closure in Ω there is a constant C such that*

$$\sup_{x\in\Omega'} |Q(D)\,u(x)|^2 \leq C(\|P^k\,u\|^2 + \|u\|^2). \tag{3.6.1}$$

THEOREM 3.7. *Let Ω be a bounded domain. If all the solutions $u \in L^2(\Omega)$ of the equation $Pu = 0$ are infinitely differentiable after correction on a null set, the operator $P(D)$ must be complete and of local type.*

PROOF. We shall prove that the first condition in Theorem 3.3 is fulfilled. This can be done my means of explicit constructions similar to those of Petrowsky [26]. However, we give a proof along the lines of this paper. Thus let Ω' be a domain with compact closure in Ω. Since P is a closed operator, the set U of all solutions u of the equation $Pu = 0$ is a closed subspace of $L^2(\Omega)$. The mapping

$$U \ni u \to \partial u/\partial x^k \in L^2(\Omega')$$

is closed, and by assumption it is defined in the whole of U. Hence it is continuous in virtue of the theorem on the closed graph, so that

$$\int_{\Omega'} \sum_{k=1}^{\nu} \left|\frac{\partial u}{\partial x^k}\right|^2 dx \leq C \int_{\Omega} |u|^2\,dx, \quad u \in U.$$

If we apply this inequality to the function $u = e^{i\langle x,\zeta\rangle}$, where $\zeta = \xi + i\eta$ is a solution of the equation $P(\zeta) = 0$, we obtain

$$\left(\sum_{k=1}^{\nu}|\zeta_k|^2\right)\int_{\Omega'} e^{-2\langle x,\eta\rangle}\,dx \leqq C\int_{\Omega} e^{-2\langle x,\eta\rangle}\,dx.$$

Hence when η is bounded, $|\eta|<A$, it follows that $|\zeta|<C'$, which proves that P is complete and satisfies the condition I of Theorem 3.3.

Theorems 3.6 and 3.7 show that *all solutions of the equation $Pu=0$ are infinitely differentiable functions if and only if P is complete and of local type.* Thus we have found the greatest class of operators, for which a generalization of Weyl's lemma holds true. We now turn to a more detailed study of the properties of the solutions of the equation $Pu=0$.

DEFINITION 3.2. *An infinitely differentiable function u, defined in a domain Ω, is said to be of class ϱ in the direction y, if to any compact set K in Ω there is a constant C, so that*

$$\sup_{x\in K}|\langle y, D\rangle^n u(x)|<C^n\,\Gamma(\varrho n). \tag{3.6.2}$$

It is well known that solutions of elliptic equations are analytic and consequently of class 1 in every direction. There is also a classical result by Holmgren, which states that the solutions of the equation of heat are of class 2 in the time variable. We now state a result of this type for any equation of local type.

THEOREM 3.8. *Let $P(D)$ be complete and of local type. Then every solution of the equation $Pu=0$ is of class $\varrho(y)$ in the direction y, $y\neq 0$, if $\varrho(y)$ is the inverse of the exponent b in Lemma 3.9, that is*

$$\varrho(y)=\overline{\lim_{|\langle y,\xi\rangle|\to\infty}}\left(\sup_{P(\zeta)=0}\frac{\log|\langle y,\xi\rangle|}{\log|\zeta-\xi|}\right). \tag{3.6.3}$$

PROOF. Let K be a compact set in Ω, and take a function $\psi\in C_0^\infty(\Omega)$, which equals 1 in a neighbourhood of K. The function $v=\psi u$ is then in $C_0^\infty(\Omega)$ and equals u in K. Furthermore, the function $\varphi=P(D)v\in C_0^\infty(\Omega)$ and vanishes in a neighbourhood of K. Denoting by E the fundamental solution given by Theorem 3.4, we have $v=E*\varphi$ in virtue of (3.4.1). Hence

$$u(x)=\int E(x')\,\varphi(x-x')\,dx',\quad x\in K, \tag{3.6.4}$$

$$\langle y, D\rangle^n u(x)=\int(\langle y, D'\rangle^n E(x'))\,\varphi(x-x')\,dx',\quad x\in K, \tag{3.6.5}$$

where D' is the operator of differentiating with respect to x'. Now we can find two positive numbers ε and A such that $\varphi(x)=0$ in any point x with distance $<\varepsilon$ or

$>A$ from a point in K. Then $\varphi(x-x')=0$ if $x\in K$ and either $|x'|<\varepsilon$ or $|x'|>A$. Hence we may integrate only over the domain $\varepsilon\leq|x'|\leq A$ in (3.6.4) and (3.6.5). In this domain we can use Theorem 3.5, which gives

$$|\langle y, D\rangle^n u(x)|\leq\Gamma\left(\frac{n}{b}\right)C^n\int|\varphi(x-x')|\,dx'=\Gamma(\varrho n)\,C^n\int|\varphi(x)|\,dx.$$

The proof is complete.

An interesting application of Theorem 3.8 concerns the growth of null solutions of P. *Suppose that u is a null solution in Ω*, so that it vanishes when $x\in\Omega$ and $\langle x,\xi\rangle<0$, where $\langle x,\xi\rangle=0$ is a characteristic plane intersecting Ω. Let y be a direction which is not contained in this plane, that is, such that $\langle y,\xi\rangle\neq 0$. *Then, if K is a compact set in Ω, we have*

$$|u(x)|\leq A\,e^{-c\langle x,\xi\rangle^{-\alpha}},\quad x\in K,\quad \langle x,\xi\rangle>0, \tag{3.6.6}$$

where α is defined by $\alpha^{-1}=\varrho(y)-1$. For in virtue of Theorem 3.8 and Taylor's formula we have for any n, if $t=\langle x,\xi\rangle$,

$$|u(x)|\leq\frac{C^n t^n}{n!}\Gamma(\varrho n),\quad x\in K. \tag{3.6.7}$$

If in (3.6.7) we let n be the smallest integer larger than $(Ct)^{-\alpha}$ and use Stirling's formula, we obtain the desired estimate (3.6.6).

Remark. We pointed out at the end of section 3.5, that all distributions u, which solve the equation $Pu=0$, are infinitely differentiable functions, if P is complete and of local type. Using our Theorem 3.6 and Théorème XXI in Schwartz [28], Chap. VI, we can also prove that a distribution u, such that $P(D)^n u$ is of bounded order when $n\to\infty$, is an infinitely differentiable function.

3.7. Spectral theory of complete self-adjoint operators of local type

We shall call the differential operator $P(D)$ (formally) self-adjoint, if $P(D)$ coincides with its algebraic adjoint, that is, if $P(\xi)$ is real for real ξ.

Lemma 3.13. *If $P(D)$ is complete, formally self-adjoint and of local type, it follows that the operator P_0 is semi-bounded for an arbitrary domain Ω, unless $P(D)$ is an ordinary differential operator of odd order.*

Proof. First suppose that $P(D)$ is not ordinary, that is, that the dimension ν of the space of ξ is greater than 1. From condition V of Theorem 3.3 it follows that $|P(\xi)|\to\infty$ when the real vector $\xi\to\infty$. If there were points where $P(\xi)$ is positive and points where $P(\xi)$ is negative outside any sphere, there would also be points where $P(\xi)=0$, since the complement of a sphere is connected. Now this is a contradiction, so that either $P(\xi)\to+\infty$ or else $P(\xi)\to-\infty$ when $\xi\to\infty$. We may restrict ourselves to the first case. Then $P(\xi)\geqq c$ for some finite real c. If $u\in C_0^\infty(\Omega)$, we have in virtue of Parseval's formula

$$(P(D)u, u)=\int P(\xi)\,|\hat u(\xi)|^2\,d\xi\geqq c\int|\hat u(\xi)|^2\,d\xi=c\,(u, u).$$

Hence $(P_0 u, u)\geqq c\,(u, u)$ when $u\in\mathcal{D}_{P_0}$. The same result is obviously valid, if $P(D)$ is an ordinary differential operator of even order.

Thus, if $P(D)$ is complete, formally self-adjoint and of local type but not an ordinary differential operator of odd order, the operator P_0 is symmetric and semi-bounded. Hence there exist self-adjoint semi-bounded extensions $\hat P$ of P_0 (see Nagy [23] or Krein [17], who gives a more detailed study). If $\hat P$ is any self-adjoint extension, we have $P_0\subset\hat P$ and consequently $\hat P=\hat P^*\subset P_0^*=P$, so that $P_0\subset\hat P\subset P$. Thus $\hat P$ is defined by a boundary problem in the sense of section 1.3. The case where $\hat P$ is the Friedrichs extension merits some comment. The degree of $P(\xi+t\mathrm{N})$ in t for fixed $\mathrm{N}\in R_\nu$ and indeterminate ξ is even, since $P(\xi)$ is semi-bounded. Denote this degree by $2m(\mathrm{N})$. Using the methods of section 2.8 we could show that the boundary conditions corresponding to $\hat P$ are, at least formally, the vanishing of $m(\mathrm{N})-1$ transversal derivatives at a point on the boundary with normal N.

For ordinary differential operators P of odd order, the situation is different. In fact, when Ω is a semi-axis, there are no self-adjoint extensions. These exceptional operators, which can be treated explicitly, will therefore be excluded in the sequel.

Thus for the rest of the section *we assume that $P(D)$ is complete, formally self-adjoint and of local type, but not an ordinary differential operator of odd order.* Let $\hat P$ be a fixed self-adjoint extension of P_0. The operator $\hat P$ gives rise to a resolution of the identity E_λ such that

$$\hat P=\int\lambda\,dE_\lambda. \tag{3.7.1}$$

We shall study certain functions of the operator $\hat P$, which will turn out to be integral operators. Let $\mathcal{B}_\infty$ be the set of all Borel measurable functions $\alpha(\lambda)$, $-\infty<\lambda<\infty$, such that the product $\alpha(\lambda)\,\lambda^k$ is bounded for every integer $k\geqq 0$. The supremum of $|\alpha(\lambda)|$ is denoted by $|\alpha|$. Now form the operator

$$\alpha(\hat P)=\int\alpha(\lambda)\,dE_\lambda,\quad \alpha\in\mathcal{B}_\infty. \tag{3.7.2}$$

Since

$$\hat{P}^k \alpha(\hat{P}) = \int \lambda^k \alpha(\lambda)\, d E_\lambda,$$

the operator $\hat{P}^k \alpha(\hat{P})$ is bounded for every integer k;

(3.7.3) $$\| \hat{P}^k \alpha(\hat{P}) \| \leqq |\lambda^k \alpha|, \quad k = 0, 1, 2, \ldots$$

Here $\lambda^k \alpha$ denotes the function $\lambda^k \alpha(\lambda)$, and $\| \ \|$ is the operator norm. Thus, if $g = \alpha(\hat{P}) f$, it follows that $g \in \mathcal{D}_{\hat{P}^k}$ for any integer k, so that g is an infinitely differentiable function in virtue of Theorem 3.6. Since

$$\| \hat{P}^k g \| \leqq |\lambda^k \alpha| \ \|f\|,$$

the second part of the following lemma also follows as a corollary of Lemma 3.12.

LEMMA 3.14. *All functions in the range of $\alpha(\hat{P})$ are infinitely differentiable, if $\alpha \in \mathcal{B}_\infty$. Moreover, for any differential operator $Q(D)$ we have, when K is a compact subset of Ω,*

$$\sup_{x \in K} |Q(D)(\alpha(\hat{P}) f(x))|^2 \leqq C(|\alpha|^2 + |\lambda^k \alpha|^2) \|f\|^2.$$

Here k is the same integer as in Lemma 3.12, and C is a constant, which may depend on K.

Applying this result to the operators $Q(D) = 1$ and $Q(D) = D_i$, we find that, for a certain integer $\varkappa$,

(3.7.4) $$\sup_{x \in K} \left(|g(x)|^2 + \sum_{i=1}^{\nu} |\partial g / \partial x^i|^2 \right) \leqq C^2 (|\alpha|^2 + |\lambda^\varkappa \alpha|^2) \|f\|^2,$$

where $g = \alpha(\hat{P}) f$, and K is a compact subset of Ω. Hence the value $g(x)$ at a fixed point is a bounded linear functional of $f \in L^2$, so that we may write

(3.7.5) $$\alpha(\hat{P}) f(x) = (f, \varphi_{x,\alpha}),$$

where $\varphi_{x,\alpha} \in L^2$. In virtue of (3.7.4) we have, if K is a compact set in Ω,

(3.7.6) $$\|\varphi_{x,\alpha}\|^2 \leqq C(|\alpha|^2 + |\lambda^\varkappa \alpha|^2), \quad x \in K.$$

Furthermore, if K is also convex, it follows from (3.7.4) that

$$|(f, (\varphi_{x,\alpha} - \varphi_{y,\alpha}))|^2 = |g(x) - g(y)|^2 \leqq |x - y|^2 \sup_{x' \in K} \sum_1^{\nu} |\partial g / \partial x'^i|^2$$
$$\leqq |x - y|^2 C^2 (|\alpha|^2 + |\lambda^\varkappa \alpha|^2) \|f\|^2.$$

Hence we have

(3.7.7) $$\|\varphi_{x,\alpha} - \varphi_{y,\alpha}\|^2 \leqq |x - y|^2 C^2 (|\alpha|^2 + |\lambda^\varkappa \alpha|^2).$$

If $f \in \mathcal{D}_{\hat{P}^k}$, we can write $(\hat{P}^k \alpha(\hat{P})) f(x) = \alpha(\hat{P})(\hat{P}^k f)(x)$, which with the notation (3.7.5) reduces to

$$(f, \varphi_{x, \lambda^k \alpha}) = (\hat{P}^k f, \varphi_{x, \alpha}), \quad f \in \mathcal{D}_{\hat{P}^k}.$$

Thus $\varphi_{x,\alpha} \in \mathcal{D}_{\hat{P}^k}$ and $\hat{P}^k \varphi_{x,\alpha} = \varphi_{x, \lambda^k \alpha}$, for any integer k. Hence it follows from Theorem 3.6 that $\varphi_{x,\alpha}$ is an infinitely differentiable function $\varphi_{x,\alpha}(x')$, $x' \in \Omega$, and if K' is a compact set in Ω, Lemma 3.12 shows that

$$|\varphi_{x,\alpha}(x')|^2 + \sum_{i=1}^{\nu} |\partial \varphi_{x,\alpha}(x')/\partial x'^i|^2 \leqq C^2 (\|\varphi_{x,\alpha}\|^2 + \|\varphi_{x, \lambda^\varkappa \alpha}\|^2), \quad x' \in K'.$$

Estimating the right-hand side of this inequality by means of (3.7.6), we thus obtain

$$(3.7.8) \qquad |\varphi_{x,\alpha}(x')|^2 + \sum_{i=1}^{\nu} |\partial \varphi_{x,\alpha}(x')/\partial x'^i|^2 \leqq C^2 (|\alpha|^2 + |\lambda^{2\varkappa} \alpha|^2), \quad x \in K,\ x' \in K'.$$

Now set $\Theta(x', x, \alpha) = \overline{\varphi_{x,\alpha}(x')}$. In virtue of the definition (3.7.5) of $\varphi_{x,\alpha}$ we have

$$(3.7.9) \qquad \alpha(\hat{P}) f(x) = \int_{\Omega} \Theta(x', x, \alpha) f(x')\, dx'.$$

We shall prove that $\Theta(x', x, \alpha)$ is a continuous function of $(x', x) \in \Omega \times \Omega$. Let x_0 and x_0' be fixed points in Ω and take compact neighbourhoods K and K' of x_0 and x_0'. From (3.7.8) it follows that, for given ε, there exists an open neighbourhood $U' \subset K'$ such that

$$|\Theta(x', x, \alpha) - \Theta(y', x, \alpha)| < \varepsilon,$$

if $x \in K$ and $x', y' \in U'$. Furthermore, (3.7.7) shows that

$$\int_{U'} |\Theta(x', x, \alpha) - \Theta(x', x_0, \alpha)|^2\, dx' < \varepsilon^2\, m U',$$

when x is in a neighbourhood U of x_0. Thus, if $x \in U$, there exists a point $y' \in U'$ so that $|\Theta(y', x, \alpha) - \Theta(y', x_0, \alpha)| < \varepsilon$. We also have

$$|\Theta(x', x, \alpha) - \Theta(y', x, \alpha)| < \varepsilon, \text{ if } x' \in U', \text{ and } |\Theta(y', x_0, \alpha) - \Theta(x_0', x_0, \alpha)| < \varepsilon.$$

Hence, if $(x', x) \in U' \times U$, we have

$$|\Theta(x', x, \alpha) - \Theta(x_0', x_0, \alpha)| < 3\varepsilon,$$

which proves the continuity of $\Theta(x', x, \alpha)$.

Let $\mathcal{B}_k$ be the set of bounded Borel functions $\alpha(\lambda)$ such that $|\lambda^k \alpha| < \infty$. Noting that we have only used the fact that $|\lambda^{2\varkappa} \alpha| < \infty$, in constructing the function $\Theta(x', x, \alpha)$ and proving its continuity, we obtain the following theorem.

THEOREM 3.9. *There exists an integer k such that $\alpha(\hat{P})$ is an integral operator with a continuous Carleman kernel, if $\alpha \in \mathcal{B}_k$. Thus the kernel $\Theta(x', x, \alpha)$ in (3.7.9) is a continuous function of $(x', x) \in \Omega \times \Omega$, and the integrals*

$$\int_\Omega |\Theta(x', x, \alpha)|^2 dx, \quad \int_\Omega |\Theta(x', x, \alpha)|^2 dx' \tag{3.7.10}$$

exist and are continuous functions of x' and x, respectively. For compact subsets K of Ω we have

$$|\Theta(x', x, \alpha)| \leqq C(|\alpha| + |\lambda^k \alpha|), \quad x, x' \in K, \ \alpha \in \mathcal{B}_k. \tag{3.7.11}$$

PROOF. With $k = 2\varkappa$ we have proved that $\Theta(x', x, \alpha)$ is continuous and that (3.7.11) is valid. Since, with our previous notations, the second integral in (3.7.10) is $\|\varphi_{x,\alpha}\|^2$, it is finite and continuous in virtue of (3.7.6) and (3.7.7). Now we have

$$\Theta(x', x, \alpha) = \overline{\Theta(x, x', \bar{\alpha})}, \tag{3.7.12}$$

which proves the existence and continuity of the first integral (3.7.10).

We now return to the original assumption that $\alpha \in \mathcal{B}_\infty$. Let J be the anti-linear operator $f \to \bar{f}$ in L^2, and set $\hat{P}' = J^{-1}\hat{P}J$. This means that $\hat{P}'f = \overline{\hat{P}\bar{f}}$, if $\bar{f} \in \mathcal{D}_{\hat{P}}$. We obviously have

$$P_0' \subset \hat{P}' \subset P',$$

where P_0' and P' are the minimal and maximal differential operators defined by $P'(D) = P(-D)$. The relation $\hat{P}\varphi_{x,\alpha} = \varphi_{x,\lambda\alpha}$, which was proved above, now gives

$$\hat{P}'\Theta(x', x, \alpha) = \Theta(x', x, \lambda\alpha), \tag{3.7.13}$$

since $\Theta(x', x, \alpha) = \overline{\varphi_{x,\alpha}(x')}$. Here $\hat{P}'$ operates on the variable x'. Using (3.7.12) we also find that

$$\hat{P}\Theta(x', x, \alpha) = \Theta(x', x, \lambda\alpha), \tag{3.7.14}$$

where $\hat{P}$ operates on x. From the last two formulas we obtain for any n

$$(P(D) + P(-D'))^n \Theta(x', x, \alpha) = 2^n \Theta(x', x, \lambda^n \alpha)$$

in the distribution sense. Here $P(D)$ operates on x and $P(-D')$ operates on x'. Now it follows from condition III of Theorem 3.3 (see also the next section) that the complete operator $P(D) + P(-D')$ is of local type in $\Omega \times \Omega$. If Ω' is a domain with compact closure in Ω, the functions $\Theta(x', x, \lambda^n \alpha)$ are square integrable in $\Omega' \times \Omega'$. Hence Theorem 3.6 proves that $\Theta(x', x, \alpha)$ is infinitely differentiable in $\Omega' \times \Omega'$ and consequently in $\Omega \times \Omega$.

If $f \in C_0^\infty(\Omega)$, we find by differentiating (3.7.9) that

$$Q(D)\,\alpha(\hat{P})\,f(x) = \int_\Omega (Q(D)\,\Theta(x', x, \alpha))\,f(x')\,dx',$$

where $Q(D)$ is a differential operator with constant coefficients. Hence the integral

$$\int_\Omega |Q(D)\,\Theta(x', x, \alpha)|^2\,dx'$$

is bounded on compact subsets of Ω, in virtue of Lemma 3.14. Since the same result is valid for the operators $D_i\,Q(D)$, the integral is in fact continuous. Summing up, we have now proved the following theorem.

THEOREM 3.10. *The kernel* $\Theta(x', x, \alpha)$ *of* $\alpha(\hat{P})$ *is infinitely differentiable, if* $\alpha \in \mathcal{B}_\infty$. *Furthermore, the integrals*

$$\int_\Omega |Q(D)\,\Theta(x', x, \alpha)|^2\,dx', \quad \int_\Omega |Q(D')\,\Theta(x', x, \alpha)|^2\,dx \tag{3.7.15}$$

exist and are continuous functions of $x \in \Omega$ *and* $x' \in \Omega'$, *respectively, if* $Q(D)$ *is any differential operator with constant coefficients.*

For self-adjoint elliptic operators with variable coefficients, Theorem 3.9 and essentially also Theorem 3.10 were proved by Browder [2, 3] and Gårding [11, 12] in studying singular eigenfunction expansions. Our statements follow Gårding's closely. Gårding [12] proved the existence of an eigenfunction expansion for any self-adjoint operator $\hat{P}$, such that a function $\alpha(\hat{P})$, where $\alpha(\lambda) \neq 0$ a.e., is a Carleman integral operator. Hence his results apply to our case in virtue of Theorem 3.9. The precise statement may be omitted, since it does not differ in any respect from the results for elliptic operators in Browder [2] and Gårding [11, 12].

3.8. Examples of operators of local type

Elliptic operators are of local type, for it is easily seen that they satisfy condition III of Theorem 3.3. Since most of our results are not new for elliptic operators, we wish to give other examples. For convenience we shall say that a polynomial $P(\xi)$ is of local type, if the operator $P(D)$ is of local type, that is, if $P(\xi)$ satisfies conditions I–V of Theorem 3.3. We first prove some necessary conditions for an operator to be of local type.

Let η be a fixed real vector and set

$$P(\xi + t\eta) = \sum t^k\,P_k(\xi, \eta). \tag{3.8.1}$$

Denote by μ the degree in t of $P(\xi+t\eta)$ for fixed η and indeterminate ξ. We shall prove that $P_\mu(\xi,\eta)$ must then be independent of ξ, if P is of local type and $\mu>0$, that is, if $\eta\notin\Lambda(P)$. In fact, if this were not true, we should have for some real ξ and some sequence α of indices with $|\alpha|\neq 0$

$$\frac{\partial^{|\alpha|} P_\mu(\xi,\eta)}{\partial \xi_\alpha}\neq 0.$$

Then we should get when $t\to\infty$

$$\frac{P^{(\alpha)}(\xi+t\eta)}{P(\xi+t\eta)}\to\frac{P_\mu^{(\alpha)}(\xi,\eta)}{P_\mu(\xi,\eta)}\neq 0,$$

which would contradict condition III of Theorem 3.3. Hence our assertion follows.

Let $p(\xi)$ be the principal part of $P(\xi)$. Denote its order by m, and form with fixed η the expansion

$$p(\xi+t\eta)=\sum_0^m t^k p_k(\xi,\eta). \tag{3.8.2}$$

We have evidently $p_m(\xi,\eta)=p(\eta)$. The polynomial $p_k(\xi,\eta)$ either vanishes for all ξ, or else it is a homogeneous polynomial of degree $m-k$ in ξ.

Now take a real vector $\eta\notin\Lambda(P)$ such that $p(\eta)=0$. Then the degree μ of $P(\xi+t\eta)$ in t is less than m, and the degree of $p(\xi+t\eta)$ in t cannot be greater than μ. Since we have proved that the polynomial $P_\mu(\xi,\eta)$ must be independent of ξ, and we have $P_\mu(\xi,\eta)=p_\mu(\xi,\eta)+$ terms of degree less than $m-\mu$ in ξ, it follows that $p_\mu(\xi,\eta)=0$ for all ξ, so that the degree of $p(\xi+t\eta)$ in t is less than μ. Thus, *if P is of local type, the polynomial $p(\xi+t\eta)$ is at most of degree $m-2$ in t, if $p(\eta)=0$.*

If $P(\xi)$ is real, we can improve this result. For we may suppose that $P(\xi)$ is not a polynomial in one variable only. Then the polynomial $P(\xi)$ is semi-bounded (Lemma 3.13), and consequently its degree m and the degrees of $P(\xi+t\eta)$ and $p(\xi+t\eta)$ in t must be even. Hence $\mu\leq m-2$, so that *the degree of $p(\xi+t\eta)$ in t is at most $m-4$, if $p(\eta)=0$.*

From these results it follows that an operator of principal type can only be of local type, if it is elliptic. We also conclude that a homogeneous complete operator of local type must be elliptic. Finally, the results suggest the examples of self-adjoint operators of local type, which we shall now give.

THEOREM 3.11. *Let $Q(\xi)$ be any real polynomial of order m, and let k be a fixed integer ≥ 2. Then the polynomial*

(3.8.3) $$P(\xi) = Q(\xi)^{2k} + R(\xi)$$

is of local type, if $R(\xi)$ *is a positive definite homogeneous polynomial of the order* $2km - 2(k-1)$.

In fact, the same result remains true, if $R(\xi)$ is an inhomogeneous polynomial of this degree and, denoting the principal parts of Q and R by q and r, we have $r(\xi) > 0$ for every $\xi \neq 0$ such that $q(\xi) = 0$. Note that the principal part of the polynomial $P(\xi)$ is $q(\xi)^{2k}$, and that $q(\xi)$ is an arbitrary real homogeneous polynomial.

Proof. We shall prove that condition III of Theorem 3.3 is fulfilled. Writing $Q(\xi)^{2k} = S(\xi)$, we have $P^{(\alpha)}(\xi) = S^{(\alpha)}(\xi) + R^{(\alpha)}(\xi)$, and since

$$\frac{|R^{(\alpha)}(\xi)|}{P(\xi)} \leqq \frac{|R^{(\alpha)}(\xi)|}{R(\xi)} \to 0, \text{ when } \xi \to \infty, \ |\alpha| \neq 0,$$

the only difficulty is to estimate $S^{(\alpha)}$. Now we can write

(3.8.4) $$S^{(\alpha)}(\xi) = \sum_{j=1}^{\min(2k, |\alpha|)} Q(\xi)^{2k-j} F_j^{\alpha}(\xi),$$

where $F_j^{\alpha}(\xi)$ is a polynomial of degree $jm - |\alpha|$ at most. In virtue of the inequality between geometric and arithmetic means we have

(3.8.5) $$|Q(\xi)|^{2k-j} R(\xi)^{j/2k} \leqq Q(\xi)^{2k} + R(\xi) = P(\xi).$$

Hence we obtain the following estimates for the terms in (3.8.4)

$$|Q(\xi)^{2k-j} F_j^{\alpha}(\xi)| \leqq C\, P(\xi)\, R(\xi)^{-j/2k}\, R(\xi)^{(jm-|\alpha|)/\mu},$$

where $\mu = 2(km - (k-1))$ is the degree of $R(\xi)$. The sum of the exponents of $R(\xi)$ is

$$\frac{j(k-1) - k|\alpha|}{k\mu} \leqq -\frac{|\alpha|}{k\mu} < 0,$$

when $j \leqq |\alpha|$ and $|\alpha| \neq 0$. Hence $S^{(\alpha)}(\xi)/P(\xi) \to 0$, when $\xi \to \infty$, if $|\alpha| \neq 0$. Thus we obtain

$$\frac{P^{(\alpha)}(\xi)}{P(\xi)} \to 0, \text{ when } \xi \to \infty, \text{ if } |\alpha| \neq 0.$$

Hence the condition III of Theorem 3.3 is fulfilled.

Finally we remark that the product of two complete operators of local type is complete and of local type, and that the sum of two self-adjoint operators of local type, which are bounded from below, is self-adjoint and of local type. The easy verification may be left to the reader. It is also an immediate consequence of condition

II of Theorem 3.3, that if P is of local type and $Q(\xi)/P(\xi) \to 0$ when $\xi \to \infty$, then $P + tQ$ is of local type for any complex number t. Combining these simple remarks with Theorem 3.11, we could construct a very wide class of differential operators of local type.

3.9. An approximation theorem

For operators of local type we shall now answer a question raised on page 169.

THEOREM 3.12. *Let $P(D)$ be of local type and let Ω be an arbitrary domain. Then the operator P is the closure of its restriction to $\mathcal{D}_P \cap C^\infty$.*

We note that the restriction of P, mentioned in the theorem, is defined for those infinitely differentiable functions u such that u and $P(D)u$ are square integrable. The value of Pu is then of course calculated in the classical way.

PROOF. Using an idea of Deny-Lions [4], p. 312, we shall for given $\varepsilon > 0$ and $u \in \mathcal{D}_P$ construct a function $v \in C^\infty$ such that

$$\|v - u\| < \varepsilon, \; \|P(D)v - Pu\| < \varepsilon. \tag{3.9.1}$$

Since these inequalities obviously imply that $v \in L^2$ and that $P(D)v \in L^2$, the theorem will then follow. Choose a locally finite covering Ω_k, $k = 1, 2, \ldots$, of Ω such that $\overline{\Omega}_k \subset \Omega$ for every k, and then take functions $\varphi_k \in C_0^\infty(\Omega_k)$ so that $\sum \varphi_k(x) = 1$ (cf. Schwartz [28], Théorème II, Chap. I). The function $u_k = \varphi_k u$ is in $\mathcal{D}_P$ in virtue of Definition 3.1, and we have

$$u = \sum u_k, \quad Pu = \sum P u_k$$

(almost everywhere); the series converge since only a finite number of terms do not vanish in a compact subset of Ω. (However, the second series is not L^2-convergent if $u \notin \mathcal{D}_{P_0}$.) Now Lemma 2.11 shows that $u_k \in \mathcal{D}_{P_0}$, so that we can find a function $v_k \in C_0^\infty(\Omega)$ such that

$$\|u_k - v_k\| < 2^{-k}\varepsilon, \quad \|Pu_k - Pv_k\| < 2^{-k}\varepsilon. \tag{3.9.2}$$

It follows from the proof of Lemma 2.11 that we may assume that v_k has also its support in Ω_k. Since the covering Ω_k is locally finite, the series $\sum v_k(x)$ converges for every x, and the sum $v(x)$ is in $C^\infty(\Omega)$. Using (3.9.2) we obtain

$$\|v - u\| \leq \sum \|v_k - u_k\| < \varepsilon, \quad \|P(D)v - Pu\| \leq \sum \|Pv_k - Pu_k\| < \varepsilon,$$

which proves (3.9.1).

CHAPTER IV

Differential Operators with Variable Coefficients

4.0. Introduction

In the two preceding chapters we have exclusively studied differential operators with constant coefficients. However, we shall see that the methods of the proof of Theorem 2.2, which is the central theorem in Chapter II, also apply when the coefficients are variable, if suitable restrictions are imposed. In order to exclude cases where the lower order terms and the variation of the coefficients may influence the strength of the operator, we shall only study operators P of principal type. This means that the characteristics have no real singular points. (When the coefficients are constant, this is equivalent to Definition 2.1 according to Theorem 2.3.) Furthermore, we shall assume that the coefficients of the principal part are real, which means that there is some self-adjoint operator with the same principal part as P, so that P is approximately self-adjoint. (It is sufficient to require that P is approximately normal in the sense that the order of $P\bar{P}-\bar{P}P$ is at least two units lower than that of $P\bar{P}$. We do not study this case here.) The minimal differential operator defined by P in a sufficiently small domain is then stronger than all operators of lower order, and has a continuous inverse. The same result is true for the algebraic adjoint $\bar{P}$. Hence, in sufficiently small domains, the equation $Pu=f$ has a square integrable solution for any square integrable function f. In the sense of section 1.3 there also exist correctly posed abstract boundary problems for the operator P. It seems that this is the first existence proof for differential operators with non-analytic coefficients, which are not of a special type.

4.1. Preliminaries

Let P be a differential operator of order m in a manifold Ω.[1] In a local coordinate system we may write

$$P=\sum_{|\alpha|\leq m} a^{\alpha}(x)\,D_{\alpha}. \tag{4.1.1}$$

Now, if φ is an infinitely differentiable function in Ω, we have for real t

$$P\,e^{it\varphi}=t^m\sum_{|\alpha|=m} a^{\alpha}\varphi_{\alpha}+O(t^{m-1})$$

[1] It is sufficient here to suppose that Ω is a domain in R^{ν}.

when $t \to \infty$, where $\varphi_k = \partial \varphi / \partial x^k$ and $\varphi_\alpha = \varphi_{\alpha_1} \cdots \varphi_{\alpha_m}$. Thus the polynomial

$$p(x, \xi) = \sum_{|\alpha| = m} a^\alpha(x)\, \xi_\alpha \tag{4.1.2}$$

is a scalar, if ξ is a covariant vector field. $p(x, \xi)$ is called the characteristic polynomial of P. The coefficients a^α ($|\alpha| = m$) form a symmetric contravariant tensor.

The differential operator P is called *elliptic* in Ω, if $p(x, \xi) \neq 0$ for every $x \in \Omega$ and every real $\xi \neq 0$, and it is said to be of *principal type* in Ω, if all the partial derivatives $\partial p(x, \xi)/\partial \xi_i$ do not vanish simultaneously for any $x \in \Omega$ and real $\xi \neq 0$.

We shall now deduce some formulas, which replace the more implicit arguments of section 2.4 in the case considered here. Let $p(x, \xi^{(1)}, \ldots, \xi^{(m)})$ be the symmetric multilinear form in the vectors $\xi^{(1)}, \ldots, \xi^{(m)}$, which is defined by $p(x, \xi)$,

$$p(x, \xi^{(1)}, \ldots, \xi^{(m)}) = \sum_\alpha a^{\alpha_1 \cdots \alpha_m}(x)\, \xi^{(1)}_{\alpha_1} \cdots \xi^{(m)}_{\alpha_m}.$$

If $k_1, \ldots, k_p$ are positive integers, $k_1 + \cdots + k_p = m$, we shall write $p(x, \xi^{(1)^{k_1}}, \ldots, \xi^{(p)^{k_p}})$ for the multilinear form where k_i arguments are equal to $\xi^{(i)}$. Sometimes we also omit the variable x. Now set for indeterminate ξ and η

$$\sum_{i,k=1}^{\nu} R^{ik}(\zeta, \bar\zeta)\, \xi_i \eta_k = m \sum_{j=0}^{m-1} p(\zeta^j, \bar\zeta^{m-1-j}, \xi)\, p(\zeta^{m-1-j}, \bar\zeta^j, \eta), \tag{4.1.3}$$

$$\sum_{i,k=1}^{\nu} S^{ik}(\zeta, \bar\zeta)\, \xi_i \eta_k = m \sum_{j=1}^{m-1} p(\zeta^j, \bar\zeta^{m-j})\, p(\zeta^{m-1-j}, \bar\zeta^{j-1}, \xi, \eta), \tag{4.1.4}$$

and $T^{ik} = R^{ik} - S^{ik}$. Evidently $T^{ik} = T^{ik}(x, \zeta, \bar\zeta)$ is a symmetric tensor which is a homogeneous polynomial of degree $m-1$ in both ζ and $\bar\zeta$. Since

$$p(\zeta^m) = p(\zeta), \quad p(\zeta^{m-1}, \xi) = \frac{1}{m} \sum_{i=1}^{\nu} \xi_i \frac{\partial p}{\partial \zeta_i},$$

it is easy to verify the following fundamental property of the tensor T^{ik}

$$\sum_{i=1}^{\nu} (\zeta_i - \bar\zeta_i)\, T^{ik}(\zeta, \bar\zeta) = p(\zeta) \frac{\partial p(\bar\zeta)}{\partial \bar\zeta_k} - \frac{\partial p(\zeta)}{\partial \zeta_k} p(\bar\zeta). \tag{4.1.5}$$

The arguments of section 2.6 were based on the fact that, in virtue of Lemma 2.2, there exist polynomials $T^{ik}(\zeta, \bar\zeta)$ satisfying the identity (4.1.5), even for a non-homogeneous polynomial p. The simple explicit formulas given above for T^{ik} in the case of a homogeneous polynomial p, have the now essential advantage that T^{ik} are homogeneous of degree $m-1$ in ζ and in $\bar{\zeta}$. For second order equations the "energy" tensor T^{ik} was given by Hörmander [15].

We shall also use the tensor $Q^{ijk}(\zeta, \bar{\zeta})$ defined by the formula

$$(4.1.6) \quad 2 \sum_{i,j,k=1}^{\nu} Q^{ijk}(\zeta, \bar{\zeta})\, \vartheta_i\, \xi_j\, \eta_k$$

$$= m \sum_{j=1}^{m-1} j\, \{p(\zeta^{m-1-j}, \bar{\zeta}^j, \vartheta)\, p(\zeta^{j-1}, \bar{\zeta}^{m-1-j}, \xi, \eta) - p(\zeta^j, \bar{\zeta}^{m-1-j}, \vartheta)\, p(\zeta^{m-1-j}, \bar{\zeta}^{j-1}, \xi, \eta)\} +$$

$$+ m \sum_{j=2}^{m-1} (j-1)\, \{p(\zeta^j, \bar{\zeta}^{m-j})\, p(\zeta^{m-1-j}, \bar{\zeta}^{j-2}, \vartheta, \xi, \eta) - p(\zeta^{m-j}, \bar{\zeta}^j)\, p(\zeta^{j-2}, \bar{\zeta}^{m-1-j}, \vartheta, \xi, \eta)\}.$$

This tensor is symmetric in the last two indices, and we have

$$(4.1.7) \qquad \sum_{i=1}^{\nu} (\zeta_i - \bar{\zeta}_i)\, Q^{ijk}(\zeta, \bar{\zeta}) = S^{jk}(\zeta, \bar{\zeta}) - \frac{1}{2}\left(p(\zeta) \frac{\partial^2 p(\bar{\zeta})}{\partial \bar{\zeta}_j\, \partial \bar{\zeta}_k} + p(\bar{\zeta}) \frac{\partial^2 p(\zeta)}{\partial \zeta_j\, \partial \zeta_k}\right).$$

4.2. Estimates of the minimal operator

We shall now prove that an analogue of Theorems 2.1 and 2.2 is valid for certain differential operators P with variable coefficients. Since our results are not valid in the large, we may assume from the outset that our operator P is defined in a neighbourhood of a sphere $|x| \leqq A$ in R^ν.

THEOREM 4.1. *Suppose that $p(x, \xi)$ is real for real ξ and of principal type, that is, that all the partial derivatives $\partial p(x, \xi)/\partial \xi_i$ do not vanish simultaneously for any real $\xi \neq 0$. Let the coefficients of $p(x, \xi)$ be continuously differentiable and the other coefficients of P be continuous. Then there exists an open neighbourhood Ω of the origin, such that*

$$(4.2.1) \qquad \sum_{|\alpha|<m} \| D_\alpha u \|^2 \leqq C \| P u \|^2, \quad u \in C_0^\infty(\Omega).$$

PROOF. It follows from (4.1.5) and the assumption that $p(x, \xi)$ is real that

$$-\sum_{l=1}^{\nu} \frac{\partial}{\partial x^l}(T^{lk}(x, D, \bar{D})\, u \bar{u}) = 2 \operatorname{Im}(p(x, D)\, u\, \overline{p^{(k)}(x, D)\, u}) + F^k(x, D, \bar{D})\, u \bar{u},$$

where $p^{(k)}(x, \xi) = \partial p(x, \xi)/\partial \xi_k$ and

$$F^k(x, \zeta, \bar{\zeta}) = -\sum_{l=1}^{\nu} \frac{\partial}{\partial x^l}(T^{lk}(x, \zeta, \bar{\zeta})).$$

Thus $F^k(x, D, \bar{D})\, u \bar{u}$ is a quadratic form in the derivatives of u of order $m-1$ and has continuous coefficients. Multiplying by x^k and integrating over an open neighbourhood Ω of the origin, we obtain, if $u \in C_0^\infty(\Omega)$,

(4.2.2)
$$\int T^{k k}(x, D, \bar{D})\, u \bar{u}\, d x$$
$$= \int 2 x^k \operatorname{Im}\left(p(x, D) u\, \overline{p^{(k)}(x, D) u}\right) d x + \int x^k F^k(x, D, \bar{D})\, u \bar{u}\, d x .$$

Denote by δ an upper bound of $|x|$ in Ω. We may suppose that $\delta \leqq A$, and shall prove that (4.2.1) is valid, if δ is sufficiently small. If we use the notations

(4.2.3)
$$|u|_n^2 = \sum_{|\alpha|=n} \int |D_\alpha u|^2\, d x, \quad \|u\|_n^2 = \sum_{|\alpha| \leqq n} \int |D_\alpha u|^2\, d x$$

and note that $p(x, D) u$ only differs from $P u$ by a sum of derivatives of u of orders $< m$, the inequality (4.2.2) and Schwarz' inequality give

(4.2.4)
$$\int T^{k k}(x, D, \bar{D})\, u \bar{u}\, d x \leqq C \delta \left(\|P u\|\, \|u\|_{m-1} + \|u\|_{m-1}^{2}\right),$$

where C is a constant. (We shall denote by C different constants, different times.) Now we have $T^{k k} = R^{k k} - S^{k k}$, so that (4.2.4) gives, after summation,

(4.2.5)
$$\int \sum_1^\nu R^{k k}(0, D, \bar{D})\, u \bar{u}\, d x \leqq \int \sum_1^\nu \left(R^{k k}(0, D, \bar{D}) - R^{k k}(x, D, \bar{D})\right) u \bar{u}\, d x +$$
$$+ \int \sum_1^\nu S^{k k}(x, D, \bar{D})\, u \bar{u}\, d x + C \delta \left(\|P u\|\, \|u\|_{m-1} + \|u\|_{m-1}^{2}\right).$$

We shall prove (4.2.1) by estimating the terms in this inequality.

The definition (4.1.3) of $R^{i k}$ shows that

$$\sum_{k=1}^{\nu} R^{k k}(0, \xi, \xi) = \sum_{k=1}^{\nu} \left(\partial p(0, \xi) / \partial \xi_k\right)^2 .$$

This is a homogeneous positive definite polynomial, since P is of principal type. Hence we have

$$\sum_{k=1}^{\nu} R^{k k}(0, \xi, \xi) \geqq c\left(\xi_1^2 + \cdots + \xi_\nu^2\right)^{m-1}$$

for some positive constant c, and using Parseval's formula (cf. formula (2.5.1)), we thus obtain

(4.2.6)
$$c\, |u|_{m-1}^{2} \leqq \int \sum_1^\nu R^{k k}(0, D, \bar{D})\, u \bar{u}\, d x .$$

It is easy to find an estimate of the first term on the right-hand side of (4.2.5). In fact, since the coefficients of $R^{k k}(0, D, \bar{D}) - R^{k k}(x, D, \bar{D})$ are continuously differentiable and vanish for $x=0$, they are $O(|x|)$. Hence

(4.2.7)
$$\int \sum_1^\nu \left(R^{k k}(0, D, \bar{D}) - R^{k k}(x, D, \bar{D})\right) u \bar{u}\, d x \leqq C \delta\, |u|_{m-1}^{2} .$$

In order to estimate the integral $\int S^{kk}(x, D, \bar{D})\, u \bar{u}\, dx$ we must first integrate it by parts. Formula (4.1.7) with $j = k$ shows that

$$S^{kk}(x, D, \bar{D})\, u \bar{u} \tag{4.2.8}$$
$$= \mathrm{Re}\,(p(x, D)\, u\, \overline{p^{(kk)}(x, D)\, u}) + G^k(x, D, \bar{D})\, u \bar{u} + \frac{1}{i} \sum_{j=1}^{\nu} \frac{\partial}{\partial x^j} (Q^{jkk}(x, D, \bar{D})\, u \bar{u}).$$

Here

$$G^k(x, \zeta, \bar{\zeta}) = -\frac{1}{i} \sum_{j=1}^{\nu} \frac{\partial}{\partial x^j} Q^{jkk}(x, \zeta, \bar{\zeta}),$$

so that $G^k(x, D, \bar{D})\, u \bar{u}$ is a sum of products of derivatives of the orders $m-2$ and $m-1$ of u. Hence Schwarz' inequality shows that

$$\int G^k(x, D, \bar{D})\, u \bar{u}\, dx \leqq C\, |u|_{m-1}\, |u|_{m-2}. \tag{4.2.9}$$

Furthermore, the integral of the last term in (4.2.8) is zero, and using again the fact that $p(x, D)\, u$ differs from $P u$ only by derivatives of order $< m$ of u, we thus obtain

$$\int S^{kk}(x, D, \bar{D})\, u \bar{u}\, dx \leqq C(\| P u \| + \| u \|_{m-1})\, |u|_{m-2} + C\, |u|_{m-1}\, |u|_{m-2}. \tag{4.2.10}$$

If the two sides of the inequality (4.2.5) are estimated by means of the inequalities (4.2.6), (4.2.7) and (4.2.10), it follows that

$$|u|_{m-1}^2 \leqq C(\| P u \| + \| u \|_{m-1})(\delta \| u \|_{m-1} + |u|_{m-2}), \quad u \in C_0^\infty(\Omega). \tag{4.2.11}$$

To prove (4.2.1) we have now only to invoke the inequality

$$|u|_{k-1} \leqq C \delta\, |u|_k, \quad u \in C_0^\infty(\Omega), \quad k = 1, \ldots, m, \tag{4.2.12}$$

which is an immediate consequence of Lemma 2.7 but also well known previously (see for example Gårding [9], p. 57). It follows from (4.2.12) that $|u|_{m-2} \leqq C\delta\, |u|_{m-1} \leqq C\delta\, \| u \|_{m-1}$, and, since $\delta \leqq A$, that $\| u \|_{m-1} \leqq C\, |u|_{m-1}$. Hence (4.2.11) gives with a constant K

$$\| u \|_{m-1}^2 \leqq K(\| P u \| + \| u \|_{m-1})\, \delta\, \| u \|_{m-1},$$

so that

$$\| u \|_{m-1} (1 - K\delta) \leqq K \delta\, \| P u \|. \tag{4.2.13}$$

Thus the inequality (4.2.1) follows, if $K\delta < 1$.

In particular, it follows from Theorem 4.1 that the operator P_0 in $L^2(\Omega)$ has a continuous inverse, if Ω is a suitable neighbourhood of the origin. Now let the coefficients of P be sufficiently differentiable, so that $\bar{P}$ also satisfies the hypotheses of Theorem 4.1. Then the operator $\bar{P}_0^{-1}$ is also continuous. Hence *the equation* $Pu = f$ *has a solution* $u \in L^2(\Omega)$ *for any* $f \in L^2(\Omega)$ in virtue of Lemma 1.7. Furthermore, using Theorem 2.15 and Theorem 4.1 it is easy to see that P_0^{-1} and $\bar{P}_0^{-1}$ are completely continuous. Thus we can apply all the results of section 1.3. In particular, it follows that *there exist completely correctly posed boundary problems for the differential operator* P.

References

[1]. M. Š. Birman, On the theory of general boundary problems for elliptic differential equations. *Doklady Akad. Nauk SSSR* (N.S.), 92 (1953), 205–208 (Russian).

[2]. F. E. Browder, The eigenfunction expansion theorem for the general self-adjoint singular elliptic partial differential operator. I. The analytical foundation. *Proc. Nat. Acad. Sci. U.S.A.*, 40 (1954), 454–459.

[3]. ——, Eigenfunction expansions for singular elliptic operators. II. The Hilbert space argument; parabolic equations on open manifolds. *Proc. Nat. Acad. Sci. U.S.A.*, 40 (1954), 459–463.

[4]. J. Deny and J. L. Lions, Les espaces du type de Beppo Levi. *Ann. Inst. Fourier Grenoble*, 5 (1955), 305–370.

[5]. K. O. Friedrichs, On the differentiability of the solutions of linear elliptic differential equations. *Comm. Pure Appl. Math.*, 6 (1953), 299–326.

[6]. K. Friedrichs and H. Lewy, Über die Eindeutigkeit und das Abhängigkeitsgebiet der Lösungen beim Anfangsproblem linearer hyperbolischer Differentialgleichungen. *Math. Ann.*, 98 (1928), 192–204.

[7]. T. Ganelius, On the remainder in a Tauberian theorem. *Kungl. Fysiografiska Sällskapets i Lund Förhandlingar*, 24, No. 20 (1954).

[8]. L. Gårding, Linear hyperbolic partial differential equations with constant coefficients. *Acta Math.*, 85 (1951), 1–62.

[9]. ——, Dirichlet's problem for linear elliptic partial differential equations. *Math. Scand.*, 1 (1953), 55–72.

[10]. ——, On the asymptotic distribution of the eigenvalues and eigenfunctions of elliptic differential operators. *Math. Scand.*, 1 (1953), 237–255.

[11]. ——, Eigenfunction expansions connected with elliptic differential operators. *Comptes rendus du Douzième Congrès des Mathématiciens Scandinaves*, Lund, 1953, 44–55.

[12]. ——, Applications of the theory of direct integrals of Hilbert spaces to some integral and differential operators. *University of Maryland, The Institute for Fluid Dynamics and Applied Mathematics, Lecture Series*, No. 11 (1954).

[13]. ——, On the asymptotic properties of the spectral function belonging to a self-adjoint semi-bounded extension of an elliptic operator. *Kungl. Fysiografiska Sällskapets i Lund förhandlingar*, 24, No. 21 (1954).

[14]. E. Hille, An abstract formulation of Cauchy's problem. *Comptes rendus du Douzième Congrès des Mathématiciens Scandinaves*, Lund, 1953, 79–89.

[15]. L. Hörmander, Uniqueness theorems and estimates for normally hyperbolic partial differential equations of the second order. *Comptes rendus du Douzième Congrès des Mathématiciens Scandinaves*, Lund, 1953, 105–115.

[16]. F. John, On linear partial differential equations with analytic coefficients. Unique continuation of data. *Comm. Pure Appl. Math.*, 2 (1949), 209–253.

[17]. M. Krein, Theory of the self-adjoint extensions of semi-bounded hermitian operators and its applications. I. *Mat. Sbornik*, 20[62] (1947), 431–495 (Russian, English summary).

[18]. O. Ladyzenskaja, On the closure of the elliptic operator. *Doklady Akad. Nauk SSSR* (N.S.), 79 (1951), 723–725 (Russian).

[19]. J. Leray, *Hyperbolic Differential Equations.* The Institute for Advanced Study, Princeton N.J. (1954).

[20]. B. Malgrange, Equations aux dérivées partielles à coefficients constants. 1. Solution élémentaire. *C. R. Acad. Sci. Paris*, 237 (1953), 1620–1622.

[21]. ——, Existence et approximation des solutions des équations aux dérivées partielles et des équations de convolution. *Thesis*, Paris, June 1955.

[22]. A. Mychkis, Sur les domaines d'unicité pour les solutions des systèmes d'équations linéaires aux dérivées partielles. *Mat. Sbornik* (N.S.), 19[61] (1946), 489–522 (Russian, French summary).

[23]. B. v. Sz. Nagy, *Spektraldarstellung linearer Transformationen des Hilbertsschen Raumes.* Berlin, 1942.

[24]. J. v. Neumann, Über adjungierte Funktionaloperatoren. *Ann. of Math.*, (2) 33 (1932), 294–310.

[25]. I. G. Petrowsky, On some problems of the theory of partial differential equations. *Amer. Math. Soc., Translation* No. 12 (translated from *Uspehi Matem. Nauk* (N.S.), 1 (1946), No. 3–4 (13–14), 44–70).

[26]. ——, Sur l'analyticité des solutions des systèmes d'équations différentielles. *Mat. Sbornik*, 5[47] (1939), 3–70 (French, Russian summary).

[27]. G. de Rham, Solution élémentaire d'équations aux dérivées partielles du second ordre à coefficients constants. *Colloque Henri Poincaré* (Octobre 1954).

[28]. L. Schwartz, *Théorie des distributions*, I–II. Paris, 1950–1951.

[29]. A. Seidenberg, A new decision method for elementary algebra. *Ann. of Math.*, (2) 60 (1954), 365–374.

[30]. S. L. Sobolev, *Some Applications of Functional Analysis in Mathematical Physics.* Leningrad, 1950 (Russian).

[31]. S. Täcklind, Sur les classes quasianalytiques des solutions des équations aux dérivées partielles du type parabolique. *Nova Acta Soc. Sci. Upsaliensis*, (4) 10 (1936), 1–57.

[32]. A. Tychonov, Théorèmes d'unicité pour l'équation du chaleur. *Mat. Sbornik*, 42 (1935), 199–216.

[33]. B. L. van der Waerden, *Moderne Algebra*, I. Berlin, 1950.

[34]. M. I. Višik, On general boundary problems for elliptic differential equations. *Trudy Moskov. Mat. Obšč.*, 1 (1952), 187–246 (Russian).

[35]. A. Zygmund, *Trigonometrical Series.* Warszawa-Lwow, 1935.

HÖRMANDER, L.
Math. Annalen 140, 169—173 (1960)

Differential Equations without Solutions

By

LARS HÖRMANDER in Stockholm

1. Introduction

Let $P(x, D)$ be a differential operator of order m with coefficients in C^∞. In [1] we have recently shown that the existence of solutions of the equation $P(x, D)u = f$ is related to the properties of the part $C_{2m-1}(x, D)$ of order $2m - 1$ in the commutator

$$\overline{p}(x, D)\, p(x, D) - p(x, D)\, \overline{p}(x, D)\,.$$

Here p is the principal part of P, and $\overline{p}$ is obtained by complex conjugation of the coefficients of p. We shall here supplement the results of [1] by proving the following theorems, which were there only given for first order operators with analytic coefficients (Theorems 3.1 and 3.2 in [1]). (For notations we refer to [1].)

Theorem 1. Suppose that the coefficients of $P(x, D)$ are in $C^\infty(\Omega)$, where Ω is an open set, and assume that the equation

$$P(x, D)u = f \tag{1.1}$$

has a solution $u \in \mathscr{D}'(\Omega)$ for every $f \in \mathscr{D}(\Omega)$. Then

$$C_{2m-1}(x, \xi) = 0 \quad \text{if} \quad p(x, \xi) = 0,\ x \in \Omega,\ \xi \in R^n. \tag{1.2}$$

Theorem 2. Suppose that the coefficients of $P(x, D)$ are in C^∞ and that (1.2) is not valid for any open non void set $\omega \subset \Omega$ when x is restricted to ω. Then there exist functions $f \in \dot{B}(\Omega)$ such that the equation (1.1) does not have a solution $u \in \mathscr{D}'(\omega)$ for any open non void set $\omega \subset \Omega$. The set of such f is of the second category.

Here $\dot{B}(\Omega)$ denotes the set of all infinitely differentiable f in Ω such that to every α and ε there is a compact set $K \subset \Omega$ so that $|D_\alpha f| < \varepsilon$ in $\complement K$. This is a complete metrizable space with the topology defined by the semi-norms $\sup_\Omega |D_\alpha f|$.

Note that if $Q(D)$ is an arbitrary non elliptic operator of order m and of principal type, there is a differential operator $P(x, D)$ of order m with coefficients in C^∞ (or even linear functions of x) such that $P(0, D) = Q(D)$ and (1.2) is not valid at 0. This shows that (1.2) is a severe restriction except in the elliptic case.

We shall prove Theorem 1 by refining the constructions used in [1]. In doing so it is sufficient to consider a fixed point $x \in \Omega$, and we may of course assume that $x = 0$.

That Theorem 2 follows from Theorem 1 may be proved by the argument used in [1] to show that Theorem 3.2 follows from Theorem 3.1. We do not repeat the proof here.

2. The basic constructions

We shall here use the following norms (the notation differs from that of [1])

$$\|u\|_k = \sup_{x\in\Omega,\,|\alpha|\leq k} |D_\alpha u(x)|\,, \qquad k = 0, 1, \dots\,. \tag{2.1}$$

As usual we denote by P^t the formal adjoint of P defined by the identity

$$\int (Pu)\, v\, dx = \int u(P^t v)\, dx;\ u, v \in \mathscr{D}(\Omega)\,.$$

Lemma 1. Let k and N be arbitrary positive integers, and τ a positive real parameter. Assume that (1.2) is not valid when $x = 0$ and let ω be an open neighbourhood of 0. Then there exist functions $f_{\tau,k} \in C_0^\infty(\omega)$ and $v_{\tau,k,N} \in C_0^\infty(\omega)$ such that for all k and N

$$\overline{\lim_{\tau\to\infty}}\, \|P^t v_{\tau,k,N}\|_N < \infty\,, \tag{2.2}$$

$$\overline{\lim_{\tau\to\infty}}\, \|f_{\tau,k}\|_k < \infty\,, \tag{2.3}$$

$$\overline{\lim_{\tau\to\infty}}\, \left|\int f_{\tau,k} v_{\tau,k,N}\, dx\right| = \infty\,. \tag{2.4}$$

This section will be devoted to the proof of Lemma 1. In section 3 we prove that Theorem 1 follows from Lemma 1.

Let ξ be a real vector such that

$$p(0, \xi) = 0\,, \quad C_{2m-1}(0, \xi) < 0\,. \tag{2.5}$$

Since C_{2m-1} is a real odd function of ξ, and (1.2) is not valid for $x = 0$, there exists a vector ξ with these properties. ξ will be kept fixed in what follows. With a function $F \in C_0^\infty(R^n)$ such that

$$\int F(x)\, e^{i\langle x,\xi\rangle}\, dx \neq 0 \tag{2.6}$$

we set

$$f_{\tau,k}(x) = \tau^{-k} F(\tau x)\,. \tag{2.7}$$

It is clear that (2.3) is valid and that $f_{\tau,k} \in C_0^\infty(\omega)$ for large τ; for smaller values of τ we may define $f_{\tau,k}$ to be identically 0.

The definition of $v_{\tau,k,N}$ is much more complex, and requires the following lemma.

Lemma 2. Let (2.5) be fulfilled and q be a positive integer. Then there is a function $u \in C^\infty(\Omega)$ which satisfies

$$p(x, \operatorname{grad} u) = O(|x|^q),\ x \to 0\,, \tag{2.8}$$

and has a Taylor expansion

$$u(x) = i \sum_1^n x^j \xi_j + \frac{1}{2} \sum_1^n \sum_1^n x^j x^k \alpha_{jk} + O(|x|^3) \tag{2.9}$$

where the matrix α_{jk} is symmetric and has negative definite real part.

Proof. The lemma is essentially a combination of Lemma 2.2 and Lemma 2.3 in [1]. The fact that $p(0, \xi) = 0$ makes (2.8) valid with $q = 1$ for all u with the expansion (2.9). Further, it was shown after the proof of Lemma 2.3 in [1], that when $C_{2m-1}(0, \xi) < 0$ it is possible to choose a matrix α_{jk} with the required property such that (2.8) holds with $q = 2$. To prove the lemma we now only have to argue as in the proof of Lemma 2.2 in [1]. First, the condition (2.8) does not change if we replace the coefficients of p by their Taylor expansions of order q at 0. Hence we may assume that the coefficients of p are analytic. The Cauchy-Kovalevsky theorem can then be applied to prove the existence of a solution of the equation $p(x, \operatorname{grad} u) = 0$ in a neighbourhood of 0 which has the Taylor expansion (2.9). This proves the lemma.

In constructing $v_{\tau, k, N}$ we use a function satisfying (2.8) with

$$q = 2(r+1),\ r = n + k + m + N\,, \tag{2.10}$$

(n is the dimension of the space and m the order of P). With functions $\varphi_0, \ldots, \varphi_{r-1}$ in $C_0^\infty(\omega)$ still to be determined we set

$$v_{\tau, k, N} = \tau^{n+1+k} e^{\tau u} \sum_0^{r-1} \varphi_j \tau^{-j}\,. \tag{2.11}$$

Our first step is to prove that (2.4) is valid if $\varphi_0(0) = 1$. To do so we note that (2.7), (2.11) give after a substitution of variables

$$\frac{1}{\tau} \int f_{\tau, k} v_{\tau, k, N}\, dx = \int F(x) e^{\tau u(x/\tau)} \left(\sum_0^{r-1} \varphi_j(x/\tau) \tau^{-j} \right) dx\,. \tag{2.12}$$

Since F has compact support and the integrand in the right hand side is uniformly convergent there, the integral converges to the limit $\int F(x) e^{i\langle x, \xi\rangle} dx$, which is $\neq 0$ in view of (2.6). This proves (2.4).

We next prove that the functions φ_j can be chosen so that (2.2) holds. To compute $P^t(x, D) v_{\tau, k, N}$ we note that the principal part of P^t is $p(x, -D)$. If ψ is a function in C^∞ we have (compare the proof of Lemma 2.1 in [1])

$$P^t(\psi e^{\tau u}) = \sum_0^m c_j \tau^j e^{\tau u}\,.$$

Here

$$c_m = i^m p(x, \operatorname{grad} u) \psi = A \psi\,, \tag{2.13}$$

so that A is a function in C^∞ which in view of (2.8) satisfies

$$A(x) = O(|x|^q),\ x \to 0\,. \tag{2.14}$$

If we write $A_j = -p^{(j)}(x, i \operatorname{grad} u)$ we have with a function $B \in C^\infty$

$$c_{m-1} = \sum_1^n A_j D_j \psi + B \psi\,. \tag{2.15}$$

We recall from [1] that it follows from (2.5) that all A_j do not vanish at 0. Using (2.13) and (2.15) with $\psi = \varphi_j$ we now obtain

$$P^t v_{\tau, k, N} = \tau^{n+1+k+m} e^{\tau u} \sum_0^{m+r-1} a_j \tau^{-j}, \tag{2.16}$$

where

$$a_0 = A\varphi_0,\ a_1 = A\varphi_1 + \sum_1^n A_j D_j \varphi_0 + B\varphi_0 . \tag{2.17}$$

If we agree to read φ_μ as 0 if $\mu \geqq r$, the general form of the coefficients is

$$a_\mu = A\varphi_\mu + \sum_1^n A_j D_j \varphi_{\mu-1} + B\varphi_{\mu-1} + L_\mu , \tag{2.18}$$

where L_μ is a linear combination of functions φ_j with $j < \mu - 1$ and their derivatives. We claim that the functions $\varphi_j \in C_0^\infty$ may be chosen successively so that $\varphi_0(0) = 1$ and

$$a_\mu(x) = O(|x|^{q-2\mu}),\ \mu \leqq r,\ x \to 0 . \tag{2.19}$$

When $\mu = 0$, the estimate (2.19) follows from (2.14). Next note that (2.14) also shows that the first term in (2.18) is irrelevant for the validity of (2.19). Thus we have to choose $\varphi_{\mu-1}$ so that

$$\sum_1^n A_j D_j \varphi_{\mu-1} + B\varphi_{\mu-1} + L_\mu = O(|x|^{q-2\mu}) . \tag{2.20}$$

Suppose that all φ_j with $j < \mu - 1$ have already been chosen and that $1 \leqq \mu \leqq r$. When choosing $\varphi_{\mu-1}$ we may assume that A_j, B and L_μ are analytic. In fact, (2.20) does not change if we replace these infinitely differentiable functions by their Taylor expansions of order q at 0. We can then find a solution of the equation

$$\sum_1^n A_j D_j \varphi_{\mu-1} + B\varphi_{\mu-1} + L_\mu = 0$$

in a neighbourhood V of 0 with given analytic values on a non characteristic plane through 0. Multiplying by a function which $\in C_0^\infty(\omega \cap V)$ and equals 1 in another neighbourhood of 0 we obtain a function $\varphi_{\mu-1} \in C_0^\infty(\omega)$ such that (2.20) is valid. When $\mu = 1$ we can also fulfill the condition $\varphi_0(0) = 1$.

To complete the proof of (2.2) and thus of Lemma 1 we only have to combine the following lemma with (2.19), (2.16) and the definition (2.10) of q.

Lemma 3. Let $\psi \in C_0^\infty(\omega)$ and assume that for some integer $s \geqq 0$

$$\psi(x) = O(|x|^{2s}),\ s \to 0 . \tag{2.21}$$

Then it follows that

$$\overline{\lim_{\tau\to\infty}} \|\tau^{s-N} \psi e^{\tau u}\|_N < \infty . \tag{2.22}$$

Proof. To prove (2.22) we only have to show that

$$\sup_\omega \tau^{s-N} \tau^{N-|\alpha|} |D_\alpha \psi|\, |e^{\tau u}| \tag{2.23}$$

is bounded as $\tau \to \infty$ when $|\alpha| \leqq N$. To do so we use for the first time that the Taylor expansion of the real part of u starts with a negative definite quadratic form. Assuming that ω has been chosen sufficiently small we thus have

$$\operatorname{Re} u(x) \leqq -a|x|^2,\quad x \in \omega , \tag{2.24}$$

where $a > 0$. Using this estimate and (2.21) we can now estimate (2.23) by a constant multiple of

$$\tau^{s-|\alpha|}|x|^{2s-|\alpha|}e^{-a\tau|x|^2} = |x|^{|\alpha|}(\tau|x|^2)^{s-|\alpha|}e^{-a\tau|x|^2},$$

and this is obviously bounded when $x \in \omega$, $\tau \geqq 0$.

3. Proof of Theorem 1

Let F be the Fréchet space of all infinitely differentiable functions with support in $\bar{\omega}$ and the topology given by the semi-norms $\|u\|_k$. Denote by E_N the set of all $f \in F$ such that the equation

$$P(x, D)u = f \tag{3.1}$$

has a solution $u \in \mathscr{D}'(\omega)$ satisfying

$$|u(\psi)| \leqq N\|\psi\|_N, \quad \psi \in \mathscr{D}(\omega). \tag{3.2}$$

If the equation (3.1) has a solution u in $\mathscr{D}'(\Omega)$ and $\bar{\omega} \subset \Omega$ it is well known that (3.2) is valid for large N. When the hypotheses of Theorem 1 are fulfilled we thus have $\bigcup_1^\infty E_N = F$. Now it is clear that E_N is closed, for the set of distributions in $\mathscr{D}'(\omega)$ satisfying (3.2) is compact. (Cf. the proof of Theorem 3.2 in [1].) Hence it follows from the category theorem, that E_N has an interior point for some N. Since E_N is convex and symmetric, the origin is an interior point, that is, there exist $\varepsilon > 0$ and k such that

$$f \in E_N \quad \text{if} \quad f \in F \quad \text{and} \quad \|f\|_k < \varepsilon. \tag{3.3}$$

We shall prove that this contradicts the hypothesis of Lemma 1 that (1.2) is not valid.

First note that by definition (3.1) means that for every $v \in \mathscr{D}(\omega)$

$$u(P^t v) = \int f v\, dx$$

and using (3.2) we get

$$\left|\int f v\, dx\right| \leqq N\|P^t v\|_N, \quad f \in E_N, \quad v \in \mathscr{D}(\omega). \tag{3.4}$$

In view of (3.3) we obtain

$$\left|\int f v\, dx\right| \leqq \frac{N}{\varepsilon}\|f\|_k\|P^t v\|_N; \ f, v \in \mathscr{D}(\omega). \tag{3.5}$$

Choosing $f = f_{\tau,k}$ and $v = v_{\tau,k,N}$ we get a contradiction with (2.2), (2.3), (2.4). This proves Theorem 1.

References

[1] Hörmander, L.: Differential operators of principal type. Math. Ann. **140**, 124—146 (1960).

(Received December 3, 1959)

HYPOELLIPTIC SECOND ORDER DIFFERENTIAL EQUATIONS

BY

LARS HÖRMANDER

The Institute for Advanced Study, Princeton, N.J., U.S.A.

1. Introduction

A linear differential operator P with C^∞ coefficients in an open set $\Omega \subset \mathbf{R}^n$ (or a manifold) is called hypoelliptic if for every distribution u in Ω we have

$$\text{sing supp } u = \text{sing supp } Pu,$$

that is, if u must be a C^∞ function in every open set where Pu is a C^∞ function. Necessary and sufficient conditions for P to be hypoelliptic have been known for quite some time when the coefficients are constant (see [3, Chap. IV]). It has also been shown that such equations remain hypoelliptic after a perturbation by a "weaker" operator with variable coefficients (see [3, Chap. VII]). Using pseudo-differential operators one can extend the class of admissible perturbations further; in particular one can obtain in that way many classes of hypoelliptic (differential) equations which are invariant under a change of variables (see [2]). Roughly speaking the sufficient condition for hypoellipticity given in [2] means that the differential equations with constant coefficients obtained by "freezing" the arguments in the coefficients at a point x shall be hypoelliptic and not vary too rapidly with x.

However, the sufficient conditions for hypoellipticity given in [2] are far from being necessary. For example, they are not satisfied by the equation

$$\frac{\partial^2 u}{\partial x^2} + x\frac{\partial u}{\partial y} - \frac{\partial u}{\partial t} = f, \tag{1.1}$$

for the operator obtained by freezing the coefficients at a point must operate along a two dimensional plane only so it cannot be hypoelliptic. But Kolmogorov [8] constructed already in 1934 an explicit fundamental solution of (1.1) which is a C^∞ function outside the diagonal, and this implies that (1.1) is hypoelliptic.

Acta mathematica 119. Imprimé le 7 février 1968.

The arguments of Kolmogorov [8] can also be applied to the more general equation

$$-\frac{\partial u}{\partial x_0}+\sum_{j,k=1}^{n} a_{jk}\frac{\partial^2 u}{\partial x_j\,\partial x_k}+\sum_{j,k=1}^{n} b_{jk}x_j\frac{\partial u}{\partial x_k}+cu=f, \tag{1.2}$$

where a_{jk}, b_{jk} and c denote real constants, and the matrix $A=(a_{jk})$ is symmetric and positive semi-definite. If we take Fourier transforms in all variables except x_0 we are led to the equation

$$-\frac{\partial U}{\partial x_0}-A(\xi,\xi)\,U-\sum_{j,k=1}^{n} b_{jk}\frac{\partial(\xi_k U)}{\partial \xi_j}+cU=-\frac{\partial U}{\partial x_0}-(A(\xi,\xi)-c')\,U-\sum_{j,k=1}^{n} b_{jk}\,\xi_k\frac{\partial U}{\partial \xi_j}=F, \tag{1.3}$$

where $c'=c-\mathrm{Tr}B$. To obtain a fundamental solution of (1.2) with pole at (y_0, y) and vanishing when $x_0<y_0$ we wish to find a solution U of (1.3) when $x_0\geqslant y_0$ such that $U=-e^{-i\langle y,\xi\rangle}$ when $x_0=y_0$. The characteristic equations for (1.3) are

$$dx_0=\frac{d\xi_j}{\Sigma\, b_{jk}\,\xi_k}=-\frac{dU}{U(A(\xi,\xi)-c')} \tag{1.4}$$

and have the solutions

$$x_0=y_0+t,\quad \xi(t)=(\exp Bt)\,\eta,\quad U=-\exp\left(-i\langle y,\eta\rangle-\int_0^t (A(\xi(s),\xi(s))-c')\,ds\right.$$

if we take the initial condition into account. Elimination of t and η gives

$$U(x_0,\xi)=-\exp\Big(-i\langle y,(\exp B(y_0-x_0))\,\xi\rangle-\int_0^{x_0-y_0} A(\exp(-Bs)\,\xi,\exp(-Bs)\,\xi)\,ds$$
$$+(x_0-y_0)\,c',\quad x_0>y_0;$$

and we set $U(x_0,\xi)=0$ when $x_0<y_0$. The quadratic form in the exponent is positive semi-definite, and it is positive definite unless for some $\xi\neq 0$ we have $A((\exp Bs)\,\xi,(\exp Bs)\,\xi)=0$ identically in s, that is, $A(B^k\xi, B^k\xi)=0$ for every k. This means that the null space of A contains a non-trivial invariant subspace for B. If this is not the case we obtain by inverting the Fourier transformation a two-sided fundamental solution which is a C^∞ function off the diagonal; for fixed x_0 and y_0 it is the exponential of a negative definite quadratic form in $(\exp{}^tBx_0)\,x-(\exp{}^tBy_0)\,y$ with eigenvalues $\to-\infty$ when $x_0-y_0\to 0$. (The eigenvalues may have different orders of magnitude so the differentiability properties of the fundamental solution may be quite different in different directions, a typical feature of the subject of this paper.) Thus it follows that (1.2) *is hypoelliptic unless the null space of A contains a non-trivial subspace which is invariant for B.*

The results of Kolmogorov have been extended by Weber [11] and Il'in [4] to the equation

$$\sum_{i,j=1}^{n} a_{ij}\frac{\partial^2 u}{\partial y_i \partial y_j} + \sum_{1}^{n} a_i \frac{\partial u}{\partial y_i} + au + \sum_{1}^{m} b_i \frac{\partial u}{\partial x_i} - \frac{\partial u}{\partial t} = f, \tag{1.5}$$

where the coefficients are C^∞ functions of x, y, t, the matrix (a_{ij}) is positive definite and the matrix $(\partial b_i/\partial y_j)$ has rank m everywhere. The hypoellipticity follows from a construction of a fundamental solution by the E. E. Levi method starting from that given above after a change of variables to make $b_i = y_i$.

In this paper we shall give a nearly complete characterization of hypoelliptic second order differential operators P with real C^∞ coefficients. First it is easy to show, as we shall do in section 2, that the principal part must be semi-definite if P is hypoelliptic. In any open set where the rank is constant we can then write locally

$$P = \sum_{1}^{r} X_j^2 + X_0 + c, \tag{1.6}$$

where $X_0, \ldots, X_r$ denote first order homogeneous differential operators in an open set $\Omega \subset \mathbf{R}^n$ with C^∞ coefficients, and $c \in C^\infty(\Omega)$. We assume from now on that P has this form but do not necessarily require that the X_j are linearly independent at every point. There is of course a large freedom in the choice of the operators X_j. In particular, we may replace X_j by

$$X_j' = \sum_{1}^{r} c_{jk} X_k, \quad j = 1, \ldots, r,$$

where (c_{jk}) is an orthogonal matrix which is a C^∞ function of $x \in \Omega$; then X_0 is replaced by an operator X_0' such that $X_0' - X_0$ is a linear combination of $X_1, \ldots, X_r$ with C^∞ coefficients.

If the Lie algebra generated by $X_0, \ldots, X_r$ has constant rank $< n$ in a neighborhood of a point $x \in \Omega$, it follows from the Frobenius theorem that there exists a local change of variables near x so that P afterwards only acts in the variables $x_1, \ldots, x_{n-1}$. If the homogeneous equation $Pu = 0$ is satisfied by some non-trivial function it follows that P is not hypoelliptic, for a new solution is obtained by changing the definition of u to 0 on one side of a hyperplane $x_n = \text{constant}$. Thus the sufficient condition for hypoellipticity in the following theorem is essentially necessary also:

THEOREM 1.1. *Let P be written in the form* (1.6) *and assume that among the operators* $X_{j_1}, [X_{j_1}, X_{j_2}], [X_{j_1}, [X_{j_2}, X_{j_3}]], \ldots, [X_{j_1}, [X_{j_2}, [X_{j_3}, \ldots, X_{j_k}]]] \ldots$ *where* $j_i = 0, 1, \ldots, r$, *there exist n which are linearly independent at any given point in Ω. Then it follows that P is hypoelliptic.*

It is a simple exercise to verify that for the equation (1.2) the condition in Theorem 1.1 is the same that we needed to construct a smooth fundamental solution by the method of Kolmogorov [8].

As mentioned above we shall discuss necessary conditions for hypoellipticity in section 2. It is proved in section 3 that Theorem 1.1 is a consequence of certain *a priori* estimates and we make some preliminary steps toward proving these. However, they are proved completely only at the end of section 5 after a preliminary study of fractional differentiability of functions along a set of non-commuting vector fields has been made in section 4.

Finally we wish to mention that there is an extensive recent literature concerning global regularity of solutions of boundary problems for second order equations with semi-definite principal part. (See Kohn–Nirenberg [6, 7], Olejnik [10] and the references in these papers.) I am very much indebted to Professor Olejnik who first called my attention to the problem studied here and to Professor W. Feller who explained to me the probabilistic meaning of equations such as (1.1) and pointed out the existence of Kolmogorov's paper [8].

2. A necessary condition for hypoellipticity

A hypoelliptic differential equation with constant coefficients must have multiple characteristics if it is not elliptic (see [3, Theorem 4.17]). It is easy to extend this result to operators with variable coefficients, and this may have been done before. However, a proof will be given here since we do not know of any reference.

THEOREM 2.1. *Let $P(x, D)$ be a differential operator in $\Omega \subset \mathbf{R}^n$ with C^∞ coefficients, and let the principal symbol $p(x, \xi)$ be real. If for some $x \in \Omega$ one can find a real $\xi \neq 0$ such that*

$$p(x, \xi) = 0, \quad \text{but} \quad \frac{\partial p(x, \xi)}{\partial \xi_j} \neq 0 \quad \text{for some } j, \tag{2.1}$$

it follows that P is not hypoelliptic.

Proof. We may assume that the point x in (2.1) is 0. The classical integration theory for the characteristic equation (cf. [3, section 1.8] shows that there exists a real valued function φ in a neighborhood of 0 such that $\operatorname{grad} \varphi(0) = \xi$ and $p(x, \operatorname{grad} \varphi) = 0$. We can replace Ω by such a neighborhood of 0, and shall then determine a *formal* solution of the equation $P(x, D)\, u = 0$ by setting

$$u = \sum_0^\infty u_j t^{-j} e^{it\varphi},$$

where $u_j \in C^\infty(\Omega)$ and t is a parameter. Now we have

$$P(v\,e^{it\varphi}) = e^{it\varphi} \sum_0^m c_j t^j,$$

where $c_m = vp(x, \operatorname{grad}\varphi) = 0$ and $c_{m-1} = \sum_1^n A_j D_j v + Bv$ with $A_j = p^{(j)}(x, \operatorname{grad}\varphi)$. Hence we obtain

$$Pu = t^m \sum_0^\infty a_j t^{-j} e^{it\varphi},$$

where $$a_0 = 0, \quad a_1 = \sum_1^n A_j D_j u_0 + Bu_0, \quad a_k = \sum_1^n A_j D_j u_{k-1} + Bu_{k-1} + L_k,$$

where L_k is a linear combination of $u_0, \ldots, u_{k-2}$ and their derivatives. Since A_j are real and not all 0 we can if Ω is conveniently chosen successively find solutions $u_0, u_1, \ldots$ of these equations with $u_0(0) = 1$.

If the equation $Pu = 0$ were hypoelliptic we would have an *a priori* estimate

$$|\operatorname{grad} u(0)| \leqslant C\{\sup|u| + \sum_{|\alpha| \leqslant N} \sup|D^\alpha Pu|\}, \quad u \in C^\infty(\Omega), \tag{2.2}$$

where $\alpha = (\alpha_1, \ldots, \alpha_n)$ is a multiorder, $|\alpha| = \alpha_1 + \ldots + \alpha_n$, and $D^\alpha = (-i\partial/\partial x_1)^{\alpha_1} \ldots (-i\partial/dx_n)^{\alpha_n}$. Indeed, the set of all continuous and bounded functions in Ω with $D^\alpha Pu$ continuous and bounded in Ω for every α is then contained in C^∞ so (2.2) follows from the closed graph theorem. However, if we apply (2.2) to

$$\sum_0^{k-1} u_j t^{-j} e^{it\varphi},$$

where $N + m \leqslant k + 1$, it follows that the right-hand side of (2.2) is bounded when $t \to +\infty$ whereas the left-hand side is not. This proves the theorem.

COROLLARY 2.2. *For a second order hypoelliptic operator with real principal part, the principal part must be a semi-definite quadratic form.*

3. Preliminaries for the proof of Theorem 1.1

Let P be a differential operator of the form (1.6) where $X_j \in T(\Omega)$, the set of all homogeneous real first order differential operators in Ω with C^∞ coefficients, and $c \in C^\infty(\Omega)$. (We shall denote by $C^\infty(\Omega)$ the space of complex valued C^∞ functions in Ω and use the notation $C^\infty(\Omega, \mathbf{R})$ for the subset of real valued functions. Clearly $T(\Omega)$ is a $C^\infty(\Omega, R)$ module.) Alternatively we may of course regard $T(\Omega)$ as the space of C^∞ sections of the tangent bundle of Ω.

The starting point for the proof of Theorem 1.1 is an *a priori* estimate which is obtained by partial integration and also occurs frequently in the work of Kohn–Nirenberg [6, 7], Olejnik [10] and others. After noting that the adjoint of X_j is $-X_j+a_j$, where $a_j \in C^\infty(\Omega, \mathbf{R})$, the inequality is obtained by taking $v \in C_0^\infty(\Omega)$ and integrating by parts as follows:

$$-\operatorname{Re}\int v\overline{Pv}\,dx = -\operatorname{Re}\sum_1^r \int v\overline{X_j^2 v}\,dx - \operatorname{Re}\int v\overline{X_0 v}\,dx - \int \operatorname{Re} c\,|v|^2\,dx$$

$$= \operatorname{Re}\sum \int (X_j v - a_j v)\,\overline{X_j v}\,dx - \tfrac{1}{2}\int X_0\,|v|^2\,dx - \int \operatorname{Re} c\,|v|^2\,dx$$

$$= \sum \int |X_j v|^2\,dx + \int d\,|v|^2\,dx,$$

where we have written

$$d = \tfrac{1}{2}\sum (X_j a_j - a_j^2) - \tfrac{1}{2}a_0 - \operatorname{Re} c.$$

Hence
$$\sum_1^r \|X_j v\|^2 + \|v\|^2 \leq C\,\|v\|^2 - \operatorname{Re}\int v\overline{Pv}\,dx, \quad v \in C_0^\infty(K), \tag{3.1}$$

if K is a compact subset of Ω and $C_0^\infty(K)$ denotes the set of elements in $C^\infty(\Omega)$ with support in K. Here we have used the notation $\|\ \|$ for the L^2 norm.

For the left-hand side of (3.1) we introduce the notation

$$|||v|||^2 = \sum_1^r \|X_j v\|^2 + \|v\|^2.$$

To have a precise estimate for the right-hand side of (3.1) we also need the dual norm

$$|||f|||' = \sup_v \left|\int fv\,dx\right| / |||v|||, \quad v \in C_0^\infty(\Omega).$$

Then we have

$$-\operatorname{Re}\int v\overline{Pv}\,dx \leq |||v|||\;|||Pv|||' \leq \tfrac{1}{2}\left(|||v|||^2 + |||Pv|||'^2\right),$$

so with a new constant C we obtain from (3.1)

$$|||v|||^2 \leq C\,\|v\|^2 + |||Pv|||'^2, \quad v \in C_0^\infty(K). \tag{3.2}$$

Noting that
$$|||X_j f|||' \leq C\,\|f\|, \quad f \in C_0^\infty(K), \quad j = 1, \dots, r,$$

we conclude that $|||X_j^2 v|||' \leq C\,\|X_j v\| \leq C\,|||v|||$, $j = 1, \dots, r$. Thus it follows from (3.2) that

$$|||v|||^2 + |||X_0 v|||'^2 \leq C\left(\|v\|^2 + |||Pv|||'^2\right), \quad v \in C_0^\infty(K). \tag{3.3}$$

Let $\|v\|_{(s)}$ denote the L^2 norm of the derivatives of v of order s (cf. [3, section 2.6]), defined by

$$\|v\|_{(s)}^2 = (2\pi)^{-n}\int |\hat{v}(\xi)|^2 (1+|\xi|^2)^s\, d\xi, \quad v \in C_0^\infty.$$

The main point of this paper is the proof given in sections 4 and 5 that for some $\varepsilon > 0$

$$\|v\|_{(\varepsilon)} \leqslant C(|||v||| + |||X_0 v|||'), \quad v \in C_0^\infty(K), \tag{3.4}$$

when the hypotheses of Theorem 1.1 are fulfilled. Combining this with (3.3) we obtain

$$\|v\|_{(\varepsilon)} \leqslant C(\|v\| + |||Pv|||'), \quad v \in C_0^\infty(K). \tag{3.5}$$

We shall now prove that it follows from (3.5) that P is hypoelliptic. The main step is the proof of the following proposition.

PROPOSITION 3.1. *Assume that* (3.5) *is valid for compact subsets K of Ω. Every $v \in L^2(\Omega) \cap \mathcal{E}'(\Omega)$ such that $|||Pv|||' < \infty$ is then in $H_{(\varepsilon)}$.*

We recall that $H_{(\varepsilon)}$ is the completion of C_0^∞ in the norm $\| \ \|_{(\varepsilon)}$ but refer to [3, section 2.6] for further discussion of this space.

Many statements of the same kind as Proposition 3.1 have been proved by Kohn and Nirenberg [6] but it seems that none of them contains Proposition 3.1 explicitly so we supply a proof here along lines similar to [3, Chap. VIII].

First note that (3.5) is valid for all $v \in H_{(2)}$ with compact support in Ω. Indeed, we can find a sequence $v_j \in C_0^\infty(K)$, where K is a compact neighborhood of supp v, such that $D^\alpha v_j - D^\alpha v \to 0$, $j \to \infty$, when $j \leqslant 2$. Hence $\|Pv_j - Pv\| \to 0$, which implies that $|||Pv_j - Pv|||' \to 0$. In particular,

$$\overline{\lim}\, |||Pv_j|||' \leqslant |||Pv|||',$$

so it follows from (3.5) applied to v_j that

$$\overline{\lim}\, \|v_j\|_{(\varepsilon)} \leqslant C(\|v\| + |||Pv|||').$$

Hence (3.5) remains valid when $v \in H_{(2)}$ and supp v is in the interior of K.

If v satisfies the hypotheses of Proposition 3.1 we choose $\chi \in C_0^\infty(\Omega)$ so that $0 \leqslant \chi \leqslant 1$ and $\chi = 1$ in a neighborhood ω of supp v, and we set

$$v_\delta = \chi(1-\delta^2\Delta)^{-1}v.$$

Here $(1-\delta^2\Delta)^{-1}v$ is defined as the inverse Fourier transform of $(1+\delta^2|\xi|^2)^{-1}\hat{v}(\xi)$. It is clear that v_δ is then in $H_{(2)}$, that supp $v_\delta \subset$ supp $\chi \subset\subset \Omega$ and that $v_\delta \to v$ in L^2 norm when $\delta \to 0$. Hence we may apply (3.5) to v_δ and conclude that $\|v\|_{(\varepsilon)} < \infty$ provided that we can show that $|||Pv_\delta|||'$ remains bounded when $\delta \to 0$. This we shall do after a few simple remarks:

1°. The inverse Fourier transform K of $(1+|\xi|^2)^{-1}$ and all derivatives of K decrease exponentially at infinity. Since

$$(1-\delta^2\Delta)^{-1}v(x) = \delta^{-n}\int K((x-y)/\delta)\,v(y)\,dy$$

it follows that any derivative of $(1-\delta^2\Delta)^{-1}v(x)$ decreases faster than any power of δ when $\delta \to 0$ if $x \notin \omega$.

2°. If Q is a differential operator of order $j \leqslant 2$ with coefficients in C_0^∞, it follows that

$$\|(1-\delta^2\Delta)^{-1}\delta^j Qu\| \leqslant C\|u\|, \quad u \in L^2. \tag{3.6}$$

Indeed, the estimate $\|Q^*\delta^j(1-\delta^2\Delta)^{-1}u\| \leqslant C\|u\|$, $u \in C_0^\infty$, for the adjoint operator is trivial since $\delta^j(1+|\xi|)^j(1+\delta^2|\xi|^2)^{-1}$ is bounded.

3°. If Q is a differential operator of order $\leqslant 1$ with coefficients in C_0^∞, then

$$\|Q(1-\delta^2\Delta)^{-1}u - (1-\delta^2\Delta)^{-1}Qu\| \leqslant C\|u\|, \quad u \in C_0^\infty. \tag{3.7}$$

For writing $w = (1-\delta^2\Delta)^{-1}u$, we have $u = (1-\delta^2\Delta)w$ and

$$Qu = (1-\delta^2\Delta)Qw + \delta^2 Rw,$$

where $R = [\Delta, Q]$ is of second order. Multiplication by $(1-\delta^2\Delta)^{-1}$ gives

$$(1-\delta^2\Delta)^{-1}Qu - Q(1-\delta^2\Delta)^{-1}u = (1-\delta^2\Delta)^{-1}\delta^2 Rw,$$

and in view of 2° the L^2 norm of the right-hand side can be estimated by $\|w\| \leqslant \|u\|$.

4°. When $\chi_j \in C_0^\infty(\Omega)$ we have

$$|||\chi_1(1-\delta^2\Delta)^{-1}\chi_2 u||| \leqslant C|||u|||, \quad u \in C_0^\infty(\Omega), \qquad |||\chi_2(1-\delta^2\Delta)^{-1}\chi_1 f|||' \leqslant C|||f|||', \quad f \in \mathcal{D}'(\Omega).$$

The second inequality follows from the first which in turn is obvious since

$$\|X_j\chi_1(1-\delta^2\Delta)^{-1}\chi_2 u - (1-\delta^2\Delta)^{-1}X_j\chi_1\chi_2 u\| \leqslant C\|u\|$$

in view of 3°.

Proof of Proposition 3.1. We recall that with the notations introduced above we have to show that $|||Pv_\delta|||'$ is bounded when $\delta \to 0$. In the neighborhood ω of supp v

we have $(1-\delta^2\Delta)v_\delta = v$. If we apply the operator P noting that $[X_j^2, \Delta] = X_j[X_j, \Delta] + [X_j, \Delta]X_j = 2X_j[X_j, \Delta] + [[X_j, \Delta], X_j]$, it follows that in ω

$$(1-\delta^2\Delta)Pv_\delta = Pv + \sum_1^r X_j\delta^2 B_j v_\delta + \delta^2 B_0 v_\delta,$$

where $B_0, \ldots, B_r$ are second order operators with compact support. In view of 1° it follows that we have everywhere

$$(1-\delta^2\Delta)Pv_\delta = Pv + \sum_1^r X_j\delta^2 B_j v_\delta + \delta^2 B_0 v_\delta + h_\delta,$$

where h_δ vanishes in ω, $\operatorname{supp} h_\delta \subset \operatorname{supp} \chi$ and $\|h_\delta\| \to 0$ when $\delta \to 0$. Hence

$$Pv_\delta = \chi_1\{(1-\delta^2\Delta)^{-1}Pv + \sum_1^r (1-\delta^2\Delta)^{-1}X_j\delta^2 B_j v_\delta + (1-\delta^2\Delta)^{-1}\delta^2 B_0 v_\delta + (1-\delta^2\Delta)^{-1}h_\delta\},$$

where χ_1 is a function in $C_0^\infty(\Omega)$ which is equal to 1 in $\operatorname{supp}\chi$. Since $v = \chi v$, it follows from 4° that

$$|||\chi_1(1-\delta^2\Delta)^{-1}Pv|||' \leqslant C|||Pv|||'.$$

The last two terms are bounded in L^2 norm in view of 2°, and since

$$(1-\delta^2\Delta)^{-1}X_j\delta^2 B_j v_\delta = X_j(1-\delta^2\Delta)^{-1}\delta^2 B_j v_\delta + [(1-\delta^2\Delta)^{-1}, X_j]\,\delta^2 B_j v_\delta,$$

we obtain using 3°

$$|||\chi_1(1-\delta^2\Delta)^{-1}X_j\delta^2 B_j v_\delta|||' \leqslant C(\|(1-\delta^2\Delta)^{-1}\delta^2 B_j v_\delta\| + \|\delta^2 B_j v_\delta\|) \leqslant C'\|\delta^2 B_j v_\delta\| \leqslant C''\|v\|.$$

This completes the proof of Proposition 3.1.

PROPOSITION 3.2. *Assume that* (3.5) *is valid for compact subsets K of Ω. If $u \in \mathcal{D}'(\Omega)$ and $Pu = f \in H_{(s)}^{\text{loc}}(\Omega)$, it follows that $u \in H_{(s+\varepsilon)}^{\text{loc}}(\Omega)$. The same is true for open subsets of Ω, so in particular P is hypoelliptic.*

Proof. Since the statement is local we may assume that $u \in H_{(t)}^{\text{loc}}(\Omega)$ for some t. It suffices to show that t can be replaced by $t+\varepsilon$ if $t \leqslant s$. Let E be a compactly supported pseudo-differential operator in Ω with symbol $e(\xi) = (1+|\xi|^2)^{t/2}$ (cf. [2]), and set $v = \chi E u$ where $\chi \in C_0^\infty(\Omega)$. If we can show that $v \in H_{(\varepsilon)}$ for every χ we will have $Eu \in H_{(\varepsilon)}^{\text{loc}}$, hence $u \in H_{(t+\varepsilon)}^{\text{loc}}$ since E is elliptic. It is clear that $v \in L^2(\Omega) \cap \mathcal{E}'(\Omega)$, so in view of Proposition 3.1 it only remains to show that $|||Pv|||' < \infty$. To do so we note that $\chi E f \in L^2$ since $t \leqslant s$, and form the difference

$$Pv - \chi E f = (P\chi E - \chi E P)\,u.$$

As in the proof of Proposition 3.1 we have $[X_j^2, \chi E] = 2X_j[X_j, \chi E] + [[X_j, \chi E], X_j]$ so it follows that

$$P\chi E - \chi E P = \sum_1^r X_j Q_j + Q_0,$$

where $Q_j = 2[X_j, \chi E]$ for $j = 1, \ldots, r$ and all Q_j are compactly supported pseudo-differential operators of order $\leqslant t$. Since $Q_j u \in L^2$ and has compact support, it follows that $|||P\chi Eu - \chi EPu|||' < \infty$, that is, $|||Pv|||' < \infty$. This completes the proof.

4. Differentiability along noncommuting vector fields

Let Ω be an open set in $\mathbf{R}^n$, K a compact subset of Ω, and let $X \in T(\Omega)$. We shall consider the one parameter (local) group of transformations in Ω defined by X. Thus let f be the solution of the initial value problem

$$\frac{df(x,t)}{dt} = X(f(x,t)), \quad f(x,0) = x. \tag{4.1}$$

It is clear that f is a C^∞ function from $K \times (-t_0, t_0)$ to Ω if t_0 is a small positive number depending on K and on X, and we have the group property

$$f(f(x,t), s) = f(x, t+s),$$

when $x \in K$ and $|t| + |s| < t_0$.

If u is a function in Ω we set

$$(e^{tX} u)(x) = u(f(x,t)).$$

When $|t| < t_0$ this defines a mapping from $C_0^\infty(K)$ to $C_0^\infty(\Omega)$ and one from $C^\infty(\Omega)$ to $C^\infty(K)$, and we have $e^{tX} e^{sX} u = e^{(t+s)X} u$ for small t and s. The differential equation for f gives

$$\frac{d(e^{tX}u)}{dt} = e^{tX} X u.$$

The left-hand side is the limit of $(e^{(t+h)X}u - e^{tX}u)/h$ when $h \to 0$, hence also equal to $X e^{tX} u$ by the same formula with u replaced by $e^{tX}u$ and t replaced by 0. Summing up, e^{tX} is a local one parameter group of transformations of functions in Ω, and

$$\frac{d(e^{tX}u)}{dt} = e^{tX} X u = X e^{tX} u.$$

When $u \in C^\infty$ we obtain the Taylor expansion at $t = 0$

$$e^{tX} u \sim \sum_0^\infty \frac{t^k X^k u}{k!}. \tag{4.2}$$

We shall be interested in Hölder continuity of functions along vector fields in the sense of L^2 norms. Thus we shall for $0<s\leqslant 1$ and $0<\varepsilon<t_0$ consider the norms

$$|u|^{\varepsilon}_{X,s}=\sup_{0<|t|<\varepsilon}\|e^{tX}u-u\|\,|t|^{-s},\quad u\in C_0^\infty(K), \tag{4.3}$$

where $\| \ \|$ denotes the L^2 norm. The norm (4.3) increases with ε, but since the difference between its values for two different choices of ε can be bounded by a constant times $\|u\|$, we shall usually omit ε from the notations below. An equivalent norm is of course obtained if we take $0<t<\varepsilon$. (Since our aim is to prove the *a priori* estimate (3.4) for $v\in C_0^\infty(K)$ we have chosen not to introduce the complete spaces corresponding to these norms and leave for the reader to state the implications for these spaces of the estimates proved below.)

LEMMA 4.1. *If $\varphi\in C^\infty(\Omega,\mathbf{R})$, it follows that*

$$|u|_{\varphi X,s}\leqslant C|u|_{X,s},\quad u\in C_0^\infty(K). \tag{4.4}$$

Proof. We keep the notation $f(x,t)$ used above so that $(e^{tX}u)(x)=u(f(x,t))$. Let $\tau(x,t)$ be the solution of the initial value problem

$$\frac{d\tau}{dt}=\varphi(f(x,\tau)),\quad \tau=0\quad\text{when}\quad t=0.$$

From the differential equation (4.1) we then obtain

$$\frac{df(x,\tau)}{dt}=(\varphi X)(f(x,\tau)),$$

Hence $e^{t\varphi X}u(x)=u(f(x,\tau))$, so that

$$\|e^{t\varphi X}u-u\|^2=\int|u(f(x,\tau(x,t)))-u(x)|^2\,dx.$$

Since τ depends on x we cannot compare this directly with (4.3), so we first note that for any σ

$$|u(f(x,\tau))-u(x)|^2\leqslant 2|u(f(x,\tau))-u(f(x,\sigma))|^2+2|u(f(x,\sigma))-u(x)|^2.$$

Integrating with respect to x and averaging over σ for $|\sigma|<|t|$, we obtain

$$\|e^{t\varphi X}u-u\|^2\leqslant|t|^{-1}\iint_{|\sigma|<|t|}|u(f(x,\tau))-u(f(x,\sigma))|^2\,dx\,d\sigma+2|u|^2_{X,s}|t|^{2s}.$$

In the integral we introduce new variables by setting

$$y = f(x, \sigma), \quad f(y, w) = f(x, \tau), \quad \text{that is,} \quad w + \sigma = \tau.$$

For fixed t and for $\sigma = 0$ we obtain

$$dy = dx + X\, d\sigma, \quad dw = d\tau - d\sigma.$$

When $t=0$ we have $\tau = 0$, hence $d\tau = 0$, so for $t = \sigma = 0$ the Jacobian $D(y, w)/D(x, \sigma)$ is equal to -1. Hence it is arbitrarily close to -1 for sufficiently small σ and t. Since $\tau = O(t)$ we have $|w| \leqslant A|t|$ for some constant A when $|\sigma| \leqslant |t|$. Thus we conclude that for sufficiently small t

$$|t|^{-1} \iint_{|\sigma|<|t|} |u(f(x,\tau)) - u(f(x,\sigma))|^2\, dx\, d\sigma$$

$$\leqslant 2|t|^{-1} \iint_{|w|<A|t|} |u(f(y,w)) - u(y)|^2\, dy\, dw \leqslant 4A(A|t|)^{2s} |u|^2_{X,s}.$$

This completes the proof of (4.4).

We shall also use a universal s-norm defined by

$$|u|_s^\varepsilon = \sup_{|h|<\varepsilon} \|\tau_h u - u\|\, |h|^{-s},$$

where $(\tau_h u)(x) = u(x+h)$. If e_j is the field of unit vectors along the jth coordinate axis, we find immediately by using the triangle inequality that

$$|u|_s^\varepsilon \leqslant \sum_1^n |u|^\varepsilon_{e_j, s}. \tag{4.5}$$

On the other hand, we can estimate $|u|^\varepsilon_{X,s}$ by a constant times $|u|_s^\varepsilon$ for an arbitrary X. This is a special case (for $N=1$) of the following

LEMMA 4.2. *Let $g(x,t)$ be a map from a neighborhood of $K \times 0$ in $\Omega \times \mathbf{R}$ to Ω such that $g(x,t) - x = O(t^N)$, $t \to 0$, where $N > 0$, and g is a C^∞ function of x which is continuous in t as well as its derivatives. Then we have for small $|t|$*

$$\int |u(g(x,t)) - u(x)|^2\, dx \leqslant C|t|^{2Ns} |u|_s^2, \quad u \in C_0^\infty(K). \tag{4.6}$$

Proof. The proof is parallel to that of Lemma 4.1. Thus we first compare with the translations τ_h and obtain

$$\int |u(g(x,t))-u(x)|^2\,dx \leqslant C|t|^{-Nn}\iint_{|h|<|t|^N} |u(g(x,t))-u(x+h)|^2\,dx\,dh + 2|u|_s^2|t|^{2Ns}.$$

We introduce new coordinates in the integral by setting $y=x+h$ and $y+w=g(x,t)$. For $t=0$ this is the linear transformation $y=x+h$, $w=-h$, which has determinant ± 1. For small t the Jacobian of the change of variables is therefore close to 1 in absolute value, and since $|w|<A|t|^N$ for some constant A in the domain of the new integral, the proof is concluded as that of Lemma 4.1.

Note that the norm $|\ |_s$ together with the L^2 norm is weaker than the usual s-norm $\|\ \|_{(s)}$ (cf. [3, section 2.6]) used in paragraph 3, but is stronger than the norm $\|\ \|_{(t)}$ when $t<s$.

If $X\in T(\Omega)$ we shall use the standard notation $\operatorname{ad} X$ for the differential operator from $T(\Omega)$ to $T(\Omega)$ defined by

$$(\operatorname{ad} X)\,Y=[X,Y],\quad Y\in T(\Omega).$$

Given elements $X_i\in T(\Omega)$, $i=0,\dots,r$, as in Theorem 1.1, and a multi-index I, that is, a sequence $(i_1,\dots,i_k)$ with $0\leqslant i_j\leqslant r$, we shall write

$$X_I=\operatorname{ad} X_{i_1}\dots\operatorname{ad} X_{i_{k-1}}X_{i_k}.$$

(Note the distinction between a multi-index and a multi-order as used in section 2.) We set $k=|I|$ and always assume that $|I|\neq 0$. The same notations will be used for other Lie algebras than $T(\Omega)$.

We can now state the main result to be proved in this section.

THEOREM 4.3. *Given $X_j\in T(\Omega)$ and $s_j\in(0,1]$, $j=0,\dots,r$, we denote by $T^s(\Omega)$ the $C^\infty(\Omega,\mathbf{R})$ submodule of $T(\Omega)$ generated by all X_I with $s(I)\geqslant s$, where*

$$\frac{1}{s(I)}=\sum_1^k\frac{1}{s_{i_j}}.$$

Assume that $T^s(\Omega)=T(\Omega)$ for some $s>0$. Then we have for every compact set $K\subset\Omega$ with C depending on X and K but not on u

$$|u|_{X,s}\leqslant C\left(\sum_0^r|u|_{X_j,s}+\|u\|\right),\quad u\in C_0^\infty(K),\quad X\in T^s(\Omega). \tag{4.7}$$

In particular,

$$|u|_s\leqslant C\left(\sum_0^r|u|_{X_j,s}+\|u\|\right),\quad u\in C_0^\infty(K),\quad \textit{if}\quad T^s(\Omega)=T(\Omega). \tag{4.8}$$

We remark that a slightly more precise version of a special case of (4.7) has been proved by Kohn [5] but his method does not seem applicable in the general case. The proof of (4.7) will be made by induction for increasing s starting from a point where we expect (4.8) to be valid. We begin with a simple lemma justifying that we have not considered more complicated commutators in Theorems 1.1 and 4.3.

LEMMA 4.4. *If* $1/t_1+1/t_2\leqslant 1/t_3$, *we have*

$$[T^{t_1}, T^{t_2}]\subset T^{t_3}.$$

Proof. Let I_1 and I_2 be multi-indices with $s(I_j)\geqslant t_j$ and let $\varphi_j\in C^\infty(\Omega)$. Then we have

$$[\varphi_1 X_{I_1}, \varphi_2 X_{I_2}]=\varphi_1(X_{I_1}\varphi_2)\, X_{I_2}-\varphi_2(X_{I_2}\varphi_1)\, X_{I_1}+\varphi_1\varphi_2[X_{I_1}, X_{I_2}].$$

Since $t_3\leqslant t_j$ for $j=1, 2$, the first two terms on the right-hand side are in T^{t_3}, and so is the third since the Jacobi identity $\operatorname{ad}[X, Y]=[\operatorname{ad} X, \operatorname{ad} Y]$ gives that $[X_{I_1}, X_{I_2}]=(\operatorname{ad} X)_{I_1} X_{I_2}$ which written explicitly is a linear combination of elements X_J where $1/s(J)=1/s(I_1)+1/s(I_2)$.

We shall have to make repeated use of the Campbell–Hausdorff formula which can be stated as follows (cf. Hochschild [1], Chap. X): If x and y are two non-commuting indeterminates, we have in the sense of formal power series in x and y that $e^x e^y=e^z$ where

$$z=\sum_1^\infty(-1)^{n+1}n^{-1}\sum_{\alpha_i+\beta_i\neq 0}(\operatorname{ad} x)^{\alpha_1}(\operatorname{ad} y)^{\beta_1}\dots(\operatorname{ad} x)^{\alpha_n}(\operatorname{ad} y)^{\beta_n-1}y/c_{\alpha,\beta},$$

where $c_{\alpha,\beta}=\alpha!\,\beta!\,|\alpha+\beta|$. (When $\beta_n=0$ the term should be modified so that the last factor is $(\operatorname{ad} x)^{\alpha_n-1}x$.) The important facts for us are that the terms of order 1 are $x+y$, that those of order two are $\frac12[x, y]$, and that all terms of higher order are (repeated) commutators of x and y.

We shall use the Campbell–Hausdorff formula to derive a product decomposition of e^{x+y}. With the notations used above we have

$$e^{-y}e^{-x}e^{x+y}=e^{-z}e^{x+y}=e^{r_2},$$

where $r_2=-z+x+y+\frac12[-z, x+y]+\dots=-\frac12[x, y]+\dots$, the dots indicating terms of order at least three which are linear combinations of commutators. Writing $z_2=-\frac12[x, y]$, we form $e^{-z_2}e^{r_2}=e^{r_3}$. The Campbell–Hausdorff formula gives $r_3=z_3+\dots$ where z_3 is a linear combination of commutators of x and of y of degree three, and the dots indicate a formal series whose terms are commutators of degree at least four. Proceeding in this way, we choose for every integer $k \geqslant 2$ a linear combination z_k of commutators of x and y of order k such that

$$e^{-z_k} e^{r_k} = e^{r_{k+1}},$$

where r_k is a formal series whose terms are commutators of x and y of order at least k. Thus we have

$$e^{-z_k} e^{-z_{k-1}} \dots e^{-z_2} e^{-y} e^{-x} e^{x+y} = e^{r_{k+1}}, \tag{4.9}$$

so that e^{x+y} is to a high degree of accuracy approximated by the product

$$e^x e^y e^{z_2} \dots e^{z_k}.$$

Lemma 4.5. *Let $X, Y \in T(\Omega)$ and denote by Z_j the linear combination of commutators of j factors X and Y obtained by replacing x and y by X and Y above. Let $0 < \sigma \leqslant 1$, and let N be an integer $\geqslant 2$. Then we have for small t and $u \in C_0^\infty(K)$*

$$\| e^{t(X+Y)} u - u \| \leqslant C (\| e^{tX} u - u \| + \| e^{tY} u - u \| + \sum_2^{N-1} \| e^{t^j Z_j} u - u \| + t^{\sigma N} |u|_\sigma). \tag{4.10}$$

Proof. The operator

$$H_N^t = \exp(-t^{N-1} Z_{N-1}) \dots \exp(-t^2 Z_2) \exp(-tY) \exp(-tX) \exp t(X+Y)$$

is induced by a mapping in Ω since every factor is. Hence there exists a C^∞ function $h_N(x,t)$ from a neighborhood of $K \times 0$ to Ω such that $H_N^t v(x) = v(h_N(x,t))$. From (4.9) and (4.2) it follows that $H_N^t v - v = O(t^N)$ if $v \in C^\infty$. Taking for v a coordinate function we conclude that

$$h_N(x,t) - x = O(t^N),$$

so Lemma 4.2 gives $$\| H_N^t v - v \| \leqslant C |t|^{\sigma N} |v|_\sigma.$$

Now we have for any bounded operators $S_1, \dots, S_k$ in L^2

$$\| S_1 \dots S_k u - u \| = \| \sum_{j=1}^k S_1 \dots S_{j-1} (S_j u - u) \| \leqslant \sum_{j=1}^k \| S_1 \| \dots \| S_{j-1} \| \, \| S_j u - u \|. \tag{4.11}$$

Since $$\exp t(X+Y) = \exp(tX) \exp(tY) \exp(t^2 Z_2) \dots \exp(t^{N-1} Z_{N-1}) H_N^t$$

and the norm of each factor is bounded uniformly in t the inequality (4.10) follows.

Lemmas 4.1 and 4.5 together will allow us to prove (4.7) for arbitrary $X \in T^s(\Omega)$ after the estimate has been established for a set of generators. We shall therefore study next some identities which give control of the commutators of the given operators X_j.

Let $x_1, \ldots, x_k$ be k non-commuting indeterminates. By the Campbell–Hausdorff formula we have

$$e^{x_{k-1}} e^{x_k} e^{-x_{k-1}} e^{-x_k} = e^{z_{k-1}},$$

where $z_{k-1} = [x_{k-1}, x_k] + \ldots$, the dots indicating a formal series all terms of which are commutators of at least three factors equal to x_k or x_{k-1}; obviously both x_k and x_{k-1} must occur at least once in every one of them. We now form successive formal power series $z_{k-1}, z_{k-2}, \ldots, z_1$ by setting

$$e^{x_j} e^{z_{j+1}} e^{-x_j} e^{-z_{j+1}} = e^{z_j}, \quad j = 1, \ldots, k-2.$$

Then e^{z_1} is a product of n_k factors $e^{\pm x_j}$, where $n_2 = 4$ and $n_{k+1} = 2 + 2n_k$, that is, $n_k = 3 \cdot 2^{k-1} - 2$, and $z_1 = c + \ldots$ where $c = \operatorname{ad} x_1 \operatorname{ad} x_2 \ldots \operatorname{ad} x_{k-1} x_k$, the dots denoting a series with terms of higher order, each of which is a commutator containing each x_j at least once. As in the discussion preceding Lemma 4.5 we can use the Campbell–Hausdorff formula again to show that for any N we can write

$$e^c = e^{z_1} e^{c_1} e^{c_2} \ldots e^{c_\nu} e^r,$$

where each c_j is a commutator formed from $x_1, \ldots, x_k$ which contains each x_j at least once and some x_j twice, and r is a formal sum of commutators of at least N factors x_j. If we recall the definition of e^{z_1} we have thus found an approximate representation of e^c by products of $e^{\pm x_j}$ and e^{c_j} where c_j are commutators of higher order than c. This allows us to prove the final lemma needed for the proof of Theorem 4.3.

Lemma 4.6. *Given X_j and s_j, $j = 0, \ldots, r$, as in Theorem 4.3, we set $m_j = 1/s_j$ and $m(I) = 1/s(I)$ when I is a multi-index. Let $\sigma > 0$. Then we have for small $t > 0$ and an arbitrary multi-index I*

$$\|\exp(t^{m(I)} X_I) u - u\| \leqslant C_1 t \sum_0^r |u|_{X_j, s_j} + C_2 t |u|_\sigma, \quad u \in C_0^\infty(K), \tag{4.12}$$

where C_1 and C_2 are constants and C_1 only depends on r and σ, not on $X_0, \ldots, X_r, s_0, \ldots, s_r$.

Proof. Since $m_j \geqslant 1$, we have $m(I) \geqslant |I|$, so (4.12) follows from Lemma 4.2 with $C_1 = 0$ if $\sigma |I| \geqslant 1$. If N is an integer with $N\sigma \geqslant 1$, we may thus prove (4.12) by induction for decreasing $|I|$, starting when $|I| = N$.

Replacing the indeterminates x_j in the discussion preceding Lemma 4.6 by $t^{m_{i_j}} X_{i_j}$, we obtain as in the proof of Lemma 4.5 an identity

$$\exp\,(t^{m(I)}X_I)=\prod \exp\,(\pm t^{m_j}X_j)\exp\,(t^{m(I_1)}X_{I_1})\ldots\exp\,(t^{m(I_\nu)}X_{I_\nu})\,H_N^t, \tag{4.13}$$

where the product contains $3\cdot 2^{|I|-1}-2$ factors as described above, the multi-indices $I_1,\ldots,I_\nu$ have greater length than I and

$$H_N^t\,v(x)=v(h_N(x,t))$$

with a C^∞ function $h_N(x,t)$ of x, depending continuously on t, such that $h_N(x,t)-x=O(t^N)$, $t\to 0$. From Lemma 4.2 we obtain

$$\|H_N^t u-u\|\leqslant C\,t^{N\sigma}\,|u|_\sigma,\quad u\in C_0^\infty(K).$$

In view of (4.11) if follows that for small t

$$\|\exp\,(t^{m(I)}X_I)\,u-u\|\leqslant 2^{|I|+1}\sum_0^r\|\exp\,(t^{m_j}X_j)\,u-u\|$$
$$+2\sum_1^\nu\|\exp\,(t^{m(I_j)}X_{I_j})\,u-u\|+C\,t^{\sigma N}\,|u|_\sigma,$$

for the norm of each factor in (4.13) is close to 1 for small t. We can apply the inductive hypothesis to the terms in the second sum, and since $\sigma N\geqslant 1$, the estimate (4.12) follows.

Remark. Since C_1 does not depend on the choice of $X_0,\ldots,X_r$, it follows that for multi-indices I containing some index $\geqslant 1$, we have for every $\varepsilon>0$

$$\|\exp\,(t^{m(I)}X_I)\,u-u\|\leqslant\varepsilon t\,|u|_{X_0,s_0}+C_\varepsilon\,t\,(\sum_1^r|u|_{X_j,s_j}+|u|_\sigma),\quad u\in C_0^\infty(K). \tag{4.12'}$$

In fact, X_I does not change if we replace X_0 by εX_0 provided that at the same time we replace X_j by $\varepsilon^{-\gamma}X_j$ for a suitable $\gamma>0$ when $j\geqslant 1$.

Proof of Theorem 4.3. Choose $\sigma>0$ so that $T^\tau(\Omega)=T(\Omega)$ for some $\tau>\sigma$. We wish to prove that

$$|u|_{X,s}\leqslant C_X\,(\sum_0^r|u|_{X_j,s_j}+|u|_\sigma),\ u\in C_0^\infty(K),\ X\in T^s(\Omega). \tag{4.14}$$

This estimate is trivial if $s\leqslant\sigma$ and it follows from Lemma 4.6 if X is any one of the commutators X_I which generate $T^s(\Omega)$. In view of Lemma 4.1 the estimate (4.14) remains valid for $X=\varphi X_I$ if $\varphi\in C^\infty(\Omega)$. By definition every $X\in T^s$ is therefore a finite sum of vector fields for which (4.14) is valid. If X, $Y\in T^s(\Omega)$ are vector fields for which (4.14) is valid, we apply Lemma 4.5 with $\sigma N\geqslant s$ noting that if follows from Lemma 4.4 that $Z_j \in T^{s/j}$. Assuming as we may that (4.14) has already been proved when s is replaced by a number $\leqslant s/2$, we conclude that (4.14) is valid when X is replaced by $X+Y$. Hence (4.14) follows.

Now recall that $T^{\tau}(\Omega)=T(\Omega)$ for some $\tau>\sigma$. Thus we obtain from (4.14) the estimate

$$|u|_\tau \leqslant C(\sum_0^r |u|_{X_j, s_j} + |u|_\sigma), \quad u \in C_0^\infty(K), \tag{4.15}$$

if we take (4.5) into account. Since $\tau>\sigma$ we have for any $\delta>0$

$$|u|_\sigma \leqslant \delta\,|u|_\tau + C_\delta\,\|u\|. \tag{4.16}$$

If we combine (4.15) and (4.16) taking $\delta C<\frac{1}{2}$, we conclude that

$$|u|_\sigma \leqslant |u|_\tau \leqslant C'(\sum_0^r |u|_{X_j, s_j} + \|u\|), \quad u \in C_0^\infty(K).$$

Using this estimate in the right-hand side of (4.14) we have proved (4.7).

5. Smoothing and estimates

In section 3 we have proved that Theorem 1.1 is a consequence of the *a priori* estimate (3.4). We recall that

$$|||v|||^2 = \sum_1^r \|X_j v\|^2 + \|v\|^2,$$

so the right-hand side of (3.4) gives us control of $|v|_{X_j,1}$ when $j=1,\ldots,r$. However, the information given about $X_0 v$ is in a weaker norm which prevents us from applying Theorem 4.3. To study the differentiability of v in the direction X_0 we consider $f(t)=\|e^{tX_0}v-v\|$. Differentiation with respect to t gives, if v is real as we may well assume

$$df(t)^2/dt = 2(e^{tX_0}X_0 v, e^{tX_0}v - v).$$

Let us assume for a moment that e^{tX_0} preserves the norms $|||\ |||$ and $|||\ |||'$, although we shall see below that this is far from true. Then we would obtain

$$df(t)^2/dt \leqslant 4\,|||X_0 v|||'\ |||v|||,$$

hence

$$f(t) \leqslant t^{\frac{1}{2}}(|||v||| + |||X_0 v|||').$$

Thus we would have control of $|v|_{X_0,\frac{1}{2}}$ and could apply Theorem 4.3 with $s_0=\frac{1}{2}$, $s_1=\ldots=s_r=1$, and this would give (3.4).

To examine the validity of the preceding arguments we must consider $|||e^{tX_0}v|||$, thus the L^2 norm of $X_j e^{tX_0}v$ for $j=1,\dots,r$. This is essentially the same as the L^2 norm of $e^{-tX_0}X_j e^{tX_0}v$. Now

$$e^{-tX_0}X_j e^{tX_0} = e^{-t\,\mathrm{ad}\,X_0}X_j = \sum_0^\infty (-t)^k(\mathrm{ad}\,X_0)^k X_j/k!,$$

where the first equality is a definition motivated by the second one which means that the two sides have the same Taylor expansion in t. This follows immediately from the fact that left and right multiplication by X_0 commute and that by definition their difference is ad X_0. Since we have no information about the differentiability of v in the direction $(\mathrm{ad}\,X_0)^k X_j$ when $k \neq 0$, the argument as given above breaks down. However, we note that the derivative in this direction occurs with a factor t^k, which indicates that we can impose sufficient smoothness on v by a regularization which does not change v too much for small t. This we shall do in the following discussion which aims at proving that $f(t^2)/t$ can be bounded by the right-hand side of (3.4). It is in fact permissible to allow in the right-hand side of the estimates any quantities which by the results of section 4 can be estimated by a small constant times $|v|_{X_0,\frac{1}{2}}$ and a large constant times $|||v|||$.

The first step is to study regularization along a vector field $X \in T(\Omega)$. Let $u \in C_0^\infty(K)$, $K \subset\subset \Omega$, and assume that $e^{\tau X}$ maps $C_0^\infty(K)$ into $C_0^\infty(\Omega)$ for $|\tau| \leqslant 1$. With $\varphi \in C_0^\infty(-1,1)$ we set

$$\varphi_X u = \int e^{\tau X} u\varphi(\tau)\,d\tau.$$

This operator is smoothing in the direction X, for

$$X\varphi_X u = \int \frac{d}{d\tau} e^{\tau X} u\varphi(\tau)\,d\tau = -\varphi'_X u = \int (u - e^{\tau X}u)\,\varphi'(\tau)\,d\tau.$$

It follows that
$$\|X\varphi_X u\| \leqslant \int |d\varphi| \sup_{|\tau|<1} \|e^{\tau X}u - u\|, \tag{5.1}$$

$$\|\varphi_X u - u\| \leqslant \sup_{|\tau|<1}\|e^{\tau X}u - u\| \quad \text{if} \quad \varphi \geqslant 0 \quad \text{and} \quad \int \varphi\,dx = 1. \tag{5.2}$$

We shall later have to consider the commutator of φ_X with operators $Y \in T(\Omega)$, so we note the formulas

$$Y\varphi_X u = \int e^{\tau X}(e^{-\tau\,\mathrm{ad}\,X}Y)\,u\varphi(\tau)\,d\tau, \tag{5.3}$$

$$\varphi_X Y u = \int (e^{\tau \operatorname{ad} X} Y) e^{\tau X} u \varphi(\tau) \, d\tau. \tag{5.4}$$

Each term in the Taylor expansion of $e^{\tau \operatorname{ad} X}$ will thus give an analogous expression with a smoothing operator defined by some other function and acting on the other side of another differential operator.

Besides the quite specific smoothing along certain vector fields using the operators φ_X we shall also employ the usual smoothing in all variables. Thus let $\Phi \in C_0^\infty(B)$, where B is the unit ball in $\mathbf{R}^n$, and set

$$\Phi_\varepsilon u(x) = \int u(x - \varepsilon h) \, \Phi(h) \, dh.$$

Then we obtain

$$\varepsilon \| D_j \Phi_\varepsilon u \| \leqslant \int |D_j \Phi| \, dx \sup_{|h|<1} \| u(x - \varepsilon h) - u(x) \|, \tag{5.1)$'$}$$

$$\| \Phi_\varepsilon u - u \| \leqslant \sup_{|h|<1} \| u(x - \varepsilon h) - u(x) \| \quad \text{if} \quad \Phi \geqslant 0 \quad \text{and} \quad \int \Phi \, dx = 1. \tag{5.2)$'$}$$

Instead of (5.3) and (5.4) we shall use Friedrichs's lemma ([3, Theorem 2.4.3]) which gives for every $Y \in T(\Omega)$

$$\| (Y \Phi_\varepsilon - \Phi_\varepsilon Y) u \| \leqslant C \| u \|, \quad u \in C_0^\infty(K), \tag{5.5}$$

where C is uniformly bounded for small ε if Y lies in a bounded set in $T(\Omega)$.

As in section 4 the notation I will stand for a multi-index and X_I for the corresponding commutator. We set $s_0 = \frac{1}{2}$, $s_1 = \ldots = s_r = 1$ and define $s(I)$ and $m(I) = 1/s(I)$ as in Theorem 4.3 and Lemma 4.6. Thus $m(I)$ is the sum of the length $|I|$ of I and the number of indices in I which are equal to 0.

Let σ be a positive number chosen so small that with the notations of Theorem 4.3 we have $T^s(\Omega) = T(\Omega)$ for some $s > \sigma$. As in section 4 we shall allow $|u|_\sigma$ to occur in the right-hand side of our estimates and use it to take care of various remainder terms in Taylor expansions. Let $\mathfrak{J}$ denote the set of all I with $\sigma m(I) \leqslant 1$ and $|I| < m(I) < 2|I|$; the latter condition means that I shall contain indices equal to 0 as well as indices $\neq 0$. Set

$$M(u) = |||u||| + |||X_0 u|||' + \sum_{I \in \mathfrak{J}} |u|_{X_I, s(I)} + |u|_\sigma.$$

Our aim is to show that

$$|u|_{X_0, \frac{1}{2}} \leqslant C M(u), \quad u \in C_0^\infty(K). \tag{5.6}$$

By (4.12)′ we can estimate $|u|_{X_I, s(I)}$ by a small constant times $|u|_{X_0, \frac{1}{2}}$ and a large constant times $|||u|||+|u|_\sigma$ when $I \in \mathfrak{J}$. Hence (5.6) implies

$$\sum_0^r |u|_{X_j, s_j} + \|u\| \leqslant C' (|||u||| + |||X_0 u|||' + |u|_\sigma), \quad u \in C_0^\infty(K). \tag{5.7}$$

Let $s > \sigma$ but $T^s(\Omega) = T(\Omega)$. Then it follows from (5.7) and Theorem 4.3 that

$$|u|_s \leqslant C'' (|||u||| + |||X_0 u|||' + |u|_\sigma), \quad u \in C_0^\infty(K),$$

and since $|u|_\sigma \leqslant \delta |u|_s + C_\delta \|u\|$ for any $\delta > 0$, we obtain with another constant C

$$|u|_s \leqslant C (|||u||| + |||X_0 u|||'), \quad u \in C_0^\infty(K),$$

hence

$$\sum_0^r |u|_{X_j, s_j} \leqslant C (|||u||| + |||X_0 u|||'), \quad u \in C_0^\infty(K). \tag{5.8}$$

In view of Theorem 4.3 we conclude

THEOREM 5.1. *Let $X_0, \ldots, X_r$ satisfy the hypotheses of Theorem 4.3 and set $s_0 = \frac{1}{2}$, $s_1 = \ldots = s_r = 1$. Then we have*

$$|u|_{X, s} \leqslant C(X) (|||u||| + |||X_0 u|||'), \quad u \in C_0^\infty(K), \quad X \in T^s(\Omega), \tag{5.9}$$

where $|||\ |||$ and $|||\ |||'$ are defined in section 3.

Clearly Theorem 5.1 completes the proof of Theorem 1.1, so all that remains now is to prove (5.6).

We give $\mathfrak{J}$ a total ordering so that $m(I)$ is an increasing function of $I \in \mathfrak{J}$ and set

$$S_t u = \prod_{I \in \mathfrak{J}} \varphi_{t^{m(I)} X_I} \Phi_{t^{1/\sigma}} u,$$

where φ and Φ are functions satisfying the hypotheses of (5.2) and (5.2)′. The factors in the product are taken from left to right in increasing order of I. If $J \in \mathfrak{J}$ we shall also write $S_t^J u$ for the similar expression with the product restricted to all I with $I \geqslant J$. We also set $\mathfrak{J}' = \mathfrak{J} \cup \infty$ and $S_t^\infty = \Phi_{t^{1/\sigma}}$, and define $\infty > I$ for every $I \in \mathfrak{J}$.

If follows immediately from (5.2), (5.2)′, the definition of the norms and (4.11) that

$$\|S_t u - u\| \leqslant C t M(u). \tag{5.10}$$

We want to estimate $\|e^{t^2 X_0} u - u\|$. Noting that

$$e^{t^2 X_0}u - u = e^{t^2 X_0}(u - S_t u) + e^{t^2 X_0} S_t u - S_t u + S_t u - u$$

and that the norm of $e^{t^2 X_0}$ as an operator in L^2 is uniformly bounded, we conclude that

$$\| e^{t^2 X_0} u - u \| \leqslant C t M(u) + \| e^{t^2 X_0} S_t u - S_t u \|. \tag{5.11}$$

The advantage of this is that the regularity built into $S_t u$ will make it possible to estimate the last term by applying the argument outlined at the beginning of the section but which was then merely heuristic. We need the following lemma which shows how differentiability is successively introduced by the regularizations in S_t. (Note that $S_t = S_t^J$ when J is the smallest element in $\mathfrak{J}$.)

LEMMA 5.2. *For every $J \in \mathfrak{J}'$ we have for small $t > 0$*

$$\sum_1^n \| t^{1/\sigma} D_j S_t^J u \| \leqslant C t M(u), \tag{5.12}$$

$$\sum_{J \leqslant I \in \mathfrak{J}} \| t^{m(I)} X_I S_t^J u \| \leqslant C t M(u), \tag{5.13}$$

$$\sum_0^r \| [t^{m_j} X_j, S_t^J] \, u \| \leqslant C t M(u). \tag{5.14}$$

Proof. For $J = \infty$ the estimate (5.12) follows from (5.1)$'$. As a superposition of compositions with C^∞ maps, each factor in the operators S_t^J is uniformly bounded in the $H_{(s)}$ norm for every s, so (5.12) is valid for all $J \in \mathfrak{J}'$. For $J = \infty$ the statement (5.13) is void and (5.14) is very much weaker than (5.5). When proving the lemma we may therefore assume that it has already been proved for larger J and arbitrary $\varphi \in C_0^\infty(-1, 1)$. In the proof of (5.13) we must then distinguish between two different cases:

1°. $I > J$. Let J' be the smallest element in $\mathfrak{J}'$ larger than J; then $I \geqslant J'$. We shall use (5.3) with Y replaced by $t^{m(I)} X_I$ and X replaced $t^{m(J)} X_J$. This allows us to let X_I pass through the first regularizer and we obtain

$$t^{m(I)} X_I S_t^J u = \int e^{\tau t^{m(J)} X_J} \left\{ \sum_{\nu < N} (\mathrm{ad} - \tau t^{m(J)} X_J)^\nu \, t^{m(I)} X_I S_t^{J'} A \, u/\nu\,! + t^{N m(J) + m(I)} Y_{t,\tau} S_t^{J'} u \right\} \varphi(\tau) \, d\tau,$$

where $Y_{t,\tau}$ belongs to a bounded set when $t \to 0$. If $N m(J) + m(I) \geqslant 1/\sigma$ if follows from (5.12) that we have the desired estimate for the term involving the remainder term $Y_{t,\tau}$. Since $(\mathrm{ad}\ X_J)^\nu X_I = X_{I'}$ for some I' with $I' \geqslant I > J$ we have $I' \geqslant J'$, so the inductive hypothesis concerning (5.13) allows us to estimate the terms in the sum to the extent that they are not taken care of by (5.12).

2°. $I=J$. With J' defined as above we have in view of (5.1)

$$\|t^{m(J)}X_J S_t^J u\| \leqslant C \sup_{|\tau|\leqslant 1} \| e^{\tau t^{m(J)}X_J} S_t^{J'} u - S_t^{J'} u\|.$$

The proof of (5.10) applies without change to prove that the right-hand side can be bounded by $CtM(u)$.

It remains to prove (5.14). We can write

$$[t^{m_j}X_j, S_t^J]\, u = \varphi_{t^{m(J)}X_J} [t^{m_j}X_j, S_t^{J'}]\, u + [t^{m_j}X_j, \varphi_{t^{m(J)}X_J}]\, S_t^{J'}\, u.$$

By the inductive hypothesis concerning (5.14) it suffices to consider the second term. Now (5.3) gives

$$[t^{m_j}X_j, \varphi_{t^{m(J)}X_J}]\, v = \int e^{\tau t^{m(J)}} X_J \left(\sum_{0<\nu<N} (\mathrm{ad} - \tau t^{m(J)}\, X_J)^\nu\, t^{m_j} X_j v/\nu! + t^{NM(J)+m_j}\, Y_{t,\tau} v \right) \varphi(\tau)\, d\tau$$

with $Y_{t,\tau}$ as above in 1°. When $\nu \neq 0$ we have $(\mathrm{ad} - X_J)^\nu X_j = X_{I'}$ for some $I' \in \mathfrak{J}'$ with $I' \geqslant J'$ or else $\sigma m(I) \geqslant 1$, so we obtain (5.14) from the inductive hypotheses concerning (5.12) and (5.13).

Proof of (5.6). Our aim is to estimate the right-hand side of (5.11) so we introduce for $0 \leqslant \tau \leqslant t^2$ the function

$$f(\tau) = \| e^{\tau X_0} S_t u - S_t u \|.$$

Differentiation of f^2 gives

$$\begin{aligned} f(\tau)\, f'(\tau) &= (e^{\tau X_0} X_0 S_t u,\, e^{\tau X_0} S_t u - S_t u) \\ &= (e^{\tau X_0} [X_0, S_t]\, u,\, e^{\tau X_0} S_t u - u) + (e^{\tau X_0} S_t X_0 u,\, e^{\tau X_0} S_t u - S_t u). \end{aligned}$$

Using the Cauchy–Schwarz inequality and (5.14), we obtain

$$f(\tau)\, f'(\tau) \leqslant Ct^{-1} M(u)\, f(\tau) + (X_0 u,\, (e^{\tau X_0} S_t)^* \,(e^{\tau X_0} S_t u - S_t u)). \tag{5.15}$$

We shall prove that the last expression can be estimated by $CM(u)^2$. Admitting this estimate for a moment, we obtain the integral inequality

$$ff' \leqslant C\left(\frac{f}{t} + M\right) M \quad \text{when} \quad 0 < \tau < t^2; \quad f(0) = 0.$$

If $g = f/Mt$, this reduces to $gg' \leqslant C(g+1)\, t^{-2}$, so

$$\int_0^g g dg/(g+1) \leqslant C\tau t^{-2} \leqslant C \quad \text{when} \quad \tau \leqslant t^2.$$

This implies that $g \leqslant C'$ when $\tau \leqslant t^2$, hence $f(t^2) \leqslant C'M(u)t$, and (5.6) follows in view of (5.11).

What remains is therefore to show that

$$|||(e^{\tau X_0} S_t)^* (e^{\tau X_0} S_t u - S_t u)||| \leqslant CM(u). \tag{5.16}$$

Write $$N_t(v) = \sum_1^r \|X_j v\| + \sum_{I \in \mathfrak{J}} \|t^{m(I)-1} X_I v\| + t^{1/\sigma - 1} \sum_1^n \|D_j v\| + \|v\|.$$

Then we have $$N_t(S_t u) \leqslant CM(u). \tag{5.17}$$

This follows immediately from Lemma 5.2 if we note that $X_j S_t u = S_t X_j u + [X_j, S_t]u$ and recall again that S_t has uniformly bounded norm in L^2. The norm N_t is quite well behaved under translations; indeed, we shall prove that

$$N_t(e^{\tau X_J} v) \leqslant CN_t(v), 0 < \tau \leqslant t^{m(J)}, \tag{5.18}$$

provided that the multi-index J contains 0.

To prove (5.18) we let $Y = X_j$, $j = 1, \ldots, r$, or $Y = X_I$ where $I \in \mathfrak{J}$, and note that

$$Y e^{\tau X_J} S_t u = e^{\tau X_J} (e^{-\operatorname{ad} \tau X_J} Y) S_t u.$$

Now $$\|(\operatorname{ad} \tau X_J)^k Y v\| \leqslant \|(\operatorname{ad} t^{m(J)} X_J)^k Y v\|$$

when $0 \leqslant \tau \leqslant t^{m(J)}$, so we obtain the desired bound for each term in the Taylor expansion of $e^{-\operatorname{ad} \tau X_J}$. The error term can be estimated by using the last sum in the definition of N_t, so (5.18) follows.

In particular, we obtain from (5.17) and (5.18)

$$N_t(e^{\tau X_0} S_t u - S_t u) \leqslant CM(u) \quad \text{when} \quad 0 \leqslant \tau \leqslant t^2. \tag{5.19}$$

Now the adjoint of a translation e^Y is equal to $J_Y e^{-Y}$ where for Y in a suitable neighborhood of 0 the Jacobian J_Y has a uniform bound together with as many derivatives as we wish. In view of (5.18) it follows that the adjoint of $e^{\tau X_0}$ for $0 \leqslant \tau \leqslant t^2$ and of $e^{\tau X_I}$ for $0 \leqslant \tau \leqslant t^{m(I)}, I \in \mathfrak{J}$, are uniformly bounded with respect to the norms N_t, as is the adjoint of the operator $\Phi_{t^{1/\sigma}}$. Since $(e^{\tau X_0} S_t)^*$ for $0 \leqslant \tau \leqslant t^2$ is a superposition with finite total mass of operators which have uniformly bounded norm with respect to N_t, it follows from (5.19) that

$$N_t((e^{\tau X_0} S_t)^* (e^{\tau X_0} S_t u - S_t u)) \leqslant CM(u),$$

and this implies (5.16). Thus we have completed the proof of (5.6) and so we have proved Theorems 5.1 and 1.1.

References

[1]. HOCHSCHILD, G., *The structure of Lie groups.* Holden–Day Inc., San Francisco, London, Amsterdam, 1965.

[2]. HÖRMANDER, L., Pseudo-differential operators and hypoelliptic equations. To appear in *Amer. Math. Soc. Proc. Symp. Pure Math.*, 10 (1967).

[3]. ——, *Linear partial differential operators.* Springer-Verlag, Berlin–Göttingen–Heidelberg, 1963.

[4]. IL'IN, A. M., On a class of ultraparabolic equations. (Russian.) *Doklady Akad. Nauk SSSR*, 159 (1964), 1214–1217. Also in *Soviet Math. Dokl.*, 5 (1964), 1673–1676.

[5]. KOHN, J. J., Boundaries of complex manifolds. *Proc. Conf. Complex Analysis (Minneapolis 1964)*, 81–94. Springer Verlag, Berlin, 1965.

[6]. KOHN, J. J. & NIRENBERG, L., Non coercive boundary value problems. *Comm. Pure Appl. Math.*, 18 (1965), 443–492.

[7]. ——, Degenerate elliptic-parabolic equations of second order. To appear in *Comm. Pure Appl. Math.*

[8]. KOLMOGOROV, A. N., Zufällige Bewegungen. *Ann. of Math.* (2), 35 (1934), 116–117.

[9]. NIRENBERG, L. & TRÈVES, F., Solvability of a first order linear partial differential equation. *Comm. Pure Appl. Math.*, 16 (1963), 331–351.

[10]. OLEJNIK, O. A., Linear second order equations with non–negative characteristic form. (Russian.) *Mat. Sb.*, 69 (1966), 111–140.

[11]. WEBER, M., The fundamental solution of a degenerate partial differential equation of parabolic type. *Trans. Amer. Math. Soc.*, 71 (1951), 24–37.

Received June 12, 1967

Vita

Kiyosi Itô

Date and Place of Birth: September 7th, 1915, Mie Prefecture, Japan

1938	B.S. Tokyo Imperial University
1940-43	Statistical officer, Statistical Bureau of Government
1943-52	Assistant Professor, Nagoya Imperial University
1945	Doctor of Science, Tokyo Imperial University
1952-79	Professor, Kyoto University
1954-56	Fellow, Institute of Advanced Study, Princeton
1961-64	Professor, Stanford University
1967-69	Professor, Aarhus University
1969-75	Professor, Cornell University
1976-79	Director of Research Institute of Mathematical Sciences, Kyoto University
1979-81	President, Mathematical Society of Japan
1979-85	Professor, Gakushūin University
1979-present	Professor Emeritus, Kyoto University

Awards:

1978	Asahi Prize
1978	Japan Academy Prize and Imperial Prize
1985	Fujiwara Prize
1987	Wolf Prize
1998	Kyoto Prize

Honorary degrees:

1981	Docteur Honoris Causa, Université Paris VI
1987	Honorary Doctor of Mathematics, ETH, Zürich
1992	Degree of Doctor of Science honoris causa, The University of Warwick

Memberships:

1989	Membre Associé Étranger, Académie des Sciences, France
1991	Member, Japan Academy
1995	Honorary Member, Moscow Mathematical Society
1998	Foreign Associate, National Academy of Sciences, USA

Foreword

As my foreword to this book I would like to outline my research development and its background, which may be of some interest to the reader.

When I was a student in the Department of Mathematics at the University of Tokyo (1935–1938), I was fascinated by the rigorous arguments and the beautiful structures seen in pure mathematics, but also I was concerned with the fact that many mathematical concepts had their origins in mechanics. Fiddling around with mathematics and mechanics, I came close to stochastic processes through statistical mechanics. At that time when probability theory was not popular in Japan, I felt myself rather isolated. But I was able to continue my work thanks to the kind encouragement of Professor Shokichi Iyanaga in whose seminar I was participating.

Having read A. Kolmogorov's *Grundbegriffe der Wahrscheinlichkeitsrechnung* (1933) I became convinced that probability theory could be developed in terms of measure theory as rigorously as in other fields of mathematics. In P. Lévy's book *Théorie de l'addition des variables aléatoires* (1937) I saw a beautiful structure of the sample paths of stochastic processes deserving the name of mathematical theory. From this book I learned stochastic processes, Wiener's Brownian motion (Wiener process), Poisson process, and processes with independent increments (differential processes). I was particularly interested in the decomposition theorem for differential processes, the core of this book. But I had a hard time following Lévy's argument because of his unique intrinsic description. Fortunately I noticed that all ambiguous points could be clarified by means of J. L. Doob's idea of regular versions presented in his paper "Stochastic processes depending on a continuous parameter" [*Trans. Amer. Math. Soc.* **42,** 1938]. Checking Lévy's argument carefully from Doob's viewpoint, I was able to introduce Poisson random measures of jumps to really understand Lévy's spirit of the decomposition theorem. In accordance with Professor Iyanaga's suggestion I published this result [1] in the *Japanese Journal of Mathematics* in 1942.

As a natural extension of Lévy's book I read A. Kolmogorov's paper on a sample continuous Markov process (diffusion process), "Über die analystische Methoden in der Wahrscheinlichkeitsrechnung" [*Math. Ann.* **104,** 1931], and W. Feller's paper on a general Markov process with jumps, "Zur Theorie der stochastichen Prozesse (Existenz-und Eindeutigkeitssätze)" [*Math. Ann.* **113,** 1936]. In these papers I saw a powerful analytic method to study the transi-

Reprinted from Kiyosi Itô: Selected Papers (Springer, 1987), pp. xiii–xvii.

tion probabilities of the process, namely Kolmogorov's parabolic equation and its extension by Feller. But I wanted to study the paths of Markov processes in the same way as Lévy observed differential processes. Observing the intuitive background in which Kolmogorov derived his equation (explained in the Introduction of the paper), I noticed that a Markovian particle would perform a time homogeneous differential process for the infinitesimal future at every instant, and arrived at the notion of a stochastic differential equation governing the paths of a Markov process that could be formulated in terms of the differentials of a single differential process. It took some years to carry out this idea. When I finished this job for a sample continuous Markov process (Kolmogorov's case), I wrote a paper on the result in Japanese (English translation [2]) and intended to write it in English after having been able to carry out my idea for a general Markov process (Feller's case).

I published this Japanese paper in a mimeographed journal (1942) of Osaka University. This journal was being published to encourage young mathematicians all over Japan to communicate their ideas at that time when there was no xeroxing.

I do not know anyone who read this paper thoroughly when it appeared except my friend G. Maruyama. As I heard from him later, he read it in the military camp where he had been drafted for World War II. Later he wrote a paper on the subject (*Cir. Mat. Palermo,* 1955). At that time Maruyama and I were the only probabilists in Japan who were really interested in the sample paths.

I wrote these two papers [1], [2] when I was working in the Statistical Bureau of the Government (1939–1943), where I was given sufficient time for my own study thanks to the kindness of Mr. T. Kawashima, Head of the Bureau.

In 1943 I obtained a teaching position at Nagoya University and stayed there until I moved to Kyoto in 1952. At Nagoya I was very happy to work with Professor Kôsaku Yosida, though this period was the dark age of World War II and its aftermath.

At that time Yosida had just completed his famous semigroup theory and was developing it further to apply it to various fields of analysis, including the Kolmogorov equation for time homogeneous Markov processes. Through discussions with him I obtained much knowledge about functional analysis that turned out to be very useful later.

I was also lucky to become acquainted with Professor S. Kakutani who returned from Princeton to Osaka because of the war breaking out. I listened with much interest to his lectures on ergodic theory, positive-definite functions on non-abelian groups, and the relation between Brownian motions and harmonic functions. I enjoyed discussions with him, because he was also interested in the sample paths and showed interest in my work.

Although I had heard much of N. Wiener's great contribution to probability theory, I had not read his work carefully until I went to Nagoya. Even his theory of Brownian motion I learned from Lévy's book and Doob's papers. Reading some of his papers I was impressed by the originality with which he initiated not only measure-theoretic probability theory but also path-theoretic process theory as early as the 1920's.

Being motivated by Kakutani's lecture on ergodic theory, Maruyama's work on ergodic properties of stationary Gaussian processes, and discussions with H. Anzai, Kakutani's student, I slightly modified Wiener's homogeneous chaos to define multiple Wiener integral [4] (1951). Later I learned that Wiener had made the same modification in a different way. In 1953 I introduced the complex multiple Wiener integral and used it to give another proof of Maruyama's result [17].

When I wrote my manuscript on stochastic differential equations for general Markov processes several years after I had planned to in 1942, I learned that there was no journal in Japan to publish such a long note because of the shortage of papers under the bad economic conditions after the war. So I sent it to Professor J. L. Doob and asked him about any possibility of its being published in the United States. He made a kind arrangement for publishing it in the *Memoirs of the American Mathematical Society* in 1951 [12].

In the later half of my Nagoya period manifold theory was beginning to draw the attention of young people. Being stimulated by such an atmosphere, I became interested in a diffusion on a compact manifold whose generator is a non-degenerate elliptic operator. The analytic theory had been initiated by Kolmogorov, and Yosida was treating it from his semi-group viewpoint. I wanted to construct the path of the diffusion by writing a stochastic differential equation in terms of local coordinates and wrote three papers [10], [11], [15]. Although my work was not complete at all, I obtained a chain rule of stochastic differentials [13] as a byproduct and noticed that this rule is useful to understand the probabilistic meaning of the generator.

In 1952 I moved to Kyoto University, from which I retired in 1979. During those 27 years I was at Kyoto for half of the period in total, because I was abroad for the rest, at Princeton (1954–1956), Stanford (1961–1964), Aarhus, Denmark (1966–1969), and Cornell (1969–1975). I am very grateful to Professors Y. Akizuki, A. Kobori, A. Komatsu, and M. Hukuhara for their kindness and generosity, thanks to which I was able to enrich my knowledge in mathematics as well as in other respects through my experience in the United States and in Europe. In particular I learned various aspects of probability theory through discussions with W. Feller, H. P. McKean, R. Getoor, J. L. Doob, S. Karlin, K. L. Chung, H. Hoffmann-Jørgensen, F. Spitzer, H. Kesten, L. Gross, H. H. Kuo, D. W. Stroock, S. R. S. Varadhan, P. Malliavin, P. Meyer, J. Neveu, H. Föllmer, E. B. Dynkin, and many others.

In spite of my frequent absence from Kyoto I was lucky to have had a number of students who contributed much to the field of my interest and are still working actively and bringing up young probabilists. N. Ikeda, T. Hida, M. Nisio, S. Watanabe, M. Fukushima and H. Kunita are among them.

In 1952 when I moved to Kyoto, Doob's famous book *Stochastic Processes* appeared. I was impressed by his beautiful theory of martingales and its applications to various problems. I was also glad to see the stochastic integral discussed in the framework of martingales.

FOREWARD

My experience at Princeton (1954–1956) was most exciting because of the presence of Professor W. Feller and his student H. P. McKean, Jr., now a Professor at the Courant Institute. When I met Feller, I was impressed by his deep insight into the paths of stochastic processes in spite of all his papers being written from the analytic side. He suggested to me and to McKean as well the problem of constructing the paths of the most general one-dimensional diffusion whose analytic theory he had completed by that time. Since I did not really understand Feller's theory at that time, I learned it from McKean; I gave him the whole set of my reprints, all of which had been written from the path-theoretic viewpoint. He read my papers very quickly. Later he wrote a nice book on stochastic integrals (1964), compact but full of interesting materials, using the relation between stochastic integrals and martingales ingeniously. After lively and often exciting discussions we succeeded in constructing the Brownian path with an elastic barrier, the problem Feller proposed to us for the first step, by using P. Lévy's local time. Thus we obtained a clue for constructing the paths of Feller's diffusion. Thereafter our joint work went on rather smoothly, even though we sometimes had technical or more serious difficulties to overcome. When we got an almost complete picture of the one-dimensional diffusion processes, we decided to write a book, which was published as *Diffusion Processes and Their Sample Paths* by Springer-Verlag in 1965. For almost ten years McKean worked very hard to collect new materials and to organize them. Without his tremendous effort our joint work would have never appeared in book form.

In 1957–1958 McKean visited Kyoto and gave a series of lectures on diffusion processes. His lectures were so inspiring that they resulted in the emergence of quite a few active probabilists in Japan. We wrote two joint papers, one for random walks and potentials [20] and the other for the construction of Brownian motion on the half-line for every possible boundary condition [21] whose analytic theory had been established under some restrictions by Feller.

Since 1960 the general theory of Markov processes represented by E. B. Dynkin's work and the potential theoretic Markov process theory represented by G. Hunt's work have developed extensively and rapidly. Doob's martingale theory played an important role in this development. Also the theory of stochastic differential equations has been completely transformed by martingale integrals (initiated by Doob and completed by H. Kunita, S. Watanabe, and P. Meyer), the Stroock–Varadhan martingale problems, the stochastic differential geometry of J. Eells, K. D. Elworthy, and P. Malliavin, and Malliavin calculus. I wrote several papers related to this development, three of which I explain below.

When I was at Stanford (1961–1964), I had a plan to construct harmonic tensor fields on a Riemannian manifold by using Brownian motion. Though I was not able to solve this problem, I obtained the idea of stochastic parallel displacement as a byproduct [23] (1962). Since this idea drew no attention except that E. B. Dynkin introduced tensor diffusions, I gave up my plan. Later I was very glad to hear that J. Eells, K. D. Elworthy, and P. Malliavin took up my idea around 1970 when they established their theory of stochastic differential geometry, in which my problem was beautifully solved. Such a solution was far beyond my reach when I introduced stochastic parallel displacement.

When I heard of the Stratonovich integral from engineering mathematicians around 1970, I was too stubborn to accept it, because their reason to use it seemed to me to be just technical. But when I learned from D. W. Stroock his construction of Brownian motion on the sphere using the Stratonovich integral, I suddenly realized the mathematical meaning of this integral. I wrote a paper [37] (1975) to clarify the relation between my integral and the Stratonovich integral in terms of martingale integrals and to illustrate some mathematical applications.

In 1970, when I was at Cornell, I wrote a paper to determine all possible behaviors of a Markov process at a recurrent point [34]. This was a generalization of my joint paper with McKean [24], but the motivation was quite different. When I was at Aarhus (1966–1969), I became interested in infinite-dimensional stochastic differential equations to deal with stochastic dynamical systems of infinite degrees of freedom and tried for the first step to study basic facts and to check special examples. After several years it became my habit to observe even finite-dimensional facts from the infinite-dimensional viewpoint. This habit led me to reduce the problem above to a Poisson point process with values in the space of excursions.

When I became free from all official duties by retiring from Gakushuin University (1979–1985), I was given a chance to spend one year in the Institute for Mathematics and Its Applications at the University of Minnesota thanks to the kind consideration of Professors H. Weinberger, S. Orey, and D. W. Stroock. Since all workshops for this year are for stochastic differential equations and their applications, I am listening to the lectures with great interest and learning that there are many interesting new theories of which I have not had the slightest idea. It would be my greatest pleasure if I could study these new theories leisurely after I returned to Kyoto.

I would like to express my sincere thanks to Professors S. R. S. Varadhan and D. W. Stroock for their arrangement of publication of my selected papers and their extremely kind Introduction, and I am also grateful to the staff at Springer-Verlag for their enthusiastic cooperation.

K. Itô

Bibliography of Kiyosi Itô[2]

(A) Papers

[1] On stochastic processes (infinitely divisible laws of probability) (Doctoral thesis). *Japan. Journ. Math.* **XVIII**, 261-301 (1942).

[2] Differential equations determining a Markoff process (original Japanese: Zenkoku Sizyo Sugaku Danwakai-si). *Journ. Pan-Japan Math. Coll.* No.1077 (1942).

[3] On the ergodicity of a certain stationary processes. In: *Proc. Imp. Acad. Tokyo* **20**, 54-55 (1944).

[4] A kinematic theory of turbulence. In: *Proc. Imp. Acad. Tokyo* **20**, 120-122 (1944).

[5] On the normal stationary process with no hysteresis. In: *Proc. Imp. Acad. Tokyo* **20**, 199-202 (1944).

[6] A screw line in Hilbert space and its application to the probability theory. *Proc. Imp. Acad. Tokyo* **20**, 203-209 (1944).

[7] Stochastic integral. In: *Proc. Imp. Acad. Tokyo* **20**, 519-524 (1944).

[8] On Student's test. *Proc. Imp. Acad. Tokyo* **20**, 694-700 (1944).

[9] On a stochastic integral equation. In: *Proc. Imp. Acad. Tokyo* **22**, 32-35 (1946).

[10] Stochastic differential equations in a differentiable manifold. *Nagoya Math. Journ.* **1**, 35-47 (1950).

[11] Brownian motions in a Lie group. In: *Proc. Imp. Acad. Tokyo* **26**, 4-10 (1950).

[12] On stochastic differential equations. *Mem. Amer. Math. Soc.* **4**, 1-51 (1951),

[13] On a formula concerning stochastic differentials. *Nagoya Math. Journ.* **3**, 55-65 (1951).

[14] Multiple Wiener integral. *Journ. Math. Soc. Japan* **3**, 157-169 (1951).

[15] Stochastic differential equations in a differentiable manifold (2). *Mem. Coll. Science, Univ. Kyoto, Ser. A.*, **28**, 81-85 (1953).

[16] Stationary random distributions. *Mem. Coll. Science, Univ. Kyoto, Ser. A.*, **28**, 209-223 (1953).

[17] Complex multiple Wiener integral. *Japan Journ. Math.* **22**, 63-86 (1952).

[18] Isotropic random current. In: *Proc. Third Berkeley Symp. Math. Statist. Prob.* **II**, 125-132 (1955).

[19] Spectral type of the shift transformation of differential processes with stationary increments. *Trans. Amer. Math. Soc.* **81**, 253-263 (1956).

[20] Potentials and random walk (with H. P. McKean, Jr.). *Illinois Journ. Math.* **4**, 119-132 (1960).

[21] Wiener integral and Feynman integral. In: *Proc. Fourth Berkeley Symp. Math. Statist. Prob.* **II**, 227-238 (1960).

[2] Those marked with * are not contained in the list in Kiyosi Itô Selected Papers

Reprinted from Itô's Stochastic Calculus and Probability Theory, eds. Ikeda, Watanabe, Fukushima and Kunita (Springer, 1996), pp. xi–xiv.

[22] Construction of diffusions. *Ann. Fac. Sci. Univ. Clermont* **2**, 23-32 (1962).
[23] The Brownian motion and tensor fields on Riemannian manifold. In: *Proc. Intern. Congr. Mathemat.* (Stockholm), 536-539 (1962).
[24] Brownian motion on a half line (with H. P. McKean, Jr.). *Illinois Journ. Math.* **7**, 181-231 (1963).
[25] The expected number of zeros of continuous stationary Gaussian processes. *Journ. Math. Kyoto Univ.* **3**, 207-216 (1964).
[26] On stationary solutions of a stochastic differential equation (with M. Nisio). *Journ. Math. Kyoto Univ.* **4**, 1-75 (1964).
[27] Transformation of Markov processes by multiplicative functionals (with S. Watanabe). *Ann. Inst. Fourier, Univ. Grenoble* **XV**, 13-30 (1965).
[28] The canonical modification of stochastic processes. *Journ. Math. Soc. Japan* **20**, 130-150 (1968).
[29] On the convergence of sums of independent Banach space valued random variables. (with M. Nisio). *Osaka Journ. Math.* **5**, 35-48 (1968).
[30] Generalized uniform complex measures in the Hilbertian metric space with their application to the Feynman integral. In: *Proc. Fifth Berkeley Symp. Math. Statist. Prob.* **II**, 145-161 (1965).
[31] On the oscillation functions of Gaussian processes (with M. Nisio). *Math. Scand.* **22**, 209-223 (1968).
[32] Canonical measurable random functions. In: *Proc. Int. Conf. Funct. Anal. Rel. Topics* (Tokyo), 369-377 (1969).
[33] The topological support of a Gaussian measure on Hilbert space. *Nagoya Math. Journ.* **38**, 181-183 (1970).
[34] Poisson point processes attached to Markov processes. In: *Proc. Sixth Berkeley Symp. Math. Statist. Prob.* **III**, 225-239 (1970).
[35] Stochastic differentials of continuous local martingales. In: *Stability of Stochastic Dynamical Systems* (Lecture Notes in Mathematics **294**), Springer-Verlag, Berlin, 1-7 (1972).
[36] Stochastic integration. In: *Vector and Operator Valued Measures and Applications*, Academic Press, New York, 141-148 (1973).
[37] Stochastic differentials, *Appl. Math. and Opt.* **1**, 374-381 (1974).
[38] Stochastic parallel displacement. In: *Probabilistic Methods in Differential Equations* (Lecture Notes in Mathematics **451**), Springer-Verlag, Berlin, 1-7 (1975).
[39] Stochastic calculus. In: *Mathematical Problems in Physics* (Lecture Notes in Physics **39**), Springer-Verlag, 218-223 (1975).
[40] Extension of stochastic integrals. In: *Proc. Int. Symp. Stochastic Differential Equations* (Kyoto), 95-109 (1976)
[41] Introduction to stochastic differential equations (with S. Watanabe). In: *Proc. Int. Symp. Stochastic Differential Equations* (Kyoto), i-xxx (1976).
[42] Continuous additive $\mathcal{S}'$-processes.

[43] Stochastic Analysis in infinite dimensions. In: *Stochastic Analysis* (A. Friedman and M. Pinsky, eds.), Academic Press, New York, 187-197 (1978).

[44] Infinite dimensional Ornstein-Uhlenbeck processes. In: *Taniguchi Symp. SA*, Katata, 197-224 (1982).

[45] Regularization of linear random functionals (with M. Nawata). In *Probability Theory and Mathematical Statistics, Fourth USSR-Japan Symposium Proceedings, 1982* (Lecture Notes in Mathematics **1021**), Springer-Verlag, Berlin, 257-267 (1983).

[46] Distribution-valued processes arising from independent Brownian motions. *Math. Zeit.* **182**, 17-33 (1983).

[47] A stochastic differential equation in infinite dimensions. In: *Contemporary Math.* **26**, 163-169 (1984).

[48]* Malliavin's $\mathcal{C}^\infty$-functionals of a centered Gaussian system. IMA preprint Series, Univ. Minnesota, No **327**(1987).

[49]* Malliavin calculus on a Segal space, in *Stochastic Analysis, Proc. of Japanese-French Seminar, Paris, 1987*, (eds. M. Métivier and S. Watanabe), Lecture Notes in Mathematics **1322**, Springer-Verlag, Berlin, 50-72 (1988).

[50]* Positive generalized functions on $(\mathbf{R}^\infty, \mathcal{B}^\infty, N^\infty)$, in *White Noise Analysis, Mathematics and Applications*, (eds. T. Hida, H.-H. Kuo, J. Potthoff and L. Streit), World Scientific, 166-179 (1990).

[51]* On Malliavin calculus, in *Proceedings of 1989 Singapore Probability Conference*, (eds. L. H. Y. Chen, K. P. Choi, K. Hu and J.-H. Lou), Walter de Gruyter, 47-72 (1992).

[52]* An elementary approach to Malliavin fields, in *Asymptotic problems in probability theory: Wiener functionals and asymptotics, Proceedings of Taniguchi Symp. Sanda and Kyoto, 1990*, (eds. K. D. Elworthy and N. Ikeda), Pitman Research Notes in Math. Series **284**, Longman, 35-89 (1993).

[53]* Semigroups in probability theory, in *Functional analysis and related topics, Proceedings of the International Conference in Memory of Professor Kôsaku Yosida, RIMS, Kyoto Univ. 1991*, (ed. H. Komatsu), Lecture Notes in Mathematics **1540**, Springer-Verlag, Berlin, 69-83 (1993).

[54]* On Malliavin tensor fields. *Communs. Pure and Appl. Math.* **XLVII**, 377-403 (1994). (The third of five special issues dedicated to Henry McKean.)

[55]* A measure-theoretic approach to Malliavin calculus, in *'New Trends in Stochastic Analysis', Proc. Taniguchi Symposium, Sept. 1994, Charingworth, (eds. K. D. Elworthy, S. Kusuoka and I. Shigekawa)*, World Scientific, 220–287 (1997).

(B) Books

[a] *Foundation of Probability Theory* (in Japanese), Iwanami, Tokyo, 1943.

[b] *Probability Theory* (in Japanese), Iwanami, Tokyo, 1952.

[c]* *Stochastic Processes* I, II (in Jpapanese), Iwanami Series of Modern Applied Mathematics A. 13, I, II, Iwanami, Tokyo, 1957.
[d] *Diffusion Processes and Their Sample Paths* (with H. P. McKean, Jr.), Springer, 1965; reprint of the 1974 Edition in the Springer Series of *Classics in Mathematics*, 1996.
[e] *Probability Theory* (in Japanese), Iwanami Series of Fundamental Mathematics, Analysis (I) vii, Iwanami, Tokyo, 1978, revised 1983.
[f] *Introduction to Probability Theory* (English translation of [e] Chapters I-IV), Cambridge Univ. Press, 1984.
[g] *Foundations of Stochastic Differential Equations in Infinite Dimensional Spaces*, (CBMS-NSF Reg. Conf. Ser. in Appl. Math. 47) SIAM, 1984.

(C) Lecture Notes

[i]* Lectures on Stochastic Processes, Tata Institute for Fundamental Research, Bombay 1961.
[ii]* Stochastic Processes, Lecture Notes Series, No. **16**, Aarhus University, 1969.

Kiyosi Itô

Born: September 7, 1915 in Mie Prefecture, Japan
Currently Professor Emeritus at Kyoto University
Doctor of Science, the Imperial University of Tokyo, 1945

Awards

Asahi Prize, 1978
Japan Academy Prize and Imperial Prize, 1978
Fujiwara Prize, 1985
Wolf Foundation Prize, 1987

Memberships

Membre Associé Étranger, Académie des Science, France, 1989
Member of Japan Academy, 1991

Honorary degrees

Docteur Honoraris Causa, Université Paris VI, 1981
Honorable Doctor of Mathematics, ETH, Zürich, 1987
Degree of Doctor of Science Honoraris Causa, The University of Warwick, 1992

ON A FORMULA CONCERNING STOCHASTIC DIFFERENTIALS

KIYOSI ITÔ

In his previous paper [1][1] the author has stated a formula[2] concering stochastic differentials with the outline of the proof. The aim of this paper is to show this formula in details in a little more general form (Theorem 6).

1. Definitions. Throughout this paper we assume that all stochastic processes[3] $\xi(t,\omega)$, $\eta(t,\omega)$, $a(t,\omega)$, $b(t,\omega)$, etc. are measurable in variables t and ω. A system of r one-dimensional Brownian motions independent of each other is called an r-dimensional Brownian motion.

Given two system of stochastic processes:

$$(1.1)\qquad \xi=\{\xi_\lambda(t,\omega),\ \lambda\in\Lambda\},\qquad \eta=\{\eta_\mu(t,\omega),\ \mu\in M\}.$$

We say that ξ has the property α with regard to η in $u\leqq t\leqq v$, if, for any t, the following two systems of random variables are independent of one another:

$$(1.2)\qquad \begin{cases}\varphi_t=\{\xi_\lambda(\tau,\omega),\ \lambda\in\Lambda,\ \eta_\mu(\tau,\omega),\ \mu\in M,\ u\leqq\tau\leqq t\}\\ \psi_t=\{\eta_\mu(\sigma,\omega)-\eta_\mu(t,\omega),\ \mu\in M,\ t\leqq\sigma\leqq v\}.\end{cases}$$

Now we shall state an outline[4] of a stochastic integral of the form:

$$(1.3)\qquad \int_s^t\xi(\tau,\omega)d\beta(\tau,\omega),\qquad u\leqq s\leqq t\leqq v,\qquad \omega\in\Omega_1,$$

where $\beta(t,\omega)$ is a one-dimensional Brownian motion and Ω_1 is a measurable subset of Ω. We shall set the two conditions on ξ;

(C.1) $\quad$ *$\xi(t,\omega)$ has the property α concerning $\beta(t,\omega)$ in $u\leqq t\leqq v$,*

(C.2) $\quad \int_u^v\xi(\tau,\omega)^2d\tau$ *for almost all* $\omega\in\Omega_1$.

Received April 16, 1951.

[1] The number in [ ] refers to the Reference at the end of this paper.

[2] Theorem 1.1 in [1].

[3] In the analytical theory of probability any stochastic process is expressed as a function of the time parameter t and the probability parameter ω which runs over a probability space $\Omega(P)$, P being the probability distribution.

[4] Cf. [2] concerning the details.

Reprinted from Nagoya Math. J., Vol. 3 (1951), pp. 55–65.

Case 1. When ξ is uniformly stepwise, that is when there exists a system of time-points:

$$u=s_0<s_1<\dots<s_n=v \tag{1.4}$$

such that

$$\xi(\tau,\omega)=\xi(s_{i-1},\omega),\quad s_{i-1}\leqq\tau<s_i,\quad i=1,2,\dots,n,$$

we define

$$\int_s^t\xi(\tau,\omega)d\beta(\tau,\omega)=\sum_{i=k+1}^{l-1}\xi(s_{i-1},\omega)(\beta(s_i,\omega)-\beta(s_{i-1},\omega)) \tag{1.5}$$
$$+\xi(s_{k-1},\omega)(\beta(s_k,\omega)-\beta(s,\omega))+\xi(s_{l-1},\omega)(\beta(t,\omega)-\beta(s_{l-1},\omega))$$
$$(s_{k-1}\leqq s<s_k,\ s_{l-1}\leqq t<s_l).$$

Case 2. When

$$\int_\Omega\int_u^v\xi(t,\omega)^2dtP(d\omega)<\infty, \tag{1.6}$$

there exists a sequence of uniformly stepwise processes $\xi_n(t,\omega)$, $n=1,2,\dots$ whose value at any time-point t is a B-measurable function of $\xi(\tau,\omega)$, $u\leqq\tau\leqq t$, such that

$$\int_\Omega\int_u^v(\xi_n(t,\omega)-\xi(t,\omega))^2dtP(dw)<8^{-n}. \tag{1.7}$$

We define

$$\int_s^t\xi(\tau,\omega)d\beta(\tau,\omega)=\lim_n\int_s^t\xi_n(\tau,\omega)d\beta(\tau,\omega). \tag{1.8}$$

As was proved in our previous paper [2], the sequence:

$$\int_s^t\xi_n(\tau,\omega)d\beta(\tau,\omega),\quad n=1,2,\dots,$$

is uniformly convergent *in* $u\leqq s\leqq t\leqq v$ for almost all ω, and the definition is independent of the special choice of the sequence $\{\xi_n(t,\omega)\}$.

Case 3. Now we shall consider the general case. Put

$$\phi_n(\lambda)=\begin{cases}1 & (|\lambda|\leqq n)\\ 0 & (|\lambda|>n)\end{cases} \tag{1.9}$$

and

$$\xi_n(t,\omega)=\phi_n\left(\int_u^t\xi(\tau,\omega)^2dt\right)\xi(t,\omega),\quad n=1,2,\dots.$$

Then $\xi_n(t,\omega)$ satisfies the conditions (1.6) and (C.1), and so their stochastic integrals are well defined. Since $\xi_n(t,\omega)=\xi(t,\omega)$, $u\leqq t\leqq v$, for a sufficiently large n for almost all $\omega\in\Omega_1$ by (C.2), we define naturally

$$\int_s^t \xi(\tau, \omega)d\beta(\tau, \omega) = \lim_n \int_s^t \xi_n(\tau, \omega)d\beta(\tau, \omega).$$

2. Preliminary properties of stochastic integrals. In the following Theorems 1, 3, and 4 we shall assume that $\xi(t, \omega)$ satisfies the conditions (C. 1) and (C. 2).

THEOREM 1. *The stochastic integral (1. 3) is continuous in s, t for almost all* $\omega \in \Omega_1$.

THEOREM 2. *If each of* $\xi(t, \omega)$ *and* $\eta(t, \omega)$ *satisfies (C. 1) and (C. 2), and if the system* $\{\xi(t, \omega), \eta(t, \omega)\}$ *has the property* α *with regard to the Brownian motion* β, *then* $\zeta(t, \omega) \equiv a\xi(t, \omega) + b\eta(t, \omega)$ *(a, b being constants) satisfies (C. 1) and (C. 2) and we obtain*

$$\int_s^t (a\xi(\tau, \omega) + b\eta(\tau, \omega))d\beta(\tau, \omega) = a\int_s^t \xi(\tau, \omega)d\beta(\tau, \omega) + b\int_s^t \eta(\tau, \omega)d\beta(\tau, \omega) \tag{2.1}$$

for $u \leqq t \leqq s \leqq v$ *for almost all* $\omega \in \Omega_1$.

THEOREM 3. *We have*

$$\int_{s_1}^{s_2} \xi(\tau, \omega)d\beta(\tau, \omega) + \int_{s_2}^{s_3} \xi(\tau, \omega)d\beta(\tau, \omega) = \int_{s_1}^{s_3} \xi(\tau, \omega)d\beta(\tau. \omega) \tag{2.2}$$

for $u \leqq s_1 \leqq s_2 \leqq s_3 \leqq v$ *for almost all* $\omega \in \Omega_1$.

THEOREM 4. *If (1. 6) is satisfied, then we have*

$$\begin{aligned} \frac{c^2}{2} Pr\Big\{ \sup_{u \leqq s \leqq t \leqq v} \Big| \int_s^t \xi(\tau, \omega)d\beta(\tau, \omega) \Big| \geqq 2c \Big\} \\ \leqq c^2 Pr\Big\{ \sup_{u \leqq t \leqq v} \Big| \int_u^t \xi(\tau, \omega)d\beta(\tau, \omega) \Big| \geqq c \Big\} \\ \leqq \int_\Omega \Big(\int_u^v \xi(t, \omega)d\beta(t, w) \Big)^2 P(d\omega) = \int_\Omega \int_u^v \xi(t, \omega)^2 dt P(d\omega). \end{aligned} \tag{2.4}$$

THEOREM 5. *If each of* $\xi_n(t, \omega)$, $n = 1, 2, \ldots,$ *satisfies (C. 1) and (C. 2) and if the system* $\{\xi_n(t, \omega), n = 1, 2, \ldots, \infty\}$ *has the property* α *with regard to* β *in* $u \leqq t \leqq v$, *and further if*

$$\int_u^v (\xi_n(t, \omega) - \xi_\infty(t, \omega))^2 dt \to 0 \tag{2.5}$$

for almost all $\omega \in \Omega_1$, *then*

$$\sup_{u \leqq s < t \leqq v} \Big| \int_s^t \xi_n(\tau, \omega)d\beta(\tau, \omega) - \int_s^t \xi_\infty(\tau, \omega)d\beta(\tau, \omega) \Big| \tag{2.6}$$

tends to 0 in probability over Ω_1.

Since Theorems 1, 2, 3 and 4 follow at once from the properties of the stochastic integral established in [2], we shall here prove Theorem 5 only.

Since we have

$$\int_u^v \xi_n(t,\omega)^2 dt \to \int_u^v \xi_\infty(t,\omega)^2 dt$$

for almost all $\omega \in \Omega_1$ by the assumption (2.5) and since there exists $M=M(\varepsilon)$ for any $\varepsilon>0$ such that

$$Pr\left\{\omega\in\Omega_1, \int_u^v \xi_\infty(t,\omega)^2 dt < M\right\} > P(\Omega_1)-\varepsilon$$

by the assumption that $\xi_\infty(t,\omega)$ satisfies (C. 2), there exists $N_1=N_1(\varepsilon)$ for any $\varepsilon>0$

$$Pr\left\{\omega\in\Omega_1, \int_u^v \xi_n(t,\omega)^2 dt < M,\ N_1<n\leqq\infty\right\} > P(\Omega_1)-2\varepsilon. \tag{2.7}$$

Put

$$\xi_n^M(t,\omega)=\phi_M\left(\sup_{n\leqq k\leqq\infty}\int_u^t \xi_k(\tau,\omega)^2 dt\right)\xi_n(t,\omega),\quad n=1, 2, \ldots, \infty,$$

where ϕ_M is defined by (1.9). Then it follows from (2.7) that

$$Pr\{\omega\in\Omega_1,\ \xi_n^M(t,\omega)=\xi_n(t,\omega),\ u\leqq t\leqq v,\ N_1<n\leqq\infty\} > P(\Omega_1)-2\varepsilon. \tag{2.8}$$

Since we have

$$\int_u^v (\xi_n^M(t,\omega)-\xi_\infty^M(t,\omega))^2 dt \leqq \int_u^v \xi_n(t,\omega)-\xi_\infty(t,\omega))^2 dt \to 0,$$

$$\int_u^v (\xi_n^M(t,\omega)-\xi_\infty^M(t,\omega))^2 dt \leqq 2\int_u^v \xi_n^M(t,\omega)^2 dt + 2\int_u^v \xi_\infty^M(t,\omega)^2 dt < 4M,$$

we obtain

$$\int_\Omega \int_u^v (\xi_n^M(t,\omega)-\xi_\infty^M(t,\omega))^2 dt P(d\omega) \to 0.$$

By Theorem 4 there exists $N_2=N_2(\varepsilon)$ for any $\varepsilon>0$ such that

$$Pr\left\{\sup_{u\leqq s\leqq t\leqq v}\left|\int_s^t \xi_n^M(\tau,\omega)d\beta(\tau,\omega)-\int_s^t \xi_\infty^M(\tau,\omega)d\beta(\tau,\omega)\right| < \varepsilon\right\} > P(\Omega_1)-3\varepsilon, \tag{2.9}$$

which, combined with (2.8), proves our theorem.

3. A formula concerning stochastic differentials. Let $\beta=(\beta^i(t,\omega), i=1, 2, \ldots, r)$ be an r-dimensional Brownian motion, and let the system:

$$\{\xi^i(t,\omega),\ a^i(t,\omega),\ b_j^i(t,\omega),\ i=1, 2, \ldots, n,\ j=1, 2, \ldots, r\} \tag{3.1}$$

have the property α with regard to β in $u\leqq t\leqq v$. When we have

$$\xi^i(s,\omega)-\xi^i(t,\omega)=\int_t^s a^i(\tau,\omega)d\tau+\int_t^s b_j^i(\tau,\omega)d\beta^j(\tau,\omega),^{5)}\quad u\leqq t\leqq s\leqq v,\ 1\leqq i\leqq n, \tag{3.2}$$

5) We omit the summation sign $\sum_{j=1}^r$ according to the usual rule of tensor calculus.

for almost all $\omega\in\Omega_1$, we write this relation in the differential form as follows:

$$d\xi^i(t, \omega)=a^i(t, \omega)dt+b_j^i(t, \omega)d\beta^j(t, \omega), \quad u\leqq t\leqq v, \quad \omega\in\Omega_1, \quad 1\leqq i\leqq n. \tag{3.3}$$

THEOREM 6. *Let $\xi^i(t, \omega)$, $i=1, 2, \ldots, n$, satisfy*

$$d\xi^i(t, \omega)=a^i(t, \omega)dt+b_j^i(t, \omega)d\beta^j(t, \omega), \qquad i=1, 2, \ldots, n, \tag{3.4}$$

and G be an open subset of the n-space R^n which contains all the points $(\xi^i(t, \omega)$, $i=1, 2, \ldots, n)$ for $u\leqq t\leqq v$, $\omega\in\Omega_1$.

Let $f(t, x^1, x^2, \ldots, x^n)$ be a continuous function defined in $u\leqq t\leqq v$, $(x^1, x^2, \ldots, x^n)\in G$, such that

$$\begin{cases} f_0(t, x^1, \ldots, x^n)=\dfrac{\partial f}{\partial t}(t, x^1, \ldots, x^n) \\ f_i(t, x^1, \ldots, x^n)=\dfrac{\partial f}{\partial x_i}(t, x^1, \ldots, x^n), \quad i=1, 2, \ldots, n, \\ f_{ij}(t, x^1, \ldots, x^n)=\dfrac{\partial^2 f}{\partial x^i\partial x^j}(t, x^1, \ldots, x^n), \quad i, j=1, 2, \ldots, n, \end{cases} \tag{3.5}$$

are all continous.

Then $\eta(t, \omega)\equiv f(t, \xi^1(t, \omega), \ldots, \xi^n(t, \omega))$ satisfies

$$\begin{aligned} d\eta(t, \omega)=(f_0(t, \xi)+f_i(t, \xi)a^i(t, \omega)+\frac{1}{2}f_{ij}(t, \xi)b_k^i(t, \omega)b_k^j(t, \omega))dt \\ +f_i(t, \xi)b_j^i(t, \omega)d\beta^j(t, \omega), \end{aligned} \tag{3.6}$$

where $\xi\equiv(\xi^1(t, \omega), \xi^2(t, \omega), \ldots, \xi^n(t, \omega))$.

LEMMA 1. *For any stochastic process $\xi(t, \omega)$ satisfying*

$$\int_u^v \xi(t, \omega)^2 dt<\infty, \qquad \omega\in\Omega_1, \tag{3.7}$$

there exists a sequence of uniformly stepwise stochastic processes $\xi_n(t, \omega)$, $n=1, 2, \ldots$ whose value at any time-point t is a B-measurable function of $\xi(\tau, \omega)$, $\tau\leqq t$, and such that

$$\int_u^v |\xi_n(t, \omega)-\xi(t, \omega)|^2 dt \to 0$$

for almost all $\omega\in\Omega_1$.

This Lemma follows immediately from Lemma 7.1 in [2].

LEMMA 2. *Let $\xi(t, \omega)$, $\eta(t, \omega)$ be stochastic processes such that the system $\{\xi(t, \omega), \eta(t, \omega)\}$ has the property α with regard to a one-dimensional Brownian motion $\beta(t, \omega)$ in $u\leqq t\leqq v$ and that*

$$\int_u^v \xi(t, \omega)^2 dt<\infty, \qquad \int_u^v \eta(t, \omega)^2 dt<\infty$$

for almost all $\omega\in\Omega_1$. Then we have

$$(3.8)\quad \int_u^v \xi(t,\omega)d\beta(t,\omega)\int_u^v \eta(s,\omega)d\beta(s,\omega)$$
$$=\int_u^v \xi(t,\omega)\int_u^t \eta(s,\omega)d\beta(s,\omega)d\beta(t,\omega)$$
$$+\int_u^v \eta(s,\omega)\int_u^s \xi(t,\omega)d\beta(t,\omega)d\beta(s,\omega)+\int_u^v \xi(t,\omega)\eta(t,\omega)dt$$

for almost all.

Proof. Firstly we shall prove (3.8) in the case that both $\xi(t,\omega)$ and $\eta(t,\omega)$ are uniformly stepwise. Then we may assume that

$$(3.9)\quad \xi(t,\omega)=\xi(u_{i-1},\omega),\quad \eta(t,\omega)=\eta(u_{i-1},\omega),\quad u_{i-1}\leqq t\leqq u_i,\ i=1,2,\ldots,n,$$

where

$$(3.10)\quad u=u_0<u_1<\ldots<u_n=v.$$

The left side of (3.8) equals the following: $(u_{ij}=u_i+\frac{j}{N}(u_{i+1}-u_i))$

$$\sum_{ijpq}\int_{u_{i,j-1}}^{u_{ij}} \xi(t,\omega)d\beta(t,\omega)\int_{u_{p,q-1}}^{u_{pq}} \eta(s,\omega)d\beta(s,\omega)$$
$$=\sum_{u_{ij}>u_{pq}}+\sum_{u_{pq}>u_{ij}}+\sum_{p=i,q=j}=\int_u^v \xi(t,\omega)\int_u^{\lambda_n(t)} \eta(s,\omega)d\beta(s,\omega)d\beta(t,\omega)$$
$$+\int_u^v \eta(s,\omega)\int_u^{\lambda_n(s)} \xi(t,\omega)d\beta(t,\omega)d\beta(s,\omega)$$
$$+\sum_i \xi(u_{i-1},\omega)\eta(u_{i-1},\omega)\sum_{j=1}^N(\beta(u_{ij},\omega)-\beta(u_{i,j-1},\omega))^2=I_1+I_2+I_3,$$

where $\lambda_n(t)$ is the maximum u_{ij} which does not exceed t.

As $N\to\infty$, I_1 tends to the first term of the right side of (3.8) in probability by virtue of Theorem 5, since we have

$$\int_u^v |\xi(t,\omega)\int_u^{\lambda_n(t)} \eta(s,\omega)d\beta(s,\omega)-\xi(t,\omega)\int_u^t \eta(s,\omega)d\beta(s,\omega)|^2dt$$
$$=\int_u^v |\xi(t,\omega)\int_{\lambda_n(t)}^t \eta(s,\omega)d\beta(s,\omega)|^2dt$$
$$=\int_u^v |\xi(t,\omega)\eta(\lambda_n(t),\omega)(\beta(t,\omega)-\beta(\lambda_n(t),\omega))|^2dt\to 0$$

for almost all ω. By the same reason we see that I_2 tends to the second term of the right side of (3.8) in probability.

Since $\sum_{j=1}^N(\beta(u_{ij},\omega)-\beta(u_{i,j-1},\omega))^2\to u_i-u_{i-1}$ (in probability), I_3 tends to the third term of the right side of (3.8).

Next we shall consider the general case. By Lemma 1 we shall construct $\xi_n(t,\omega)$, $n=1,2,\ldots$, and $\eta_n(t,\omega)$, $n=1,2,\ldots$ for $\xi(t,\omega)$ and $\eta(t,\omega)$ respectively. Since our Lemma 2 holds for uniformly stepwise processes as is proved above, we have

$$(3.11)\quad \int_u^v \xi_n(t, \omega)d\beta(t, \omega)\int_u^v \eta_n(t, \omega)d\beta(s, \omega)$$
$$= \int_u^v \xi_n(t, \omega)\int_u^t \eta_n(s, \omega)d\beta(s, \omega)d\beta(t, \omega)$$
$$+ \int_u^v \eta_n(s, \omega)\int_u^s \xi_n(t, \omega)d\beta(t, \omega)d\beta(s, \omega) + \int_u^v \xi_n(t, \omega)\eta_n(t, \omega)dt.$$

Put

$$\zeta_n(t, \omega) = \int_u^t \eta_n(s, \omega)d\beta(s, \omega), \qquad \zeta(t, w) = \int_u^t \eta(s, \omega)d\beta(s, \omega),$$
$$\rho_n(t, \omega) = \int_u^t \xi_n(s, \omega)d\beta(s, \omega), \qquad \rho(t, \omega) = \int_u^t \eta(s, \omega)d\beta(s, \omega).$$

By taking adequate subsequences we see, by Theorem 5, that $\zeta_n(t, \omega)$ and $\rho_n(t, \omega)$ converge uniformly in t to $\zeta(t, \omega)$ and $\rho(t, \omega)$ respectively for almost all $\omega \in \Omega_1$. Therefore we have

$$\int_u^v |\xi_n(t, \omega)\zeta_n(t, \omega) - \xi(t, \omega)\zeta(t, \omega)|^2 dt$$
$$\leqq 2\int_u^v |\xi_n(t, \omega) - \xi(t, \omega)|^2 |\zeta_n(t, \omega)|^2 dt$$
$$+ 2\int_u^v |\xi(t, \omega)|^2 |\zeta_n(t, \omega) - \zeta(t, \omega)|^2 dt \to 0$$

for almost all $\omega \in \Omega_1$, from which follows by Theorem 5

$$\int_u^v \xi_n(t, \omega)\zeta_n(t, \omega)d\beta(t, \omega) \to \int_u^v \xi(t, \omega)\zeta(t, \omega)d\beta(t, \omega) \quad \text{(in probability)}.$$

Similaly we have

$$\int_u^v \eta_n(s, \omega)\rho_n(s, \omega)d\beta(s, \omega) \to \int_u^v \eta(s, \omega)\rho(s, \omega)d\beta(s, \omega) \quad \text{(in probability)}.$$

Further we have

$$\left|\int_u^v \xi_n(t, \omega)\eta_n(t, \omega)dt - \int_u^v \xi(t, \omega)\eta(t, \omega)dt\right.$$
$$\leqq \sqrt{\int_u^v \eta_n(t, \omega)^2 dt \int_u^v (\xi_n(t, \omega) - \xi(t, \omega))^2 dt}$$
$$+ \sqrt{\int_u^v \xi(t, \omega)dt \int_u^v (\eta_n(t, \omega) - \eta(t, \omega))^2 dt} \to 0.$$

Thus our Lemma 2 is completely proved.

By the same way as above, we obtain the following Lemmas 3 and 4.

LEMMA 3. *Let* $\xi(t, \omega)$ *and* $\eta(t, \omega)$ *be stochastic processes such that the system* $\{\xi(t, \omega), \eta(t, \omega)\}$ *has the property* α *with regard to the two-dimensional Brownian motion* $(\beta(t, \omega), \gamma(t, \omega))$ *and that*

$$\int_u^v \xi(t, \omega)^2 dt < \infty, \qquad \int_u^v \eta(t, \omega)^2 dt < \infty$$

for almost all $\omega \in \Omega_1$.

Then we have

$$\int_u^v \xi(t, \omega) d\beta(t, \omega) \int_u^t \eta(s, \omega) d\gamma(s, \omega) \tag{3.12}$$

$$= \int_u^v \xi(t, \omega) \int_u^t \eta(s, \omega) d\gamma(s, \omega) d\beta(t, \omega)$$

$$+ \int_u^v \eta(s, \omega) \int_u^s \xi(t, \omega) d\beta(t, \omega) d\gamma(s, \omega)$$

for almost all $\omega \in \Omega_1$.

LEMMA 4. *Let* $a(t, \omega)$ *and* $b(t, \omega)$ *be stochastic processes such that the system* $a(t, \omega)$, $b(t, \omega)$ *has the property* α *with regard to a one-dimensional Brownian motion* $\beta(t, \omega)$ *and that*

$$\int_u^v |a(t, \omega)| dt < \infty, \qquad \int_u^v |b(t, \omega)|^2 dt < \infty$$

for almost all $\omega \in \Omega_1$. *Then we have*

$$\int_u^v a(t, \omega) dt \int_u^v b(s, \omega) d\beta(s, \omega) \tag{3.13}$$

$$= \int_u^v a(t, \omega) \int_u^t b(s, \omega) d\beta(s, \omega) dt + \int_u^v b(s, \omega) \int_u^s a(t, \omega) dt \, d\beta(s, \omega).$$

LEMMA 5. *Let* $\xi^i(t, \omega)$, $i = 1, 2, \ldots, n$, *be determined as in Theorem 6. Then we have*

$$(\xi^i(s, \omega) - \xi^i(t, \omega))(\xi^j(s, \omega) - \xi^j(t, \omega)) \tag{3.14}$$

$$= \int_t^s \{(\xi^i(\tau, \omega) - \xi^i(t, \omega)) a^j(\tau, \omega) + (\xi^j(\tau, \omega) - \xi^j(t, \omega)) a^i(\tau, \omega) + b_k^i(\tau, \omega) b_k^j(\tau, \omega)\} d\tau$$

$$+ \int_t^s \{(\xi^i(\tau, \omega) - \xi^i(t, \omega)) b_k^j(\tau, \omega) + (\xi^j(\tau, \omega) - \xi^j(t, \omega)) b_k^i(\tau, \omega)\} d\beta^k(\tau, \omega)$$

for almost all $\omega \in \Omega_1$ *and for* $u \leqq t \leqq s \leqq v$.

Proof. By the assumption we have

$$(\xi^i(s, \omega)-\xi^i(t, \omega))(\xi^j(s, \omega)-\xi^j(t, \omega))$$
$$=\int_t^s a^i(\tau, \omega)d\tau\int_t^s a^j(\sigma, \omega)d\sigma+\int_t^s b_k{}^i(\tau, \omega)d\beta^k(\tau, \omega)\int_t^s a^j(\sigma, \omega)d\sigma$$
$$+\int_t^s a(\tau, \omega)d\tau\int_t^s b_l{}^i(\sigma, \omega)d\beta^l(\sigma, \omega)$$
$$+\int_t^s b_k{}^i(\tau, \omega)d\beta(\tau, \omega)\int_t^s b_l{}^i(\sigma, \omega)d\beta^l(\sigma, \omega),$$

from which follows (3.14) at once by virtue of Lemmas 2, 3 and 4.

LEMMA 6. *Let* $\xi^i(t, \omega)$, $i=1, 2, \ldots, n$, *be determined as in Theorem 6. For the point-system:*

$$\Delta : t=t_0<t_1<\ldots<t_m=s$$

we put

$$S(\Delta, \omega)=\sum_{\mu=1}^{m}|\xi^i(t_\mu, \omega)-\xi^i(t_{\mu-1}, \omega)|\,|\xi^j(t_\mu, \omega)-\xi^j(t_{\mu-1}, \omega)|.$$

Then there exists $M=M(\varepsilon)$ *independent of* Δ *for any* $\varepsilon>0$ *such that*

$$Pr\{\omega\in\Omega_1,\ S(\Delta, \omega)>M\}<\varepsilon. \tag{3.15}$$

Proof. We may consider the case that

$$\int_\Omega\int_t^s b_k^p(\tau, \omega)dtP(d\omega)<\infty, \qquad p=i, j, \qquad k=1, 2, \ldots, n,$$

since, if our Lemma is established in this case, we can easily deduce our Lemma in the general case by the definition of stochastic integral. When there is no confusion, we omit the time parameter and the probability parameter ω in the following.

$$(\xi^i(t_\mu)-\xi^i(t_{\mu-1}))(\xi^j(t_\mu)-\xi^j(t_{\mu-1}))$$
$$=\int_{t_{\mu-1}}^{t_\mu} a^i d\tau\int_{t_{\mu-1}}^{t_\mu} a^j d\sigma+\int_{t_{\mu-1}}^{t_\mu} b_k{}^i d\beta^k\int_{t_{\mu-1}}^{t_\mu} a^j d\sigma$$
$$+\int_{t_{\mu-1}}^{t_\mu} a^i d\tau\int_{t_{\mu-1}}^{t_\mu} b_k{}^i d\beta^k+\int_{t_{\mu-1}}^{t_\mu} b_k{}^i d\beta^k\int_{t_{\mu-1}}^{t_\mu} b_k{}^j d\beta^k.$$

Since

$$S_1(\Delta, \omega)\equiv\sum_\mu\left|\int_{t_{\mu-1}}^{t_\mu} a^i d\tau\right|\left|\int_{t_{\mu-1}}^{t_\mu} a^j d\sigma\right|\leqq\max_{t\leqq t'\leqq s'\leqq s}\left|\int_{t'}^{s'} a^i d\tau\right|\int_t^s|a^j|\,d\sigma$$

for almost all $\omega\in\Omega_1$ we may find $M_1(\varepsilon)$ independent of Δ for any $\varepsilon<0$ such that $M>M_1(\varepsilon)$ implies

$$Pr\{\omega\in\Omega_1,\ S_1(\Delta, \omega)>M\}<\varepsilon/4. \tag{3.16}$$

By the same way we may find $M_{2k}(\varepsilon)$ and $M_{3k}(\varepsilon)$ independent of Δ for any

$\varepsilon<0$ such that $M>M_{2k}(\varepsilon)$ or $M>M_{3k}(\varepsilon)$ implies

(3.17) $$Pr\{\omega\in\Omega_1,\ S_{2k}(\Delta,\ \omega)>M\}<\varepsilon/4r \quad \text{or} \quad Pr\{\omega\in\Omega_2,\ S_{3k}(\Delta,\ \omega)>M\}<\varepsilon/4r$$

respectively, where

$$S_{2k}(\Delta,\ \omega)=\sum_{\mu}\left|\int_{t_{\mu-1}}^{t_\mu} b_k{}^i d\beta^k\right|\left|\int_{t_{\mu-1}}^{t_\mu} a^j d\sigma\right|,$$

$$S_{3k}(\Delta,\ \omega)=\sum_{\mu}\left|\int_{t_{\mu-1}}^{t_\mu} a^i d\tau\right|\left|\int_{t_{\mu-1}}^{t_\mu} b_k{}^j d\beta^k\right|.$$

Put

$$S_{4kl}(\Delta,\ \omega)\equiv\sum_{\mu}\left|\int_{t_{\mu-1}}^{t_\mu} b_k{}^i d\beta^k\right|\left|\int_{t_{\mu-1}}^{t_\mu} b_l{}^j d\beta^l\right|=\sum_{\mu}A_\mu B_\mu.$$

$$\int S_{4kl}(\Delta,\ \omega)P(d\omega)=\sum_{\mu}\int_\Omega |A_\mu|^2P(d\omega)+\sum_{\mu}\int_\Omega |B_\mu|^2P(d\omega)$$

$$\leqq\sum_{\mu}\int_\Omega\int_{t_{\mu-1}}^{t_\mu}(b_k{}^i)^2d\tau P(d\omega)+\sum_{\mu}\int_\Omega\int_{t_{\mu-1}}^{t_\mu}(b_l{}^j)^2d\tau P(d\omega)$$

$$\leqq\int_\Omega\int_t^s(b_k{}^i)^2d\tau P(d\omega)+\int_\Omega\int_t^s(b_l{}^j)^2d\sigma P(d\omega).$$

Thus we may find $M_{4kl}(\varepsilon)$ independent of Δ for any $\varepsilon>0$ such that $M>M_{4kl}(\varepsilon)$ implies

$$Pr\{\omega\in\Omega_1,\ S_{4kl}(\Delta,\ \omega)>M\}<\varepsilon/4r^2.$$

Since

$$S(\Delta,\ \omega)\leqq S_1(\Delta,\ \omega)+\sum_k S_{2k}(\Delta,\ \omega)+\sum_k S_{3k}(\Delta,\ \omega)+\sum_{k,l} S_{4kl}(\Delta,\ \omega),$$

we have (3.15) by putting

$$M=M(\varepsilon)=M_1(\varepsilon)+\sum_k M_{2k}(\varepsilon)+\sum_k M_{3k}(\varepsilon)+\sum_{k,l} M_{4kl}(\varepsilon).$$

Proof of THEOREM 6. By Taylor expansion of $f(t, x^1, \ldots, x^n)$ we have (the probability parameter ω being omitted)

$$\eta(s,\ \omega)-\eta(t,\ \omega)$$

$$=\sum_{k=1}^{m}(\eta(t_k^m,\ \omega)-\eta(t_{k-1}^m,\ \omega)) \qquad \left(t_k^m=t+\frac{k}{m}(s-t)\right)$$

$$=\sum_{k=1}^{m}[f(*k)(t_k^m-t_{k-1}^m)+\sum_{i=1}^{n}f_i(*k)(\xi^i(t_k^m)-\xi^i(t_{k-1}^m))$$

$$+\frac{1}{2}\sum_{ij}f_{ij}(*k)(\xi^i(t_k^m)-\xi^i(t_{k-1}^m))(\xi^j(t_k^m)-\xi^j(t_{k-1}^m))$$

$$+\sum_{ij}\theta_{ijk}^m(\xi^i(t_k^m)-\xi^i(t_{k-1}^m))(\xi^j(t_k^m)-\xi^j(t_{k-1}^m))],$$

where

$$*k=(t_{k-1}^m,\ \xi^1(t_{k-1}^m),\ \ldots,\ \xi^n(t_{k-1}^m)).$$

Since $f_{ij}(t, x^1, \ldots, x^n)$ are continuous and $\xi^i(t, \omega)$, $i=1, 2, \ldots, n$ are all continuous in t for almost all $\omega \in \Omega_1$, $\theta^m_{i,j,k}$ tends to 0 uniformly in m and k as $n \to \infty$ for almost all $\omega \in \Omega_1$. Therefore we have

$$\sum_{kij} \theta^m_{ijk} (\text{''}) (\text{''}) \to 0 \text{ (in probability on } \Omega_1)$$

by virtue of Lemma 6.

By Lemma 5 the remainder equals the following expression:

$$\begin{aligned}(3.18) \quad & \int_t^s (f_0(*) + f_i(*) a^i(\tau) + \frac{1}{2} f_{ij}(*) b_k{}^i(\tau) b_k{}^j(\tau)) d\tau + \int_t^s f_i(*) b_j{}^i(\tau) d\beta^j(\tau) \\ & + \frac{1}{2} \int_t^s f_{ij}(*) [(\xi^i(\tau) - \xi^i(\lambda_m(\tau))) a^j(\tau) + (\xi^j(\tau) - \xi^j(\lambda_m(\tau))) a^i(\tau)] d\tau \\ & + \frac{1}{2} \int_t^s f_{ij}(*) [(\xi^i(\tau) - \xi^i(\lambda_m(\tau))) b_k{}^j(\tau) + (\xi^j(\tau) - \xi^j(\lambda_m(\tau))) b_k{}^i(\tau)] d\beta^k(\tau),\end{aligned}$$

where $\lambda_m(\tau)$ denotes the maximum t^m_k which dose not exceed τ and $*$ denotes $(\lambda_m(\tau), \xi^1(\lambda_m(\tau)), \ldots, \xi^n(\lambda_m(\tau)))$.

But $\xi^i(\lambda_n(\tau), \omega) \to \xi^i(\tau, \omega)$ uniformly in τ for almost all $\omega \in \Omega_1$ as $n \to \infty$. Therefore, by letting $n \to \infty$ in (3.18) we obtain

$$\begin{aligned}\eta(s) - \eta(t) = & \int_t^s (f_0(\tau, \xi) + f_i(\tau, \xi) a^i(\tau) + \frac{1}{2} f_{ij}(\tau, \xi) b_k{}^i(\tau) b_k{}^j(\tau)) d\tau \\ & + \int_t^s f_i(\tau, \xi) b_j{}^i(\tau) d\beta^j(\tau),\end{aligned}$$

which proves our Theorem 5.

References

[1] K. Itô: On stochastic differential equations in a differentiable manifold, this Journal Vol. 1, 1950.

[2] K. Itô: Stochastic differential equations, Memoris of the American Mathematical Society, 4, 1951.

Mathematical Institute,
Nagoya University

Journal of the Mathematical Society of Japan Vol. 3, No. 1, May, 1951.

Multiple Wiener Integral

Kiyosi Itô

The notion of *multiple Wiener integral* was introduced first by N. Wiener[1] who termed it *polynomial chaos*. Our definition in the present paper is obtained by a slight modification of Wiener's one, and seems to be more convenient in the point that our integrals of different degrees are orthogonal to each other while Wiener's polynomial chaos has not such a property.

In § 1 we shall define a normal random measure as a generalization of a brownian motion process. In § 2 we shall define multiple Wiener integral and show its fundamental property. In § 3 we shall establish a close relation between our integrals and Hermite polynomials. By making use of this relation we shall give, in § 4, an orthogonal expansion of any L_2-functional of the normal random measure, which proves to be coincident with the expansion given by S. Kakutani[2] for the purpose of the spectral resolution of the shift operator in the L_2 over the brownian motion process. In § 5 we shall treat the case of a brownian motion process, and in this case we shall show that we can define the multiple Wiener integral by the iteration of stochastic integrals.[3]

§ 1. Normal random measure

A system of real random veriables $\xi_\alpha(\omega)$, $\alpha \in A$, ω being a probability parameter, is called normal when the joint destribution of $\xi_{\alpha_1}, \ldots, \xi_{\alpha_n}$; $\alpha_1, \ldots, \alpha_n \in A$, is always a multivariate Gaussian distribution (including degenerate cases) with the mean vector $(0, \cdots, 0)$.

By making use of Kolmogoroff's theorem[4] of introducing a probability distribution in R^A, we can easily prove the following

---

1) N. Wiener: The homogeneous chaos, Amer. Journ. Math. Vol. *LV*, No. 4, 1938.

2) S. Kakntani: Determination of the spectrum of the flow of Brownian motion, Proc. Nat. Acad. Sci., U.S.A. **36** (1950), 319–323.

3) K. Itô: Stochastic integral, Proc. Imp. Acad. Tokyo, Vol. *XX*, No. 8, 1944.

4) A. Kolmogoroff: Grundbegriffe der Wahrscheinlichkeitsrechnung, Berlin, 1933. The consistency-condition is well satisfied by virtue of the property of multivariate Gaussian distribution.

Theorem 1. 1 *If $v_{\alpha\beta}$; $\alpha, \beta \in A$, satisfies the following two conditions:*

symmetric : $v_{\alpha\beta}=v_{\beta\alpha}$; (1.1)

positive-definite : $\sum x_i \bar{x}_j v_{\alpha_i \alpha_j} \geqq 0$ *(for any $\alpha_1,\cdots,\alpha_n \in A$ and for any complex numbers $x_1, x_2,\cdots, x_n$), then there exists a normal random system ξ_α, $\alpha \in A$, which satisfies* (1.2)

$$v_{\alpha\beta}=\mathfrak{E}(\xi_\alpha\xi_\beta)=\int \xi_\alpha(\omega)\xi_\beta(\omega)d\omega. \tag{1.3}$$

Definition. Let $(T, \boldsymbol{B}, m)$ be a measure space. We denote by $\boldsymbol{B}^*$ the system $\{E\,; E \in \boldsymbol{B}, m(E) < \infty\}$. A normal system $\beta(E, \omega)$, $E \in \boldsymbol{B}^*$, is called a *normal random measure* on $(T, \boldsymbol{B}, m)$, if

$$\mathfrak{E}(\beta(E)\ \beta(E'))=m(E\cap E') \text{ for any } E, E' \in \boldsymbol{B}^*. \tag{1.4}$$

Remark. Since we have $m(E\cap E')=m(E'\cap E)$ and $\sum x_i\bar{x}_j m(E_i\cap E_j) =\int|\sum_i x_i C_i(t)|^2\ m(dt) \geqq 0$, $C_i(t)$ being the characteristic function of the set E_i, we can see, by Theorem 1.1, the existence of a normal random measure on any measure space $(T, \boldsymbol{B}, m)$.

The following theerem, which can be easily shown, justifies the name of normal random " measure."

Theorem 1. 2 *Let $\beta(E)$ be a normal random measure on $(T, \boldsymbol{B}, m)$. If $E_1, E_2,\cdots$ are disjoint, then $\beta(E_1), \beta(E_2),\cdots$ are independent. Furthermore if $E=E_1+E_2+\cdots \in \boldsymbol{B}^*$, then $\beta(E)=\sum_n \beta(E_n)$ (in mean convergence).*

Remark. Since $\beta(E_1)$, $\beta(E_2),\cdots$, are independent, then the mean convergence of $\sum\beta(E_n)$ implies the almost certain convergence by virtue of Levy's theorem.[5)]

Hereafter we set the following restriction on the measure m.

Continuity. For any $E \in \boldsymbol{B}^*$ and $\varepsilon>0$ there exists a decomposition of E:

$$E=\sum_{i=1}^n E_i \tag{1.5}$$

such that

$$m(E_i) < \varepsilon,\ i=1, 2,\cdots, n. \tag{1.6}$$

5) P. Lévy: Théorie de l'addition des variariables aléatoires, Paris, 1937.

§ 2. Definition of multiple Wiener integral

By $L^2(T^p)$ we denote the totality of square-summable complex-valued functions defined on the product measure space $(T, \boldsymbol{B}, m)^p$. An elementary function[6] $f(t_1,\cdots,t_p)$ is called *special* if $f(t_1,\cdots,t_p)$ vanishes except for the case that $t_1,\cdots,t_p$ are all different. We shall denote by S_p the totality of special elementary functions.

Theorem 2. 1. *S_p is a linear manifold dense in $L^2(T^p)$.*

Proof. It suffices to show that the characteristic function $c(t_1,\cdots,t_p)$ of any set E of the form:

$$E=E_1\times E_2\times\cdots\times E_p \quad (E_i \in \boldsymbol{B}^*,\ i=1,2,\cdots,p) \tag{2.1}$$

can be approximated (in the L_2-norm) by a special elementary function.

For any $\varepsilon>0$ we can determine, by the *continuity condition*, a set-system $\boldsymbol{F}=\{F_1,\cdots,F_n\}\in\boldsymbol{B}^*$ which satisfies

$\boldsymbol{F}$. 1. $F_1, F_2,\cdots,F_n$ are disjoint,

$\boldsymbol{F}$. 2. $m(F_i)<\varepsilon_1\equiv\varepsilon/\binom{p}{2}\cdot(\sum m(E_i))^{p-1}$, $\binom{p}{2}=\dfrac{p(p-1)}{2\cdot 1}$,

$\boldsymbol{F}$. 3. each E_i is expressible as the sum of a subsystem of $\boldsymbol{F}$.

Then $c(t_1,\cdots,t_p)$ is expressible in the form:

$$c(t_1,\cdots,t_p)=\sum \varepsilon_{i_1,\ldots,i_p} c_{i_1}(t_1)\ldots c_{i_1}(t_p) \tag{2.2}$$

where $\varepsilon_{i_1,\ldots,i_p}=0$ or 1 and $c_i(t)$ is the characteirstic function of F_i, $i=1,2,\ldots,n$. We devide $\sum$ into two parts: $\sum'$ and $\sum''$: $\sum'$ corresponds to the indices $\{i_1,\ldots,i_p\}$ which are all different, while $\sum''$ corresponds to the others.

We put

$$f(t_1,\ldots,t_p)=\sum{}' \varepsilon_{i_1,\ldots,i_p} c_{i_1}(t_1)\ldots c_{i_p}(t_p). \tag{2.3}$$

Then $f\in S_p$ and

$$\|c-f\|^2=\int\cdots\int|c(t_1,\ldots,t_p)-f(t_1,\cdots,t_p)|^2\, m(dt_1)\ldots m(dt_p)$$
$$=\sum{}'' \varepsilon_{i_1\ldots i_p} m(F_{i_1})\ldots m(F_{i_p})$$

6) An elementary function of $(t_1,\ldots,t_p)$ is defined as a linear combination of the characteristic functions of the sets of the form $E_1\times\ldots\times E_p$, $E_i\in\boldsymbol{B}^*$, $i=1, 2,\ldots,n$.

$$\leqq \binom{p}{2} \sum m(F_i)^2 (\sum m(F_i))^{p-2}$$

$$\leqq \binom{p}{2} \varepsilon_1 (\sum m(F_i))^{p-1} = \binom{p}{2} \varepsilon_1 (\sum m(E_i))^{p-1} < \varepsilon.$$

Now we shall define the multiple wiener integral of $f \in L^2(T^p)$, which we denote by

$$I_p(f) \text{ or } \int \dots \int f(t_1, \dots, t_p) d\beta(t_1) \dots d\beta(t_p)$$

Let f be a special elementary function. Then f can be expressible as follows:

$$f(t_1, \dots, t_p) = a_{i_1 \dots i_p} \text{ for } (t_1, \dots, t_p) \in T_{i_1} \times \dots \times T_{i_p}, \tag{2.4}$$
$$= 0 \qquad \text{elsewhere,}$$

where $T_1, T_2, \dots, T_n$ are disjoint and $m(T_i) < \infty$, $i = 1, 2, \dots, n$, and $a_{i_1 \dots i_p} = 0$ if any two of $i_1, \dots, i_p$ are equal. We define $I_p(f)$ for such f by

$$I_p(f) = \sum a_{i_1 \dots i_p} \beta(T_{i_1}) \dots \beta(T_{i_p}). \tag{2.5}$$

Then we obtain

$$I_p(af + bg) = aI_p(f) + bI_p(g) \tag{I.1}$$

$$I_p(f) = I_p(\tilde{f}), \tag{I.2}$$

where

$\tilde{f}(t_1, \dots, t_p) = \frac{1}{\underline{|p}} \sum_{(\pi)} f(t_{\pi_1}, \dots, t_{\pi_p})$, $(\pi) = (\pi_1, \dots, \pi_p)$ running over all permutations of $(1, 2, \dots, p)$ $(\underline{|p} = 1 \cdot 2 \cdot \dots \cdot p)$.

$$(I_p(f), I_p(g)) = \underline{|p} \, (\tilde{f}, \tilde{g}), \tag{I.3}$$

where $(I_p(f), I_p(g)) \equiv \mathfrak{E}(I_p(f) \overline{I_p(g)}) \equiv \int I_p(f) \cdot \overline{I_p(g)} d\omega$, and

$$(\tilde{f}, \tilde{g}) \equiv \int \dots \int \tilde{f}(t_1, \dots, t_p) \overline{\tilde{g}(t_1, \dots, t_p)} m(dt_1) \dots m(dt_p).$$

$$(I_p(f), I_p(g)) = 0, \text{ if } p \neq q. \tag{I.4}$$

$(I.1)$ is clear. In order to show $(I.2)$ and $(I.3)$ we may assume that f and g are expressible as follows:

$$f(t_1,\dots,t_p)=a_{i_1,\dots,i_p},\ g(t_1,\dots,t_p)=b_{i_1,\dots,i_p}$$
$$\text{for}\quad (t_1,\dots,t_p)\in T_{i_1}\times\cdots\times T_{i_p}$$

and

$$f(t_1,\dots,t_p)=0,\ g(t_1,\dots,t_p)=0\quad\text{elsewhere.}$$

Then we have

$$I_p(f)=\sum_{i_1<\dots<i_p}\Big(\sum_{(j)\sim(i)}a_{j_1,\dots,j_p}\Big)\,\beta(T_{i_1})\dots\beta(T_{i_p})^{7)}$$
$$=\underline{|p}\sum_{i_1<\dots<i_p}\Big(\frac{1}{\underline{|p}}\sum_{(j)\sim(i)}a_{j_1,\dots,j_p}\Big)\,\beta(T_{i_1})\dots\beta(T_{i_p})$$
$$=\sum_{i_1,\dots,i_p}\Big(\frac{1}{\underline{|p}}\sum_{(j)\sim(i)}a_{j_1,\dots,j_p}\Big)\,\beta(T_{i_1})\dots\beta(T_{i_p})$$
$$=I(\tilde f),\ (\underline{|p}=1\cdot2\cdot\ldots\cdot p)$$

which proves $(I.2)$.

$$(I_p(f),\ I_p(g))=\Big(\sum_{i_1<\dots<i_p}\Big(\sum_{(j)\sim(i)}a_{j_1,\dots,j_p}\Big)\,\beta(T_{i_1})\dots\beta(T_{i_p}),$$
$$\sum_{i_1<\dots<i_p}\Big(\sum_{(j)\sim(i)}b_{j_1,\dots,j_p}\Big)\,\beta(T_{i_1})\dots\beta(T_{i_p})\Big)$$
$$=\sum_{i_1<\dots<i_p}\Big(\sum_{(j)\sim(i)}a_{j_1,\dots,j_p}\Big)\cdot\Big(\sum_{(j)\sim(i)}\bar b_{j_1,\dots,j_p}\Big)\ m(T_{i_1})\dots m(T_{i_p})$$
$$=\frac{1}{\underline{|p}}\sum_{i_1,\dots,i_p}\Big(\sum_{(j)\sim(i)}a_{j_1,\dots,j_p}\Big)\Big(\sum_{(j)\sim(i)}\bar b_{j_1,\dots,j_p}\Big)\ m(T_{i_1})\dots m(T_{i_p})$$
$$=\underline{|p}\sum_{i_1,\dots,i_p}\Big(\frac{1}{\underline{|p}}\sum_{(j)\sim(i)}a_{j_1,\dots,j_p}\Big)\Big(\frac{1}{\underline{|p}}\sum_{(j)\sim(i)}b_{j_1,\dots,j_p}\Big)\ m(T_{i_1})\dots m(T_{i_p})$$
$$=\underline{|p}\int\dots\int\tilde f(t_1,\dots,t_p)\cdot\overline{\tilde g(t_1,\dots,t_p)}\ m(dt_1)\dots m(dt_p)$$
$$=\underline{|p}\ (\tilde f,\tilde g).$$

---

7) $(j)\sim(i)$ means that $(j)\equiv(j_1,\dots j_q)$ is a permutation of $(i)\equiv(i_1,\dots,i_p)$.

Thus $(I.3)$ is proved.

By the similar computations we can prove $(I.4)$.

By putting $f=g$ in $(I.3)$, we obtain

$$\|I_p(f)\|^2=\lfloor p \,\|\tilde{f}\|^2 \leqq \lfloor p \,\|f\|^2, \tag{$I.3'$}$$

the last inequality being obtained by virtue of Schwarz' inequality.

Therefore I_p can be considered as a bounded linear operator from S_p into $L_2(\omega)$, and so it can be extended to an operator from the closure of $S_p(=L^2(T^p)$ by Theorem 2.1) into $L_2(\omega)$ which satisfies also $(I.1)$, $(I.2)$, $(I.3)$, $(I.4)$ and $(I.3')$.

For the later use we denote by L^2_0 the totality of complex numbers and we define as $I_0(c)=c$. Thus $(I.1)$, $(I.2)$, $(I.3)$, $(I.4)$ and $(I.3')$ are true for $p, q=0, 1, 2, \ldots$.

§ 3. Relation between multiple Wiener integrals and Hermite polynomials.

Theorem 3.1. *Let $\varphi_1(t)$, $\varphi_2(t), \ldots, \varphi_n(t)$ be an orthogonal system of real-valued functions in $L^2(T)$ and $H_p(x)$ be the Hermite polynomial of degree p. Then we have*

$$\int\cdots\int \varphi_1(t_1)\cdots\varphi_1(t_{p_1})\cdot\varphi_2(t_{p_1+1})\cdots\varphi_2(t_{p_1+p_2})\cdots$$

$$\times\varphi_n(t_{p_1+\cdots+p_{n-1}+1})\cdots\varphi_n(t_{p_1+\cdots+p_n})\,d\beta(t_1)\ldots d\beta(t_{p_1+\cdots+p_n})$$

$$=\prod_{\nu=1}^{n}\frac{H_{p_\nu}\left(\frac{1}{\sqrt{2}}\int\varphi_\nu(t)\,d\beta(t)\right)}{\sqrt{2}^{\,p_\nu}}.$$

For the proof of this theorem we prepare the following

Theorem 2.2. I, *If $\varphi(t_1,\ldots,t_p)\in L^2(T^p)$ and $\psi(t)\in L^2(T)$, then*

$$\int\cdots\left[|\varphi(t_1,\ldots,t_p)\psi(t_k)|\,m(dt_k)\right]^2 m(dt_1)\ldots m(dt_{k-1})\,m(dt_{k-1})\ldots m(dt_p)$$

$$\leqq \|\psi\|^2\cdot\|\varphi\|^2<\infty.$$

Therefore

II. $$\varphi\underset{(k)}{\times}\psi(t_1\ldots t_{k-1}t_{k+1}\ldots t_p)\equiv\int\varphi(t_1,\ldots,t_p)\psi(t_k)\,m(dt_k)$$

is a square-summable function of $t_1, \cdots, t_{k-1}, t_{k+1}, \cdots, t_p$, *and it holds*

$$\|\varphi \underset{(k)}{\times} \psi\| \leqq \|\varphi\| \cdot \|\psi\|. \tag{3.3}$$

III. *We have*

$$I_{p+1}(\varphi\psi) = I_p(\varphi) \cdot I_1(\psi) - \sum_{k=1}^{p} I_{p-1}(\varphi \underset{(k)}{\times} \psi) \tag{3.4}$$

Proof. (3.2) is clear by virtue of Schwarz' inequality and (3.3) is also true by the definition of the norm $\|.\|$ in L_2. For the proof of (3.4) we consider firstly the case when φ and ψ are special elementary functions. Then we may express φ and ψ in the form

$$\varphi(t_1, \ldots, t_p) = a_{i_1 \ldots i_p} \text{ for } (t_1, \ldots, t_p) \in T_{i_1} \times \ldots \times T_{i_p},$$
$$= 0 \qquad \text{elsewhere},$$
$$\psi(t) = b_i \qquad \text{for} \qquad t_i \in T_i$$
$$= 0 \qquad \text{elsewhere},$$

where $T_1, T_2, \ldots, T_N$ are disjoint and $m(T_i) < \infty$, $i = 1, 2, \ldots, N$ and $a_{i_1 \ldots i_p} = 0$ if any two of $i, \ldots, i_p$ are equal.

Put $S = T_1 + \ldots + T_N$, $A = \max |a_i|$, and $B = \max |b_i|$. Then

$$m(S),\ A,\ B < \infty$$

On account of the continuity-condition of m we may assume that

$$m(T_i) < \varepsilon,\ i = 1, 2, \ldots, N,$$

for any asigned $\varepsilon > 0$, by subdividing each T_i, if necessary. S, A and B remain invariant by this subdivision.

Now we define a special elementary function χ_ε by

$$\chi_\varepsilon(t_1, \ldots, t_p, t) = a_{i_1 \ldots i_p} b_i, \text{ if } (t_1, \ldots, t_p, t) \in T_{i_1} \times \ldots \dot{\times} T_{i_p} \times T_i,$$
$$\text{and } i \neq i_1, \ldots, i_p.$$
$$= 0, \text{ if otherwise.}$$

Then we have

$$I_p(\varphi) \cdot I_1(\psi) = \sum a_{i_1 \ldots i_p} \beta(T_{i_1}) \ldots \beta(T_{i_p}) \sum b_i \beta(T_i)$$

$$= \sum_{i \neq i_1, \ldots, i_p} a_{i_1 \ldots i_p} b_i \beta(T_{i_1}) \ldots \beta(T_{i_p}) \beta(T_i)$$

$$+ \sum_{k=1}^{p} \sum a_{i_1 \ldots i_p} b_{i_k} \beta(T_{i_1}) \ldots \beta(T_{i_{k-1}}) \beta(T_{i_k})^2 \beta(T_{i_{k+1}}) \ldots \beta(T_{i_p})$$

$$= I_{p+1}(\chi_\varepsilon) + \textstyle\sum_{k=1}^{p} \sum a_{i_1 \ldots i_p} b_{i_k} \beta(T_{i_1}) \ldots \beta(T_{i_{k-1}}) m(T_{i_k}) \beta(T_{i_{k+1}}) \ldots \beta(T_{i_p})$$

$$+ \textstyle\sum_{k=1}^{p} \sum a_{i_1 \ldots i_p} b_{i_k} \beta(T_{i_1}) \ldots \beta(T_{i_{k-1}}) (\beta(T_{i_k})^2 - m(T_{i_k})) \beta(T_{i_{k+1}}) \ldots \beta(T_{i_p})$$

$$= I_{p+1}(\chi_\varepsilon) + \textstyle\sum_{k=1}^{p} I_{p-1}(\varphi \underset{(k)}{\times} \psi) + \sum R_k$$

$$\|I_{p+1}(\chi_\varepsilon) - I_{p+1}(\varphi\psi)\|^2 = \underline{|p} \; \|\chi_\varepsilon - \varphi \cdot \psi\|^2$$

$$= \textstyle\sum_{k=1}^{p} \sum a^2_{i_1 \ldots i_p} b^2_{i_k} m(T_{i_1}) \ldots m(T_{i_{k-1}}) m(T_{i_k})^2 \ldots m(T_{i_p})$$

$$\leqq pA^2B^2(\textstyle\sum m(T_i))^{p-1} \cdot (\sum m(T_i))^2$$

$$\leqq \varepsilon p A^2 B^2 (\textstyle\sum m(T_i))^p = \varepsilon p A^2 B^2 m(S)^p$$

$$\|R_k\|^2 = c \sum a^2_{i_1 \ldots i_p} b^2_{i_k} m(T_{i_1}) \ldots m(T_{i_{k-1}}) m(T_{i_k})^2 \ldots m(T_{i_p})$$

$$\left(c = \frac{1}{2\pi} \int_{-\infty}^{\infty} (x^2-1)^2 e^{-x^2/2} dx\right), \quad \leqq \varepsilon \, c \, A^2 B^2 m(S)^p.$$

Thus we obtain, as $\varepsilon \longrightarrow 0$, $\quad I_p(\varphi) \cdot I_1(\psi) = I_{p+1}(\varphi\psi) + \sum_{k=1}^{p} I_{p-1}(\varphi k \psi)$.

Let φ and ψ any functions respectively in $L^2(T^p)$ and $L^2(T)$. By virtue of Theorem 2.1 we can find special elementary functions $\varphi_n \in L^2(T^p)$ and $\psi_n \in L^2(T)$ such that $\quad \|\varphi_n - \varphi\| \to 0, \quad \|\psi_n - \psi\| \to 0.$

By the above argument we have

$$I_{p+1}(\varphi_n \psi_n) = I_p(\varphi_n) I_1(\psi_n) - \sum_{k=1}^{p} I_{p-1}(\varphi_n \underset{(k)}{\times} \psi_n). \tag{3.5}$$

By making use of (3.2), (3.3) and (*I*.3′) (§ 2) we obtain

$$\|I_{p+1}(\varphi_n\psi_n) - I_{p+1}(\varphi\psi)\|_1 = \|I_{p+1}(\varphi_n\psi_n - \varphi\psi)\|_1 (\|\cdot\|_1 \text{ being the } L_1\text{-norm})$$

$$\leqq \|I_{p+1}(\varphi_n\psi_n - \varphi\psi)\| \leqq \sqrt{\underline{|p+1}} \; \|\varphi_n\psi_n - \varphi\psi\|$$

$$\leqq \sqrt{\underline{|p+1}} \; \|\varphi_n(\psi_n - \psi)\| + \sqrt{\underline{|p+1}} \; \|(\varphi_n - \varphi)\psi\|$$

$$= \sqrt{\underline{|p+1}}\,\|\varphi_n\|\cdot\|\psi_n-\psi\| + \sqrt{\underline{|p+1}}\,\|\varphi_n-\varphi\|\cdot\|\psi\|.$$

$$\|I_p(\varphi_n)\cdot I_1(\psi_n)-I_p(\varphi)\cdot I_1(\psi)\|_1 \leqq \|I_p(\varphi_n)I_1(\psi_n-\psi)\|_1$$
$$+\|I_p(\varphi_n-\varphi)I_1(\psi)\|_1$$
$$\leqq \|I_p(\varphi_n)\|\cdot\|I_1(\psi_n-\psi)\|+\|I_p(\varphi_n-\varphi)\|\cdot\|I_1(\psi)\|$$
$$\leqq \sqrt{\underline{|p}}\,\|\varphi_n\|\cdot\|\psi_n-\psi\|+\sqrt{\underline{|p}}\,\|\varphi_n-\varphi\|\cdot\|\psi\|.$$

$$\|I_{p-1}(\varphi_n \underset{(k)}{\times} \psi)-I_{p-1}(\varphi\underset{(k)}{\times}\psi)\|_1 \leqq \|I_{p-1}(\varphi_n\underset{(k)}{\times}\psi_n-\varphi\underset{(k)}{\times}\psi)\|$$
$$\leqq \sqrt{\underline{|p-1}}\,\|\varphi_n\underset{(k)}{\times}\psi_n-\varphi\underset{(k)}{\times}\psi\|$$
$$\leqq \sqrt{\underline{|p-1}}\,\|(\varphi_n-\varphi)\underset{(k)}{\times}\psi_n\|+\sqrt{\underline{|p-1}}\,\|\varphi\underset{(k)}{\times}(\psi_n-\psi)\|$$
$$\leqq \sqrt{\underline{|p-1}}\,\|\varphi_n-\varphi\|\cdot\|\psi_n\|+\sqrt{\underline{|p-1}}\,\|\varphi\|\cdot\|\psi_n-\psi\|.$$

Thus we see that (3.4) is true in the general case by letting n tend to ∞ in (3.5).

Proof of Theorem 3.1. We make use of the mathematical induction with regard to $p_1+\dots+p_n$. The theorem is trivially true in case $p_1+\dots+p_n=0$ or 1. It suffices to show that (3.1) is also true for $p_1+\dots+p_n=p+1$ under the assumption that (3.1) is valid for $p_1+\dots+p_n=p-1$, p. We may suppose that $p_1\geqq 1$ with no loss of generality.

If we put, in Theorem 3.2,

$$\varphi(t_1,\dots,t_p)=\varphi_1(t_1)\dots\varphi_1(t_{p_1-1})\varphi_2(t_{p_1})\dots\varphi_2(t_{p_1+p_2-1})\dots$$
$$\times\varphi_n(t_{p_1+\dots+p_{n-1}})\dots\psi_n(t_{p_1+\dots+p_n-1})$$
$$\psi(t)=\varphi_1(t),$$

we obtain, by the assumption of induction

$$\int\dots\int\varphi(t_1,\dots,t_p)\varphi_1(t)d\beta(t_1)\dots d\beta(t_p)d\beta(t)$$
$$=\int\dots\int\varphi(t_1,\dots,t_p)d\beta(t_1)\dots d\beta(t_p)\int\varphi_1(t)d\beta(t)$$
$$-\textstyle\sum_{k=1}^{p}\int\dots\int(\varphi\underset{(k)}{\times}\varphi_1)d\beta(t_1)\dots d\beta(t_{k-1})d\beta(t_{k+1})\dots d\beta(t_p)$$

$$= \prod_{\nu=2}^{n} \frac{H_{p_\nu}\left(\frac{1}{\sqrt{2}}\int \varphi_\nu(t)d\beta(t)\right)}{\sqrt{2}^{p_\nu}} \cdot \frac{H_{p_1-1}\left(\frac{1}{\sqrt{2}}\int \varphi_1(t)d\beta(t)\right)}{\sqrt{2}^{p_1-1}}$$

$$\cdot \int \varphi_1(t)d\beta(t) + (p_1-1)$$

$$\times \prod_{\nu=2}^{n} \frac{H_{p_\nu}\left(\frac{1}{\sqrt{2}}\int \varphi_\nu(t)d\beta(t)\right)}{\sqrt{2}^{p_\nu}} \cdot \frac{H_{p_1-2}\left(\frac{1}{\sqrt{2}}\int \varphi_1(t)d\beta(t)\right)}{\sqrt{2}^{p_\nu-2}}$$

$$= \prod_{\nu=1}^{n} \frac{H_{p_\nu}\left(\frac{1}{\sqrt{2}}\int \varphi_\nu(t)d\beta(t)\right)}{\sqrt{2}^{p_\nu}}$$

in considering

$$\int \varphi(t_1,\dots,t_p)\varphi_1(t_k)m(dt_k) = \begin{cases} \varphi_1(t_1)\cdots\varphi_1(t_{k-1})\varphi_1(t_{k+1})\cdots\varphi_1(t_{p_1-1})\varphi_2(t_{p_1}) \\ \quad \cdots\varphi_n(t_{p_1+\dots+p_n-1}) \quad (1 \leqq k \leqq p_1-1) \\ 0 \qquad (k \geqq p_1) \end{cases}$$

(which follows from the orthonormality of $\{\varphi_1,\dots,\varphi_n\}$)

and

$$H_{p_1}\left(\frac{x}{\sqrt{2}}\right) = \sqrt{2}\, H_{p_1}\left(\frac{x}{\sqrt{2}}\right) - 2(p_1-1)H_{p_1-2}\left(\frac{x}{\sqrt{2}}\right)$$

(which is the recursion formula of Hermite polynomials).

§ 4. Orthogonal development of L_2-functionals of β by multiple Wiener integrals.

A mapping from R^{B^*} into the complex number space K is called *B-measurable* if the inverse image of any Borel set in K is a B-measurable set in R^{B^*}, which is a set belonging to the least complete additive class that contains all the Borel cylinder sets in R^{B^*}. A complex-valued random variable $\xi(\omega)$ is a B-measurable function of β if it is expressible in the from :

$$\xi(\omega) = f(\beta(E, \omega),\ E \in \boldsymbol{B}^*), \tag{4.1}$$

for any ω, f being a B-meassurable mapping from R^{B^*} into K.

Furthermore if

$$\|\xi\|^2 = \int |\xi(\omega)|^2 d\omega < \infty, \tag{4.2}$$

then we say that $\xi(\omega)$ is an L_2-functional of β.

Theorem 4.1 (R. H. Cameron and W. T. Martin)[8] *Let* $\{\varphi_\alpha(t)\}$ *be a complete orthonormal system. Then any* L_2*-functional* $\xi(\omega)$ *of* β *can be developed as follows:*

$$\xi = \sum_p \sum_{p_1+\cdots+p_n=p} \sum_{\alpha_1,\ldots,\alpha_n} a_{p_1\ldots p_n}^{\alpha_1\ldots\alpha_n} \prod_{\nu=1}^n H_{p_\nu}\left(\frac{1}{\sqrt{2}} \int \varphi_{\alpha_\nu}(t) d\beta(t)\right).$$

Cameron and Martin has shown this theorem in the case when β is a normal random measure derived from a brownian motion process, but their proof is available for our general case.

Theorem 4.2 *Any* L_2*-functional* ξ *of* β *can be expressible in the form:*

$$\xi = \sum I_p(f_p) = \sum I_p(\tilde{f}_p), \tag{4.3}$$

where f *is given by the following orthogonal development*

$$f_p(t_1,\ldots,t_p) = \sqrt{2} \sum_{p_1+\cdots+p_n=p} \sum_{\alpha_1,\ldots,\alpha_n} a_{p_1\ldots p_n}^{\alpha_1\ldots\alpha_n} \varphi_{\alpha_1}(t_1)\cdots\varphi_{\alpha_1}(t_{p_1})$$

$$\times \varphi_{\alpha_2}(t_{p_1+1})\cdots\varphi_{\alpha_2}(t_{p_1+p_2})\cdots\varphi_{\alpha_n}(t_{p_1+\cdots+p_n+1})\cdots\varphi_{\alpha_n}(t_{p_1+\cdots+p_n}),$$

$\{\varphi_\alpha\}$ *and* $\{a_{p_1\ldots p_n}^{\alpha_1\ldots\alpha_n}\}$ *being the same as those appearing in Theorem* 4.1.

Since $I_p(f_p)$ (or $I_p(\tilde{f}_p)$), $p=0, 1, 2,\ldots$, are orthogonal to each other, (4.3) may be considered as an orthogonal development,

We shall give another method of defining the symmetric functions $\{\tilde{S}\}$ which satisfy $\xi = \sum I_p(\tilde{S}_p)$. Put

$$F(\tilde{h}_p) = \frac{1}{\underline{|p}}(\xi, I_p(\tilde{h}_p)), \tilde{h} \in \tilde{L}^2(T^p),$$

where $\tilde{L}_\sigma(T^p)$ is the totality of symmetric functions in $L^2(T^p)$ which forms a closed linear subspace of $L^2(T^p)$.

8) R. H. Cameron and W. T. Martin: The orthogonal development of non-linear functionals in series of Fourier-Hermite functions,

Then F_p is a bounded linear functional on $\tilde{L}^2(T^p)$, since

$$F_p(a\tilde{h}_p+b\tilde{g}_p)=aF_p(\tilde{h}_p)+bF_p(\tilde{g}_p)$$

$$|F_p(\tilde{h}_p)|\leqq\frac{1}{\underline{|p}}\ \|\xi\|\cdot\|I_p(\tilde{h}_p)\|=\|\xi\|\cdot\|\tilde{h}_p\|.$$

By Riesz-Fischer's theorem in Hilbert space, we can find $\tilde{s}_p\in\tilde{L}^2(T^p)$ such that

$$F_p(\tilde{h}_p)=(\tilde{s}_p,\ \tilde{h}_p).$$

By (4.3) we have $F_p(\tilde{h}_p)=\frac{1}{\underline{|p}}(I_p(\tilde{f}_p),\ I_p(\tilde{h}_p))=(\tilde{f}_p,\tilde{h}_p).$

Thus we have

$$(\tilde{s}_p,\tilde{h}_p)=(\tilde{f}_p,\tilde{h}_p)\ \text{for}\ h_p\in\tilde{L}^2(T^p),$$

which proves $\tilde{s}_p=\tilde{f}_p$.

From the above argument follows at once

Theorem 4.3. $\xi=\sum I_p(f_p)=\sum I_p(g_p)$ *implies* $\tilde{f}_p=\tilde{g}_p$.

§ 5. The case of a brownian motion process.

Let $\beta(t)$, $a<t<b$, be a brownian motion process.
If we put

$$\beta(E)=\int c_E(t)d\beta(t),$$

where $c_E(t)$ is the characteristic function of the set E and the integral is the so-called Wiener integral. Then $\beta(E)$ is a normal random measure on $T=(a,b)$, the measure m on T being the so-called Lebesgue measure, which clearly fulfills the continuity-condition.

Let $f(t_1,\dots,t_p)\in L^2(T^p)$. Then we can consider

$$I=\int\cdots\int f(t_1,\dots,t_p)d\beta(t_1)\dots d\beta(t_p).$$

Theorem 5.1. *The above multiple Wiener integral I is expressible in the form of iterated stochastic intgrals*[9]

9) loc. cit. 2).

$$I=\underline{|p}\int_a^b\Big(\int_a^{t_p}\Big(\cdots\int_a^{t_3}\Big(\int_a^{t_2} f(t_1,\ldots,t_p)\,d\beta(t_1)\Big)\,d\beta(t_2)\Big)\cdots\Big)\,d\beta(t_{p-1})\Big)\,d\beta(t_p)$$

Proof. If f is a special elementary function, this theorem is easily verified by the definitions. In the general case we can show it by approximating f with a special elementary function and making use of the properties of multiple Wiener integrals and stochastic integral.

Any Wiener functional of the brownian motion process is an L_2-functional of the normal random measure derived from it, and vice versa. Therefore we see that Theorem 4.2 gives an orthogonal development of Wiener functionals.

I express my hearty thanks to Mr. H. Anzai for his friendly aid and valuable suggestions.

Mathematical Institute, Nagoya University.

THE BROWNIAN MOTION AND TENSOR FIELDS ON RIEMANNIAN MANIFOLD

By KIYOSI ITÔ

1. Introduction

The Brownian motion $\xi(t)$ on an r-dimensional Riemannian manifold M^r is defined as a diffusion process with the generator

$$g = \tfrac{1}{2} g^{ij} \nabla_i \nabla_j, \tag{1.1}$$

where ∇_i denotes covariant differentiation. The paths of the Brownian motion can be constructed by means of the stochastic differential equation [1,2,3]. We shall denote with P_a the probability law of the Brownian motion starting at a point $a \in M^r$ and the corresponding expectation with E_a.

Let f be a scalar field on M^r. It is clear by the definition that

$$u(t,a) = E_a[f(\xi(t))] \tag{1.2}$$

satisfies a heat equation.

$$\frac{\partial}{\partial t} u = \tfrac{1}{2} g^{ij} \nabla_i \nabla_j u \tag{1.3 a}$$

with the initial condition

$$u(0+,a) = f(a). \tag{1.3}$$

Our problem is to establish a similar fact in case f is a vector field f_k or more generally a tensor field $f_{k_1 k_2 \ldots k_p}$. We cannot replace f in (1.2) with $f_{k_1 k_2 \ldots k_p}$, because $f_{k_1 k_2 \ldots k_p}(\xi(t))$ is a tensor attached to the point $\xi(t)$ which varies with t, while $u(t,a)$ should be attached to $a = \xi(0)$. Therefore we shall shift $f_{k_1 \ldots k_p}(\xi(t))$ back to $a = \xi(0)$ along the path ξ by parallel displacement. Denoting with $T_{\xi,t}$ the parallel displacement along ξ from $a = \xi(0)$ to $\xi(t)$, we can prove that

$$u_{k_1 \ldots k_p}(t,a) = E_a[T_{\xi,t}^{-1} f_{k_1 \ldots k_p}(\xi(t))] \tag{1.4}$$

satisfies a differential equation of the same form as (1.3):

$$\frac{\partial}{\partial t} u_{k_1 \ldots k_p} = \tfrac{1}{2} g^{ij} \nabla_i \nabla_j u_{k_1 \ldots k_p}. \tag{1.5}$$

In order to define the parallel displacement $T_{\xi,t}$ used above, we cannot use Levi-Civita's definition in its usual form because of the non-differentiability of the Brownian paths. We shall define $T_{\xi,t}$ as follows. Let

$$\Delta : 0 = t_0 < t_1 < \ldots < t_n = t \tag{1.6}$$

Reprinted from Proc. Int. Congr. Mathematicians, pp. 536–539 (1962).

be a division of the time interval $[0,t]$, make a polygonal curve $\xi_\Delta(t)$ which consists of the geodesic curves connecting $\xi(t_{i-1})$ with $\xi(t_i)$, $i=1,2,\dots,n$, and denote with $T_{\xi,t}$ the Levi-Civita parallel displacement along ξ_Δ. Then we can prove that $T_{\xi_\Delta,t}$ tends to a limit in probability as $|\Delta| = \max_i(t_i - t_{i-1})$ tends to 0. This limit is defined to be $T_{\xi,t}$.

Noticing that the operator $\Delta = d\delta + \delta d$ introduced in the Hodge theory of harmonic tensor fields by Kodaira and also by Bidal and de Rham is expressed in the form [5]

$$\Delta = -g^{ij}\nabla_i\nabla_j + \text{linear transformation} \tag{1.7}$$

and modifying the transformation $T_{\xi,t}$ slightly in the above discussion, we can also get a solution of

$$\frac{\partial}{\partial t}u_{k_1\dots k_p} = -\tfrac{1}{2}\Delta u_{k_1\dots k_p} \tag{1.8 a}$$

with the initial condition

$$u_{k_1\dots k_p}(0+, a) = f_{k_1\dots k_p}(a) \tag{1.8 b}$$

As was proved by Milgram and Rosenbloom [4], $u_{k_1\dots k_p}(+\infty, a)$ is the harmonic tensor field with the same periods as $f_{k_1\dots k_p}(a)$ in case M^r is compact. We shall discuss this in a separate paper.

2. The paths of the Brownian motion on a Riemannian manifold

The Brownian motion on a Riemannian manifold M^r is a diffusion (a strict Markov process with continuous paths) with the generator

$$\begin{aligned} g &= \tfrac{1}{2}g^{ij}\nabla_i\nabla_j \\ &= \tfrac{1}{2}g^{ij}\frac{\partial^2}{\partial x^i \partial x^j} - \tfrac{1}{2}g^{ij}\Gamma^k_{ij}\frac{\partial}{\partial x^k}. \end{aligned} \tag{2.1}$$

Introducing σ^i_k and m^k by

$$\sum_k \sigma^i_k \sigma^j_k = g^{ij}, \tag{2.2 a}$$

$$m^k = -\tfrac{1}{2}g^{ij}\Gamma^k_{ij} \tag{2.2 b}$$

and solving a stochactic differential equation

$$d\xi^i(t) = \sigma^i_k(\xi(t))\,d\beta^k(t) + m^i(\xi(t))\,dt, \tag{2.3 a}$$

$$\xi^i(0) = a^i \tag{2.3 b}$$

(β is the n-dimensional Wiener process), we can construct the paths of the Brownian notion starting at $a \in M^r$. See [1, 2 and 3] for the details.

3. The parallel displacement along the Brownian paths

Since almost all paths are non-differentiable, we shall define the parallel displacement along the Brownian paths as follows.

Let $(\xi^k(t),\ k=1,2,\dots,r)$ be the Brownian notion on M^r starting at $a=(a^k)$ and $\alpha_{k_1 k_2 \dots k_p}$ be a tensor attached to a. Let

$$\Delta : 0=t_0<t_1<t_2<\dots<t_n=t \tag{3.1}$$

be a division of the time interval $[0,t]$, make a polygonal curve $\xi_\Delta(s),\ 0\leqslant s\leqslant t$, which consists of the geodesic curves connecting $\xi(t_{i-1})$ with $\xi(t_i)$, $i=1, 2,\dots,n$, and denote with $\alpha^\Delta_{k_1\dots k_p}(t)$ be the tensor obtained from $\alpha_{k_1\dots k_p}$ by the Levi-Civita parallel displacement along ξ_Δ. Then we have

THEOREM 1. *As* $|\Delta| = \max_i(t_i - t_{i-1})$ *tends to* 0, $\alpha^\Delta_{i_1\dots i_p}(t)$ *tends in probability for each fixed* t *to the solution* $\alpha_{i_1\dots i_p}(t)$ *of the stochastic differential equation*

$$\begin{aligned} d\alpha_{k_1\dots k_p}(t) &\\ = \sum_\nu \Gamma^k_{ik_\nu}(\xi(t))\,\alpha_{k_1\dots k_{\nu-1}kk_{\nu+1}\dots k_p}(t)\,d\xi^i(t) &\\ + \tfrac{1}{2}\sum_\nu g^{mi}(\xi(t))\left[\frac{\partial\Gamma^k_{ik_\nu}}{\partial x^m}(\xi(t) + \Gamma^l_{ik_\nu}(\xi(t))\,\Gamma^k_{ml}(\xi(t))\right] &\\ \times \alpha_{k_1\dots k_{\nu-1}kk_{\nu+1}\dots k_p}(\xi(t))\,dt &\\ + \tfrac{1}{2}\sum_{\mu\neq\nu} g^{mi}(\xi(t))\,\Gamma^k_{mk_\mu}(\xi(t))\,\Gamma^l_{ik_\nu}(\xi(t)) &\\ \times \alpha_{k_1\dots k_{\mu-1}kk_{\mu+1}\dots k_{\nu-1}lk_{\nu+1}\dots k_p}(\xi(t))\,dt, & \end{aligned} \tag{3.2 a}$$

with the initial condition

$$\alpha_{k_1\dots k_p}(0) = \alpha_{k_1\dots k_p}. \tag{3.2 b}$$

See [1,2] for the meaning of the this equation.

Definition. $\alpha_{k_1\dots k_p}(t)$ is called the parallel displacement of $\alpha_{k_1\dots k_p}$ along ξ.

We shall sketch the proof of this theorem.

Noticing the symbolic estimation

$$\Delta\xi^i(t)\,\Delta\xi^j(t) \sim g^{ij}(\xi(t))\,\Delta t \tag{3.3}$$

(see [2] for the exact meaning), we can ignore $o((\Delta\xi^i(t))^2)$ as well as $o(\Delta t)$. Therefore the geodesic curve C connecting $a=\xi(t)$ with $b=\xi(t+\Delta t)$ is approximately

$$x^i(t) = a^i + (b^i - a^i)t + \frac{t(1-t)}{2}\Gamma^i_{jk}(a)(b^j-a^j)(b^k-a^k). \tag{3.4}$$

Therefore the tensor $\beta_{k_1\dots k_p}$ at b obtained from $\alpha_{k_1\dots k_p}$ at a by the Levi-Civita parallel displacement along C is approximately given by

$$
\begin{aligned}
\beta_{k_1\dots k_p} \sim \alpha_{k_1\dots k_p} &+ \sum_{\nu} \Gamma^{k}_{ik_\nu}(a)\alpha_{k_1\dots k_{\nu-1}kk_{\nu+1}\dots k_p}(b^i - a^i) \\
&+ \frac{1}{2}\frac{\partial \Gamma^k_{ik_\nu}}{\partial x^m}(a) + \Gamma^l_{ik_\nu}(a)\,\Gamma^k_{ml}(a) \\
&\qquad \times \alpha_{k_1\dots k_{\nu-1}kk_{\nu+1}\dots k_p}(a)\,(b^m - a^m)\,(b^i - a^i) \\
&+ \tfrac{1}{2}\sum_{\mu\neq\nu} \Gamma^l_{ik_\nu}(a)\,\Gamma^l_{mk_\mu}(a)\,\alpha_{i_1\dots i_{\mu-1}ki_{\mu+1}\dots i_{\nu-1}l_{i_{\nu+1}}\dots i_p}(a) \\
&\qquad \times (b^m - a^m)\,(b^i - a^i).
\end{aligned}
\tag{3.5}
$$

using (3.3) we shall get the stochastic differential equation (3.2 a) for $\alpha_{i_1\dots i_p}(t)$.

4. Heat equation

Using the notations as in section 3, consider the mapping

$$T_{\xi,t} : \alpha_{k_1\dots k_p} \rightarrow \alpha_{k_1\dots k_p}(t) \tag{4.1}$$

and the tensor

$$u_{k_1\dots k_p}(t, a) = E_a(T^{-1}_{\xi,t} f_{k_1\dots k_p}(\xi(t))) \tag{4.2}$$

Then we have

THEOREM 2. $u_{k_1\dots k_p}(t, a)$ satisfies the heat equation

$$\frac{\partial}{\partial t} u_{k_1\dots k_p} = \tfrac{1}{2} g^{ij}\nabla_i\nabla_j u_{k_1\dots k_p}. \tag{4.3}$$

The Markov property of the Brownian motion implies that

$$\frac{\partial}{\partial t} u_{k_1\dots k_p} = A u_{k_1\dots k_p}, \tag{4.4}$$

where
$$A f_{k_1\dots k_p} = \lim_{t\to 0} \frac{f_{k_1\dots k_p}(t, a) - f_{k_1\dots k_p}(a)}{t}. \tag{4.5}$$

To identify A with $\frac{1}{2}g^{ij}\nabla_i\nabla_j$, it is enough to compute $E(T^{-1}_{\xi,t} f_{i_1\dots i_p}(\xi(t))$ neglecting $o((d\xi^i)^2)$ as well as $o(dt)$. Though the computation is complicated, the technique is essentially the same as in [1].

REFERENCES

[1]. ITÔ, K., Stochastic differential equations in a differentiable manifold. *Nagoya Math. J.*, 1 (1950), 35–47.

[2]. —— On a formula concerning stochastic differentials. *Nagoya Math. J.*, 3 (1951), 55–65.

[3]. —— Stochastic differential equations in a differentiable manifold. *Mem. Coll. Sci. Univ. Kyoto, Ser. A. Math.*, 28, (1953), 81–85.

[4]. MILGRAM, A. N. & ROSENBLOOM, P. C., Harmonic forms and heat conduction. I. Closed Riemannian manifolds. *Proc. Nat. Acad. Sci. U.S.A.*, 37 (1951), 180–184.

[5]. DE RHAM, G., Variétés différentiables. *Actualités Sci. Ind.*, 1222 (1955), Paris.

Stochastic Differentials

K. Itô*

Department of Mathematics
Cornell University
Ithaca, New York 14850

Communicated by A. V. Balakrishnan

By a stochastic differential we understand a random interval function induced by a continuous local quasi-martingale [11]. Hence the stochastic integral in the usual sense, the stochastic integral in the Stratonovich sense and the quadratic variation in the Kunita-Watanabe sense are interpreted as operations in the space of stochastic differentials. In this note we will present a simple relation among these operations and its application.

1. The Space of Stochastic Differentials.

Let $\{\mathscr{F}_t\}_{t\in[0,\infty)}$ be an increasing family of σ-algebras of measurable events and $\mathscr{B}$ denote the measurable processes adapted to $\{\mathscr{F}_t\}$ and having locally bounded sample functions. Among the sub-classes of $\mathscr{B}$ the following classes are important for our purpose.

$$\mathscr{C} = \{X \in \mathscr{B}\colon \text{ the sample function of } X \text{ is continuous}\}$$

$$\mathscr{M} = \{X \in \mathscr{C}\colon\ X \text{ is a local martingale relative to } \{\mathscr{F}_t\}\}$$

$$\mathscr{A} = \{X \in \mathscr{C}\colon \text{ the sample function of } X \text{ is locally bounded variation and } X(0) = 0\}$$

$$\mathscr{Q} = \{X = M+A\colon\ M \in \mathscr{M} \text{ and } A \in \mathscr{A}\}$$

$$= \{X \in \mathscr{C}\colon\ X \text{ is a local quasi-martingale relative to } \{\mathscr{F}_t\}\}.$$

Every $X \in \mathscr{Q}$ is expressed uniquely as $X = M_X + A_X$, $M_X \in \mathscr{M}$ and $A_X \in \mathscr{A}$. M_X and A_X are called the *martingale part* and the *bounded variation part* of X respectively.

With each $X \in \mathscr{Q}$ we associate a random interval function

$$dX(I) = X(t) - X(s), \ I = (s.t],$$

which is continuous in the sense that

$$dX(I) \to 0 \text{ as } I\downarrow t \text{ for every } t.$$

---

* Supported by NSF GP-33136X, Cornell University.

Applied Mathematics & Optimization, Vol. 1, No. 4

Let $d\mathscr{Q}$, $d\mathscr{M}$ and $d\mathscr{A}$ denote the classes

$$\{dX : X \in \mathscr{Q}\},\ \{dM : M \in \mathscr{M}\} \text{ and } \{dA : A \in \mathscr{A}\}$$

respectively.

Now we introduce two operations in $d\mathscr{Q}$.

A. Addition:

$$dX + dY = d(X+Y)$$

M. $\mathscr{B}$-Multiplication:

$$\Phi \cdot dX(I) = \int_I \Phi \cdot dX \text{ (\textit{Stochastic integral} [1], [2], [3], [7], [8])}.$$

In case $\Phi \in \mathscr{C}$, we have

$$(\Phi \cdot dX)(I) = \underset{|\Delta| \to 0}{\text{l.i.p.}} \sum_{i=1}^{n} \Phi(t_{i-1}) dX(I_i) \tag{1.1}$$

where $\Delta = \{I_i \equiv (t_{i-1}, t_i]\}_{i=1}^n$ is a subdivision of I and $|\Delta| = \max_i (t_i - t_{i-1})$.

P. Product:

$$(dX \cdot dY)(I) = d[M_X, M_Y],\quad X, Y \in \mathscr{Q}$$

where $[M_X, M_Y]$ is the *quadratic variation* of the pair (M_X, M_Y). For each I we have

$$(dX \cdot dY)(I) = \underset{|\Delta| \to 0}{\text{l.i.p.}} \sum_{i=1}^{n} dX(I_i) dY(I_i).$$

where Δ and $|\Delta|$ are the same as above.

Observing

$$dX(I_i) \cdot dY(I_i) = d(XY)(I_i) - Y(t_{i-1}) dX(I_i) - X(t_{i-1}) dY(I_i),$$

we can show that

$$dX \cdot dY = d(X \cdot Y) - YdX - XdY. \tag{1.2}$$

Hence

$$d(XY) = Y \cdot dX + X \cdot dY + dX \cdot dY, \text{ i.e. } XY \in \mathscr{Q}.$$

Therefore $\mathscr{Q}$ *is a commutative ring*.

It is easy to prove the following:

Theorem 1. *The space $d\mathscr{Q}$ with the operations A, M and P is a commutative $\mathscr{B}$-ring. $d\mathscr{A}$ is a subring of $d\mathscr{Q}$, and $d\mathscr{M}$ is a submodule of the $\mathscr{B}$-module $d\mathscr{Q}$ with A and M. Also we have*

$$d\mathscr{Q} \cdot d\mathscr{Q} \subset d\mathscr{A},\ d\mathscr{A} \cdot d\mathscr{Q} = 0 \text{ and } d\mathscr{Q} \cdot d\mathscr{Q} \cdot d\mathscr{Q} = 0.$$

Let us consider a third operation:

SM. Symmetric $\mathscr{Q}$-multiplication.

$$Y \circ dX = Y \cdot dX + \tfrac{1}{2} dX \cdot dY$$

For each I we have

$$Y \circ dX(I) = \underset{|\Delta| \to 0}{\text{l.i.p.}} \sum_{i=1}^{n} \frac{Y(t_{i-1}) + Y(t_i)}{2} dX(I_i), \quad Y \in \mathcal{Q}, dX \in d\mathcal{Q} \tag{1.3}$$

where Δ and $|\Delta|$ are the same as above. This is obvious by

$$\frac{Y(t_{i-1}) + Y(t_i)}{2} \cdot dX(I_i) = Y(t_{i-1}) \cdot dX(I_i) + \tfrac{1}{2} dY(I_i) \cdot dX(I_i)$$

and the formula (1.1). $Y \circ dX(I)$ coincides with the *symmetric stochastic integral of Stratonovich*: $(s) \int_I Y \cdot \delta X$ in the special case discussed by him [9]. The stochastic integral of this type was also discussed by Wong and Zakai [4] and most extensively by McShane [5]. It is trivial that

$$Y \circ dX = Y \cdot dX \text{ for } Y \in \mathcal{A} \text{ or } dX \in d\mathcal{A}$$

We can easily prove the following:

Theorem 2. *The space $d\mathcal{Q}$ with operations A, SM and P is a $\mathcal{Q}$-ring.*

Since the product $dX \cdot dY$ is expressed by (2) in terms of $\mathcal{B}$-multiplication, the symmetric $\mathcal{Q}$-multiplication can be expressed in terms of $\mathcal{B}$-multiplication as follows:

$$Y \circ dX = Y \cdot dX + \tfrac{1}{2}(d(XY) - X \cdot dY - Y \cdot dX).$$

However, there is no way of expressing $\mathcal{B}$-multiplication in terms of the symmetric $\mathcal{Q}$-multiplication even in case $Y \in \mathcal{Q}$. The only conceivable way is

$$Y \cdot dX = p_{d\mathcal{M}}(Y \circ dX) + Y \cdot (dX - p_{d\mathcal{M}} dX),$$

where $p_{d\mathcal{M}}$ is the projection from $d\mathcal{Q}$ to $d\mathcal{M}$, i.e.

$$p_{d\mathcal{M}} dZ = dM_Z,$$

but this needs an extra operation $p_{d\mathcal{M}}$.

In the stochastic calculus the following transformation formula plays an important role. Let $f: \mathbf{R}^n \to \mathbf{R}$ be of class C^2. If $X_1, X_2, \ldots, X_n \in \mathcal{Q}$, then

$$f(X) = f(X_1, X_2, \ldots, X_n) \in \mathcal{Q} \text{ and } \partial_i f(X), \partial_i \partial_j f(X) \in \mathcal{C}$$

and

$$(T) \quad df(X) = \sum_{i=1}^{n} \partial_i f(X) \cdot dX_i + \tfrac{1}{2} \sum_{i,j=1}^{n} \partial_i \partial_j f(X) \cdot dX_i \cdot dX_j.$$

For the index i with $X_i \in A$, we need only assume $\partial_i f \in C^1$ by removing the terms involving all such indexes from the second summation.

If f is of class C^3, then $\partial_i f(X) \in \mathcal{Q}$ and

$$d(\partial_i f(X)) = \sum_j \partial_j \partial_i f(X) \cdot dX_j + \tfrac{1}{2} \sum_{j,k} \partial_k \partial_j \partial_i f(X) dX_j dX_k.$$

Since $dX_k \cdot dX_j \cdot dX_i = 0$ by Theorem 1, the formula (T) implies that

$$\sum_{i=1}^{n} \partial_i f(X) \circ dX_i = \sum_{i=1}^{n} \partial_i f(X) \cdot dX_i + \tfrac{1}{2} \sum_{i=1}^{n} d(\partial_i f(X)) \cdot dX_i$$

$$= \sum_{i=1}^{n} \partial_i f(X)\cdot dX_i + \tfrac{1}{2}\sum_{i=1}^{n} \partial_i\partial_j f(X)\cdot dX_j dX_i,$$

$$= df(X),$$

i.e.

$$(T_s)\quad df(X) = \sum_{i=1}^{n} \partial_i f(X)\circ dX_i.$$

This is the *Stratonovich transformation formula* [9]. This formula can be proved for $f \in C^2$ if we extend the definition of the symmetric multiplication but we will not get into it, because in practical application the function f can be assumed to be continuously differentiable as many times as we need.

Because of simplicity of the formula (T_s), the symmetric multiplication is very convenient for some purposes, as will be illustrated in Section 3.

2. *Stochastic Differential Equations.*

Let $B(t) = (B_1(t),B_2(t),\ldots,B_n(t))$, $t \in [0,1)$, be an n-dimensional Brownian motion and let $\{\mathscr{F}_t\}$ be an increasing family of σ-algebras of measurable events. We assume that (1) $(B(t))$ is adapted to $\{\mathscr{F}_t\}$ and (2) for every t $(B(s+t)-B(t))_{s\in[0,\infty)}$ and $\mathscr{F}_t$ are independent; for example, the σ-algebras

$$\mathscr{F}_t = \sigma[X_u, u \le t],\ t \in [0,\infty)$$

have these properties.

Let us consider the space $d\mathscr{Q}$ of stochastic differentials relative to this increasing family $\{\mathscr{F}_t\}$ and its subspaces $d\mathscr{M}$ and $d\mathscr{A}$. Then

$$dB_i \in d\mathscr{M},\ dt \in d\mathscr{A} \tag{2.1}$$

$$dB_i dB_j = \delta_{ij}dt,\ dB_i dt = 0 \text{ and } dt\cdot dt = 0. \tag{2.2}$$

Now we consider two stochastic differential equations:

$$(E)\quad dX_\alpha = a_\alpha(t,X)dt + \sum_{i=1}^{n} \sigma_{\alpha i}(t,X)dB_i,\ \alpha = 1,2,\ldots,m,$$

$$(E_s)\quad dY_\alpha = a_\alpha(t,Y)\circ dt + \sum_{i=1}^{n} \sigma_{\alpha i}(t,Y)\circ dB_i,\ \alpha = 1,2,\ldots,m.$$

where $a_\alpha(t,\xi)$ and $\sigma_{\alpha i}(t,\xi)$ are smooth.

From now on we often omit the summation sign $\sum$ on the indexes appearing twice as in differential geometry.

It is known that (E) determines a continuous Markov process (i.e. a diffusion) with generator at t:

$$\mathscr{G}_t f = a_\alpha \partial_\alpha f + \tfrac{1}{2}\sigma_{\alpha i}\sigma_{\beta i}\partial_\alpha\partial_\beta f.$$

We can derive this by using the transformation formula (T) and (2) as follows.

$$\begin{aligned} df(X) &= \partial_\alpha f dX + \tfrac{1}{2}\partial_\alpha\partial_\beta f\, dX_\alpha dX_\beta \\ &= \partial_\alpha f(a_\alpha dt + \sigma_{\alpha i} dB_i) + \tfrac{1}{2}\partial_\alpha\partial_\beta f(a_\alpha dt + \sigma_{\alpha i} dB_i)(a_\beta dt + \sigma_{\beta i} dB_i) \\ &= (\partial_\alpha f a_\alpha + \tfrac{1}{2} a_{\alpha i} a_{\beta i}\partial_\alpha\partial_\beta f)dt + \partial_\alpha f \sigma_{\alpha i} dB_i, \end{aligned}$$

$$\begin{aligned} \text{so } \mathscr{G}_t f(X(t))) &= \lim_{\epsilon\downarrow 0} \frac{1}{\epsilon} E(df(X)(t,t+\epsilon]|\mathscr{F}_t) \\ &= \lim_{\epsilon\downarrow 0} E\left(\frac{1}{\epsilon}\int_t^{t+\epsilon} (a_\alpha\partial_\alpha f + \tfrac{1}{2}\sigma_{\alpha i}\sigma_{\beta i}\partial_\alpha\partial_\beta f + \partial_\alpha f\sigma_{\alpha i} dB_i)d\tau|\mathscr{F}_t\right) \\ &= (a_\alpha\partial_\alpha f + \tfrac{1}{2}\sigma_{\alpha i}\sigma_{\beta i} f)(X(t)). \end{aligned}$$

The equation (E_s) is the *symmetric stochastic differential equation due to Stratonovich* [4]. By the definition (E_s) is written as

$$\begin{aligned} dY_\alpha &= a_\alpha dt + \sigma_{\alpha i} dB_i + \tfrac{1}{2} d\sigma_{\alpha i} dB_i \\ &= a_\alpha dt + \sigma_{\alpha i} dB_i + \tfrac{1}{2}\partial_\beta\sigma_{\alpha i} dY_\beta dB_i \text{ (note } d\mathscr{Q}\cdot d\mathscr{Q}\cdot d\mathscr{Q} = 0) \\ &= (a_\alpha + \tfrac{1}{2}(\partial_\beta\sigma_{\alpha i})\sigma_{\beta i})dt + \sigma_{\alpha i} dB_i \text{ by (2.2)} \end{aligned}$$

Therefore (E_s) determines a Markov process Y different from the process X determined by (E). The generator of Y is given by

$$\begin{aligned} \mathscr{G}_t f &= a_\alpha\partial_\alpha f + \tfrac{1}{2}\partial_\beta(\sigma_{\alpha i})\sigma_{\beta i}\partial_\alpha\partial_\beta f + \tfrac{1}{2}\sigma_{\alpha i}\sigma_{\beta i}\partial_\alpha\partial_\beta f \\ &= a_\alpha\partial_\alpha f + \tfrac{1}{2}\sigma_{\beta i}\partial_\beta(\sigma_{\alpha i}\partial_\alpha f). \end{aligned}$$

We can derive this more directly from (E_s) by using the Stratonovich transformation formula (T_s) as follows:

$$\begin{aligned} df(Y) &= \partial_\alpha f \circ dY \\ &= \partial_\alpha f \circ (a_\alpha \circ dt + \sigma_{\alpha i} \circ dB_i) \\ &= a_\alpha\partial_\alpha f dt + \sigma_{\alpha i}\partial_\alpha f dB_i + \tfrac{1}{2}\partial_\beta(\sigma_{\alpha i}\partial_\alpha f) dY_\beta dB_i \\ &= (a_\alpha\partial_\alpha f + \tfrac{1}{2}\partial_\beta(\sigma_{\alpha i}\partial_\alpha f)\sigma_{\beta i})dt + \sigma_{\alpha i}\partial_\alpha f dB_i, \end{aligned}$$

so $\mathscr{G}_t f(Y(t)) = (a_\alpha\partial_\alpha f + \tfrac{1}{2}\sigma_{\beta i}\partial_\beta(\sigma_{\alpha i}\partial_\alpha f))(Y(t))$.

Consider the ordinary differential equation

$$(C) \quad \frac{dx_\alpha}{dt} = a_\alpha(t,x) + \sigma_{\alpha i}(t,x)u_i(t), \quad \alpha = 1,2,\ldots,m,$$

where $u = (u_1(t), u_2(t),\ldots,u_n(t))$ is a square integrable vector function. We can regard (C) as a control equation of the equation

$$\frac{dx_\alpha}{dt} = a_\alpha(t,x), \quad \alpha = 1,2,\ldots,m,$$

under the control u. The *Stroock-Varadhan support theorem* [6] claims that the solutions of (C), with initial vector $X(0) = (X(0),\ldots,X_n(0))$ given, is dense in the topological support of the probability law of the sample function of the solution of the Stratonovich symmetric stochastic differential equation (E_s) with the same initial vector. In this sense (E_s) *is more closely connected to* (C) *than* (E).

3. *Examples of Stochastic Calculus.*

Example 1. The ordinary differential equation

$$dy = ydx$$

has a unique solution $y = y(0)e^x$. We have two stochastic versions of this equation:

$$(E_s) \qquad dY = Y \circ dX,$$
$$(E) \qquad dY = Y \cdot dX.$$

Using the transformation formula (T_s) we can prove that (E_s) has unique solution:

$$Y(t) = Y(0) \exp (X(t) - X(0)) \left(= Y(0) \exp (dX(0,t]) = Y(0) \exp \int_0^t dX\right). \quad (1)$$

The equation (E) can be written as

$$\begin{aligned} dY &= Y \circ dX - \tfrac{1}{2} dY \cdot dX \\ &= Y \circ dX - \tfrac{1}{2} Y \cdot (dX)^2 \\ &= Y \circ (dX - \tfrac{1}{2}(dX)^2) \quad (\text{Note: } Y \cdot (dX)^2 = Y \circ (dX)^2) \\ &= Y \circ dZ, \; Z(t) = \int_0^t (dX - \tfrac{1}{2}(dX)^2). \end{aligned}$$

Hence we have

$$Y = Y(0)e^{Z(t)} = Y(0) \exp \left\{\int_0^t (dX - \tfrac{1}{2}(dX)^2\right\} \quad (2)$$

by (1).

As a special case: $dX = \Phi \cdot dB$ (B: Brownian motion), we obtain the following results:

$$dY = Y \circ (\Phi \cdot dB) \Rightarrow Y(t) = Y(0) \exp \left(\int_0^t \Phi dB\right) \quad (1')$$

$$dY = Y \cdot \Phi \cdot dB \Rightarrow Y(t) = Y(0) \exp \left(\int_0^t \Phi dB - \tfrac{1}{2}\Phi^2 ds)\right). \quad (2')$$

Example 2. Consider the Stroock equation [10] determining the spherical Brownian motion:

$$dX = a(X) \circ dB, \; a(x) = I - \frac{x \cdot x'}{|x|^2}, \quad (3)$$
$$\alpha, \; i = 1,2,\ldots,n,$$

where $B = (B_1, B_2, \ldots, B_n)$ is an n-dimensional Brownian motion, all n-vectors in (1) are column vectors and x' denotes the transpose of the vector x (regarded as an $n \times 1$ matrix). The matrix $a(x)$ gives the projection to the hyperplane normal to the direction x; in fact,

$$x' \cdot a(x) = 0.$$

By applying the formula (T_s) component-wise and using Theorem 2, we have

$$\begin{aligned} d|X|^2 &= 2X' \circ dX (= 2 \textstyle\sum_\alpha X_\alpha \circ dX_\alpha) \\ &= 2X' \cdot a(X) \circ dB \\ &= 0. \end{aligned}$$

Hence the solution of (3) lies on a sphere with center 0. This fact is also regarded as an example of the Stroock-Varadhan support theorem. Thus the solution X of (3) is a diffusion process on a sphere with centre 0. Since dB is rotation invariant, and since

$$a(0x) = 0a(x)0' \qquad (0 = \text{an orthogonal matrix}),$$

we have

$$d(0X) = 0 \cdot dX = 0a(X) \cdot dB = 0a(X)0'0dB = a(0X)0B,$$

showing that X is a rotation invariant diffusion on a sphere with centre 0, i.e. a Brownian motion on the sphere.

Writing (3) (dimension $n = 3$) in polar coordinates we have

$$\begin{pmatrix} \sin\Theta\cos\Phi & R\cos\Theta\cos\Phi & -R\sin\Theta\sin\Phi \\ \sin\Theta\sin\Phi & R\cos\Theta\sin\Phi & R\sin\Theta\cos\Phi \\ \cos\Theta & -R\sin\Theta & 0 \end{pmatrix} \circ \begin{pmatrix} dR \\ d\Theta \\ d\Phi \end{pmatrix}$$

$$= \begin{pmatrix} 1-\sin^2\Theta\cos^2\Phi & -\sin^2\Theta\cos\Phi\sin\Phi & -\cos\Theta\sin\Theta\cos\Phi \\ -\sin^2\Theta\cos\Phi\sin\Phi & 1-\sin^2\Theta\sin^2\Phi & -\cos\Theta\sin\Theta\sin\Phi \\ -\cos\Theta\sin\Theta\cos\Phi & -\cos\Theta\sin\Theta\sin\Phi & \sin^2\Theta \end{pmatrix} \circ \begin{pmatrix} dB_1 \\ dB_2 \\ dB_3 \end{pmatrix}$$

By Theorem 2 in Section 1 we can use the Cramer formula to solve this for dR, $d\Theta$ and $d\Phi$ and we obtain

$$dR = 0$$

$$d\Theta = \frac{1}{R}\cos\Theta\cos\Phi \circ dB_1 + \frac{1}{R}\cos\Theta\sin\Phi \circ dB_2 - \frac{1}{R}\sin\Theta \circ dB_3,$$

$$d\Phi = -\frac{1}{R}\frac{\sin\Phi}{\sin\Theta} \circ dB_1 + \frac{1}{R}\frac{\cos\Phi}{\sin\Theta} \circ dB_2.$$

Thus the generator of the diffusion X in polar coordinates is

$$\frac{1}{r^2}\left[\frac{1}{\sin\theta}\,\partial_\theta(\sin\theta\,\partial_\theta) + \frac{1}{\sin^2\theta}\,\partial_\varphi^2\right];$$

note that the expression in the square bracket equals the spherical Laplacian on the unit sphere.

References

[1] D. L. Fisk, Quasi-martingales, *Trans. Amer. Math. Soc.*, **120** (1965), 369–389.

[2] P. Courrege, Intégrals stochastiques et martingales de carré integrable, *Sem. Th. Potent.* 7. Inst. Henri Poincaré, Paris, 1963.

[3] H. Kunita and S. Watanabe, On square integrable martingales, *Nagoya Math. J.* **30** (1967), 209–245.

[4] E. Wong and M. Zakai, On the relation between ordinary and stochastic differential equations, *Internat. J. Engrg. Sci.* 3 (1965), 213–229.

[5] E. J. McShane, Stochastic differential equations and models of random processes, *Prc. 6th Berkeley Symp. on Math. Stat. and probability* 3 (1970), 263–294.

[6] D. W. Stroock and S. R. S. Varadhan, On the support of diffusion processes with applications to strong maximum principle, *Proc. 6th Berkeley Symp. on Math. Stat. and probability* 3 (1970), 333–360.

[7] P. A. Meyer, *Intégrals stochastiques I-IV*. Lecture Notes in Math. (Springer) 39 Semi. Prob. (1967), 72–162.

[8] P. W. Millar, Martingale intégrals, *Trans. Amer. Math. Soc.* **133** (1968), 145–168.

[9] R. L. Stratonovich, *Conditioned Markov processes and their application to optimal control*, R. Elsevier, New York (1968).

[10] D. W. Stroock, On the growth of stochastic integrals, *Z. Wakr. and Vew. Geb.* **18** (1971), 340–344.

[11] K. Itô, Stochastic differentials of continuous local quasi-martingales, Stability of stochastic dynamical systems. Lecture Notes in Math. (Springer), **294** (1972), 1–7.

Curriculum Vitae of Joseph B. Keller

Position, Address, Personal Data

Professor of Mathematics and Mechanical Engineering, Emeritus

Stanford University, Department of Mathematics, Stanford, CA 94305-2125

Date of Birth: July 31, 1923 Birthplace: Paterson, NJ

Education

Ph.D., 1948, New York University, New York, NY

M.S., 1946, New York University, New York, NY

B.A., 1943, New York University, New York, NY

Positions

Stanford University
- Visiting Professor of Mathematics, 1969-1970, 1976-1978
- Professor of Mathematics and Mechanical Engineering, 1978-Present

Courant Institute of Mathematical Sciences, New York University
- Assistant, Associate, and Full Professor of Mathematics 1948–1979
- Chairman, Department of Mathematics, University College, School of Engineering and Science, and Graduate School of Engineering and Science, New York University, 1967-1973
- Director, Division of Wave Propagation and Applied Mathematics, 1967-1979

Office of Naval Research, Washington DC
- Head, Mathematics Branch, September 1953–September 1954

Columbia University
- Research Assistant, March 1944–October 1945

Princeton University
- Instructor in Physics, July 1943–February 1944

Selected Honors

1997 Wolf Prize, Wolf Foundation, Jerusalem, Israel

1996 Frederic Esser Nemmers Prize in Mathematics, Northwestern University, Evanston, IL

1995 Honorary Doctor of Science, New Jersey Institute of Technology, Newark, NJ

1995 Boeing Chair, Applied Mathematics Department, University of Washington

1995 National Academy of Sciences Award in Applied Mathematics and Numerical Analysis

1993–1994 Professorship Carlos III, Universidad Carlos III de Madrid

1993 Honorary Doctor of Philosophy, University of Crete

1993 Rouse Ball Lecturer, University of Cambridge

1990 Lewis M. Terman Professor and Scholar, Stanford University

1989–1993 Honorary Professor of Mathematical Sciences, University of Cambridge

1988 National Medal of Science

1988 Doctor of Science, Northwestern University

1984 Timoshenko Medal, American Society of Mechanical Engineers

1983 von Neumann Lecturer, Society of Industrial and Applied Mechanics

1981 Eringen Medal, Society of Engineering Sciences

1979 von Karman Prize, Society of Industrial and Applied Mathematics

1979 Doctor Technices Honoris Causa, Technical University of Denmark

1977 Gibbs Lecturer, American Mathematical Society

1977 Hedrick Lecturer of Mathematical Association of America

1977 Lester R. Ford Award of the Mathematical Association of America for outstanding expository writing, "The Feynman Integral"

1976 Lester R. Ford Award of the Mathematical Association of America for outstanding writing, "Inverse Problems"

Biographical Sketch of Joseph B. Keller

Joseph B. Keller was born in Paterson, New Jersey in 1923 and educated at New York University (Ph. D. 1948). He remained there as a Professor of Mathematics in the Courant Institute of Mathematical Sciences until 1979. Then he moved to Stanford University where he was Professor of Mathematics and Mechanical Engineering until 1993, when he became Professor Emeritus.

His research concerns the use of mathematics to solve problems of science and engineering. For example, he developed the Geometrical Theory of Diffraction to describe the propagation of waves. It is widely used to analyze radar reflection from objects, to calculate elastic wave scattering from flaws in solids, to study acoustic wave propagation in the ocean, etc. Another example is his formulation of the EBK method of quantization to determine energy levels of atoms and molecules in quantum mechanics and to solve characteristic value problems in other fields. These two examples are described in the papers reproduced here. Other problems he has worked on are described in his list of publications.

Publication List of Joseph B. Keller

Diffraction of Sound Around a Circular Disk, (with H. Primakoff, M.J. Klein and E. Carstensen) *J.A.S.A.*, **19**, 132–142, Jan. 1947; *Math. Revs.* **8**, 545, 1947.

Reflection and Transmission of Sound by Thin Curved Shells, (with H. Primakoff), *J.A.S.A.*, **19**, 820–831, Sept. 1947; *Math. Revs.*, **9**, 315, 1948.

Reflection and Transmission of Sound by a Spherical Shell, (with H.B. Keller), *J.A.S.A.*, **20**, 310–313, May 1948; *Math. Revs.*, **9**, 635, 1948.

On the Solution of the Boltzmann Equation for Rarefied Gases, *Comm. Pure Appl. Math.*, **1**, 275–285, Sept. 1948; *Math. Revs.* **10**, 639, 1949.

The Solitary Wave and Periodic Waves in Shallow Water, *Comm. Pure Appl. Math.*, **1**, 323–340, Dec. 1948; *Math. Revs.* **11**, 227, 1950. Also in Ocean Surface Waves, Annals of the N.Y. Acad. of Sci., **51**, 345-350, May 1949.

Reflection and Transmission of Electromagnetic Waves by a Spherical Shell, (with H.B. Keller), *J.Appl. Phys.*, **20**, 393–396, April 1949; *Math. Revs.*, **10**, 659, 1949.

Determination of Reflected and Transmitted Fields by Geometrical Optics, (with H.B. Keller), *J. Opt. Soc. Am.*, **40**, 48–52, Jan. 1950; *Math. Revs.*, **11**, 561, 1950.

Reflection and Transmission of Electromagnetic Waves by Thin Curved Shells, *J. Appl. Phys.*, **21**, 896–901, Sept. 1950; *Math. Revs.*, **12**, 305, 1951.

Reflection of Waves from Floating Ice in Water of Finite Depth, (with M.L. Weitz), *Comm. Pure Appl. Math.*, **3**, 305–318, Sept. 1950; *Math. Revs.*, **12**, 762, 1951.

Diffraction and Reflection of Pulses by Wedges and Corners, (with A. Blank), *Comm. Pure Appl. Math.*, **4**, 75–94, June 1951; *Math. Revs.*, **12**, 564, 1951; **13**, 304, 1952.

Bowing of Violin Strings, *Proc. of the Eighth Internat'l Cong. on Theor. and Appl. Mech.*, Istanbul, Turkey, Sept. 1951.

Comments on Channels of Communication in Small Groups, *Am. Soc. Rev.*, **16**, 842–843, Dec. 1951.

Diffraction of a Shock or an Electromagnetic Pulse by a Right-Angled Wedge, *J. Appl. Phys.*, **23**, 1267–1268, Nov. 1952.

Scattering of Water Waves Treated by the Variational Method, (abstract), *Gravity Waves, Proc. NBS Sesquicentennial Symp. on Gravity Waves, NBS Circular 521*, Nov. 28, 1952, p. 127.

Water Wave Reflection Due to Surface Tension and Floating Ice, (with E. Goldstein), *Trans. Am. Geophys. Union*, **34**, 43-48, Feb 1953; *Math. Revs.*, **14**, 810, 1953.

Finite Amplitude Sound Waves, *J.A.S.A.*, **25**, 212–216, March 1953; *Math. Revs.*, **14**, 923, 1953.

The Geometrical Theory of Diffraction, *Proc. of the Symp. on Microwave Optics*, Eaton Electronics Laboratory, McGill University, Montreal, Canada, June 1953, **Vol. 1** (4 pages); Reprinted as, The Geometric Optics Theory of Diffraction, The McGill Symp. on Microwave Optics, B.S. Karasik and F.J. Zucker, eds., AFCRC, Bedford, MA, 1959, Vol. 2, 207–210.

Asymptotic Evaluation of the Field at a Caustic, (with I. Kay), *J. Appl. Phys.*, **25**, 876–883, 1954.

Bohm's Interpretation of the Quantum Theory in Terms of Hidden Variables, *Phys. Rev.*, **89**, 1040–1041, 1953; *Math. Revs.*, **16**, 984, 1954.

Reflection and Transmission Coefficients for Water Waves Entering or Leaving an Icefield, (with M.L. Weitz), *Comm. Pure Appl. Math.*, **6**, 415–417, July 1953; *Math. Revs.*, **17**, 571, 1954.

Parallel Reflection of Light by Plane Mirrors, *Quart. Appl. Math.*, **11**, 216–219, July 1953; *Math. Revs.*, **14**, 1042, 1953.

Bowing of Violin Strings, *Comm. Pure Appl. Math.*, **6**, 483–495, Nov. 1953; *Math. Revs.*, **15**, 707, 1954.

The Scope of the Image Method, *Comm. Pure Appl. Math.*, **6**, 505–512, Nov. 1953; *Math. Revs.*, **14**, 877, 1953.

Decay of Spherical Sound Pulses due to Viscosity and Heat Conduction, *J.A.S.A.*, **26**, 58, Jan. 1954; *Math. Revs.*, **15**, 757, 1954.

Finite Amplitude Sound Produced by a Piston in a Closed Tube, *J.A.S.A.*, **26**, 253-254, 1954.

Multiple Shock Reflection in Corners, *J. Appl. Phys.*, **25**, 588-590, May 1954; *Math. Revs.*, **16**, 85, 1955.

Asymptotic Evaluation of the Field at a Caustic, (with I. Kay), *J. Appl. Phys.*, **25**, 876–883, July 1954; *Math. Revs.*, **16**, 199, 1955.

Instability of Liquid Surfaces and the Formation of Drops, (with I.I. Kolodner), *J. Appl. Phys.*, **25**, 918–921, July 1954; *Math. Revs.*, **16**, 638, 1955.

Lowest Eigenvalues of Nearly Circular Regions, (with H.B. Keller), *Quart. Appl. Math.*, **12**, 141–150, July 1954; *Math. Revs.*, **15**, 959, 1954.

Geometrical Acoustics I. The Theory of Weak Shock Waves, *J. Appl. Phys.*, **25**, 938–947, Aug. 1954; *Math. Revs.*, **16**, 761, 1955.

Asymptotic Expansion of Solutions of $(\nabla^2+k^2)u=0$, (with F.G. Friedlander), *Comm. Pure Appl. Math.*, **8**, 387–394, Aug. 1955; *Math. Revs.*, **16**, 482, 1955; **17**, 41, 1956.

Geometrical Acoustics II. Diffraction, Reflection and Refraction of a Weak Spherical or Cylindrical Shock at a Plane Interface, (with K.O. Friedrichs), *J. Appl. Phys.*, **26**, 961–966, Aug. 1955; *Math. Revs.*, **17**, 553, 1956.

Reflection and Transmission of Sound by a Moving Medium,, *J.A.S.A.*, **27**, 1044–1047, Nov. 1955; *Math. Revs.*, **17**, 553, 1956.

Determination of the Potential from Scattering Data, (with I. Kay and J. Shmoys), *Phys. Rev.*, **102**, 557–559, April 1956; *Math. Revs.*, **18**, 204, 1957.

Asymptotic Solution of Some Diffraction Problems, (with R.M. Lewis and B.D. Seckler), *Comm. Pure Appl. Math.*, **9**, 207–265, June 1956; *Math. Revs.*, **17**, 41, 1956; **18**, 43, 1957.

Spherical, Cylindrical and One-Dimensional Gas Flows, *Quart. Appl. Math.*, **14**, 171–184, July 1956; *Math. Revs.*, **18**, 253, 1957.

Electrohydrodynamics I. The Equilibrium of a Charged Gas in a Container, *J. Rat. Mech. Anal.*, **5**, 715–724, July 1956; *Math. Revs.*, **18**, 442, 1957.

Diffraction by a Convex Cylinder, IRE Trans. on Antennas and Prop., Symp. on Electromagnetic Wave Theory, AP–4, 312–321, July 1956; *Math. Revs.*, **19**, 103, 1959.

Damping of Underwater Explosion Bubble Oscillations, (with I.I. Kolodner), *J. Appl. Phys.*, **27**, 1152-1161, Oct. 1956.

Upward Falling Jets and Surface Tension, (with M.L. Weitz), *J. Fluid Mech.*, **2**, 201–203, March 1957.

Diffraction by an Aperture, *J. Appl. Phys.*, **28**, 426- 444, April 1957; *Math. Revs.*, **20**, 835, 1959; **21**, 105, 1960.

A Theory of Thin Jets, (with M.L. Weitz), *Proc. Ninth Internt'l Cong. Appl. Mech.*, **1**, 316–323, Brussels, Belgium, 1957.

Diffraction by an Aperture II, (with R.M. Lewis and B.D. Seckler), *J. Appl. Phys.*, **28**, 570–579, May 1957; *Math. Revs.* **20**, 835, 1959; **21**, 105, 1960.

Bounds on Phase Shifts, Il Nuovo Cimento, Series X, **5**, 1122– 1127, May 1957.

Teapot Effect, *J. Appl. Phys.*, **28**, 859–864, Aug. 1957; *Math. Revs.* **19**, 348, 1958.

Acoustic Torques and Forces on Disks, *J.A.S.A.*, **29**, 1085–1090, Oct. 1957; *Math. Revs.* **19**, 707, 1958; **20**, 248, 1959.

On Solutions of $\Delta u = f(u)$, *Comm. Pure Appl. Math.*, **10**, 503-510, Nov. 1957; *Math. Revs.* **19**, 964, 1958.

On Solutions of Nonlinear Wave Equations, *Comm. Pure Appl. Math.*, **10**, 523-530, Nov. 1957; *Math. Revs.* **20**, 558, 1959.

Propagation of Electromagnetic Pulses Around the Earth, (with B.R. Levy), IRE Trans. on Antennas and Prop., **AP–6**, 56-61, Jan. 1958; *Math. Revs.* **19**, 1011, 1958.

Errata: Diffraction by an Aperture, *J. Appl. Phys.*, **29**, 744, April 1958.

A Geometrical Theory of Diffraction, *Calculus of Variations and Its Applications, Proceedings of Symposia in Applied Mathematics*, **8**, 27- -52, McGraw–Hill, New York, 1958; *Math. Revs.* **20**, 103, 1959.

Corrected Bohr-Sommerfeld Quantum Conditions for Nonseparable Systems, *Annals of Phys.*, **4**, 180–188, June 1958; *Math. Revs.* **20**, 934, 1959.

Propagation of a Magnetic Field into a Superconductor, *Phys. Rev.*, **111**, 1497–1499, Sept. 1958; *Math. Revs.* **20**, 730, 1959.

Surface Waves on Water of Nonuniform Depth, *J. Fluid Mech.*, **4**, 607–614, Nov. 1958; *Math. Revs.*, **21**, 201, 1960.

The Geometrical Theory of Diffraction in Inhomogeneous Media, (with B.D. Seckler), *J.A.S.A.*, **31**, 192–205, Feb. 1959; *Math. Revs.* **20**, 836 and 1132, 1959.

Asymptotic Theory of Diffraction in Inhomogeneous Media, (with B.D. Seckler), *J.A.S.A.*, **31**, 206–216, Feb. 1959; *Math. Revs.* **20**, 836 and 1132, 1959.

Diffraction by a Smooth Object, (with B.R. Levy), *Comm. Pure Appl. Math.*, **12**, 159–209, Feb. 1959; *Math. Revs.*, **21**, 212, 1960.

Water Waves Produced by Explosions, (with H.C. Kranzer), *J. Appl. Phys.*, **30**, 398–407, March 1959; *Math. Revs.*, **21**, 202, 1960.

The Inverse Scattering Problem in Geometrical Optics and the Design of Reflectors, *IRE Trans. on Antennas and Prop.*, **AP–7**, 146–149, April 1959.

Determination of the Intermolecular Potentials from Thermodynamic Data and the Law of Corresponding States, (with B. Zumino), *J. Chem. Phys.*, **30**, 1351–1353, May 1959.

Elastic Wave Propagation in Homogeneous and Inhomogeneous Media, (with F.C. Karal, Jr.), *J.A.S.A.*, **31**, 694–705, June 1959; *Math. Revs.*, **21**, 457, 1960.
How Dark is the Shadow of a Round-Ended Screen?, *J. Appl. Phys.*, **30**, 1452–1454, Sept. 1959.
Large Amplitude Motion of a String, *Am. J. Phys.*, **27**, 584-586, Nov. 1959; *Math. Revs.* **21**, 733, 1960.
Decay Exponents and Diffraction Coefficients for Surface Waves on Surfaces of Non-constant Curvature, (with B.R. Levy), IRE *Trans. on Antennas and Prop.*, **AP–7**, S52–S61, Dec. 1959.
The Stefan Problem for a Nonlinear Equation, (with W.L. Miranker), *J. Math. Mech.*, **9**, 67–70, Jan. 1960; *Math. Revs.* **22** (2B), 231, 1961.
Asymptotic Solution of Eigenvalue Problems, (with S.I. Rubinow), *Annals of Phys.*, **9**, 24–75, Jan. 1960; Errata, No. 72.
Diffraction by a Spheroid, (with B.R. Levy), *Canadian J. of Phys.*, **38**, 128–144, Jan. 1960; *Math. Revs.* **22** (2B), 228, 1961.
Boundary Layer Problems in Diffraction Theory, (with R.N. Buchal), *Comm. Pure Appl. Math.*, **13**, 85–114, Feb. 1960; *Math. Revs.* **22** (10B), 1797, 1961.
Backscattering from a Finite Cone, *IRE Trans. on Antennas and Prop.*, **AP–8**, 175–182, Mar. 1960.
The Shape of the Strongest Column, *Archiv. Rat. Mech. Anal.*, **5**, 275–285, Mar. 1960; *Math. Revs.*, **23**, (3B), 217, 1962.
Surface Wave Excitation and Propagation, (with F.C. Karal, Jr.), *J. Applied Phys.*, **31**, 1039–1046, June 1960; *Math. Revs.* **25**, 2753, 1963.
Errata: Asymptotic Solution of Eigenvalue Problems, (with S.I. Rubinow), *Annals of Phys.*, **10**, 303–305, 1960.
Standing Surface Waves of Finite Amplitude, (with I. Tadjbakhsh), *J. Fluid Mech.*, **8**, 442–451, July 1960; *Math. Revs.*, **22** (8B), 1270, 1961.
Equation of State and Phase Transition of the Spherical Lattice Gas, (with W. Pressman), *Phys. Rev.*, **120**, 22–32, Oct. 1960; *Math. Revs.*, **22** (7B), 1044, 1961.
Multiple Diffraction by an Aperture in a Hard Screen, (with S.N. Karp), Optica Acta, **8**, 61–72, Jan. 1961.
Solution of the Functional Differential Equation for the Statistical Equilibrium of a Crystal, (with R.M. Lewis), *Phys. Rev.*, **121**, 1022–1037, Feb. 1961: *Math. Revs.* **23** (1B), 1962.
Simple Proofs of the Theorems of J.S. Lomont and H.E. Moses on the Decomposition and Representation of Vector Fields, *Comm. Pure Appl. Math.*, **14**, 77–80, 1961; *Math. Revs.*, **23** (1B), 1962.
Lower Bounds and Isoperimetric Inequalities for Eigenvalues of the Schröinger Equation, *J. Math. Phys.*, **2**, 262–266, Mar. 1961; *Math. Revs.*, **22**, (11B), 2017, 1961.
Shift of the Shadow Boundary and Scattering Cross Section of an Opaque Object, (with S.I. Rubinow), *J. Appl. Phys.*, **32**, 814–820, May 1961; *Math. Revs.* **22** (10B), 1797, 1961.
Asymptotic Solution of Systems of Linear Ordinary Differential Equations with Discontinuous Coefficients, (with C.R. Chester), *J. Math. Mech.*, **10**, 557–567, July 1961; *Math. Revs.*, **23**, A2569, 1962.

Backscattering from a Finite Cone–Comparison of Theory and Experiment, *IRE Trans. Antennas and Prop.*, **AP–9**, 411-412, July 1961.

Diffraction by a Semi-Infinite Screen with a Rounded End, (with D.G. Magiros), *Comm. Pure Appl. Math.*, **14**, 475–471, 1961.

Quantization of the Fluxoid in Superconductivity, (with B.Zumino), *Phys. Rev. Letters*, **7**, 164–165, Sept. 1961.

Diffraction of Polygonal Cylinders, *Electromagnetic Waves*, R.E. Langer, ed., University of Wisconsin Press, Madison, 1962; 129–137; *Math. Revs.* **24**, B212, 1962.

Current on and Input Impedance of a Cylindrical Antenna, (with Y.M. Chen), Radio Prop., *J. Res. Nat. Bur. Stan.*, **66D**, 15–21, Jan. 1962; *Math. Revs.* **24**, B1518, 1962.

Three-Dimensional Standing Surface Waves of Finite Amplitude, (with G.R. Verma), *Phys. Fluids*, **5**, 52–56, Jan. 1962.

Determination of a Potential from its Energy Levels and Undetectability of Quantization at High Energy, *Am. J. Phys.*, **30**, 22–26, Jan. 1962.

Geometrical Theory of Diffraction, *J. Opt. Soc. Am.*, **52**, 116–130, Feb. 1962; *Math. Revs.* **24**, B1115, 1962.

Strongest Columns and Isoperimetric Inequalities for Eigenvalues, (with I. Tadjbakhsh), *J. Appl. Mech.*, **29E** 159–164, Mar. 1962; *Appl. Mech. Revs.*, **15**, 5774, 1962.

Buckled States of Circular Plates, (with H.B. Keller and E.L. Reiss), *Quart. Appl. Math.*, **20**, 55–65, April 1962.

The Transverse Force on a Spinning Sphere Moving in a Viscous Fluid, (with S.I. Rubinow), *J. Fluid Mech.*, **11**, 447–459, 1961; *Appl. Mech. Rev.*, **15**, 3407, 1962.

Wave Propagation in Random Media, *Proc. Symp. in Appl. Math.*, **13**, *Hydrodynamic Instability*, Am. Math. Soc., 227–246, 1962.

Factorization of Matrices by Least Squares, *Biometrika*, **49**, 239–242, 1962.

Exponential-Like Solutions of Systems of Linear Ordinary Differential Equations, (with H.B. Keller), *SIAM J. Appl. Math.*, **10**, 246–259, 1962.

A Survey of Short Wavelength Diffraction Theory, *Symposium on Electromagnetic Theory and Antennas, Copenhagen*, June 1962, Pergamon Press, New York, 1963, 3–9.

Conductivity Tensor and Dispersion Equation for a Plasma, (with R.M. Lewis), *Phys. Fluids*, **5**, 1248–1263, 1962.

The Field of a Pulsed Dipole in an Interface, (with C.S. Gardner), *Comm. Pure Appl. Math.*, **15**, 99–108, 1962.

Reaction Kinetics of a Long Chain Molecule, *J. Chem. Phys.* **37**, 2584–2586, 1962.

Reaction Kinetics of a Long Chain Molecule II. Arends' Solution, *J. Chem. Phys.*, **38**, 325–326, 1963.

Low-Energy Expansion of Scattering Phase Shifts for Long-Range Potentials, (with B.R. Levy), *J. Math. Phys.*, **4**, 54–64, 1963.

Geometrical Methods and Asymptotic Expansions in Wave Propagation, *J. Geophys. Res.*, **68**, 1182–1183, 1963.

Small Vibrations of a Slightly Stiff Pendulum, (with G.H. Handleman), *Proc. Fourth. U.S. Nat. Cong. Appl. Mech.*, Am. Soc. Mech. Eng., New York, 1963, 195–202.

Conductivity of a Medium Containing a Dense Array of Perfectly Conducting Spheres or Cylinders or Nonconducting Cylinders, *J. Appl. Phys.*, **34**, 991–993, 1963.

Zeros of Hankel Functions and Poles of Scattering Amplitudes, (with S.I. Rubinow and M. Goldstein), *J. Math. Phys.*, **4**, 829–832, 1963.

Scattering of Short Waves, (with B.R. Levy), *Interdisciplinary Conference on Electromagnetic Scattering*, August 1962, Milton Kerker, ed. Clarkson College of Technology, MacMillan, New York, 1963, 3–24.

Asymptotic Solution of the Dirac Equation, (with S.I. Rubinow), *Phys. Rev.*, **131**, 2789–2796, 1963.

Tsunamis–Water Waves Produced by Earthquakes, *Proceedings of the Conference on Tsunami Hydrodynamics*, Institute of Geophysics, University of Hawaii, **24**, 154–166, 1961.

The High-Frequency Asymptotic Field of a Point Source in an Inhomogeneous Medium, (with G.S.S. Avila), *Comm. Pure Appl. Math.*, **16**, 363–381, 1963.

Instability Intervals of Hill's Equation, (with D.M. Levy), *Comm. Pure Appl. Math.*, **16**, 469–476, 1963.

Impedance Between Perfect Conductors in a Finitely Conducting Medium With Application to Composite Media, (with R.N. Buchal), *J. Appl. Phys.*, **34**, 3414, 1963.

The Field of an Antenna Near the Center of a Large Circular Disk, *J. Soc. Indust. Appl. Math.*, **11**, 1110–1112, 1963.

The Steepest Minimal Surface Over the Unit Circle, (Appendix to Robert Finn, New Estimates for Equations of Minimal Surface Type), *Arch. Rat. Mech. Anal.*, **14**, 337–375, 1963.

Geometrical Theory of Elastic Surface–Wave Excitation and Propagation, (with F.C. Karal, Jr.), *J.A.S.A.*, **36**, 32–40, 1964.

Viscous Flow Through a Grating or Lattice of Cylinders, *J. Fluid Mech.*, **18**, 94–96, 1964.

Elastic, Electromagnetic, and Other Waves in a Random Medium, (with F.C. Karal, Jr.), *J. Math. Phys.*, **5**, 537–547, 1964.

A Theorem on the Conductivity of a Composite Medium, *J. Math. Phys.*, **5**, 548–549, April 1964.

Growth and Decay of Gas Bubbles in Liquids, *Cavitation in Real Liquids*, Robert Davies, ed., Elsevier, Amsterdam, 1964, 19–29.

Partial Differential Equations with Periodic Coefficients and Bloch Waves in Crystals, (with F. Odeh), *J. Math. Phys.*, **5**, 1499– 1504, 1964.

Stochastic Equations and Wave Propagation in Random Media, *Proc. Symp. Appl. Math.*, **16**, 145–170, Am. Math. Soc., McGraw–Hill, New York, 1964.

Survey of the Theory of Diffraction of Short Waves by Edges, (with E.B. Hansen), *Acta Physica Polonica*, **27**, 217–234, 1965.

The Inverse Problem of Electromagnetic Scattering by a Metallic Object, *Proc.First GISAT Symp.*, Mitre Corp., Bedford, MA., 1965, 13–21.

Statistical Mechanics of the Moment Stress Tensor, (with E.F. Keller), *Phys. Fluids*, **9**, 3–7, 1966.

Statistical Mechanics of a Fluid in an External Potential, (with E.F. Keller), *Phys. Rev.*, **142**, 90–99, 1966.

Effective Dielectric Constant, Permeability, and Conductivity of a Random Medium and the Velocity and Attenuation Coefficient of Coherent Waves, (with F.C. Karal, Jr.), *J. Math. Phys.*, **7**, 661–670, 1966.

Nonlinear Vibrations Governed by Partial Differential Equations, *Proceedings of 5th U.S. National Congress on Applied Mech.*, 1966, Minneapolis, MN., 15–20.

Quantum-Mechanical Second Virial Coefficient of a Hard Sphere Gas at High Temperature, (with R.A. Handelsman), *Phys. Rev.*, **148**, 94–97, Aug. 1966.

Some Recent Developments in Diffraction and Scattering Theory, Proceedings of the URSI General Assembly, Munich, 1966.

Periodic Oscillations in Model of Thermal Convection, *J. Fluid Mech.*, **26**, 599–606, 1966.

The Tallest Column, (with F.I. Niordson), *J. Math. Mech.*, **16**, 433–446, 1966.

Periodic Vibrations of Systems Governed by Nonlinear Partial Differential Equations, (with L. Ting), *Comm. Pure Appl. Math.*, **19**, 371–420, 1966.

Axially Symmetric Potential Flow Around a Slender Body, (with R.A. Handelsman), *J. Fluid Mech.* **28**, 131–147, 1967.

The Electrostatic Field Around a Slender Conducting Body of Revolution, (with R.A. Handelsman), *SIAM J. Appl. Math.*, **15**, 824–841, 1967.

The Velocity and Attenuation of Waves in a Random Medium, Electromagnetic Scattering, R.L. Rowell and R.S. Stein, eds., Proceedings of ICES II, Gordon and Breach Science Publishers, New York, 1967, 823–834.

Extremum Principles for Slow Viscous Flows with Applications to Suspensions, (with L.A. Rubenfeld and J.E. Molyneux), *J. Fluid Mech.*, **30**, 97–125, 1967.

Refractive Index, Attenuation, Dielectric Constant, and Permeability for Waves in a Polarized Medium, (with D.J. Vezzeti), *J. Math. Phys.*, **8**, 1861–1870, 1967

Uniform Asymptotic Solutions for Potential Flow Around a Thin Airfoil and the Electrostatic Potential About a Thin Conductor, (with J.F. Geer), *SIAM J. Appl. Math.*, **16**, 75–101, 1968.

Ray Theory of Reflection from the Open End of A Waveguide, (with H.Y. Yee and L.B. Felsen), *SIAM J. Appl. Math.*, **16**, 268–300, 1968.

Loss of Boundary Conditions in the Asymptotic Solution of Linear Ordinary Differential Equations, I Eigenvalue Problems, (with G.H. Handelman and R.E. O'Malley, Jr.), *Comm. Pure Appl. Math.*, **21**, 243–261, 1968.

Loss of boundary Conditions in the Asymptotic Solution of Linear Ordinary Differential Equations, II Boundary Value Problems, (with R.E. O'Malley, Jr.), *Comm. Pure Appl. Math.*, **21**, 263–270, 1968.

Hydrodynamic Aspects of the Circulatory System, (with S.I. Rubinow), *Proceedings of the 1st International Congress of Hemorheology*, Reykjavik, Iceland, 1968, 149–155.

A Survey of the Theory of Wave Propagation in Continuous Random Media, *Symposium on Turbulence of Fluids and Plasmas*, Polytechnic Institute of Brooklyn, N.Y., 1968, 131–142.

Spectra of Water Waves in Channels and Around Islands, (with M.C. Shen and R.E. Meyer), *Phys. Fluids*, **11**, 2289–2304, 1968.

Perturbation Theory of Nonlinear Boundary-Value Problems, (with M.H. Millman), *J. Math. Phys.*, **10**, 342–361, 1969.

Bifurcation Theory and Nonlinear Eigenvalue Problems, J.B. Keller and S. Antman, eds., Benjamin, New York, 1969.

Bifurcation Theory for Ordinary Differential Equations, *Bifurcation Theory and Nonlinear Eigenvalue Problems*, J.B. Keller and S. Antman, eds., Benjamin, New York, 1969, 17–48.

Rossby Waves in the Presence of Random Currents, (with G. Veronis), *J. Geophys. Res.*, **74**, 1941–1951, 1969.

Perturbation Theory of Nonlinear Electromagnetic Wave Propagation, (with M.H. Millman), *Phys. Rev.*, **181**, 1730–1747, 1969.

Nonlinearity in Electromagnetic Wave Propagation, *Proceedings of the URSI Symposium on Electromagnetic Waves*, Stresa, Italy, 1968; *Alta Frequenza*, **38** Special Number, 198–203, May 1969.

Survey of the Theory of Turbulence, *Contemporary Physics*, Vol. I, International Atomic Energy Agency, Vienna, 1969, 257–272.

Reflection of Elastic Waves from Cylindrical Surfaces, (with D.S. Ahluwalia and E. Resende), *J. Math. Mech.*, **19**, 93–105, 1969.

Accuracy and Validity of the Born and Rytov Approximations, *J. Opt. Soc. Amer.*, **59**, 1003–1004, 1969.

Internal Wave Propagation in an Inhomogeneous Fluid of Non-Uniform Depth, (with V.C. Mow), *J. Fluid Mech.*, **38**, 365–374, 1969.

Bounds on Elastic Moduli of Composite Media, (with L.A. Rubenfeld), *SIAM J. Appl. Math.*, **17**, 495–510, 1969.

Reflection and Transmission by a Random Medium, (with P.Chow, I. Kupiec, L.B. Felsen and S. Rosenbaum), *Radio Science*, **4**, 1067–1077, 1969.

Progressing Waves Diffracted by Smooth Surfaces, (with D.S. Ahluwalia), *J. Math. Mech.*, **19**, 515–530, 1969.

Uniform Asymptotic Solution of Second Order Linear Ordinary Differential Equations with Turning Points, (with R.Y.S. Lynn), *Comm. Pure Appl. Math.*, **23**, 379–408, 1970.

Internal Wave Wakes of a Body Moving in a Stratified Fluid, (with W.H. Munk), *Phys. Fluids*, **13**, 1425–1431, 1970.

Asympotic Solutions of Initial Value Problems of Nonlinear Partial Differential Equations, (with S. Kogelman), *SIAM J. Appl. Math.*, **18**, 748–758, 1970.

Classical and Quantum Mechanical Correlation Functions of Fields in Thermal Equilibrium, *J. Math. Phys.*, **11**, 2286–2296, 1970.

Extremum Principals for Irreversible Processes, *J. Math. Phys.*, **11**, 2919–2931, 1970.

Complex Rays with an Application to Gaussian Beams, (with W. Streifer), *J. Opt. Soc. Amer.*, **61**, 40–43, 1971.

Finite Amplitude Sound Wave Propagation in a Waveguide, (with M.H. Millman), *J. Acoust. Soc. Amer.*, **49**, 329–333, 1971.

Diffraction by a Curved Wire, (with D.S. Ahluwalia), *SIAM J. Appl. Math.*, **20**, 390–405, 1971.

Transient Behavior of Unstable Nonlinear Systems with Applications to the Benard and Taylor Problems, (with S. Kogelman), *SIAM J. Appl. Math.*, **20**, 619–637, 1971.

Mean Power Transmission Through a Slab of Random Medium, (with J.A. Morrison and G.C. Papanicolaou), *Comm. Pure Appl. Math.*, **24**, 473–489, 1971.

Force on a Rigid Sphere in an Incompressible Inviscid Fluid, (with S.I. Rubinow), *Phys. Fluids*, **14**, 1302–1304, 1971.

Stochastic Differential Equations with Applications to Random Harmonic Oscillators and Wave Propagation in Random Media, (with G. Papanicolaou), *SIAM J. Appl. Math.*, **21**, 287–305, 1971.

Propagation of Acoustic Waves in a Turbulent Medium, (with A.R. Wenzel), *J. Acous. Soc. Amer.*, **50**, 911–920, 1971.

Wave Propagation in a Fluid-Filled Tube, (with S.I. Rubinow), *J. Acous. Soc. Amer.*, **50**, 198–223, 1971.

Diffraction Coefficients for Higher Order Edges and Vertices, (with L. Kaminetzky), *SIAM J. Appl. Math.*, **22**, 109–134, 1972.

Temperature of a Nonlinearly Radiating Semi-Infinite Solid, (with W.E. Olmstead), *Q. Appl. Math.*, **30**, 559–566, 1972.

Short Time Asymptotic Expansions of Solutions of Parabolic Equations, (with J.K. Cohen and F.G. Hagin), *J. Math. Anal. Appl.*, **38**, 82–91, 1972.

Dipole Moments in Rayleigh Scattering, (with R.E. Kleinman and T.B.A. Senior), *J. Inst. Math. Appl.*, **9** 14–22, 1972.

Flow of a Viscous Fluid Through an Elastic Tube with Applications to Blood Flow, (with S.I. Rubinow), *J. Theor. Bio.*, **35**, 299–313, 1972.

Quantum Mechanical Cross Sections for Small Wavelengths, *Am. J. Phys.*, **40**, 1035–1036, 1972.

Wave Propagation in a Random Lattice I, (with P.-L. Chow), *J. Math. Phys.*, **13**, 1404–1411, 1972.

Nonlinear Stability Theory, *Proceedings of the Conference on Mathematical Topics in Stability Theory*, March 1972, Univ. of Washington, Pullman, Washington, 85–98.

Expansion and Contraction of Planar, Cylindrical, and Spherical Underwater Gas Bubbles, (with D. Epstein), *J. Acous. Soc. Am.*, **52**, 975-980, 1972.

Post Buckling Behavior of Elastic Tubes and Rings with Opposite Sides in Contact, (with J.E. Flaherty and S.I. Rubinow), *SIAM J. Appl. Math.*, **23**, 446–455, 1972.

Backscattering from a Circular Loop of Wire, (with D.S. Ahluwalia), *Proc. IEEE*, **60**, 1552–1554, 1972.

Some Trends in Applied Mechanics, *Proc. First Iranian Congress of Civil Engineering and Appl. Mech.*, Pahlavi University, Shiraz, Iran, May 1972, 102–104.

Contact Problems Involving a Buckled Elastica, (with J.E. Flaherty), *SIAM J. Appl. Math.* **24**, 215–225, 1973.

Asymptotic Theory of Nonlinear Wave Propagation, (with S. Kogelman), *SIAM J. Appl. Math.* **24**, 352–361, 1973. (Addendum *SIAM J. Appl. Math.* **47**, 941–958, 1987.)

A Theory of Competitive Running, *Physics Today*, **26**, No. 9, 42–47, 1973. Also abridged in *Optimal Strategies in Sports*, S. Ladany and R. Machol, eds., American Elsevier, New York, 1977, 172–178.

Uniform Asymptotic Solution of Eigenvalue Problems for Convex Plane Domains, (with D.S. Ahluwalia), *SIAM J. Appl. Math.*, **25**, 583–591, 1973.

Flows of Thin Streams with Free Boundaries, (with J. Geer), *J. Fluid Mech.*, **59**, 417–432, 1973.

Stochastic Differential Equations, J. B. Keller and H. P. McKean, eds., Am. Math. Soc., Providence, RI, 1973.

Elastic Behavior of Composite Media, (with J.E. Flaherty), *Comm. Pure Appl. Math.*, **26**, 565–580, 1973.

Ray Method for Nonlinear Wave Propagation in a Rotating Fluid of Variable Depth, (with M.C. Shen), *Phys. Fluids*, **16**, 1565–1572, 1973. (erratum: *Phys. Fluids* **24**, 786 (1981)).

Traveling Wave Solutions of a Nerve Conduction Equation, (with J. Rinzel), *Biophys. J.*, **13**, 1313–1337, 1973.

Spatial Instability of a Jet, (with S.I. Rubinow and Y.O. Tu), *Phys. Fluids*, **16**, 2052–2055, 1973.

Elastic Waves Produced by Surface Displacements, (with D.S. Ahluwalia and R. Jarvis), *SIAM J. Appl. Math.*, **26**, 108–119, 1974.

Asymptotic Solution of Neutron Transport Problems for Small Mean Free Paths, (with E. W. Larsen), *J. Math. Phys.*, **15**, 75–81, 1974.

Contact of Inflated Membranes with Rigid Surfaces, (with A.J. Callegari), *J. Appl. Mech.*, **41**, 189–191, 1974.

Nonlinear Forced and Free Vibrations in Acoustic Waveguides, *J. Acoust. Soc. Am.*, **55**, 524–527, 1974.

Wave Propagation in Elastic Rods of Arbitrary Cross Section, (with G. Rosenfeld), *J. Acoust. Soc. Am.*, **55**, 555–561, 1974.

Optimal Velolicty in a Race, *Am. Math. Monthly* **81**, 474–480, 1974.

Asymptotic Theory of Propagation in Curved and Nonuniform Waveguides, (with D.S. Ahluwalia and B.J. Matkowsky), *J. Acoust. Soc. Am.*, **55**, 7–12, 1974.

Planing of a Flat Plate at High Froude Number, (with L. Ting), *Phys. Fluids*, **17**, 1080–1086, 1974.

Effect of Viscosity on Swimming Velocity of Bacteria, *Proc. Nat. Acad. Sci.*, **71**, 3253–3254, 1974.

Optimum Checking Schedules for Systems Subject to Random Failure, *Management Sci.* **21**, 256–260, 1974.

Mechanical Aspects of Athletics, *Proc. Seventh U.S. Nat. Cong. Appl. Mech.*, Boulder, CO, June 1974, 22–26, The American Society of Mechanical Engineers, New York, N.Y. Also abridged in *Optimal Strategies in Sports*, S. Ladany and R. Machol, eds., American Elsevier, New York, NY, 1977, 186–187.

Wave Propagation in Nonuniform Elastic Rods, (with G. Rosenfeld), *J. Acoust. Soc. Am.* **57**, 1094–1096, 1975.

The Feyman Integral, (with D.W. McLaughlin), *Am. Math. Monthly,* **82**, 451–465, 1975.

Diffraction by Edges and Vertices of Interfaces, (with L. Kaminetzky), *SIAM J. Appl. Math.*, **28**, 839–856, 1975.

Uniform Ray Theory of Surface, Internal and Acoustic Wave Propagation in a Rotating Ocean or Atmosphere, (with M.C. Shen), *SIAM J. Appl. Math.*, **28**, 857–875, 1975.

Effective Conductivity, Dielectric Constant and Permeability of a Dilute Suspension, *Philips Res. Repts.*, **30**, 83–90, 1975.

Closest Unitary, Orthogonal and Hermitian Operators to a Given Operator, *Mathematics Magazine*, **48**, 192–197, 1975.

Asymptotic Analysis of Stochastic Models in Population Genetics, (with R. Voronka), *Math. Biosci.,* **25**, 331–362, 1975.

Asymptotic solution of Eigenvalue Problems for Second Order Ordinary Differential Equations, (with D.U. Anyanwu), *Comm. Pure Appl. Math.*, **28**, 753–763, 1975.

Inverse Problems, *Am. Math. Monthly,* **83**, 107–118, 1976.

Swimming of Flagellated Microorganisms, (with S.I. Rubinow), *Biophys. J.*, **16**, 151–170, 1976.

Wave Patterns of Non-Thin or Full-Bodied Ships, *Proc. Tenth Symp. Naval Hydro.*, Office of Naval Research, Department of the Navy, Arlington, VA, June 1974, 543–545.

Wave Resistance and Wave Patterns of Thin Ships, (with D.S. Ahluwalia), *J. Ship Res.*, **20**, 1–6, 1976.

Slender-Body Theory for Slow Viscous Flow, (with S.I. Rubinow), *J. Fluid Mech.* **75**, 705–714, 1976.

Synchronization of Periodical Cicada Emergences, (with F.C. Hoppensteadt), *Science*, **194**, 335–337, 1976.

The Minimum Ratio of Two Eigenvalues, *SIAM J. Appl. Math.*, **31**, 485–491, 1976.

Optimal Shape of a Planing Surface at High Froude Number, (with L. Ting), *J. Ship Res.*, **21**, 40–43, 1977.

Green's Function with the Singularity Near the Boundary, (with C.H. Wu), *Recent Adv. Eng. Sci.*, **8**, 169–174, 1977.

A Theory of Transformed Cell Growth with Applications to Initiation-Promotion Data, (with A. Whittemore), *Environmental Health: Quantitative Methods*, Alice Whittemore, ed., SIAM, Philadelphia, PA, 1977, 183–193.

Radiation from the Open End of a Cylindrical or Conical Pipe and Scattering from the End of a Rod or Slab, (with L. Ting), *J. Acoust. Soc. Am.*, **61**, 1438–1444, 1977.

Effective Behavior of Heterogeneous Media, *Statistical Mechanics and Statistical Methods in Theory and Application*, U. Landman, ed., Plenum, New York, 1977, 631–644.

Lecture Notes on Wave Propagation and Underwater Acoustics, J.B. Keller and J.S. Papadakis, eds., Springer, New York, 1977.

Ray and Asymptotic Methods in Underwater Sound Propagation, (with D.S. Ahluwalia), *Lecture Notes on Wave Propagation and Underwater Acoustics*, J.B. Keller and J.S. Papadakis, eds., Springer, New York, 1977, 14–85.

Survey of Wave Propagation and Underwater Acoustics, *Lecture Notes on Wave Propagation and Underwater Acoustics*, J.B. Keller and J.S. Papadakis, eds., Springer, New York, 1977, 1–12.

Peeling, Slipping and Cracking–Some One-Dimensional Free-Boundary Problems in Mechanics, (with R. Burridge), *SIAM Rev.*, **20**, 31–61, 1978.

Quantitative Theories of Carcinogenesis, (with A. Whittemore), *SIAM Rev.*, **20**, 1-30, 1978.

Asymptotic Solution of Higher-Order Differential Equations with Several Turning Points, and Application to Wave Propagation in Slowly Varying Waveguides, (with D.U. Anyanwu), *Comm. Pure Appl. Math.*, **31**, 107– 121, 1978.

Asymptotic Analysis of Diffusion Equations in Population Genetics, (with C. Tier), *SIAM J. Appl. Math.*, **34**, 549–576, 1978.

Heat Conduction in a One-Dimensional Random Medium, (with C.G. Papanicolaou and J. Weilenmann), *Comm. Pure Appl. Math.*, **32**, 583–592, 1978.

Modification of Underwater Bubble Oscillations by Nearby Objects, *J. Acoust. Soc. Am.*, **64**, 937, 1978.

Rays, Waves and Asymptotics, *Bull. Am. Math. Soc.*, **84**, 727–750, 1978.

Wave Propagation in a Viscoelastic Tube Containing a Viscous Fluid, (with S.I. Rubinow), *J. Fluid Mech.*, **88**, 181–203, 1978.

Elastic Waveguides, *Modern Problems in Elastic Wave Propagation*, J. Miklowitz and J.D. Achenbach, eds., Wiley, New York, 1978, 401–415.

A Tri-Allelic Diffusion Model with Selection, (with C. Tier), *SIAM J. Appl. Math.*, **35**, 521–535, 1978.

Stochastic Theories of Carcinogenesis and Population Genetics, *Lecture Notes on Mathematics in the Life Sciences*, Vol. 11, S. Levin, ed., Am. Math. Soc., Providence, 1979, 1–19.

The Ray Theory of Ship Waves and the Class of Streamlined Ships, *J. Fluid Mech.*, **91**, 465–488, 1979.

Progress and Prospects in the Theory of Linear Wave Propagation, *SIAM Rev.* **21**, 229–245, 1979.

Slender Streams, (with J. Geer), *J. Fluid Mech.*, **93**, 97–115, 1979.

Training in Applied Mathematics, *Proc. of the Conf. on Graduate Training in Math.*, T.L. Sherman, ed., Rocky Mountain Mathematics Consortium, Tempe, Arizona, 1979, 110–113.

Lichen Growth, (with S. Childress), *J. Theor. Biol.*, **82**, 157–165, 1980.

A New Family of Capillary Waves, (with J.-M. Vanden-Broeck), *J. Fluid Mech.*, **98**, 161–169, 1980.

Plate Failure Under Pressure, *SIAM Rev.*, **22**, 227–228, 1980.

Darcy's Law for Flow in Porous Media and the Two-Space Method, *Nonlinear Partial Differential Equations in Engineering and Applied Science*, R.L. Sternberg, A.J. Kalinowski and J.S. Papadakis, eds., Marcel Dekker, New York, 1980, 429–443.

Bubble Oscillations of Large Amplitude, (with M. Miksis), *J. Acoust. Soc. Am.*, **68**, 628–633, 1980.

Bubble or Drop Distortion in a Straining Flow in Two Dimensions, (with J.-M. Vanden-Broeck), *Phys. Fluids*, **23**, 1491–1495, 1980.

Some Bubble and Contact Problems, *SIAM Rev.*, **22**, 442– 458, 1980.

Deformation of a Bubble or Drop in a Uniform Flow, (with J.-M. Vanden-Broeck), *J. Fluid Mech..*, **101**, 673–686, 1980.

Liesegang Rings and a Theory of Fast Reaction and Slow Diffusion, *Dynamics and Modeling of Reactive Systems*, W. Stewart, ed., Academic Press, New York, 1980, 221–224.

Tendril Shape and Lichen Growth, *Some Mathematical Questions in Biology, Lectures on Mathematics in the Life Sciences*, **Vol. 13**, Am. Math. Soc., Providence, 1980, 257–274.

Shape of a Sail in a Flow, (with J.-M. Vanden-Broeck), *Phys. Fluids*, **24**, 552–553, 1981.

Temperley's Model of Gas Condensation, *J. Chem. Phys.*, **74**, 4203–4204, 1981.

Kelvin Wave Production, (with J.G. Watson), *J. Phys. Ocean*, **11**, 284–285, 1981.

Recurrent Precipitation and Liesegang Rings, (with S.I. Rubinow), *J. Chem, Phys.* **74**, 5000–5007, 1981.

Axisymmetric Bubble or Drop in a Uniform Flow, (with M.J. Miksis and J.-M. Vanden-Broek), *J. Fluid Mech.*, **108**, 89–100, 1981.

Quench Front Propagation, (with R.E. Caflisch), *Nucl. Eng. Design*, **65**, 97–102, 1981.

Internal and Surface Wave Production in a Stratified Fluid, (with D.M. Levy and D.S. Ahluwalia), *Wave Motion*, **3**, 215–229, 1981.

Oblique Derivative Boundary Conditions and the Image Method, *SIAM J. Appl. Math.*, **41**, 294–300, 1981.

Parabolic Approximations for Ship Waves and Wave Resistance, (with J.-M. Vanden-Broek), *Proc. of the Third International Conference on Numerical Ship Hydrodynamics*, Paris, France, Bassin d'Essais des Carenes, 1982, 1–12.

Poroelasticity Equations Derived from Microstructure, (with R. Burridge), *J. Acoust. Soc. Am.*, **70**, 1140–1146, 1981.

Optimum Inspection Policies, *Management Sci.*, **28**, 447–450, 1982.

Biot's Poroelasticity Equations by Homogenization, (with R. Burridge), *Macroscopic Properties of Disordered Media*, ed. by R. Burridge, S. Childress and G. Papanicolau, Springer, New York, 1982, 51–57.

Rising Bubbles, (with M.J. Miksis and J.-M. Vanden-Broeck), *J. Fluid Mech.*, **123**, 31–41, 1982.

Jets Rising and Falling Under Gravity, (with J.-M. Vanden-Broeck), *J. Fluid Mech.*, **124**, 335–345, 1982.

Time-dependent Queues, *SIAM Rev.*, **24**, 401–412, 1982.

Surface Tension Driven Flows, (with M.J. Miksis), *SIAM J. Appl. Math.*, **43**, 268–277, 1983.

Weakly Nonlinear High Frequency Waves, (with J.K. Hunter) *Comm. Pure Appl. Math.* **36**, 547–569, 1983.

Capillary Waves on a Vertical Jet, *J. Fluid. Mech.*, **135**, 171–173, 1983.

Crawling of Worms, (with M.S. Falkovitz), *J. Theor. Bio.*, **104**, 417–442, 1983.

Reflection, Scattering, and Absorption of Acoustic Waves by Rough Surfaces, (with J.G. Watson), *J. Acoust. Soc. Amer.*, **74**, 1887– 1894, 1983.

Eigenvalues of Slender Cavities and waves in Slender Tubes, (with J.F. Geer), *J. Acoust. Soc. Am.*, **74**, 1895–1904, 1983.

Breaking of Liquid Films and Thread, *Phys. Fluids*, **26**, 3451–3453, 1983.

Weak Shock Diffraction (with J.K. Hunter), *Wave Motion*, **6**, 79–89, 1984.

Optimal Catalyst Distribution in a Membrane (with M.S. Falkovitz and H. Frisch), *Chem. Eng. Sci.* **39**, 601–604, 1984.

Probability of a Shutout in Raquetball, *SIAM Rev.*, **26**, 267–268, 1984.

Effective Viscosity of a Periodic Suspension, (with K.C. Nunan), *J. Fluid Mech.*, **142**, 269–287, 1984.

Effective Elasticity Tensor of a Periodic Composite, (with K.C. Nunan), *J. Mech. Phys. Solids*, **32**, 259–280, 1984.

Rough Surface Scattering via the Smoothing Method, (with J.G. Watson), *J. Acoust. Soc. Am.*, **75**, 1705–1708, 1984.

Hanging Rope of Minimum Elongation, (with G.R. Verma), *SIAM Rev.*, **26**, 569–571, 1984.

Discriminant, Transmisison Coefficient, and Stability Bands of Hill's Equation, *J. Math. Phys.*, **25**, 2903–2904, 1984.

Genetic variability due to Geographical Inhomogeneity, *J. Math. Biol.* **20**, 223–230, 1984.

Free Boundary Problems in Mechanics, *Seminar in Nonlinear Partial Differential Equations*, S.S. Chern, ed., Springer-Verlag, New York, 99–115, 1984.

Macroscopic Modelling of Turbulent Flows, J.B. Keller, U. Frisch, G. Papanicolaou and O. Pironneau, eds., Lecture Notes in Physics, No. 230, Springer-Verlag, Berlin Heidelberg, 1985.

One Hundred Years of Diffraction Theory, *IEEE Trans. Antennas Prop.*, **AP–33**, 123–126, 1985.

Acoustoelastic Effect and Wave Propagation in Heterogeneous Weakly Anisotropic Materials, (with Luis L. Bonilla), *J. Mech. Phys. Solids*, **33**, 241–261, 1985.

Hill's Equation with a Large Potential, (with M. I. Weinstein), *SIAM J. Appl. Math.*, **45**, 200–214, 1985.

Computers and Chaos in Mechanics, *Theoretical and Applied Mechanics*, F.I. Nioordson and N. Olhoff, eds. Elsevier, 31–41, 1985.

Asymptotic Analysis of a Viscous Cochlear Model, (with J.C. Neu), *J. Acoust. Soc. Am.*, **77**, 2107–2110, 1985.

Soliton Generation and Nonlinear Wave Propagation, *Phil. Trans. R. Soc. Lond. A.* **315**, 367–377, 1985.

Reciprical Relations for Effective Conductivities of Anisotropic Media, (with J. Nevard), *J. Math. Phys.* **26**, 2761–2765, 1985.

Semiclassical Mechanics, *SIAM Rev.*, **27**, 485–504, 1985.

Irrevesibilty and Nonrecurrence, (with L.L. Bonilla), *J. Statistical Phys.*, **42**, 1115–1125, 1986.

Inverse Elastic Scattering in Three Dimensions, (with W.E. Boyse), *J. Acoust. Soc. Am.*, **79**, 215–218, 1986.

Reaction Kinetics on a Lattice, *J. Chem. Phys.* **84**, 4108–4109, 1986.

Impact with Friction, *ASME J. Appl. Mech.*, **53**, 1–4, 1986.

Uniform Solutions for Scattering by a Potential Barrier and Bound States of a Potential Well, *Am. J. Phys.*, **54**, 546–550, 1986.

The Probability of Heads, *Am. Math. Monthly*, **93**, 191– 197, 1986.

Melting or Freezing at Constant Speed, *Phys. Fluids*, **92**, 2013, 1986.

On Tango's Index forDisease Clustering in Time, (with A. Whittemore), *Biometrics*, **42**, 218, 1986.

Survival Estimation Using Splines, (with A.S. Whittemore) *Biometrics*, **42**, 495–506, 1986.

Review of *Stochastic Wave Propagation* by K. Sobcyzk, *SIAM Review* **28**, 593–594, 1986.

Scattering by a slender body, (with D.S. Ahluwalia) *J. Acoust. Soc. Am.* **80**, 1782–1792, 1986.

Finite elastic deformation governed by linear equations, *J. Appl. Mech.* **53**, 819–820, 1986.

Pouring Flows, (with J.-M. Vanden-Broeck) *Phys. Fluids* **29**, 3958–3961, 1986.

Finite amplitude vortices in curved channel flow, (with W.H. Finlay and J.H. Ferziger), Proc. of the 25th AIAA Aerospace Sciences Meeting, Reno, Jan. 1987, 1–9.

Free surface flow due to a sink, (with J.-M. Vanden-Broeck) *J. Fluid Mech.* **175**, 109–117, 1987.

Weir Flows, (with J.-M. Vanden-Broeck) *J. Fluid Mech.* **176**, 283–293, 1987.

Acoustoelasticity, *Dynamical problems in continuum physics*, eds. J.L. Bona, C. Dafermos, J.L. Ericksen and D. Kinderlehrer, Springer-Verlag, New York, 1987, pp. 193–203.

Impact with an impulsive frictional moment, *ASME J. Appl. Mech.* **54**, 239–240, 1987.

Effective conductivities of reciprocal media, *Random Media*, ed. G. Papanicolaou, Springer-Verlag, New York, 1987, pp. 183–188.

Addendum: Asymptotic theory of nonlinear wave propagation, (with S. Kogelman), *SIAM J. Appl. Math.* **47**, 454, 1987.

Caustics of nonlinear waves (with J. K. Hunter), *Wave Motion* **9**, 429–443, 1987.

Asymptotic behavior of stability regions for Hill's equation, (with M.I. Weinstein), *SIAM J. Appl. Math.* **47**, 941–958, 1987.

Stability of periodic plane waves, (with P.K.Newton) *SIAM J.Appl. Math.* **47**, 959–964, 1987.

Sound waves in a periodic medium containing rigid spheres, (with Dov Bai) *J. Acoust. Soc. Am.* **82**, 1436–1441, 1987.

Effective conductivity of periodic composites composed of two very unequal conductors, *J. Math. Phys.* **28**, 2516–2520, 1987.

Lower bounds on permeability (with J. Rubinstein) *Phys. Fluids* **30**, 2919–2921, 1987.

Ropes in Equilibrium, (with J.H. Maddocks), *SIAM J. Appl. Math.* **47**, 1185–1200, 1987.

Misuse of game theory, *J. Chronic Diseases*, **40**, 1147–1148, 1987.

Newton's Second Law, *Am. J. Phys*, **55**, 1145–1146, 1987.

Precipitation Pattern Formation, (with M.S. Falkowitz), *J.Chem.Phys.*, **88**, 416–421, 1988.

Stability of plane wave solutions of nonlinear systems (with P.K. Newton), *Wave Motion*, **10**, 183-191, 1988.

Resonantly interacting water waves, *J. Fluid Mech.*, **191**, 529-534, 1988.

Nonlinear hyperbolic waves, (with J. K. Hunter), Proc. Royal Society Lond., **A 417**, 299-308, 1988.

Instability and transition in curved channel flow, (with W.H. Finlay and J.H. Ferziger), *J. Fluid Mech.*, **194**, 417-456, 1988.

Flows over rectangular weirs, (with F. Dias and J.-M. Vanden-Broeck), *Phys. Fluids*, **31**, 2071-2076, 1988.

Spilling, *But the Crackling is Superb*, N. and G. Kurti, eds., Hilger, Bristol, 1988, 25-26.

Instability and transition in nonaxisymmetric curved channel flow (with W.H. Finlay and J.H. Ferziger), AIAA Paper No. 88-3761, First National Fluid Dynamics Congress, July 1988, 1–8.

Approximations for regression with covariate measurement error, (with A.S. Whittemore), *J. Am. Stat. Assoc.*, **83**, 1057-1066, 1988.

Fast reaction, slow diffusion and curve shortening (with J. Rubinstein and P. Sternberg), *SIAM J. Appl. Math.*, **49**, 116-133, 1989.

Surfing on solitary waves, (with J.-M. Vanden-Broeck), *J. Fluid Mech.*, **198**, 115-125, 1989.

Pouring flows with separation, (with J.-M. Vanden-Broeck), *Phys. Fluids A* **1**, 156–158, 1989.

Fair Dice, (with P. Diaconis) *Am. Math. Monthly*, **96**, 337-339, 1989.

Sedimentation of a dilute suspension, (with J. Rubinstein), *Phys. Fluids A* **1**, 637-643, 1989.

Exact non-reflecting boundary conditions, (with D. Givoli), *J. Comp. Phys.*, **82**, 172-192, 1989.

Ocular dominance column development: analysis and simulation, (with K.D. Miller and M.P. Stryker), *Science*, **245**, 605-615, 11 August 1989.

Particle distribution functions in suspensions (with J. Rubinstein), *Phys. Fluids A* **1**, 1632–1641, 1989.

Reaction-diffusion processes and evolution to harmonic maps, (with J. Rubinstein and P. Sternberg), *SIAM J. Appl. Math*, **49**, 1722-1733, 1989.

A Finite Element Method for Large Domains, (with D. Givoli), *Comp. Meth. Appl. Mech. and Eng.*, **76**, 41–66, 1989.

Heat transport into a shear flow at high Peclet number, *Proc. Roy. Soc. Lond. A*, **427**, 25-30, 1990.

On unsymmetrically impinging jets, *J. Fluid Mech.*, **211**, 653–655, 1990.

Partition asymptotics from recursion equations, (with C. Knessl), *SIAM J. Appl. Math.*, **50**, 323–338, 1990.

Stability of crystals that grow or evaporate by step propagation, (with R. Ghez and H.G. Cohen) *Appl. Phys. Lett.*, **56**, 1977–1979, 1990.

Non-reflecting boundary conditions for elastic waves (with D. Givoli), *Wave Motion*, **12**, 261–279, 1990.

Collapse of wavefunctions and probability densities, *Am. J. Phys.*, **58**, 768–770, 1990.

Slender jets and thin sheets with surface tension (with L. Ting), *SIAM J. Appl. Math.*, **50**, 1533–1546, 1990.

Stirling number asymptotics from recursion equations using the ray method, (with C. Knessl), *Studies in Applied Math.* **84**, 43–56, 1991.

Diffusively coupled dynamical systems, *Applied and Industrial Mathematics*, R. Spigler, ed., Kluwer, Amsterdam 1991, 49–56.

Changes in adiabatic invariants (with Ye Mu), *Annals of Physics*, **205**, 219–227, 1991.

Asymptotic properties of eigenvalues of integral equations, (with C. Knessl), *SIAM J. Appl. Math.*, **51**, 214–232, 1991.

Nonlinear wave motion in a strong potential, (with J. Rubinstein), *Wave Motion*, **13**, 291–302, 1991.

Nonlinear eigenvalue problems under strong localized perturbations with applications to chemical reactors, (with M.J. Ward), *Studies in Applied Math.* **85**, 1–28, 1991.

Mathematical model of granulocytopoiesis and chronic myelogenous leukemia, (with A.S. Fokas and B.D. Clarkson), *Cancer Research*, **51**, 2084–2091, 1991.

Flexural rigidity of a liquid surface, (with G.J. Merchant), *J. Statistical Phys.*, **63**, 1039–1051, 1991.

Family data determine all parameters in Mendelian incomplete penetrance models (with A.S. Whittemore and M.J. Ward), *Ann. Hum. Genet.* **55**, 175–177, 1991.

Low-Grade, Latent Prostate Cancer Volume: Predictor of Clinical Cancer Incidence?, (with A.S. Whittemore and R. Betensky), *J. Natl. Cancer Inst.*, **83**, 1231–1235, 1991.

Asymptotic behavior of high order differences of the partition function, (with C. Knessl), *Comm. Pure Appl. Math.*, **44**, 1033– 1045, 1991.

Surface tension (with A. King and G.J. Merchant), *Of Fluid Mechanics and Related Matters, Proceedings of a Symposium Honoring John Miles on his Seventieth Birthday*, R. Salmon and D. Betts, eds., Scripps Institute of Oceanography, U.C.S.D., 1991, 161–168.

Free surface flow around a ship, (with J.-M. Vanden-Broeck), *Mathematical Approaches in Hydrodynamics*, Touvia Miloh, ed., SIAM, Philadelphia, 1991, 289–299.

Contact Angles, (with G.J. Merchant), *Phys. Fluids A* **4**, 477–485, 1992.

A finite element method for domains with corners, (with D. Givoli and L. Rivkin), *Int. J. Num. Meth. Eng.*, **35**, 1329–1345, 1992.

Stability of rotating shear flows in shallow water, (with C. Knessl), *J. Fluid Mech.*, **244**, 605–614, 1992.

Diffraction of Acoustic Waves from Material Discontinuities, (with P. Barbone) Flow-Structure and Flow-Sound Interactions, Proceedings of the 1992 Symposium on Flow Induced Vibration and Noise, ASME Press, New York, T.M. Farabee and M.P Païdoussis, eds.

The shape of a Möbius band, (with L. Mahadevan), *Proc. Roy. Soc. Lond. A* **440**, 149–162, 1993.

Drop evaporation through a thin membrane, (with H.L. Frisch), *J. Colloid and Interface Science,* **155**, 262–263, 1993.

Phase fronts in reaction-diffusion problems, *Emerging applications in free boundary problems*, J. Chadam and H. Rasmussen, eds., Wiley, New York, 24–28, 1993.

Singularities of semilinear waves, (with L. Ting), *Comm. Pure Appl. Math.*, **46**, 341–352, 1993.

The stability of growing or evaporating crystals, (with R. Ghez and H.G. Cohen), *J. Appl. Phys.*, **73**, 3685–3693, 1993.

The stability of rapidly growing or evaporating crystals, (with G.J. Merchant and H.G. Cohen), *J. Appl. Phys.*, **73**, 3694–3697, 1993.

Strong localized perturbations of eigenvalue problems, (with M. J. Ward), *SIAM J. Appl. Math.*, **53**, 770–798, 1993.

Summing logarithmic expansions for singularly perturbed eigenvalue problems, (with M.J. Ward and W.D. Henshaw), *SIAM J. Appl. Math.* **53**, 799–828, 1993.

Stresses in narrow regions, *J. Appl. Mech.*, **60**, 1054–1056, 1993.

Front interaction and nonhomogeneous equilibria for tri-stable reaction-diffusion equations, (with J. Rubinstein and P. Sternberg), *SIAM J. Appl. Math.*, **53**, 1669–1685, 1993.

Asymptotic evaluation of oscillatory sums, (with C. Knessl), *Euro. J. Appl. Math.*, **4**, 361–380, 1993.

Asymptotic and numerical results for blowing-up solutions to semilinear heat equations, (with J. Lowengrub), *Singularities in Fluids, Plasmas and Optics*, R.E. Caflisch and G.C. Papanicolaou, eds., Kluwer 1993, 111–129.

Removing small features from computational domains, *J. Comp. Phys.*, **113**, 148–150, 1994.

Eulerian number asymptotics (with E. Giladi). *Proc. Royal Soc. A*, **445**, 291–303, 1994.

Nonreflecting boundary conditions (with M. Grote), Anniversary Volume, DCAMM, Technical Univ. of Denmark, Lyngby, Denmark, October 1994, 47–54.

Optimal Pricing of Scarce Natural Resources, (with P.S. Hagan, D.E. Woodward and R.E. Caflish), *Appl. Math. of Finance*, **1**, 87–108, 1994.

A Characterization of the Poisson Distribution and the Probability of Winning a Game, *Am. Stat.*, **48**, pages 294–298, 1994.

Special finite elements for use with high order boundary conditions (with D. Givoli), *Comp. Meth. Appl. Mech. and Eng.*, **199**, 199–213, 1994.

Range of the first two eigenvalues of the Laplacian, (with S. A. Wolf), *Proc. Royal Soc. A*, **447**, 397–412, 1994.

Blob Formation, (with A. King and L. Ting), *Phys. of Fluids*, **7**, 226–228, 1995.

Exact non-reflecting boundary condition for the time dependent wave equation, (with Marcus Grote), *SIAM J. Appl. Math.* **55**, 280–297, 1995.

How many shuffles to mix a deck?, *SIAM Review*, **37**, 88–89, 1995.

Rossby Waves, (with C. Knessl), *Studies in Appl. Math.*, **94**, 359–376, 1995.

Short acoustic, electromagnetic and elastic waves in random media, (with W. Boyse), *J. Opt. Soc. Amer. A*, **12**, 380–389, 1995.

Wave propagation, *Proc. Int'l. Cong. Math.*, Zurich 1994.

Asymptotic methods for partial differential equations: the reduced wave equation and Maxwell's equations, (with R.M. Lewis), *Surveys in Applied Mathematics*, edited by J.B. Keller, G. Papanicolau, and D. McLaughlin, Plenum Publishing, NY, 1995.

Stability of linear shear flows in shallow water, (with C. Knessl), *J. Fluid Mech.*, **303**, 203–214, 1995.

A Hybrid Asymptotic-Numerical Method for Calculating Low Reynolds Number Flows Past Symmetric Cylindrical Bodies, (with M.C.A. Kropinski and M.J. Ward), *SIAM J. Appl. Math.***55**,1484–1510, 1995.

Periodic Folding of Thin Sheets, (with L. Mahadevan), *SIAM J. Appl. Math.*,**55**, 1609–1624, 1995.

On nonreflecting boundary conditions, (with M. Grote), *J. Comp. Phys.*, **122**, 231–243, 1995.

Asymptotics beyond all orders for a low Reynolds number flow, (with M. Ward), *J. Eng. Math.*, **30**, 253–265, 1996.

Coiling of Flexible Ropes, (with L. Mahadevan), *Proc. Royal Soc.* A, **452**, 1679–1694, 1996.

Nonreflecting Boundary Conditions for Time Dependent Scattering, (with M. Grote), *J. Comp. Phys.*, **127**, 52–65, 1996.

Stability of the P to S energy ratio in the diffusive regime (with G. Papanicolaou and L. Ryzhik), *Bulletin of the Seismological Soc. of Amer.*, **86**, 1107–1115, 1996.

Turbulent diffusion in a Gaussian velocity field, 1996 Summer Study Program in Geophysical Fluid Dynamics, Woods Hole Oceanographic Institution, 1996.

Exact boundary conditions on artificial boundaries, *Proc. 3rd Hellenic-European Conf. on Math. and Informatics*, Sept. '96, Athens, Greece, 75–88.

Three dimensional water waves (with P. Milewski), *Studies in Appl. Math.*, **97**, 149–166, 1996.

Transport equations for elastic and other waves in random media (with G. Papanicolaou, L. Ryzhik), *Wave Motion*, **24**, 327–370, 1996

Advection-diffusion past a strip I. Normal Incidence. (with C. Knessl), *J. Math. Phys.*, **38**, 267–282, 1997.

Advection-diffusion past a strip II. Oblique Incidence. (with C. Knessl), *J. Math. Phys.*, **38**, 902–925, 1997.

High Order Boundary Conditions and Finite Elements for Infinite Domains, (with D. Givoli and I. Patlashenko), *Comp. Meth. Appl. Mech. Eng.*, **143**, 13–39, 1997.

Axisymmetric free surface with a 120° angle along a circle, (with J.-M. Vanden Broeck), *J. of Fluid Mech.*, **342**, 403–409, 1997.

Boundary and initial boundary-value problems for separable backward-forward parabolic problems, (with H.F.Weinberger), *J. Math. Phys.*, **38**, p. 4343–53, 1997.

Iterative solution of elliptic problems by approximate factorization, (with E. Giladi), *J. Comp. and Appl. Math.*, **85**, 287–313, 1997.

Transport Equations for Waves in a Half Space, (with L. Ryzhik and G. Papanicolaou), *Comm. P.D.E.*, **22**, 1869–1910, 1997.

Homogenization of rough boundaries and interfaces (with J. Nevard), *SIAM J. Appl. Math.* , **57**, 1660–1686, 1997.

Large Deviation Theory for Stochastic Difference Equations, (with R. Kuske), *Euro. J. Appl. Math.*, **8**, 567–580, 1997.

Nonreflecting Boundary Conditions for Maxwell's Equations (with M. Grote), *J. Comp. Phys.* **139**, 327–342, 1998.

Inner and outer iterations for the Chebyshev algorithm (with E. Giladi and G. Golub), *SIAM J. Num. Anal.*, **35**, 300, 1998.

Singularities on free surfaces of fluid flows (with P. Milewski and J.-M. Vanden-Broeck), *Studies in Appl. Math.*, **100**, 245–267, 1998.

Gravity waves on ice-covered water, *J. Geophys. Res.- Oceans*, vol. 103, **C-4**, 7663–7669, April 15, 1998.

Advection-diffusion around a curved obstacle (with D. Ahluwalia and C. Knessl) *J. Math. Phys.*, **39**, 3694–3710, 1998.

Optimal exercise boundary for an American put option, (with R. Kuske), *Appl. Math. Fin.*, **5**, 107–116, 1998.

Weak Shock Diffraction and Singular Rays (Extended Abstract), (with L. Ting), *ZAMM*, **78**, Suppl. 2, pp. S767–770, 1998.

Discrete Dirichlet-to-Neumann maps for unbounded domains, (with D. Givoli and I. Patlashenko), *Comp. Meth. Appl. Mech. & Eng..*, Vol. 164, 173–185, 1998.

Surface tension force on a partially submerged body, *Phys. of Fluids*, **10**, 3009-3010, 1998.

Singularities at the tip of a plane angular sector, *J. Math. Phys.*, **40**, 1087–1092, 1999.

Ray solution of a backward-forward parabolic problem for data handling systems (with C. Knessl) *Euro. J. Appl. Math.* (in press).

Exact nonreflecting boundary conditions for elastic waves, (with M. Grote) *SIAM J. Appl. Math.* (in press).

Probability of Brownian motion hitting an obstacle (with C. Knessl) *SIAM J. Appl. Math.* (in press).

Transport Theory for acoustic waves with reflection and transmissio at interfaces (with G. Papanicolaou, G. Bal, L. Ryzhik) *Wave Motion* (in press).

JOURNAL OF THE OPTICAL SOCIETY OF AMERICA VOLUME 52, NUMBER 2 FEBRUARY, 1962

Geometrical Theory of Diffraction*

JOSEPH B. KELLER
Institute of Mathematical Sciences, New York University, New York, New York
(Received September 13, 1961)

The geometrical theory of diffraction is an extension of geometrical optics which accounts for diffraction. It introduces diffracted rays in addition to the usual rays of geometrical optics. These rays are produced by incident rays which hit edges, corners, or vertices of boundary surfaces, or which graze such surfaces. Various laws of diffraction, analogous to the laws of reflection and refraction, are employed to characterize the diffracted rays. A modified form of Fermat's principle, equivalent to these laws, can also be used. Diffracted wave fronts are defined, which can be found by a Huygens wavelet construction. There is an associated phase or eikonal function which satisfies the eikonal equation. In addition complex or imaginary rays are introduced. A field is associated with each ray and the total field at a point is the sum of the fields on all rays through the point. The phase of the field on a ray is proportional to the optical length of the ray from some reference point. The amplitude varies in accordance with the principle of conservation of energy in a narrow tube of rays. The initial value of the field on a diffracted ray is determined from the incident field with the aid of an appropriate diffraction coefficient. These diffraction coefficients are determined from certain canonical problems. They all vanish as the wavelength tends to zero. The theory is applied to diffraction by an aperture in a thin screen diffraction by a disk, etc., to illustrate it. Agreement is shown between the predictions of the theory and various other theoretical analyses of some of these problems. Experimental confirmation of the theory is also presented. The mathematical justification of the theory on the basis of electromagnetic theory is described. Finally, the applicability of this theory, or a modification of it, to other branches of physics is explained.

1. INTRODUCTION

GEOMETRICAL optics, the oldest and most widely used theory of light propagation, fails to account for certain optical phenomena called diffraction. We shall describe an extension of geometrical optics which does account for these phenomena. It is called the geometrical theory of diffraction.[1,2] Like geometrical optics, it assumes that light travels along certain straight or curved lines called rays. But it introduces various new ones, called diffracted rays, in addition to the usual rays. Some of them enter the shadow regions and account for the light there while others go into the illuminated regions.

Diffracted rays are produced by incident rays which hit edges, corners, or vertices of boundary surfaces, or which graze such surfaces. Ordinary geometrical optics does not describe what happens in these cases but the new theory does. It does so by means of several laws of diffraction which are analogous to the laws of reflection and refraction. Like them, the new laws are deducible from Fermat's principle, appropriately modified. Away from the diffracting surfaces, diffracted rays behave just like ordinary rays.

In terms of the new rays, diffracted wave fronts can be defined. A Huygens wavelet construction can also be devised to determine them. It is also possible to introduce an eikonal or phase function which is constant on these wave fronts and which satisfies the usual eikonal equation. Thus all the fundamental principles of ordinary geometrical optics can be extended to the geometrical theory of diffraction.

Ordinary geometrical optics is often used to determine the distribution of light intensity, polarization, and phase throughout space. This is accomplished by assigning a field value to each ray and letting the total field at a point be the sum of the fields on all the rays through that point. The phase of the field on a ray is assumed to be proportional to the optical length of the ray from some reference point where the phase is zero. The amplitude is assumed to vary in accordance with the principle of conservation of energy in a narrow tube of rays. The direction of the field, when it is a vector, is given by a unit vector perpendicular to the ray. This vector slides parallel to itself along the ray in a homogeneous medium, and rotates around the ray in a specific way as it slides along it in an inhomogeneous medium.

Exactly the same principles as those just described can be used to assign a field to each diffracted ray. The only difficulty occurs in obtaining the initial value of the field at the point of diffraction. In the case of the ordinary rays, the field on a ray emerging from a source is specified at the source. But on a reflected or transmitted ray, the initial value is obtained by multiplying the field on the incident ray by a reflection or transmission coefficient. By analogy the initial value of the field on a diffracted ray is obtained by multiplying the field on the incident ray by a diffraction coefficient, which is a matrix for a vector field. There are different coefficients for edge diffraction, vertex diffraction, etc.

All the diffraction coefficients vanish as the wavelength λ of the field tends to zero. Then the sum of the fields on all diffracted rays, which we call the diffracted field, also vanishes. The geometrical optics field alone remains in this case, as we should expect because diffraction is usually attributed to the fact that λ is not zero.

* This paper was presented as an invited address at the Seminar on Recent Developments in Optics and Related Fields at the October 13, 1960 meeting of the Optical Society of America, Boston, Massachusetts. The research on which it is based was sponsored in part by the Air Force Cambridge Research Laboratories, Office of Aerospace Research.

[1] J. B. Keller, "The geometrical theory of diffraction," Proceedings of the Symposium on Microwave Optics, Eaton Electronics Research Laboratory, McGill University, Montreal, Canada (June, 1953).

[2] See J. B. Keller, in *Calculus of Variations and its Applications, Proceedings of Symposia in Applied Math*, edited by L. M. Graves (McGraw-Hill Book Company, Inc., New York and American Mathematical Society, Providence, Rhode Island, 1958), Vol. 8.

Dimensional considerations show that edge-diffraction coefficients are proportional to $\lambda^{\frac{1}{2}}$ and tip- or vertex-diffraction coefficients to λ. The field diffracted around a curved surface decreases exponentially with λ, and is consequently weaker than the field diffracted by a tip which is in turn weaker than that diffracted by an edge.

Diffraction coefficients can be characterized by recognizing that only the immediate neighborhood of the point of diffraction can affect their values. Thus the directions of incidence and diffraction, the wavelength, and the geometrical and physical properties of the media at the point of diffraction determine them. This suggests that they are determined by certain simpler problems in which only the local geometrical and physical properties enter. These "canonical" problems must be solved in order to determine the diffraction coefficients mathematically. Alternatively, experimental measurements on these canonical configurations can yield the coefficients.

The theory outlined above suffices for the analysis of a large class of situations involving diffraction. However, there remain phenomena which can be analyzed only by the inclusion of still another type of ray—the complex or imaginary ray. In a homogeneous medium such a ray is a complex straight line, while in an inhomogeneous medium it is a complex-valued solution of the differential equations for rays. Such rays occur as transmitted rays whenever total internal reflection occurs, and also in many other situations. Fields can be associated with these rays just as well as with the other kinds of rays.

A different kind of diffraction effect, not covered by the theory as explained so far, occurs at a caustic or focus of the ordinary or the diffracted rays. At such places neighboring rays intersect and the cross-sectional area of a tube of rays becomes zero. Consequently the principle of energy conservation in a tube of rays leads to an infinite amplitude for the field there. In order to obtain a finite value for the field at such places the present theory introduces a caustic correction factor. When the field on a ray passing through a caustic is multiplied by the appropriate factor, it becomes finite at the caustic. The caustic correction factors are determined by the wavelength and the local-ray geometry near the caustic, and are obtained from canonical problems.

In the subsequent sections this theory will be explained more fully and applied to some typical illustrative examples. Other applications of the theory will also be described. The mathematical interpretation of the theory in terms of asymptotic expansions will be explained in order to relate it to electromagnetic theory. The construction of similar theories in other branches of physics will also be commented upon.

2. EDGE-DIFFRACTED RAYS

The fundamental premise underlying the geometrical theory of diffraction is that light propagation is entirely a local phenomenon because the wavelength of light is so small. By this it is meant that the manner of propagation at a given point is determined solely by the properties of the medium and the structure of the field in an arbitrarily small neighborhood of the point. Furthermore all fields, no matter how they are produced, must have the same local structure, namely, that of a plane wave. Therefore, all fields must propagate in the same way. In particular, then, diffracted fields must travel along rays just like the ordinary geometrical optics field does, and, in fact, these rays must obey the ordinary geometrical-optics laws. The rays along which the diffracted field propagates are the diffracted rays.

Where do diffracted rays come from? It seems clear that they must be produced by some of the ordinary optical rays, but, which ones? The laws of propagation, reflection, and refraction enable us to follow the usual rays from the source outward, and determine where they go and what rays they produce, with some exceptions. The usual laws fail to specify what happens to a ray which hits an edge or a vertex, or grazes a boundary surface. Therefore, such rays must give rise to diffracted rays. We hypothesize that they do. In the case of edges, this hypothesis is related to Thomas Young's idea that diffraction is an edge effect.

This hypothesis can be tested mathematically in those cases where diffraction problems have been solved by other means. One such case is the diffraction of a plane wave by a semi-infinite screen with a straight edge. The solution of this problem obtained by Sommerfeld[3] consists of incident and reflected plane waves plus a third wave which is called a diffracted wave. When the incident wave is propagating in a direction normal to the edge of the screen, the diffracted wave is cylindrical with the edge as its axis. The straight lines orthogonal to the cylindrical wave fronts of the diffracted wave appear to come from the edge and can be identified with our diffracted rays. This example also suggests that an incident ray normal to the edge produces diffracted rays which are also normal to the edge and which leave it in all directions.

When the incident rays in the direction of propagation of the incident wave are oblique to the edge of the screen, the diffracted wave in Sommerfeld's solution is conical. This means that the diffracted wave fronts are parallel cones with the edge as their common axis. The straight lines orthogonal to these cones also appear to come from the edge, and can be identified with our diffracted rays. This example suggests the *law of edge diffraction.* A diffracted ray and the corresponding incident ray make equal angles with the edge at the point of diffraction, provided that they are both in the same medium. They lie on opposite sides of the plane normal to the edge at the point of diffraction. When the two rays lie in different media, the ratio of the sines of the

[3] A. J. W. Sommerfeld, *Optics* (Academic Press, Inc., New York, 1954).

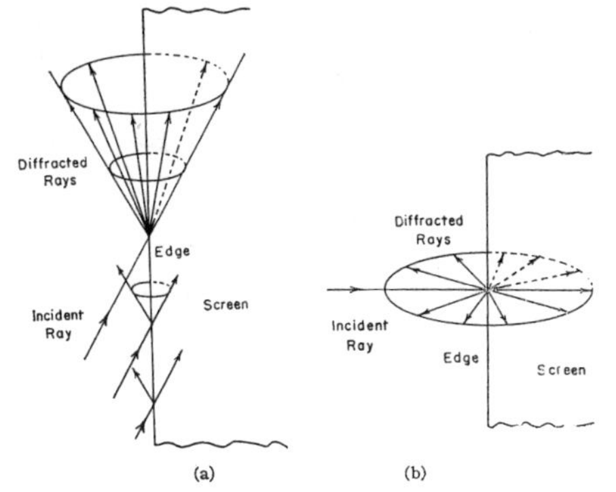

FIG. 1. (a) The cone of diffracted rays produced by an incident ray which hits the edge of a thin screen obliquely. (b) The plane of diffracted rays produced by a ray normally incident on the edge of a thin screen.

angles between the incident and diffracted rays and the normal plane is the reciprocal of the ratio of the indices of refraction of the two media. See Fig. 1.

The second part of this law, pertaining to diffraction into a different medium, is not suggested by the above example but is suggested by Snell's law of refraction. Both parts of the law are consequences of the following modified form of Fermat's principle which we call *Fermat's principle for edge diffraction.* An edge-diffracted ray from a point P to a point Q is a curve which has stationary optical length among all curves from P to Q with one point on the edge.

The derivation of the law of edge diffraction from this principle is particularly simple when the edge is straight and both rays lie in the same homogeneous medium. Then it is obvious that the ray consists of two straight-line segments meeting at a point on the edge. Let us rotate the plane containing the edge and the point Q around the edge until it contains P. In doing so the length of the segment from Q to the axis is unchanged, and the angle between the segment and the axis is unchanged. After the rotation, P, Q, and the edge lie in one plane and the stationary optical path is that given by the law of reflection. Thus the two segments must make equal angles with the edge and lie on opposite sides of the plane normal to the edge at the point of diffraction —but this is the law of edge diffraction for rays in the same medium. A similar argument, using Snell's law, yields the law of edge diffraction for rays in different media.

The law of edge diffraction is also confirmed by several approximate solutions of the problem of diffraction by an aperture in a thin screen. The Kirchhoff method, sometimes called the method of physical optics, represents the field diffracted through an aperture as an integral over the aperture. Rubinowicz[4] evaluated this integral asymptotically for short wavelengths when the incident field was a spherical wave, by using the Maggi transformation. He showed that the diffracted field at a point Q consisted of contributions from a small number of points on the edge. If we draw straight lines from these points to Q and call them diffracted rays we find that all of them at smooth parts of the edge satisfy the law of edge diffraction. The integrals of the Kirchhoff and modified Kirchhoff method, which employs Rayleigh's formulas, have been evaluated asymptotically for short wavelengths by van Kampen[5] and by Keller *et al.*[6] for arbitrary incident fields. The latter authors also evaluated Braunbek's[7] improved version of the Kirchhoff integrals. In all cases the points on smooth parts of the edge which contribute to the diffracted field correspond to diffracted rays satisfying the law of edge diffraction.

An indirect experimental verification of the existence of edge diffracted rays and of the law of edge diffraction is contained in the results of Coulson and Becknell.[8] They photographed the cross sections of the shadows of thin opaque disks of various shapes and detected bright lines in the shadows. When the incident field was normally incident on a disk, the bright line was found to be the evolute of the edge of the disk. In the special case of a circle the evolute is just the center, which appears as the well-known bright spot on the axis of the shadow. According to the law of edge diffraction, the diffracted rays lie in planes normal to the edge when the incident rays are normal to the edge. Therefore the caustic of these rays is a cylinder normal to the disk. Its cross section is the envelope of the normals to the edge in the plane of the disk, which is just the evolute of the edge. Thus the cross section of the caustic of the diffracted rays coincides with the bright lines which were observed. A similar interpretation applies to the bright lines observed at oblique incidence. Similar bright lines were observed by Nienhuis[9] in the diffraction patterns of apertures. He found that they were exactly the caustics of diffracted rays which he assumed to emanate normally from the edge.

3. FIELDS DIFFRACTED BY STRAIGHT EDGES

Let us now consider the field u_e on a ray diffracted from an edge.[10] For simplicity let us suppose that the ray is in a homogeneous medium so that it is a straight line. Let us begin with the two-dimensional case in which the edge is a straight line and the incident rays all lie in planes normal to the edge. Then the diffracted rays are also normal to the edge, and emanate from it in all

[4] A. Rubinowicz, Ann. Physik **53**, 257 (1917); **73**, 339 (1924).

[5] N. G. van Kampen, Physica **14**, 575 (1949).

[6] J. B. Keller, R. M. Lewis, and B. D. Seckler, J. Appl. Phys. **28**, 570 (1957).

[7] W. Braunbek, Z. Physik **127**, 381 (1950); **127**, 405 (1950).

[8] J. Coulson and G. G. Becknell, Phys. Rev. **20**, 594 (1922); **20**, 607 (1922).

[9] K. Nienhuis, Thesis, Groningen, 1948.

[10] J. B. Keller, J. Appl. Phys. **28**, 426 (1957).

directions. Thus it suffices to consider only the rays in one plane normal to the edge.

If λ denotes the wavelength of the incident field u_i, it is convenient to define the propagation constant $k=2\pi/\lambda$. We also denote by r the distance from the edge. Then the phase on a diffracted ray is just kr plus the phase ψ_i of the incident ray at the edge. To find the amplitude $A(r)$, which we assume to be a scalar, we consider as a tube of rays two neighboring rays in the same plane normal to the edge. Actually the tube is a cylinder of unit height. The cross-sectional area of this tube is proportional to r and the flux through it is proportional to rA^2. Since the flux must be constant, $A(r)$ is proportional to $r^{-\frac{1}{2}}$. The amplitude is also proportional to the incident amplitude A_i at the edge so we write $A(r)=DA_ir^{-\frac{1}{2}}$, where D is a diffraction coefficient. Thus the diffracted field is

$$u_e=DA_ir^{-\frac{1}{2}}e^{i(kr+\psi_i)}=Du_ir^{-\frac{1}{2}}e^{ikr}. \quad (1)$$

Let us compare our result (1) with Sommerfeld's[3] exact solution for diffraction of a plane scalar wave by a half-plane. When his result is asymptotically expanded for large values of kr, it agrees perfectly with (1) provided that

$$D=-\frac{e^{i\pi/4}}{2(2\pi k)^{\frac{1}{2}}\sin\beta}[\sec\tfrac{1}{2}(\theta-\alpha)\pm\csc\tfrac{1}{2}(\theta+\alpha)]. \quad (2)$$

Here β is the angle between the incident ray and the edge which is $\pi/2$ in this case. The angles between the incident and diffracted rays and the normal to the screen are θ and α, respectively (see Fig. 2). The upper sign applies when the boundary condition on the half-plane is $u=0$ while the lower sign applies if it is $\partial u/\partial n=0$.

The agreement between (1) and the exact solution confirms our theory and also determines the edge diffraction coefficient D. Similar agreement occurs for oblique incidence on a half-plane when (1) is replaced by the appropriate expression and the denominator $\sin\beta$ is included in (2). In this case θ and α are defined as above after first projecting the rays into the plane normal to the edge. In case the half-plane is replaced by a wedge of angle $(2-n)\pi$, comparison of (1), and its modified form for $\beta\neq\pi/2$, with Sommerfeld's exact solution for a wedge yields agreement when

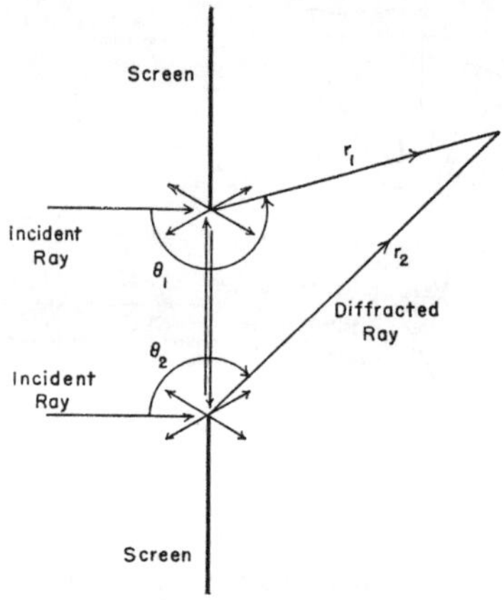

FIG. 3. The diffracted rays produced by a plane wave normally incident on a slit in a thin screen. The two incident rays which hit the slit edges are shown, with some of the singly diffracted rays they produce. One diffracted ray from each edge is shown crossing the slit and hitting the opposite edge, producing doubly diffracted rays.

$$D=\frac{e^{i\pi/4}\sin\frac{\pi}{n}}{n(2\pi k)^{\frac{1}{2}}\sin\beta}\left[\left(\cos\frac{\pi}{n}-\cos\frac{\theta-\alpha}{n}\right)^{-1}\mp\left(\cos\frac{\pi}{n}-\cos\frac{\theta+\alpha+\pi}{n}\right)^{-1}\right]. \quad (3)$$

For $n=2$ the wedge becomes a half-plane and (3) reduces to (2). [Equation (3) is misprinted in Eq. (A10) of reference 10.] In the electromagnetic case D is a matrix which has been determined in reference 10 for a perfectly conducting thin screen.

We shall now apply (1) and (2) to determine the field diffracted through an infinitely long slit of width $2a$ in a thin screen. By Babinet's principle, this will also be the field diffracted by a thin strip of width $2a$. For simplicity we shall assume that the incident field is a plane wave propagating in a direction normal to the edges of the slit. Then the problem is two dimensional and we can confine our attention to a plane normal to the edges. In this plane let the screen lie in the y axis of a rectangular coordinate system with the edges of the slit at $x=0$, $y=\pm a$. Let the incident field be $e^{ik(x\cos\alpha-y\sin\alpha)}$. Two singly diffracted rays, one from each edge, pass through any point P. Thus the singly diffracted field at P, $u_{e1}(P)$, is the sum of two terms of the form (1).

$$u_{e1}(P)=-\frac{e^{ik(r_1-a\sin\alpha)+i\pi/4}}{2(2\pi kr_1)^{\frac{1}{2}}}[\sec\tfrac{1}{2}(\theta_1+\alpha)\pm\csc\tfrac{1}{2}(\theta_1-\alpha)]$$
$$-\frac{e^{ik(r_2+a\sin\alpha)+i\pi/4}}{2(2\pi kr_2)^{\frac{1}{2}}}+[\sec\tfrac{1}{2}(\theta_2-\alpha)\pm\csc\tfrac{1}{2}(\theta_2+\alpha)]. \quad (4)$$

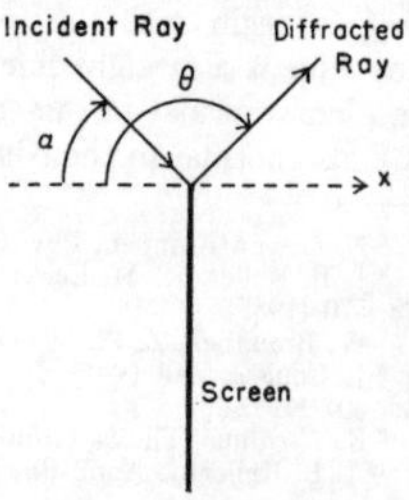

FIG. 2. The projection of incident and diffracted rays into a plane normal to the edge of a screen. The angles α and θ are those between the projections and the normal to the screen, measured as shown in the figure. The edge is normal to the plane of the figure.

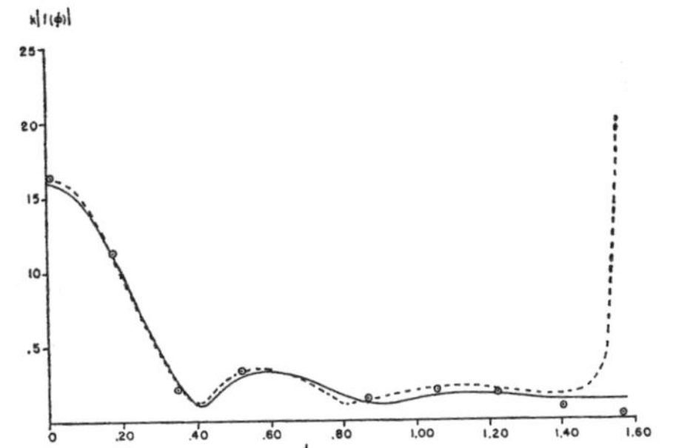

FIG. 4. The far-field diffraction pattern of a slit of width $2a$ hit normally by a plane wave; $ka=8$. The solid curve based upon Eq. (6) results from single diffraction, and applies to a screen on which $u=0$ or $\partial u/\partial n=0$. The dashed curve includes the effects of multiple diffraction for a screen on which $u=0$. The dots are based upon the exact solution of the reduced wave equation for a screen on which $u=0$.[11] The ordinate is $k|f(\varphi)|$ and the abscissa is φ in radians.

In (4) r_1 and r_2 denote the distances from P to the upper and lower edges, and the angles θ_1 and θ_2 are determined by the corresponding rays, as shown in Fig. 3.

The diffraction pattern of the slit, $f_s(\varphi)$, can be obtained from (4) by introducing the polar coordinates r,φ of P. When P is far from the slit $(r\gg a)$ we have $r_1\sim r-a\sin\varphi$, $r_2\sim r+a\sin\varphi$, $\theta_1\sim\pi+\varphi$, and $\theta_2\sim\pi-\varphi$. Then (4) becomes

$$u_{el}(P)=-(k/2\pi r)^{\frac{1}{2}}e^{ikr+i\pi/4}f_s(\varphi). \tag{5}$$

The diffraction pattern $f_s(\varphi)$ due to single diffraction is found to be

$$f_s(\varphi)=i\frac{\sin[ka(\sin\varphi+\sin\alpha)]}{k\sin\frac{1}{2}(\varphi+\alpha)}\pm\frac{\cos[ka(\sin\varphi+\sin\alpha)]}{k\cos\frac{1}{2}(\varphi-\alpha)}. \tag{6}$$

This equation shows that $|f_s(\varphi)|$ is not zero for any value of φ and that it is the same for the two types of boundary conditions on the screen.

In Figs. 4 and 5, $|kf_s(\varphi)|$ is shown as a function of φ for normal incidence $(\alpha=0)$. The range of φ is 0 to $\pi/2$ which is the region behind the slit. In Fig. 4 for which $ka=8$, the exact values of the diffraction pattern are shown. They are based upon numerical evaluation of the exact solution of the reduced wave equation with $u=0$ on the screen, and are given by Karp and Russek.[11] The agreement between our theory and the exact values is seen to be quite good even though the slit is only about 3 wavelengths wide.

The transmission cross section σ of the slit per unit length can be obtained simply by employing the cross-section theorem. According to this theorem it is equal to the imaginary part of the diffraction pattern in the forward direction $\varphi=-\alpha$. By applying the theorem to $f_s(\varphi)$ we obtain $\sigma=2a\cos\alpha$, which is just the geometrical optics result.

[11] S. N. Karp and A. Russek, J. Appl. Phys. 27, 886 (1956).

To obtain a better value of σ, let us consider the two doubly diffracted rays which go in the direction $\varphi=-\alpha$. They are produced by two singly diffracted rays which cross the slit and hit the opposite edges. The field at the upper edge on the singly diffracted ray from the lower edge is given by the second term in (4) with $\theta_2=\pi/2$ and $r_2=2a$. If we choose the upper sign, appropriate to a screen on which $u=0$, we obtain

$$u_{e1}=-\frac{e^{ika(2+\sin\alpha)+i\pi/4}}{2(\pi ka)^{\frac{1}{2}}}\sec\frac{1}{2}\left(\frac{\pi}{2}-\alpha\right). \tag{7}$$

We now use (7) as the incident field in (1) to obtain the field on the doubly diffracted rays. We proceed similarly to find the doubly diffracted field from the lower edge and add the two results to obtain the doubly diffracted field u_{e2}. Far from the slit we can write it in the form (5) with $f_d(\varphi)$ instead of $f_s(\varphi)$. In the direction $\varphi=-\alpha$, f_d has the value

$$f_d(-\alpha)=-\frac{1}{k(\pi ka)^{\frac{1}{2}}}\left[\frac{e^{i2ka(1+\sin\alpha)+i\pi/4}}{1+\sin\alpha}+\frac{e^{i2ka(1-\sin\alpha)+i\pi/4}}{1-\sin\alpha}\right]. \tag{8}$$

To obtain σ we apply the cross-section theorem which requires taking the imaginary part of $f_s(-\alpha)+f_d(-\alpha)$. The result is

$$\sigma=2a\cos\alpha-\frac{1}{k(\pi ka)^{\frac{1}{2}}}\left[\frac{\cos[2ka(1+\sin\alpha)-\pi/4]}{1+\sin\alpha}+\frac{\cos[2ka(1-\sin\alpha)-\pi/4]}{1-\sin\alpha}\right]. \tag{9}$$

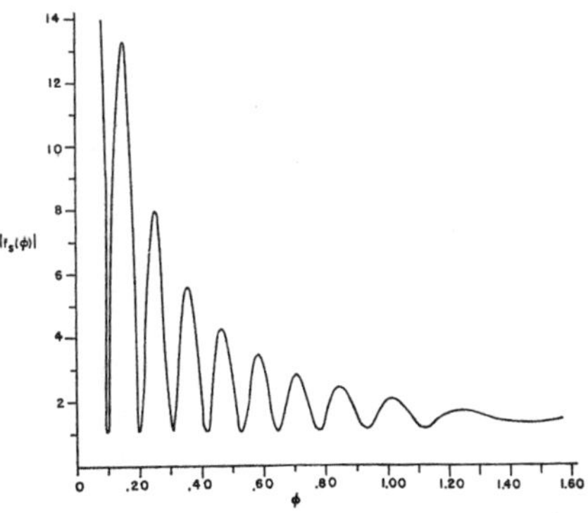

FIG. 5. The far-field diffraction pattern of a slit of width $2a$ hit normally by a plane wave; $ka=10\pi$. The curve, based on Eq. (6), results from single diffraction for a screen on which $u=0$ or $\partial u/\partial n=0$. The ordinate is $k|f_s(\varphi)|$ and the abscissa is φ in radians. The value at $\varphi=0$ is finite but too high to be shown.

This expression has also been obtained by H. Levine by a different method. For normal incidence, $\alpha=0$, and (9) becomes

$$\frac{\sigma}{2a}=1-\frac{\cos(2ka-\pi/4)}{\pi^{1/2}(ka)^{3/2}}. \tag{10}$$

In Fig. 6 $\sigma/2a$ is shown as a function of ka based on (10), and points obtained from the exact solution of the boundary-value problem are also shown. The agreement is quite good even when the wavelength is several times the slit width. When additional multiply diffracted rays are taken into account, even better agreement is obtained, as the dashed curve in the figure shows. The expression for the field obtained by summing all the multiply diffracted fields coincides with that obtained by Karp and Russek[11] in a different way.

The above calculation of the doubly diffracted field was performed for a screen on which $u=0$. For a screen on which $\partial u/\partial n=0$ no diffracted field is produced when a plane wave travels toward the edge in a direction parallel to the screen. As a consequence the diffraction coefficient D, given by (2) with the lower sign, vanishes when $\alpha=\pi/2$. In this case the diffracted field is proportional to $\partial u_i/\partial n$, the normal derivative of the incident field at the edge. Thus (1) must be replaced by

$$u_e=D'\frac{\partial u_i}{\partial n}\cdot r^{-1/2}e^{ikr}. \tag{11}$$

Here D' is a new diffraction coefficient which can be obtained by solving the problem of diffraction of the wave xe^{-iky} by a half-plane lying on the negative y axis. This was done by Karp and Keller[12] with the result

$$D'(\theta)=-\frac{1}{ik}\frac{\partial}{\partial\alpha}D(\theta,\pi/2)$$

$$=-\frac{e^{-i\pi/4}}{2(2\pi)^{1/2}k^{3/2}\sin^2\beta}\cdot\frac{\sin\left(\frac{\pi}{4}-\frac{\theta}{2}\right)}{\cos^2\left(\frac{\pi}{4}-\frac{\theta}{2}\right)}. \tag{12}$$

Let us use (11) and (12) to determine the doubly diffracted field for a screen on which $\partial u/\partial n=0$. First we obtain from the second term in (4) at $\theta_2=\pi/2$, $r_2=2a$ the result

$$\frac{\partial u_{e1}}{\partial n}=-\frac{e^{ika(2+\sin\alpha)+i\pi/4}}{8(\pi ka^3)^{1/2}}\cdot\frac{\sin\left(\frac{\pi}{4}-\frac{\alpha}{2}\right)}{\cos^2\left(\frac{\pi}{4}-\frac{\alpha}{2}\right)}. \tag{13}$$

We now use (13) and (12) in (11) to obtain the doubly diffracted field from the upper edge, and do a similar calculation for the lower edge. Then we add the two doubly diffracted fields to obtain

$$u_{e2}=-\frac{e^{ika(2+\sin\alpha)+ikr_1}}{16\pi(ka)^{3/2}(2kr_1)^{1/2}}\cdot\frac{\sin\left(\frac{\pi}{4}-\frac{\alpha}{2}\right)}{\cos^2\left(\frac{\pi}{4}-\frac{\alpha}{2}\right)}\cdot\frac{\sin\left(\frac{\pi}{4}-\frac{\theta_1}{2}\right)}{\cos^2\left(\frac{\pi}{4}-\frac{\theta_1}{2}\right)},$$

$$-\frac{e^{ika(2-\sin\alpha)+ikr_2}}{16\pi(ka)^{3/2}(2kr_2)^{1/2}}\cdot\frac{\sin\left(\frac{\pi}{4}+\frac{\alpha}{2}\right)}{\cos^2\left(\frac{\pi}{4}+\frac{\alpha}{2}\right)}\cdot\frac{\sin\left(\frac{\pi}{4}-\frac{\theta_2}{2}\right)}{\cos^2\left(\frac{\pi}{4}-\frac{\theta_2}{2}\right)}. \tag{14}$$

Far from the slit we write u_{e2} in the form (5) and obtain $f_d(\varphi)$. Then we apply the cross-section theorem to $f_s(-\alpha)+f_d(-\alpha)$ and find

$$\sigma=2a\cos\alpha-\frac{2a}{32\pi^{1/2}(ka)^{3/2}}\times\left\{\frac{\sin^2\left(\frac{\pi}{4}-\frac{\alpha}{2}\right)}{\cos^4\left(\frac{\pi}{4}-\frac{\alpha}{2}\right)}\sin[2ka(1+\sin\alpha)-\pi/4]+\frac{\sin^2\left(\frac{\pi}{4}+\frac{\alpha}{2}\right)}{\cos^4\left(\frac{\pi}{4}+\frac{\alpha}{2}\right)}\sin[2ka(1-\sin\alpha)-\pi/4]\right\}. \tag{15}$$

For normal incidence $\alpha=0$ and (15) becomes

$$\frac{\sigma}{2a}=1-\frac{\sin(2ka-\pi/4)}{8\pi^{1/2}(ka)^{3/2}}. \tag{16}$$

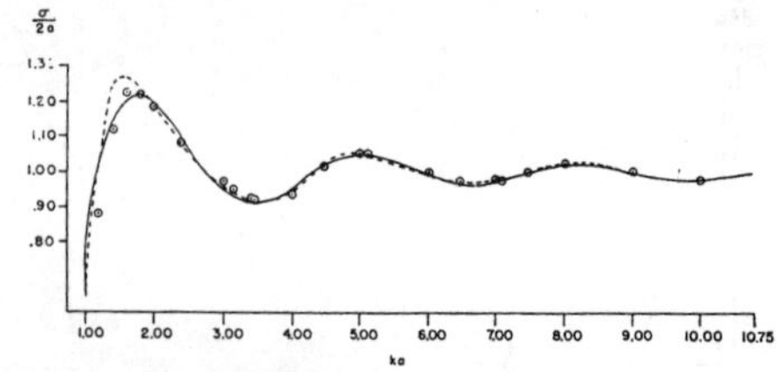

FIG. 6. The transmission cross section of a slit of width $2a$ as a function of ka, for normal incidence with $u=0$ on the screen. The solid curve, based on Eq. (10), results from single and double diffraction; the dashed curve includes single and all multiple diffraction. The dots are based upon the exact solution of the reduced wave equation with $u=0$ on the screen.[11] The ordinate is $\sigma/2a$ and the abscissa is ka.

[12] S. N. Karp and J. B. Keller, Optica Acta (Paris) 8, 61 (1961).

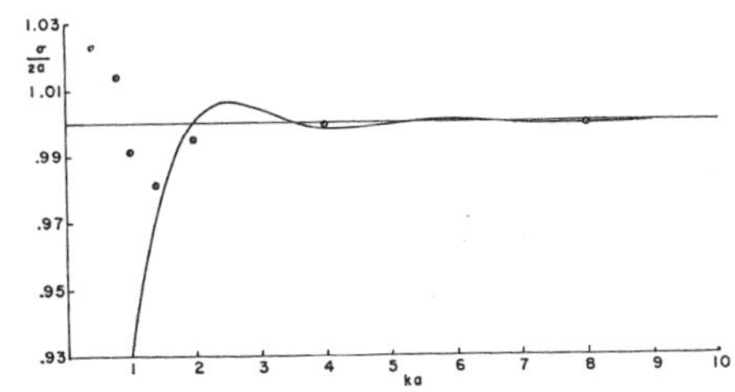

FIG. 7. The transmission cross section of a slit of width $2a$ as a function of ka, for normal incidence with $\partial u/\partial n=0$ on the screen. The curve, based on Eq. (16), results from single and double diffraction. The dots are based upon the exact solution of the reduced wave equation.[13] The ordinate is $\sigma/2a$ and the abscissa is ka.

A graph of $\sigma/2a$ versus ka based on (16) is shown in Fig. 7 together with some values of $\sigma/2a$ computed by Skavlem[13] from the exact solution of the boundary value problem.

Let us now consider a diffraction grating consisting of $2N+1$ parallel slits each of width $2a$ with centers a distance b apart. Let us again consider the two-dimensional case in which a plane wave is incident at angle α with its wavefronts parallel to the edges. Then one singly diffracted ray from each edge will pass through every point and the total field is the sum of $2(2N+1)$ terms of the form (1). Let us number the slits from $t=-N$ to $t=N$ with the center of slit t at $y=tb$ and let r_t,φ_t be polar coordinates with their origin at this center. We now add together the fields on the two rays from the edges of slit t. If $r_t\gg a$ we obtain the result (5) for that slit, with r_t,φ_t instead of r,φ and multiplied by the factor $e^{-iktb\sin\alpha}$ to account for the phase of the incident wave. Thus far from the grating we have

$$u_{el}=-\sum_{t=-N}^{t=N}\left(\frac{k}{2\pi r_t}\right)^{\frac{1}{2}}e^{ik(r_t-tb\sin\alpha)+i\pi/4}f_s(\varphi). \quad (17)$$

To simplify (17), let r,φ be polar coordinates with their origin at the center of the middle slit. At points far from the grating compared to its length $2Nb+2a$ we have $\varphi_t\sim\varphi$ and $r_t\sim r-tb\sin\varphi$ and (17) becomes

$$u_{el}=-\left(\frac{k}{2\pi r}\right)^{\frac{1}{2}}e^{ikr+i\pi/4}f_s(\varphi)\sum_{t=-N}^{t=N}e^{-iktb(\sin\varphi+\sin\alpha)}$$

$$=-\left(\frac{k}{2\pi r}\right)^{\frac{1}{2}}e^{ikr+i\pi/4}f_s(\varphi)\times\frac{\sin[(N+\frac{1}{2})kb(\sin\varphi+\sin\alpha)]}{\sin[\frac{1}{2}kb(\sin\varphi+\sin\alpha)]}. \quad (18)$$

The same result applies with N instead of $N+\frac{1}{2}$ if there are $2N$ slits and the origin is at the midpoint of the grating. The result expresses the diffraction pattern as the product of the pattern of a single slit and a grating factor which is the quotient of sines.

The method of this section has been applied to diffraction by a semi-infinite thick screen with a flat end and by a truncated wedge by Burke and Keller.[14] In computing the doubly diffracted field for a screen upon which $u=0$ it was found that $D=0$. This is related to the fact that the singly diffracted field vanishes at the edge. Here again the doubly diffracted field is proportional to $\partial u_i/\partial n$ at the edge. The corresponding diffraction coefficient $D'(\theta,n)$ was found by solving a canonical problem, as was done to obtain (12). If $D(\alpha,\theta,n)$ denotes the coefficient in (3), the result is

$$D'(\theta,n)=\frac{1}{ik}\frac{\partial}{\partial\alpha}D(-\pi/2,\theta,n)=\frac{2e^{-i\pi/4}\sin(\pi/n)}{n^2k^{\frac{1}{2}}(2\pi)^{\frac{1}{2}}}\times\frac{\sin\left[\left(\theta+\frac{\pi}{2}\right)\Big/n\right]}{\left\{\cos\frac{\pi}{n}-\cos\left[\left(\theta+\frac{\pi}{2}\right)\Big/n\right]\right\}}. \quad (19)$$

In the case of grazing incidence additional considerations which we shall not explain are required because the incident ray and a diffracted ray both continue along the surface together.

4. FIELDS DIFFRACTED BY CURVED EDGES

The field on an edge-diffracted ray is given by (1) only in the special case when the diffracted wave is cylindrical. In general (1) must be replaced by

$$u_e=Du_i[r(1+\rho_1^{-1}r)]^{-\frac{1}{2}}e^{ikr}. \quad (20)$$

Here ρ_1 denotes the distance from the edge to the caustic of the diffracted rays, measured negatively in the direction of propagation. (See Fig. 8.) We obtain (20) by employing the energy principle according to which A varies inversely as the square root of the cross-sectional area of a tube of rays. If ρ_1 and ρ_2 denote the radii of curvature of the wave front which is orthogonal to the tube at one cross section, then ρ_1+r and ρ_2+r are the corresponding radii a distance r further along the tube. (See Fig. 9.) Thus the cross-sectional area is proportional to $(\rho_1+r)(\rho_2+r)$ and A is proportional to $[(\rho_1+r)(\rho_2+r)]^{-\frac{1}{2}}$. When $\rho_2=0$, A varies as $[r(1+\rho_1^{-1}r)]^{-\frac{1}{2}}$. Since the diffracted rays form a caustic at the edge, $\rho_2=0$ there. If r denotes distance along a ray from the edge, the last expression for A applies. By using it together with the phase factor and the factors Du_i previously explained, we obtain (20).

The distance ρ_1 in (20) can be found by geometrical considerations. When the edge is a plane curve, it is

[13] C. J. Bouwkamp, Repts. Progr. in Phys. 17, 35 (1954).

[14] J. E. Burke and J. B. Keller, Research Rept. EDL-E48, Electronic Defense Laboratories, Sylvania Electronic Systems, Mountain View, California (March, 1960).

given by the following formula, which is analogous to the lens and mirror laws

$$\frac{1}{\rho_1}=-\frac{\dot{\beta}}{\sin\beta}-\frac{\cos\delta}{\rho\,\sin^2\beta}. \tag{21}$$

Here $\rho\geqq 0$ denotes the radius of curvature of the edge, β is the angle between the incident ray and the (positive) tangent to the edge, $\dot{\beta}$ is the derivative of β with respect to arclength s along the edge, and δ is the angle between the diffracted ray and the normal to the edge. Since both β and $\partial/\partial s$ change sign when the direction of increasing arclength is reversed, this direction can be chosen arbitrarily without effecting the value of ρ_1. The normal, which lies in the plane of the edge, is assumed to point towards its center of curvature.

As an application of (21) let us consider a spherical wave diffracted by a straight edge. In this case $\rho=\infty$ so the focal length of the edge is infinite and the last term in (21) vanishes. A simple calculation shows that $\dot{\beta}=-R^{-1}\sin\beta$ at a point on the edge a distance R from the source. Thus (21) yields $\rho_1=R$ as we should have expected. If $u_i=e^{ikR}/R$ then (20) yields

$$u_e=-\frac{e^{ik(R+r)+i\pi/4}}{2\sin\beta(2\pi k)^{\frac{1}{2}}[rR(r+R)]^{\frac{1}{2}}}\times[\sec\tfrac{1}{2}(\theta-\alpha)\pm\csc\tfrac{1}{2}(\theta+\alpha)]. \tag{22}$$

This result coincides with the asympotic expansion of the diffracted field given by the exact solution of the reduced wave equation for this problem, providing another confirmation of our theory.

When a plane wave is normally incident upon a screen containing an aperture of any shape, $\beta=\pi/2$ for every ray and (21) yields $\rho_1=-\rho/\cos\delta$. If $\beta=\pi/2$ it follows from the definitions of θ and δ that $\delta=\theta-\pi/2$ so $\rho_1=-\rho/\sin\theta$. For a circular aperture of radius a, $\rho=a$ and $\rho_1=-a/\sin\theta$. In this case two singly diffracted rays

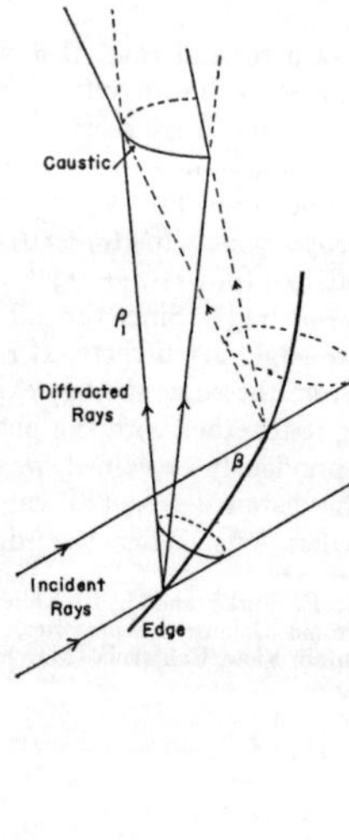

FIG. 8. A pair of neighboring incident rays hitting a curved edge, and some of the resulting diffracted rays. The two cones of diffracted rays intersect at the caustic, which is at the distance ρ_1 from the edge along the rays.

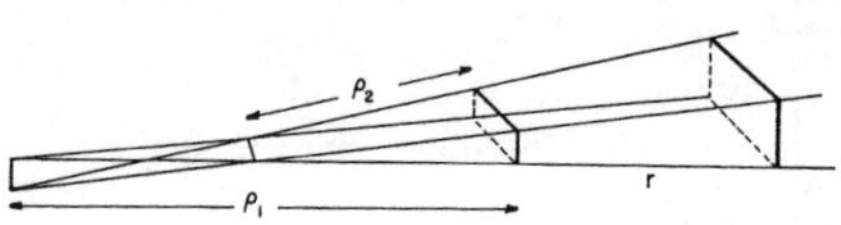

FIG. 9. A tube of rays and two small portions of wave fronts normal to them a distance r apart. The neighboring rays intersect at the two centers of curvature of the wave fronts. They are at the distances ρ_1 and ρ_2 from one wave front and at ρ_1+r and ρ_2+r from the other. The ratio of the areas of the wavefront sections is seen to be $\rho_1\rho_2/(\rho_1+r)(\rho_2+r)$.

pass through each point P not on the axis. They come from the nearest and farthest points on the edge. If the incident field is $u_i=e^{ikz}$ and the screen is in the plane $x=0$, we obtain by adding two terms of the form (20) the result

$$u_{el}(P)=-\frac{e^{ikr+i\pi/4}}{2(2\pi k)^{\frac{1}{2}}}\left[\sec\frac{\theta_1}{2}\pm\csc\frac{\theta_1}{2}\right][r_1(1-a^{-1}r_1\sin\theta_1)]^{-\frac{1}{2}}-\frac{e^{ikr_2+i\pi/4}}{2(2\pi k)^{\frac{1}{2}}}\left[\sec\frac{\theta_2}{2}\pm\csc\frac{\theta_2}{2}\right]\times[r_2(1-a^{-1}r_2\sin\theta_2)]^{-\frac{1}{2}}. \tag{23}$$

The angles and distances in (23) are as shown in Fig. 3.

Far from the aperture (23) simplifies to

$$u_{el}=-\frac{ke^{ikr}}{2\pi r}f_s(\varphi). \tag{24}$$

The diffraction pattern $f_s(\varphi)$ due to single diffraction is from (23)

$$f_s(\varphi)=k^{-\frac{3}{2}}(2\pi a/\sin\varphi)^{\frac{1}{2}}\left\{i\frac{\sin[ka\sin\varphi-(\pi/4)]}{\sin(\varphi/2)}\pm\frac{\cos[ka\sin\varphi-(\pi/4)]}{\cos(\varphi/2)}\right\}. \tag{25}$$

A graph of $|kf_s(\varphi)|$ based on (25) is given in Fig. 10 for $ka=3\pi$. It shows that $f_s(\varphi)$ is infinite at $\varphi=0$, which is a consequence of the fact that the axis, a caustic of the diffracted rays, and the shadow boundary $r=a$ both extend in the direction $\varphi=0$.

Let us examine (23) near the axis by introducing ρ and x, the distances from P to the axis and to the plane of the screen, and $\delta=\tan^{-1}(x/a)$. When $\rho\ll a$, (23) can be simplified to

$$u_{el}=-\frac{a^{\frac{1}{2}}e^{ik(x\sin\delta+a\cos\delta)}}{(2\pi k\rho)^{\frac{1}{2}}(x^2+a^2)^{\frac{1}{4}}}\left[\sec\left(\frac{\delta}{2}+\frac{\pi}{4}\right)\pm\csc\left(\frac{\delta}{2}+\frac{\pi}{4}\right)\right]\times\cos\left(k\rho\cos\delta-\frac{\pi}{4}\right). \tag{26}$$

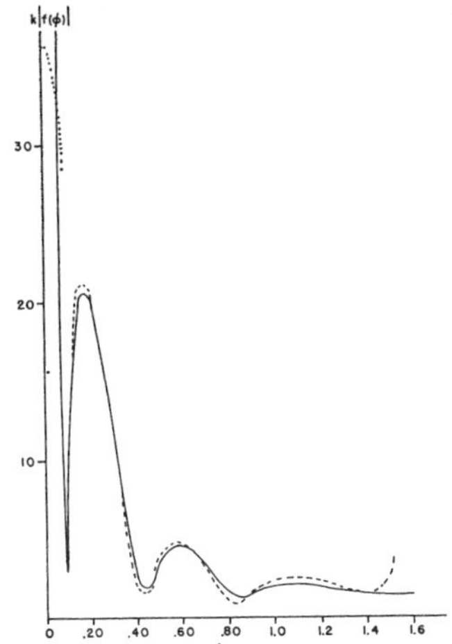

FIG. 10. The diffraction pattern of a circular hole of radius a hit normally by a plane wave; $ka=3\pi$. The solid curve, based on Eq. (25) results from single diffraction with either $u=0$ or $\partial u/\partial n=0$ on the screen. The dashed curve, for the case $u=0$ on the screen, also includes multiple diffraction. The dotted curve near $\varphi=0$ includes the caustic and shadow-boundary corrections given by (30).

This equation shows that u_{e1} is singular like $\rho^{-\frac{1}{2}}$ on the axial caustic $\rho=0$. To modify it so that u_{e1} is finite there, we consider the following exact solution of the reduced wave equation, in which δ is a constant and J_0 is the zero-order Bessel function

$$e^{ikz\sin\delta}J_0(k\rho\cos\delta)\sim\frac{2e^{ikz\sin\delta}}{(2\pi k\rho\cos\delta)^{\frac{1}{2}}}\cos(k\rho\cos\delta-\pi/4). \quad (27)$$

The right side of (27) is the asymptotic expansion of the solution on the left side when $k\rho$ is large. It shows that the solution corresponds to rays converging on an axial caustic, making the angle $\pi/2-\delta$ with the axis, just as is the case in (26). Near the axis the right side of (27) is not applicable and the same is true of (26). Therefore we assume that the right side of (26) can be converted to its correct value on and near the axis by multiplying it by the ratio of the left side of (27) to the right side, and we call this ratio the axial caustic correction factor F

$$F=\tfrac{1}{2}(2\pi k\rho\cos\delta)^{\frac{1}{2}}\sec(k\rho\cos\delta-\pi/4)J_0(k\rho\cos\delta). \quad (28)$$

[The factor $\frac{1}{2}$ was omitted in (A16) of reference 10]. When we multiply the right side of (26) by F, we obtain on and near the axis

$$u_{e1}=-\frac{(a\cos\delta)^{\frac{1}{2}}}{2(x^2+a^2)^{\frac{1}{4}}}\left[\sec\left(\frac{\delta}{2}+\frac{\pi}{4}\right)\pm\csc\left(\frac{\delta}{2}+\frac{\pi}{4}\right)\right]$$
$$\times J_0(k\rho\cos\delta)\exp[ik(x^2+a^2)^{\frac{1}{2}}]. \quad (29)$$

Far from the screen δ approaches $\pi/2$ and $\sec[(\delta/2)+(\pi/4)]\approx 2x/a$, so u_{e1} does not decrease with x along the axis. This is because the shadow boundary $r=a$ extends in the direction parallel to the boundary. When we combine u_{e1} with the geometrical optics field $u_g=e^{ikx}$ in the region $r<a$, the plane wave terms cancel and we obtain a spherical wave

$$u_g+u_{e1}\approx e^{ikx}-\frac{a}{2x}e^{ikx}\left(1+\frac{ika^2}{2x}\right)\frac{2x}{a}J_0(ka\sin\varphi)$$
$$=-\frac{ke^{ikx}}{2\pi x}i\pi a^2J_0(ka\sin\varphi). \quad (30)$$

Thus $f_s(\varphi)$ has the finite value $i\pi a^2J_0(ka\sin\varphi)$ near $\varphi=0$, rather than the singular value given by (25). The transmission cross section σ is the imaginary part of $f(0)$ according to the cross-section theorem, so (30) yields $\sigma=\pi a^2$, the geometrical optics value.

The result (30) applies only in the immediate vicinity of $\varphi=0$ since its derivation employed the geometrical optics field in the region $r<a$. To obtain an improved form of the diffraction pattern (25) which is valid for a larger range of φ, we shall introduce a caustic and shadow boundary correction directly into (25). We do this by observing that $(\sin\varphi)^{-\frac{1}{2}}\sin(ka\sin\varphi-\pi/4)$ and $(\sin\varphi)^{-\frac{1}{2}}\cos(ka\sin\varphi-\pi/4)$ are proportional to the asymptotic forms of $J_1(ka\sin\varphi)$ and $J_0(ka\sin\varphi)$, respectively. Upon using these functions in (25) in place of their asymptotic forms, we obtain

$$f_s(\varphi)=\frac{\pi a}{k}\left[\frac{iJ_1(ka\sin\varphi)}{\sin(\varphi/2)}\pm\frac{J_0(ka\sin\varphi)}{\cos(\varphi/2)}\right]. \quad (31)$$

Let us now consider the doubly diffracted field in order to obtain a more accurate value of σ. For a screen on which $u=0$ the first term in (23) becomes at $r_1=2a$, $\theta=\pi/2$

$$u_{e1}=-\frac{e^{2ika+i\pi/4}}{(2\pi ka)^{\frac{1}{2}}}. \quad (32)$$

This is the field on a ray which has crossed the aperture to the opposite edge. By treating it as the incident field in (20) we obtain for the doubly diffracted field

$$u_{e2}=\frac{e^{ik(r_1+2a)}}{2\pi ka^{\frac{1}{2}}}[r_1(1-a^{-1}r_1\cos\delta_1)]^{-\frac{1}{2}}\sec\frac{\delta_1}{2}$$
$$+\frac{e^{ik(r_2+2a)}}{2\pi ka^{\frac{1}{2}}}[r_2(1-a^{-1}r_2\cos\delta_2)]^{-\frac{1}{2}}\sec\frac{\delta_2}{2}. \quad (33)$$

Near the axis we find by comparing (33) with (26) that

$$u_{e2}=-\frac{2e^{2ika+\pi/4}}{(2\pi ka)^{\frac{1}{2}}}\sec\frac{\delta}{2}\left[\sec\left(\frac{\delta}{2}+\frac{\pi}{4}\right)+\csc\left(\frac{\delta}{2}+\frac{\pi}{4}\right)\right]^{-1}u_{e1}. \quad (34)$$

From (34) we see that u_{e2} becomes infinite on the axis like u_{e1} does and it can be made regular by the same caustic correction. The result can be obtained by using (29) in (34). In this way we obtain at points far behind the screen and near the axis

$$u_{e2} = -\frac{ke^{ikx}}{2\pi x}\cdot\left[-\frac{2}{k}\left(\frac{\pi a}{k}\right)^{\frac{1}{2}} e^{2ika-i\pi/4}J_0(ka\sin\varphi)\right]. \quad (35)$$

The bracketed quantity in (35) is $f_d(\varphi)$, the diffraction-pattern contribution from double diffraction. Upon applying the cross-section theorem to it, and adding the result to πa^2, which was obtained from (30), we find

$$\frac{\sigma}{\pi a^2} = 1 - \frac{2\sin[2ka-\pi/4]}{\pi^{\frac{1}{2}}(ka)^{\frac{3}{2}}}. \quad (36)$$

This result, which was also obtained in a different way by Levine,[15] is shown in Fig. 11. Values of $\sigma/\pi a^2$ obtained from the exact solution of the corresponding boundary value problem for the reduced wave equation are also shown. The agreement can be seen to be quite good, even when the wavelength is twice the diameter.

For an aperture in a screen on which $\partial u/\partial n = 0$ we must use (11) with a factor $(1+\rho_1^{-1}r)$ on the right side, to find the doubly diffracted field. We begin by obtaining from the first term in (23) at $r_1 = 2a$, $\theta_2 = \pi/2$ the result

$$\frac{\partial u_{e1}}{\partial n} = -\frac{e^{2ika-i\pi/4}}{4(2\pi k a^3)^{\frac{1}{2}}}. \quad (37)$$

This is the normal derivative of the singly diffracted field on a ray which has crossed the aperture to the opposite edge. We now use (37) in (11) with the extra

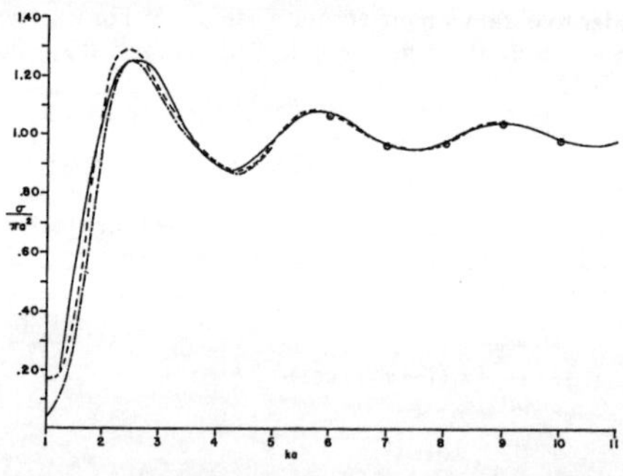

FIG. 11. The transmission cross section σ of a circular aperture of radius a in a thin screen on which $u=0$. The wave is normally incident. The solid curve, based on Eq. (36), results from single and double diffraction; the dashed curve also includes all multiple diffractions. The dots and the broken curve up to $ka=5$ are based upon the exact solution of the reduced wave equation.[13] The ordinate is $\sigma/\pi a^2$ and the abscissa ka.

[15] H. Levine, Institute of Mathematical Sciences, New York University, New York, Research Rept. EM-84 (1955).

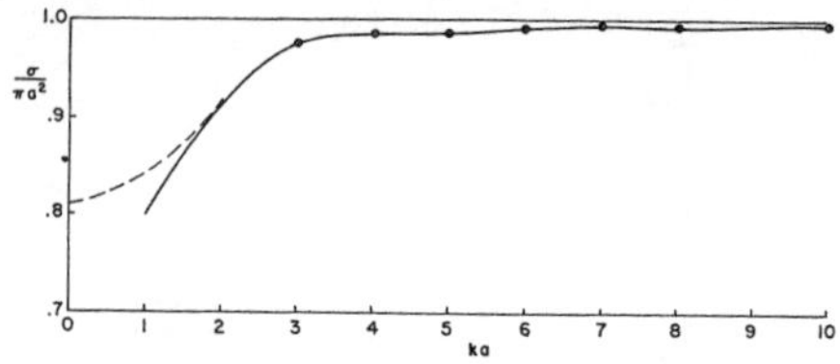

FIG. 12. The transmission cross section σ for normal incidence on a circular hole in a screen on which $\partial u/\partial n = 0$. The ordinate is $\partial/\pi a^2$ and the abscissa is ka. The solid curve is based upon Eq. (40) which results from single and double diffraction. The encircled points and the dashed curve are the exact values computed by Bouwkamp.[13]

factor to obtain the doubly diffracted field which is

$$u_{e2} = -\frac{ie^{ik(r_1+2a)}}{16\pi(k^4a^3)^{\frac{1}{2}}}[r_1(1-a^{-1}r_1\cos\delta_1)]^{-\frac{1}{2}}\frac{\sin(\delta_1/2)}{\cos^2(\delta_1/2)}$$
$$-\frac{ie^{ik(r_2+2a)}}{16\pi(k^4a^3)^{\frac{1}{2}}}[r_2(1-a^{-1}r_2\cos\delta_2)]^{-\frac{1}{2}}\frac{\sin(\delta_2/2)}{\cos^2(\delta_2/2)}. \quad (38)$$

Proceeding as before, we evaluate (38) at points near the axis, then apply the caustic correction factor (28), and finally evaluate the result far from the screen. In this way we obtain

$$u_{e2} = -\frac{ke^{ikx}}{2\pi x}\left[\frac{\pi^{\frac{1}{2}}J_0(ka\sin\varphi)}{4k^{\frac{3}{2}}a^{\frac{1}{2}}}e^{2ika+i\pi/4}\right]. \quad (39)$$

To compute σ we apply the cross-section theorem which requires us to take the imaginary part of the bracketed expression in (39) at $\varphi=0$. We must add this contribution to σ from double diffraction to πa^2, obtained from (30). We also add the term $-\pi/4ka^2$ which was shown by Buchal and Keller[16] to come from the second term in the singly diffracted field u_{e1}. In this way we obtain

$$\frac{\sigma}{\pi a^2} = 1 - \frac{1}{4(ka)^2} + \frac{\sin(2ka+\pi/4)}{4\pi^{\frac{1}{2}}(ka)^{\frac{5}{2}}}. \quad (40)$$

This result was also obtained by Levine and Wu.[17] It is shown in Fig. 12 together with values based upon the exact solution of the reduced wave equation with $\partial u/\partial n = 0$ on the screen. The agreement appears to be quite good for $ka \geqq 2$.

De Vore and Kouyoumjian[18] have used the method of this section to calculate the singly and doubly diffracted fields produced by a plane electromagnetic wave inci-

[16] R. N. Buchal and J. B. Keller, Commun. Pure Appl. Math. 8, 85 (1960).

[17] H. Levine and T. T. Wu, Tech. Rept. 71, Applied Mathematics and Statistics Laboratory, Stanford University, Stanford, California, (July, 1957).

[18] R. DeVore and R. Kouyoumjian, "The back scattering from a circular disk," URSI-IRE Spring meeting, Washington D. C. (May, 1961).

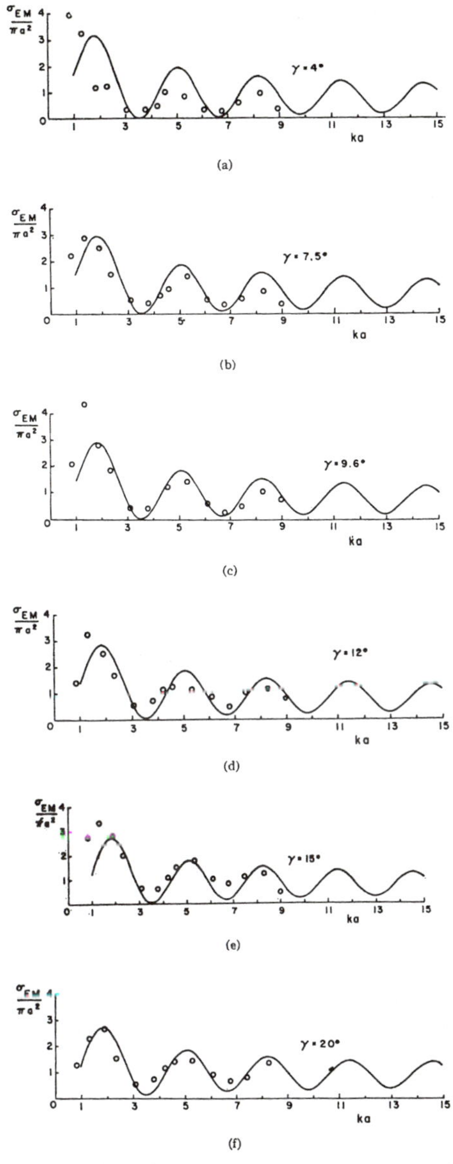

FIG. 13. Experimental and theoretical values of the electromagnetic back scattering cross section σ_{EM} of a finite circular metal cone for axial incidence. The ordinate is $\sigma_{EM}/\pi a^2$ and the abscissa is ka where a is the radius of the base of the cone. The circles are measured values and the curves are computed using the singly and doubly diffracted fields from the edge. The half-angle γ of the cone is indicated on each curve.

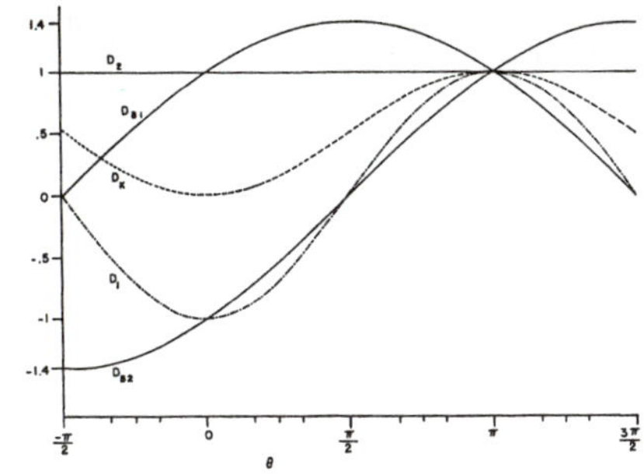

FIG. 14. Comparison of the Kirchhoff edge-diffraction coefficients D_1, D_2, and D_k with the Braunbek coefficients D_{B1} and D_{B2} for normal incidence ($\alpha=0$). The latter coincide with our D, given by Eq. (2) with the upper and lower signs, respectively. D_1 and D_2 result from the two usual modifications of the Kirchhoff theory and $D_k=\frac{1}{2}(D_1+D_2)$ comes from the usual form of it. The ordinate is $-2(2\pi k)^{\frac{1}{2}}e^{-i\pi/4}\sin\theta D$ and the abscissa is θ. All the coefficients are singular on the shadow boundary $\theta=\pi$ but the singular factor $\sin\theta$ has been removed. The dark side of the screen is at $\theta=3\pi/2$ and the illuminated side at $\theta=-\pi/2$.

dent from any direction on a perfectly conducting thin disk. They also performed accurate measurements of the diffracted field and found excellent agreement with the theory. Keller[19] has also used it to calculate the singly and doubly diffracted fields scattered by a perfectly conducting finite circular cone with a flat base hit by an axially incident plane wave. The back-scattered field comes primarily from the sharp edge at the rear of the cone and may be calculated by using the diffraction coefficient D given by (3) with the sign depending upon the orientation of the incident electric field. The results have been compared[20] with the measured values of Keys and Primich.[21] The comparison, shown in Fig. 13, indicates that the theory is fairly accurate even when ka is as small as unity. The theory has also been applied to objects of other shapes by Burke and Keller.[22]

It has been shown[6] by asymptotic evaluation, that the Kirchhoff theory, its two usual modifications and Braunbek's modification all lead to expressions for the field diffracted by an aperture in a thin screen, which can be interpreted as sums of fields on diffracted rays. The expression for the field on each ray was found to be exactly of the form (21), which is an additional corroboration of our theory. The Braunbek modification yielded exactly the diffraction coefficients (2) given by our theory, but the various other Kirchhoff methods gave different coefficients. The various coefficients are compared in Fig. 14 from which it can be seen that the

[19] J. B. Keller, IRE Trans. Antennas and Prop. **AP-8**, 175 (1960).
[20] J. B. Keller, IRE Trans. Antennas and Prop. **AP-9**, 411 (1961).
[21] J. E. Keys and R. I. Primich, Defense Research Telecommunications Establishment, Ottawa, Canada, Rept. 1010 (May, 1959).
[22] J. E. Burke and J. B. Keller, Research Rept. EDL-E49, Electronic Defense Laboratories, Sylvania Electronic Systems, Mountain View, California (April, 1960).

Kirchhoff results agree with ours in the forward direction, but differ considerably at other angles. Thus we may expect the Kirchhoff theory to lead to incorrect results at large diffraction angles.

The present method has been used by Keller[23] to determine the force and torque exerted on a rigid immobile thin disk or strip by an incident plane acoustic wave. In this case, if u denotes the acoustic pressure, the force on the disk is proportional to its total scattering cross section. By Babinet's principle this is found to be twice the transmission cross section of the complementary aperture. Similarly the torque on the disk is proportional to the angular derivative of the diffraction pattern of the disk, evaluated in the forward direction, and this pattern is the negative of that of the complementary aperture. For strips the same statements apply to the force and torque per unit length. Thus the results obtained above for σ yield the force directly. By differentiating the pattern $f(\varphi)$ with respect to φ at $\varphi=-\alpha$, we can also obtain the torque. Graphs of some of the results for the torque are shown in Figs. 15–17.

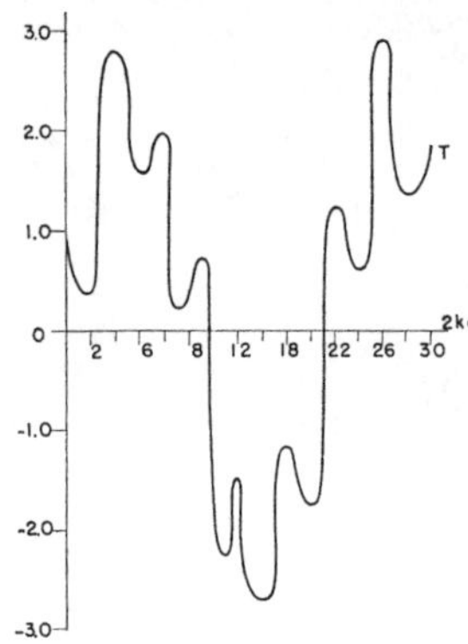

FIG. 16. The acoustic torque T per unit length on a strip of width $2a$ as a function of ka for $\alpha=\pi/4$ (45°). The vertical scale is the same as in Fig. 15.

5. CORNER OR TIP DIFFRACTION

Let us now consider diffraction by special points such as the corner of an edge of a boundary surface or the tip of a conical boundary surface. We call such points vertices and assume that an incident ray which hits one produces infinitely many diffracted rays which satisfy the *Law of vertex diffraction:* A diffracted ray and the corresponding incident ray may meet at any angle at a vertex of a boundary surface. This law follows at once from *Fermat's principle for vertex diffraction:* A vertex-diffracted ray from a point P to a point Q is a curve which has stationary optical length among all curves from P to Q passing through the vertex.

Both of these formulations show that a ray incident on a vertex produces a two-parameter family of diffracted rays which leave the vertex in all directions. In a homogeneous medium they are straight lines and the corresponding diffracted wave fronts are spheres with the vertex as their center. Thus the diffracted wave is spherical, as we might expect. This expectation is verified by the exact solution of the boundary-value problem for the reduced wave equation, corresponding to diffraction by a cone. This confirmation lends support to the law of vertex diffraction. Additional support is provided by the asymptotic evaluation[6] for short wavelengths of the various forms of the Kirchhoff approximation for diffraction through an aperture in a thin screen. This evaluation shows that the field at any point contains a sum of terms, one for each vertex on the edge of the aperture. Each of these terms corresponds to a vertex-diffracted ray, in agreement with our theory.

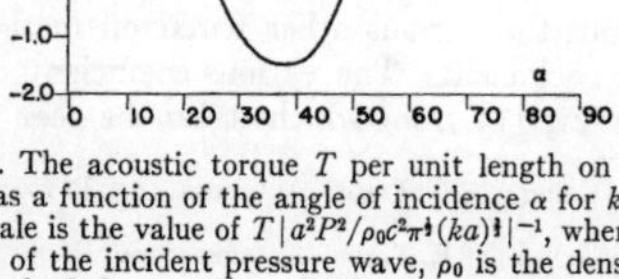

FIG. 15. The acoustic torque T per unit length on a strip of width $2a$ as a function of the angle of incidence α for $ka=5$. The vertical scale is the value of $T|a^2P^2/\rho_0c^2\pi^{\frac{1}{2}}(ka)^{\frac{1}{2}}|^{-1}$, where P is the amplitude of the incident pressure wave, ρ_0 is the density of the medium and c is its sound speed. The values of α at which $T=0$ are equilibrium angles which are stable if $T_\alpha<0$ and unstable if $T_\alpha>0$. The curve does not apply near $\alpha=\pi/2$, which corresponds to grazing incidence, at which $T=0$. T is an odd function of α.

To determine the field on a vertex-diffracted ray let us still consider a homogeneous medium. Then the phase on such a ray at the distance r from the vertex is $kr+\Psi_i$ where Ψ_i is the phase of the incident field at the vertex. The amplitude varies as r^{-1} since the cross-sectional area of a tube of rays is proportional to r^2. Therefore we write the field on the diffracted ray as

$$u=Cu_i(e^{ikr}/r). \tag{41}$$

Here C is the appropriate vertex or corner diffraction coefficient. It depends upon the directions of the incident and diffracted rays, the local geometry of the vertex and the local properties of the media at it. Dimensional considerations show that C is proportional to a length, so it must be proportional to k^{-1}.

The field diffracted by any cone when a plane wave is incident is of the form (41), as we find by analyzing the boundary value problem for the reduced wave equation. This not only confirms our theory but enables us to determine C when the boundary-value problem can be solved. It has been solved for elliptic cones, including the plane angular sector, by Kraus and Levine,[24] but C

[23] J. B. Keller, J. Acoust. Soc. Am. 29, 1085 (1957).

[24] L. Kraus and L. Levine, Commun. Pure Appl. Math. 14, 49 (1961).

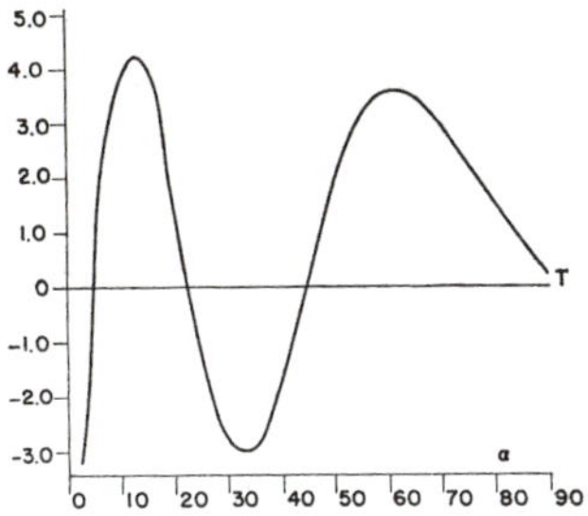

FIG. 17. The acoustic torque T on a circular disk of radius a as a function of the angle of incidence α for $ka=5$. The vertical scale is the value of $T|P^2a^3/\rho_0c(ka)^2|^{-1}$. P, ρ_0, and c are defined in the caption of Fig. 15. The values of x at which $T=0$ are equilibrium values, stable if $T_\alpha<0$ and unstable if $T_\alpha>0$. The curve does not apply near $\alpha=0$ (normal incidence) nor $\alpha=\pi/2$ (grazing incidence).

has not been evaluated. For circular cones, it has been evaluated by Felsen[25] and by Siegel *et al.*[26] The various forms of the Kirchhoff integral for diffraction through an aperture, when evaluated asymptotically,[6] yield fields of the form (41) for each corner on the aperture edge. Here again the resulting expressions for C are different for the different versions of the Kirchhoff method, and none can be expected to coincide with the exact value contained in the solution of Kraus and Levine. We shall not consider any applications of vertex diffraction.

6. SURFACE-DIFFRACTED RAYS

Let us now consider an incident ray which grazes a boundary surface, i.e., is tangent to the surface. We assume that such a ray gives rise to a surface-diffracted ray in accordance with the *law of surface diffraction:* An incident ray and the resulting surface-diffracted ray in the same medium are parallel to each other at the point of diffraction and lie on opposite sides of the plane normal to the ray at this point. When the two rays lie in different media, they obey the law of refraction.

The surface ray travels along the surface in a manner determined by the usual differential equations for rays on a surface. Therefore in a homogeneous medium it is an arc of a geodesic or shortest path on the surface. At every point the surface ray sheds a diffracted ray satisfying the law of surface diffraction. All of these properties of surface-diffracted rays follow from *Fermat's principle for surface diffraction:* A surface-diffracted ray from a point P to a point Q is a curve which makes stationary the optical length among all curves from P to Q having an arc on the boundary surface.

To illustrate these ideas, let us consider a boundary surface which is a cylinder parallel to the z axis. Let the intersection of the cylinder with the xy plane be a smooth closed convex curve C and suppose that the medium surrounding the cylinder is homogeneous. To find a surface-diffracted ray from P to Q, both of which lie outside C, we observe that the optical length of a curve is proportional to its geometric length. Therefore to utilize Fermat's principle we imagine a string from P to Q, and consider it to be pulled taut. Then it will consist of two straight-line segments from P and Q to the cylinder and a geodesic arc along the cylinder. When P and Q are both in the xy plane the arc is just an arc of C; otherwise it is a helical arc along the cylinder. It may wrap around the cylinder any number of times in either direction, so there are countably many surface diffracted rays from P to Q.

We define a surface-diffracted wavefront to be a surface orthogonal to a family (i.e., normal congruence) of surface-diffracted rays. In the example just described let us suppose that P is a line source parallel to the z axis. Then the surface-diffracted wavefronts which it produces are cylinders with generators parallel to the z axis. Their intersections with the xy plane are the involutes of the curve C.

Two examples of surface-diffracted rays are shown in Figs. 18 and 19. In both cases the grazing ray is incident horizontally from the left. One of the shed surface-diffracted rays is shown in Fig. 18 and two are shown in Fig. 19. In both cases the figures show cross sections of opaque screens surrounded by homogeneous media.

To construct the field on a surface diffracted ray we assume that the phase of the field increases in proportion to optical length along the ray. The amplitude is assumed to be proportional to the amplitude of the incident field at the point of diffraction. The coefficient of proportionality involves a new surface-diffraction coefficient B which depends upon local properties of the boundary surface and the media at the point of diffraction. Along the portion of the diffracted ray on the surface, the amplitude is assumed to vary in accordance with the principle of energy conservation in a narrow strip of rays on the surface. Thus it varies inversely as the square root of the width of this strip because of

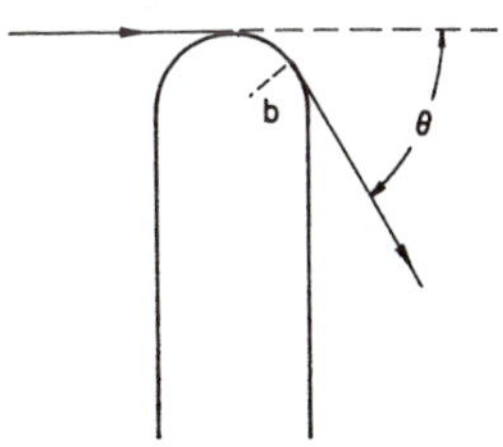

FIG. 18. Cross section of a screen of width $2b$ with a rounded end of radius b. A plane wave is normally incident upon it from the left. The dashed line is the shadow boundary. An incident ray which grazes the end of the screen is shown together with one of the surface diffracted rays which it produces in the shadow region. The angle between this ray and the shadow boundary is θ.

[25] L. B. Felsen, J. Appl. Phys. 26, 138 (1955).
[26] K. M. Siegel, J. W. Crispin, and C. E. Schensted, J. Appl. Phys. 26, 309 (1955).

convergence or divergence. However, it also decays because of radiation from the surface along the shed diffracted rays. The decay rate is assumed to be determined by a decay exponent α which depends upon local properties of the surface and media. The amplitude on a shed diffracted ray varies in the usual manner and is proportional to the field on the surface diffracted ray at the point of shedding. The proportionality factor involves another diffraction coefficient which can be shown to be the same as the coefficient B introduced above as a consequence of the principle of reciprocity. The details of this construction of the field are given for cylinders by Keller[27] and for three-dimensional surfaces by Levy and Keller.[28]

The theory has been tested by applying it to diffraction of a cylindrical wave by a circular cylinder and comparing the predicted field with the asymptotic expansion of the exact solution. The two results agreed exactly when appropriate expressions were used for the diffraction coefficient B and the decay rate α. Thus the theory was confirmed and B and α were determined. The theory was further confirmed by making similar comparisons for diffraction by a parabolic cylinder, an elliptic cylinder (Levy[29]), a spheroid (Levy and Keller[30]) and the screen of Fig. 19 (Magiros and Keller[31]). Of course, in all cases the same values of B and α were employed.

A numerical test of the accuracy of the theory was made by Keller[32] for the screen of Fig. 19. The field u far from the end of the screen in the shadow region can be written as $u=f(\theta,kb)e^{ikr+i\pi/4}(kr)^{-\frac{1}{2}}$. The function $f(\theta,kb)$ was evaluated for $\theta=\pi/4$ as a function of kb from the exact solution of the boundary-value problem and from the formula given by our theory. The results are shown in Fig. 20. The upper curves and points apply to a screen on which $\partial u/\partial n=0$; the lower ones to a screen on which $u=0$. The solid curves are obtained when a simple expression is used for the decay exponent α and the dashed curve when a more accurate expression is used for α. The agreement between the curves given by our theory and the points from the exact solution appears to be quite good for $kb>2$. Similar curves for the screen of Fig. 18 are shown in Fig. 21, but there is no exact solution with which to compare them.

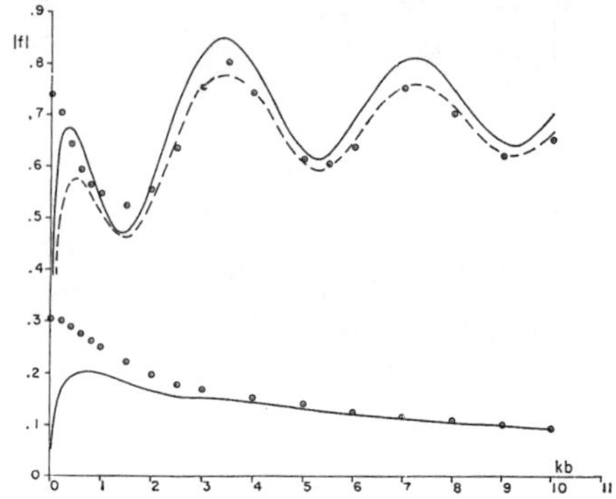

FIG. 20. The far-field amplitude $|f(\pi/4,kb)|$ in the shadow of the screen of Fig. 19 in the direction $\theta=\pi/4$ as a function of kb. The upper curves and points apply to a screen on which $\partial u/\partial n=0$ while the lower ones pertain to a screen on which $u=0$. The encircled points were obtained by laborious numerical computation of the series solution of the boundary-value problem; the curves were obtained from the formulas given by the geometrical theory of diffraction. The dashed curve was obtained from a formula in which an improved expression for α was used.

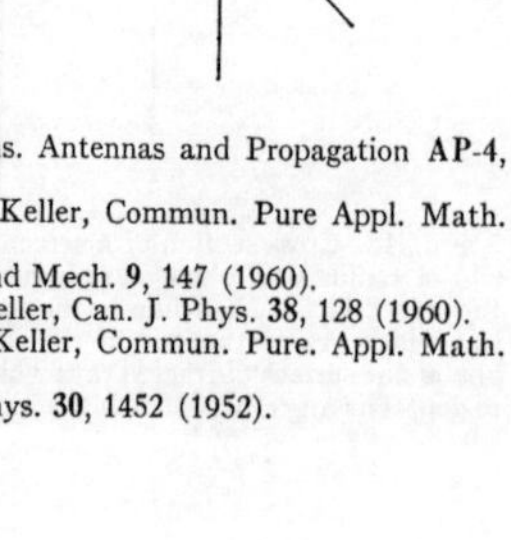

FIG. 19. Cross section of an infinitely thin screen with a cylindrical tip of radius b. A plane wave is normally incident upon it from the left. The dotted line is the shadow boundary. An incident ray which grazes the tip is shown together with two of the surface diffracted rays which it produces. One of them is reflected from the screen, and both ultimately make the angle θ with the shadow boundary.

Diffracted rays have also been introduced by Franz and Depperman[33] who called the associated waves "creeping waves." They applied them to explain oscillations in the measurements of the radar back-scattering cross section of metallic circular cylinders. A refinement of the method described above for determining B and α was given by Levy and Keller.[34] They showed how variation of the curvature of the diffracting surface modifies the values of B and α determined from a circular cylinder. The determination of the field near the diffracting surface requires special considerations, described in references 27 and 28, because the surface is a caustic of the diffracted rays. Another special treatment described by Buchal and Keller[16] is required near the shadow boundary. Near the point of diffraction, where the shadow boundary meets the diffracting surface, a still different special analysis due to V. Fock and to C. L. Pekeris is required. A uniform expression for the field in these various regions has been obtained, in two-dimensional cases, by Logan and Yee[35] by combining

[27] J. B. Keller, IRE Trans. Antennas and Propagation **AP-4**, 243 (1956).
[28] B. R. Levy and J. B. Keller, Commun. Pure Appl. Math. **12**, 159 (1959).
[29] B. R. Levy, J. Math. and Mech. **9**, 147 (1960).
[30] B. R. Levy and J. B. Keller, Can. J. Phys. **38**, 128 (1960).
[31] D. Magiros and J. B. Keller, Commun. Pure. Appl. Math. **14**, 457 (1961).
[32] J. B. Keller, J. Appl. Phys. **30**, 1452 (1952).

[33] W. Franz and K. Depperman, Ann. Physik **10**, 361 (1952).
[34] B. R. Levy and J. B. Keller, IRE Trans. Antennas and Propagation, **AP-7**, 552 (1959).
[35] N. A. Logan and K. S. Yee, Symposium on Electromagnetic Theory, U. S. Army Mathematics Research Center, University of Wisconsin, Madison, Wisconsin (April, 1961).

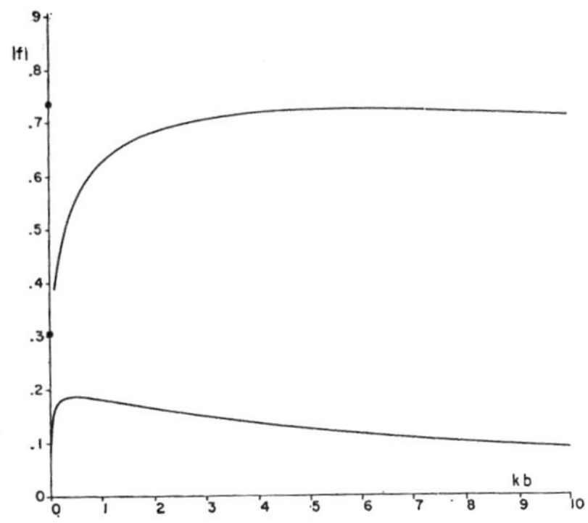

FIG. 21. The far-field amplitude $|f(\pi/4,kb)|$ in the shadow of the screen of Fig. 18 in the direction $\theta=\pi/4$ as a function of kb. The curves are based upon the geometrical theory of diffraction while the points at $kb=0$ are determined from the solution for diffraction by a half-plane. The upper curve and point apply to a screen on which $\partial u/\partial n=0$; the lower ones to a screen on which $u=0$. Probably only the decreasing portions of the curves are correct, and most likely the amplitude decreases monotonically for all values of kb.

the method of Fock and Pekeris with the geometrical considerations we have described.

7. FURTHER DEVELOPMENTS

The geometrical theory of diffraction which we have described has been applied to inhomogeneous media by Friedrichs and Keller[36] and by Seckler and Keller.[37] It has also been extended by the introduction of complex or imaginary rays.[2]

A similar theory can be constructed to describe any kind of wave propagation and this has been done to some extent for water waves, elastic waves, quantum-mechanical waves, surface waves, etc. When anisotropic media are present, or when more than one propagation velocity exist, there are more kinds of rays and the theory is correspondingly more complicated. However the principles are essentially unchanged.

From a mathematical point of view, the field constructed by means of the present theory is the leading part of the asymptotic expansion of the exact field for small values of λ or large values of k. The full asymptotic expansion consists of additional terms in the amplitude of the field on each ray. These terms are smaller than the first term by factors of k^{-n}, $n=1, 2, \cdots$. These statements have been proved in special cases, but not in general.

[36] K. O. Friedrichs and J. B. Keller, J. Appl. Phys. **26**, 961 (1955).

[37] B. D. Seckler and J. B. Keller, J. Acoust. Soc. Am. **31**, 192 (1958).

Reprinted from ANNALS OF PHYSICS, Vol. 4, No. 2, June, 1958
Academic Press Inc.

ANNALS OF PHYSICS: 4, 180–188 (1958)

Corrected Bohr-Sommerfeld Quantum Conditions for Nonseparable Systems*

JOSEPH B. KELLER

Institute of Mathematical Sciences, New York University, New York, New York

For a separable or nonseparable system an approximate solution of the Schrödinger equation is constructed of the form $Ae^{i\hbar^{-1}S}$. From the single-valuedness of the solution, assuming that A is single-valued, a condition on S is obtained from which follows A. Einstein's generalized form of the Bohr-Sommerfeld-Wilson quantum conditions. This derivation, essentially due to L. Brillouin, yields only integer quantum numbers. We extend the considerations to multiple valued functions A and to approximate solutions of the form

$$\sum A_k \exp (i\hbar^{-1}S_k).$$

In this way we deduce the corrected form of the quantum conditions with the appropriate integer, half-integer or other quantum number (generally a quarter integer). Our result yields a classical mechanical principle for determining the type of quantum number to be used in any particular instance. This fills a gap in the formulation of the "quantum theory", since the only other method for deciding upon the type of quantum number—that of Kramers—applies only to separable systems, whereas the present result also applies to nonseparable systems.

In addition to yielding this result, the approximate solution of the Schrödinger equation—which can be constructed by classical mechanics—may itself prove to be useful.

INTRODUCTION

In the "quantum theory" the motion of a system is described by classical mechanics but certain constants of the motion are restricted to be integers. These restrictions are called the quantum conditions and the integers occurring in them are called the quantum numbers. In many cases better agreement between theory and observation is obtained if half-integer quantum numbers are employed instead of integers. However no theoretical principle is available to determine whether an integer, half-integer or other quantum number is to be used in any particular case. It is the purpose of this article to provide such a principle.

* This article is based upon a report (*1*) sponsored by the Geophysics Research Directorate, Air Force Cambridge Research Center, Air Research and Development Command, under Contract No. AF19(122)-463.

Historically the problem of deciding between integer and half-integer quantum numbers was circumvented by the invention of quantum mechanics, which replaced the "quantum theory". Therefore the discovery of a procedure for deciding between the two kinds of quantum numbers might now be considered to be of purely academic interest. However this is not necessarily the case because the procedure can still be used in the approximate solution of quantum mechanical problems.

A way of deciding between integer and half-integer quantum numbers for separable systems was found by Kramers (*2*) by means of an approximate solution of the Schrödinger equation. His result also showed that for such systems these conditions are consequences of quantum mechanics in the limit as Planck's constant h tends to zero. Another derivation of the quantum conditions from the Schrödinger equation was given by Brillouin (*3*). It was more general than Kramers' since it applied to both separable and nonseparable systems, but it was incorrect since it yielded only integer quantum numbers. Brillouin's argument is essentially the following:

Consider a quantum mechanical system with N coordinate operators q_r, N conjugate momentum operators $p_r \equiv -i\hbar^{-1}(\partial/\partial q_r)$, $r = 1, \cdots, N$, and Hamiltonian operator $H(q_r, p_r, t)$. Let the Schrödinger representor of the state of the system be $\Psi(q_r, t)$. Suppose that Ψ is approximately equal to Ψ_0 defined by

$$\Psi_0 = A(q_r, t, \hbar) \exp [i\hbar^{-1}S(q_r, t, \hbar)]. \tag{1}$$

The function Ψ and hence Ψ_0 must be single valued. Therefore if S is multiple valued and if ΔS denotes the difference between any two of its values, it is necessary that

$$\Delta S = nh. \tag{2}$$

Here n is an integer. This condition guarantees that Ψ_0 will be single valued even if S is not. The quantum conditions with integer quantum numbers follow from (2), as will be shown below. We will also show that (2) is equivalent to the quantum condition postulated by Einstein (*4*) in generalizing the Bohr-Sommerfeld-Wilson conditions to nonseparable systems.

We note that Brillouin's argument assumes that A is single valued. However if A is not single valued then in order for Ψ_0 to be single valued (2) must be replaced by

$$\Delta S = h\left[n + i\frac{\Delta \log A}{2\pi}\right]. \tag{3}$$

Suppose, for example, that two values of A differ only in sign. Then $\Delta \log A = -i\pi$ and (3) yields

$$\Delta S = h[n + \tfrac{1}{2}]. \tag{4}$$

From (4) the quantum condition with a half-integer quantum number follows.

We have just indicated how to modify Brillouin's argument in order to obtain the appropriate integer, half-integer or other quantum number. To complete the argument we must examine the amplitude A. This will be done in the next sections. However we must first make another change in the method. Instead of (1), we will assume that Ψ_0 is a sum of terms given by

$$\Psi_0 = \sum_{k=1}^{M} A_k(q_r, t, \hbar) \exp [i\hbar^{-1} S_k(q_r, t, \hbar)]. \tag{5}$$

This more general form of approximate solution is required in almost all problems in which S is multiple valued. The various functions S_k and A_k will be considered to be different branches of multiple valued functions S and A. We will also make our considerations more definite by assuming that Ψ is asymptotic to Ψ_0 as Planck's constant $\hbar$ tends to zero. This viewpoint was introduced by Birkhoff (*5*).

ASYMPTOTIC SOLUTION OF THE SCHRÖDINGER EQUATION

The function Ψ satisfies the Schrödinger equation

$$H(q_r, p_r, t)\Psi(q_r, t) = i\hbar \frac{\partial \Psi}{\partial t} (q_r, t). \tag{6}$$

Upon inserting the expression (5) for Ψ_0 into (6) and considering the leading terms in $\hbar$ we obtain equations for the A_k and S_k. These equations were derived by Dirac (*6*) when Ψ_0 consists of a single term and the same analysis applies when Ψ_0 is a sum of terms. The result is that each S_k satisfies the classical Hamilton-Jacobi equation

$$H\left(q_r, \frac{\partial S_k}{\partial q_r}, t\right) = -\frac{\partial S_k}{\partial t}. \tag{7}$$

The equation for each A_k, written in terms of $P_k = A_k^2$ and

$$v_r = \frac{\partial H}{\partial p_r} (q_r, \partial S/\partial q_r, t),$$

is

$$\frac{\partial P_k}{\partial t} + \sum_{r=1}^{N} \frac{\partial}{\partial q_r} [v_r P_k] = 0. \tag{8}$$

Equation (8) is the Liouville equation of classical statistical mechanics for the probability distribution P_k of a classical mechanical system with Hamiltonian H. It is a special form because P_k depends only upon q_r and t, but not upon v_r or p_r as is usual in classical mechanics. This is a consequence of the fact that quantum mechanics does not yield joint probability distributions of conjugate variables. It is to be noted that P_k is not necessarily positive, nor even real, in the present case. We will also permit S_k to be complex.

Let us solve (7) when H is independent of t. Then $H = E$, where the constant E is the total energy of the system. Now if $q_r(t)$, $p_r(t)$ denote a trajectory, then the function S_k is given at any point of the trajectory in terms of its value $S_k(0)$ at some fixed point on the trajectory by

$$\begin{aligned} S_k(q_r, t, \hbar) &= S_k(0) + \int_0^t \sum_{r=1}^N p_r(\tau) \frac{dq_r(\tau)}{d\tau} d\tau - Et \\ &= S_k(0) + \int_0^t \sum_{r=1}^N p_r \, dq_r - Et. \end{aligned} \tag{9}$$

Equation (8) can also be solved at once if P_k is independent of t, for then it becomes

$$\sum_{r=1}^N \frac{\partial}{\partial q_r} [v_r P_k] = 0. \tag{10}$$

Equation (10) asserts that the probability flux is divergenceless, and therefore by applying Gauss' theorem to a tube of trajectories we obtain

$$P_k v = P_k(0) v_0 \frac{d\sigma_0}{d\sigma}. \tag{11}$$

In (11)

$$v = \left[\sum_{r=1}^N v_r^2 \right]^{1/2}$$

and $d\sigma$ is the normal cross sectional area of the tube of trajectories, both evaluated at the same point at which P_k is evaluated. The corresponding quantities v_0, $d\sigma_0$, and $P_k(0)$ are evaluated at some other point on the same trajectory, and thus (11) merely asserts the conservation of "probability". Actually (11) holds in the limit as $d\sigma_0$ tends to zero, in which case $d\sigma/d\sigma_0$ denotes the Jacobian of the mapping of one cross section onto another by means of trajectories. The problem of Rutherford scattering has been treated by Gordon (*7*) and by the present author (*1*) by means of Eqs. (9) and (11).

THE QUANTUM CONDITIONS

Quantum mechanics requires that Ψ be a single valued function of the q_r and therefore Ψ_0 must satisfy the same condition. Consequently each term in the expression (5) for Ψ_0 must be single valued. As we have shown in the Introduction, this requires that each S_k and the corresponding A_k must satisfy the condition (3). If any A_k is single valued, this condition becomes the simpler condition (2) for the corresponding S_k.

The differences $\Delta S_k(q_r, t)$ and $\Delta \log A_k(q_r, t)$ are expressible as line integrals over some closed curve in q_r space beginning and ending at q_r. In termś of these integrals, (3) becomes, for each value of k,

$$\oint \nabla S \cdot ds = h\left[n + \frac{i}{2\pi}\oint \nabla \log A \cdot ds\right]. \tag{12}$$

Equation (12) must hold for every closed curve in the q_r space, since only then will Ψ_0 be single valued at every point. However the line integrals in (12) have the same value for every two closed curves which are deformable into one another without crossing a singularity of the integrand. For example, they are zero for a curve which can be deformed into a point. There are, in general, only a finite number of classes of independent curves which cannot be deformed into points. Any other curve is deformable into a linear combination of such curves with integer coefficients. Therefore (12) will be satisfied for all curves if it is satisfied by one curve in each of the independent classes of curves. Thus we have, in general, a finite number of quantum conditions, in each of which the integer n is arbitrary.

Since $\nabla S \cdot ds = \sum_r p_r \, dq_r$, (12) can be rewritten as

$$\oint \sum_r p_r \, dq_r = h\left[n + \frac{i}{2\pi}\oint \nabla \log A \cdot ds\right]. \tag{13}$$

When A_k is single valued, the conditions (13) become exactly the quantum conditions postulated by Einstein (*4*) for a system in a steady state of constant energy. He pointed out that these quantum conditions are invariant under a contact transformation of variables because $\sum p_r \, dq_r$ is invariant. If the variables are separable in the Hamilton-Jacobi equation, so that each p_r can be expressed in terms of the corresponding q_r alone, and if A_k is single valued, then these conditions reduce to the well known Bohr-Sommerfeld-Wilson quantum conditions for a separable system in a steady state of constant energy.

In order to clarify the conditions (12) and (13) let us consider a multiple valued solution S of the Hamilton-Jacobi equation. Such a solution generally has infinite multiplicity, i.e., an infinite number of different values or branches. However only a finite number of its branches, say M of them, are essentially distinct. Every other branch differs from one of these branches by an additive constant. Therefore the function ∇S will have only the finite multiplicity M since any two branches differing by a constant yield the same value for ∇S. Let us introduce an M-sheeted q_r space and associate one branch of the function ∇S with each sheet. We will denote each sheet by an integer k and the corresponding branch of ∇S by ∇S_k, with k ranging from one to M. Any two sheets—say sheets j and k—are to be joined together at all points where $\nabla S_j = \nabla S_k$. Furthermore if ∇S is defined in only part of the q_r space then only that part is covered by additional sheets. The M sheeted q_r space so constructed is called the covering space for the function ∇S. Its main property is that on it ∇S is a single valued function. The Riemann surfaces of function theory are examples of such spaces.

The same considerations may be applied to the multiple valued function $\log A$. Its gradient has the same multiplicity M as does ∇S as we see from (8) and its different branches become equal where those of ∇S do. Therefore the same covering space on which ∇S is single valued also serves as the covering space for $\nabla \log A$. Consequently the line integrals in (12) and (13) may be thought of as being evaluated along a closed path on this covering space. Then the omission of the subscript k in (12) and (13) is appropriate since any number of branches may be involved in each integral. Also the question of whether one closed curve is deformable into another becomes clear in this space. Furthermore, the independent closed curves can be recognized as the basis of the fundamental group of the covering space. In this way we see that the topology of the covering space determines the number of quantum conditions. This number is just the number of closed curves in the basis of the fundamental group.

Let us now consider the evaluation of

$$\Delta \log A = \oint \nabla \log A \cdot ds.$$

We will restrict our attention to steady states since then $A^2 = P$ is explicitly given by (11). From this equation we see that P becomes infinite whenever $v d\sigma$ becomes zero. We will call points at which this occurs caustic points, in analogy with optics where points at which $d\sigma = 0$ are so named. A locus of caustic points is called a caustic of the family of trajectories associated with the S function under consideration. Those caustics which correspond to the vanishing of $d\sigma$ are envelopes of the family of trajectories. Therefore ∇S is multiple valued near these surfaces. Consequently such caustics are the loci of points at which two different branches of ∇S become equal. Thus these caustics form the boundaries at which different sheets of the covering space for ∇S are joined together. Those caustics at which $v = 0$ also form part of these boundaries, assuming that each p_r either changes sign along each trajectory on which v vanishes or is identically zero near the caustic. For then ∇S, which has the p_r as components, reverses its direction at the caustic. Thus this type of caustic is also a boundary on which two branches of ∇S join.

We have seen that A becomes infinite on a caustic and that a caustic must be crossed by a path which goes from one sheet of the covering space to another. It is well known in optics that the phase of A is retarded by $\pi/2$ (i.e., A is multiplied by $e^{-i\pi/2}$) on a ray which passes through a caustic on which $d\sigma$ vanishes simply. (The positive direction along a ray is the direction of ∇S.) Furthermore the phase is retarded by π on a ray passing through a focus, which is a caustic point at which $d\sigma$ vanishes to the second order. The usual method for proving these facts is based upon the asymptotic evaluation of certain double integrals which represent the wave function. These integral representations are deduced from Green's theorem, and the integrals are evaluated by the method of stationary phase. Both of these considerations can be immediately extended to problems such as the present one in which the number of dimensions is N. The result is this: The phase of A is retarded by $m\pi/2$ on a trajectory which passes through a caustic on which $d\sigma$ vanishes to the mth order. We may replace the statement "$d\sigma$ vanishes to the mth order" by the equivalent statement "the dimensionality of the cross section of a tube of trajectories is reduced by m". This result is an analogue for partial differential equations of the Kramers connection formulas which are employed in the WKB treatment of ordinary differential equations. In the present case the caustics play the role of the turning points.

From the foregoing analysis we see that log A changes by $-im\pi/2$ along a path which passes from one sheet of the covering space to another in the direction of ∇S. Here the positive integer m is the number of dimensions "lost" by a tube of trajectories at the caustic. Obviously m must be replaced by $-m$ if the path is traversed in the opposite direction. Considerations similar to those outlined above show that at the caustics on which v vanishes the phase of A is also retarded by $m\pi/2$, so log A changes by $-im\pi/2$ where m is the number of p_r which change sign at the caustic.

The total change Δ log A along a closed curve is generally just the change associated with the various caustics through which the curve passes. Therefore, in general, we have

$$\frac{i}{2\pi}\,\Delta \log A = \frac{i}{2\pi} \oint \nabla \log A \cdot ds = \frac{m}{4}. \tag{14}$$

Here m denotes the total number of dimensions "lost" by the trajectories at the caustics through which the curve passes plus the number of p_r which change sign at the $v = 0$ caustics through which the curve passes. In evaluating m account must be taken of whether the curve traverses the caustic in the direction of increasing or decreasing S. When (14) is used the quantum conditions (13) finally become

$$\oint \sum p_r \, dq_r = h\left[n + \frac{m}{4}\right]. \tag{15}$$

These are the corrected quantum conditions for separable or nonseparable systems. In each quantum condition the positive integer n is arbitrary but the integer m is determined by the considerations described above.

THE "CLASSICAL Ψ FUNCTION"

We will call the function Ψ_0 given by (5) the "classical Ψ function" because it can be constructed by classical mechanical considerations alone. In spite of this the probabilities computed from $|\Psi_0|^2$ still show quantum mechanical interference effects if the sum in (5) contains more than one term.

If only one term occurs in (5) then $|\Psi_0|^2 = |A|^2 \exp(-2h^{-1}ImS)$, and the probability, as well as Ψ_0 itself, is exponentially damped in regions where $ImS > 0$. The probability in such regions vanishes as h tends to zero, corresponding to the fact that these regions are excluded in classical mechanics. This can be seen from the fact that the solution of the Hamilton-Jacobi equation is not real there. The exponential tail shows that Ψ_0 describes such quantum mechanical effects as "tunneling".

If $ImS = 0$ and $|A|^2 = A^2 = P$, then $|\Psi_0|^2 = P$. Thus as h tends to zero the quantum mechanical probability distribution approaches that given by the Liouville equation of classical statistical mechanics, and we may say that quantum mechanics approaches classical statistical mechanics, as h tends to zero. The customary statement that quantum mechanics approaches classical mechanics is thus not strictly correct, but holds only when the initial data are such that $P = 0$ except on one trajectory, in which case classical statistical mechanics reduces to classical mechanics. Since the classical method of computing differential scattering cross sections is actually based on classical statistical mechanics, the preceding considerations show that the quantum mechanical cross sections will approach them as h tends to zero when only one term occurs in (5).

Finally it is to be noted from (13) or (15) that h and n enter the solution only in the combination $(n + \frac{1}{4}m)h$. In some problems the solution for fixed n does not have the asymptotic behavior assumed in the derivation. However when the limit in which n becomes infinite while h becomes zero and $(n + \frac{1}{4}m)h$ is constant is considered, the assumed asymptotic behavior may result. In such cases the asymptotic solution applies only for high quantum numbers n.

AN EXAMPLE—THE HARMONIC OSCILLATOR

To exemplify the preceding results, let us consider the steady state of a one-dimensional harmonic oscillator of mass m, energy E, frequency ν_0, and momentum p. Recalling that $S_x = p$, we have from the definition of momentum

$$S_x = p = \pm(2m\,E - m\,\nu_0 x^2)^{1/2}. \tag{16}$$

We see that $\nabla S = S_x$ is real and double valued in the interval $-x_0 \leqq x \leqq x_0$ where $x_0 = (2E\nu_0^{-1})^{1/2}$. The two branches of S_x become equal at the endpoints of this interval. Thus the covering space for S_x consists of two line segments joined together at their two ends. This space is topologically equivalent to a circle, and there is only one basic closed curve on it. Therefore there is only one quantum condition. The closed curve passes through the two caustics $x = \pm x_0$ at both of which $v = m^{-1}p$ vanishes and p changes sign. Thus in (15) we have $m = 2$ so (15) becomes

$$\int_{-x_0}^{x_0} (2mE - m\nu_0 x^2)^{1/2}\, dx - \int_{x_0}^{-x_0} (2mE - m\nu_0 x^2)^{1/2}\, dx = h\left(n + \frac{1}{2}\right). \quad (17)$$

Equation (17) is just the result given by the usual WKB method in this case. A similar analysis holds for a particle in any one dimensional potential well.

To construct Ψ_0 we note that $d\sigma_0/d\sigma = 1$ in the present case so (11) yields, with A_0 a constant,

$$A = A_0\, v^{-1/2}. \quad (18)$$

Inserting $v = m^{-1}p$ in (18) with p given by (16) we obtain the two results

$$A_+ = A_0 m^{1/4}(2E - \nu_0 x^2)^{-1/4} \quad (19)$$

$$A_- = e^{-i\pi/2} A_0 m^{1/4}(2E - \nu_0 x^2)^{-1/4}. \quad (20)$$

The phase retardation represented by the factor $e^{-i\pi/2}$ in (20) accounts for the phase shift which occurs upon passing through either caustic. In the present example it arises formally when p is negative and the square root of p^{-1} is taken. Of course our previous considerations are necessary to ensure that we take the correct root.

Upon inserting (19), (20), and (16) into (5) and setting $S(x_0) = 0$, we obtain for $\Psi_0(x)$ in the region $|x| \leqq x_0$,

$$\Psi_0(x) = A_0 m^{1/4}(2E - \nu_0 x^2)^{-1/4}\left\{\exp\left[-i\hbar^{-1}\int_x^{x_0}(2mE - m\nu_0 x^2)^{1/2}\, dx\right]\right.$$

$$\left. + \exp\left[-i\frac{\pi}{2} + i\hbar^{-1}\int_x^{x_0}(2mE - m\nu_0 x^2)^{1/2}\, dx\right]\right\} \quad (21)$$

$$= e^{-i(\pi/4)} A_0 m^{1/4}(2E - \nu_0 x^2)^{-1/4}\cos\left[\hbar^{-1}\int_x^{x_0}(2mE - m\nu_0 x^2)^{1/2}\, dx - \frac{\pi}{4}\right].$$

This is the usual WKB result. We may obtain the result for a particle in any one dimensional potential $V(x)$ by replacing $\nu_0 x^2$ by $2V(x)$ in (21).

RECEIVED: January 23, 1958

REFERENCES

1. J. B. KELLER, New York University, Mathematics Research Group, Research Report No. CX-10, July, 1953 (unpublished).
2. H. A. KRAMERS, *Z. Physik* **39**, 828–840 (1926).
3. L. BRILLOUIN, *J. phys. radium* **7**, 353–368 (1926).
4. A. EINSTEIN, *Verhandl. deut. physik. Ges.* (1917). References to prior work are given in this paper.
5. G. D. BIRKHOFF, *Amer. Math. Soc. Bull.* **39**, 681–700 (1933).
6. P. A. M. DIRAC, "The Principles of Quantum Mechanics," 3rd ed., pp. 121–123. Oxford Univ. Press, London and New York, 1947.
7. W. GORDON, *Z. Physik* **48**, 180 (1928).

LIFE OF KUNIHIKO KODAIRA

YOICHI MIYAOKA AND KENJI UENO

FIRST SIGNS OF INTEREST IN MATHEMATICS AND MUSIC

Kunihiko Kodaira was born on March 16, 1915 in Yodobashi-ku (now Shinjuku-ku), Tokyo as the first son of Gon-ichi and Ichi Kodaira.

Both parents originated from Suwa district, Nagano Prefecture, in a mountainous area of central Japan.

Kodaira's father, Gon-ichi (1884–1976), majored in agriculture and politics at Tokyo Imperial University and he made his career at the Ministry of Agriculture, after which he was promoted to Vice-Minister (1938–39). He played an active role in Japan's commitment to the agricultural development program in South America, to which a considerable percentage of the population emigrated due to overpopulation and poverty caused by the Great Depression of the '30s. After his retirement from the Ministry of Agriculture, he was elected member of the Lower House of Parliament during World War II. Although Gon-ichi was a liberal and was not really involved in wartime militarism, he was purged from public office as soon as the Allied Power occupied Japan.

As a specialist in agronomics, Gon-ichi wrote as many as 40 books and 350 research papers, a principal work being his doctorate thesis "Treatise on Agricultural Finance", which was over 1,000 pages. The manuscript of this book was once lost in a tram, but he rewrote the book from scratch at once. Kodaira was unaware of his father's scholarly contributions until two biographies of Gon-ichi were published in the '80s.

His mother, Ichi (1894–1993), is remembered by D.C. Spencer as a "remarkable woman". She spoke English and "will be remembered by all of us who visited the Kodairas in Tokyo as a warm and generous hostess"[1]. She solemnly declared to her sons immediately after the surrender of Imperial Japan on August 15, 1945 that "Now all of us must learn English." During the boyhood of her sons, she and the boys would spend summer vacations at her parents' house in Kami-Suwa Town. Her father, Kyuji Kanai, was a school master, mayor of Kami-Suwa and member of the prefectural assembly. He studied Chinese classical literature and philosophy and was, at the same time, interested in modern European natural sciences, especially zoology.

[1] D.C. Spencer: *Kunihiko Kodaira*, Notices of the AMS, **45** (1998), 388–389.

Kodaira had a younger brother Nobuhiko (1919–), who worked at meteorological observatories and undertook the charge of launching the first Japan-made meteorological satellite.

From his infancy Kodaira was keenly interested in numbers. His mother used to recall that he had had great fun in counting the beans in their pods.

When Kodaira was ten years old, he made an experiment on whether dogs have the concept of quantity when his dog gave birth to six puppies. First Kodaira hid all of them. The dog searched for her puppies, crying frantically, until she found them. Then he hid five of them. Only one remained with her. Since she seemed completely happy and gave the impression that she did not care about the absence of the missing five, Kodaira confidently concluded that dogs did not have any concept of numbers.

When Kodaira began elementary school in 1921, he did not fit well into school life because of his shyness and his habit of stammering when he was under stress. Since he was a slow runner, he was particularly unhappy during physical education class. Naturally, he was far more comfortable in mathematics class.

Kodaira's interest in mathematics vigorously developed in his middle school years (1927–32). The mathematics classes given at middle schools consisted of arithmetic, algebra, plane geometry, and space geometry. Students were supposed to master computing skills and logical thinking by solving exercises of algebra and plane geometry in three years. Having solved all the exercises in the textbooks in one and a half years, Kodaira bought the two-volume, 1400-page thick "Advanced Algebra" by M. Fujiwara. The book was not easy to understand. Kodaira copied some complicated proofs several times until he could clearly see the structure of them. In this manner, he studied matrices, determinants, continued fractions, reciprocal law of quadratic residues and so on, just short of Galois theory.

Kodaira was now conscious of the fact that he loved mathematics. However, he was not yet aware of the existence of a strange profession called mathematical research. The profession he had in his mind was engineering.

From 1921 to 1922, Kodaira's father stayed in Germany, where inflation was astronomical. With the strong yen, he was able to purchase quantities of goods to bring back to Japan, some of which were presents for the two boys. The first thing which interested young Kodaira was a German assembly kit box containing steel plates, bolts, nuts, axes, wheels and gears. He would construct cranes and trains by assembling various parts, the process of which he enjoyed so much that he wanted to design and construct machines as an engineer.

This kit, by the way, also taught him that a triangle with edges of lengths 3, 4 and 5 is a right triangle.

Among the presents his father brought home from Germany, the piano was also of great importance for Kodaira. When he was 15 years old, he started playing the piano. His progress was quick. Everyday he practised for at least two hours, and when he entered Tokyo Imperial University, he was able to play Beethoven's Tempest sonata, Schumann's Humoreske, and Chopin's Ballade No. 1.

Kodaira was now interested in classical music, and his love for music as well as his love for mathematics continued throughout his life. He regularly attended concerts given in Tokyo and was deeply impressed by maestri such as Arthur Rubinstein, Wilhelm Kempf, and Leonid Kreutzer. As for artists he knew through records, he particularly liked Josef Lhevine.

In 1932, Kodaira took the entrance examination of Dai-Ichi High School.

Before World War II, the standard of Japanese high schools was high, comparable to that of the undergraduate courses of American universities. The entrance examinations of high schools were indeed the highest hurdle for the to-be elites, as only one out of eight or nine applicants was admitted. Dai-Ichi High School was known as the best and, accordingly, the hardest to enter.

The questions of the examination Kodaira took were as difficult as ever. Kodaira believed that he had failed in all subjects (English, Chinese classics, and Japanese classics) but mathematics. Feeling disappointed, he went to stay at a seaside villa of his best friend's father where he received a telegram from his mother: "You passed the exam! Come home!" The reality was that he had passed with the highest score of all the applicants.

Dai-Ichi High School had a humanities course and a science course. Kodaira did not hesitate to choose the science course and selected German as the second foreign language after English.

Students in high schools had to study foreign languages and mathematics extensively. There were ten German classes every week. In half a year, Kodaira finished basic grammar and was required to read novels and other literature written for native Germans.

Mathematics classes consisted of trigonometry, plane geometry, algebra and calculus. In parallel to the regular lectures at school, he continued reading Fujiwara's "Advanced Algebra" and began to study "Lectures on Elementary Number Theory", by T. Takagi.

Dai-Ichi High School admitted only about 400 students each year, and the students lived in dormitories, which gave the entire school the ambience of an extended family. Kodaira made many friends who later became scientists, medical doctors, bureaucrats, and businessmen. This network of close friendship proved extremely useful later when, for example, he tried to raise funds for the International Congress of Mathematicians at Kyoto, 1990.

Kodaira often visited Professor Aramata who was in charge of Kodaira's class and taught him calculus for two years. Aramata's wife would always serve beer and dinner for her husband's students. Although he had wanted to be an engineer during middle school, he now hoped to become a professor in a high school. Apparently he was attracted by the way of life Professor Aramata led, who looked so happy to be with his students.

Student at Tokyo Imperial University

In 1935, Kodaira was admitted to Tokyo Imperial University, the most prestigious academic institution in Japan.

For three years, he studied mathematics. The Mathematics Institute had only five professors (T. Takagi, S. Nakagawa, S. Kakeya, T. Takeuchi and J. Suetsuna), two associate professors (M. Tsuji and S. Iyanaga) and an assistant (S. Kametani). Among the 15 classmates in the Mathematics Institute, there were K. Ito (the second Japanese winner of the Wolf Prize in mathematics), Y. Kawada and S. Furuya, the latter two of whom were destined to be colleagues at the University of Tokyo.

In the first two years, Kodaira skipped most of the lectures. He read books by himself on Lebesgue integrals, topology (Alexandroff–Hopf), algebra (Deuring) and so forth. Shortly before the term examinations, he would borrow Kawada's notebooks and copy them out by hand, which was enough for him to thoroughly understand the lectures.

There were no lectures for third-year students. Instead they were required to take a seminar. Kodaira read "Topologie" by Alexandroff–Hopf under the supervision of Shokichi Iyanaga, his future brother-in-law.

In 1938 Kodaira graduated from the Mathematics Institute and then re-entered the Physics Institute of the university. The primary reason was that he realized the close relationship between mathematics and developing quantum physics, as was illustrated in the publication of H. Weyl's "Gruppentheorie und Quantenmechanik" and J. von Neumann's "Mathematische Grundlagen der Quantenmechanik". A second reason was that he needed a rest during the tensest period of Japanese history.

Kodaira's college life was parallel to the period of Japan's aggression into China, from Manchuria through North China to the Yang-Tse River Region. The army and navy, in alliance with combative bureaucrats, gradually deprived the establishment of political power through blackmail and, sometimes, sheer violence (they launched two coup d'etats and plotted several aborted ones). Japan was then dominated by militarism and nationalistic fanaticism.

Although the special police ruthlessly cracked down on liberalism, universities were comparatively protected from militarism and fanaticism. College students were regarded as elites in every sense and were exempted from military service until 1943. On February 26, 1937, a fanatical group in the army launched the second coup, killing several politicians and businessmen. An examination of algebra was scheduled on that date, which was naturally canceled. Kodaira and his classmates were very happy due to the canceled examination, and visited a zoo nearby.

Kodaira graduated from the Physics Institute in 1941, having written 10 research articles. His earliest papers up to this stage dealt with several scattered topics: ring theory, homology theory, dimension theory, operators on Hilbert spaces, metrized spaces, and especially, topological groups. His interest in topological/Lie groups and in Hilbert spaces is apparently related to the fact that he studied quantum mechanics. However, the papers he wrote in this period were purely mathematical in nature and most of them convey the author's affiliation to the Mathematics Institute instead of the Physics Institute. His 10th paper "Über die Beziehung zwischen den Massen und den Topologien in einer Gruppe" dated November 16, 1940, shortly before his graduation from the Physics Institute, was the proof of a theorem of A. Weil, which was announced in *Comptes Rendus* in 1936.

A young researcher in a troubled period

Immediately after his graduation from the Physics Institute of Tokyo Imperial University, Kodaira started his professional career by being appointed a research fellow at the Physics Institute, and then, in 1942, he became an associate professor at the Mathematical Institute, Tokyo Bun-Rika Daigaku (Tokyo College of Humanities and Sciences).

In 1941, Japan was completely isolated and there was no hope to win the long war with China. The embargo of strategic materials from the U.S. seriously damaged the life of the Japanese. On December 8 (Japan time; December 7 in the U.S.), Japan initiated the desperate, fatal war with the Allied Powers by attacking Pearl Harbor, Hawaii. This exacerbated the isolation of Japan and the devastation of life for the Japanese to an unheard-of extent. It was then impossible to import new journals and books from Europe or America, and talented researchers had nothing to resort to but their own ideas. Under these miserable circumstances, some researchers were still able to develop their original ideas, such as the renormalization theory of Shin-Ichiro Tomonaga, one of Kodaira's colleagues at Tokyo Bun-Rika Daigaku.

It was fortunate for the mathematical world that Kodaira was one of the few people who successfully discovered his originality during these difficult years. In this period, he began the study of harmonic forms on manifolds, which constituted the first series of his major papers and culminated in his Ph.D. thesis.

In 1944, Kodaira became associate professor at the Physics Institute of Tokyo Imperial University, still retaining the same position at Tokyo Bun-Rika Daigaku.

The situation of the War on the Pacific was abominable and the impending defeat was obvious to intellectuals in Japan. In 1943, the Japanese government decided that university and college students, who had been exempted from military service, were to be immediately conscripted into the military service. (It was almost unbelievable that the government still had reason to let young researchers in natural sciences continue their own work even if it was purely theoretical.) Furthermore, Tokyo was under the imminent threat of air raid by American aircraft, making it too dangerous to conduct researches in Tokyo. Evacuation to a safer place was considered and Kodaira played a leading role in organizing the evacuation of the departments of mathematics and physics to Suwa district, from where the Kodairas originated.

In 1945, at the faculty meeting, Kodaira proposed that the Physics Institute be relocated to a rural district and he explained that his father could find a suitable place for the relocation. He was astonished that the participants of the meeting immediately and unanimously agreed to the idea. He had believed that a consensus on such a major issue would be reached only after long, heated discussions. Moreover, the Mathematics Institute decided to relocate also. Since the evacuation was his idea, Kodaira was responsible to the entire plan. The connection of Kodaira's father to the people of Suwa helped smooth the negotiations with town offices, school masters and local hotels, and thus the two institutes, together with the students, successfully moved to Chino town and Osachi village.

Without doubt, the new residents had to endure various inconveniences in exchange for a secure place free from bombs. Kodaira lived in the birthplace of his father, from where he had to walk 7 km to reach the Physics Institute. However, obtaining food was a far more serious problem. Although the institutes were surrounded by rice fields and vegetable plots, food was controlled by a strict quota which was not sufficient. The only way to obtain additional ration was to barter clothes or jewelry illegally.

Marriage

On May 30, 1943, Kodaira married Seiko Iyanaga, the Iyanagas being another distinguished family.

Her brother, Shokichi Iyanaga, is a well-known mathematician and was, as mentioned above, the advisor of Kodaira at Tokyo Imperial University. As a student and later as the successor of Teiji Takagi of Tokyo Imperial University (currently the University of Tokyo), he studied mainly class field theory and number theory of local fields. He was also by far the most influential figure among the Japanese mathematicians in the '50s and '60s by producing astounding range of students, including K. Ito, K. Iwasawa, T. Tamagawa, G. Shimura, Y. Taniyama, I. Satake and Y. Ihara. Kodaira is one of the earliest and brightest students of Iyanaga.

Seiko's second brother, Kyojiro, was president of Nihon Optics, the company renowned for Nikon cameras, and her third brother, Teizo, was a professor of Japanese history at the University of Tokyo.

Seiko had heard about Kodaira as one of the brightest students of her brother, but personally they became acquainted through music.

Kodaira's second piano teacher, Tazuko Nakajima, was a professional violinist. When he was a student of physics, he was asked to be the accompanist for the annual recitals of her violin pupils. (He had special talent for sight reading.) Seiko was one of her violin pupils.

The young couple honeymooned in Gôra, a hot spring resort in Hakone (a volcano renowned for beautiful scenery and hot springs). Since food quotas were so severe, they were forced to bring rice to stay at the hotel.

Their first child, Kazuhiko, was born in March the following year. In the autumn, Seiko and Kazuhiko, and other female members of the Iyanagas with children, moved to their summer villa in Karuizawa, located far north of Tokyo and considered to be safe. During the winter holidays, the men also stayed there.

Since Karuizawa is located on a plateau 1,000 m above sea level, it was extremely cold during the winter, where temperatures sometimes dropped to −20°C. They bathed in water heated by firewood, but the wall of the bathroom was covered by ice all day long.

After the evacuation of the Physics and Mathematics Institutes was completed, the Kodairas, except Gon-Ichi who had to stay in Tokyo as a responsible statesman, moved to Yonezawa village, Gon-Ichi's birthplace. This was because their house in Tokyo, in which Goin-ichi, together with his secretary and a maid, resided in, was reduced to ashes by the air raid on April 13, 1945.

On August 15, Japan surrendered to the Allied Powers. The two institutes returned to Tokyo in early autumn and so did Kodaira slightly later, leaving Seiko and the boy in Yonezawa village.

Kodaira lived with his father in a room of a dormitory in the southwest part of Tokyo until a new small house was built on the estate of their old house.

Despite the miserable economic conditions, the academic activities at the university were quite vigorous. One of the many topics taken up at Kodaira's seminar was a paper of Heisenberg on S-matrices, which was brought to Japan during the War crossing the three oceans in a submarine. The theory of S-matrices was naturally related to the spectral theory of ordinary linear differential operators of second order, and Kodaira found a general result on eigenfunction expansions such that it implied the original theorem of Heisenberg if applied to the Schrödinger equation. The paper "The eigenvalue problem for ordinary differential equations of the second order and Heisenberg's theory of S-matrices" was brought to Hermann Weyl by H. Yukawa, who was invited to the Institute for Advanced Study in 1948. Weyl wrote to Kodaira that the same result had been independently obtained by E.C. Titchmarsh but that Kodaira's paper, based on a completely different method, was worth being published. This was the first personal contact between Kodaira and Weyl, his second mentor after Iyanaga. The paper was eventually published in the *American Journal of Mathematics* in 1950.

Another topic Kodaira was working on during this period was harmonic analysis on Riemannian manifolds, which he had started as soon as he graduated from the Physics Institute. The objective was to generalize Weyl's construction of meromorphic functions on Riemann surfaces to that on higher dimensional manifolds. He tried to generalize Weyl's theory of harmonic tensor fields, and found that it would pass over to general manifolds with the aid of the De Rham theory, Hadamard's fundamental solutions of partial differential equations, and Weyl's orthogonal projections in Hilbert spaces. The resumé of this major result was published in 1944 under the title "Über die harmonischen Tensorfelder in Riemannschen Mannigfaltigkeiten", although it took him three years to work out the details.

The year 1947 was a year of tragedy for the Kodairas.

In January, their young son Kazuhiko, suffering from nephrosis, was hospitalized in the nearby town of Kami-Suwa. Seiko, then pregnant, lived in the hospital room to care for her son. In the dire conditions of Japanese society after the war, they had to prepare their own food in the hospital ward. On weekends, the mother and child were joined by Kodaira. In May, a daughter, Yasuko was born, but Kazuhiko was in critical condition. In November, the Kodairas could not afford the medical care at the hospital any more. Kazuhiko returned to Yonezawa village, only to pass away on the 13th of that month.

Beside the sick boy's bed, Kodaira completed the final pages of "Harmonic fields in Riemannian manifolds (generalized potential theory)". The paper was too long to be published in Japan, as paper itself was in shortage. In 1948, S. Kakutani, who had an acquaintance in the occupying American army, arranged for the manuscript to be submitted to the *Annals of Mathematics*, of which Weyl was one of the editors. In October, Kodaira received a letter of acceptance from Lefschetz, the editor-in-chief, and the article was published in 1949 and became his Ph.D. thesis at the University of Tokyo (formerly Tokyo Imperial University).

The paper was a catalyst to his future research activities in two senses. First, the theory of harmonic forms and projection method on compact manifolds provided a basis to the great series of papers to come, including the well-known articles on the Kodaira vanishing theorem and the Kodaira embedding theorem. Secondly, together with the paper on S-matrices, it caught the attention of Weyl, who brought Kodaira to the Institute for Advanced Study in Princeton, New Jersey.

The first three years in the United States

When his second daughter, Mariko, was to be born, Kodaira received Weyl's invitation to come to the Institute for Advanced Study. The period from late '40s and to the '50s was an era when massive emigration of prominent Japanese mathematicians and scientists (H. Yukawa, S. Kakutani, K. Ito, Y. Nambu, K. Iwasawa, G. Shimura, etc.) to the United States took place. This was later called the *brain drain.*

Japanese air companies were still banned from operating aircraft and it was impossible to buy flight tickets from an American company in foreign currencies. Traveling via ship was the only transportation available, and it took two weeks to travel from Yokohama, a harbor city south of Tokyo, to San Francisco via Honolulu.

The ship *President Wilson* left Yokohama on August 9, 1949. Shin-Ichiro Tomonaga, his former colleague at Tokyo Bun-Rika Daigaku and a future Nobel Laureate in Physics, 1965, was also on board. He was invited to the Institute by Oppenheimer. Kodaira was seasick and was unable to digest any food when he arrived at Oahu Island. But due to a dispute among the sailors, the ship was anchored in Honolulu Harbor for three days, giving the two scientists the opportunity to see the beautiful tropical sights of the island. In Honolulu, Kodaira was particularly impressed by the rich flavor of ice cream and the number of cars traversing the city, which was in sharp contrast to the poverty in Japan.

During their long trip to Princeton, Kodaira and Tomonaga met many people.

The two arrived in San Francisco on August 24, and flew to Chicago on the 27th. At Chicago Airport, Kakutani met them and took them to the University of Chicago, where Kodaira was introduced to André Weil and Tomonaga to E. Fermi. Kodaira was shocked when Weil waved from his office window, his face covered by a mask of satan. He still had difficulty following conversations in English, so Weil proposed that he gave talks in Chicago the following year when his English had improved. During their one-week stay in New York City, they visited H. Yukawa, then a professor at Columbia University.

On September 9, the two Japanese arrived at the Institute and were introduced to Weyl and Oppenheimer.

For people who came from the devastated condition of Japan, the affluence in America must have been a wonderland. Everything American, from ice cream to cars, from refrigerators to houses with lawns, from beef steaks to operas, was inconceivable in Japan still under the occupation of the Allied Powers. And, above all, eminent mathematicians from all over the world lived in this continent, and Princeton, *par excellence*, was seemingly the center of gravity. The staff of the School of Mathematics of the Institute as of 1949 consisted of five professors (Weyl, Veblen, Morse, Siegel and von Neumann) and four fellows (Gödel, Alexander, Montgomerry and Selberg).

Kodaira describes his first days at Princeton as follows:

> Every morning at around 10, I went to the Institute. In my office I would read mathematical books and write research papers; at noon I would go to the cafeteria on the fourth floor. Usually Professor Weyl took lunch there with the young visiting fellows. It was embarrassing that I could not follow the conversations. Everybody would laugh at the jokes of Weyl, while I sat vacantly silent. For Europeans, conversing was much easier than writing, and they could not imagine that the reverse was the case for me. Somebody even asked me, "Did you really write your papers?" Weyl enjoyed all this and said, "You shall give talks at our seminar next year."
>
> However, everything worked well *because of* my poor English. The secretary, Miss Eigleheart, kindly did all my everyday business, from banking to typing business letters. She was born in Karuizawa, Japan, and had taught music to Seiko at her high school. I was deeply impressed by how people's lives are so closely entwined in the world.
>
> In late September, lectures and seminars started. I attended the lectures of Siegel on three-body problems. Every week he gave three one-hour lectures without using a single memo. He memorized by heart every formula, however complicated it was. Later when I was at Johns Hopkins, he gave a colloquium talk, which was followed by a dinner at a Chinese restaurant. He said "I often work on mathematics from 9 in the morning till midnight, skipping meals. When I have to eat food I skipped during the day after midnight, I have some trouble in the stomach." I thought, "Nobody could defeat such a person, unless he is a superman."
>
> In early October, I received a message from Professor Spencer of Princeton University expressing his desire to meet me. When I went see him, he proposed that I give seminar talks on my paper on harmonic tensor fields. I refused the request, saying that I did not speak English. Then he said, "You spoke English when you said you did not speak English!". So I had to promise to give talks at his seminar. I had no idea that this would be the beginning of our collaboration which lasted for twelve years.

The seminar organized by Weyl and Siegel started in February 1950. Speakers were Weyl (on historical introduction), de Rham (on harmonic differential forms based on currents), and Kodaira (on applications of harmonic forms to complex manifolds).

The summer of 1950 was devoted to a long trip across the United States. Tomonaga was returning home, and Kodaira joined him on the long drive to San Francisco. In San Francisco, Kodaira met K. Iwasawa, a new visitor to IAS, and the two mathematicians stayed in Chicago for one and a half months to discuss mathematics with Weil. From late August to early September, Kodaira attended the ICM held at Harvard as well as satellite conferences on several complex variables and on algebraic geometry. At the conference, he was requested by W.-L. Chow to be a one-year visiting professor at Johns Hopkins University, Baltimore. His salary at Johns Hopkins was to be $ 6,000, while that of Princeton was $ 4,000. He accepted the offer after consulting Weyl, who was in Switzerland on vacation at the time.

In June 1951, Kodaira returned to IAS, and in the same month Seiko arrived with the two little daughters. The younger daughter, Mariko, was born after Kodaira left Japan.

The family was happy with their new life in America. At home Kodaira was completely protected by Seiko, who was not only his private secretary but his cook and chauffeur. His favorite place for research was at the dining table in his home.

Now, with the arrival of a charming hostess, Kodaira's residence became a clubhouse for many Japanese visitors to Princeton including K. Ito, H. Yamabe and Y. Nambu. The Kodairas played music together again, which was their common hobby. Seiko ordered a 10-dollar violin at Sear's, while Kodaira bought a 60-dollar old piano, whose pitch was a half-note lower than normal. They played chamber music with the neighbors, and Kodaira mastered his special intriguing talent to play pieces a half-note higher than they were originally written.

In September 1952, Kodaira became an associate professor at Princeton University, due to the recommendation of Spencer. The chairman of the mathematics department was Lefschetz, whose first comment about Kodaira was: "You (Seiko) are taller than your husband." The new visitors to IAS that year included F. Hirzebruch.

Harmonic integrals on compact complex manifolds

The first joint program for Kodaira and Spencer at Princeton was to construct the theory of complex manifolds in the language of sheaves and cohomology.

The notion of sheaves grew out from the ideas of J. Leray and of K. Oka during World War II. In order to connect local data to global ones, Leray introduced cohomology with coefficients in sheaves and spectral sequences, while Oka independently introduced the notion of *idéaux de domaines indéterminés*. H. Cartan, who knew both of them well, unified their ideas to powerful machinery to investigate holomorphic functions in several variables. This theory was brought to America by French mathematicians, and sheaves were first called *faisceaux*, and later *stacks* as seen in Kodaira's 1953 papers. The term *sheaf* first appears in "On Kähler varieties of restricted type", which was accepted by *Proceedings of National Academy of Sciences U.S.A.* on February 23, 1954.

It was Spencer who, in 1952, proposed that a seminar on sheaves should be organized at Princeton. The textbook was *Séminaire Cartan* and was read by a young student whose hobby was to repair old used cars. Kodaira's first impression on sheaves was that sheaves are odd abstracts without reality. He was yet to use this new notion in the papers on the Riemann–Roch theorem on Kähler surfaces in 1953. However, by the spring of 1953, he realized that they were really useful to study complex manifolds.

Other new notions which became useful were those of fiber bundles and their characteristic classes. Fiber bundles were defined by Whitney in around 1935. Since that time and throughout the World War period, they were studied by Stiefel, Hopf and others. Chern forms were introduced by S.-S. Chern in 1944.

With this newly tuned machinery, Kodaira and Spencer started joint works on several topics which were related to the Riemann–Roch theorem in dimension two or more. Every day at lunch time, they fervently discussed mathematics. The first fruit of their lunch-time discussions was the affirmative answer to the Severi conjecture (1949) to the effect that two arithmetic genera, defined in two manners, should be identical. The result was published as "On arithmetic genera of algebraic varieties" in 1953. In late autumn of that year, one of the principal problems of the theory of compact complex manifolds was conquered by the young F. Hirzebruch, who was staying at the Institute for Advanced Study.

The Hirzebruch Riemann–Roch states that the alternate sum of the dimensions of the cohomology groups is a topological invariant. This, however, falls short of computing each dimension. To get the exact computation, the condition for higher cohomology to vanish is of great importance. The Kodaira vanishing theorem (1953) gives such a (sufficient) condition.

The theorem was proved in the paper "On a differential-geometric method in the theory of analytic stacks". Having completely mastered sheaves and other new notions through the seminar supervised by Spencer and the joint research with him, Kodaira combined these new powerful tools with the theory of harmonic integrals he had been studying since the years of World War II. He applied S. Bochner's differential-geometric method to prove the vanishing of $H^q(M, L(K_M))$, $q > 0$ for a compact Kähler manifold M and a positive line bundle L.

The implication of the Kodaira vanishing theorem is immense. It lies in the foundation of the so-called "minimal model program", basically implying important theorems of Grauert–Riemenschneider vanishing, Kawamata-Viehweg vanishing, Shokurov non-vanishing, Kawamata freeness, and cones of curves.

As an important application of the Kodaira vanishing, Kodaira also succeeded in giving a differential geometric characterization of complex projective manifolds ("On Kähler varieties of restricted type", 1954). Namely, a compact Kähler manifold is projective if and only if it carries a Kähler metric of which the Kähler form represents an integral cohomology class.

The astounding productivity of Kodaira naturally placed him under international spotlight. The International Congress of Mathematicians in 1954 was to be held in Amsterdam. One morning that year, Spencer brought Kodaira a message from Weyl, which stated that Kodaira was nominated as one of the two Fields medalists. It came as a complete surprise to him, as he had not made any plans to attend the ICM.

Together with Seiko, Kodaira left New York for Europe in mid August, to attend the ICM and to do some sightseeing before the conference. They first went to Rome, Naples, Pompeii and the Isle of Capri. Then they went to Switzerland to meet De Rham in Lausanne and the Weyls in Zurich.

On September 2 at the Opening Ceremony of the ICM, Kodaira was awarded the Fields Medal along with Jean-Pierre Serre. The President of the Fields Medal Committee was H. Weyl, Kodaira's mentor at Princeton, and the other members were E. Bompiani, F. Bureau, H. Cartan, A. Ostrowski, A. Pleijel, G. Szegö, and E.C. Tichmarsh. Weyl's beautiful address at the International Congress of Mathematicians[2] summarizes the work of Kodaira on harmonic integrals. To Kodaira the ceremony was so impressive that he remembered all the piano pieces played there (an *impromptu*, a *nocturne*, and a *scherzo*, of Chopin).

Kodaira's vivid report on this congress would be worth quoting:

> Following the program prepared beforehand, on the third day of the Congress I gave a lecture entitled "Some results in the transcendental theory of algebraic varieties". On the second day Serre, the other medalist, gave a talk on cohomology and algebraic geometry.
> I met Severi, the great master of algebraic geometry, with his "niece". Long ago when I was a student at Tokyo Imperial University, he gave a series of lectures as a cultural envoy from Italy. I attended the lectures, but did not remember what he had talked about. The only thing I remembered was the beautifully bold head of Professor Nakagawa, who had been sitting in front of me. So I was aghast when I heard Severi assert that I was one of his students for the reason that I had been among the audience of his lectures.
> On September 8, the Fields medalists were together with a dozen or so great mathematicians such as Weyl, Severi, Hodge and von Neumann. We were invited to the Queen's tea party which was held in the garden of a villa in the suburbs of Amsterdam. The enormous garden was

[2] H. Weyl: *Address of the President of the Fields Medal Committee 1954*, in: Proceedings of the International Congress of Mathematicians 1954, Erven P. Noordhoff, Groningen and North-Holland, Amsterdam, 1957, pp. 161–174.

covered by clean pebbles, and tables and chairs were standing in a corner. Serre carelessly started smoking and asked me, "How should I extinguish this cigarette?" I replied, "Hide it under the pebbles." He chuckled, "As a cat would do?"

After the Congress, the Kodairas visited Paris accompanied by Iyanaga to meet C. Chevalley, and then Cambridge to visit Hodge. The Hodges invited the Kodairas to tea, where Mrs. Hodge whispered to Seiko, "Aren't mathematicians odd?"

Prime years at Princeton

In 1955, Kodaira became a professor at Princeton University and a fellow at the Institute. During the fall, he gave three classes for graduate students at the university, while during the spring he belonged to the Institute and had no teaching duties. Atiyah, who was a 1955–56 visiting member of the Institute, has commented about Kodaira's lectures at Princeton with the following:

> (...) the front rows were filled by the new generation of young geometers: Hirzebruch, Serre, Bott, Singer. The front rows were actually rather crowded since Kodaira's voice rarely rose above a whisper. Fortunately he wrote very clearly, very slowly and in very large handwriting so his lecture notes were impeccable.[3]

In the same year, Kodaira resigned from the University of Tokyo where he had been promoted to professor at the Department of Mathematics from associate professor at the Department of Physics.

From the autumn of 1956, Kodaira started a joint research on deformations with Spencer.

Theory of deformation of complex manifolds is one of the most important achievements of Kodaira and was done in close collaboration with Donald C. Spencer, who had been working on deformation of the complex structure of Riemann surfaces with A.C. Schaeffer at Stanford. Since Kodaira never learned to drive, Spencer usually offered him a lift to and from the university. Their mathematical discussions during the drive home would not stop by the time they arrived at Kodaira's house and the discussions would go on in the car parked in front of the gate. The neighbors of the Kodairas were soon familiar with Spencer.

M.F. Atiyah summarizes their mutual relationship as follows:

[3] Obituary of Kunihiko Kodaira for the London Mathematical Society, *Bull. London Math. Soc.* **31** (1999) 489–493.

The Kodaira–Spencer collaboration was more than just a working relationship. The two had very different personalities which were complementary. Kodaira's shyness and reticence were balanced by Spencer's dynamism. In the world of university politics Spencer was able to exercise his talents on Kodaira's behalf, providing a protective environment in which Kodaira's mathematical talents could flourish.[4]

In the preface to his book "Complex Manifolds and Deformation of Complex Structures", Kodaira vividly described how the joint program on the deformation theory started and how it developed.

Deformation of the complex structure of Riemann surfaces is an idea which goes back to Riemann, who, in his famous memoir on Abelian functions published in 1857, calculated the number of effective parameters on which the deformation depends. (...) The deformation of algebraic surfaces seems to have been considered first by Max Noether in 1888. However, the deformation of higher dimensional complex manifolds had been curiously neglected for 100 years. In 1957, exactly 100 years after Riemann's memoir, Fröhlicher and Nijhenhuis published a paper in which they studied deformation of higher dimensional complex manifolds by a differential geometric method and obtained an important result. Inspired by their result, D.C. Spencer and I conceived a theory of deformation of compact complex manifolds which is based on the primitive idea that, since a compact complex manifold M is composed of a finite number of coordinate neighborhoods patched together, its deformation would be a shift in the patches. Quite naturally, it follows from this idea that an infinitesimal deformation of M should be represented by an element of the cohomology group $\mathrm{H}^1(M,\Theta)$ of M with coefficients in the sheaf Θ of germs of holomorphic vector fields. However, there seemed to be no reason that any given element of $\mathrm{H}^1(M,\Theta)$ represents an infinitesimal deformation of M. In spite of this, examination of familiar examples of compact complex manifolds M revealed a mysterious phenomenon that $\dim \mathrm{H}^1(M,\Theta)$ coincides with the number of effective parameters involved in the definition of M. In order to clarify this mystery, Spencer and I developed the theory of deformation of compact complex manifolds. The process of the development was the most interesting experience in my entire mathematical life. It was similar to an experimental science developed by the interaction between experiments (examination of examples) and theory. In this book I have tried to reproduce this interesting

[4] *Ibid.*

experience; however I could not fully convey it. Such an experience may be a passing phenomenon which cannot be reproduced.[5]

In the great series of their joint papers from 1957 to 1962, Kodaira and Spencer introduced characteristic maps (Kodaira–Spencer maps), and proved the fundamental existence theorem, completeness theorem and stability theorem. The analytic theory of deformation was finally completed by a theorem of M. Kuranishi on the existence of local universal deformation space (1962). The formal aspects of deformation theory were later developed by A. Grothendieck and were formulated in the functorial language in M. Schlessinger, "Functors of Artin rings" (1968).

In the late '50s, along with the collaboration with Spencer, Kodaira started a second program toward the structure theory of elliptic surfaces. He wrote:

> It was great fun to closely study the structure of elliptic surfaces by applying the general theory of complex manifolds. Classical theory of elliptic functions is so miraculously applied to it that the investigation smoothly progressed with surprising ease. ... My impression was that I simply carved out a statue called the theory of elliptic surfaces which had been embedded in the mass of marble called mathematics.

The work was to be incorporated into the overall classification theory of complex analytic surfaces.

The classification of algebraic surfaces was established by the Italian school of algebraic geometry, represented in particular by Castelnuovo and Enriques. However, their classical language was so hard to follow and some notions were so badly founded that the classification table must be rewritten in the modern, rigorous language of sheaves and cohomology groups.

The re-classification of algebraic surfaces was independently conducted by Kodaira in Princeton and by I.R. Shafarevich and his students in Moscow.

Shafarevich and his group did this in the framework of the classical pluri-canonical linear systems, introducing the symbol κ, which is an invariant that extracts the most essential property of the (algebraic) surfaces. The symbol now means the most important birational invariant of arbitrary dimensional algebraic varieties, and is called the *Kodaira dimension.* (This generalized invariant was first defined by S. Iitaka in 1972, inspired by Kodaira's plurigenera formula for surfaces. Eager to name the invariant after Kodaira, Iitaka invited Kodaira to a tea shop to get permission to use

[5]Intoduction to: K. Kodaira, *Complex Manifolds and Deformation of Complex Structures*, translated by K. Akao, Grundlehren der Mathematischen Wissenschaften 283, Springer Verlag, 1986.

his name. Kodaira said nothing and Iitaka concluded that the term was authorized.)

On the other hand, Kodaira, with the powerful general theory of complex manifolds at hand, extended the classification to that of compact complex surfaces. Only from this extended point of view, the structures of abelian surfaces, K3 surfaces and elliptic surfaces were well understood. Furthermore, he found a mysterious group of surfaces, which are never diffeomorphic to algebraic surfaces, and he named them surfaces of class VII_0. Kodaira's classification of complex surfaces appeared in another great series which occupies the main body of the third volume of his collected papers from 1960 through 1970.

In the late '50s, Kodaira was under the protection of the chairman Lefschetz, and his family fully enjoyed life in Princeton. He bought a house in 1956, and was immediately relieved from playing the piano a half-note lower by purchasing a cheap second-hand middle-grand Steinway. Now all the family members played an instrument: Kunihiko and the elder daughter Yasuko the piano, Seiko and the younger daughter Mariko the violin. They often had home concerts, sometimes with several guest musicians. They also regularly attended a small theater in Princeton, which held concerts given by musicians such as Rudolph Serkin, Isaac Stern, Andreas Segovia, Maira Hess, Budapest Quartet, Philadelphia Symphony Orchestra.

Departure from Princeton

In 1957, Kodaira received the Japan Academy Prize, and then The Cultural Medal, the highest level of recognition in Japan, for his mathematical achievement. He was the second mathematician to receive this honor, the first being T. Takagi and the third K. Oka.

Though the research activities of Kodaira and Spencer were at their peak, they were to experience a very unhappy departure from Princeton.

Kodaira wrote:

> One day after the colloquium, the participants went to dinner at a restaurant in the suburbs. Some twenty mathematicians sat around a rectangular table. Spencer and I were discussing mathematics, when we heard a sarcastic remark from Feller who sat opposite us: "The two rarely see each other so they started mathematical discussions as soon as they meet!" Spencer was surprised and said, "I did not realize we attracted so much attention." Borel, who sat at the left end of the table exclaimed: "Ah! Jealousy, jealousy! Another paper will appear soon!"
> In the same period, I visited Teruhisa Matsusaka of Brandeis who was staying at the Institute. The topic of salaries at universities came up. When Matsusaka knew my salary, he said "Your salary is so low!

> Princeton University must be a terrible place." Since Lefschetz had retired, I gradually realized that the older professors at Princeton hated me.

O. Zariski invited Kodaira to be a visiting professor at Harvard for the 1961/2 academic year. Unhappy at Princeton, he willingly accepted the offer. Harvard at that time was growing as the flourishing center of algebraic geometry. On the second day of his stay at Harvard, he saw a strange man with a completely bald head and short pants in a nearby supermarket. This was A. Grothendieck. Zariski gave parties every two weeks, with the participants being only men. Regular participants were Kodaira, Hironaka, Grothendieck, Tate, Mumford, and M. Artin, and they went on discussing mathematics from 9 p.m. to midnight with a glass of wine in their hands. In winter 1961, Hironaka proved the resolution of singularities which made him the second Fields Medalist from Japan.

In 1962, Chow, the chairman of the mathematics department of Johns Hopkins University, offered Kodaira professorship with an 18,000-dollar annual salary. Princeton did not try to retain him by raising his salary. After discussing the options with Spencer over the telephone, he decided to leave Princeton for Johns Hopkins. Enraged by the fact that the mathematics department did not stop Kodaira from leaving, Spencer also resigned from Princeton and moved to Stanford.

At Johns Hopkins, Kodaira gave a three-hour course on complex analysis and another three-hour course for advanced students. The teaching duty made Kodaira aware that there were some very strange students. For example Kodaira wrote:

> Since the results of the fall term exam on complex analysis was so terrible, I assigned exercises to one of the three classes of the spring term. After a few weeks, a student came to my office and complained that it was unfair to give questions which could not be solved without nice ideas. I was a little bit amazed, but after some pause I asked, "But you are not a computer, are you?"

The students at Johns Hopkins had to take certifying examinations two or three years after entering the university.

> I kicked a student out after the certifying examination. He had a special kind of memory. Whenever I asked a question, he knew that the answer was written on which page of which book, but could not tell what the answer was. The level of the questions became more and more elementary, and eventually I asked how he could prove that a polynomial equation of degree n has at most n roots. I received no answer from him. It was not possible to pass a graduate student

> who could not answer so basic a question. The examination committee unanimously decided that he be expelled and I informed him of the result. He showed up later at my office, saying, "I will not give up; Weierstrass also failed his exam." He was totally oblivious to the fact that he did not understand mathematics.

In 1964 the Kodairas spent July and August at the AMS Summer Institute on algebraic geometry at Woodshole, Massachusetts, and then in Stanford, California where Spencer lived. In October, he received an offer from Stanford for a position which was to begin in 1965. His decided to move immediately because he wanted to work with Spencer again.

Stanford was beautiful. He enjoyed the mild climate and beautiful scenery. The Kodairas lived in Palo Alto, the town next to Stanford. In the same year, Yasuko, his elder daughter, graduated from high school and was admitted to the University of California, Berkeley and to the International Christian University, Tokyo. She decided to return to Tokyo to become more aware of her own roots.

Return to Japan

Japanese mathematicians had long been asking Kodaira to return to Japan. They informally offered him professorship at the University of Tokyo.

In the summer of 1966, Kodaira and Spencer visited Japan to attend an international conference in Kyoto. After a 17-year stay in America, Japan had completely changed. Kodaira felt as if he were in America with everything being on a smaller scale. Yasuko guided him in Tokyo. Even the Japanese language was different from what they used to speak.

During his stay in Japan, Kodaira was formally requested to return to the University of Tokyo, which offered him professorship without the burden of administrative duties.

To reach a decision was extraordinarily difficult.

On the one hand, the prosperity of America seemed to be endless; the society was safe and the economy was stable with no inflation. For instance the price of a cup of coffee remained at 10 cents throughout the 18 years of his stay. His salary was now $ 24,000 after being raised during his second year at Stanford. He lived in a spacious house in beautiful Palo Alto, where the climate was mild year round with sunny skies. His daughters had been raised in America and felt more at home in English than in Japanese.

On the other hand, his personality was not at all American. He was too shy to easily survive in the assertive society of the United States. The offer from Tokyo with no administrative duties was attractive. The University of Tokyo, his *alma mater*, would present him a larger number of bright students.

The final decision was made by his daughter Yasuko, who said "Dad would be happier in Japan." He returned to Japan in mid-August 1967, to become professor at the Department of Mathematics of University of Tokyo. Major Japanese newspapers celebrated the return of Kodaira.

Prior to his departure to Japan, an informal meeting of some twenty participants took place at Stanford during which it was agreed to collect papers dedicated to Kodaira. The papers were published in 1969 under the title "Global Analysis" and was edited by S. Iyanaga and D.C. Spencer, two mathematicians who had played important roles in Kodaira's mathematical life. The contributors included Kodaira's old friends from Princeton years such as Spencer, Nirenberg, de Rham, Hirzebruch, Atiyah, Thom and Grauert; a new generation of colleagues in algebraic geometry, such as Artin, Mumford, Griffiths, Satake, Kuranishi; and his students Baily, Kas, Wavrik and Morrow.

Unfortunately, Kodaira did not have many students in the States. At Princeton, however, Kodaira had one outstanding student, Walter. L. Baily Jr., who has been a close friend of the Kodaira family since his student days. With his Japanese wife, Yaeko, Baily regularly visited Japan and the Kodairas. His most important work is the one on natural compactifications of arithmetic quotients of the hermitian bounded symmetric domains, known as the Baily–Borel compactifications. In his Johns Hopkins years, Kodaira had two students, A. Kas and J. Wavrik, who followed Kodaira to Stanford, and they received their Ph.D.s there. His student at Stanford, J. Morrow, wrote a book "Complex Manifolds" (Holt, Rinehart and Winston, Inc., 1971) together with his teacher.

As soon as Kodaira returned to Japan, he attracted a corps of young mathematicians of talent.

Even during his years in America, his influence was clearly visible in Japan. In this sphere of influence, we could count Tetsuji Shioda, Soichi Kawai, Shigeru Iitaka and Nobuo Sasakura who started their early works under the impact of Kodaira's classification theory of compact complex surfaces.

Having returned to Tokyo, Kodaira had an impressive number of students of high quality between 1968 and 1974: Tatsuo Suwa, Kenji Ueno, Yukihiko Namikawa, Eiji Horikawa, Masahisa Inoue, Masaki Kashiwara, Kazuo Akao, Iku Nakamura, Masahide Kato, Fumio Sakai, Takao Fujita, Toshiki Mabuchi, Kazuhisa Maehara, Yoichi Miyaoka, and Tohru Tsujishita. One of his first students, M. Kashiwara, proved a nice theorem while he attended Kodaira's undergraduate seminar. On his first visit to Japan in 1972, F. Hirzebruch praised K. Ueno's master thesis on the degeneration of curves of genus two, saying that it was worthy of three Ph.D. theses.

In Tokyo, Kodaira chaired a weekly seminar on complex manifolds with twenty to thirty participants on Saturday afternoons. The talks were organized by several young lecturers and assistants around him, specifically T. Shioda, S. Iitaka, K. Ueno, E. Horikawa and K. Akao. The topics were centered on the classification theory of complex surfaces and related subjects, including Shioda's elliptic modular surfaces, Iitaka's program for general classification theory, degeneration theory by Ueno and Namikawa, Horikawa's deformation theory of morphisms, and Inoue's new surfaces of class VII_0.

After the seminar, he would always have a piece of cake at a nearby coffee shop with the participants. Though not talkative, he did speak of new interesting topics he had read about in *Time, Newsweek* and *Scientific American*; he sometimes made satirical jokes and ironical observations about the Japanese society, the future of the civilization and new math. But more often he talked about music and old maestri who he loved so much. On the contrary, *avant garde* music represented by John Cage was sheer absurdity to him.

Meanwhile, the 1968 student riots which had started in France in 1968 quickly propagated to Japan in the same year. The universities were closed and collective bargainings between faculty and students were conducted. To Kodaira, the rioting was senseless and absurd. Students blamed faculty members of being idiotic specialists, implying that they knew nothing but their specialized areas. Kodaira received a questionnaire from the Faculty of Sciences on what should be said against such accusations during the bargaining session. He wrote, "If one is not an idiotic specialist, he is simply an idiot" a phrase which was, to his surprise, adopted by the faculty.

In 1971, Kodaira was elected Dean of the Faculty of Science. He, who believed that he was exempted from administrative duties, was miserably depressed as the promise of being exempted had been agreed upon only among professors at the mathematics department and not among the staff of the Faculty of Sciences as a whole.

The aftermath of the 1968 student riots was yet to die out, and being dean was not an easy task. He performed this difficult job effectively and conscientiously. However, the position caused him only agony.

In mid-March 1972, Kodaira was invited to a three-day conference in Princeton to celebrate the 60th birthday of Spencer. The participants included Bott, Griffiths, Mumford, Hirzebruch and Atiyah. On this occasion, he made a two-month visit to the United States along with Seiko and Mariko. In Baltimore, after giving a colloquium talk, he met Evans, who was also Dean of Johns Hopkins. Evans said, "I can't believe you are a dean" twice, first when they shook hands and later when they said goodbye. Kodaira's conclusion was "The dean of Johns Hopkins knew better than the faculty members of the University of Tokyo that I am not the person appropriate to become a dean."

Kodaira quit the position of dean on April 1, 1973. However, eventually, the pressure of being dean put an end to Kodaira's astonishing productivity in research. His last research paper was "Holomorphic mappings of polydiscs into compact complex manifolds" based on his lecture on Nevanlinna theory given in 1970.

Later Years

Kodaira, a member of the Japan Academy of Science since 1965, became a corresponding member of Göttingen Akademie (1974), a foreign associate of the National Academy of Sciences (1975), an honorary member of the Academy of Arts and Sciences (1978), and an honorary member of the London Mathematical Society (1979).

In 1975, he received another prestigious award, the Fujiwara Prize, for his achievement which was not covered by the Fields Medal, particularly those in the theory of complex surfaces and in deformation theory.

In that year Kodaira became 60 years old and retired from the University of Tokyo. On this occasion, a conference on complex analysis and algebraic geometry was held at the university in which Spencer naturally participated. Afterwards "Complex Analysis and Algebraic Geometry", a second collection of papers dedicated to Kodaira, was published with contributions from old friends and students.

In April, he moved to Gakushuin University, a highly esteemed private university and the former college for the imperial and aristocratic families.

When he was at the University of Tokyo, Kodaira did not really know the intellectual level of ordinary university students. Only at Gakushuin did he realize that intellectual competence of students deteriorated year by year. During the ten years at Gakushuin, he was so alarmed by the decline that he began to criticize the post-War educational system which he believed caused it.

His approach was of course not ideological but practical. The principal points of his argument were that (1) basic subjects like Japanese and mathematics should be more extensively taught in the early stages of education, because (2) less basic subjects, such as social studies or natural sciences, should be taught later after students master enough skills in basics to efficiently study such subjects. He also claimed that (3) for young students, classical concrete mathematics represented by plane geometry would best develop logical thinking and mathematical intuition.

Though Kodaira clearly addressed his opinions at public hearings held by the Ministry of Education, the educational system remained essentially unchanged. The system depends in essence on the complex power balance of many contradicting interest groups of politicians, beaurocrats and teachers.

However, to Kodaira's surprise, the Ministry of Education decided that elementary set theory for high school students be removed from textbooks following his third point that abstract approach to mathematics was no good for high school students. On hearing this piece of news, he decided to write high school mathematics textbooks in order to improve the mathematical education in Japan.

At the same time, he wrote a series of essays on mathematical education and on his own mathematical life which was compiled into two books,

"A Memoir of a Lazy Mathematician"(1986) and "I Could Not do Anything but Mathematics"(1987). These books are interesting and humorous, sometimes even hilarious. Although shy when conversing, he could express his impressions and philosophy effortlessly in his writings. The second book is a concise but consistent autobiography, including vivid episodes out of his mathematical and aesthetical life, while the first book contains his letters to Seiko in Tokyo from Princeton as well as his essays on education and on his own mathematical life.

To Kodaira, mathematics was something very concrete and real. He asserted that there was the sense of mathematics as well as the sense of sight or of hearing. He wrote:

> Though general perception believes that mathematics is akin to logic, mathematics has in fact little to do with logic. Of course mathematics must follow logic. However, logic to mathematics is similar to grammar to literature. Everybody understands ordinary logic, while many high school students do not understand mathematics. Therefore I believe mathematics is fundamentally different from logic.
> Mathematics is a science which studies mathematical phenomena in nature. To understand mathematics is to "see" mathematical phenomena. I do not mean to visually "see" something, but to perceive something by a kind of sense. This sense is hard to explain; it is a pure sense, quite different from logic or reasoning. For me this sense is close to that of sight.
> One might say that it is absurd that mathematical objects are natural phenomena. But the fact that mathematical phenomena are as real as physical phenomena is explicitly stated in the expression that a mathematician "discovered" a theorem instead of "invented" one. I proved several new theorems, but I never thought that I invented those. Rather I felt like I accidentally found theorems which had long been there.[6]

In March 1985, Kodaira retired from Gakushuin at age 70, followed by a birthday party at the mathematics department of Gakushuin, and a smaller party given by his students at the University of Tokyo.

In December of the previous year, he received a wonderful piece of news that he would be awarded the Wolf Prize along with Hans Levi. In May, Kodaira flew to Israel, accompanied by Yasuko in place of Seiko. One of the speeches delivered at the Wolf Prize Ceremony on May 12 was very impressive to him. "Japanese are good at technology while Jews are good at mathematics. But Professor Kazil, a Jew, won the Japan International

[6] An extract from "A Memoir of a Lazy Mathematician"

Prize in biotechnology, and Professor Kodaira, a Japanese, won the Wolf Prize in mathematics. What happened?"

From the early '80s Kodaira suffered from weak health. At Stanford, the first symptom of asthma started which afflicted Kodaira in the years to come. In the early years of his stay in America, he loved to drink wine, but later he developed an allergy against alcohol. After he returned to Japan, he did not drink any alcohol at all.

What was even more devastating for the enthusiastic music lover was a disorder which developed in his ears.

In the late '70s, he bought expensive audio equipment and a vast number of LP discs. He particularly loved piano music as he was an excellent piano player. His favorite pianists were mostly old virtuosi like Rachmaninov, Godowsky or Rubinstein, but he had high opinions of younger pianists such as Argerich and Kisin. As for composers, perhaps he liked Chopin the best, but in his later years he loved to listen to Mozart.

The first symptom he suffered from was difficulty in hearing high notes, followed by a painful ringing in his ears. He had to gradually abandon listening to music and by 1988 he quit playing the Steinway which had been at his side since his early years in Princeton.

In 1986, Kodaira was nominated to be the president of the ICM 90 organizing committee. He was the most natural candidate for the president of the first ICM which was to be held in Asia.

Despite his weak health, he made great effort to raise funds for the Congress. He signed thousands of letters of appreciation to the donators. He visited businessmen for corporal donations.

In order to collect donations from industry and financial institutions, he had a valuable resource of the network of his old friends, mostly classmates during his high school years. H. Tanimura, Chairman of the Tokyo Stock Exchange, was particularly influential, as he was also former Vice Minister of Finance and former Chairman of the Fair Trade Commission. He was the key person for helping raise funds along with Dr. K. Miyairi, another close friend and classmate.

The Kyoto Congress was a great success, but his weak health and asthma prevented Kodaira from chairing the ceremonies.

From the mid '90s, Kodaira was semi-bedridden due not only to asthma, but also to cancer of the prostrate and cerebral infraction. As his illnesses grew progressively worse, he was sadly hospitalized until he passed away on July 26, 1997.

Hundreds of people attended Kodaira's funeral which was held on July 29, 1997. He was even honored with flowers from the Emperor which were placed directly above the coffin among the other flowers which surrounded him. At the bequest of his family's wishes, and because he was not a religious person, a simple ceremony was held. Chopin's piano nocturnes and ballades which he so dearly loved echoed softly in the background as each mourner presented him with a white carnation to pay his last respect to Kodaira.

Bibliography

Research Papers

(1) **Über die Struktur des endlichen, vollständig primären Ringes mit verschwindendum Radikalquadrat**
Japan J. Math., 14(1937), 15–21.

(2) **Über den allgemeinen Zellenbegriff und die Zellenzerspaltungen der Komplexe**
Proc. Imp. Acad. Tokyo, 14(1938), 49–52.

(3) **Eine Bemerkung zur Dimensionstheorie**
Proc. Imp. Acad. Tokyo, 15(1939), 174–176.

(4) **On some fundamental theorems in the theory of operators in Hilbert space**
Proc. Imp. Acad. Tokyo, 15(1939), 207–210.

(5) **On the theory of almost periodic functions in a group** (collaborated with S. Iyanaga)
Proc. Imp. Acad. Tokyo, 16(1940), 136–140.

(6) **Über die Differenzierbarkeit der einparametrigen Untergruppe Liescher Gruppen**
Proc. Imp. Acad. Tokyo, 16(1940), 165–166.

(7) **Über zusammenhängende kompakte abelsche Gruppen** (collaborated with M. Abe)
Proc. Imp. Acad. Tokyo, 16(1940), 167–172.

(8) **Die Kuratowskische Abbildung und der Hopfsche Erweiterungssatz**
Compositio Math., 7(1940), 177–184.

(9) **Über die Gruppe der messbaren Abbildungen**
Proc. Imp. Acad. Tokyo, 17(1941), 18–23.

(10) **Über die Beziehung zwischen den Massen und den Topologien in einer Gruppe**
Proc. Phys.-Math. Soc. Japan(3), 23(1941), 67–119.

(11) **Normed ring of a locally compact abelian group** (collaborated with S. Kakutani)
Proc. Imp. Acad. Tokyo, 19(1943), 360–365.

(12) **Über die Harmonischen Tensorfelder in Riemannschen**

Mannigfaltigkeiten, (I), (II), (III)
Proc. Imp. Acad. Tokyo, 20(1944), 186–198, 257–261, 353–358.

(13) **Über die Rand- und Eigenwertprobleme der linearen elliptischen Differentialgleichungen zweiter Ordnung**
Proc. Imp Acad. Tokyo, 20(1944), 262–268.

(14) **Über das Haarsche Mass in der lokal bikompakten Gruppe** (collaborated with S. Kakutani)
Proc. Imp. Acad. Tokyo, 20(1944), 444–450.

(15) **Relations between harmonic fields in Riemannian manifolds**
Math. Japonicae, 1(1948), 6–23.

(16) **On the existence of analytic functions on closed analytic surfaces**
Kōdai Math. Sem. Reports, 1(1949), 21–26.

(17) **Harmonic fields in Riemannian manifolds (generalized potential theory)**
Ann. of Math., 50(1949), 587–665.

(18) **The eigenvalue problem for ordinary differential equations of the second order and Heisenberg's theory of *S*-matrices**
Amer. J. Math., 71(1949), 921–945.

(19) **On ordinary differential equations of any even order and the corresponding eigenfunction expansions**
Amer. J. Math., 72(1950), 502–544.

(20) **A non-separable translation invariant extension of the Lebesgue measure space** (collaborated with S. Kakutani)
Ann. of Math., 52(1950), 574–579.

(21) **Harmonic integrals,** Part II
Lectures delivered in a seminar conducted by Professors H. Weyl and C. L. Siegel at the Institute for Advanced Study (1950).

(22) **The theorem of Riemann-Roch on compact analytic surfaces**
Amer. J. Math., 73(1951), 813–875.

(23) **Green's forms and meromorphic functions on compact analytic varieties**
Canad. J. Math., 3(1951), 108–128.

(24) **The theorem of Riemann-Roch for adjoint systems on 3-dimensional algebraic varieties**
Ann. of Math., 56(1952), 298–342.

(25) **On analytic surfaces with two independent meromorphic functions** (collaborated with W.-L. Chow)
Proc. Nat. Acad. Sci. U.S.A., 38(1952), 319–325.

(26) **On the theorem of Riemann-Roch for adjoint systems on**

Kählerian varieties
Proc. Nat. Acad. Sci. U.S.A., 38(1952), 522–527.

(27) **Arithmetic genera of algebraic varieties**
Proc. Nat. Acad. Sci. U.S.A., 38(1952), 527–533.

(28) **The theory of harmonic integrals and their applications to algebraic geometry**
Work done at Princeton University, 1952.

(29) **The theorem of Riemann-Roch for adjoint systems on Kählerian varieties**
Contributions to the Theory of Riemann Surfaces, Annals of Math. Studies, No. 30 (1953), 247–264.

(30) **Some results in the transcendental theory of algebraic varieties**
Ann. of Math., 59(1954), 86–134.

(31) **On arithmetic genera of algebraic varieties** (collaborated with D. C. Spencer)
Proc. Nat. Acad. Sci. U.S.A., 39(1953), 641–649.

(32) **On cohomology groups of compact analytic varieties with coefficients in some analytic faisceaux**
Proc. Nat. Acad. Sci. U.S.A., 39(1953), 865–868.

(33) **Groups of complex line bundles over compact Kähler varieties** (collaborated with D. C. Spencer)
Proc. Nat. Acad. Sci. U.S.A., 39(1953), 868–872.

(34) **Divisor class groups on algebraic varieties** (collaborated with D. C. Spencer)
Proc. Nat. Acad. Sci. U.S.A., 39(1953), 872–877.

(35) **On a differential-geometric method in the theory of analytic stacks**
Proc. Nat. Acad. Sci. U.S.A., 39(1953), 1268–1273.

(36) **On a theorem of Lefschetz and the lemma of Enriques-Severi-Zariski** (collaborated with D. C. Spencer)
Proc. Nat. Acad. Sci. U.S.A., 39(1953), 1273–1278.

(37) **On Kähler varieties of restricted type**
Proc. Nat. Acad. Sci. U.S.A., 40(1954), 313–316.

(38) **On Kähler varieties of restricted type (an intrinsic characterization of algebraic varieties)**
Ann. of Math., 60(1954), 28–48.

(39) **Some results in the transcendental theory of algebraic varieties**
Proc. Intern. Congress of Mathematicians, 1954, Vol. III, 474–480.

(40) **Characteristic linear systems of complete continuous systems**
Amer. J. Math., 78(1956), 716–744.

(41) **On the complex projective spaces** (collaborated with F. Hirzebruch)
J. Math. Pures Appl., 36(1957), 201–216.

(42) **On the variation of almost-complex structure** (collaborated with D. C. Spencer)
Algebraic Geometry and Topology, Princeton Univ. Press, 1957, pp. 139–150.

(43) **On deformations of complex analytic structures, I–II** (collaborated with D. C. Spencer)
Ann. of Math., 67(1958), 328–466.

(44) **On the existence of deformations of complex analytic structures** (collaborated with L. Nirenberg and D. C. Spencer)
Ann. of Math., 68(1958), 450–459.

(45) **A theorem of completeness for complex analytic fibre spaces** (collaborated with D. C. Spencer)
Acta Math., 100(1958), 281–294.

(46) **Existence of complex structure on a differentiable family of deformations of compact complex manifolds** (collaborated with D. C. Spencer)
Ann. of Math., 70(1959), 145–166.

(47) **A theorem of completeness of characteristic systems of complete continuous systems** (collaborated with D. C. Spencer)
Amer. J. Math., 81(1959), 477–500.

(48) **On deformations of complex analytic structures, III, Stability theorems for complex structures** (collaborated with D. C. Spencer)
Ann. of Math., 71(1960), 43–76.

(49) **On deformations of some complex pseudo-group structures**
Ann. of Math., 71(1960), 224–302.

(50) **Multifoliate structures** (collaborated with D. C. Spencer)
Ann. of Math., 74(1961), 52–100.

(51) **On compact analytic surfaces**
Analytic Functions, Princeton Univ. Press, 1960, pp. 121–135.

(52) **On compact complex analytic surfaces, I**
Ann. of Math., 71(1960), 111–152.

(53) **A theorem of completeness for analytic systems of surfaces, with ordinary singularities**
Ann. of Math., 74(1961), 591–627.

(54) **A theorem of completeness of characteristic systems for analytic families of compact submanifolds of complex**

manifolds
Ann. of Math., 75(1962), 146–162.

(55) **On stability of compact submanifolds of complex manifolds**
Amer. J. Math., 85(1963), 79–94.

(56) **On compact analytic surfaces, II–III**
Ann. of Math., 77(1963), 563–626, 78(1963), 1–40.

(57) **On the structure of compact complex analytic surfaces**
Proc. Nat. Acad. Sci. U.S.A., 50 (1963), 218–221.

(58) **On the structure of compact complex analytic surfaces, II**
Proc. Nat. Acad. Sci. U.S.A., 51(1964), 1100–1104.

(59) **On the structure of compact complex analytic surfaces**
Lecture Notes prepared in connection with the A.M.S. Summer Institute on Algebraic Geometry held at the Whitney Estate, Woods Hole, Mass. July 6–July 31, 1964.

(60) **On the structure of compact complex analytic surfaces, I**
Amer. J. Math., 86(1964), 751–798.

(61) **On characteristic systems of families of surfaces with ordinary singularities in a projective space**
Amer. J. Math., 87(1965), 227–256.

(62) **Complex structures on $S^1 \times S^3$**
Proc. Nat. Acad. Sci. U.S.A., 55(1966), 240–243.

(63) **On the structure of compact complex analytic surfaces, II**
Amer. J. Math., 88(1966), 682–721.

(64) **A certain type of irregular algebraic surfaces**
J. Anal. Math., 19(1967), 207–215.

(65) **Pluricanonical systems on algebraic surfaces of general type**
Proc. Nat. Acad. Sci. U.S.A., 58(1967), 911–915.

(66) **On the structure of compact complex analytic surfaces, III**
Amer. J. Math., 90(1968), 55–83.

(67) **Pluricanonical systems on algebraic surfaces of general type**
J. Math. Soc. Japan, 20(1968), 170–192.

(68) **On the structure of complex analytic surfaces, IV**
Amer. J. Math., 90(1968), 1048–1066.

(69) **On homotopy $K3$ surfaces**
Essays on Topology and Related Topics, Mémoires dédiés à Georges de Rham, Springer, 1970, pp. 58–69.

(70) **Holomorphic mappings of polydiscs into compact complex manifolds**
J. Differential Geometry, 6(1971), 33–46.

Books

[B1] (with J. Morrow) *Complex manifolds*, Holt, Rinehart and Winston, Inc., 1971.

[B2] *Introduction to Calculus* (in Japanese), Iwanami-Shoten Publishers, Tokyo, 1982.

[B3] *Complex Manifolds* (in Japanese), Iwanami-Shoten Publishers, Tokyo, 1982.

[B4] *Complex Analysis* (in Japanese), Iwanami-Shoten Publishers, Tokyo, 1983.

[B5] *Introduction to Complex Analysis*, translated from [B4] by E. Horikawa, Cambridge University Press, Cambridge-New York, 1984.

[B6] *Complex Manidfolds and Deformation of Complex Structures*, translated from [B3] by K. Akao, Grundlehren der Mathematischen Wissenschaften 283, Springer-Verlag, Berlin-Heidelberg-New York, 1986.

[B7] *A Memoir of a Lazy Mathematician* (in Japanese), Iwanami-Shoten Publishers, Tokyo, 1986.

[B8] *I Could Not do Anything but Mathematics* (in Japanese), Nikkei-Science, Tokyo, 1987.

[B9] *Joy of Geometry* (in Japanese), Iwanami-Shoten Publishers, Tokyo, 1987.

[B10] *An Invitation to Geometry* (in Japanese), Iwanami-Shoten Publishers, Tokyo, 1991.

Acknowledgement

This short biography essentially follows Kodaira's own account in his two books "A Memoir of a Lazy Mathematician" and "I Could Not do Anything but Mathematics". Unless otherwise mentioned, the citations are taken from the second book. Some additional important data came from interviews with Mrs. Seiko Kodaira, Mrs. Yasuko Hashimoto and Mrs. Mariko Oka, to whom the authors are heartily grateful.

*ON A DIFFERENTIAL-GEOMETRIC METHOD IN THE THEORY OF ANALYTIC STACKS**

By K. Kodaira

Department of Mathematics, Princeton University

Communicated by S. Lefschetz, September 9, 1953

1. *Introduction.*—Let V be a compact Kähler variety of complex dimension n and let F be a complex line bundle[1] over V whose structure group is the multiplicative group of complex numbers. Moreover let $\Omega^p(F)$ be the stack (faisceau) over V of germs of holomorphic p-forms with coefficients in F and let $H^q(V; \Omega^p(F))$ be the qth cohomology group of V with coefficients in $\Omega^p(F)$. It is important for applications to determine the circumstances under which the cohomology group $H^q(V; \Omega^p(F))$ vanishes. In the present note we shall prove by a differential-geometric method due to Bochner[2] some sufficient conditions for the vanishing of $H^q(V; \Omega^p(F))$ in terms of the characteristic class of the bundle F.

2. *Harmonic Forms with Coefficients in Complex Line Bundles.*[3] We denote by $ds^2 = 2\sum g_{\alpha\beta^*}(dz^\alpha d\bar{z}^\beta)$ the Kähler metric on V. Take a sufficiently fine finite covering $\mathfrak{U} = \{U_j\}$ of V. Then the bundle F is determined by the system $\{f_{jk}\}$ of non-vanishing holomorphic functions f_{jk} defined, respectively, in $U_j \cap U_k$ and satisfying $f_{jk}f_{kl}f_{lj} = 1$ in $U_j \cap U_k \cap U_l$. Clearly there exists a system $\{a_j\}$ of real positive functions a_j of class C^∞ defined, respectively, in U_j satisfying

$$\frac{a_j}{a_k} = |f_{jk}|^2, \quad \text{in } U_j \cap U_k. \tag{1}$$

Reprinted from Proc. Nat. Acad. Sci., Vol. 39 (1953), pp. 1268–1273.

A form φ with coefficients in F is, by definition, a system $\{\varphi_j\}$ of exterior differential forms φ_j defined, respectively, in U_j such that

$$\varphi_j = f_{jk}\cdot\varphi_k, \quad \text{in } U_j \cap U_k.$$

For a form $\varphi = \{\varphi_j\}$ with coefficients in F, $\bar{\partial}\varphi = \{(\bar{\partial}\varphi)_j\}$ and $\mathfrak{b}\varphi = \{(\mathfrak{b}\varphi)_j\}$ are defined, respectively, by

$$(\bar{\partial}\varphi)_j = \bar{\partial}\varphi_j, \qquad (\mathfrak{b}\varphi)_j = -*a_j\partial\left(\frac{1}{a_j} *\varphi_j\right),$$

provided that φ is differentiable, where $*\varphi_j$ denotes the dual form of φ_j with respect to the preassigned Kähler metric ds^2 on V. The form φ is called *harmonic* if $\bar{\partial}\varphi = 0$ and $\mathfrak{b}\varphi = 0$.

Let $\Omega^p(F)$ be the stack (faisceau) over V of germs of holomorphic p-forms with coefficients in F. Then, as was shown by the author,[4] *the cohomology group $H^q(V; \Omega^p(F))$ of V with coefficients in $\Omega^p(F)$ is isomorphic to the space $H^{p,q}(F)$ consisting of all harmonic forms of type (p, q) with coefficients in F:*

$$H^q(V; \Omega^p(F)) \cong H^{p,q}(F). \tag{2}$$

As one readily infers, the mapping

$$\varphi = \{\varphi_j\} \to \varphi^\dagger = \{\varphi_j^\dagger\}, \quad \varphi_j^\dagger = \frac{1}{a_j} *\bar{\varphi}_j$$

maps $H^{p,q}(F)$ isomorphically onto $H^{n-p,n-q}(-F)$, where $-F$ denotes the complex line bundle defined by the system $\{f_{jk}^{-1}\}$. Combining this with (2), we obtain therefore the isomorphism[5]

$$H^q(V; \Omega^p(F)) \cong H^{n-q}(V; \Omega^{n-p}(-F)). \tag{3}$$

3. An Inequality.—The equations $\bar{\partial}\varphi = 0$, $\mathfrak{b}\varphi = 0$ for $\varphi = \{\varphi_j\}$,

$$\varphi_j = (1/p!q!)\sum \varphi_{j\tau_1\ldots\tau_p\alpha_1^*\ldots\alpha_q^*}\, dz^{\tau_1}\ldots dz^{\tau_p} d\bar{z}^{\alpha_1}\ldots d\bar{z}^{\alpha_q},$$

can be written explicitly in the forms

$$\sum_{m=0}^{q} (-1)^m \nabla_{\alpha_m^*}\varphi_{j\tau_1\ldots\tau_p\alpha_0^*\ldots\alpha_{m-1}^*\alpha_{m+1}^*\ldots\alpha_q^*} = 0, \tag{4$_1$}$$

$$\sum g^{\lambda\mu^*}(\nabla_\lambda + \rho_{j\lambda})\varphi_{j\tau_1\ldots\tau_p\mu^*\alpha_2^*\ldots\alpha_q^*} = 0, \tag{4$_2$}$$

where ∇_λ, ∇_{α^*} denote the covariant differentiations with respect to the local coordinates z^λ, $\bar{z}^\alpha$ and where

$$\rho_{j\lambda} = -\partial_\lambda \log a_j, \qquad \partial_\lambda = \frac{\partial}{\partial z^\lambda}.$$

Let $R^\sigma{}_{\tau\alpha^*\beta} = \bar{\partial}_\alpha(\sum g^{\sigma\nu^*}\partial_\beta g_{\tau\nu^*})$ be the curvature tensor and let $R_{\alpha^*\beta} = \sum_\tau R^\tau{}_{\tau\alpha^*\beta}$ be the Ricci tensor.[6] Moreover we set

$$X_{\lambda\alpha^*} = -\bar{\partial}_\alpha \rho_{j\lambda} = \partial_\lambda \bar{\partial}_\alpha \log a_j. \tag{5}$$

Then, by a simple computation, we infer from (4) that, for any harmonic form $\varphi = \{\varphi_j\} \epsilon H^{p,q}(F)$, the following relation holds:

$$\sum g^{\lambda\mu^*}(\nabla_\lambda + \rho_{j\lambda})\nabla_{\mu^*}\varphi_{j\tau_1\ldots\tau_p\alpha_1^*\ldots\alpha_q^*} = \sum_m \sum_h \sum R^\sigma{}_{\tau_h\alpha_m^*}{}^{\mu^*} \times$$

$$\varphi_{j\tau_1\ldots(\sigma)_h\ldots\tau_p\alpha_1^*\ldots(\mu^*)_m\ldots\alpha_q^*} + \sum_m \sum [X^{\mu^*}{}_{\alpha_m^*} - R^{\mu^*}{}_{\alpha_m^*}] \times$$

$$\varphi_{j\tau_1\ldots\tau_p\alpha_1^*\ldots(\mu^*)_m\ldots\alpha_q^*}, \tag{6}$$

where $X^{\mu^*}{}_{\alpha^*} = \sum g^{\lambda\mu^*}X_{\lambda\alpha^*}$, $R^{\mu^*}{}_{\alpha^*} = \sum g^{\lambda\mu^*}\bar{R}_{\lambda^*\alpha}$ and where $(\sigma)_h$ or $(\mu^*)_m$ indicates that τ_h or α_m^* is to be replaced by σ or μ^* respectively. We note here that $\partial\bar{\partial} \log a_j = \partial\bar{\partial} \log a_k$ in $U_j \cap U_k$, so

$$\sum X_{\lambda\alpha^*}dz^\lambda d\bar{z}^\alpha = \partial\bar{\partial} \log a_j \tag{7}$$

is a well defined 2-form on the whole of V. Now we set

$$\xi_{\mu^*} = \frac{1}{a_j}\sum \nabla_{\mu^*}\varphi_{j\tau_1\ldots\tau_p\alpha_1^*\ldots\alpha_q^*}\cdot\bar{\varphi}_j^{\tau_1\ldots\tau_p\alpha_1^*\ldots\alpha_q^*}$$

where

$$\bar{\varphi}_j^{\tau_1\ldots\tau_p\alpha_1^*\ldots\alpha_q^*} = \sum g^{\tau_1\sigma_1^*}\ldots g^{\tau_p\sigma_p^*}g^{\beta_1\alpha_1^*}\ldots g^{\beta_q\alpha_q^*}\overline{\varphi_{j\sigma_1\ldots\sigma_p\beta_1^*\ldots\beta_q^*}}.$$

Then $\xi = \sum \xi_{\tau^*} d\bar{z}^\tau$ is a well defined 1-form on V and

$$-\delta\xi = \sum g^{\lambda\mu^*}\nabla_\lambda\xi_{\mu^*}$$

$$= \# + \frac{1}{a_j}\sum g^{\lambda\mu^*}(\nabla_\lambda + \rho_{j\lambda})\nabla_{\mu^*}\varphi_{j\tau_1\ldots\tau_p\alpha_1^*\ldots\alpha_q^*}\cdot\bar{\varphi}_j^{\tau_1\ldots\tau_p\alpha_1^*\ldots\alpha_q^*} \tag{8}$$

where

$$\# = \frac{1}{a_j}\sum g^{\lambda\mu^*}\nabla_{\mu^*}\varphi_{j\tau_1\ldots\tau_p\alpha_1^*\ldots\alpha_q^*}\cdot\overline{\nabla_{\lambda^*}\varphi_j}^{\tau_1\ldots\tau_p\alpha_1^*\ldots\alpha_q^*} \geqq 0.$$

Denoting by g the determinant $g = |g_{\alpha\beta^*}|$, we have $\int_V \delta\xi\cdot g dV = 0$, where $dV = (i)^n dz^1 d\bar{z}^1\ldots dz^n d\bar{z}^n$. Combining (6) with (8), we obtain therefore the inequality

$$0 \geqq \int_V \Big\{ q\sum(\delta^\sigma{}_\tau[X^{\beta^*}{}_{\alpha^*} - R^{\beta^*}{}_{\alpha^*}] + pR^\sigma{}_{\tau\alpha^*}{}^{\beta^*})\cdot$$

$$\frac{1}{a_j}\varphi_{j\sigma\tau_2\ldots\tau_p\beta^*\alpha_2^*\ldots\alpha_q^*}\bar{\varphi}_j^{\tau\tau_2\ldots\tau_p\alpha^*\alpha_2^*\ldots\alpha_q^*}\Big\} g dV. \tag{9}$$

By a (p, q)-form we shall mean a $(p + q)$-form of type (p, q). A real (1,1)-form ψ on V can be written in the form $\psi = i\sum \psi_{\alpha\beta^*}(z, \bar{z})dz^\alpha d\bar{z}^\beta$, where $\psi_{\alpha\beta^*} = \psi_{\alpha\beta^*}(z, \bar{z})$ satisfy $\bar{\psi}_{\alpha\beta^*} = \psi_{\beta\alpha^*}$. *We say that such a form* ψ

is positive and write $\psi > 0$ *if the Hermitian form* $\sum \psi_{\alpha\beta^*}(z, \bar{z})\bar{u}^{\alpha}u^{\beta}$ *in* n *variables* $u^1, u^2, \ldots, u^n$ *is positive definite at each point* z *on* V.

Now, let $c(F)$ be the characteristic class of the principal bundle associated with F.

LEMMA. *Let* $\{a_j\}$ *be a system of functions satisfying* (1). *Then the real d-closed* (1,1)*-form* $\gamma = (i/2\pi)\partial\bar{\partial} \log a_j$ *belongs*[7] *to the characteristic class* $c(F)$ *of the bundle* F. *Conversely, given a d-closed real* (1,1)*-form* γ *of class* C^{∞} *belonging to the characteristic class* $c(F)$, *there exists a system* $\{a_j\}$ *of positive functions* a_j *of class* C^{∞} *satisfying* (1) *such that* $(i/2\pi)\partial\bar{\partial} \log a_j = \gamma$.

Proof: Let π be the canonical projection of F onto V and let ζ_j be the linear coordinate on the fibre $\pi^{-1}(z)$, $z \in U_j$, such that $\zeta_j = f_{jk}(z)\zeta_k$ for $z \in U_j \cap U_k$. The principal bundle F^* associated with F is obtained from F by removing the origin $\zeta_j = 0$ of each fibre $\pi^{-1}(z)$. Now

$$\Phi = \frac{1}{2\pi i} (d \log \zeta_j - \partial \log a_j)$$

is a well defined 1-form of class C^{∞} on the principal bundle F^* and

$$-d\Phi = \frac{i}{2\pi} \partial\bar{\partial} \log a_j,$$

while the restriction of Φ to the fibre $\pi^{-1}(z)^* = \pi^{-1}(z) - (0)$ of F^* equals the d-closed 1-form $(1/2\pi i)d \log \zeta_j$ which represents the basic cohomology class of $\pi^{-1}(z)^*$. This shows that $\gamma = (i/2\pi)\partial\bar{\partial} \log a_j$ belongs to the characteristic class $c(F)$ of F^*.

To prove that every γ belonging to $c(F)$ can be written in the form $\gamma = (i/2\pi)\partial\bar{\partial} \log a_j$, we take a system $\{a_j'\}$ satisfying (1) and consider the difference $\gamma_0 = \gamma - (i/2\pi)\partial\bar{\partial} \log a_j'$. Obviously the harmonic part $H\gamma_0$ of γ_0 vanishes and consequently $\gamma_0 = \Delta G\gamma_0 = 2\partial\mathfrak{d}G\gamma_0$, where G denotes the Green's operator. Using the formula[8] $\Lambda\partial - \partial\Lambda = -i\mathfrak{d}$, we get therefore $\gamma_0 = i2\partial\bar{\partial}\Lambda G\gamma_0$. Now $r = 2\Lambda G\gamma_0$ is a real scalar function of class C^{∞} defined on the whole of V and, hence, setting $a_j = a_j' \cdot e^{2\pi r}$ we obtain a system $\{a_j\}$ of positive functions a_j of class C^{∞} satisfying (1) and also $(i/2\pi)\partial\bar{\partial} \log a_j = \gamma$.

Now, letting $\gamma = (i/2\pi) \sum X_{\alpha\beta^*}dz^{\alpha}d\bar{z}^{\beta}$ be an arbitrary real (1,1)-form, we introduce the Hermitian form

$$\Theta^{p,q}(\gamma, u; z) = \sum (\delta^{\sigma}{}_{\tau}[X^{\beta^*}{}_{\alpha^*} - R^{\beta^*}{}_{\alpha^*}] + pR^{\sigma}{}_{\tau\alpha^*}{}^{\beta^*})\times u_{\sigma\tau_2\ldots\tau_p\beta^*\alpha_2^*\ldots\alpha_q^*}\bar{u}^{\tau\tau_2\ldots\tau_p\alpha^*\alpha_2^*\ldots\alpha_q^*}$$

in a skew-symmetric tensor $u_{\tau_1\tau_2\ldots\tau_p\alpha_1^*\ldots\alpha_q^*}$ at each point z on V, provided that $q \geq 1$. Then the inequality (9) can be written in the following form:

$$\int_V \frac{g}{a_j} \Theta^{p,q}(\gamma, \varphi_j; z)dV \leq 0, \quad \text{for } q \geq 1. \tag{10}$$

With the help of the above lemma, we derive from (10) the following theorem:

THEOREM 1. *If the characteristic class $c(F)$ of F contains a real d-closed (1,1)-form $\gamma = (i/2\pi)\sum X_{\alpha\beta^*}dz^\alpha d\bar{z}^\beta$ which is sufficiently large with respect to the order $>$ in the sense that the Hermitian form $\Theta^{p,q}(\gamma, u; z)$ is positive definite at each point z on V, then the cohomology groups $H^q(V, \Omega^p(F))$ and $H^{n-q}(V, \Omega^{n-p}(-F))$ both vanish, provided that $1 \leqq q \leqq n$.*

Proof: In view of (2) and (3), it is sufficient to show that $H^{p,q}(F) = 0$. By virtue of the above lemma, we may choose the system $\{a_j\}$ satisfying (1) such that $(i/2\pi)\partial\bar{\partial} \log a_j = \gamma$. Then the inequality (10) holds for any $\varphi = \{\varphi_j\} \in H^{p,q}(F)$, while, by hypothesis, $\Theta^{p,q}(\gamma, \varphi_j; z) > 0$ unless $\varphi_j = 0$. This proves that $H^{p,q}(F) = 0$, q. e. d.

In the case where $p = n$, we have

$$\Theta^{n,q}(\gamma, u; z) = n!\sum X^{\beta^*}{}_{\alpha^*} u_{12\ldots n\beta^*\alpha_2^*\ldots\alpha_q^*}\bar{u}^{12\ldots n\alpha^*\alpha_2^*\ldots\alpha_q^*}.$$

Hence $\Theta^{n,q}(\gamma, u; z)$ is positive definite if and only if $\gamma > 0$. As a corollary to the above theorem, we therefore obtain the following:

THEOREM 2. *If $c(F)$ contains a real d-closed (1,1)-form $\gamma > 0$, then the cohomology groups $H^q(V, \Omega^n(F))$ and $H^{n-q}(V, \Omega^0(-F))$ both vanish for $1 \leqq q \leqq n$.*

By the *canonical bundle* over V we shall mean the complex line bundle K defined by the system $\{J_{jk}\}$ of Jacobians

$$J_{jk} = \frac{\partial(z_k^1, \ldots, z_k^n)}{\partial(z_j^1, \ldots, z_j^n)},$$

where $(z_j^1, \ldots, z_j^n)$ is the local coordinates in U_j. We note that $-c(K)$ equals the first Chern class of the variety V. Since the isomorphism

$$\Omega^0(F) \cong \Omega^n(F - K)$$

holds in an obvious manner, we get from the above theorem the following:

THEOREM 3. *If $c(F) - c(K)$ contains a real d-closed (1,1)-form $\gamma > 0$, then the cohomology group $H^q(V, \Omega^0(F))$ vanishes for $1 \leqq q \leqq n$.*

* This work was supported by a research project at Princeton University sponsored by the Office of Ordnance Research, U. S. Army.

[1] By a complex line bundle over V will be meant an analytic fibre bundle over V whose fibre is the complex number field C and whose structure group is the mutiplicative group C^* of complex numbers acting on C. Cf. Kodaira, K., and Spencer, D. C., "Groups of Complex Line Bundles Over Compact Kähler Varieties," these PROCEEDINGS, 39, 868–872 (1953).

[2] Bochner, S., "Curvature and Betti Numbers" (I) and (II), *Ann. Math.*, 49, 379–390 (1948); 50, 77–93 (1949).

[3] See Kodaira, K., "On Cohomology Groups of Compact Analytic Varieties with Coefficients in Some Analytic Faisceaux," these PROCEEDINGS, 39, 865–868 (1953).

[4] Kodaira, *loc. cit.*

[5] This result reduces to a special case of Serre's duality theorem.

[6] See Garabedian, P. R., and Spencer, D. C., "A Complex Tensor Calculus for Kähler Manifolds," *Acta Math.*, 89, 279–331 (1953).

[7] In view of de Rham's theorem, any cohomology class may be regarded as a class of d-closed forms. By a d-closed form we shall mean a form which is closed under d.

[8] Garabedian and Spencer, *loc. cit.*, p. 290.

On Compact Analytic Surfaces

Kunihiko Kodaira
PRINCETON UNIVERSITY AND
INSTITUTE FOR ADVANCED STUDY

THE present note is a preliminary report on a study of structures of compact analytic surfaces.

1. Let V be a compact analytic surface, i.e. a compact complex manifold of complex dimension 2. Let $\mathscr{M}(V)$ be the field of all meromorphic functions on V and denote by $\dim \mathscr{M}(V)$ the degree of transcendency of $\mathscr{M}(V)$ over the field $\mathbf{C}$ of all complex numbers. By a result due to Chow [2] and Siegel [9], we have

$$\dim \mathscr{M}(V) \leqq 2.$$

We denote by $p_g(V)$ the *geometric genus* of V, by c_1 the first Chern class of V and by $c_1^2(V)$ the value of $c_1^2 = c_1 \cdot c_1$ on the fundamental 4-cycle V. The geometric genus $p_g(V)$ is, by definition, the number of linearly independent holomorphic 2-forms on V. By the *canonical bundle* K of V we mean the complex line bundle over V of forms of type (2, 0). We note that the characteristic class $c(K)$ of K is equal to $-c_1$.

THEOREM 1. (Chow and Kodaira [3], Kodaira [6].) *A compact analytic surface V is a non-singular algebraic surface imbedded in a projective space if and only if* $\dim \mathscr{M}(V) = 2$.

Let Δ be a non-singular algebraic curve (i.e. a compact Riemann surface). We say that V is an *analytic fibre space of curves of genus π over* Δ if there exists a holomorphic map Φ of V *onto* Δ such that the inverse image $C_u = \Phi^{-1}(u)$ of any "general point" u of Δ is a non-singular irreducible curve of genus π, and we call Φ the canonical projection of V onto Δ, the inverse image $\Phi^{-1}(u)$ of each point $u \in \Delta$ a *fibre* of V. In what follows we are mainly concerned with analytic fibre spaces of *elliptic* curves i.e. curves of genus 1.

THEOREM 2. (Kodaira [6].) *If* $\dim \mathscr{M}(V) = 1$, *then V is an analytic fibre space of elliptic curves over a curve Δ and the canonical projection $\Phi : V \to \Delta$ induces an isomorphism*

$$\Phi^* : \mathscr{M}(V) \cong \mathscr{M}(\Delta),$$

where $\mathscr{M}(\Delta)$ denotes the field of meromorphic functions on Δ. Moreover V contains no irreducible curve other than the components of the fibres of V.

Kodaira, K. and Nevanlinna, R., Analytic Functions, © 1960 by Princeton Univ. Press, pp. 121–135.
Reprinted with permission of Princeton University Press.

We denote by (CD) the *intersection multiplicity* of two curves C and D on V and write (C^2) for (CC). A curve C on V will be called an *exceptional curve* (of the the first kind) if C is a non-singular rational curve with $(C^2) = -1$ (compare Zariski [10], pp. 36–41. By an exceptional curve we mean always an exceptional curve of the first kind). For an arbitrary point $p \in V$ we denote by Q_p the *quadratic transformation* with the center p (see Hopf [5]). Moreover we call any surface $W = \dots Q_{p_3}Q_{p_2}Q_{p_1}(V)$ obtained from V by applying a finite number of quadratic transformations a *quadratic transform* of V. We recall that $S = Q_p(p)$ is an exceptional curve on the quadratic transform $\hat{V} = Q_p(V)$ and that Q_p^{-1} is a holomorphic map of $\hat{V}$ onto V which is biregular between $\hat{V} - S$ and $V - p$. Moreover Q_p^{-1} induces an isomorphism: $\mathcal{M}(Q_p(V)) \simeq \mathcal{M}(V)$.

THEOREM 3. *Assume that* $\dim \mathcal{M}(V) \geqq 1$. *If V contains an exceptional curve S, then there exists a compact analytic surface W and a point $p \in W$ such that $V = Q_p(W)$ and $S = Q_p(p)$.*

This theorem has been established for algebraic surfaces by Castelnuovo and Enriques [1] (compare also Kodaira [7]). In view of Theorem 1, it suffices therefore to consider the case in which $\dim \mathcal{M}(V) = 1$. Now, if $\dim \mathcal{M}(V) = 1$, V is an analytic fibre space of elliptic curves over a curve Δ and the canonical projection Φ maps S onto a single point u on Δ. Let U be a small neighborhood of u on Δ. A detailed analysis of the structure of the fibre space V shows that the neighborhood $\Phi^{-1}(U)$ of S can be imbedded in an algebraic surface. Hence the theorem is reduced to the case of algebraic surfaces.

As to compact analytic surfaces with no meromorphic functions except constants, we have the following

THEOREM 4. (Kodaira [6].) *Let V be a compact Kähler surface with* $\dim \mathcal{M}(V) = 0$. *The geometric genus $p_g(V)$ of V is equal to* 1 *and the first Betti number $b_1(V)$ of V is equal to either* 4 *or* 0. *If $b_1(V) = 4$, then V is a quadratic transform of a complex torus (of complex dimension* 2*). If $b_1(V) = 0$ and if V contains no exceptional curve, then the canonical bundle K of V is trivial and the second Betti number $b_2(V)$ of V is equal to* 22.

It can be shown that $c_1^2(V) \leqq 0$ if V is an analytic fibre space of elliptic curves over a curve, while, with the help of Riemann-Roch's theorem, we infer readily that $\dim \mathcal{M}(V) \geqq 1$ if $p_g(V) \geqq 1$ and if $c_1^2(V) > 0$. Hence we obtain from Theorems 1, 2, and 4 the following

THEOREM 5. *If V is a compact Kähler surface and if $c_1^2(V) > 0$, then V is an algebraic surface.*

2. Let V be an analytic fibre space of elliptic curves over a curve Δ having no exceptional curve and let Φ be the canonical projection of V onto Δ. For an arbitrary point $a \in \Delta$, we denote by τ_a the local uniformization variable on Δ with the center a and by $\tau_a(u)$ the value of τ_a at a point u in a neighborhood of a on Δ. Moreover we denote by (z_1, z_2) a system of local coordinates on V. Now, if

$$(1) \qquad \left|\frac{\partial\tau_a(\Phi(z))}{\partial z_1}\right|^2 + \left|\frac{\partial\tau_a(\Phi(z))}{\partial z_2}\right|^2 > 0$$

at each point z on $\Phi^{-1}(u)$, we call $C_u = \Phi^{-1}(u)$ a *regular fibre* of V. Clearly each regular fibre C_u is a non-singular elliptic curve. By a theorem of Bertini, there exists a finite set $\{a_\rho\}$ of points a_ρ, $\rho = 1, 2, 3, \ldots$, on Δ such that $C_u = \Phi^{-1}(u)$ is a regular fibre for each point $u \in \Delta - \{a_\rho\}$. In this section we assume that the fibres $\Phi^{-1}(a_\rho)$ are *not* regular. Clearly $\tau_{a_\rho}(\Phi) : z \to \tau_{a_\rho}(\Phi(z))$ is a holomorphic function defined on a neighborhood of $\Phi^{-1}(a_\rho)$ which vanishes at each point on $\Phi^{-1}(a_\rho)$. By the *singular fibre* C_{a_ρ} of V over a_ρ we mean the divisor of the holomorphic function $\tau_{a_\rho}(\Phi)$. We write each singular fibre C_{a_ρ} in the form

$$C_{a_\rho} = \textstyle\sum_s n_{\rho s}\Theta_{\rho s},$$

where $\Theta_{\rho s}$ are irreducible curves and $n_{\rho s}$ are positive integers. Clearly $\Phi^{-1}(a_\rho)$ coincides with the union of the curves $\Theta_{\rho s}$; thus C_{a_ρ} may be considered as the inverse image of a_ρ attached with the proper multiplicities $n_{\rho s}$.

The expression $\sum_s n_{\rho s}\Theta_{\rho s}$ may be regarded as a 2-*cycle* on the polyhedron $\Phi^{-1}(a_\rho) = \bigcup_s \Theta_{\rho s}$. By the *type* of the singular fibre $C_{a_\rho} = \sum_s n_{\rho s}\Theta_{\rho s}$ we mean the topological structure of the polyhedron $\Phi^{-1}(a_\rho)$ together with the homology class of the 2-cycle $\sum n_{\rho s}\Theta_{\rho s}$ on $\Phi^{-1}(a_\rho)$. It is not difficult to determine all possible types of singular fibres of V. We denote these types by the symbols ${}_mI_b$, I_b^*, II, II*, III, III*, IV, IV*. In order to describe the characteristic properties of each type (see Figure 1), we write $\sum n_s\Theta_s$ for $\sum n_{\rho s}\Theta_{\rho s}$ and denote by $\Theta_s \cdot \Theta_t$ the *intersection cycle* of Θ_s and Θ_t. For instance the formula $\Theta_s \cdot \Theta_t = 2p_1 + 3p_2$ indicates that Θ_s and Θ_t meet at two points p_1 and p_2 with intersection multiplicities 2 and 3, respectively.

${}_mI_0$: $C_{a_\rho} = m\Theta_0$, $m = 2, 3, 4, \ldots$, where Θ_0 is a non-singular elliptic curve.

${}_mI_1$: $C_{a_\rho} = m\Theta_0$, $m = 1, 2, 3, \ldots$, where Θ_0 is a rational curve with one ordinary double point.

${}_mI_2$: $C_{a_\rho} = m\Theta_0 + m\Theta_1$, $m = 1, 2, 3, \ldots$, where Θ_0 and Θ_1 are non-singular rational curves with $\Theta_0 \cdot \Theta_1 = p_1 + p_2$.

II : $C_{a_\rho} = \Theta_0$ is a rational curve with one cusp.

III : $C_{a_\rho} = \Theta_0 + \Theta_1$, where Θ_0 and Θ_1 are non-singular rational curves with $\Theta_0 \cdot \Theta_1 = 2p$.

IV : $C_{a_\rho} = \Theta_0 + \Theta_1 + \Theta_2$, where Θ_0, Θ_1, Θ_2 are non-singular rational curves and $\Theta_0 \cdot \Theta_1 = \Theta_1 \cdot \Theta_2 = \Theta_2 \cdot \Theta_0 = p$.

The singular fibres of the types ${}_mI_b$, $b \geqq 3$, I_b^*, II*, III*, IV* are composed of non-singular rational curves $\Theta_0, \Theta_1, \ldots, \Theta_s, \ldots$ in such a way that $(\Theta_s\Theta_t) \leqq 1$ (i.e. Θ_s and Θ_t have at most one simple intersection point) for $s < t$

and $\Theta_r \cap \Theta_s \cap \Theta_t$ is empty for $r < s < t$. These types are therefore described completely by showing *all pairs* Θ_s, Θ_t *with* $(\Theta_s\Theta_t) = 1$ besides $\sum n_s\Theta_s$.

${}_m\mathrm{I}_b$: $C_{a_\rho} = m\Theta_0 + m\Theta_1 + \ldots + m\Theta_{b-1}, m = 1, 2, 3, \ldots, b = 3, 4, 5, \ldots,$ $(\Theta_0\Theta_1) = (\Theta_1\Theta_2) = \ldots = (\Theta_s\Theta_{s+1}) = \ldots = (\Theta_{b-2}\Theta_{b-1}) = (\Theta_{b-1}\Theta_0) = 1.$

I_b^* : $C_{a_\rho} = \Theta_0 + \Theta_1 + \Theta_2 + \Theta_3 + 2\Theta_4 + 2\Theta_5 + \ldots + 2\Theta_{4+b}, b = 0, 1, 2, \ldots, (\Theta_0\Theta_4) = (\Theta_1\Theta_4) = (\Theta_2\Theta_{4+b}) = (\Theta_3\Theta_{4+b}) = (\Theta_4\Theta_5) = (\Theta_5\Theta_6) = \ldots = (\Theta_{3+b}\Theta_{4+b}) = 1.$

II* : $C_{a_\rho} = \Theta_0 + 2\Theta_1 + 3\Theta_2 + 4\Theta_3 + 5\Theta_4 + 2\Theta_5 + 4\Theta_6 + 3\Theta_7 + 6\Theta_8,$ $(\Theta_0\Theta_1) = (\Theta_1\Theta_2) = (\Theta_2\Theta_3) = (\Theta_3\Theta_4) = (\Theta_5\Theta_6) = (\Theta_4\Theta_8) = (\Theta_6\,\Theta_8) = (\Theta_7\Theta_8) = 1.$

III* : $C_{a_\rho} = \Theta_0 + 2\Theta_1 + 3\Theta_2 + \Theta_3 + 2\Theta_4 + 3\Theta_5 + 2\Theta_6 + 4\Theta_7, (\Theta_0\Theta_1) = (\Theta_1\Theta_2) = (\Theta_3\Theta_4) = (\Theta_4\Theta_5) = (\Theta_2\Theta_7) = (\Theta_5\Theta_7) = (\Theta_6\Theta_7) = 1.$

IV* : $C_{a_\rho} = \Theta_0 + 2\Theta_1 + \Theta_2 + 2\Theta_3 + \Theta_4 + 2\Theta_5 + 3\Theta_6, (\Theta_0\Theta_1) = (\Theta_2\Theta_3) = (\Theta_4\Theta_5) = (\Theta_1\Theta_6) = (\Theta_3\Theta_6) = (\Theta_5\Theta_6) = 1.$

A singular fibre $C_{a_\rho} = \sum_s n_{\rho s}\Theta_{\rho s}$ will be called *simple* or *multiple* according as min $\{n_{\rho s}\} = 1$ or $\geqq 2$. It is clear that C_{a_ρ} is a multiple singular fibre if and only if C_{a_ρ} is of type ${}_m\mathrm{I}_b$, $m \geqq 2$. Suppose that C_{a_ρ}, $1 \leqq \rho \leqq l$, are of types ${}_{m_\rho}\mathrm{I}_{b_\rho}$, $m_\rho \geqq 2$, respectively, and that C_{a_ρ}, $\rho \geqq l + 1$, are simple. Let m_0 be the l.c.m. of $m_1, \ldots, m_\rho, \ldots, m_l$, and let $d = m_1m_2 \ldots m_l$. Moreover let a_0 be an arbitrary point on $\Delta - \{a_\rho\}$. Then there exists a d-fold abelian covering surface $\tilde{\Delta}$ of Δ which is unramified over $\Delta - \{a_0, a_1, \ldots, a_l\}$ and has d/m_ρ branch points $b_{\rho k}$, $k = 1, 2, \ldots, d/m_\rho$ of order $m_\rho - 1$ over each point a_ρ, $\rho = 0, 1, 2, \ldots, l$. Letting ϖ be the canonical projection of $\tilde{\Delta}$ onto Δ, we define the analytic fibre space $\tilde{V}$ over $\tilde{\Delta}$ *induced from V by the map* ϖ to be the *minimal non-singular model* of the subvariety of $V \times \tilde{\Delta}$ consisting of all points $(z, \tilde{u}) \in V \times \tilde{\Delta}$ satisfying $\Phi(z) = \varpi(\tilde{u})$. The canonical projection $\tilde{\Phi} : \tilde{V} \to \tilde{\Delta}$ is induced from the projection $V \times \tilde{\Delta} \to \tilde{\Delta}$. $\tilde{V}$ is also an analytic fibre space of elliptic curves. It is clear that, for $\tilde{u} \neq b_{\rho k}$, the fibre $\tilde{C}_{\tilde{u}}$ of $\tilde{V}$ over $\tilde{u}$ has the same type as the fibre C_u of V over $u = \varpi(\tilde{u})$ and, in particular, $\tilde{C}_{\tilde{u}}$ is regular if $\varpi(\tilde{u}) \notin \{a_\rho\}$, while the fibre $\tilde{C}_{b_{\rho k}}$, $1 \leqq \rho \leqq \ell$, is of type ${}_1\mathrm{I}_{m_\rho b_\rho}$ and $\tilde{C}_{b_{0k}}$ is regular. Thus the induced fibre space $\tilde{V}$ is free from multiple singular fibres. On the other hand $\tilde{V}$ is a d-fold abelian covering manifold of V which is unramified over $V - C_{a_0}$ and has d/m_0 branch curves $\tilde{C}_{b_{0k}}$, $k = 1, 2, \ldots, d/m_0$, of order $m_0 - 1$ over C_{a_0}. The covering map $\Pi : \tilde{V} \to V$ is induced from the projection $V \times \tilde{\Delta} \to V$. Thus we obtain the following

THEOREM 6. *An analytic fibre space V of elliptic curves over a curve* Δ *having no exceptional curve induces over a suitably chosen finite ramified* *abelian covering surface $\tilde{\Delta}$ of Δ an analytic fibre space $\tilde{V}$ of elliptic curves free from multiple singular fibres. $\tilde{V}$ is a finite abelian covering of V having branch curves over a regular fibre C_{a_0} of V and contains no exceptional curve. V is represented as the factor space: $V = \tilde{V}/\mathfrak{G}$, where $\mathfrak{G}$ is the covering transformation group of $\tilde{V}$ over V.*

3. Now let V be an analytic fibre space of elliptic curves over Δ having neither exceptional curve nor multiple singular fibre and let $\Phi : V \rightarrow \Delta$ be

FIGURE 1

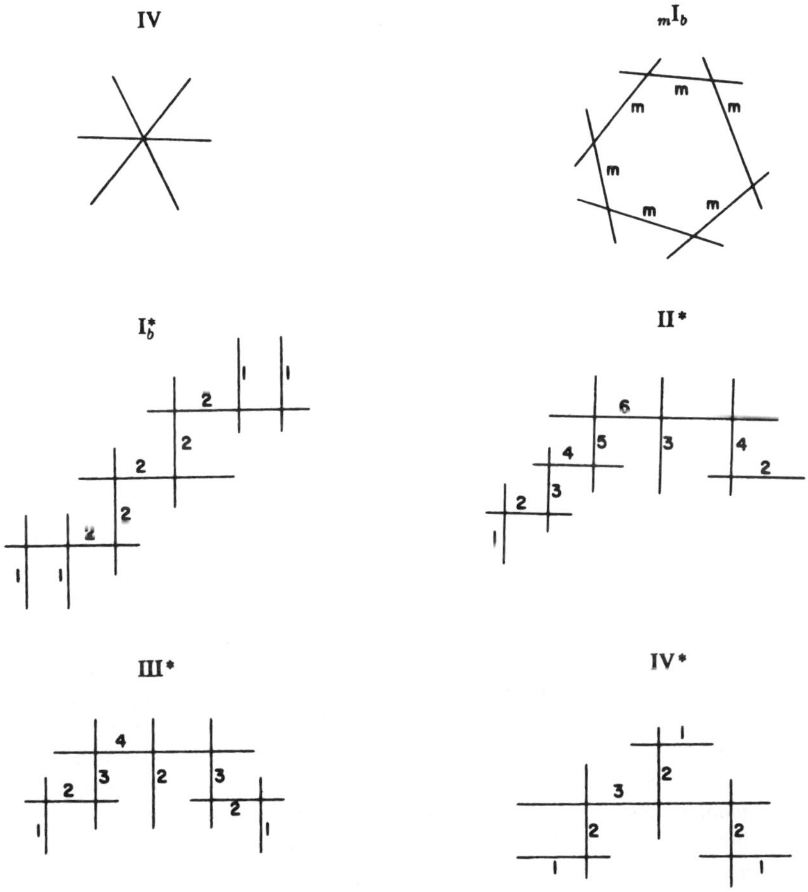

Each line represents $\Theta_{\rho s}$; the integers attached to the line gives $n_{\rho s}$.

the canonical projection. Moreover let $\{a_\rho\}$ be a finite set of points $a_1, \ldots, a_\rho, \ldots, a_r$ on Δ such that the fibre $C_u = \Phi^{-1}(u)$ is regular for $u \neq a_\rho$ and let $\Delta' = \Delta - \{a_\rho\}$, where, for convenience, we do *not* assume that each fibre C_{a_ρ} is singular. C_{a_ρ} is therefore either a regular fibre or a singular fibre of one of the types I_b, I_b^*, II, II*, III, III*, IV, IV*, where we write I_b for ${}_1\mathrm{I}_b$.

The restriction $V \mid \Delta' = \Phi^{-1}(\Delta') = \bigcup_{u \in \Delta'} C_u$ of V to Δ' is a differentiable fibre bundle of tori over Δ' and therefore the first homology groups $H_1(C_u, \mathbf{Z})(\cong \mathbf{Z} \oplus \mathbf{Z})$ of the fibres C_u with coefficients in the integers $\mathbf{Z}$ forms a locally constant sheaf

$$G' = \bigcup_{u \in \Delta'} H_1(C_u, \mathbf{Z})$$

over Δ' in a canonical manner. Let E_ρ be a circular neighborhood of a_ρ on Δ and let $E'_\rho = E_\rho - a_\rho$. Then the group $\Gamma(G' \mid E'_\rho)$ of sections of G' over E'_ρ is independent of the size of E'_ρ. We extend G' to a sheaf G over Δ by defining $\Gamma(G' \mid E'_\rho)$ to be the stalk G_{a_ρ} of G over a_ρ:

$$G = \bigcup_\rho G_{a_\rho} \cup G', \quad G_{a_\rho} = \Gamma(G' \mid E'_\rho)$$

and we call G the *homological invariant* of the fibre space V over Δ.

We denote by $J(\omega)$ the *elliptic modular function* defined on the upper half plane $\mathbf{C}^+ = \{\omega \mid \Im\omega > 0\}$. As is well known, $\omega \to J(\omega)$ is a holomorphic map of $\mathbf{C}^+$ onto the whole plane $\mathbf{C}$ and the equality $J(\omega) = J(\omega')$ holds if and only if ω' is obtained from ω by a *modular transformation*

$$S : \omega \to \omega' = \frac{a\omega + b}{c\omega + d}, \qquad ad - bc = 1,$$

where a, b, c, d are rational integers. For each point $u \in \Delta'$, we represent the elliptic curve C_u as a complex torus with periods $(\omega(u), 1)$, $\Im\omega(u) > 0$, and set

$$\mathcal{J}(u) = J(\omega(u)).$$

$\omega(u)$ is determined by C_u uniquely up to a modular transformation and *depends holomorphically* on $u \in \Delta'$. Hence $\mathcal{J}(u)$ is a single-valued holomorphic function of u defined on $\Delta' = \Delta - \{a_\rho\}$. Moreover it can be shown that each point a_ρ is either a pole or a removable singular point of $\mathcal{J}(u)$. Thus *$\mathcal{J}(u)$ is a meromorphic function on Δ*. We call $\mathcal{J}(u)$ the *functional invariant* of the fibre space V of elliptic curves over Δ. By a suitable choice of the finite set $\{a_\rho\}$, we may assume that $\mathcal{J}(u) \neq 0, 1, \infty$ for $u \in \Delta' = \Delta - \{a_\rho\}$.

Suppose conversely that an algebraic curve Δ of genus p and a meromorphic function $\mathcal{J}(u)$ on Δ are given. Take a finite set $\{a_\rho\}$ of points on Δ such that $\mathcal{J}(u) \neq 0, 1, \infty$ for $u \in \Delta' = \Delta - \{a_\rho\}$. Then there exists one and only one multi-valued holomorphic function $\omega(u)$ with

$\Im\omega(u) > 0$ on Δ' satisfying $J(\omega(u)) = \mathcal{J}(u)$. Take a point o on Δ' and suppose that each element of the fundamental group $\pi_1(\Delta')$ of Δ' is represented by a closed arc β on Δ' starting and ending at o. By the analytic continuation along the arc β, $\omega(u)$ is transformed into $S_\beta\omega(u)$, where S_β is a modular transformation depending only on the homotopy class of β relative to o. We indicate this fact by the formula

$$\omega(u) \xrightarrow{\beta} S_\beta\omega(u) = \frac{a_\beta\omega(u) + b_\beta}{c_\beta\omega(u) + d_\beta}. \tag{2}$$

The correspondence $\beta \to S_\beta$ gives a representation of $\pi_1(\Delta')$, provided that we take a fixed branch of $\omega(u)$ at the starting point o of β. In the particular case in which $\omega(u) = \omega_0$ is a constant, the condition (2), which states simply that ω_0 is a fixed point of S_β, may not be sufficient to determine the modular transformation S_β. Then we choose S_β arbitrarily, provided that S_β satisfies (2) and that $\beta \to S_\beta$ gives a representation of $\pi_1(\Delta')$. For each modular transformation S_β, $\beta \in \pi_1(\Delta')$, we choose

$$(S_\beta) = \pm\begin{pmatrix} a_\beta & b_\beta \\ c_\beta & d_\beta \end{pmatrix}$$

in such a way that $\beta \to (S_\beta)$ gives a representation: $\pi_1(\Delta') \to SL(2, \mathbf{Z})$. This is possible in $2p + r - 1$ different manners, provided that $r \geqq 1$, since $\pi_1(\Delta')$ is generated by $2p + r$ generators $\beta_1, \beta_2, \ldots, \beta_{2p}, \alpha_1, \ldots, \alpha_r$ with the single relation:

$$\beta_1\beta_2\beta_1^{-1}\beta_2^{-1}\beta_3\beta_4 \ldots \beta_{2p-1}^{-1}\beta_{2p}^{-1}\alpha_1\alpha_2 \ldots \alpha_r = 1.$$

The representation $\beta \to (S_\beta)$ defines a locally constant sheaf G' over Δ' whose stalks are isomorphic to $\mathbf{Z} \oplus \mathbf{Z}$ and G' can be extended uniquely to the sheaf $G = \bigcup_\rho G_{a_\rho} \cup G'$ over Δ, where $G_{a_\rho} = \Gamma(G' \mid E'_\rho)$. Under these circumstances we say that *the sheaf G belongs to the meromorphic function $\mathcal{J}(u)$*. It can be shown that, given an analytic fibre space V of elliptic curves over Δ free from multiple singular fibres, *the homological invariant G of V belongs to the functional invariant $\mathcal{J}(u)$ of V.*

DEFINITION 1. *Given an algebraic curve Δ, a meromorphic function $\mathcal{J} = \mathcal{J}(u)$ on Δ and a sheaf G over Δ belonging to $\mathcal{J}$, we denote by $\mathcal{F}(\mathcal{J}, G)$ the family of all fibre space V of elliptic curves over Δ free from multiple singular fibres and from exceptional curves having the given invariants $\mathcal{J}$ and G.*

We say that the analytic fibre spaces V_1 and V_2 of (elliptic) curves over Δ are *analytically equivalent* if and only if there exists an analytic homeomorphism $f: V_1 \to V_2$ such that $\Phi_2 f = \Phi_1$, where Φ_1, Φ_2 are the canonical projections of V_1, V_2 onto Δ, respectively. In the above definition of the family $\mathcal{F}(\mathcal{J}, G)$, analytically equivalent fibre spaces may be considered as the *same* fibre space. More precise classifications of analytic fibre spaces of elliptic curves will be introduced later.

Let α_ρ be a small circle with positive orientation around a_ρ. The restriction $G \mid E_\rho$ of G to a circular neighborhood E_ρ of a_ρ is determined by the representation (S_{α_ρ}) of α_ρ.

THEOREM 7. *The type of the fibre* C_{a_ρ} *of any fibre space* $V \in \mathcal{F}(\mathcal{J}, G)$ *is determined uniquely by* (S_{α_ρ}) (see Table I).

By a holomorphic section of the fibre space V over an open subset

TABLE I

Normal form of (S_{α_ρ})	Type of C_{a_ρ}	$C^\sharp_{a_\rho}$	$\Theta^\sharp_{\rho 0}$	Behavior of $\mathcal{J}(u)$ at a_ρ
1	regular	torus	torus	regular
$\begin{pmatrix} -1 & 0 \\ 0 & -1 \end{pmatrix}$	I_0^*	$C \times Z_2 \times Z_2$	C	
$\begin{pmatrix} 1 & b \\ 0 & 1 \end{pmatrix}$	I_b	$C^* \times Z_b$	C^*	pole of order b
$\begin{pmatrix} -1 & -b \\ 0 & -1 \end{pmatrix}$	I_b^*	$C \times Z_2 \times Z_2$	C	
$\begin{pmatrix} 1 & 1 \\ -1 & 0 \end{pmatrix}$	II	C	C	$\mathcal{J}(a_\rho) = 0$
$\begin{pmatrix} 0 & -1 \\ 1 & 1 \end{pmatrix}$	II*	C	C	
$\begin{pmatrix} 0 & 1 \\ -1 & 0 \end{pmatrix}$	III	$C \times Z_2$	C	$\mathcal{J}(a_\rho) = 1$
$\begin{pmatrix} 0 & -1 \\ 1 & 0 \end{pmatrix}$	III*	$C \times Z_2$	C	
$\begin{pmatrix} 0 & 1 \\ -1 & -1 \end{pmatrix}$	IV	$C \times Z_3$	C	$\mathcal{J}(a_\rho) = 0$
$\begin{pmatrix} -1 & -1 \\ 1 & 0 \end{pmatrix}$	IV*	$C \times Z_3$	C	

Z_b denotes the cyclic group of order b.

$E \subset \Delta$ we mean a holomorphic map $\varphi : u \to \varphi(u)$ of E into V such that $\Phi(\varphi(u)) = u$.

THEOREM 8. *The family $\mathscr{F}(\mathscr{J}, G)$ contains one and only one fibre space B which has a global holomorphic section over Δ.*

We note that the fibre space B is an *algebraic surface*.

Let $\mathscr{G}$ and M be complex manifolds. We say that $\mathscr{G}$ is an analytic fibre space over M free from singular fibres if there is a holomorphic map Ψ of $\mathscr{G}$ onto M such that, at each point on $\mathscr{G}$, the rank of the Jacobian of Ψ is equal to the dimension of M.

DEFINITION 2. *Let $\mathscr{G}$ be an analytic fibre space over M free from singular fibres such that each fibre $\mathscr{G}_u = \Psi^{-1}(u)$, $u \in M$, is a complex Lie group whose complex structure is that of the complex submanifold $\mathscr{G}_u$ of $\mathscr{G}$. Let $\mathscr{D}$ be the complex submanifold of $\mathscr{G} \times \mathscr{G}$ consisting of all points $(g_1, g_2) \in \mathscr{G} \times \mathscr{G}$ satisfying $\Psi(g_1) = \Psi(g_2)$. $\mathscr{G}$ will be called an analytic fibre system of complex Lie groups over M if the following two conditions are satisfied:* (i) *$u \to 1_u$ is a holomorphic map of M into $\mathscr{G}$, where 1_u denotes the unit of $\mathscr{G}_u$,* (ii) *$(g_1, g_2) \to g_1 \cdot g_2^{-1}$ is a holomorphic map of $\mathscr{D}$ onto $\mathscr{G}$.*

Now we consider the fibre space $B \in \mathscr{F}(\mathscr{J}, G)$. We denote the canonical projection $B \to \Delta$ by Ψ and one of the global holomorphic sections of B by $o : u \to o(u)$. Each fibre $C_u = \Psi^{-1}(u)$, $u \in \Delta'$; of B is a complex torus. The canonical additive group structure on a complex torus is determined uniquely by giving the position of the unit 0. We choose the additive group structure on each complex torus C_u such that the unit 0 coincides with $o(u)$. Then the subspace $B' = B \mid \Delta' = \Psi^{-1}(\Delta')$ of B forms an *analytic fibre system of complex Lie groups over* Δ' in the sense of the above Definition 2.

We denote by $B^\sharp$ the open subset of B consisting of all points z satisfying

$$\left|\frac{\partial \tau_a(\Psi(z))}{\partial z_1}\right|^2 + \left|\frac{\partial \tau_a(\Psi(z))}{\partial z_2}\right|^2 > 0$$

(compare (1)). Clearly we have

$$B^\sharp = \textstyle\bigcup_\rho C^\sharp_{a_\rho} \cup B', \qquad B' = B \mid \Delta',$$

where

$$C^\sharp_{a_\rho} = B^\sharp \cap \Psi^{-1}(a_\rho).$$

The fibre $C^\sharp_{a_\rho}$ is written in the form

$$C^\sharp_{a_\rho} = \bigcup_{n_{\rho s}=1} \Theta^\sharp_{\rho s}, \qquad \Theta^\sharp_{\rho s} = B^\sharp \cap \Theta_{\rho s},$$

where $\bigcup_{n_{\rho s}=1}$ denotes the sum extended over all s for which $n_{\rho s} = 1$. If C_{a_ρ} is regular, then $C^\sharp_{a_\rho} = C_{a_\rho}$; otherwise each component $\Theta^\sharp_{\rho s}$ of $C^\sharp_{a_\rho}$ is (analytically homeomorphic to) either $\mathbf{C}$ or $\mathbf{C}^*$.

PROPOSITION 1. *The analytic fibre space $B^\sharp$ over Δ has a unique structure of analytic fibre system of complex abelian Lie groups which is an extension of the structure of analytic fibre system of complex Lie groups on B'. The group structures of the fibres $C^\sharp_{a_\rho}$ of $B^\sharp$ are given in* Table I.

Let E be an arbitrary open domain on Δ and let $\varphi_\nu : u \to \varphi_\nu(u)$, $\nu = 1, 2$, be holomorphic sections of $B^\sharp$ over E. Then, by Proposition 1, $u \to \varphi_1(u) \pm \varphi_2(u)$ is also a holomorphic section of $B^\sharp$ over E, where $+$ or $-$ denotes the group operation on each fibre of $B^\sharp$. Thus the space $\Gamma(B^\sharp \mid E)$ of holomorphic sections of $B^\sharp$ over E forms an additive group in a canonical manner and therefore *the sheaf over Δ of germs of holomorphic sections of $B^\sharp$ is defined. We denote this sheaf by $\Omega(B^\sharp)$.*

For any holomorphic section $\varphi : u \to \varphi(u)$ of $B^\sharp$ over E,

$$L^\sharp(\varphi) : z \to z + \varphi(\Psi(z))$$

is a fibre preserving analytic homeomorphism of $B^\sharp \mid E$ onto $B^\sharp \mid E$.

PROPOSITION 2. $L^\sharp(\varphi) : B^\sharp \mid E \to B^\sharp \mid E$ can be extended uniquely to a fibre preserving analytic homeomorphism $L(\varphi)$ of $B \mid E$ onto $B \mid E$.

In view of this proposition, $\Omega(B^\sharp)$ may be regarded as a *sheaf of structure groups* acting on the fibre space B (compare Grothendieck [4]).

We consider the first cohomology $H^1(\Delta, \Omega(B^\sharp))$ of Δ with coefficients in the sheaf $\Omega(B^\sharp)$. Given an element $\sigma \in H^1(\Delta, \Omega(B^\sharp))$, let $\{E_j\}$ be a sufficiently fine covering of Δ and let $\{\sigma_{jk}\}$, $\sigma_{jk} \in \Gamma(B^\sharp \mid E_j \cap E_k)$, be a 1-cocycle on the nerve of the covering $\{E_j\}$ representing the cohomology class σ.

DEFINITION 3. *B^σ is the fibre space of elliptic curves over Δ obtained from the collection $\{B \mid E_j\}$ of the pieces $B \mid E_j$ of the fibre space B by identifying $z_j \in B \mid E_j$ and $z_k \in B \mid E_k$ if and only if $z_j = L(\sigma_{jk}) \cdot z_k$.*

It is clear that B^σ is determined uniquely (up to an analytic equivalence) by the cohomology class σ of the 1-cocycle $\{\sigma_{jk}\}$.

THEOREM 9. *The family $\mathscr{F}(\mathscr{J}, G)$ consists of all fibre spaces B^σ, $\sigma \in H^1(\Delta, \Omega(B^\sharp))$.*

We denote by $\Theta^\sharp_{\rho 0}$ the component of $C^\sharp_{a_\rho}$ containing the unit $o(a_\rho)$ of $C^\sharp_{a_\rho}$ and let

$$B^\sharp_0 = \textstyle\bigcup_\rho \Theta^\sharp_{\rho 0} \cup B'.$$

$B^\sharp_0$ has also a structure of an analytic fibre system of complex Lie groups which is the restriction of the structure of $B^\sharp$ and therefore $\Omega(B^\sharp_0)$ is a subsheaf of $\Omega(B^\sharp)$. The factor sheaf $F = \Omega(B^\sharp)/\Omega(B^\sharp_0)$ has the following property:

$$F_u = \begin{cases} 0, & \text{for } u \neq a_\rho, \\ \text{finite abelian group}, & \text{for } u = a_\rho, \end{cases}$$

where F_u denotes the stalk of F over u. Hence we obtain the exact sequence

$$(3) \qquad \ldots \to H^0(\Delta, F) \to H^1(\Delta, \Omega(B_0^\sharp)) \to H^1(\Delta, \Omega(B^\sharp)) \to 0.$$

For an element σ of $H^1(\Delta, \Omega(B_0^\sharp))$ we define B^σ in the same manner as in Definition 3. In view of (3), we obtain from Theorem 9 the following

THEOREM 10. *The family* $\mathscr{F}(\mathscr{J}, G)$ *consists of all fibre spaces* B^σ, $\sigma \in H^1(\Delta, \Omega(B_0^\sharp))$.

The notions of fibre spaces and their equivalences depend on the sheaf of structure groups (compare Grothendieck [4]). We mean by a Ξ-fibre space a fibre space with the sheaf of structure groups Ξ and we say that two fibre spaces are Ξ-equivalent if they are equivalent as Ξ-fibre spaces. $V = B^\sigma$, $\sigma \in H^1(\Delta, \Omega(B_0^\sharp))$, may be considered as an analytic fibre space, as an $\Omega(B^\sharp)$-fibre space, or as an $\Omega(B_0^\sharp)$-fibre space, and the element $\sigma \in H^1(\Delta\ \Omega(B_0^\sharp))$ represents the $\Omega(B_0^\sharp)$-equivalence class of B^σ.

4. Clearly the fibre $C_{0u}^\sharp$ of $B_0^\sharp$ over u is given by

$$C_{0u}^\sharp = \begin{cases} C_u, & \text{for } u \in \Delta', \\ \Theta_{\rho 0}^\sharp, & \text{for } u = a_\rho, \end{cases}$$

where $\Theta_{\rho 0}^\sharp$ is isomorphic to $\mathbf{C}$, $\mathbf{C}^*$, or a complex torus if C_{a_ρ} is regular (compare Table I). Let $\mathfrak{f}_u$ be the tangent space of $C_{0u}^\sharp$ at $o(u)$. Then

$$\mathfrak{f} = \bigcup_{u \in \Delta} \mathfrak{f}_u$$

forms a *complex line bundle* over Δ in a canonical manner. Since $\mathfrak{f}_u$ may be regarded as the infinitesimal group of the complex Lie group $C_{0u}^\sharp$ which is isomorphic to a complex torus, $\mathbf{C}$, or $\mathbf{C}^*$, we have a canonical homomorphism

$$h_u : \mathfrak{f}_u \to C_{0u}^\sharp.$$

It follows that there is a locally biregular holomorphic map

$$h : \mathfrak{f} \to B_0^\sharp$$

whose restriction to each fibre $\mathfrak{f}_u$ coincides with h_u. Clearly h induces a homomorphism: $\Omega(\mathfrak{f}) \to \Omega(B_0^\sharp)$ of the sheaf $\Omega(\mathfrak{f})$ of holomorphic sections of $\mathfrak{f}$ onto the sheaf $\Omega(B_0^\sharp)$. We denote this homomorphism by the same symbol h.

PROPOSITION 3. *We have the exact sequence*

$$(4) \qquad 0 \to G \to \Omega(\mathfrak{f}) \xrightarrow{h} \Omega(B_0^\sharp) \to 0,$$

where G is the homological invariant of the fibre space B.

We obtain from (4) the corresponding *exact cohomology sequence*

$$(5) \quad \ldots \to H^1(\Delta, G) \to H^1(\Delta, \Omega(\mathfrak{f})) \xrightarrow{h^*} H^1(\Delta, \Omega(B_0^\sharp)) \xrightarrow{\delta^*} H^2(\Delta, G) \to 0.$$

We define

$$c(\sigma) = \delta^*(\sigma), \qquad \text{for } \sigma \in H^1(\Delta, \Omega(B_0^\sharp)).$$

Let $A(\mathfrak{f})$ and $A(B_0^\sharp)$ be the sheaves over Δ of differentiable sections of $\mathfrak{f}$ and $B_0^\sharp$, respectively. $A(B_0^\sharp)$ is a sheaf of structure groups acting on B considered as a *differentiable* fibre space. From the exact commutative diagram

$$\begin{array}{ccccccccc} 0 & \to & G & \to & \Omega(\mathfrak{f}) & \to & \Omega(B_0^\sharp) & \to & 0 \\ & & \| & & \downarrow{\scriptstyle \iota} & & \downarrow{\scriptstyle \iota} & & \\ 0 & \to & G & \to & A(\mathfrak{f}) & \to & A(B_0^\sharp) & \to & 0, \end{array}$$

where ι denotes the inclusion map, we obtain the exact commutative diagram

$$\begin{array}{ccccccc} \to H^1(\Delta, \Omega(\mathfrak{f})) & \to & H^1(\Delta, \Omega(B_0^\sharp)) & \to & H^2(\Delta, G) & \to & 0 \\ \downarrow & & \downarrow{\scriptstyle \iota^*} & & \| & & \\ 0 & \longrightarrow & H^1(\Delta, A(B_0^\sharp)) & \to & H^2(\Delta, G) & \to & 0 \end{array}$$

This shows that $c(\sigma)$ represents the $A(B_0^\sharp)$-equivalence class of the fibre space B^σ over Δ. In particular we have

THEOREM 11. *If $c(\sigma) = c(\eta)$, then the fibre spaces B^σ and B^η are differentiably equivalent.*

We call $c(\sigma)$ the *characteristic class* of the $\Omega(B_0^\sharp)$-fibre space B^σ over Δ.

By a *complex analytic family* of compact complex manifolds we mean the family $\{V_t \mid t \in M\}$ of fibres $V_t = \pi^{-1}(t)$ of an analytic fibre space $\mathscr{V}$ free from singular fibres over a *connected* complex manifold M whose fibres are compact, where π is the canonical projection $\mathscr{V} \to M$ (see Kodaira and Spencer [8]). Moreover we say that a compact complex manifold W is a (complex analytic) *deformation* of V if W and V belong to one and the same complex analytic family. It is clear that any deformation W of V is differentiably homeomorphic to V.

THEOREM 12. *$\{B^{\sigma+h^*(t)} \mid t \in H^1(\Delta, \Omega(\mathfrak{f}))\}$ forms a complex analytic family of compact complex manifolds, where σ is an element of $H^1(\Delta, \Omega(B_0^\sharp))$.*

We remark here that $H^1(\Delta, \Omega(\mathfrak{f}))$ has a canonical structure of complex vector space. We set

$$\mathscr{V}^{(c)} = \{B^{\sigma+h^*(t)} \mid t \in H^1(\Delta, \Omega(\mathfrak{f}))\}, \qquad c = c(\sigma).$$

In view of (5), the family $\mathscr{V}^{(c)}$ is composed of all fibre spaces B^η, $c(\eta) = c$, with repetitions. Hence we have

THEOREM 13. *If $c(\eta) = c(\sigma)$, then B^η is a deformation of B^σ.*

It is clear that the projection $\Phi : B^\sigma \to \Delta$ induces an isomorphism Φ^* of $\mathscr{M}(\Delta)$ into $\mathscr{M}(B^\sigma)$ and therefore $\dim \mathscr{M}(B^\sigma) \geqq 1$. Using the fact that, if $\dim \mathscr{M}(B^\sigma) = 1$, B^σ contains no irreducible curve other than the components of the fibres (see Theorem 2), we obtain the following

THEOREM 14. *B^σ is an algebraic surface if and only if σ is an element of finite order of $H^1(\Delta, \Omega(B_0^\sharp))$.*

Combining this with Theorem 13, we get

THEOREM 15. *If $c(\sigma)$ is an element of finite order of $H^2(\Delta, G)$, then B^σ is a deformation of an algebraic surface.*

5. We denote by $c(\mathfrak{f})$ the characteristic class of the complex line bundle $\mathfrak{f}$ over Δ. In view of the canonical isomorphism: $H^2(\Delta, \mathbf{Z}) \cong \mathbf{Z}$, we may suppose that $c(\mathfrak{f})$ is an integer. Moreover we denote by $\nu(T)$ the number of the singular fibres of B of type T and by j the order of the meromorphic function $\mathcal{J}(u)$, i.e. the total multiplicities of the zeros of $\mathcal{J}(u)$.

PROPOSITION 4. *$\delta = -c(\mathfrak{f})$ is given by*

$$12\delta = j + 6\sum_b \nu(\mathrm{I}_b^*) + 2\nu(\mathrm{II}) + 10\nu(\mathrm{II}^*) + 3\nu(\mathrm{III}) + 9\nu(\mathrm{III}^*) + 4\nu(\mathrm{IV}) + 8\nu(\mathrm{IV}^*).$$

This proposition shows, in particular, that $c(\mathfrak{f}) \leqq 0$ and that $c(\mathfrak{f}) = 0$ if and only if B has no singular fibre and $\mathcal{J}(u)$ is reduced to a constant. By using Riemann-Roch's theorem we obtain

$$\dim H^1(\Delta, \Omega(\mathfrak{f})) = \begin{cases} p - 1 + \delta, & \text{if } \mathfrak{f} \neq 0, \\ p, & \text{if } \mathfrak{f} = 0, \end{cases} \tag{6}$$

where p is the genus of Δ.

PROPOSITION 5. *$H^2(\Delta, G)$ is a finite abelian group, provided that G is not trivial.*

It is clear that, if G is the trivial sheaf $\mathbf{Z} \oplus \mathbf{Z}$, $H^2(\Delta, G)$ is isomorphic to $\mathbf{Z} \oplus \mathbf{Z}$.

Now we consider an arbitrary fibre space $V = B^\sigma$, $\sigma \in H^1(\Delta, \Omega(B_0^\sharp))$, belonging to $\mathcal{F}(\mathcal{J}, G)$. Let $\mathfrak{k}$ be the canonical bundle of Δ.

THEOREM 16. *The canonical bundle K of V is induced from $\mathfrak{k} - \mathfrak{f}$ by the projection Φ of V onto Δ : $K = \Phi^*(\mathfrak{k} - \mathfrak{f})$.*

Combining this with (6), we infer that the geometric genus $p_g(V)$ is given by

$$p_g(V) = \begin{cases} p - 1 + \delta, & \text{if } \mathfrak{f} \neq 0, \\ p, & \text{if } \mathfrak{f} = 0. \end{cases}$$

THEOREM 17. *The Euler number $c_2(V)$ of V is given by $c_2(V) = 12\delta$.*

An analytic fibre space V of (elliptic) curves over Δ is called an analytic *fibre bundle* if each point $u \in \Delta$ has a neighborhood E_u such that $V \mid E_u = E_u \times C_u$ in the *complex analytic sense*. It can be shown that, if G is trivial, B is analytically trivial i.e. $B = \Delta \times C_0$ in the complex analytic sense and therefore $V = B^\sigma$ is an analytic principal bundle of elliptic curves, i.e. an analytic fibre bundle over Δ whose fibre is a non-singular elliptic curve C_0 and whose structure group is the translation group acting on C_0. We infer from Proposition 5 and Theorem 14 that, in case G is not trivial, *each complex analytic family $\mathcal{V}^{(r)}$ contains an algebraic surface.* Hence we obtain

THEOREM 18. *An analytic fibre space V of elliptic curves over a curve Δ*

free from exceptional curves and from multiple singular fibres is a deformation of an algebraic surface unless V is an analytic principal bundle of elliptic curves.

Since the group $H^1(\Delta, G)$ is finitely generated, we infer from (5) and Theorem 14 the following

THEOREM 19. *If G is not trivial and if* $\dim H^1(\Delta, \Omega(\mathfrak{f})) \geqq 1$, *the general member of each complex analytic family* $\mathscr{V}^{(c)}$ *is not algebraic, or, more precisely, the member* $B^{\sigma+h^*(t)}$ *of* $\mathscr{V}^{(c)}$ *is algebraic if and only if t belongs to a countable subset of* $H^1(\Delta, \Omega(\mathfrak{f}))$.

In case G is trivial, each family $\mathscr{V}^{(c)}$, $c \neq 0$, contains no algebraic surface. The family $\mathscr{V}^{(0)}$ contains the algebraic surface $B = \Delta \times C_0$, but, if $\dim H^1(\Delta, \Omega) \geqq 1$, the general member of $\mathscr{V}^{(0)}$ is not algebraic.

6. Now we consider an arbitrary compact analytic surface V with $\dim \mathscr{M}(V) = 1$. By Theorems 3 and 2, V is a quadratic transform of an analytic fibre space W of elliptic curves over a curve Δ_0 containing no exceptional curve. By Theorem 6, W is represented in the form $W = \tilde{W}/\mathfrak{G}$, where $\tilde{W}$ is an analytic fibre space of elliptic curves over a finite ramified covering $\tilde{\Delta}_0$ of Δ_0 free from multiple singular fibres. Let $\Delta = \tilde{\Delta}_0$. Then, by Theorem 10, W has the form B^σ, $\sigma \in H^1(\Delta, \Omega(B_0^\sharp))$. Thus we obtain

THEOREM 20. *Every compact analytic surface V with* $\dim \mathscr{M}(V) = 1$ *is represented in the form* $V = \dots Q_{p_3} Q_{p_2} Q_{p_1}(B^\sigma/\mathfrak{G})$, $\sigma \in H^1(\Delta, \Omega(B_0^\sharp))$, *where* $\mathfrak{G}$ *is a finite abelian group of analytic automorphisms of* B^σ *introduced in Theorem* 6.

An analytic fibre space V of elliptic curves over Δ containing no exceptional curve will be called an *analytic quasi-bundle* of elliptic curves if each singular fibre C_{a_ρ} of V is of type ${}_m\mathrm{I}_0$, m depending on ρ, and if $V \mid \Delta'$ is an analytic principal bundle of elliptic curves over $\Delta' = \Delta - \{a_\rho\}$. It can be shown that $W = \tilde{W}/\mathfrak{G}$ is an analytic quasi-bundle of elliptic curves over Δ_0 if and only if $\tilde{W} = B^\sigma$ is an analytic principal bundle of elliptic curves over $\tilde{\Delta}_0 = \Delta$.

Suppose that $\tilde{W} = B^\sigma$ is not an analytic principal bundle of elliptic curves. Let X be the underlying differentiable manifold on which the complex structure B^σ is defined and consider $\mathfrak{G}$ as a group of differentiable automorphisms of X. By a detailed analysis of the structures of the automorphisms of $\mathfrak{G}$, we infer that there exists a linear subspace M of $H^1(\Delta, \Omega(\mathfrak{f}))$ satisfying the following two conditions: (i) The complex analytic family $\{B^{\sigma+h^*(t)} \mid t \in M\}$ contains an algebraic surface, (ii) The automorphisms of $\mathfrak{G}$ are biregular with respect to each complex analytic structure $B^{\sigma+h^*(t)}$, $t \in M$, defined on X. Letting

$$V_t = \dots Q_{p_3} Q_{p_2} Q_{p_1}(B^{\sigma+h^*(t)}/\mathfrak{G}),$$

we obtain a complex analytic family $\{V_t \mid t \in M\}$ of deformations V_t of $V_0 = V$ which contains an algebraic surface. Thus we obtain

THEOREM 21. *A compact analytic surface V with* $\dim \mathcal{M}(V) \geqq 1$ *is a deformation of an algebraic surface unless V is a quadratic transform of an analytic quasi-bundle of elliptic curves over a curve.*

In case $\tilde{W} = B^{\sigma}$ is an analytic principal bundle of elliptic curves, it can be shown that, if $V = \dots Q_{p_2} Q_{p_1}(B^{\sigma}/\mathfrak{G})$ has a Kähler metric, V is a deformation of an algebraic surface. Combining this with Theorem 21, we obtain

THEOREM 22. *Every compact Kähler surface having at least one non-constant meromorphic function is a deformation of an algebraic surface.*

Now we consider a compact Kähler surface V with $\dim \mathcal{M}(V) = 0$. If the first Betti number $b_1(V)$ of V is positive, then, by Theorem 4, V is a quadratic transform of a complex torus, and therefore V is a deformation of an algebraic surface. Combining this with the above Theorem 22, we obtain

THEOREM 23. *Every compact Kähler surface V with the first Betti number $b_1(V) > 0$ is a deformation of an algebraic surface.*

REFERENCES

[1] G. CASTELNUOVO and F. ENRIQUES, Sopra alcune questioni fondamentali nella teoria delle superficie algebriche, *Annali di Math., Ser. III, 6* (1901), pp. 165–225.

[2] W. L. CHOW, On complex analytic varieties, to appear in the *Amer. J. Math.*

[3] W. L. CHOW and K. KODAIRA, On analytic surfaces with two independent meromorphic functions, *Proc. Nat. Acad. Sciences 38* (1952), pp. 319–325.

[4] A. GROTHENDIECK, *A general theory of fibre spaces with structure sheaf*, University of Kansas, 1955.

[5] H. HOPF, Schlichte Abbildungen und lokale Modificationen 4-dimensionaler komplexer Mannigfaltigkeiten, *Comm. Math. Helv. 29* (1955), pp. 132–156.

[6] K. KODAIRA, *On compact complex analytic surfaces I*, to appear in *Ann. of Math.*

[7] ———, On Kähler varieties of restricted type, *Ann. of Math. 60* (1954), pp. 28–48.

[8] K. KODAIRA and D. C. SPENCER, On deformations of complex analytic structures, I–II, *Ann. of Math.* 67 (1958) pp. 328–466.

[9] C. L. SIEGEL, Meromorphe Funktionen auf kompakten analytischen Mannigfaltigkeiten, *Nachr. Akad. Wiss.* Göttingen (1955), pp. 71–77.

[10] O. ZARISKI, Algebraic surfaces, *Ergeb. Math. Berlin* (1935).

ROBERT LANGLANDS

BIBLIOGRAPHY

(1) *Some holomorphic semi-groups*, Proc. Nat. Acad. Sci. 46 (1960), 361-363.

(2) *On Lie semi-groups*, Can J. Math. 12 (1960), 686-693.

(3) *Dirichlet series associated with quadratic forms*, Princeton University, 1961 (preprint).

(4) *The dimension of spaces of automorphic forms*, Am. J. Math. 85 (1963), 99-125.

(5) *The volume of the fundamental domain for some arithmetical subgroups of Chevalley groups*, in Algebraic Subgroups and Discontinuous Subgroups, Amer. Math. Soc. 1966, pp. 143-148.

(6) *Eisenstein series*, loc. cit., pp. 235-252.

(7) *Dimension of spaces of automorphic forms*, loc. cit., pp. 253-257.

(8) *Representations of abelian algebraic groups*, Yale University, 1968 (preprint).

(9) *Problems in the theory of automorphic forms*, in Lectures in Modern Analysis and Applications III, Springer-Verlag Lecture Notes 170, 1970, pp. 18-61.

(10) *On the functional equation of the Artin L-functions*, Yale University, 1970 (preprint).

(11) (with H. Jacquet), *Automorphic Forms on GL(2)*, Springer-Verlag Lecture Notes 114, 1970, pp. vii + 548.

(12) *On Artin's L-function*, Rice University Studies 56 (2) (1970), pp. 23-28.

(13) *Automorphic Forms on GL(2)*, Proc. Int. Congr. Math., Nice, 1970, Vol. 2, pp. 327-329.

(14) *Euler Products*, Yale University Press, 1971, v + 53 pp.

(15) *Modular forms and ℓ-adic representations, in Modular Functions of One Variable III*, Springer-Verlag Lecture Notes 349, 1973, pp. 361-500.

(16) *On the classification of irreducible representations of real algebraic groups*, Institute for Advanced Study, 1973 (preprint).

(17) *Some contemporary problems with origins in the Jugendtraum*, in Mathematical developments arising from Hilbert problems, Proc. Sympos. Pure Math. 28 (1976), pp. 401-418.

(18) *Shimura varieties and the Selberg trace formula*, U.S.-Japan Seminar on Number

Theory, Ann Arbor, 1975; Can. J. Math. 29 (1977), pp. 1292-1299.

(19) *Base Change for GL(2)*, Ann. Math. Studies 96, 1980, pp. vii + 237.

(20) *On the functional equations satisfied by Eisenstein series*, Springer Lecture Notes in Math. 544 (1976). (Contains [3] above and other supplementary material.)

(21) *Stable conjugacy: Definitions and lemmas*, Can. J. Math. 31 (1979), pp. 700-725.

(22) (with J.-P. Labesse), *L-indistinguishability for SL(2)*, Can. J. Math. 31 (1979), pp. 726-785.

(23) *L-functions and automorphic representations*, Proc. Int. Congr. Math., Helsinki, 1978, pp. 165-175.

(24) *On the zeta functions of some simple Shimura varieties*, Can. J. Math. 31 (1979), pp. 1121-1216.

(25) *Sur la mauvaise réduction d'une variété de Shimura*, Astérique 65 (1979), pp. 125-153.

(26) *Automorphic representations, Shimura varieties, and motives*, Proc. Sympos. Pure Math. 33 (1979), I, pp. 205-246.

(27) *On the notion of an automorphic representation*, loc. cit. II, pp. 203-206.

(28) *Les débuts d'une formule des traces stable*, Université de Paris (1983).

(29) *Orbital integrals on forms of SL(3)*, Am. Jour. Math. 105 (1983), pp. 465-506.

(30) (with D. Shelstad), *On principal values on p-adic manifolds, in Lie group representations II*, Springer-Verlag Lecture Notes in Math. 1041 (1984), pp. 250-279.

(31) (with G. Harder and M. Rapoport), *Algebraische Zyklen auf Hilbert-Blumenthal-Flächen*, Crelle's Journal 366 (1986) pp. 53-120.

(32) *Dirac monopoles and induced representations*, Pac. Jour. of Math. 126 (1987), pp. 145-151.

(33) (with M. Rapoport), *Shimuravarietäten und Gerben*, Crelle's Journal, 378 (1987), pp. 113-220.

(34) (with D. Shelstad), *On the definition of transfer factors*, Math. Ann. 278 (1987) pp. 219-271.

(35) *On unitary representations of the Virasoro algebra* in Infinite-dimensional Lie alge-

bras and their applications, ed. S. Kass, World Scientific (1988) pp. 141-159.

(36) *Representation theory and arithmetic*, in The Mathematical Heritage of Hermann Weyl, Amer. Math. Soc. (1988) pp. 25-33.

(37) *Eisenstein series, the trace formula, and the modern theory of automorphic forms*, in Number theory, trace formulas, and discrete groups, ed. K.R. Aubert et al, Academic Press, 1988, pp. 125-155.

(38) (with D. Shelstad), *Orbital integrals on forms of SL(3), II*, Can. Jour. Math., Vol. XLI, 1989, pp. 480-507.

(39) *On the classification of irreducible representations of real algebraic groups*, in Representation Theory and Harmonic Analysis on Semisimple Lie Groups, ed. P. Sally and D. Vogan (this is [16] above), 1989, pp. 101-170.

(40) *The factorization of a polynomial defined by partitions*, in Comm. Math. Physics, Vol. 124 (1989) pp. 251-284.

(41) (with D. Shelstad), *Descent for transfer factors*, in the Grothendieck Festschrift, Vol. II, Birkhäuser (1990) pp. 485-563.

(42) *Representation Theory: Its rise and its role in number theory*, Proceedings of the Gibbs Symposium, AMS (1990) pp. 181-210.

(43) *Rank-one residues of Eisenstein series*, in Festschrift in Honor of I.I. Piatetski-Shapiro, Vol. 2, Israel Mathematical Conference Proceedings (1990) pp. 111-126.

(44) *Remarks on Igusa Theory and real orbital integrals*, in The zeta functions of Picard modular surfaces, ed. Robert P. Langlands and Dinakar Ramakrishnan, Les Publications CRM, Montréal (1992).

(45) (with C. Pichet, Ph. Pouliot, and Y. Saint-Aubin), *On the universality of crossing probabilities in two-dimensional percolation*, Jour. Stat. Phys. Vol. 67 (1992), pp. 553-574.

(46) *Dualität bei endlichen Modellen der Perkolation*, Math. Nach. Bd. 160 (1993) pp. 7-58.

(47) (with Ph. Pouliot and Y. Saint-Aubin), *Conformal invariance in two-dimensional percolation*, B.A.M.S., Vol. 30 (1994) pp. 1-61.

(48) (with M.-A. Lafortune), *Finite models for percolation*, Cont. Math., No. 177, Amer. Math. Soc. (1994), pp. 227-246.

(49) (with Y. Saint-Aubin), *Algebro-geometric aspects of the Bethe equations, in Strings and Symmetries*, Lect. Notes in Physics, Springer (1995), pp. 40-53.

(50) Reprinting of [18] with commentary in Can. Math. Soc. 1945-1995, Vol. 2, Selecta, (1996).

(51) (with Y. Saint-Aubin), *Aspects combinatoires des équations de Bethe*, in Advances in Mathematical Sciences: CRM's 25 years, AMS, (1997).

(52) (with Fan Chung), *A combinatorial Laplacian with vertex weights*, Jour. of Comb. Theory, Ser. A, Vol. 75, pp. 316-327, (1996).

(53) An essay on the dynamics and statistics of critical field theories, in Can. Math. Soc. 1945-1995, Vol. 3, Invited Papers, pp. 173-210, (1996).

(54) *Where stands functoriality today*, in Proc. of Symp. in Pure Mathematics, Vol. 61, 1997, pp. 457-471.

(55) Representations of abelian algebraic groups, (this is [8] above), Pac. Jour. of Math. 1998, pp. 231-250 (Olga Taussky-Todd memorial issue).

(56) (with M.-A. Lewis and Y. Saint-Aubin), *Universality and conformal invariance for the Ising model in domains with boundary*, to appear in Jour. Stat. Phys.

Miscellaneous Publications

(1) Review of $SL2(R)$ by Serge Lang, B.A.M.S., Vol. 82 (1976).

(2) Review of *The Theory of Eisenstein Systems* by M. Scott Osborne and Garth Warner, B.A.M.S., Vol. 9 (1983).

(3) Review of *Harish-Chandra, Collected Papers*, Bull. Lon. Math. Soc. Vol. 17 (1985).

(4) *Pure Mathematics* in The Canadian Encyclopedia, Edmonton (1985).

(5) Speech printed in *Harish-Chandra, 1923-1983*, Princeton (1985).

(6) *Harish-Chandra* in Biographical Memoirs of Fellows of the Royal Society. London (1985).

(7) Review of *Elliptic Curves* by Anthony W. Knapp, B.A.M.S. (new series), Vol. 30, (1994).

Editor

(1) (with D. Ramakrishnan), *The zeta functions of Picard modular surfaces*, Les Publications CRM, Montréal, (1992).

Problems in the Theory of Automorphic Forms†

To Salomon Bochner
In Gratitude

1. There has recently been much interest, if not a tremendous amount of progress, in the arithmetic theory of automorphic forms. In this lecture I would like to present the views not of a number theorist but of a student of group representations on those of its problems that he finds most fascinating. To be more precise, I want to formulate a series of questions which the reader may, if he likes, take as conjectures. I prefer to regard them as working hypotheses. They have already led to some interesting facts. Although they have stood up for a fair length of time to the most careful scrutiny I could give, I am still not entirely easy about them. Indeed even at the beginning in the course of the definitions, which I want to make in complete generality, I am forced, for lack of time and technical competence, to make various assumptions.

I should perhaps apologize for such a speculative lecture. However, there are some interesting facts scattered amongst the questions. Moreover, the unsolved problems in group representations arising from the theory of automorphic forms are much less technical than the solved ones, and their significance can perhaps be more easily appreciated by the outsider.

Suppose G is a connected reductive algebraic group defined over a global field F. F is then an algebraic number field or a function field in one variable over a finite field. Let $\mathbb{A}(F)$ be the adèle ring of F. $G_{\mathbb{A}(F)}$ is a locally compact topological group with G_F as a discrete subgroup. The group $G_{\mathbb{A}(F)}$ acts on the functions on $G_F \setminus G_{\mathbb{A}(F)}$. In particular, it acts on $L^2(G_F \setminus G_{\mathbb{A}(F)})$. It should be possible, although I have not done so and it is not important at this stage, to attach a precise meaning to the assertion that a given irreducible representation π of $G_{\mathbb{A}(F)}$ occurs in $L^2(G_F \setminus G_{\mathbb{A}(F)})$. If G is abelian it would mean that π is a character of $G_F \setminus G_{\mathbb{A}(F)}$. if G is not abelian it would be true for at least those representations which act on an irreducible invariant subspace of $L^2(G_F) \setminus G_{\mathbb{A}(F)}$.

If G is $GL(1)$ then to each such π one, following Hecke, associates an L-function. If G is $GL(2)$ then Hecke has also introduced, without explicitly mentioning group representations, some L-functions. The problems I want to discuss center about the possibility of defining L-functions for all such π and proving that they have the analytic properties we have grown used to expecting of such functions. I

† Appears in *Lectures in Modern Analysis and Applications* III, C. T. Taam, ed., Springer Lecture Notes in Mathematics 170, 1970, pp. 18–61. This is the postscript file downloaded from http://www.sunsite.ubc.ca/DigitalMathArchive/Langlands/intro.html/#problems

shall also comment on the possible relations of these new functions to the Artin L-functions and the L-functions attached to algebraic varieties.

Given G, I am going to introduce a complex analytic group $\hat{G}_F$. To each complex analytic representation σ of $\hat{G}_F$ and each π I want to attach an L-function $L(s,\sigma,\pi)$. Let me say a few words about the general way in which I want to form the function. $G_{\mathbb{A}(F)}$ is a restricted direct product $\prod_{\mathfrak{p}} G_{F_\mathfrak{p}}$. The product is taken over the primes, finite and infinite, of F. It is reasonable to expect, although to my knowledge it has not yet been proved in general, that π can be represented as $\prod_{\mathfrak{p}} \otimes \pi_\mathfrak{p}$ where $\pi_\mathfrak{p}$ is a unitary representation of $G_{F_\mathfrak{p}}$.

I would like to have first associated to any algebraic group G defined over $F_\mathfrak{p}$ a complex analytic group $\hat{G}_{F_\mathfrak{p}}$ and to any complex analytic representation $\sigma_\mathfrak{p}$ of $\hat{G}_{F_\mathfrak{p}}$ and any unitary representation $\pi_\mathfrak{p}$ of $G_{F_\mathfrak{p}}$ a local L-function $L(s,\sigma_\mathfrak{p},\pi_\mathfrak{p})$ which, when $\mathfrak{p}$ is non-archimedean, would be of the form

$$\prod_{i=1}^{n} \frac{1}{1-\alpha_i|\varpi_\mathfrak{p}|^s}$$

where n is the degree of $\sigma_\mathfrak{p}$. Some of the α_i may be zero. For $\mathfrak{p}$ infinite it would be, basically, a product of Γ-functions. $L(s,\sigma_\mathfrak{p},\pi_\mathfrak{p})$ would depend only on the equivalence classes of $\sigma_\mathfrak{p}$ and $\pi_\mathfrak{p}$. I would also like to have defined for every non-trivial additive character $\psi_{F_\mathfrak{p}}$ of $F_\mathfrak{p}$ a factor $\varepsilon(s,\sigma_\mathfrak{p},\pi_\mathfrak{p},\psi_{F_\mathfrak{p}})$ which, as a function of s, has the form ae^{bs}.

There would be a complex analytic homomorphism of $\hat{G}_{F_\mathfrak{p}}$ into $\hat{G}_F$ determined up to an inner automorphism of $\hat{G}_F$. Thus σ determines for each $\mathfrak{p}$ a representation of $\sigma_\mathfrak{p}$ of $\hat{G}_{F_\mathfrak{p}}$. I want to define

$$L(s,\sigma,\pi) = \prod_{\mathfrak{p}} L(s,\sigma_\mathfrak{p},\pi_\mathfrak{p}) \,. \tag{A}$$

Of course it has to be shown that the product converges in a half-plane. We shall see how to do this. Then we will want to prove that the function can be analytically continued to a function meromorphic in the whole complex plane. Let ψ_F be a non-trivial character of $F\setminus\mathbb{A}(F)$ and let $\psi_{F_\mathfrak{p}}$ be the restriction of ψ_F to $F_\mathfrak{p}$. We will want $\varepsilon(s,\sigma_\mathfrak{p},\pi_\mathfrak{p},\psi_{F_\mathfrak{p}})$ to be 1 for all but finitely many $\mathfrak{p}$. We will also want

$$\varepsilon(s,\sigma,\pi) = \prod_{\mathfrak{p}} \varepsilon(s,\sigma_\mathfrak{p},\pi_\mathfrak{p},\psi_{F_\mathfrak{p}})$$

to be independent of ψ_F. The functional equation should be

$$L(s,\sigma,\pi) = \varepsilon(s,\sigma,\pi)L(1-s,\tilde{\sigma},\pi)$$

if $\tilde{\sigma}$ is the representation contragredient to σ.

We are asking for too much too soon. What we should try to do is to define the $L(s,\sigma_{\mathfrak{p}},\pi_{\mathfrak{p}})$ and the $\varepsilon(s,\sigma_{\mathfrak{p}},\pi_{\mathfrak{p}},\psi_{F_{\mathfrak{p}}})$ when there is no ramification, verify that there is ramification at only a finite number of primes, and show that if the product in (A) is taken only over the unramified primes it converges for $Re\, s$ sufficiently large. As we learn how to prove the functional equation we shall be able to make the definitions at the unramified primes. By the way, we introduce the additive characters, whose appearance must appear rather mysterious, only because we can indeed prove some things and know better than to leave them out.

What does unramified mean in our context? First of all for $\mathfrak{p}$ to be unramified G will have to be quasi-split over $F_{\mathfrak{p}}$ and split over an unramified extension. In that case there is, as we shall see, a canonical conjugacy class of maximal compact subgroups of $G_{F_{\mathfrak{p}}}$. For $\mathfrak{p}$ to be unramified, the restriction of $\pi_{\mathfrak{p}}$ to any one of these groups will have to contain the identity representation. There is also a condition to be imposed on $\psi_{F_{\mathfrak{p}}}$. Although it is not very important I would like to mention it explicitly. If $\mathfrak{p}$ is non-archimedean the largest ideal of $F_{\mathfrak{p}}$ on which $\psi_{F_{\mathfrak{p}}}$ is trivial will have to be $O_{F_{\mathfrak{p}}}$, the ring of integers in $F_{\mathfrak{p}}$. If $F_{\mathfrak{p}}$ is $\mathbb{R}$ then $\psi_{F_{\mathfrak{p}}}(x)$ will have to be $e^{2\pi i x}$ and if $F_{\mathfrak{p}}$ is $\mathbb{C}$ then $\psi_{F_{\mathfrak{p}}}(z)$ will have to be $e^{4\pi i\, Re\, z}$. We want $\varepsilon(s,\sigma_{\mathfrak{p}},\pi_{\mathfrak{p}},\psi_{F_{\mathfrak{p}}})$ to be 1 if $\mathfrak{p}$ is unramified.

2. $\hat{G}_F$ can be identified for a connected reductive group over any field F. Take first a quasi-split group G over F which splits over the Galois extension K. Choose a Borel subgroup B of G which is defined over F and let T be a maximal torus of B which is also defined over F. Let L be the group of rational characters of T. Write G as G^0G^1 where G^0 is abelian and G^1 is semi-simple. Then $G^0\cap G^1$ is finite. If $T^0=G^0$ and $T^1=T\cap G^1$ then $T=T^0T^1$. Let L_+^0 be the group of rational characters of T^0 and let L_-^0 be the elements of L_+^0 which are 1 on $T^0\cap T^1$. Let L_-^+ be the group generated by the roots of T^1. If R is any field let $E_R^1=L_-^1\otimes_{\mathbb{Z}}R$. The Weyl group Ω acts on L_-^1 and therefore on E_R^1. Let $(\ \cdot\ ,\ \cdot\ )$ be a non-degenerate bilinear form on $E_{\mathbb{C}}^1$ which is invariant under Ω. Suppose also that its restriction to $E_{\mathbb{R}}^1$ is positive definite. Let

$$L_+^1=\left\{\lambda\in E_{\mathbb{C}}^1\ \middle|\ 2\frac{(\lambda,\alpha)}{(\alpha,\alpha)}\in\mathbb{Z}\text{ for all roots }\alpha\right\}.$$

Set $L_-=L_-^0\oplus L_-^1$ and $L_+=L_+^0\oplus L_+^1$. We may regard L as a sublattice of L_+. It will contain L_-.

Let $\alpha_1,\ldots,\alpha_\ell$ be the simple roots of T^1 with respect to B and let

$$(A_{ij})=2\frac{(\alpha_i,\alpha_j)}{(\alpha_i,\alpha_i)}$$

be the Cartan matrix. If σ belongs to $\mathfrak{G}(K/F)$ and λ belongs to L then $\sigma\lambda$, where $\sigma\lambda(t)=\sigma(\lambda(\sigma^{-1}t))$, also belongs to L. Thus $\mathfrak{G}(K/F)$ acts on L. It also acts on L_- and L_+ and the actions on these three lattices are consistent. Moreover the roots $\alpha_1, \ldots, \alpha_\ell$ are permuted amongst themselves and the Cartan matrix is left invariant.

If R is any field containing $\mathbb{Q}$ let $E_R = L \otimes_{\mathbb{Z}} R$ and let $\hat{E}_R = \mathrm{Hom}_R(E_R, R)$. The lattices

$$\begin{aligned}\hat{L}_+ &= \mathrm{Hom}(L_-, \mathbb{Z}) = \mathrm{Hom}(L_-^0, \mathbb{Z}) \oplus \mathrm{Hom}(L_-^1, \mathbb{Z}) = L_+^0 \oplus \hat{L}_+^1 \\ \hat{L} &= \mathrm{Hom}(L, \mathbb{Z}) \\ \hat{L}_- &= \mathrm{Hom}(L_+, \mathbb{Z}) = \mathrm{Hom}(L_+^0, \mathbb{Z}) \oplus \mathrm{Hom}(L_+^1, \mathbb{Z}) = \hat{L}_-^0 \oplus \hat{L}_-^1\end{aligned}$$

may be regarded as subgroups of $\hat{E}_{\mathbb{C}}$. IF $E_R^0 = L_-^0 \otimes_{\mathbb{Z}} R$ then $E_R = E_R^0 \otimes E_R^1$. With the obvious definitions of $\hat{E}_R^0$ and $\hat{E}_R^1$ we have $\hat{E}_R = \hat{E}_R^0 \oplus \hat{E}_R^1$. Let $(\,\cdot\,,\,\cdot\,)$ also denote the form on $\hat{E}_{\mathbb{C}}^1$ adjoint to the given form on $E_{\mathbb{C}}^1$. To be precise if λ and μ belong to $E_{\mathbb{C}}^1$, if $\hat{\lambda}$ and $\hat{\mu}$ belong to $\hat{E}_{\mathbb{C}}^1$, and if $\left\langle \eta, \hat{\lambda} \right\rangle = (\eta, \lambda)$ and $\langle \eta, \hat{\mu} \rangle = (\eta, \mu)$ for all η in $E_{\mathbb{C}}^1$ then $(\lambda, \mu) = (\hat{\lambda}, \hat{\mu})$.

If α is a root define its coroot $\hat{\alpha}$ in $\hat{E}_{\mathbb{C}}^1$ by the condition:

$$\langle \lambda, \hat{\alpha} \rangle = 2\,\frac{(\lambda, \alpha)}{(\alpha, \alpha)}$$

for all λ in $E_{\mathbb{C}}^1$. The coroots generate $\hat{L}_-^1$. Moreover

$$(\hat{\alpha}, \hat{\beta}) = 4\,\frac{(\alpha, \beta)}{(\alpha, \alpha)(\beta, \beta)}$$

and

$$2\,\frac{(\hat{\alpha}, \hat{\beta})}{(\hat{\alpha}, \hat{\alpha})} = 2\,\frac{(\alpha, \beta)}{(\beta, \beta)}\ .$$

Thus the matrix

$$(\hat{A}_{ij}) = \left(2\,\frac{(\hat{\alpha}_i, \hat{\alpha}_j)}{(\hat{\alpha}_i, \hat{\alpha}_i)}\right)$$

is the transpose of (A_{ij}). The linear transformation $\hat{S}_i$ of $\hat{E}_{\mathbb{C}}^1$ defined by

$$\hat{S}_i(\hat{\alpha}_j) = \hat{\alpha}_j - \hat{A}_{ij}\hat{\alpha}_i = \hat{\alpha}_j - A_{ji}\hat{\alpha}_i$$

is contragredient to the linear transformation S_i of $\hat{E}_{\mathbb{C}}^1$ defined by

$$S_i(\alpha_j) = \alpha_j - A_{ij}\alpha_i\ .$$

Thus the group $\hat{\Omega}$ generated by $\{\hat{S}_i | 1 \le i \le \ell\}$ is canonically isomorphic to the finite group Ω and, by a well-known theorem (cf. Chapter VII of [7]) $(\hat{A}_{ij})$ is the Cartan matrix of a simply-connected complex group $\hat{G}_+^1$. Let $\hat{B}_+^1$ be a Borel subgroup of $\hat{G}_+^1$ and let $\hat{T}_+^1$ be a Cartan subgroup in $\hat{B}_+^1$. We identify the simple roots of $\hat{T}_+^1$ with respect to $\hat{B}_+^1$ with $\hat{\alpha}_1, \ldots, \hat{\alpha}_\ell$ and the free vector space over $\mathbb{C}$ with basis

$\{\hat{\alpha}_1, \ldots, \hat{\alpha}_\ell\}$ with $\hat{E}^1_{\mathbb{C}}$. We may also identify Ω and $\hat{\Omega}$. The roots of $\hat{T}^1_+$ are the vectors $\omega\hat{\alpha}_i$, $\omega \in \Omega$, $1 \le i \le \ell$. If $\omega\alpha_i = \alpha$ then $\omega\hat{\alpha}_i = \hat{\alpha}$ because

$$\langle \lambda, \omega\hat{\alpha}_i \rangle = \langle \omega^{-1}\lambda, \hat{\alpha}_i \rangle = 2\,\frac{(\omega^{-1}\lambda, \alpha_i)}{(\alpha_i, \alpha_i)} = 2\,\frac{(\lambda, \omega\alpha_i)}{(\omega\alpha_i, \omega\alpha_i)} = 2\,\frac{(\lambda, \alpha)}{(\alpha, \alpha)}$$

Thus the roots of $\hat{T}^1_+$ are just the coroots. If λ belongs to $\hat{E}^1_{\mathbb{C}}$ then

$$2\,\frac{(\lambda, \hat{\alpha})}{(\hat{\alpha}, \hat{\alpha})} = \langle \alpha, \lambda \rangle$$

so that

$$\hat{L}^1_+ = \left\{ \lambda \in \hat{E}^1_{\mathbb{C}} \;\middle|\; 2\,\frac{(\lambda, \hat{\alpha})}{(\hat{\alpha}, \hat{\alpha})} \in \mathbb{Z} \text{ for all coroots } \hat{\alpha} \right\}$$

and is therefore just the set of weights of $\hat{T}^1_+$.

Let

$$\hat{G}^0_+ = \mathrm{Hom}_{\mathbb{Z}}(\hat{L}^0_+, \mathbb{C}^*) \ .$$

$\hat{G}^0_+$ is a reductive complex Lie group. Set $\hat{G}_+ = \hat{G}^0_+ \times \hat{G}^1_+$. If $\hat{T}^0_+ = \hat{G}^0_+$ and $\hat{T}_+ = \hat{T}^0_+ \times \hat{T}^1_+$ then $\hat{L}_+$ is the set of complex analytic characters of $\hat{T}_+$. If

$$\hat{Z} = \left\{ t \in \hat{T}_+ \;\middle|\; \lambda(t) = 1 \text{ for all } \lambda \text{ in } \hat{L} \right\}$$

then $\hat{Z}$ is a normal subgroup of $\hat{G}_+$ and $G = \hat{G}_+/\hat{Z}$ is also a complex Lie group. $\mathfrak{G}(K/F)$ acts in a natural fashion on $\hat{L}_-$, $\hat{L}$, and $\hat{L}_+$. The action leaves the set $\{\hat{\alpha}_1, \ldots \hat{\alpha}_\ell\}$ invariant. $\mathfrak{G}(K/F)$ acts naturally on $\hat{G}^0_+$. I want to define an action on $\hat{G}^1_+$ and therefore an action on $\hat{G}_+$. Choose $H_1, \ldots, H_\ell$ in the Lie algebra of $\hat{T}^1_+$ so that

$$\lambda(H_i) = \langle \alpha_i, \lambda \rangle$$

for all λ in $\hat{L}^1_+$. Choose root vectors $X_1, \ldots, X_\ell$ belonging to the coroots $\hat{\alpha}_1, \ldots, \hat{\alpha}_\ell$ and root vectors $Y_1, \ldots, Y_\ell$ belonging to their negatives. Suppose $[X_i, Y_i] = H_i$. If σ belongs to $\mathfrak{G}(K/F)$ let $\sigma(\hat{\alpha}_i) = \hat{\alpha}_{\sigma(i)}$. There is (cf. Chapter VII of [7]) a unique isomorphism σ of the Lie algebra of $\hat{G}^1_+$ such that

$$\sigma(H_i) = H_{\sigma(i)},\ \sigma(X_i) = X_{\sigma(i)},\ \sigma(Y_i) = Y_{\sigma(i)} \ .$$

These isomorphisms clearly determine an action of $\mathfrak{G}(K/F)$ on the Lie algebra and therefore one on $\hat{G}^1_+$ itself. Since $\mathfrak{G}(K/F)$ leaves L invariant its action on $\hat{G}_+$ can be transferred to $\hat{G}$. If $\hat{B}$ is the image of $\hat{B}_+ = \hat{T}^0_+ \times \hat{B}^1_+$ and $\hat{T}$ the image of $\hat{T}_+$ in $\hat{G}$ the action leaves $\hat{B}$ and $\hat{T}$ invariant. I want to define $\hat{G}_F$ to be the semi-direct product $\hat{G} \times \mathfrak{G}(K/F)$.

However $\hat{G}_F$ as defined depends upon the choice of B, T and $X_1 \ldots, X_\ell$ and $\hat{G}_F$ comes provided with a Borel subgroup $\hat{B}$ of its connected component, a Cartan subgroup $\hat{T}$ of $\hat{B}$, and a one-to-one correspondence between the simple roots of T with respect to B and those of $\hat{T}$ with respect to $\hat{B}$. Suppose G' is another quasi-split group over F which is isomorphic to G over K by means of an isomorphism φ such that $\varphi^{-1}\sigma(\varphi)$ is inner for all σ in $\mathfrak{G}(K/F)$, B' is a Borel subgroup of G' defined over F, and T' is a Cartan subgroup of B' also defined over F. There is an inner automorphism ψ of G which is defined over K so that $\varphi\psi$ takes B to B' and T to T'. Then $\varphi\psi$ determines an isomorphism of $\hat{L}$ and $\hat{L}'$ and a one-to-one correspondence between $\{\alpha_1, \ldots, \alpha_\ell\}$ and $\{\alpha'_1, \ldots, \alpha'_\ell\}$ both of which depend only on φ and, as is easily verified, commute with the action of $\mathfrak{G}(K/F)$. There is then a natural isomorphism of $\hat{G}^0_+$ with $(\hat{G}^0_+)'$ associated to φ. Moreover there is a unique isomorphism of $\hat{G}^1_+$ with $(\hat{G}^1_+)'$ whose action on the Lie algebra takes H_i to H'_i, X_i to X'_i, and Y_i to Y'_i. The two together define an isomorphism of $\hat{G}_+$ with $\hat{G}'_+$. If we assume that α_i corresponds to α'_i, $1 \leq i \leq \ell$ this isomorphism takes $\hat{Z}$ to $\hat{Z}'$ and determines an isomorphism of $\hat{G}$ with $\hat{G}'$ which commutes with $\mathfrak{G}(K/F)$. This in turn determines an isomorphism $\hat{\varphi}$ of $\hat{G}'_F$ with $\hat{G}_F$. In particular taking $G' = G$ and φ to be the identity we see that $\hat{G}_F$ is determined up to a canonical isomorphism.

Suppose G is any reductive group over F, K is a Galois extension of F, G' and G'' are quasi-split groups over F which split over K, and φ: $G' \to G$, ψ: $G'' \to G$ are isomorphisms defined over K such that $\varphi^{-1}\sigma(\varphi)$ and $\psi^{-1}\sigma(\psi)$ are inner for all σ in $\mathfrak{G}(K/F)$. Then $(\psi^{-1}\varphi)^{-1}\sigma(\psi^{-1}\varphi)$ is also inner so that there is a canonical isomorphism of $\hat{G}'_F$ and $\hat{G}''_F$. We are thus free to set $\hat{G}_F = \hat{G}'_F$. $\hat{G}_F$ depends on K but there is no need to stress this. However we shall sometimes write $\hat{G}_{K/F}$ instead of $\hat{G}_F$.

3. Although it is a rather simple case, it may be worthwhile to carry out the previous construction when G is $GL(n)$ and $K = F$. We take T to be the diagonal and B to be the upper triangular matrices. G^0 is the group of non-zero scalar matrices and G^1 is $SL(n)$. If λ belongs to L and

$$\lambda: \begin{pmatrix} t_1 & & 0 \\ & \ddots & \\ 0 & & t_n \end{pmatrix} \longrightarrow t_1^{m_1} \ldots t_n^{m_n}$$

with $m_1, \ldots, m_n$ in $\mathbb{Z}$ we write $\lambda = (m_1, \ldots, m_n)$. Thus L is identified with $\mathbb{Z}^n$. We may identify $E_{\mathbb{R}}$ with $\mathbb{R}^n$ and $E_{\mathbb{C}}$ with $\mathbb{C}^n$. If λ belongs to L^0_+ and

$$\lambda: tI \to t^m$$

with m in $\mathbb{Z}$ we write $\lambda = \left(\frac{m}{n}, \ldots, \frac{m}{n}\right)$. Then L^0_- which is a subgroup of both L and L^0_+ consists of the elements $(m, \ldots, m)$ with m in $\mathbb{Z}$. The rank ℓ is $n-1$ and

$$\begin{aligned} \alpha_1 &= (1, -1, 0, \ldots, 0) \\ \alpha_2 &= (0, 1, -1, 0, \ldots, 0) \\ &\vdots \\ \alpha_\ell &= (0, \ldots, 0, 1, -1) \end{aligned}$$

Thus

$$L^1_- = \left\{ (m_1, \ldots, m_n) \in L \;\Big|\; \sum_{i=1}^n m_i = 0 \right\} .$$

$E^1_{\mathbb{C}}$ is the set of all $(z_1, \ldots, z_n)$ in $E_{\mathbb{C}}$ for which

$$\sum_{i=1}^n z_i = 0 .$$

The bilinear form on $E^1_{\mathbb{C}}$ may be taken as the restriction of the form

$$(z, w) = \sum_{i=1}^n z_i w_i$$

on $E_{\mathbb{C}}$. Thus

$$L^1_+ = \left\{ (m_1, \ldots, m_n) \;\Big|\; \sum_{i=1}^n m_i = 0 \text{ and } m_i - m_j \in \mathbb{Z} \right\} .$$

We may use the given bilinear form to identify $\hat{E}_{\mathbb{C}}$ with $E_{\mathbb{C}}$. Then the operation " $\hat{}$ " leaves all lattices and all roots fixed. Thus $\hat{G}^0_+ = \mathrm{Hom}(L^0_+, \mathbb{C})$. Any non-singular complex scalar matrix tI defines an element of $\hat{G}^0_+$, namely, the homomorphism

$$\left(\frac{m}{n}, \ldots, \frac{m}{n}\right) \longrightarrow t^m .$$

We identify $\hat{G}^0_+$ with the group of scalar matrices. $\hat{G}^1_+$ is $SL(n, \mathbb{C})$. There is a natural map of $\hat{G}^0_+ \times \hat{G}^1_+$ onto $GL(n, \mathbb{C})$. It sends $tI \times A$ to tA. The kernel is easily seen to be $\hat{Z}$ so that $\hat{G}_F$ is $GL(n, \mathbb{C})$.

4. To define the local L-functions, to prove that almost all primes are unramified, and to prove that the product of the local L-functions over the unramified primes converges for $Re\, s$ sufficiently large we need some facts from the reduction theory for groups over local fields (cf. [1]). Much progress has been made in that theory, but it is still incomplete. Unfortunately, the particular facts we need do not

seem to be in the literature. Very little is lost at this stage if we just assume them. For the groups about which something definite can be said, they are easily verified.

Suppose K is an unramified extension of the non-archimedean local field F and G is a quasi-split group over F which splits over K. Let B be a Borel subgroup of G and T a Cartan subgroup of B, both of which are defined over F. Let v be the valuation on K. It is a homomorphism from K^*, the multiplicative group of K, onto $\mathbb{Z}$ whose kernel is the group of units. If t belongs to T_F, let $v(t)$ in $\hat{L}$ be defined by $\langle \lambda, v(t) \rangle = v(\lambda(t))$ for all λ in L. If σ belongs to $\mathfrak{G}(K/F)$, then

$$\langle \lambda, \sigma v(t) \rangle = \langle \sigma^{-1}\lambda, v(t) \rangle = v(\sigma^{-1}(\lambda(\sigma t))) = v(\lambda(t))$$

because $\sigma t = t$ and $v(\sigma^{-1}a) = v(a)$ for all a in K^*. Thus v is a homomorphism of T_F into $\hat{M}$, the groups of invariants of $\mathfrak{G}(K/F)$ in $\hat{L}$. It is in fact easily seen that it takes T_F onto $\hat{M}$.

We assume the following lemma.

Lemma 1. *There is a Chevalley lattice in the Lie algebra of G whose stabilizer U_K is invariant under $\mathfrak{G}(K/F)$. U_K is its own normalizer. Moreover, $G_K = B_K U_K$, $H^1(\mathfrak{G}(K/F), U_K) = 1$, and $H^1(\mathfrak{G}(K/F), B_K \cap U_K) = 1$. If we choose two such Chevalley lattices with stabilizers U_K and U'_K, respectively, then U'_K is conjugate to U_K in G_K.*

If g belongs to G_K and σ belongs to $\mathfrak{G}(K/F)$, let $g^\sigma = \sigma^{-1}(g)$. If g belongs to G_F, we may write it as $= bu$ with b in B_K and u in U_K. Then $g^\sigma = b^\sigma u^\sigma$ and $u^\sigma u^{-1} = b^{-\sigma} b$. By the lemma, there is a v in $B_K \cap U_K$ such that $u^\sigma u^{-1} = b^{-\sigma} b = v^\sigma v^{-1}$. Then $b' = bv$ belongs to B_F, $u' = v^{-1}u$ belongs to $U_F = G_F \cap U_K$, and $g = b'u'$. Thus, $G_F = B_F U_F$.

If $g U_K g^{-1} = U'_K$ for some g in G_K, then $g^\sigma U_K g^{-\sigma} = U'_K$ so that $g^{-\sigma} g$ belongs to U_K, which is its own normalizer. By the lemma, there is u in U_K such that $g^{-\sigma} g = u^\sigma u^{-1}$. Then $g_1 = gu$ lies in G_F and $g_1 U_K g_1^{-1} = U'_K$. Thus, U_F and U'_F are conjugate in G_F.

Let $C_c(G_F, U_F)$ be the set of all compactly supported functions for G_F such that $f(gu) = f(ug) = f(g)$ for all u in U_F and all g in G_F. $C_c(G_F, U_F)$ is an algebra under convolution. It is called the Hecke algebra. If N is the unipotent radical of B let dn be a Haar measure on N_F and let $\frac{d(bnb^{-1})}{dn} = \delta(b)$ if b belongs to B_F. If λ belongs to $\hat{M}$, choose t in T_F such that $v(t) = \lambda$. If f belongs to $C_c(G_F, U_F)$, set

$$\hat{f}(\lambda) = \delta^{1/2}(t) \left\{ \int_{N_F \cap U_F} dn \right\}^{-1} \int_{N_F} f(tn) dn \, .$$

The group $\mathfrak{G}(K/F)$ acts on Ω. Let Ω^0 be the group of invariant elements. Ω^0 acts on $\hat{M}$. Let $A(\hat{M})$ be the group algebra of $\hat{M}$ over $\mathbb{C}$, and let $\Lambda^0(\hat{M})$ be the invariants of Ω^0 in $\Lambda(\hat{M})$. We also assume the following lemma (cf. [12]).

Lemma 2. *The map $f \to \hat{f}$ is an isomorphism of $C_c(G_F, U_F)$ and $\Lambda^0(M)$.*

Suppose B is replaced by B_1 and T by T_1. Observe that $T \simeq B/N$ and $T_1 \simeq B_1/N_1$. If u in G_F takes B to B_1, it takes N to N_1 and defines a map from T to T_1. This map does not depend on u. It determines $\mathfrak{G}(K/F)$ invariant maps from L_1 to L and from $\hat{L}$ to $\hat{L}_1$ and thus maps from $\hat{M}$ to $\hat{M}_1$ and from $\Lambda^0(\hat{M})$ to $\Lambda^0(\hat{M}_1)$. Suppose $\hat{f}$ goes to $\hat{f}_1$ and $\hat{\lambda}$ goes to $\hat{\lambda}_1$. If we choose, as we may, u in U_F, then

$$\hat{f}_1(\hat{\lambda}_1) = \hat{f}(\hat{\lambda}) = \delta^{1/2}(t)\left\{\int_{N_F \cap U_F} dn\right\}^{-1} \int_{N_F} f(tn)dn\ .$$

Let $N_F \cap U_F = V$. Denote the corresponding group associated to N_1 by V_1. Then $uVu^{-1} = V_1$. Choose $d(unu^{-1}) = dn_1$. Since $f(ugu^{-1}) = f(g)$, the expression on the right equals

$$\delta^{1/2}(utu^{-1})\left\{\int_{V_1} dn_1\right\}^{-1} \int_{N_F} f(utu^{-1}unu^{-1})dn\ .$$

If utu^{-1} projects on t_1 in T_1, then $\delta(utu^{-1}) = \delta(t_1)$ and $v(t_1) = \hat{\lambda}_1$. Moreover,

$$\int f(utu^{-1}unu^{-1})dn = \int f(t_1n_1)dn_1$$

and the diagram

$$\begin{array}{ccc} & C_c(G_F, U_F) & \\ \swarrow & & \searrow \\ \Lambda^0(M) & \longrightarrow & \Lambda^0(M_1) \end{array}$$

is commutative.

If $gU_Fg^{-1} = U'_F$, the map $f \to f'$ with $f'(h) = f(g^{-1}hg)$ is an isomorphism of $C_c(G_F, U_F)$ with $C_c(G_F, U'_F)$. It does not depend on g. We can take g in B_F. Then

$$\hat{f}'(\lambda) = \delta^{1/2}(t)\left\{\int_{N_F \cap U'_F} dn\right\}^{-1} \int_{N_F} f(g^{-1}tng)dn\ .$$

Since $g^{-1}tng = t(t^{-1}g^{-1}tg)g^{-1}ng$, the second integral is equal to

$$\int_{N_F} f(tg^{-1}ng)dn\ .$$

Since

$$\frac{d(g^{-1}ng)}{dn} = \left\{\int_{N_F \cap U'_F} dn\right\}^{-1} \int_{N_F \cap U_F} dn$$

we conclude that $\hat{f}'(\hat{\lambda}) = \hat{f}(\hat{\lambda})$ and that the diagram

$$\begin{array}{ccc} C_c(G_F,U_F) & \longrightarrow & C_c(G_F,U'_F) \\ & \searrow \quad \swarrow & \\ & \Lambda^0(\hat{M}) & \end{array}$$

is commutative.

I shall not explicitly mention the commutativity of these diagrams again. However, they are important because they imply that the definitions to follow have the invariance properties which are required if they are to have any sense.

If π is an irreducible unitary representation of G_F on H whose restriction to U_F contains the identity representation, then

$$H_0 = \{x \in H \mid \pi(u)x = x \text{ for all } u \text{ in } U_F\}$$

is a one-dimensional subspace. If f belongs to $C_c(G_F, U_F)$, then

$$\pi(f) = \int_G f(g)\pi(g)dg$$

maps H_0 into itself. The representation of $C_c(G_F, U_F)$ on H_0 determines a homomorphism χ of $C_c(G_F, U_F)$ or of $\Lambda^0(\hat{M})$ into the ring of complex numbers. π is determined by χ. To define the local L-functions, we study such homomorphisms. First of all, observe that, if χ is associated to a unitary representation, then

$$|\chi(f)| \le \int_{G_F} |f(g)|dg\ .$$

Since $\Lambda(\hat{M})$ is a finitely generated module over $\Lambda^0(\hat{M})$, any homomorphism of $\Lambda^0(\hat{M})$ into $\mathbb{C}$ may be extended to a homomorphism of $\Lambda(\hat{M})$ into $\mathbb{C}$ which will necessarily be of the form

$$\sum \hat{f}(\lambda)\lambda \longrightarrow \sum \hat{f}(\lambda)\lambda(t) \tag{B}$$

for some t in $\hat{T}$. Conversely, given t the formula (B) determines a homomorphism χ_t of $\Lambda^0(\hat{M})$ into $\mathbb{C}$. We shall show that $\chi_{t_1} = \chi_{t_2}$ if and only if $t_1 \times \sigma_F$ and $t_2 \times \sigma_F$, where σ_F is the Frobenius subsitution, are conjugate in $\hat{G}_F$. If t belongs to $\hat{G}$ and σ belongs to $\mathfrak{G}(K/F)$, we shall abbreviate $t \times \sigma$ to $t\sigma$. It is known [4] that every semi-simple element of $\hat{G}_F$ whose projection on $\mathfrak{G}(K/F)$ is σ_F is conjugate to some $t\sigma_F$ with t in $\hat{T}$. Thus, there is a one-to-one correspondence between homomorphisms of the Hecke algebra into $\mathbb{C}$ and semi-simple conjugacy classes in $\hat{G}_F$ whose projection on $\mathfrak{G}(K/F)$ is σ_F.

If ρ is a complex analytic representation of $\hat{G}_F$ and χ_t is the homomorphism of $\Lambda^0(\hat{M})$ into $\mathbb{C}$ associated to π, we define the local L-function to be

$$L(s, \rho, \pi) = \frac{1}{\det(I - \rho(t\sigma_F)|\pi_F|^2)}$$

if π_F generates the maximal ideal of O_F.

$\hat{T}$ may be identified with $\mathrm{Hom}_{\mathbb{Z}}(\hat{L}, \mathbb{C}^*)$. The exact sequence

$$0 \longrightarrow \mathbb{Z} \xrightarrow{\varphi} \mathbb{C} \xrightarrow{\psi} \mathbb{C}^* \longrightarrow 0$$

with $\varphi(z) = \frac{2\pi i}{\log|\pi_F|} z$ and $\psi(z) = |\pi_F|^{-z}$ leads to the exact sequence

$$0 \longrightarrow L = \mathrm{Hom}_{\mathbb{Z}}(\hat{L}, \mathbb{Z}) \xrightarrow{\varphi} E_{\mathbb{C}} = \mathrm{Hom}_{\mathbb{Z}}(\hat{L}, \mathbb{C}) \xrightarrow{\psi} \hat{T} \longrightarrow 0 .$$

Let $V_{\mathbb{C}}$ be the invariants of $\mathfrak{G}(K/F)$ in $E_{\mathbb{C}}$ and let $W_{\mathbb{C}}$ be the range of $\sigma_F - 1$. Then $E_{\mathbb{C}} = V_{\mathbb{C}} \oplus W_{\mathbb{C}}$. If w belongs to $W_{\mathbb{C}}$ and λ belongs to $\hat{M}$, then $\langle w, \lambda \rangle = 0$ and, replacing t by $t\psi(w)$ does not change χ_t. If $w = \sigma_F v - v$ and $\psi(v) = s$, then

$$t\psi(w)\sigma_F = ts^{-1}\sigma_F(s)\sigma_F = s^{-1}(t\sigma_F)s$$

is conjugate to $t\sigma_F$. Thus, we have to show that if $t_1 = \psi(v_1)$ and $t_2 = \psi(v_2)$ with v_1 and v_2 in $\mathbb{C}$, then $t_1\sigma_F$ and $t_2\sigma_F$ are conjugate if and only if $\chi_{t_1} = \chi_{t_2}$.

Some preliminary remarks are necessary. We also have a decomposition $\hat{E}_{\mathbb{C}} = \hat{V}_{\mathbb{C}} \oplus \hat{W}_{\mathbb{C}}$ and $\hat{M} = \hat{L} \cap \hat{V}_{\mathbb{C}}$. Let $\hat{Q}$ be the elements of $\hat{V}_{\mathbb{C}}$ obtained by projecting the positive coroots on $\hat{V}_{\mathbb{C}}$. If S is an orbit of $\mathfrak{G}(K/F)$ in the set of positive coroots, every element in S has the same projection on $\hat{V}_{\mathbb{C}}$. Since $\sum_{\hat{\alpha} \in S} \hat{\alpha}$ belongs to $\hat{V}_{\mathbb{C}}$, the projection must be

$$\frac{1}{n(S)} \sum_{\hat{\alpha} \in S} \hat{\alpha}$$

if $n(S)$ is the number of elements in S. Let $S_1, \dots, S_m$ be the orbits of $\mathfrak{G}(K/F)$ in $\{\hat{\alpha}_1, \dots, \hat{\alpha}_\ell\}$ and set

$$\hat{\beta}_i = \frac{1}{n(S_i)} \sum_{\hat{\alpha} \in S_i} \hat{\alpha} .$$

Every element of $\hat{Q}$ is a linear combination of $\hat{\beta}_1, \dots, \hat{\beta}_m$ with non-negative coefficients. Notice that if ω belongs to Ω^0 and ω acts trivially on $\hat{M}$, then ω leaves each β_i fixed and therefore takes positive roots to positive roots. Thus, it is 1. If we extend the inner product in any way from $\hat{E}^1_{\mathbb{R}}$ to $\hat{E}_{\mathbb{R}}$ and set

$$\hat{C} = \left\{ x \in \hat{V}_{\mathbb{R}} \mid (\hat{\beta}_i, x) \geq 0, \ 1 \leq i \leq m \right\}$$

and

$$\hat{D} = \left\{ x \in \hat{E}_{\mathbb{R}} \mid (\hat{\alpha}_i, x) \geq 0, \ 1 \leq i \leq \ell \right\} ,$$

then $\hat{C} = \hat{D} \cap \hat{V}_{\mathbb{R}}$. Consequently, no two elements of $\hat{C}$ belong to the same orbit of Ω^0.

Let $\widehat{\mathfrak{g}}_i$ be the subalgebra of the Lie algebra of $\hat{G}$ generated by the root vectors belonging to the coroots in S_i and their negatives. $\widehat{\mathfrak{g}}_i$ is fixed by $\mathfrak{G}(K/F)$. Let $\hat{G}_i$ be the corresponding analytic group and let $\hat{T}_i = \hat{T} \cap \hat{G}_i$. Let μ_i be the unique element of the Weyl group of $\hat{T}_i$ which takes every positive root to a negative root. If σ belongs to $\mathfrak{G}(K/F)$, then $\sigma(\mu_i)$ has the same property, so that $\sigma(\mu_i) = \mu_i$. Let w be any element in the normalizer of $\hat{T}$ whose image in $\hat{\Omega}$ is μ_i. Then $w\sigma_F(w^{-1})$ lies in $\hat{T}$. Its image in $\hat{T}/\psi(W_{\mathbb{C}})$ is independent of w. I claim that this image is 1. To see this write $\widehat{\mathfrak{g}}_i$ as a direct sum $\sum_{k=1}^{n} \widehat{\mathfrak{g}}_{ik}$ of simple algebras. If $[K\colon F] = n$ the stabilizer of $\widehat{\mathfrak{g}}_{i1}$ is $\left\{\sigma_F^{jn_i} \mid 0 \le j \le \frac{n}{n_i}\right\}$. We may suppose that

$$\widehat{\mathfrak{g}}_{ik} = \sigma_F^{k-1}(\widehat{\mathfrak{g}}_{i1}) \ .$$

If $\hat{G}_{ik}$ is the analytic subgroup of $\hat{G}$ with Lie aglebra $\widehat{\mathfrak{g}}_{ik}$, choose w_1 in the normalizer of $\hat{T} \cap \hat{G}_{i1}$ so that w_1 takes the positive roots of $\widehat{\mathfrak{g}}_{i1}$ to the negative roots. We may choose w to be $\prod_{k=0}^{n_i-1} \sigma_F^k(w_1)$. Then

$$\begin{aligned} w\sigma_F(w^{-1}) &= (w_1\sigma_F(w_1^{-1}))(\sigma_F(w_1)\sigma_F^2(w_1^{-1}))\dots(\sigma_F^{n_i-1}(w_1)\sigma_F^{n_i}(w_1^{-1})) \\ &= w_1\sigma_F^{n_i}(w_1^{-1}) \ . \end{aligned}$$

The Dynkin diagram of $\widehat{\mathfrak{g}}_{i1}$ is connected and the stabilizer of $\widehat{\mathfrak{g}}_{i1}$ in $\mathfrak{G}(K/F)$ acts transitively on it. This means that it is of type A_1 or A_2.

In the first case the diagram reduces to a point and the action of the stabilizer must be trivial, so that $w_1 = \sigma_F^{n_i}(w_1)$. In the second case $SL(3,\mathbb{C})$ is the simply-connected covering group of G_{i1}; we may choose the covering map to be such that $\hat{T} \cap \hat{G}_{i1}$ is the image of the diagonal matrices and $\sigma_F^{n_i}$ corresponds to the automorphism

$$A \to \begin{pmatrix} 0 & 0 & 1 \\ 0 & -1 & 0 \\ 1 & 0 & 0 \end{pmatrix} {}^tA^{-1} \begin{pmatrix} 0 & 0 & 1 \\ 0 & -1 & 0 \\ 1 & 0 & 0 \end{pmatrix}$$

of $SL(3,\mathbb{C})$. We may take w_1 to be the image of

$$\begin{pmatrix} 0 & 0 & 1 \\ 0 & -1 & 0 \\ 1 & 0 & 0 \end{pmatrix} .$$

Then $\sigma_F^{n_i}(w_1) = w_1$.

μ_i acts on $\hat{V}$ as the reflection in the hyperlane perpendicular to β_i. Thus $\mu_1, \dots, \mu_m$ generate Ω^0. If ω belongs to Ω^0, choose w in the normalizer of $\hat{T}$ whose image in Ω is ω. The image of $w\sigma_F(w^{-1})$ in $\hat{T}/\psi(W_{\mathbb{C}})$ depends only on ω. Call it δ_ω. Then

$$\delta_{\omega_1\omega_2} = w_1w_2\sigma_F(w_2^{-1}w_1^{-1}) = w_1(w_2\sigma_F(w_2^{-1}))w_1^{-1}(w_1\sigma_F(w_1^{-1})) = \omega_1(\delta_{\omega_1})\delta_{\omega_1} \ .$$

Since δ_ω is 1 on a set of generators, this relation shows that it is identically 1.

Returning to the original problem, we show first that if $\chi_{t_1} = \chi_{t_2}$ there is an ω in Ω^0 such that $\omega(w_1) = t_2$. Then, if w lies in the normalizer of $\hat{T}$ in $\hat{G}$ and its image in Ω is ω, we will have $w(t_1\sigma_F)w^{-1} = t_2 w\sigma_F(w^{-1})\sigma_F$. Since $w\sigma_F(w^{-1})$ lies in $\psi(W_\mathbb{C})$, the element on the right is conjugate to $t_2\sigma_F$.

If t belongs to $\hat{T}$, let χ_t also denote the homomorphism

$$\sum \hat{f}(\lambda)\lambda \to \sum \hat{f}(\lambda)\lambda(t)$$

of $\Lambda(\hat{M})$ into $\mathbb{C}$. If there were no ω such that $\omega(t_1) = t_2$, there would be an $\hat{f}$ in $\Lambda(\hat{M})$ such that

$$\chi_{t_2}(\hat{f}) \neq \chi_{\omega(t_1)}(\hat{f})$$

for all ω in Ω^0. Let

$$\prod (X - \omega(\hat{f})) = \sum_{k=0}^{n} \hat{f}_k X^k \, .$$

Each $\hat{f}_k$ belongs to $\Lambda^0(\hat{M})$. Applying χ_{t_1} and χ_{t_2}, we find that

$$\prod_\omega (X - \chi_{\omega(t_1)}(\hat{f})) = \sum_{k=0}^{n} \chi_{t_1}(\hat{f}_k) X^k = \sum_{k=0}^{n} \chi_{t_2}(\hat{f}_k) X^k = \prod_\omega (X - \chi_{\omega(t_2)}(\hat{f})) \, .$$

The polynomial on the right has $\chi_{t_2}(\hat{f})$ as a root, but that on the left does not. This is a contradiction.

If $t_1\sigma_F$ and $t_2\sigma_F$ are conjugate, then for every representation ρ of $\hat{G}_F$

$$\text{trace } \rho(t_1\sigma_F) = \text{trace } \rho(t_2\sigma_F) \, .$$

Let ρ act on X and if λ belongs to $\hat{M}$, let t_λ be the trace of $\rho(\sigma_F)$ on

$$X_\lambda = \left\{ x \in X \mid \rho(t)x = \lambda(t)x \text{ for all } t \text{ in } \hat{T} \right\} \, .$$

If t belongs to $\psi(W_\mathbb{C})$, then $\lambda(t) = 1$. If ω belongs to Ω^0 and w in the normalizer of T has image ω in Ω, the $X_{\omega\lambda} = \rho(w)X_\lambda$. Then $t_{\omega\lambda}$ is the trace of $w^{-1}\sigma_F w = w^{-1}\sigma_F(w)\sigma_F$ on X_λ. Since $\lambda(w^{-1}\sigma_F(w)) = 1$, we have $t_{\omega\lambda} = t_\lambda$ and

$$\text{trace } \rho(t\sigma_F) = \sum_{\lambda \in \hat{C}} t_\lambda \left(\sum_{\mu \in S(\lambda)} \mu(t) \right)$$

if $S(\lambda)$ is the orbit of λ. If

$$\hat{f}_\rho = \sum_{\lambda \in \hat{C}} t_\lambda \sum_{\mu \in S(\lambda)} \mu$$

then $\hat{f}_\rho$ belongs to $\Lambda^0(\hat{M})$ and

$$\text{trace}\ \rho(t\sigma_F) = \chi_t(\hat{f}_\rho)\ .$$

All we need to show is that the elements $\hat{f}_\rho$ generate $\Lambda^0(\hat{M})$ as a vector space. This is an easy induction argument because every λ in $\hat{C}$ is the highest weight of a representation of $\hat{G}_F$ whose restriction to $\hat{G}$ is irreducible.

5. If t belongs to $\hat{T}$, there is a unique function ϕ_t on G_F which satisfies $\phi_t(ug) = \phi_t(gu) = \phi_t(g)$ or all u in U_F and all g in $_F$, such that

$$\chi_t(f) = \int_{G_F} \phi_t(g) f(g) dg$$

for all f in $C_c(G_F, U_F)$. A formula for ϕ_t, valid under very general assumptions, has been found by I. G. MacDonald. However, because of the present state of reduction theory, his assumptions do not cover the cases in which we are interested. I am going to assume that the obvious generalization of his theorem is valid. In stating it we may as well suppose that t belongs ot $\psi(V_{\mathbb{C}})$.

Let $\hat{N}$ be the unipotent radical of $\hat{B}$, let $\hat{\mathfrak{n}}$ be its Lie algebra, and let τ be the representation of $\hat{T} \times \mathfrak{G}(K/F)$ on $\hat{\mathfrak{n}}$. If t belongs to $\psi(V_{\mathbb{C}})$, consider the function θ_t on $\hat{M}$ defined by

$$\theta_t(\lambda) = c|\pi_F|^{-\langle\rho,\lambda\rangle} \sigma_{\omega\in\Omega^0} \frac{\det(I - |\pi_F|\tau^{-1}(\omega(t)\sigma_F))}{\det(I - \tau^{-1}(\omega(t)\sigma_F))} \lambda^{-1}(\omega(t))\ .$$

If $n(\hat{B})$ is the number of positive roots projection onto $\hat{\beta}$ in $\hat{Q}$,

$$c = \prod_{\beta\in Q} \left\{ \frac{1 - |\pi_F|^{n(\hat{\beta})\langle\rho,\hat{\beta}\rangle}}{1 - |\pi_F|^{n(\beta)(\langle\rho,\beta\rangle+1)}} \right\}\ .$$

As it stands, $\theta_t(\lambda)$ makes sense only when none of the eigenvalues of $\tau(\omega(t)\sigma)F$ are 1 for any ω in Ω^0. However, using the results of Kostant [8], we can write it in a form which makes sense for all t. Let $\hat{\rho}$ be one-half the sum of the positive coroots. $\hat{\rho}$ belongs to $\hat{V}$. If λ belongs to $\hat{M}$ and $\lambda + \hat{\rho}$ is non-singular, that is $(\lambda + \hat{\rho}, \hat{\beta}) \neq 0$ for all $\hat{\beta}$ in $\hat{Q}$, let ω in Ω^0 take $\lambda + \hat{\rho}$ to $\hat{C}$ and let χ_λ be sgn ω times the character of the representation of $\hat{G}_F$ with highest weight $\omega(\lambda + \hat{\rho}) - \hat{\rho}$. If $\lambda + \hat{\rho}$ is singular, let $\chi_\lambda \equiv 0$. If

$$\det(I - |\pi_F|\tau^{-1}(t\sigma_F)) = \sum_{\mu\in\hat{M}} b_\mu \mu(t)$$

then

$$\theta_t(\lambda) = c|\pi_F|^{-\langle\rho,\lambda\rangle} \sum_{\mu\in\hat{M}} b_\mu \chi_{\mu-\lambda}((t\sigma_F))\ .$$

Clearly b_μ is 0 unless

$$\mu = -\sum_{\hat\alpha\in S}\hat\alpha$$

where S is a subset of the set of positive coroots invariant under $\mathfrak{G}(K/F)$. If U is the collection of such μ, then $\{\hat\rho+\mu|\mu\in M\}$ is invariant under Ω^0. Suppose $\hat\rho+\mu$ is non-singular and belongs to $\hat C$. Since $\langle\alpha_i,\hat\rho\rangle = 1$ and $\langle\alpha_i,\mu\rangle$ is integral, for $1\le i\le \ell$, μ itself must belong to $\hat C$. This can only happen if μ is 0. Thus if $b_\mu\ne 0$ either $\hat\rho+\mu$ is singular or $\hat\rho+\mu$ belongs to the orbit of $\hat\rho$ and $\chi_\mu(g)\equiv\pm1$ on $\hat G_F$. As a consequence $\theta_t(0)$ is independent of t. Choose t_0 such that $\hat\beta_i(t_0) = |\pi|^{-\langle\rho,\hat\beta_i\rangle}$ for $1\le i\le m$. The eigenvalues of $\tau(\omega(t_0)\sigma_F)$ are the numbers $\zeta|\pi_F|^{-\langle\rho,\omega^{-1}\hat\beta\rangle}$ where $\hat\beta$ belongs to Q and ζ is an $n(\hat\beta)^{\mathrm{th}}$ root of unity. If $\omega\ne1$ there is a $\hat\beta_i$ such that $\omega^{-1}\hat\beta = -\hat\beta_i$ for some β in Q. Then $\left\langle\rho,\omega^{-1}\hat\beta\right\rangle = -\left\langle\rho,\hat\beta_i\right\rangle = -1$ and $\tau(\omega(t_0)\sigma_F)$ has $|\pi_F|$ is an eigenvalue. Thus

$$\theta_{t_0}(0) = c\,\frac{\det(I-|\pi_f|\tau^{-1}(t_0\sigma_F))}{\det(I-\tau^{-1}(t_0\sigma_F))} = 1\ .$$

We are going to assume that if t belongs to $\psi(V_{\mathbb C})$, a belongs to T_F, and $\lambda = v(a)$, then

$$\phi_t(a) = \theta_t(\lambda)\ .$$

If

$$|\chi_t(f)| \le \int_{G_F}|f(g)|dg$$

for all f in $C_c(G_F,U_F)$ then ϕ_t is bounded. I want to show that if ϕ_t is bounded, λ belongs to $\hat L$, $\bar\lambda$ in $\hat D$ belongs to the orbit of λ under Ω, and t lies in $\psi(V_{\mathbb C})$, then

$$|\lambda(t)| \le |\pi_F|^{-\langle\rho,\bar\lambda\rangle}\ .$$

Let $t=\psi(v)$. Then v is not determined by t but $Re\,v$ is and

$$|\lambda(t)| = |\pi_F|^{-Re\langle v,\lambda\rangle}\ .$$

We will show that if ϕ_t is bounded then $Re\,\langle v,\lambda\rangle\le\langle\rho,\bar\lambda\rangle$ for all λ in $\hat E_{\mathbb R}$. If ω belongs to Ω^0 and $Re\,\omega v$ lies in $\hat C$ then $Re\,\langle\omega v,\omega\lambda\rangle = Re\,\langle v,\lambda\rangle$. With no loss of generality, we may suppose that v lies in C, the analogue of $\hat C$. Then, as is well-known,

$$Re\,\langle v,\lambda\rangle \le Re\,\langle v,\bar\lambda\rangle$$

and we may as well assume that $\lambda=\bar\lambda$. We want to show that $Re\,\langle v,\lambda\rangle\le\langle\rho,\lambda\rangle$ for all λ in $\hat D$. Since ρ and v both belong to $V_{\mathbb C}$, it is sufficient to verify it for λ in $\hat C$. Let $\hat C^0$ be the interior of $\hat C$. The set of

λ in $\hat{C}$ for which the assertion is true is closed, convex, and positively homogeneous. Therefore, if it contains $\hat{M} \cap \hat{C}^0$, it is $\hat{C}$.

Let S be the set of simple coroots $\hat{\alpha}$ for which $Re\,\langle v, \hat{\alpha} \rangle = 0$. Let Σ_0 be the positive coroots which are linear combinations of the elements of S and let Σ_+ be the other positive coroots. If $\hat{\mathfrak{n}}_0$ is the span of the root vectors associated to the coroots in Σ_0 and $\hat{\mathfrak{n}}_+$ is the span of the root vectors associated to the coroots in Σ_+, then τ breaks up into the direct sum of a representation τ_0 on $\hat{\mathfrak{n}}_0$ and a representation τ_+ on $\hat{\mathfrak{n}}_+$. Let $\hat{H}$ be the analytic subgroup of $\hat{G}_F$ whose Lie algebra is generated by the root vectors associated to the coroots of Σ_0 and their negatives and let Θ^0 be the subgroup of Ω^0 consisting of those elements with representatives in $\hat{H}$. If ω belongs to Ω^0 and $Re\,\omega v = Re\,v$, then ω belongs to Θ^0. If $Re\,\omega v \neq Re\,v$, then $Re\,\langle \omega v, \lambda \rangle < Re\,\langle v, \lambda \rangle$ for λ in $\hat{M} \cap \hat{C}^0$. Write $\lambda = \lambda_1 + \lambda_2$ where λ_1 is a linear combination of the coroots in S and λ_2 is orthogonal to these roots. If $s = \psi(u)$ with u in $V_{\mathbb{C}}$, consider

$$\theta'_s(\lambda) = c|\pi_F|^{\langle u-\rho, \lambda_2 \rangle} \frac{\det(I - |\pi_F|\tau_+^{-1}(s\sigma_F))}{\det(I - \tau_+^{-1}(s\sigma_F))} \left\{ \sum_{\Theta^0} \frac{\det(I - |\pi_F|\tau_0^{-1}(s\sigma_F))}{\det(I - \tau_0^{-1}(s\sigma_F))} |\pi_F|^{\langle \omega u - \rho, \lambda_1 \rangle} \right\} .$$

The function θ'_s is not necessarily defined for all s. However, the preceding discussion, applied to $\hat{H}$ rather than $\hat{G}$, shows that it is defined at t and that $\theta'_t(0) \neq 0$. A simple application of l'Hospital's rule shows that, as a function of λ, θ'_t is the product of $|\pi_F|^{\langle v-\rho, \lambda \rangle}$ and a linear combination of products of polynomials and purely imaginary exponentials in λ_1. Thus, it does not vanish identically in any open cone.

Set $\theta''_t = \theta_t - \theta'_t$. θ''_t is a linear combination of products of polynomials in λ and an exponential $|\pi_F|^{\langle \omega v - \rho, \lambda \rangle}$ with $Re\,\omega v \neq Re\,v$. Thus, if λ belongs to the interior of $\hat{C}$,

$$\lim_{n \to -\infty} |\pi_F|^{\langle \rho - v, n\lambda \rangle} \theta''_t(n\lambda) = 0$$

and

$$\lim_{n \to -\infty} |\pi_F|^{\langle \rho - v, n\lambda \rangle} \theta_t(n\lambda) = \lim_{n \to -\infty} |\pi_F|^{\langle \rho - v, n\lambda \rangle} \theta'_t(n\lambda) .$$

If $\langle \rho, \lambda \rangle$ is less than $Re\,\langle v, \lambda \rangle$ for some λ in $\hat{C}$, then $\langle \rho, \lambda \rangle$ is less than $Re\,\langle v, \lambda \rangle$ for a λ in $\hat{C}$ for which $\theta'_t(n\lambda)$ does not vanish identically as a function of n. Since ϕ_t is bounded,

$$\lim_{n \to -\infty} |\pi_F|^{\langle \rho - v, n\lambda \rangle} \theta'_t(n\lambda) = 0 .$$

But $|\pi_F|^{\langle \rho - v, n\lambda \rangle} \theta'_t(n\lambda)$ is a function of the form

$$\sum_{k=0}^{q} \varphi_k(n) n^k$$

where $\varphi_k(n)$ is a linear combination of purely imaginary exponentials e^{ixn}. It is easy to see that it cannot approach 0 as n approaches $-\infty$.

6. Suppose G is a group defined over the global field F. There is a quasi-split group G' over F and an isomorphism φ: $G \to G'$ defined over a Galois extension K of F such that, for every σ in $\mathfrak{G}(K/F)$, $a_\sigma = \varphi^\sigma\varphi^{-1}$ is an inner automorphism of G'. We assume that there is a lattice $\mathfrak{g}_{O_F}$ over O_F in the Lie algebra of G' such that $O_K\mathfrak{g}_{O_F}$ is a Chevalley lattice.

If $\mathfrak{p}$ is a finite prime of K and $\mathfrak{P}$ is a prime of K dividing $\mathfrak{p}$, the group G over $F_\mathfrak{p}$ is obtained from G' by twisting by the restriction $\bar{a}$ of the cocycle $\{a_\sigma\}$ to $\mathfrak{G}(K_\mathfrak{P}/F_\mathfrak{p})$. Let $\overline{G}'$ be the adjoint group of G'. If $\bar{U}'_{K_\mathfrak{P}}$ is the stabilizer of the lattice $O_{K_\mathfrak{P}}\mathfrak{G}_{0_F}$ then, for almost all $\mathfrak{p}$, $\bar{a}$ takes values in $\bar{U}'_{K_\mathfrak{P}}$. If $K_\mathfrak{P}/F_\mathfrak{p}$ is also unramified, then G is quasi-split over $F_\mathfrak{p}$ because $H^1(\mathfrak{G}(K_\mathfrak{P}/F_\mathfrak{p}), \bar{U}'_{K_\mathfrak{p}}) = \{1\}$. Let S be the set of those $\mathfrak{p}$, unramified in K, for which $\bar{a}$ takes values in $\bar{U}'_{K_\mathfrak{P}}$. Let G act on a vector space X over F and let X_{O_F} be a lattice in X_F. Let $U_{F_\mathfrak{p}}$ be the stabilizer of $O_{F_\mathfrak{p}}X_{O_F}$ in $G_{F_\mathfrak{p}}$ and let $U'_{F_\mathfrak{p}}$ be the stabilizer of $O_{F_\mathfrak{p}}\mathfrak{g}_{O_F}$ in $G'_{F_\mathfrak{p}}$. Then $\varphi(U_{F_\mathfrak{p}}) = U'_{F_\mathfrak{p}}$ for almost all $\mathfrak{p}$. If $\mathfrak{p}$ is also in S, choose u in $\bar{U}'_{F_\mathfrak{p}}$ so that $\varphi^\sigma\varphi^{-1} = \operatorname{Ad} u^\sigma u^{-1}$ for all σ in $\mathfrak{G}(K_\mathfrak{P}/F_\mathfrak{p})$. Then $\varphi^{-1}\operatorname{Ad} u$ is defined over F and $\varphi^{-1}\operatorname{Ad} u(U'_{F_\mathfrak{p}}) = U_{F_\mathfrak{p}}$. Consequently, $U_{F_\mathfrak{p}}$ is one of the compact subgroups of the fourth paragraph.

To show that almost all $\mathfrak{p}$ are unramified, all we need do is observe that if π occurs in $L^2(G_F \setminus G_{\mathbb{Z}(F)})$, whatever the precise meaning of this is to be, and $\pi = \prod_\mathfrak{p} \otimes\pi_\mathfrak{p}$, then for almost all $\mathfrak{p}$, the restriction of $\pi_\mathfrak{p}$ to $U_{F_\mathfrak{p}}$ contains the trivial representation.

If $\mathfrak{p}$ is unramified let the homomorphism of $C_c(G_{F_\mathfrak{p}}, U_{F_\mathfrak{p}})$ associated to $\pi_\mathfrak{p}$ be $\chi_{t_\mathfrak{p}}$. To show that the product of the local L-function converges in a half plane it would be enough to show that there is a positive constant a such that for all unramified $\mathfrak{p}$ every eigenvalue of $\rho(t_\mathfrak{p}\sigma_{F_\mathfrak{p}})$ is bounded by $|\pi_\mathfrak{p}|^{-a}$. We may suppose that $\sigma_{F_\mathfrak{p}}(t_\mathfrak{p}) = t_\mathfrak{p}$. If $n = [K : F]$, then $(t_\mathfrak{p}\sigma_{F_\mathfrak{p}})^n = t_\mathfrak{p}^n$ so that we need only show that the eigenvalues of $\rho(t_\mathfrak{p})$ are bounded by $|\pi_\mathfrak{p}|^{-a}$. This we did in the previous paragraph.

7. Once the definitions are made we can begin to pose questions. My hope is that these questions have affirmative answers. The first question is the one initially posed.

Question 1. *Is it possible to define the local L-functions $L(s, \rho, \pi)$ and the local factors $\varepsilon(s, \rho, \pi, \psi_F)$ at the ramified primes so that if F is a global field $\pi = \prod \otimes\pi_\mathfrak{p}$, and*

$$L(s, \rho, \pi) = \prod_\mathfrak{p} L(s, \rho_\mathfrak{p}, \pi_\mathfrak{p})$$

then $L(s,\rho,\pi)$ is meromorphic in the entire complex plane with only a finite number of poles and satisfies the functional equation

$$L(s,\rho,\pi)=\varepsilon(s,\rho,\pi)L(1-s,\tilde{\rho},\pi)$$

and

$$\varepsilon(s,\rho,\pi)=\prod_{\mathfrak{p}}\varepsilon\left(s,\rho_{\mathfrak{p}},\pi_{\mathfrak{p}},\psi_{F_{\mathfrak{p}}}\right) .$$

The theory of Eisenstein series can be used [9] to give some novel instances in which this question has, in part, an affirmative answer. However, that theory does not suggest any method of attacking the general problem. If $G = GL(n)$ then $\hat{G}_F = GL(n,\mathbb{C})$. The work of Godement and earlier writers allows one to hope that the methods of Hecke and Tate can, once the representation theory of the general linear group over a local field is understood, be used to answer the first question when $G = GL(n)$ and ρ is the standard representation of $GL(n,\mathbb{C})$. The idea which led Artin to the general reciprocity law suggests that we try to answer it in general by answering a further series of questions. For the sake of precision, but not clarity, I write them down in an order opposite to that in which they suggest themselves. If G is defined over the local field F let $\Omega(G_F)$ be the set of equivalence classes of irreducible unitary representations of G_F.

Question 2. *Suppose G and G' are defined over the local field F, G is quasi-split and G' is obtained from G by an inner twisting. Then $\hat{G}_F = \hat{G}'_F$. Is there a correspondence F whose domain is $\Omega(G'_F)$ and whose range is contained in $\Omega(G_F)$ such that if $\pi = R(\pi')$ then $L(s,\rho,\pi) = L(s,\rho,\pi')$ for every representation ρ of $\hat{G}_F$?*

Notice that R is not required to be a function. I do not know whether or not to expect that

$$\varepsilon(s,\rho,\pi,\psi_F)=\varepsilon(s,\rho,\pi',\psi_F) .$$

One should, but I have not yet done so, look carefully at this question when F is the field of real numbers. For this one will of course need the work of Harish-Chandra.

Supposing that the second question has an affirmative answer, one can formulate a global version.

Question 3.* *Suppose that G and G' are defined over the global field F, G is quasi-split, and G' is obtained from G by an inner twisting. Suppose $\pi' = \prod_{\mathfrak{p}} \otimes \pi'_{\mathfrak{p}}$ occurs in $L^2(G'_F \setminus G'_{\mathbb{A}(F)})$. Choose for each $\mathfrak{p}$ a representation $\pi_{\mathfrak{p}}$ of $G_{F_{\mathfrak{p}}}$ such that $\pi_{\mathfrak{p}} = R(\pi'_{\mathfrak{p}})$. Does $\pi = \prod_{\mathfrak{p}} \otimes \pi_{\mathfrak{p}}$ occur in $L^2(G_F \setminus G_{\mathbb{A}(F)})$?*

---

* The question, in this crude form, does not always have an affirmative answer (cf. [6]). The proper question is certainly more subtle but not basically different.

Affirmative evidence is contained in papers of Eichler [3] and Shimizu [16] when $G = GL(2)$ and G' is the group of invertible elements in a quaternion algebra. Jacquet [16], whose work is not yet complete, is obtaining very general results for these groups.

Question 4. *Suppose G and G' are two quasi-split groups over the local field F. Let G split over K and let G' split over K' with $K \subseteq K'$. Let ψ be the natural map $\mathfrak{G}(K'/F) \to \mathfrak{G}(K/F)$. Suppse φ is a complex analytic homomorphism from $\hat{G}'_{K'/F}$ to $\hat{G}_{K/F}$ which makes*

$$\begin{array}{ccc} \hat{G}'_{K'/F} & \longrightarrow & \mathfrak{G}(K'/F) \\ \downarrow{\scriptstyle\varphi} & & \downarrow{\scriptstyle\psi} \\ \hat{G}_{K/F} & \longrightarrow & \mathfrak{G}(K/F) \end{array}$$

commutative. Is there a correspondence R_φ with domain $\Omega(G'_F)$ whose range is contained in $\Omega(G_F)$ such that if $\pi = R_\varphi \pi'$, then, for every representation ρ of $\hat{G}_F$ and every non-trivial additive character ψ_F, $L(s,\rho,\pi) = L(s,\rho\circ\varphi,\pi')$ and $\varepsilon(s,\rho,\pi,\psi_F) = \varepsilon(s,\rho\circ\varphi,\pi',\psi_F)$?

R_φ should of course be functorial and, in an unramified situation, if π' is associated to the conjugacy class $t' \times \sigma'_F$, then π should be associated to $\varphi(t' \times \sigma'_F)$. I have not yet had a chance to look carefully at this question when F is the field of real numbers.

The question has a global form.

Question 5. *Suppose G and G' are two quasi-split groups over the global field F. Let G split over K and let G' split over K' with $K \subseteq K'$. Suppose φ is a complex analytic homomorphism from $\hat{G}'_{K'/F}$ to $\hat{G}_{K/F}$ which makes*

$$\begin{array}{ccc} \hat{G}'_{K'/F} & \longrightarrow & \mathfrak{G}(K'/F) \\ \downarrow{\scriptstyle\varphi} & & \downarrow \\ \hat{G}_{K/F} & \longrightarrow & \mathfrak{G}(K/F) \end{array}$$

commutative. If $\mathfrak{P}'$ is a prime of K', let $\mathfrak{P} = \mathfrak{P}' \cap K$ and let $\mathfrak{p} = \mathfrak{P}' \cap F$. φ determines a homomorphism $\varphi_\mathfrak{p}$: $\hat{G}'_{K'_{\mathfrak{P}'}/F_\mathfrak{p}} \to \hat{G}_{K_\mathfrak{P}/F_\mathfrak{p}}$ which makes

$$\begin{array}{ccc} \hat{G}'_{K'_{\mathfrak{P}'}/F_\mathfrak{p}} & \longrightarrow & \mathfrak{G}(K'_{\mathfrak{P}'}/F_\mathfrak{p}) \\ \downarrow & & \downarrow \\ \hat{G}_{K_\mathfrak{P}/F_\mathfrak{p}} & \longrightarrow & \mathfrak{G}(K_\mathfrak{P}/F_\mathfrak{p}) \end{array}$$

commutative. If $\pi' = \prod \otimes \pi'_{\mathfrak{p}}$ *occurs in* $L^2(G'_F \setminus F'_{\mathbb{A}(F)})$ *choose for each* $\mathfrak{p}$ *a* $\pi_{\mathfrak{p}} = R_{\varphi_{\mathfrak{p}}}(\pi'_{\mathfrak{p}})$. *If* $\pi = \prod_{\mathfrak{p}} \otimes \pi_{\mathfrak{p}}$ *does* π *occur in* $L^2(G_F \setminus G_{\mathbb{A}(F)})$ *?*

An affirmative answer to the third and fifth questions would allow us to solve the first question by examining automorphic forms on the general linear groups.

It is probably worthwhile to point out the difficulty of the fifth question by giving some examples. Take $G' = \{1\}$ $G = GL(1)$, K' any Galois extension of F and $K = F$. The assertion that, in this case, the last two questions have affirmative answers is the Artin reciprocity law.

Suppose G is quasi-split and $G' = T$. We may identify $\hat{G}'_F$ with $\hat{T} \times \mathfrak{G}(K/F)$ which is contained in $\hat{G}_F$. Thus we take $K' = K$. Let φ be the imbedding. In this case π' is a character of $G'_F \setminus G'_{\mathbb{A}(F)}$. The fourth question is, with certain reservations, answered affirmatively by the theory of induced representations. The fifth question is, with similar reservations, answered by the theory of Eisenstein series. The reservations are not important. I only want to point out that the theory of Eisenstein series is a prerequisite to the solution of these problems. with G as before, take $G'' = \{1\}$ and $K'' = K$ so that $\hat{G}''_F = \mathfrak{G}(K/F)$. Let ψ take σ in $\mathfrak{G}(K/F)$ to σ to $\hat{G}_F$. There is only one choice of π''. The associated space of automorphic forms on $G_F \setminus F_{\mathbb{A}(F)}$ should be the space of automorphic forms associated to the trivial character of $G'_F \setminus G'_{\mathbb{A}(F)}$. For this character all the reservations apply. I point out that the space associated to π'' is not the obvious one. It is not the space of constant functions. To prove its existence will require the theory of Eisenstein series.

Take $G = GL(2)$ and let G' be the multiplicative group of a separable quadratic extension K' of F. Take $K = F$. Then $\hat{G}'_F$ is a semi-direct product $(\mathbb{C}^* \times \mathbb{C}^*) \times \mathfrak{G}(K'/F)$. If σ is the non-trivial element of $\mathfrak{G}(K'/F)$, then $\sigma((t_1, t_2)) = (t_2, t_1)$. Let φ be defined by

$$\varphi\colon\ (t_1, t_2) \longrightarrow \begin{pmatrix} t_1 & 0 \\ 0 & t_2 \end{pmatrix}$$

$$\varphi\colon\ \sigma \longrightarrow \begin{pmatrix} 0 & 1 \\ 1 & 0 \end{pmatrix}$$

The existence of R_φ in the local case is a known fact (see for example [6]) in the theory of representations of $GL(2, F)$. An affirmative answer to the fifth question can be given by means of the Hecke theory [6] and by other means [15].

Let E be a separable extension of F and let G be the group over F obtained from $GL(2)$ over D by restriction of scalars. Let G' be $GL(2)$ over F and let $K' = K$ be any Galois extension containing

E. Let X be the homogeneous space $\mathfrak{G}(K/E) \setminus \mathfrak{G}(K/F)$. Then $\hat{G}_F$ is the semi-direct product of $\prod_{x \in X} GL(2, \mathbb{C})$ and $\mathfrak{G}(K/F)$. If σ belongs to $\mathfrak{G}(K/F)$, then

$$\sigma\Big(\prod_{x \in X} A_x\Big)\sigma^{-1} = \prod_{x \in X} B_x$$

with $B_x = A_{x\sigma}$. Define φ by

$$\varphi(A \times \sigma)\Big(\prod_{x \in X} A\Big) \times \sigma\ .$$

Although not much is known about the fifth question in this case, the paper [2] of Doi and Naganuma is encouraging.

Suppose G and K are given. Let $G' = \{1\}$ and let K' be any Galois extension of F containing K. If F is a local field, the fourth question asks that, to every homomorphism φ of $\mathfrak{G}(K'/F)$ into $\hat{G}_F$ which makes

$$\begin{array}{ccc} \mathfrak{G}(K'/F) & \xrightarrow{\varphi} & \hat{G}_F \\ & \searrow & \downarrow \\ & & \mathfrak{G}(K/F) \end{array}$$

commutative, there be associated at least one irreducible unitary representation of G_F. If F is global, the fifth question asks that to φ there be associated a presentation of $G_{\mathbb{A}(F)}$ occurring in $L^2(G_F \backslash G_{\mathbb{A}(F)})$.

The L-functions we have introduced have been so defined that they include the Artin L-functions. However, Weil [17] has generalized the notion of an Artin L-function. The preceding observations suggest a relation between the generalized Artin L-function and the L-functions of this paper. Weil's definition requires the introduction of some locally compact groups – the Weil groups. If F is a local field, let C_F be the multiplicative group of F. If F is a global field, let C_F be the idèle class group. If K is a Galois extension of F, the Weil group $W_{K/F}$ is an extension

$$1 \to C_K \to W_{K/F} \to \mathfrak{G}(K/F) \to 1$$

of $\mathfrak{G}(K/F)$ by C_K. There is a canonical homomorphism $\tau_{K/F}$ of $W_{K/F}$ onto C_F. If F is a global field, $\mathfrak{P}$ a prime of K, and $\mathfrak{p} = F \cap \mathfrak{P}$, there is a homomorphism $\alpha_{\mathfrak{p}}$: $W_{K_{\mathfrak{P}}/F_{\mathfrak{p}}} \to W_{K/F}$. $\alpha_{\mathfrak{p}}$ is determined up to an inner automorphism. If σ is a representation of $W_{K/F}$, the class of $\sigma_{\mathfrak{p}} = \sigma \circ \alpha_{\mathfrak{p}}$ is independent of $\alpha_{\mathfrak{p}}$. By a representation σ of $W_{K/F}$ we understand a finite dimensional complex representation such that $\sigma(w)$ is semi-simple for all w in $W_{K/F}$.

If F is a local field and ψ_F a non-trivial additive character of F, then for any representation σ of $W_{K/F}$ we can define (cf. [11]) a local L-function $L(s,\sigma)$ and a factor $\varepsilon(s,\sigma,\psi_F)$. If F is a global field and σ is a representation of $W_{K/F}$, the associated L-function is

$$L(s,\sigma)=\prod_{\mathfrak{p}} L(s,\sigma_{\mathfrak{p}})\ .$$

The product is taken over all primes, including the archimedean ones. If ψ_F is a non-trivial character of $F\setminus\mathbb{A}(F)$, then $\varepsilon(s,\sigma_{\mathfrak{p}},\psi_{F_{\mathfrak{p}}})$ is 1 for almost all $\mathfrak{p}$,

$$\varepsilon(s,\sigma)=\prod_{\mathfrak{p}}\varepsilon(s,\sigma_{\mathfrak{p}},\psi_{F_{\mathfrak{p}}})$$

is independent of ψ_F, and

$$L(s,\sigma)=\varepsilon(s,\sigma)L(1-s,\tilde{\sigma})$$

if $\tilde{\sigma}$ is contragredient to σ.

Question 6. *Suppose G is quasi-split over the local field F and splits over the Galois extension K. Let $\hat{U}_F$ be a maximal compact subgroup of $\hat{G}_F$. Let K' be a Galois extension of F which contains K and let φ be a homomorphism of $W_{K'/F}$ onto $\hat{U}_F$ which makes*

$$\begin{array}{ccc} W_{K'/F} & \longrightarrow & \mathfrak{G}(K'/F)\\ \downarrow{\scriptstyle\varphi} & & \downarrow\\ \hat{U}_F & \longrightarrow & \mathfrak{G}(K/F)\end{array}$$

commutative. Is there an irreducible unitary representation $\pi(\varphi)$ of G_F such that, for every representation σ of $\hat{G}_F$, $L(s,\sigma,\pi(\varphi))=L(s,\sigma\circ\varphi)$ and $\varepsilon(s,\sigma,\pi(\varphi),\psi_F)=\varepsilon(s,\sigma\circ\varphi,\psi_F)$?

Changing φ by an inner automorphism $\hat{U}_F$ will not change $\pi(\varphi)$, or at least not its equivalence class. If F is non-archimedean and K'/F is unramified, the composition of v, the valuation on F, and $\tau_{K'/F}$ defines a homomorphism ω of $W_{K/F}$ onto $\mathbb{Z}$. If $u=t\times\sigma_F$ belongs to $\hat{U}_F$, we could define φ by

$$\varphi(w)=u^{\omega(w)}\ .$$

Then $\pi(\varphi)$ would be the representation associated to the homomorphism χ_t of the Hecke algebra into $\mathbb{C}$.

We can also ask the question globally.

Question 7. *Suppose G is quasi-split over the global field F and splits over K. Let K' be a Galois extension of F containing K and let φ be a homomorphism of $W_{K'/F}$ into $\hat{U}_F$ which makes*

$$\begin{array}{ccc} W_{K'/F} & \longrightarrow & \mathfrak{G}(K'/F) \\ \downarrow{\scriptstyle\varphi} & & \downarrow \\ \hat{U}_F & \longrightarrow & \mathfrak{G}(K/F) \end{array}$$

commutative. If $\mathfrak{P}'$ is a prime of K' and $\mathfrak{p} = \mathfrak{P}' \cap F$, then $\varphi_{\mathfrak{p}} = \varphi \circ \alpha_{\mathfrak{p}}$ takes $W_{K'_{\mathfrak{P}'}/F_{\mathfrak{p}}}$ into $\hat{U}_{F_{\mathfrak{p}}}$. If $\pi(\varphi) = \prod_{\mathfrak{p}} \otimes \pi(\varphi_{\mathfrak{p}})$, does $\pi(\varphi)$ occur in $L^2(G_F \setminus G_{\mathbb{A}(F)})$?

Both questions have affirmative answers if G is abelian [10] and the correspondence $\varphi \to \pi(\varphi)$ is surjective. In this case our L-functions are all generalized Artin L-functions. If $G = GL(2)$ and $K = F$, it appears that the Hecke theory can be used to give an affirmative answer to both questions if it is assumed that certain of the generalized Artin L-functions have the expected analytic properties. If all goes well, the details will appear in [6].

I would like very much to end this series of questions with some reasonably precise questions about the relation of the L-functions of this paper to those associated to non-singular algebraic varieties. Unfortunately, I am not competent to do so. Since it may be of interest, I would like to ask one question about the L-functions associated to elliptic curves. If C is defined over a local field F of characteristic zero, I am going to associate to it a representation $\pi(C/F)$ of $GL(2,F)$. If C is defined over a global field F which is also characteristic zero, then for each prime $\mathfrak{p}$, $\pi(C/F_{\mathfrak{p}})$ is defined. Does $\pi = \prod_{\mathfrak{p}} \otimes \pi(C/F_{\mathfrak{p}})$ occur in $L^2(GL(2,F)\setminus GL(2,\mathbb{A}(F)))$? If so, $L(s,\sigma,\pi)$, with σ the standard representation of $GL(2,\mathbb{C})$, whose analytic properties are known [6] will be one of the L-functions associated to the elliptic curve. There are examples on which the question can be tested. I hope to comment on them in [6].

To define $\pi(C/F)$, I use the result of Serre [14]. Suppose that F is non-archimedean and the j-invariant of C is integral. Take any prime ℓ different from the characteristic of the residue field and consider the ℓ-adic representation. There is a finite Galois extension K of F such that, if A is the maximal unramified extension of K, the ℓ-adic representation can be regarded as a representation of $\mathfrak{G}(A/F)$. There is a homomorphism of $W_{K/F}$ into $\mathfrak{G}(A/F)$. The ℓ-adic representation of $\mathfrak{G}(A/F)$ determines a representation φ of $W_{K/F}$ in $GL(2,R)$, where R is a finitely generated subfield of the ℓ-adic field $\mathbb{Q}_\ell$. Let σ be an isomorphism of R with a subfield fo $\mathbb{C}$. Then

$$\psi\colon\ w \to |\tau_{K/F}(w)|^{1/2}\varphi^{\sigma}(w)$$

is a representation of $W_{K/F}$ in a maximal compact subgroup of $GL(2,\mathbb{C})$. Let $\pi(C/F)$ be the representation $\pi(\psi)$ of Question 6. If C has good reduction, the class of ψ is independent of ℓ and σ. I do not know if this is so in general. It does not matter, because we do not demand that $\pi(C/F)$ be uniquely determined by C.

If the j-invariant is not integral, the ℓ-adic representation can be put in the form

$$\sigma \rightarrow \begin{pmatrix} \chi_1(\sigma) & * \\ 0 & \chi_2(\sigma) \end{pmatrix}$$

where χ_1 and χ_2 are two representations of the Galois group of the algebraic closure of F in the multiplicative group of $\mathbb{Q}_\ell$. If A is the maximal abelian extension of F, then χ_1 and χ_2 may be regarded as representations of $\mathfrak{G}(A/F)$. There is a canonical map of F^*, the multiplicative group of F, into $\mathfrak{G}(A/F)$. χ_1 and χ_2 thus define characters μ_1 and μ_2 of F^*. μ_1 and μ_2 take values in $\mathbb{Q}^*$ and $\mu_1\mu_2(x) = \mu_1\mu_2^{-1}(x) = |x|^{-1}$. In, for example, [6], there is associated to the pair of generalized characters $x \rightarrow |x|^{1/2}\mu_1(x)$ and $x \rightarrow |x|^{1/2}\mu_2(x)$ a unitary representation of $GL(2,F)$, a so-called special representation. This we take as $\pi(C/F)$.

If F is $\mathbb{C}$, take $\pi(C/F)$ to be the representation of $GL(2,\mathbb{C})$ associated to the map

$$s \rightarrow \begin{pmatrix} \frac{z}{|z|} & 0 \\ 0 & \frac{\bar{z}}{|z|} \end{pmatrix}$$

of $\mathbb{C}^* = W_{\mathbb{C}/\mathbb{C}}$ into $GL(2,\mathbb{C})$ by Question 6. $\mathbb{C}^*$ is of index two in $W_{\mathbb{C}/\mathbb{R}}$. The representation of $W_{\mathbb{C}/\mathbb{R}}$ induced from the character $z \rightarrow \frac{1}{|z|}$ of $\mathbb{C}^*$ has degree 2. If $F = \mathbb{R}$, let $\pi(C/F)$ be the representation of $GL(2,\mathbb{R})$ associated to the induced representation by Question 6.

8. I would like to finish up with some comments on the relation of the L-functions of this paper to Ramanujan's conjecture and its generalizations. Suppose $\pi = \prod \otimes \pi_\mathfrak{p}$ occurs in the space of cusp forms. The most general form of Ramanujan's conjecture would be that for all $\mathfrak{p}$ the character of $\pi_\mathfrak{p}$ is a tempered distribution [5]. However, neither the notion of a character nor that of a tempered distribution has been defined for non-archimedean fields. A weaker question is whether or not at all unramified non-archimedean primes the conjugacy class in $\hat{G}_F$ associated to $\pi_\mathfrak{p}$ meets $\hat{U}_F$ (cf. [13]). If this is so, it should be reflected in the behavior of the L-functions.

Suppose, to remove all ramification, that G is a Chevalley group and that $K = F = \mathbb{Q}$. Suppose also that each π_p is unramified. If p is non-archimedean, there is associated to π_p a conjugacy class $\{t_p\}$ in $G_\mathbb{Q}$. We may take t_p in $\hat{T}$. The conjecture is that, for all λ in $\hat{L}$,

$$|\lambda(t_p)| = 1 \,.$$

Since there is no ramification at ∞, one can, as in [9], associate to π_∞ a semi-simple conjugacy class $\{X_\infty\}$ in the Lie algebra of $\hat{G}_{\mathbb{Q}}$. We may take X_∞ in the Lie algebra of $\hat{T}$. The conjecture at ∞ is that, for λ in $\hat{L}$,

$$Re\,\lambda(X_\infty) = 0\ .$$

If σ is a complex analytic representation of $\hat{G}_{\mathbb{Q}}$, let $m(\lambda)$ be the multiplicity with which λ occurs in σ. Then

$$L(s,\sigma,\pi) = \prod_\lambda \left\{ \pi^{\frac{-s(s+\lambda(\infty))}{2}} \Gamma\left(\frac{2+\lambda(X_\infty)}{2}\right) \prod_p \frac{1}{1-\frac{\lambda(t_p)}{p^s}} \right\}^{m(\lambda)} .$$

If the conjecture is true, $L(s,\sigma,\pi)$ is analytic to the right of $Re\, s = 1$ for all σ.

Let F be any non-archimedean local field and G any quasi-split group over F which splits over an unramified extension field. If f belongs to $C_c(G_F,U_F)$, let $f^*(g) = \bar{f}(g^{-1})$. If $\hat{f}$ and $\hat{f}^*$ are the images of f and f^* in $\Lambda^0(M)$, then $\hat{f}^*(\lambda)$ is the complex conjugate of $\hat{f}(-\lambda)$. If t belongs to $\hat{T}$, define t^* by the condition that $\lambda(t^*) = \overline{\lambda(t^{-1})}$ for all λ in $\hat{L}$. The complex conjugate of $\chi_t(t^*)$ is

$$\sum \hat{f}(-\lambda)\overline{\lambda(t)} = \sum \hat{f}(\lambda)\lambda(t^*) = \chi_{t^*}(f)\ .$$

If χ_t is the homomorphism associated to a unitary representation, then $\chi_t(f^*)$ is the complex conjugate of $\chi_t(t)$ for all f so that $t \times \sigma_F$ is conjugate to $t^* \times \sigma_F$ and for any representation ρ of $\hat{G}_F$, the complex conjugate of trace $\rho(t\times\sigma_F)$ is trace $\tilde{\rho}(t\times\sigma_F)$ if $\tilde{\rho}$ is the contragredient of ρ. In the case under consideration, when $K = F$ this means that trace $\rho(t_p)$ is the complex conjugate of trace $\tilde{\rho}(t_p)$. A similar argument can be applied at the infinite prime so show that the eigenvalues of $\rho(X_\infty)$ are the complex conjugates of the eigenvalues of $\tilde{\rho}(X_\infty)$.

Suppose $L(s,\sigma,\pi)$ is analytic to the right of $Re\, s = 1$ for all σ. Since the Γ-function has no zeros,

$$\prod_\lambda \left\{ \prod_p \frac{1}{1-\frac{\lambda(t_p)}{p^s}} \right\}^{m(\lambda)} \tag{C}$$

is also. Let σ be $\rho\otimes\tilde{\rho}$. Then the logarithm of this Dirichlet series is

$$\sum_p \sum_{n=1}^{\infty} \frac{1}{n} \frac{\text{trace}\ \sigma^n(t_p)}{p^{ns}} .$$

Since

$$\text{trace}\ \sigma^n(t_p) = \text{trace}\ \rho^n(t_p)\text{trace}\ \tilde{\rho}^n(t_p) = |\text{trace}\ \rho^n(t_p)|^2$$

the series for the logarithm has positive coefficients. Thus, the original series does too. By Landau's theorem, it converges absolutely for $Re\, s > 1$ and so does the series for its logarithm. In particular,

$$\det\left(1 - \frac{\sigma(t_p)}{p^s}\right)$$

does not vanish for $Re\, s > 1$ so that the eigenvalues of $\sigma(t_p)$ are all less than or equal to p in absolute value. If λ is a weight, choose ρ such that $m\lambda$ occurs in ρ. Then $(m\lambda)(t_p) = \lambda(t_p)^m$ is an eigenvalue of $\rho(t_p)$ and $\overline{\lambda(t_p)}^m$ is an eigenvalue of $\tilde{\rho}$, so that $|\lambda(t_p)|^{2m}$ is an eigenvalue of σ and

$$|\lambda(t_p)| \leq p^{\frac{1}{2m}}$$

for all m and all λ. Thus, $|\lambda(t_p)| \leq 1$ for all λ. Replacing λ by $-\lambda$, we see that $|\lambda(t_p)| = 1$ for all λ. Since the function defined by (C) cannot vanish for $Re\, s > 1$ when $\sigma = \rho \otimes \tilde{\rho}$, the function

$$\prod_{\lambda} \Gamma\left(\frac{2 + \lambda(X_\infty)}{2}\right)^{m(\lambda)}$$

must be analytic for $Re\, s > 1$. This implies that

$$Re\, \lambda(X_\infty) \geq -1$$

if $m(\lambda) > 0$. The same argument as before leads to the conclusion that $Re\, \lambda(X_\infty) = 0$ for all λ.

Granted the generalizations of Ramanujan's conjecture, one can ask about the asymptotic distribution of the conjugacy classes $\{t_p\}$. I can make no guesses about the answer. In general, it is not possible to compute the eigenvalues of the Hecke operators in an elementary fashion. Thus, Question 7 cannot be expected to lead by itself to elementary reciprocity laws. However, when the groups G_{F_p} at the infinite primes are abelian or compact, these eigenvalues should have an elementary meaning. Thus, Question 7, together with some information on the range of the correspondences of Question 3, may eventually lead to elementary, but extremely complicated, reciprocity laws. At the present it is impossible even to speculate.

References

1. F. Bruhat and J. Tits, *Groupes algébriques simples sur un corps local*, Driebergen Conference on Local Fields, Springer-Verlag, 1967.
2. K. Doi and H. Naganuma, *On the algebraic curves uniformized by arithmetical automorphic functions*, Ann. of Math. vol. 86 (1967).
3. M. Eichler, *Quadratische Formen und Modulfunktion*, Acta Arith. vol. 4 (1958).
4. F. Gantmacher, *Canonical representation of automorphisms of a complex semi-simple Lie group*, Mat. Sb. vol. 47 (1939).
5. Harish-Chandra, *Discrete Series for Semi-Simple Lie Groups* II, Acta Math. vol. 116 (1966).
6. H. Jacquet and R. P. Langlands, *Automorphic Forms on* $GL(2)$, in preparation.
7. N. Jacobson, *Lie Algebras*, Interscience, 1962.
8. B. Kostant, *Lie Algebraic Cohomology and the Generalized Borel-Weil Theorem*, Ann. of Math. vol. 74 (1961).
9. R. P. Langlands, *Euler Products*, Lecture Notes, Yale University (1967).
10. ———, *Representations of Abelian Algebraic Groups*, Notes, Yale University (1968).
11. ———, *On the Functional Equation of the Artin L-functions*, in preparation.
12. I. Satake, *Theory of Spherical Functions on Reductive Algebraic Groups over* $\mathfrak{p}$*-adic Fields*, Publ. Math. No. 18, I.H.E.S (1963).
13. ———, *Spherical Functions and Ramanujan Conjecture*, in *Algebraic Groups and Discontinuous Subgroups*, Amer. Math. Soc. (1966).
14. J. P. Serre, *Groupes de Lie* ℓ*-adic attachés aux courbes elliptiques*, Colloque de Clermont-Ferrand (1964).
15. J. A. Shalika and S. Tanaka, *On an Explicit Construction of a Certain Class of Automorphic Forms*, Notes, Institute for Advanced Study (1968).
16. H. Shimizu, *On Zeta functions of quaternion algebras*, Ann. of Math. vol. 81 (1966).
17. A. Weil, *Sur la Theorie du Corps de Classes*, Jour. Math. Soc. Japan vol. 3 (1951).

Jean Leray Works

Leray's Selected Papers were published in 1997 jointly by Springer Verlag and the Société Mathématique de France in three volumes; each volume is devoted to a single field and starts with an important analysis of Leray's influence written by a leading mathematician in the field: Armand Borel for the volume on Topology, Peter Lax for the volume on PDE, Guennadi Henkin for the volume on Complex Analysis. The deepness and the wideness of the scope of Leray's works, are amazing; in order to convey these facts we have selected a large sample of papers spanning more than fifty years, at the price of restricting three papers to their flat introduction; also each of the nine selected Comptes Rendus Notes has been prolongated by a subsequent large research paper.

The paper [1933*a*] shows that initial value problem for the Navier–Stokes evolution of a viscous fluid has a weak solution, called by Leray *turbulent solution*, defined for all time, this turbulent solution being smooth for a set of regular time of full Lebesgue measure; at the irregular time branching in the evolution can appeared. The paper is based on a virtuose manipulation of energy dissipation inequalities together with the modern ideas which will constitute fifteen years later the theory of distributions.

[1933*b*] is an epoch making paper for the existence problems in Functional Analysis: the Mathematical Review enumerates nowadays more than six hundred articles based on this Leray–Schauder theory. Before this work the single general tool available for existence was the Picard theory of contracting map which gives at the same time existence and uniqueness; Leray–Schauder gives existence in case where there is no uniqueness. The underlying idea is to define a *Brouwer type degree* in functional spaces and to show the invariance of the degree by a continuous deformation which do not interfere with the boundary of a sufficient large ball; this non-interference can be insured by simple uniform majoration of the family of deformed maps: then *the existence of a solution inside the ball for a single value of the deformation parameter implies the existence of a solution for all values of this parameter.* By using Leray–Schauder Theorem, Leray solves the free boundary problem associated with the determination of the *wake* of a perfect fluid around an obstacle (see in particular [1934]); he also proves existence of solutions for the Dirichlet problem for the Monge–Ampère equation.

During the World War, Jean Leray was kept in a prisoner camp located in Austria for five years; without any contact with the scientific world, without any library available, suffering from severe starving conditions, and being cut for months from any news from his family, Leray assigned himself to the task of constructing a theory of fixed point valid for topological space free from any linearity assumptions. In such adverse conditions, almost every mathematician would have failed to produce lasting results. On the contrary, Leray took advantage and rebuilt by himself the Algebraic Topology from scratch and produced a monument of shinning originality. Before the war, Elie Cartan commissioned Leray in writing one of his main books: *Le Repère mobile et la théorie des groupes continus* (*Paris 1935*)*;* from this time Leray became proefficient in exterior calculus and in its use in differential geometry. Leray took as prerequisite to his reconstruction of Algebraic Topology that the *wedge operation must be preserved*, prerequisite which brought him to work on *cohomology ring with the cup product.* He wanted to localize the cohomology ring by taking its values in a local ring of coefficients and this prerequisite led him to the concept *sheave.* Then he associated [1946*a*] to a map the cohomology of the target space with values in the sheave generated by the cohomology of the fiber. Leray discovered *spectral sequences* [1946*b*] which gave a construction making possible the recovery of the cohomology of the source. Applying the spectral sequence to the projection map, Leray computed in [1946*c*] the *cohomology of a Fiber Bundle.* Using his proefficiency in Lie group theory, Leray computed the cohomology ring of homogeneous spaces [1949]. These ideas started what some German colleagues kindly called the "French Revolution of 1950": Henri Cartan introduced sheaves in the theory of functions of several complex variables; Jean-Pierre Serre evaluated the homotopy groups of spheres by using an extended spectral sequence.

The wave equation in R^3 has a fundamental solution which is carried by the light cone; this mathematical statement has for consequence the possibility to hear at distance without interference the sound of an orchestra. Hadamard raised the interesting problem: *characterize the constant coefficient operators for which the fundamental solution satisfies such lacuna property.* Leray obtained in [1962] an explicit integral formula representing the fundamental solution of any constant coefficient operator, formula which led in 1970 Bott–Gärding to a solution of the Hadamard problem. To obtain this result Leray constructed [1958] from scratch a *Residue theory in several complex variables,* theory based on cohomological considerations. He acted also as an analyst by obtaining [1959] a wide range of *Cauchy–Leray integral formulae.* Ten years after the sheave revolution, Leray ideas renewed once more the theory of functions of several complex variables by making available integral formulae leading to quantitative estimates out of the realm of sheave theory.

In [1976] propagation problem motivated in particular by magnetohydrodynamics was discussed. At the age of seventy-nine Leray contributed to [1985], which described with an incredible precision the domain of analytic continuation of the solution of a linear Cauchy problem.

Receiving at the Lincei Academy the Feltrinelli Prize, Jean Leray expressed himself in the following terms: "*all the different fields of mathematics are as inseparable as the different part of a living organism; as a living organism mathematics has to be permanently recreated; each generation must reconstruct it wider, larger and more beautiful. The death of mathematical research would be the death of mathematical thinking which constitutes the structure of scientific language itself and by consequence the death of our scientific civilization. Therefore we must transmit to our children strength of character, moral values and drive towards an endeavouring life.*"

Paul Malliavin

Bibliography

[1933*a*] *Sur le mouvement d'un fluide visqueux emplissant l'espace: Acta Mathematica 63* (*Introduction only*).

[1933*b*] (Avec J. Schauder), *Topologie et équations fonctionnelles: C.R. Acad. Sci. Paris 197, 115–117.*

[1934] (Avec A. Weinstein), *Sur un problème de représentation conforme posé par la théorie de Helmholtz: C.R. Acad. Sci. Paris 198, 430–432.*

[1946*a*] *L'anneau d'homologie d'une représentation: C.R. Acad. Sci. Paris 222, 1366–1368.*

[1946*b*] *Structure de l'anneau d'homologie d'une représentation: C.R. Acad. Sci. Paris 222, 1419–1422.*

[1946*c*] *Propriétés de l'anneau d'homologie de la projection d'un espace fibré sur sa base: C.R. Acad. Sci. Paris 223, 395–397.*

[1949] *Détermination, dans les cas non exceptionnels, de l'anneau de cohomologie de l'espace homogène quotient d'un groupe de Lie compact par un sous-groupe de même rang: C.R. Acad. Sci. Paris 228, 1902–1904.*

[1962] *Prolongement de la transformation de Laplace: Proc. Int. Cong. of Mathematicians, Stockholm, 360–367.*

[1958] *La théorie des résidus sur une variété analytique complexe: C.R. Acad. Sci. Paris 247, 2253–2257.*

[1959] *Le calcul différentiel et intégral sur une variété analytique complexe: C.R. Acad. Sci. Paris 248, 22–28.*

[1976] (en collaboration avec Y. Hamada et C. Wagschal), *Systèmes d'équations aux dérivées partielles à caractéristiques multiples: probléme de Cauchy ramifié; hyperbolicité partielle: J. Math. Pures App. 55 (Introduction only).*

[1985] (en collaboration avec Y. Hamada et A. Takeuchi), *Prolongements analytiques de la solution du problème de Cauchy linéaire: J. Math. Pures App. 64 (Introduction only).*

Biographie de Jean Leray

• Né le 7 novembre 1906 à Chantenay (Loire-Atlantique). Épouse en 1932 Marguerite Trumier ; ses parents ainsi que ceux de sa femme étaient tous les quatre instituteurs à l'école publique de Chantenay. Marguerite Leray, tout en menant une carrière complète de professeur de mathématiques dans les lycées, éduquera trois enfants : Jean-Claude (1933) ingénieur au Corps des Ponts ; Françoise (1947) directeur de recherches en biologie à l'Hôpital Henri Mondor de Créteil ; Denis (1949) médecin.

• Ecole normale supérieure (1926-1929) ; Docteur ès sciences (1933) ; Chargé de recherche (1933) ; Professeur : Université de Nancy (1938-1939) ; Université de Paris (1945-1947) ; Collège de France (1947-1978).

• Prix internationaux : Malaxa (Roumanie) 1938, partagé avec J. Schauder ; Prix Feltrinelli (Accademia dei Lincei) 1971 ; Prix Wolf (Israël) 1979 ; Médaille Lomonosov (Académie des Sciences d'U.R.S.S.) 1985.

• Académie des Sciences de Paris : présenté par Henri Lebesgue pour un poste de Correspondant en Mathématiques Pures (1938) ; élu Membre en section de Mécanique sur rapport de Henri Villat (1953).

• Académies étrangères : Accademia delle Scienze di Torino (1958) ; American Academy of Arts and Sciences (1959) ; American Philosophical Society (1959) ; Membre d'honneur de la Société Mathématique Suisse (1960) ; Académie Royale de Belgique (1962) ; Akademie der Wissenschaften in Göttingen (1963) ; National Academy of Sciences, Washington (1965) ; Académie des Sciences d'URSS (1966) ; Accademia di Scienze, Lettere e Arti di Palermo (1967) ; Istituto Lombardo, Accademia di Scienze e Lettere (1974) ; Accademia Nazionale delle Scienze Detta dei XL (1975) ; Académie Polonaise des Sciences (1977) ; Accademia Nazionale dei Lincei (1980) ; The Royal Society of London (1983).

• Officier de réserve, il rejoint à la mobilisation son affectation dans l'artillerie antiaérienne. Après avoir combattu jusqu'au bout dans la sanglante bataille de mai-juin 1940, il est fait prisonnier de guerre et restera interné dans un camp en Autriche jusqu'en avril 1945. Voulant éviter à tout prix de contribuer à l'effort industriel ennemi, il abandonne la mécanique des fluides pour se lancer dans la topologie algébrique, ceci malgré un environnement matériel très précaire. Il continue sa collaboration commencée avant guerre au Zentralblatt für Mathematik.
Commandeur de la Légion d'Honneur.

• Retrouve en 1945, dans un camp de réfugiés, la fille unique de son ami Juliusz Schauder, orpheline à neuf ans à la suite des massacres nazis ; la fait guérir dans un hôpital parisien de la grave affection pulmonaire qu'elle avait contractée en se cachant dans les égouts de Varsovie.

• Professeur à temps partiel à l'Institute for Advanced Study, Princeton, USA (1952-1961) ; propose pendant cette période à Marston Morse, Director of the School of Mathematics, de nombreuses invitations de jeunes mathématiciens français.

Paul Malliavin

Une autobiographie de Jean Leray a paru dans "Hommes de Science" Hermann ed., Paris 1990, pages 160–169.

Sur le mouvement d'un fluide visqueux emplissant l'espace

Acta Math. **63** (1934) 193–248

Introduction.[2]

I. *La théorie de la viscosité* conduit à admettre que les mouvements des liquides visqueux sont régis par les équations de Navier; il est nécessaire de justifier a posteriori cette hypothèse en établissant *le théorème d'existence suivant:* il existe une solution des équations de Navier qui correspond à un état de vitesse donné arbitrairement à l'instant initial. C'est ce qu'a cherché à démontrer M. Oseen[3]; il n'a réussi à établir l'existence d'une telle solution que pour une durée peut-être très brève succédant à l'instant initial. On peut vérifier en outre que l'énergie cinétique totale du liquide reste bornée[4]; mais il ne semble pas possible de déduire de ce fait que le mouvement lui-même reste régulier; j'ai même indiqué une raison qui me fait croire à l'existence de mouvements devenant irréguliers au bout d'un temps fini[5]; je n'ai malheureusement pas réussi à forger un exemple d'une telle singularité.

---

[1] Ce mémoire a été résumé dans une note parue aux Comptes rendus de l'Académie des Sciences, le 20 février 1933, T. 196 p. 527.

[2] Les pages 59—63 de ma Thèse (Journ. de Math. 12, 1933) annoncent ce mémoire et en complètent l'introduction.

[3] Voir Hydrodynamik (Leipzig, 1927), § 7, p. 66. Acta mathematica T. 34. Arkiv för matematik, astronomi och fysik. Bd. 6, 1910. Nova acta reg. soc. scient. Upsaliensis Ser. IV, Vol. 4, 1917.

[4] l. c. 2, p. 59—60.

[5] l. c. 2, p. 60—61. Je reviens sur ce sujet au § 20 du présent travail (p. 224).

Il n'est pas paradoxal de supposer en effet que la cause qui régularise le mouvement — la dissipation de l'énergie — ne suffise pas à maintenir bornées et continues les dérivées secondes des composantes de la vitesse par rapport aux coordonnées; or la théorie de Navier suppose ces dérivées secondes bornées et continues; M. Oseen lui-même a déjà insisté sur le caractère peu naturel de cette hypothèse; il a montré en même temps comment le fait que le mouvement obéit aux lois de la mécanique peut s'exprimer à l'aide d'équations intégro-différentielles[1], où figurent seulement les composantes de la vitesse et leurs dérivées premières par rapport aux coordonnées spatiales. Au cours du présent travail je considère justement un système de relations[2] qui équivalent aux équations intégro-différentielles de M. Oseen, complétées par une inégalité exprimant la dissipation de l'énergie. Ces relations se déduisent d'ailleurs des équations de Navier à l'aide d'intégrations par parties qui font disparaître les dérivées d'ordres les plus élevés. Et, si je n'ai pu réussir à établir le théorème d'existence énoncé plus haut, j'ai pu néammoins démontrer le suivant[3]: les relations en question possèdent toujours *au moins une solution* qui correspond à un état de vitesse donné initialement et *qui est définie pour une durée illimitée*, dont l'origine est l'instant initial. Peut-être cette solution est-elle trop peu régulière pour posséder à tout instant des dérivées secondes bornées; alors elle n'est pas, au sens propre du terme, une solution des équations de Navier; je propose de dire qu'elle en constitue »*une solution turbulente*«.[4]

Il est d'ailleurs bien remarquable que chaque solution turbulente satisfait effectivement les équations de Navier proprement dites, sauf à certaines époques d'irrégularité; ces époques constituent un ensemble fermé de mesure nulle; à ces époques sont seules vérifiées certaines conditions de continuité extrêmement

[1] Oseen, Hydrodynamik, § 6, équation (1).

[2] Voir relations (5. 15), p. 240.

[3] Voir p. 241.

[4] Je me permets de citer le passage suivant de M. Oseen (Hydrodynamik): »A un autre point de vue encore il semble valoir la peine de soumettre à une étude attentive les singularités du mouvement d'un liquide visqueux. S'il peut surgir des singularités, il nous faut manifestement distinguer deux espèces de mouvements d'un liquide visqueux, les mouvements réguliers, c'est-à-dire les mouvements sans singularité, et les mouvements irréguliers, c'est-à-dire les mouvements avec singularité. Or on distingue d'autre part en Hydraulique deux sortes de mouvements, les mouvements laminaires et les mouvements turbulents. On est dès lors tenté de présumer que les mouvements laminaires fournis par les expériences sont identiques aux mouvements réguliers théoriques et que les mouvements turbulents expérimentaux s'identifient aux mouvements irréguliers théoriques. Cette présomption répond-elle à la réalité? Seules des recherches ultérieures pourront en décider«.

larges. Une *solution turbulente* a donc *la structure* suivante: elle se compose *d'une succession de solutions régulières*.

Si j'avais réussi à construire des solutions des équations de Navier qui deviennent irrégulières, j'aurais le droit[1] d'affirmer qu'il existe effectivement des solutions turbulentes ne se réduisant pas, tout simplement, à des solutions régulières. Même si cette proposition était fausse, la notion de solution turbulente, qui n'aurait dès lors plus à jouer aucun rôle dans l'étude des liquides visqueux, ne perdrait pas son intérêt: il doit bien se présenter *des problèmes* de Physique mathématique pour lesquels *les causes physiques de régularité* ne suffisent pas à justifier *les hypothèses de régularité faites lors de la mise en équation;* à ces problèmes peuvent alors s'appliquer des considérations semblables à celles que j'expose ici.

Signalons enfin les deux faits suivants:

Rien ne permet d'affirmer l'unicité de la solution turbulente qui correspond à un état initial donné. (Voir toutefois Compléments 1°, p. 245; § 33).

La solution qui correspond à un état initial suffisamment voisin du repos ne devient jamais irrégulière. (Voir les cas de régularité que signalent les § 21 et 22, p. 226 et 227)

II. Le travail présent concerne les liquides visqueux illimités. Les conclusions en sont extrêmement analogues à celles d'un autre mémoire[2] que j'ai consacré aux mouvements plans des liquides visqueux enfermés dans des parois fixes convexes; ceci autorise à croire que ces conclusions s'étendent au cas général d'un liquide visqueux à deux ou trois dimensions que limitent des parois quelconques (même variables).

L'absence de parois introduit certes quelques complications concernant l'allure à l'infini des fonctions inconnues[3], mais simplifie beaucoup l'exposé et met mieux en lumière les difficultés essentielles; le rôle important que joue l'homogénéité des formules est plus évident; (les équations aux dimensions permettent de prévoir a priori presque toutes les inégalités que nous écrirons).

---

[1] En vertu du théorème d'existence du § 31 (p. 241) et du théorème d'unicité du § 18 (p. 222).

[2] Journal de Mathématiques, T. 13, 1934.

[3] Les conditions à l'infini par lesquelles nous caractérisons celles des équations de Navier que nous nommons régulières diffèrent essentiellement des conditions qu'emploie M. Oseen.

Rappelons que nous avons déjà traité le cas des mouvements plans illimités[1]: il est assez spécial[2]; la régularité du mouvement est alors assurée.

Sommaire du mémoire.

Le chapitre I rappelle au Lecteur une série de propositions d'Analyse, qui sont importantes, mais qu'on ne peut pas toutes considérer comme classiques.

Le chapitre II établit diverses inégalités préliminaires, aisément déduites des propriétés que possède la solution fondamentale de M. Oseen.

Le chapitre III applique ces inégalités à l'étude des solutions régulières des équations de Navier.

Le chapitre IV énonce diverses propriétés des solutions régulières, dont fera usage le chapitre VI.

Le chapitre V établit qu'à tout état initial correspond au moins une solution turbulente, qui est définie pendant une durée illimitée. La démonstration de ce *théorème d'existence* repose sur le *principe* suivant: On n'aborde pas directement le problème posé, qui est de résoudre les équations de Navier; mais on traite d'abord un problème voisin dont on peut s'assurer qu'il admet toujours une solution régulière, définie pendant une durée illimitée; on fait tendre ce problème voisin vers le problème posé et l'on construit la limite (ou les limites) de sa solution. Il existe bien une façon élémentaire d'appliquer ce principe: c'est celle qu'utilise mon étude des mouvements plans des liquides visqueux limités par des parois; mais elle est intimement liée à cette structure des solutions turbulentes que nous avons précédemment signalée; elle ne s'appliquerait pas si cette structure n'était pas assurée. Nous procèderons ici d'une autre façon, dont la portée est vraisemblablement plus grande, qui justifie mieux la notion de solution turbulente, mais qui fait appel à quelques théorèmes peu usuels cités au chapitre I.

Le chapitre VI étudie la structure des solutions turbulentes.

---

[1] These, Journal de Mathématiques 12, 1933; chapitre IV p. 64—82. (On peut donner une variante intéressante au procédé que nous y employons en utilisant la notion d'état initial semi-régulier qu'introduit le mémoire présent.)

[2] On peut dans ce cas baser l'étude du problème sur la propriété que possede alors le maximum du tourbillon à un instant donné d'être une fonction décroissante du temps. (Voir: Comptes rendus de l'Académie des Sciences, T. 194; p. 1893; 30 mai 1932). — M. Wolibner a lui aussi fait cette remarque.

ANALYSE MATHÉMATIQUE. — *Topologie et équations fonctionnelles.* Note de MM. **Jean Leray** et **Jules Schauder**, présentée par M. Henri Villat.

1. Soit un espace abstrait, $\mathcal{E}$, normé, linéaire et complet ([1]); un ensemble ouvert ([2]), ω de $\mathcal{E}$; enfin une transformation fonctionnelle,

$$y = x - \mathcal{F}(x) = \Phi(x) \tag{1}$$

définie sur $\overline{\omega}$; $\mathcal{F}(x)$ *n'est pas supposée linéaire*, mais est supposée *complètement continue* ([3]) (vollstetig), et ne doit prendre que des valeurs appartenant à $\mathcal{E}$. Nous avons pu définir, en accord avec les célèbres travaux de M. Brouwer ([4]), le degré topologique, $d[\Phi, \omega, b]$ de la transformation Φ en un point b étranger à $\Phi(\omega')$, en sorte que les propriétés bien connues de ce degré continuent à être exactes. Indiquons brièvement notre définition : nous supposons b en o; nous posons

$$h = \text{minimum de la distance de o à } \Phi(\omega');$$

([1]) Au sens de M. Banach (cf. *Fund. Math.*, 3, 1922, p. 133-181).

([2]) Nous désignerons par ω' la frontière de ω; par $\overline{\omega} = \omega + \omega'$ son ensemble de fermeture.

([3]) C'est-à-dire $\mathcal{F}(x)$ est continue et $\mathcal{F}(x)$ transforme tout ensemble borné en un ensemble compact.

([4]) *Math. Annalen*, 71, 1912, p. 97-115.

Reprinted from C. R. Acad. Sci. Paris, Vol. 197 (1933), pp. 115–117.

nous envisageons l'une des transformations fonctionnelles $\mathcal{F}_h(x)$ qui approchent $\mathcal{F}(x)$ à h près et dont toutes les valeurs font partie d'un même sous-ensemble linéaire de $\mathcal{E}$, à nombre fini de dimensions n; soit ω_n l'intersection, supposée non vide, de ω par ce sous-ensemble; soit $\Phi_h(x) = x - \mathcal{F}_h(x)$; nous posons

$$d[\Phi, \omega, o] = d[\Phi_h, \omega_n, o]$$

(la définition de ce deuxième degré résulte des travaux de M. Brouwer), en démontrant que le choix de Φ_h et ω_n n'affecte pas la valeur de $d[\Phi, \omega, o]$.

2. Nous nommons *indice total* des solutions de l'équation

$$x - \mathcal{F}(x) = o \tag{2}$$

qui sont contenues dans ω le degré en o de la transformation (1).

Théorème. — Supposons que l'équation (2) dépende continûment ([1]) d'un paramètre k variable sur un segment K de l'axe réel. Soit un ensemble ouvert et borné Ω de l'espace ([2]) $[\mathcal{E}_x \mathrm{K}]$, tel que $\mathcal{F}$ soit définie sur $\overline{\Omega}$ et que Ω' ne contienne aucune solution de (1). Alors l'indice total ([3]) des solutions intérieures à Ω est le même pour toutes les valeurs de k. Il en résulte que *l'existence d'au moins une solution de* (2) est assurée pour chaque valeur de k *quand on connaît un point de* K *où cet indice total diffère de zéro.*

3. Le champ *d'applications de ce théorème d'existence* est vaste, il contient en particulier les problèmes aux limites relatifs aux équations aux dérivées partielles du second ordre et du type elliptique : nous avons réussi à généraliser des théorèmes d'existence bien connus de M. S. Bernstein ([4]). Bornons-nous ici au problème de Dirichlet concernant l'équation quasi linéaire

---

([1]) $\mathcal{F}(x)$, qui est complètement continue en x, est supposée uniformément continue en k.

([2]) $[\mathcal{E} \times \mathrm{K}]$, espace produit des espaces $\mathcal{E}$ et K, est l'ensemble des couples (x, k) d'éléments x de $\mathcal{E}$, k de K. Un point γ intérieur à un ensemble E de $[\mathcal{E} \times \mathrm{K}]$ est un point tel que tout point de $[\mathcal{E} \times \mathrm{K}]$ suffisamment voisin de γ appartient à E : l'ensemble ouvert Ω peut contenir des points (x, k) tels que k soit une extrémité de K.

([3]) Si à une valeur de k ne correspond aucun point intérieur à Ω, alors l'indice total est par définition zéro.

([4]) Voir *Encyclopädie der Math. Wissenschaft*, *Analysis*, II, C.12, p. 1327-1328. M. Bernstein fait des hypothèses entraînant l'unicité de la solution des problèmes qu'il se pose; notre théorie permet de s'en libérer; c'est ainsi que M. Bernstein étudie l'équation (5) dans le cas où z n'y figure pas.

du type elliptique :

$$(3)\qquad \sum_{\substack{i=1,2,\ldots,n\\ j=1,2,\ldots,n}} A_{ij}\left[x_1,\ldots,x_n;z;\frac{\partial z}{\partial x_1},\ldots,\frac{\partial z}{\partial x_n};k\right]\frac{\partial^2 z}{\partial x_i\,\partial x_j} = D\left[x_1,\ldots,x_n;z;\frac{\partial z}{\partial x_1},\ldots,\frac{\partial z}{dx_n};k\right];$$

les valeurs frontières de z sont supposées données en fonction de k. Nous avons réussi (¹) comme suit à transformer le problème en une équation fonctionnelle, $z = Z(z, h)$, *qui soit du type* (1) : nous avons introduit la fonction Z qui satisfait les conditions aux limites données et l'équation

$$(4)\qquad \sum_{\substack{i=1,2,\ldots,n\\ j=1,2,\ldots,n}} A_{ij}\left[x_1,\ldots,x_n;z;\frac{\partial z}{\partial x_1},\ldots,\frac{\partial z}{\partial x_n};k\right]\frac{\partial^2 Z}{\partial x_i\,\partial x_j} = D\left[x_1,\ldots,x_n;z;\frac{\partial z}{\partial x_1},\ldots,\frac{\partial z}{\partial x_n};k\right].$$

D'où, en particulier, les résultats suivants :

THÉORÈME. — *Si* K *contient un point* k_0 *en lequel*

$$D(x_1, x_2, \ldots, x_n; z; p_1, \ldots, p_n; k_0] \equiv 0$$

si les dérivées $\partial z/\partial x_i$ *de toutes les solutions de* (3) *qui correspondent aux divers points de* k *restent bornées et satisfont* (²) *dans leur ensemble à une même condition de Hölder; alors le problème de Dirichlet considéré possède une solution quelle que soit la valeur de* k *dans* K (³).

Corollaire. — Le problème de Dirichlet concernant le cercle et l'équation du type elliptique :

$$(5)\qquad A\left(x,y,z,\frac{\partial z}{\partial x},\frac{\partial z}{\partial y}\right)\frac{\partial^2 z}{\partial x^2} + 2B(\ldots)\frac{\partial^2 z}{\partial x\,\partial y} + C(\ldots)\frac{\partial z^2}{\partial y^2} = 0$$

admet toujours au moins une solution.

Les considérations s'appliquent aussi aux équations elliptiques les plus générales.

(¹) Nous avons eu besoin de théorèmes que M. Schauder publiera incessamment dans la *Math. Zeitschrift*.

(²) Dans le cas où $n = 2$ nous avons seulement besoin de savoir que les dérivées $\partial z/\partial x_i$ sont bornées dans leur ensemble.

(³) Nous utilisons le théorème de M. Rado sur les Sattelfunktionen (*Act. Lit. Ac. Scient.*, 4, 1924-1926, p. 228-253; Von NEUMANN, *Abh. d. Math. Sem. Hamburg*, 8, 1931, p. 28-31).

(avec A. Weinstein)

Sur un problème de représentation conforme posé par la théorie de Helmholtz

C. R. Acad. Sci., Paris **198** (1934) 430–432

ANALYSE MATHÉMATIQUE. — *Sur un problème de représentation conforme posé par la théorie de Helmholtz*. Note de MM. **Jean Leray** et **Alexandre Weinstein**, présentée par M. Henri Villat.

Considérons le mouvement plan suivant : un jet liquide jaillissant hors de parois données, symétriques et polygonales; il pose un problème de représentation conforme bien distinct de celui dont l'étude est devenue classique :

Problème. — Transformer conformément la bande $0 < \psi < \pi/2$ du plan $f = \varphi + i\psi$ en un domaine D du plan $z = x + iy$, de sorte que soient réalisées les conditions suivantes : D doit être limité par l'axe des x, par une ligne polygonale *donnée* ϖ (de sommets : $z_0 = iy_0$, $z_1 = x_1 + iy_1$, ..., $z_n = x_n + iy_n$, $z_{n+1} = -\infty + iy_n$; $y_0 > 0, y_1 > 0, \ldots, y_n > 0$) et par une ligne *non donnée* λ qui se détache de z_0, le long de laquelle la transformation cherchée multiplie les longueurs par une constante positive inconnue μ (c'est-à-dire $|dz/df| = \mu$ sur λ) (1).

Nous désignerons par l_k la longueur du segment $z_k z_{k-1}$ ($k = 1, \ldots n$) et par θ_k sa direction; nous supposerons $\theta_1 \leqq \theta_2 \leqq \ldots \leqq \theta_n \leqq \theta_{n+1} = 0$, nous poserons $\pi\beta_k = \theta_{k+1} - \theta_k$ et nous nommerons « courbure totale » de ϖ la quantité $\pi \sum_{k=1}^{n} \beta_k = -\theta_1$.

La formule bien connue de Schwarz-Villat (2) permet d'affirmer que toute solution éventuelle du problème, $z(f)$, est donnée par une formule de M. Cisotti [*loc. cit.* (2)], laquelle exprime tous les éléments de la solution au moyen de $n+1$ paramètres. D'après cette formule le problème ci-dessus se réduit à la résolution, par rapport aux $n+1$ paramètres $\mu, \sigma_1, \ldots, \sigma_n$ ($\mu > 0$; $0 < \sigma_1 < \ldots < \sigma_n < \pi/2$) des $n+1$ équations

$$(1) \quad \begin{cases} y_0 = \mu\pi\left(1 - \dfrac{\mu\pi}{y_n}\right) + \mu \displaystyle\sum_{k=1}^{n} \sin\theta_k \int_{\sigma_{k-1}}^{\sigma_k} \prod_{h=1}^{n} \left|\frac{\sin\sigma_h - \sin\sigma}{\sin\sigma_h + \sin\sigma}\right|^{\beta_h} \operatorname{tang}\sigma\, d\sigma, \\ l_k = \mu \displaystyle\int_{\sigma_{k-1}}^{\sigma_k} \prod_{h=1}^{n} \left|\frac{\sin\sigma_h + \sin\sigma}{\sin\sigma_h - \sin\sigma}\right|^{\beta_h} \operatorname{tang}\sigma\, d\sigma \qquad (k = 1, \ldots n;\ \sigma_0 = 0). \end{cases}$$

Cette question ne fut étudiée jusqu'à présent que dans le cas où la courbure totale de ϖ est inférieure à $\pi/2$ (3). Un travail important de M. Friedrichs (4) nous permet de donner une extension considérable de cette théorie : les résultats ci-dessous (5) valent pour toutes les lignes ϖ de courbure totale inférieure à π.

I. Théorème 1 (*unicité locale*). — *Le déterminant*

$$\frac{D(y_0, l_1, \ldots, l_n)}{D(\mu, \sigma_1, \ldots, \sigma_n)}$$

n'est jamais nul quand $\mu > 0$, $0 < \sigma_1 < \ldots < \sigma_n < \pi/2$.

---

(1) Nous ferons correspondre aux points $f = -\infty$, $i\pi/2$, $+\infty$ les points $z = -\infty$, z_0, $+\infty$ et nous admettrons que dz/df tend vers des valeurs réelles pour $f = \pm\infty$.

(2) H. Villat, *Leçons sur l'Hydrodynamique*, p. 11 (Gauthier-Villars, 1929); U. Cisotti, *Idromeccanica piana*, **1**, 1921, p. 18; **2**, 1921, p. 254.

(3) A. Weinstein, *Math. Zeitschrift*, **31**, 1929, p. 424.

(4) K. Friedrichs, *Ueber ein Minimumproblem* (*Math. Annalen*, **109**, 1933, p. 60).

(5) Les démonstrations seront développées dans un Mémoire ultérieur.

Cette proposition a été ramenée ([1]) à la proposition auxiliaire suivante, indépendamment de toute hypothèse concernant la valeur de la courbure totale :

II. Théorème 2 (*unicité locale au sens restreint*). — *Il ne peut exister deux solutions infiniment voisines de notre problème qui correspondent à une même valeur de* μ.

La démonstration de ce théorème pour les courbures totales inférieures à π se déduit immédiatement de l'Appendice II du Mémoire de M. Friedrichs.

III. Les quantités μ, $\pi/2-\sigma_n$, σ_1, $\sigma_{k+1}-\sigma_k (k=1, \ldots, n-1)$ possèdent des *bornes inférieures et supérieures* positives, fonctions continues de la courbure totale et des données : $y_0, l_1, \ldots, l_n$.

IV. Théorème d'existence. — *La méthode de continuité classique déduit de* I *et* III *que le problème posé admet au moins une solution.*

V. *Unicité absolue*. — Appliquons la méthode de continuité en faisant varier les quantités $\beta_1, \ldots, \beta_n$; considérons-les à cet effet comme étant n nouveaux paramètres et n nouvelles inconnues; et adjoignons aux équations (1) les suivantes : $\beta_k=\beta_k\ (k=1, \ldots, n)$. Nous avons, d'après I,

$$\frac{D(y_0, l_1, \ldots, l_n; \beta_1, \ldots, \beta_n)}{D(\mu, \sigma_1, \ldots, \sigma_n; \beta_1, \ldots, \beta_n)} \neq 0.$$

D'autre part l'unicité de la solution est assurée pour $\beta_1=\ldots=\beta_n=0$. Donc le problème étudié possède *une et une seule solution*.

Compléments. — Les théorèmes topologiques de M. Brouwer permettent de baser le théorème d'existence IV seulement sur les propriétés III et sur le fait que l'unicité est évidente pour $\beta_1=\ldots=\beta_n=0$. Le cas des parois courbes et non plus polygonales pose un autre problème dont l'inconnue est une fonction et non plus le système des $n+1$ nombres $\mu, \sigma_1, \ldots, \sigma_n$. Une extension récente des théories de M. Brouwer au domaine fonctionnel ([2]) permettra à l'un de nous d'appliquer la méthode de continuité à cet autre problème; en même temps sera indiquée une nouvelle réduction de I à II.

([1]) A. Weinstein, *Rend. d. Accad. d. Lincei*, 3, série 6ª, 1927, p. 157.

([2]) J. Leray et J. Schauder, *Comptes rendus*, 197, 1933, p. 115.

L'anneau d'homologie d'une représentation

C. R. Acad. Sci., Paris 222 (1946) 1366–1368

Nous nous proposons d'indiquer sommairement comment les méthodes par lesquelles nous avons étudié la topologie d'un espace (²) peuvent être adaptées à l'étude de la topologie d'une représentation.

1. *Définitions préliminaires.* — Un *faisceau* $\mathcal{B}$ de modules (ou d'anneaux) sera défini sur un espace topologique E par les données que voici : 1° à chaque ensemble fermé F de points de E est associé un module (ou un anneau) $\mathcal{B}_F$, qui est nul quand F est vide; 2° à chaque couple d'ensembles fermés, f et F, de points de E, tels que $f \subset F$, est associé un homomorphisme de $\mathcal{B}_F$ dans $\mathcal{B}_f$, qui transforme un élément b_F de $\mathcal{B}_F$ en son *intersection* $b_F.f$ par f; si $f' \subset f \subset F$ et si $b_F \in \mathcal{B}_F$, on doit avoir $(b_F.f).f' = b_F.f'$. Le faisceau $\mathcal{B}$ est dit *normal* quand il possède les deux propriétés suivantes : 1° si $b_F \in \mathcal{B}_F$, il existe un voisinage fermé V de F et un élément b_V de $\mathcal{B}_V$ tel que $b_F = b_V.F$; 2° si $b_F \in \mathcal{B}_F$, si $f \subset F$ et si $b_F.f = 0$, alors f possède dans F un voisinage fermé v tel que $b_F.v = 0$. Exemple : Les classes d'homologie (³) à p dimensions des ensembles fermés de points d'un espace E constituent un faisceau que nous nommerons $p^{\text{ième}}$ *faisceau d'homologie* de E; si E est normal, ce faisceau est normal (T. A., lemmes 22 et 23).

2. Nous nommerons formes de E les expressions du type $\sum_\alpha b_\alpha X^{q;\alpha}$; les $X^{q;\alpha}$ sont les éléments à q dimensions d'une couverture de E; b_α, au lieu d'être

(¹) Séance du 27 mai 1946.

(²) Trois articles sur la Topologie algébrique, *Journ. de Math.*, 24, 1945, pp. 95-248; nous désignerons ces articles par T. A.

(³) Au sens de T. A., il s'agit donc, avec la terminologie la plus usuelle, des classes de cohomologie.

comme dans T. A. un élément d'un module indépendant de α, sera un élément de $\mathcal{B}_{|X^{q;\alpha}|}$; plus généralement b_α pourra être un élément de $\mathcal{B}_F$, si $|X^{q;\alpha}|\subset F$, étant convenu que $b_\alpha X^{q;\alpha}=(b_\alpha.|X^{q;\alpha}|)X^{q;\alpha}$; la dérivée de $\sum_\alpha b_\alpha X^{q;\alpha}$ sera $\sum_\alpha b_\alpha \dot{X}^{q;\alpha}$; le quotient du module des formes à q dimensions dont la dérivée est nulle par le module que constituent les dérivées des formes à $q-1$ dimensions sera nommé $q^{\text{ième}}$ *module d'homologie de* E *relatif au faisceau* $\mathcal{B}$ (4). L'étude de cette définition suppose E et $\mathcal{B}$ normaux et exige un examen approfondi des propriétés des simplexes. Les propriétés des classes d'homologie relatives à l'intersection, au produit topologique, aux représentations et à l'homotopie qu'établit T. A. se généralisent aisément. Si E possède une couverture C à supports simples relativement à $\mathcal{B}$, E a mêmes modules d'homologie que C (5) et ces modules peuvent être effectivement déterminés. Quand $\mathcal{B}$ est un faisceau d'anneaux, les classes d'homologie de E relatives à $\mathcal{B}$ sont nilpotentes. Soit $\mathcal{B}^p$ le $p^{\text{ième}}$ faisceau d'homologie de E relatif à $\mathcal{B}$; le faisceau d'homologie de E relatif à $\mathcal{B}^p$ est nul si $p>0$ (6), est $\mathcal{B}^0$ lui-même si $p=0$.

3. Soit π une *représentation fermée* (c'est-à-dire transformant tout ensemble fermé en un ensemble fermé) d'un espace normal E dans un espace normal E*; soit $\mathcal{B}$ un faisceau normal de modules défini sur E; nous désignerons par $\pi(\mathcal{B})$ le faisceau de modules défini sur E* comme suit : le module associé à l'ensemble fermé F* de points de E* est $\mathcal{B}_{\overset{-1}{\pi}(F^*)}$; l'intersection par F* d'un élément de $\pi(\mathcal{B})$ est l'intersection par $\overset{-1}{\pi}(F^*)$ de l'élément correspondant de $\mathcal{B}$. Le faisceau $\pi(\mathcal{B})$ est normal.

Soit $\mathfrak{A}$ un anneau; soit $\mathcal{B}^p$ le $p^{\text{ième}}$ faisceau d'homologie de E relatif à $\mathfrak{A}$; le $q^{\text{ième}}$ module d'homologie de E* relatif à $\pi(\mathcal{B}^p)$ sera nommé $(p, q)^{\text{ième}}$ *module d'homologie* de π relatif à $\mathfrak{A}$; une classe d'homologie $Z^{p,q}$ à q dimensions de E* relative à $\pi(\mathcal{B}^p)$ sera nommée classe d'homologie à (p, q) dimensions de π relative à $\mathfrak{A}$. $Z^{p,q}$ est nulle quand q dépasse la dimension de E* ou quand p dépasse la dimension maximum des images inverses $\overset{-1}{\pi}(x^*)$ des points x^* de E*, E* étant bicompact. L'intersection de deux classes d'homologie $Z^{p,q}$ et $Z^{r,s}$ de π est définie; c'est une classe $Z^{p+r,q+s}\sim Z^{p,q}.Z^{r,s}\sim(-1)^{(p+q)(r+s)}Z^{r,s}.Z^{p,q}$ de π; il convient donc de parler de *l'anneau d'homologie d'une représentation.*

---

(4) Un cas très particulier de cette notion a été envisagé par M. Steenrod, *Homology with local coefficients* (*Ann. of Math.*, 44, 1943, p. 610).

(5) Dans cet énoncé on doit envisager les modules d'homologie de C relatifs au *faisceau réduit* de $\mathcal{B}$; ce faisceau réduit est le quotient de $\mathcal{B}$ par le sous-faisceau que constituent les éléments réductibles de $\mathcal{B}$, un élément b de $\mathcal{B}_F$ étant dit réductible quand on peut recouvrir F avec un nombre fini d'ensembles fermés f_α tels que $b.f_\alpha=0$. Signalons que les modules d'homologie de E relatifs à $\mathcal{B}$ et au faisceau réduit de $\mathcal{B}$ sont les mêmes.

(6) Car le faisceau réduit de $\mathcal{B}^p$ est nul si $p>0$.

Les classes d'homologie des représentations sont nilpotentes, comme celles d'un espace. L'anneau d'homologie de π s'identifie à celui de E quand π est une représentation topologique de E dans E^*. Mais l'anneau d'homologie d'une représentation quelconque a une structure particulière, que nous expliciterons ultérieurement.

4. Soit F^* un ensemble fermé de points de E^*; nous nommerons *intersection* de π par F^* et nous désignerons par $\pi.F^*$ la représentation dont le champ de définition est $\overset{-1}{\pi}(F^*)$ et qui y est égale à π; cette intersection définit un homomorphisme de l'anneau d'homologie de π dans celui de $\pi.F^*$; cet homomorphisme est un isomorphisme quand $F^* = \pi(E)$. Soit une seconde représentation fermée π' d'un espace normal E' dans un espace normal E'^*. Nous nommerons *produit topologique* $\pi \times \pi'$ de π et π' la representation $\pi(x) \times \pi'(x')$ de $E \times E'$ dans $E^* \times E'^*$; quand $\mathfrak{A}$ est un corps, l'anneau d'homologie de $\pi \times \pi'$ est le produit direct des anneaux d'homologie de π et de π'. Nous dirons qu'une représentation φ de E' dans E constitue une *représentation* de π' dans π quand il existe une représentation φ^* de E'^* dans E^* telle que $\pi\varphi(x') = \varphi^*\pi'(x')$; $\overset{-1}{\varphi}$ définit un homomorphisme de l'anneau d'homologie de π dans celui de π'; cet homomorphisme est un isomorphisme quand les deux représentations φ et φ^* sont topologiques. Deux représentations θ et ψ de π' dans π seront dites *homotopes* dans π quand il existe une représentation φ de π' dans π qui a les propriétés suivantes : φ dépend continûment d'un paramètre qui décrit un espace connexe; pour deux valeurs particulières de ce paramètre, φ s'identifie à θ et à φ. Si θ et ψ sont homotopes dans π et si $Z^{p,q}$ est une classe d'homologie de π, alors $\overset{-1}{\theta}(Z^{p,q})$ est identique à $\overset{-1}{\psi}(Z^{p,q})$. Nous dirons que π est *homotope* a $\pi.F^*$ dans π quand il existe une représentation de π dans $\pi.F^*$ qui est l'identité sur $\overset{-1}{\pi}(F^*)$ et qui est homotope dans π à la représentation identique; alors l'intersection par F^* constitue un isomorphisme de l'anneau d'homologie de π sur celui de $\pi.F^*$. En particulier si π est homotope dans π à son intersection par l'un des points de $\pi(E)$, toutes les classes d'homologie de π sont du type $Z^p E^{*0}$, Z^p étant une classe d'homologie de E relative à $\mathfrak{A}$ et E^{*0} étant la classe d'homologie unité de E^*; ce que nous exprimerons en disant que π est *simple*. Si les intersections de π par les supports d'une couverture C^* de E^* sont des représentations simples, alors le $(p, q)^{\text{ième}}$ module d'homologie de π s'identifie au $q^{\text{ième}}$ module d'homologie de C^* relatif au faisceau réduit de $\pi(\mathcal{B}^p)$, et peut donc être effectivement déterminé.

Dépôt légal d'éditeur. — 1946. — N° d'ordre 64.
Dépôt légal d'imprimeur. — 1946. — N° d'ordre 144.

GAUTHIER-VILLARS, IMPRIMEUR-LIBRAIRE DES COMPTES RENDUS DES SÉANCES DE L'ACADÉMIE DES SCIENCES.
Paris — Quai des Grands-Augustins, 55.

Structure de l'anneau d'homologie d'une représentation

C. R. Acad. Sci., Paris 222 (1946) 1419–1422

1. Étant donnés un anneau $\mathcal{A}$ et une représentation fermée π d'un espace normal E dans un espace normal E*, nous avons défini récemment ([2]) l'anneau d'homologie de π relatif à $\mathcal{A}$. *Cet anneau a la structure suivante :*

Le $(p, q)^{\text{ième}}$ module d'homologie $\mathfrak{L}_1^{p,q}$ de π possède les *sous-modules*

$$0 = \mathfrak{Q}_{-1}^{p,q} = \mathfrak{Q}_0^{p,q} \subset \mathfrak{Q}_1^{p,q} \subset \mathfrak{Q}_2^{p,q} \subset \ldots \subset \mathfrak{Q}_{q-2}^{p,q} \subset \mathfrak{Q}_{q-1}^{p,q} \subset \mathfrak{L}_{p+1}^{p,q} \subset \mathfrak{L}_p^{p,q} \subset \ldots \subset \mathfrak{L}_2^{p,q} \subset \mathfrak{L}_1^{p,q};$$

le $p^{\text{ième}}$ module d'homologie $\mathcal{E}^{p,0}$ de E relatif à $\mathcal{A}$ possède les *sous-modules*

$$0 = \mathcal{E}^{-1,p+1} \subset \mathcal{E}^{0,p} \subset \mathcal{E}^{1,p-1} \subset \ldots \subset \mathcal{E}^{p-1,1} \subset \mathcal{E}^{p,0};$$

il existe des *homomorphismes* Δ_r de $\mathfrak{L}_r^{p,q}$ sur $\mathfrak{Q}_r^{p-r,q+r+1} / \mathfrak{Q}_{r-1}^{p-r,q+r+1}$ ayant pour noyau $\mathfrak{L}_{r+1}^{p,q}$ $(1 \leqq r \leqq p)$;

il existe des *homomorphismes* Γ^* de $\mathfrak{L}_{p+1}^{p,q}$ sur $\mathcal{E}^{p,q} / \mathcal{E}^{p-1,q+1}$ ayant pour noyau $\mathfrak{Q}_{q-1}^{p,q}$.

2. *Les définitions* de ces sous-modules et de ces homomorphismes sont les suivantes :

1° Soient $Z^{*p,q}$ et $Z^{*p-r,q+r+1}$ des classes d'homologie de E* relatives respectivement à $\pi(\mathcal{B}^p)$ et à $\pi(\mathcal{B}^{p-r})$; la condition

$$Z^{*p,q} \in \mathfrak{L}_r^{p,q}, \qquad Z^{*p-r,q+r+1} \in \mathfrak{Q}_r^{p-r,q+r+1}, \qquad \Delta_r(Z^{*p,q}) \sim Z^{*p-r,q+r+1} \bmod \mathfrak{Q}_{r-1}^{p-r,q+r+1}$$

équivaut à celle-ci : il existe une couverture C* de E, dont nous nommerons les éléments $X^{*r;\gamma}$, un cycle $\sum_\alpha z^{p;\alpha} X^{*q;\alpha}$ appartenant à la classe $Z^{*p,q}$, un cycle $\sum_\beta z^{p-r;\beta} X^{*q+r+1;\beta}$ appartenant à la classe $Z^{*p-r,q+r+1}$ et une forme $L^{p,q}$ de E, à

(1) Séance du 27 mai 1946.
(2) *Comptes rendus*, 222, 1946, p. 1366.

coefficients pris dans $\mathcal{A}$, tels que

$$L^{p,q}=\sum_{\alpha} L'^{p;\alpha}.\overset{-1}{\pi}(X^{*q;\alpha})+\sum_{s>0,\lambda} L'^{p-s;\lambda}.\overset{-1}{\pi}(X^{*q+s;\lambda}),$$

$$\dot{L}^{p,q}=\sum_{\beta} L'^{p-r;\beta}.\overset{-1}{\pi}(X^{*q+r+1;\beta})+\sum_{t>0,\mu} L'^{p-r-t;\mu}.\overset{-1}{\pi}(X^{*q+r+1+t;\mu}),$$

$$L'^{p;\alpha}.\overset{-1}{\pi}(|X^{*q;\alpha}|)\sim z^{p;\alpha} \qquad \text{et} \qquad L'^{p-r;\beta}.\overset{-1}{\pi}(|X^{*q+r+1;\beta}|)\sim z^{p-r;\beta};$$

dans ces formules $z^{p;\alpha}$ représente une classe d'homologie à p dimensions de $\overset{-1}{\pi}(|X^{*q;\alpha}|)$ relative à $\mathcal{A}$ et les L′ représentent des formes, à coefficients pris dans $\mathcal{A}$, d'une couverture C′ de E.

2° Soit $Z^{*p,q}$ une classe d'homologie de E* relatvie à $\pi(\mathcal{B}^p)$; soit Z^{p+q} une classe d'homologie de E relative à $\mathcal{A}$; la condition $Z^{*p,q}\in\mathcal{X}^{p,q}_{p+1}$, $Z^{p+q}\in\mathcal{E}^{p,q}$, $\Gamma(Z^{*p,q})\sim Z^{p+q} \bmod \mathcal{E}^{p-1,q+1}$ équivaut à la suivante : il existe un cycle $\sum_{\alpha} z^{p;\alpha}X^{*q;\alpha}$ de la classe $Z^{*p,q}$ et un cycle $L^{p,q}$ de la classe Z^{p+q} tels que

$$L^{p,q}=\sum_{\alpha} L'^{p;\alpha}.\overset{-1}{\pi}(X^{*q;\alpha})+\sum_{s>0,\lambda} L'^{p-s;\lambda}.\overset{-1}{\pi}(X^{*q+s;\lambda}),$$

$$L'^{p;\alpha}.\overset{-1}{\pi}(|X^{*q;\alpha}|)\sim z^{p;\alpha}.$$

3. *En particulier,* si Z^p et Z^{*q} sont des classes d'homologie de E et de E* relatives à $\mathcal{A}$, Z^pZ^{*q} est une classe d'homologie de π

$$Z^pZ^{*q}\in\mathcal{X}^{p,q}_{p+1}; \qquad \Gamma(Z^pZ^{*q})\sim Z^p.\overset{-1}{\pi}(Z^{*q}) \qquad \bmod \mathcal{E}^{p-1,q+1}.$$

L'homomorphisme $\overset{-1}{\Gamma}$ de $\mathcal{E}^{p,0}$ sur $\mathcal{X}^{p,0}_{p+1}$ est donc $\overset{-1}{\Gamma}(Z^p)\sim Z^pE^{*0}$; il en résulte que $\mathcal{X}^{p,0}_{p+1}$ est l'ensemble des classes d'homologie de π du type Z^pE^{*0} et que $\mathcal{E}^{p-1,1}$ est l'ensemble des classes d'homologie Z^p de E ayant la propriété que voici : on peut recouvrir E* avec un nombre fini d'ensembles fermés F^*_λ tels que $Z^p.\overset{-1}{\pi}(F^*_\lambda)\sim 0$ quel que soit λ; quand E* est bicompact, on peut définir $\mathcal{E}^{p-1,1}$ comme l'ensemble des classes d'homologie Z^p de E telles que $Z^p.\overset{-1}{\pi}(x^*)\sim 0$ quel que soit le point x^* de E*.

4. Les propriétés de l'*intersection* sont les suivantes : désignons par $\mathcal{X}^{p,q}_r.\mathcal{X}^{s,t}_r$ le plus petit module contenant les intersections des divers éléments de $\mathcal{X}^{p,q}_r$ par les divers éléments de $\mathcal{X}^{s,t}_r$; convenons de poser $\mathcal{X}^{p,q}_{p+1}=\mathcal{X}^{p,q}_{p+2}=\ldots$ et $\Delta_r(Z^{p,q})\sim 0$ quand $Z^{p,q}\in\mathcal{X}^{p,q}_{p+1}$ et $r\geqq p+1$; on a

$$\mathcal{X}^{p,q}_r.\mathcal{X}^{s,t}_r\subset\mathcal{X}^{p+s,q+t}_r; \qquad \mathcal{X}^{p,q}_r.\mathcal{D}^{s,t}_{r-1}\subset\mathcal{D}^{p+s,q+t}_{r-1}; \qquad \mathcal{E}^{p,q}.\mathcal{E}^{r,s}\subset\mathcal{E}^{p+r,q+s};$$

$$\Delta_r(Z^{p,q}.Z^{s,t})\sim\Delta_r(Z^{p,q}).Z^{s,t}+(-1)^{p+q}Z^{p,q}.\Delta_r(Z^{s,t}) \qquad \bmod \mathcal{D}^{p+s-r,q+t+r+1}_{r-1},$$

quand $Z^{p,q}\in\mathcal{X}^{p,q}_r$ et $Z^{s,t}\in\mathcal{X}^{s,t}_r$;

$$\Gamma(Z^{p,q}.Z^{r,s})\sim\Gamma(Z^{p,q}).\Gamma(Z^{r,s}) \qquad \bmod \mathcal{E}^{p+r-1,q+s+1},$$

quand $Z^{p,q}\in\mathcal{X}^{p,q}_{p+1}$ et $Z^{r,s}\in\mathcal{X}^{r,s}_{r+1}$.

Les homomorphismes Δ_r et Γ du *produit* de deux représentations se rattachent aux homomorphismes Δ_r et Γ de ces deux représentations par des formules analogues aux précédentes. Les homomorphismes de l'anneau d'homologie d'une représentation π que définissent l'intersection de π par un ensemble ou la transformation de π par l'inverse d'une représentation [*loc. cit.* (1), 4], respectent les homomorphismes Δ_r et Γ.

5. Les propriétés que nous avons énoncées de l'anneau d'homologie d'une représentation peuvent servir à l'étude de l'anneau d'homologie d'un espace et à l'étude de la transformation de cet anneau par l'inverse d'une représentation. Par exemple supposons que E^* soit bicompact et que, quel que soit le point x^* de E^*, $\overset{-1}{\pi}(x^*)$ soit simple (c'est-à-dire ait pour seules classes d'homologie les produits par les éléments de $\mathfrak{A}$ de la classe unité); alors $\overset{-1}{\pi}$ est un isomorphisme de l'anneau d'homologie de E^* relatif à $\mathfrak{A}$ sur l'anneau d'homologie de E. Supposons que E^* soit bicompact et que $\overset{-1}{\pi}(x^*)$ soit connexe quel que soit le point x^* de E^*; alors $\overset{-1}{\pi}$ est un isomorphisme du premier module d'homologie de E^* dans (c'est-à-dire sur un sous-module de) celui de E. Supposons que π soit la projection d'un espace fibré E sur son espace de base E^*, que cet espace de base E^* soit simplement connexe et que E, E^* et la fibre F soient des multiplicités; alors le $(p, q)^{\text{ième}}$ module d'homologie de π est le $q^{\text{ième}}$ module d'homologie de E^* relatif au $p^{\text{ième}}$ module d'homologie de F et, $\mathfrak{A}$ étant supposé un corps, l'application aux homomorphismes Δ_r et Γ du théorème sur le rang d'un quotient de modules fournit le résultat suivant : soient $\mathcal{E}(t)$, $\mathcal{E}^*(t)$ et $\mathcal{F}(t)$ les polynomes de Poincaré de E, E^* et F; il existe un polynome $\mathcal{B}(t)$ à coefficients entiers non négatifs tel que $\mathcal{E}(t) = \mathcal{F}(t)\mathcal{E}^*(t) - (1+t)\mathcal{B}(t)$. Soient un groupe bicompact, simplement connexe, E, et un sous-groupe fermé, connexe, à un paramètre, F, de E; soit E^* l'espace homogène défini par E et F; soit π la projection de chaque classe de E suivant F sur le point correspondant de E^*; supposons que $\mathfrak{A}$ soit le corps des nombres rationnels; alors l'anneau d'homologie de E^* relatif à $\mathfrak{A}$ est engendré par des classes d'homologie Z^{*2}, $Z^{*2p+1;\alpha}$ et la seule relation $(Z^{*2})^{n+1} \sim 0$; on a $\overset{-1}{\pi}(Z^{*2}) \sim 0$; l'anneau d'homologie de E est engendré par une classe hypermaximale Z^{2n+1} et par les classes hypermaximales $\overset{-1}{\pi}(Z^{*2p+1;\alpha})$. Les propriétés de l'anneau d'homologie d'une représentation permettent également de retrouver les théorèmes de M. Gysin sur les espaces fibrés dont les fibres sont des sphères et le théorème de M. Samelson sur les groupes bicompacts transformant transitivement une sphère (3).

(3) GYSIN, *Commentarii Helv.*, **14**, 1941, p. 61; SAMELSON, *Ann. of Math.*, **42**, 1941, p. 1000.

Dépôt légal d'éditeur. — 1946. — N° d'ordre 64.
Dépôt légal d'imprimeur. — 1946. — N° d'ordre 144.

GAUTHIER-VILLARS, IMPRIMEUR-LIBRAIRE DES COMPTES RENDUS DES SÉANCES DE L'ACADÉMIE DES SCIENCES.
Paris. — Quai des Grands-Augustins, 55.

Propriétés de l'anneau d'homologie de la projection d'un espace fibré sur sa base

C. R. Acad. Sci., Paris **223** (1946) 395–397

1. *Le polynome de Poincaré* $\mathfrak{P}(t, t^*)$ *d'une représentation* π. — Soit π une représentation d'un espace bicompact E dans un espace bicompact E*; soit un corps $\mathfrak{A}$; soient $\mathfrak{P}_1^{p,q}$ les modules d'homologie (²) de π relatifs à $\mathfrak{A}$; supposons qu'ils aient des bases finies; désignons par $\rho(\mathfrak{M})$ le rang d'un $\mathfrak{A}$-module $\mathfrak{M}$; posons

$$\mathfrak{P}(t, t^*) = \sum_{p,q} t^p t^{*q} \rho(\mathfrak{P}_1^{p,q}); \qquad \mathcal{E}(t, t^*) = \sum_{p,q} t^p t^{*q} \rho(\mathcal{E}^{p,q}/\mathcal{E}^{p-1,q+1});$$

$$\mathfrak{D}_r(t, t^*) = \sum_{p \geqq r, q} t^{p-r} t^{*q} \rho(\mathfrak{P}_r^{p,q}/\mathfrak{P}_{r-1}^{p,q});$$

l'application aux homomorphismes Δ_r et Γ du théorème sur le rang d'un quotient de modules donne

$$\mathcal{E}(t, t^*) = \mathfrak{P}(t, t^*) - \sum_{r \geqq 1} (t^r + t^{*r+1}) \mathfrak{D}_r(t, t^*); \tag{1}$$

$$\mathcal{E}(t, t) = \sum_p t^p \rho(\mathcal{E}^{p,0}) \tag{2}$$

est le polynome de Poincaré de E relatif à $\mathfrak{A}$;

$$\mathcal{E}(t, 0) = \sum_p t^p \rho(\mathcal{E}^{p,0}/\mathcal{E}^{p-1,1}); \tag{3}$$

(¹) Séance du 26 août 1946.

(²) Deux Notes antérieures, dont nous conservons les notations, définissent l'anneau d'homologie d'une représentation et précisent sa structure (*Comptes rendus*, 222, 1946, pp. 1366 et 1419).

nous désignons par $\mathcal{E}^{p,0}$ le $p^{\text{ième}}$ module d'homologie de E relatif à $\mathcal{A}$ et par $\mathcal{E}^{p-1,1}$ le sous-module de $\mathcal{E}^{p,0}$ que constituent les classes d'homologie Z^p de E, telles que $Z^p . \overset{-1}{\pi}(x^*) \sim 0$, quel que soit le point x^* de E^*; si $\overset{-1}{\pi}(x^*)$ est connexe quel que soit x^*, on a, $\mathcal{E}^{*p}$ représentant le $p^{\text{ième}}$ module d'homologie de E^* relatif à $\mathcal{A}$,

$$\mathcal{E}(0, t^*) = \sum_p t^p \rho\left(\overset{-1}{\pi}(\mathcal{E}^{*p})\right), \tag{4}$$

2. *Extension du théorème de dualité de H. Poincaré à la projection π d'un espace fibré* E *sur sa base* E^*. — Supposons *que* E *et* E^* *soient des multiplicités fermées, connexes, orientables* à $l+m$ et m dimensions, *que la fibre* F *soit une multiplicité fermée orientable à l dimensions* et *que $\mathcal{A}$ soit le corps des rationnels ou le corps des entiers calculés mod. n, n étant premier*. Les modules d'homologie $\mathcal{L}_1^{p,q}$ de π relatifs à $\mathcal{A}$ ont des bases finies et une généralisation du théorème de dualité de Poincaré prouve que $\mathcal{L}_1^{p,q}$ et $\mathcal{L}_1^{l-p,m-q}$ sont duals ([3]) relativement à leur intersection qui est définie dans $\mathcal{L}_1^{l,m}$, module isomorphe à $\mathcal{A}$; les propriétés des homomorphismes Δ_r et Γ permettent de déduire de cette dualité les conclusions suivantes : *Les modules d'homologie $\mathcal{L}_1^{p,q}$ et $\mathcal{L}_1^{l-p,m-q}$ de π relatifs à $\mathcal{A}$ sont duals, leurs sous-modules $\mathcal{L}_r^{p,q}$ et $\mathcal{Q}_r^{l-p,m-q}$ étant annulateurs l'un de l'autre; les modules d'homologie $\mathcal{E}^{p,0}$ et $\mathcal{E}^{m+n-p,0}$ de* E *relatifs à $\mathcal{A}$ sont duals (H. Poincaré), leurs sous-modules $\mathcal{E}^{p-q,q}$ et $\mathcal{E}^{m-p+q-1,n-q+1}$ étant annulateurs l'un de l'autre*. D'où

$$\mathcal{L}(t, t^*) = t^l t^{*m} \mathcal{L}\left(\frac{1}{t}, \frac{1}{t^*}\right); \quad \mathcal{E}(t, t^*) = t^l t^{*m} \mathcal{E}\left(\frac{1}{t}, \frac{1}{t^*}\right); \quad \mathcal{D}_r(t, t^*) = t^{l-r} t^{*m-r-1} \mathcal{D}_r\left(\frac{1}{t}, \frac{1}{t^*}\right).$$

3. *Cas où l'anneau d'homologie de π est le produit direct des anneaux d'homologie de* F *et de* F^*. — Ce cas se présente quand, F *étant connexe* et $\mathcal{A}$ *étant un corps*, on peut établir entre l'anneau d'homologie d'une fibre fixe et l'anneau d'homologie de la fibre la plus générale F un isomorphisme qui varie continûment avec F, car le $(p, q)^{\text{ième}}$ module d'homologie de π est alors le $q^{\text{ième}}$ module d'homologie de E^* relatif au $p^{\text{ième}}$ module d'homologie de F; il en est en particulier ainsi quand E^* est simplement connexe ou quand F est une sphère qu'on peut orienter continûment. Soient $\mathcal{E}(t)$, $\mathcal{F}(t)$ et $\mathcal{E}^*(t)$ les polynomes de Poincaré de E, F et E^*; soit $\mathcal{E}_\pi(t)$ [et $\mathcal{F}_E(t)$] le polynome dont le coefficient de t^p est le rang du module que constituent les transformées par $\overset{-1}{\pi}$ des classes d'homologie de E^* [les intersections par F des classes d'homologie de E] : avec les notations du n° 1

$$\mathcal{L}(t, t^*) = \mathcal{F}(t)\, \mathcal{E}^*(t^*); \quad \mathcal{E}(t, t) = \mathcal{E}(t); \quad \mathcal{E}(0, t) = \mathcal{E}_\pi(t); \quad \mathcal{E}(t, 0) = \mathcal{F}_E(t);$$

posons $\mathcal{D}(t) = \sum_{r \geq 1} t^r \mathcal{D}_r(t, t)$; *le symbole* $0 \leq \mathcal{A}(t)$ exprimera que le développe-

([3]) C'est-à-dire : chacun de ces modules est le groupe des caractères de l'autre; ils sont isomorphes.

ment de $\mathcal{A}(t)$ suivant les puissances croissantes de t a des coefficients positifs ou nuls; les formules du n° 1 et les relations

$$\mathfrak{P}_r^{p,0} . \mathfrak{P}_1^{0,q} \subset \mathfrak{P}_r^{p,q} \quad \text{et} \quad \mathfrak{P}_r^{p,0} . \mathfrak{Q}_{r-1}^{0,q} \subset \mathfrak{Q}_{r-1}^{p,q}$$

donnent

$$(5) \qquad \mathcal{E}(t) = \mathcal{F}(t)\,\mathcal{E}^*(t) - (1+t)\,\mathcal{D}(t), \qquad \text{où } 0 \leqq \mathcal{D}(t);$$

$$(6) \quad 1 \leqq \mathcal{E}_\pi(t) \leqq \mathcal{E}^*(t); \qquad \mathcal{E}_\pi(t) \leqq \mathcal{E}(t); \qquad 1 \leqq \mathcal{F}_E(t) \leqq \mathcal{F}(t); \qquad \mathcal{F}_E(t) \leqq \mathcal{E}(t);$$

$$(7) \qquad \mathcal{F}(t) - \mathcal{F}_E(t) \leqq \mathcal{D}(t) \leqq [\mathcal{F}(t) - \mathcal{F}_E(t)]\,\mathcal{E}^*(t);$$

$$(8) \qquad \mathcal{F}_E(t)\,[\mathcal{E}^*(t) - \mathcal{E}_\pi(t)] \leqq t\mathcal{D}(t).$$

D'où les inégalités suivantes entre $\mathcal{E}(t)$, $\mathcal{F}(t)$ et $\mathcal{E}^*(t)$

$$(9) \qquad \frac{1 - t[\mathcal{F}(t) - 1]}{1+t}\,\mathcal{E}^*(t) \leqq \frac{\mathcal{E}(t)}{1+t} \leqq \frac{\mathcal{F}(t)\,\mathcal{E}^*(t)}{1+t}$$

et plus particulièrement

$$(10) \qquad \{1 - t[\mathcal{F}(t) - 1]\}\,\mathcal{E}^*(t) \leqq \mathcal{E}(t) \leqq \mathcal{F}(t)\,\mathcal{E}^*(t);$$

$$(11) \qquad \mathcal{E}^*(t) \leqq \frac{\mathcal{E}(t)}{1 - t[\mathcal{F}(t) - 1]}.$$

M. G. Hirsch m'a signalé qu'il a obtenu (5) par un autre procédé. Dans le cas particulier où F est une sphère, (5) et un corollaire de (11) furent établis par M. W. Gysin; le raisonnement de M. Gysin utilise un homomorphisme qui généralise l'invariant que M. H. Hopf a attaché à une représentation d'une sphère à $4k-1$ dimensions dans une multiplicité à $2k$ dimensions [4]; l'invariant de M. Hopf et l'homomorphisme de M. Gysin peuvent être rattachés à notre homomorphisme Δ_l.

[4] H. Hopf, *Fund. math.*, **25**, 1935, p. 427; W. Gysin, *Comm. math. helv.*, **14**, 1941, p. 112, th. 35 et 34.

Dépôt légal d'éditeur. — 1946. — N° d'ordre 64.
Dépôt légal d'imprimeur. — 1946. — N° d'ordre 144.

GAUTHIER-VILLARS, IMPRIMEUR-LIBRAIRE DES COMPTES RENDUS DES SÉANCES DE L'ACADÉMIE DES SCIENCES
Paris. — Quai des Grands-Augustins, 55.

Détermination, dans les cas non exceptionnels, de l'anneau de cohomologie de l'espace homogène quotient d'un groupe de Lie compact par un sous-groupe de même rang

C. R. Acad. Sci., Paris 228 (1949) 1902–1904

Notations. — X est un groupe de Lie compact et connexe; Y est un sous-groupe de X; T est un sous-groupe abélien de Y :

$$T \subset Y \subset X;$$

X, Y et T ont le même rang l. Soit N_T le normalisateur de T dans X; si $n \in N_T$ l'automorphisme $t \to n^{-1}tn (t \in T)$ de T et l'application $xT \to xnT (x \in X)$ de l'espace homogène $U = X/T$ sur lui-même ne dépendent que de nT : si nous posons

$$\Phi_Y = (Y \cap N_T)/T \subset \Phi_X = N_T/T, \qquad V = Y/T \subset U = X/T, \qquad W = X/Y = U/V,$$

le groupe fini Φ_X opère sur les espace T et U; son sous-groupe Φ_Y opère en outre sur V et applique identiquement W sur lui-même. $\mathcal{H}_X$, $\mathcal{H}_U$, ... sont les algèbres de cohomologie de X, U, ... relatives à un même corps commutatif, de caractéristique nulle; $X(s)$ désigne le polynome de Poincaré de X.

Nous utiliserons les propriétés [1] de T, U, Φ_X découvertes par Killing, E. Cartan, H. Weyl, A. Weil, H. Hopf, H. Samelson, E. Stiefel et les invariants topologiques [2] que nous avons attachés aux applications. Tous les groupes envisagés, étant des groupes de Lie compacts, sont des produits de

---

(1) *Voir* H. Hopf, *Comm. math. helv.*, **15**, 1942, p. 59-70.

(2) *Journ. Math.* (à paraître); *Comptes rendus*, **222**, 1946, p. 1366, 1419; **223**, 1946, p. 395, 412; **228**, 1949, p. 1545, 1784. Ces Notes seront désignées par $(N_1)\dots(N_6)$. Le n° 1 de (N_5) énonce avec la terminologie actuelle les conclusions de (N_1) et (N_2); la différentielle Δ_1 de (N_4) est actuellement notée δ_2. Signalons que dans la formule (8) de (N_3) $\mathcal{D}(t)$ doit être remplacé par $t\mathcal{D}(t)$.

groupes simples, *que nous supposerons appartenir aux quatre grandes classes;* pour lever cette restriction il suffirait d'étendre le lemme aux cinq groupes simples *exceptionnels.*

LEMME. — *a. Les éléments de $\mathcal{H}_T$ ayant pour degré 1 constituent une représentation linéaire fidèle $\mathfrak{T}$ de Φ_X; $\mathcal{H}_T$ est l'algèbre extérieure $\bigwedge \mathfrak{T}$ de $\mathfrak{T}$, c'est-à-dire l'anneau des polynomes ayant pour arguments les éléments de $\mathfrak{T}$, la multiplication de ces éléments étant anticommutative; soit $\mathfrak{Q}$ l'anneau des polynomes ayant pour arguments les éléments de $\mathfrak{T}$, la multiplication de ces éléments étant commutative; soit $\mathcal{R}_X$ l'anneau que constituent les éléments de $\mathfrak{Q}$ invariants par Φ_X (c'est-à-dire par chaque opération de Φ_X); soit $\mathfrak{S}_X$ l'idéal de $\mathfrak{Q}$ qu'engendrent les éléments de $\mathcal{R}_X$ de degré >0 : il existe un isomorphisme canonique, doublant le degré, de $\mathfrak{Q}/\mathfrak{S}_X$ sur $\mathcal{H}_U$; $\mathfrak{Q}/\mathfrak{S}_X$ et $\mathcal{H}_U$ sont des représentations de Φ_X équivalentes à l'algèbre de ce groupe; l'algèbre $\mathcal{H}_U$ est engendrée par son unité et l'ensemble de ses éléments de degré 2; cet ensemble sera noté $\mathfrak{T}_U$.*

b. $\mathcal{H}_U \otimes \mathcal{H}_T$ a une différentielle δ_2; son algèbre d'homologie est $\mathcal{H}_X$; δ_2 applique isomorphiquement $1 \otimes \mathfrak{T}$ sur $\mathfrak{T}_U \otimes 1$.

c.

$$X(s) = \prod_{\lambda=1}^{l} (1 + s^{2m_\lambda - 1}); \qquad U(s) = \prod_{\lambda=1}^{l} \frac{1 - s^{2m_\lambda}}{1 - s^2};$$

$\mathcal{R}_X$ est engendré par l éléments indépendants de degrés m_λ.

Preuve. — Si ce lemme est vrai pour deux groupes X, il est vrai pour leur produit; or il est vrai pour les groupes simples des quatre grandes classes d'après (N_4) et le n° 5*b* de (N_6).

THÉORÈME. — *a. L'application canonique η de $U = X/T$ sur $W = X/Y$ a pour réciproque un isomorphisme $\overset{-1}{\eta}$ de $\mathcal{H}_W$ dans $\mathcal{H}_U$; $\overset{-1}{\eta}\mathcal{H}_W$ est l'ensemble des éléments de $\mathcal{H}_U$ invariants par Φ_Y. L'isomorphisme canonique de $\mathcal{H}_U$ sur $\mathfrak{Q}/\mathfrak{S}_X$ applique $\overset{-1}{\eta}\mathcal{H}_W$ sur $\mathcal{R}_Y\mathfrak{S}_X/\mathfrak{S}_X \simeq \mathcal{R}_Y/(\mathcal{R}_Y \cap \mathfrak{S}_X)$. La caractéristique d'Euler de W est* (³) *l'indice de Φ_Y dans Φ_X.*

b. Si Y *est connexe, on a, conformément à une hypothèse de G. Hirsch, vérifiée par J.-L. Koszul quand* W *est symétrique :*

$$W(s) = \prod_{\lambda=1}^{l} \frac{1 - s^{2m_\lambda}}{1 - s^{2n_\lambda}}; \qquad X(s) = \prod_{\lambda=1}^{l} (1 + s^{2m_\lambda - 1}); \qquad Y(s) = \prod_{\lambda=1}^{l} (1 + s^{2n_\lambda - 1}).$$

c. Soit Z *la composante de* Y *contenant l'unité; le groupe fondamental de* W *est $Y/Z \simeq \Phi_Y/\Phi_Z$; ce groupe opère sur X/Z, qui est le revêtement simplement connexe de* W*; ce groupe opère sur $\mathcal{H}_{X/Z}$ comme Φ_Y/Φ_Z opère sur l'ensemble des éléments de $\mathcal{H}_U$ invariants par Φ_Z.*

Preuve, quand Y *est connexe.* — L'isomorphisme que définit δ_2 de $\mathfrak{T}$ sur $\mathfrak{T}_U$

(³) H. HOPF et H. SAMELSON, *Comm. math. helv.*, **13**, 1940, p. 240.

(lemme *b*) s'obtient en composant celui de $\mathfrak{T}$ sur $\mathfrak{T}_U$ et l'homomorphisme de $\mathcal{H}_U$ dans $\mathcal{H}_V$ nommé section par V ; cette section est donc un isomorphisme de $\mathfrak{T}_U$ sur $\mathfrak{T}_V$ et, puisque $\mathfrak{T}_V$ engendre $\mathcal{H}_V$, un homomorphisme de $\mathcal{H}_U$ sur $\mathcal{H}_V$. Or W est simplement connexe; donc, d'après les formules (5), (7), (8) de (N_3):

l'application canonique de U sur $W = U/V$ a pour réciproque un isomorphisme de $\mathcal{H}_W$ dans $\mathcal{H}_U$;

le polynome de Poincaré de U est le produit de ceux de V et W.

Plus précisément, $\mathcal{H}_W \otimes \mathcal{H}_V$ et $\mathcal{H}_U$ sont des représentations linéaires équivalentes de Φ_Y; or $\mathcal{H}_W$ est l'ensemble des éléments de $\mathcal{H}_W \otimes \mathcal{H}_Y$ invariants par Φ_Y; l'image de $\mathcal{H}_W$ dans $\mathcal{H}_U$ est donc l'ensemble des éléments de $\mathcal{H}_V$ invariants par Φ_Y.

Preuve quand Y *n'est pas connexe.* — Soient ζ et θ les applications canoniques de X/T dans X/Z et de X/Z dans W : $\eta = \theta\zeta$. Le groupe Y/Z opère sur l'espace X/Z; en identifiant les transformés par ce groupe de chaque point de X/Z on obtient l'espace W ; donc $\overset{-1}{\theta}$ est un isomorphisme de $\mathcal{H}_W$ sur l'ensemble des éléments de $\mathcal{H}_{X/Z}$ invariants par Y/Z. Il en résulte que $\overset{-1}{\eta} = \overset{-1}{\zeta}\overset{-1}{\theta}$ est un isomorphisme de $\mathcal{H}_W$ sur l'ensemble des éléments de $\overset{-1}{\zeta}\mathcal{H}_W$ invariants par Φ_Y/Φ_Z; cet ensemble est celui des éléments de $\mathcal{H}_U$ invariants par Φ_Y.

Corollaire 1. — *Soient* Y *et* Z *deux sous-groupes de* X *ayant même rang que* X *et tels que* $Z \subset Y \subset X$: X/Z *a pour fibre* Y/Z *et pour base* X/Y. *L'application de* X/Z *sur* X/Y *a pour réciproque un isomorphisme de* $\mathcal{H}_{X/Y}$ *dans* $\mathcal{H}_{X/Z}$. *Supposons* Y *connexe : l'application topologique de* Y/Z *dans* X/Z *a pour réciproque un homomorphisme de* $\mathcal{H}_{X/Z}$ *sur* $\mathcal{H}_{Y/Z}$; $\mathcal{H}_{X/Z}$ *et* $\mathcal{H}_{X/Y} \otimes \mathcal{H}_{Y/Z}$ *sont des modules isomorphes,* mais non des algèbres isomorphes : si Z est abélien, $\mathcal{H}_{X/Z}$ et $\mathcal{H}_{Y/Z}$ sont engendrés par des éléments de degré 2 ; $\mathcal{H}_{X/Y}$ ne l'est pas.

Corollaire 2. — *a.* $\mathcal{H}_W$ *ne dépend que de* X *et* $Y \cap N_T$. *On obtient donc l'ensemble* $\mathcal{E}(X)$ *des algèbres de cohomologie des espaces homogènes quotients de* X *par un sous-groupe* Y *de même rang en choisissant pour* Y *tous les sous-groupes de* N_T *contenant* T.

b. Les éléments de $\mathcal{H}_U$ *invariants par un sous-groupe de* Φ_X *constituent un anneau; l'ensemble de ces anneaux est* $\mathcal{E}(X)$.

c. $\mathcal{E}(X_1 \times X_2)$ *est l'ensemble des produits tensoriels des éléments de* $\mathcal{E}(X_1)$ *par ceux de* $\mathcal{E}(X_2)$.

GAUTHIER-VILLARS IMPRIMEUR-LIBRAIRE DES COMPTES RENDUS DES SÉANCES DE L'ACADÉMIE DES SCIENCES.
Paris. — Quai des Grands-Augustins 55.

Prolongement de la transformation de Laplace

International Congress of Mathematicans, Stockholm 1962, pp. 360–367

La primitive $I(x,y)$ d'ordre $n-1$, s'annulant $n-1$ fois en y, d'une fonction $F(x)$ de la variable x est donnée par la formule de Cauchy :

$$I(x, y) = \int_y^x \frac{(x-t)^{n-2}}{(n-2)!} F(t)\, dt.$$

Introduisons dans cette formule les fonctions linéaires de x

$$\xi \cdot x = \xi_0 + \xi_1 x, \quad \text{où } \xi = (\xi_0, \xi_1);$$

notons f la fonction homogène de degré $-n$

$$f(\xi) = \frac{1}{\xi_1^n} F\left(-\frac{\xi_0}{\xi_1}\right),$$

$\omega^*(\xi)$ la forme différentielle

$$\omega^*(\xi) = \xi_0\, d\xi_1 - \xi_1\, d\xi_0$$

et h un arc joignant les ensembles de points ξ d'équations respectives

$$y^* : \ \xi \cdot y = 0, \quad x^* : \ \xi \cdot x = 0;$$

il vient
$$I(x, y) = \int_h \frac{(\xi \cdot x)^{n-2}}{(n-2)!} f(\xi)\, \omega^*(\xi).$$

Nous allons étudier l'intégrale analogue en un nombre quelconque de variables; nous obtiendrons ainsi un prolongement $\mathfrak{L}$ de la transformation de Laplace, qui permet de construire la solution élémentaire d'opérateurs différentiels.

1. *La fonction* $\mathfrak{L}[f]$. — Soit $\tilde{\Psi}$ une multiplicité analytique complexe, de dimension complexe

$$\dim_c \tilde{\Psi} = l;$$

ses points sont notés $\tilde{\varphi}$. Soient $\tilde{\Sigma}(x,y)$ un sous-variété analytique de $\tilde{\Psi}$, de codimension 1, d'équation réelle, dépendant *linéairement* de x et *holomorphiquement* de y; x est un point d'un espace affine réel X, ayant même dimension que $\tilde{\Psi}$:

$$\dim X = l;$$

y est un point d'un domaine Y de X.

Pour écrire l'équation de $\tilde{\Sigma}$, notons $(x_1, \ldots, x_l)$ les coordonnées de $x \in X$ et

$$\xi \cdot x = \xi_0 + \xi_1 x_1 + \ldots + \xi_l x_l$$

une fonction linéaire, numérique complexe, sur X; ξ est donc un vecteur d'un espace vectoriel complexe Ξ;

$$\dim_c \Xi = l+1.$$

Notons ξ^* l'image de $\xi \neq 0$ dans l'espace projectif

$$\Xi^* = (\Xi - 0)/\Lambda$$

quotient de $\Xi - 0$ par le groupe Λ de ses homothéties; autrement dit : ξ^* est l'hyperplan de X d'équation $\xi \cdot x = 0$.

L'équation locale de $\tilde{\Sigma}(x, y)$ est donc

$$\tilde{\Sigma}(x, y) : \ \xi(\tilde{\varphi}, y) \cdot x = 0,$$

$\xi(\tilde{\varphi}, y)$ étant une application holomorphe d'un domaine de $\tilde{\Psi} \times Y$ dans Ξ; deux choix différents de $\xi(\tilde{\varphi}, y)$ sont proportionnels; nous supposerons que

$$\xi(\tilde{\varphi}, y) \neq 0;$$

ces divers $\xi(\tilde{\varphi}, y)$ définissent donc une application holomorphe de $\tilde{\Psi} \times Y$ tout entier :

$$\xi^*(\tilde{\varphi}, y) : \tilde{\Psi} \times Y \to \Xi^*;$$

nous la nommerons *projection*; elle projette $\tilde{\Sigma}(x, y)$ (et $\tilde{\Sigma}(y, y)$) dans les hyperplans x^* (et y^*) de Ξ^* d'équations :

$$x^* : \xi \cdot x = 0, \quad y^* : \xi \cdot y = 0.$$

Nous notons $K(y)$ l'ensemble des points x de Y tels que $\tilde{\Sigma}(x, y)$ ait une singularité.

Nous notons $\omega^*(\xi)$ la forme différentielle extérieure

$$\omega^*(\xi) = \textstyle\sum_{j=0}^{l} (-1)^j \xi_j \, d\xi_0 \wedge \ldots \wedge d\xi_{j-1} \wedge d\xi_{j+1} \wedge \ldots \wedge d\xi_l;$$

son produit par une fonction de ξ, homogène de degré $-l-1$, est évidemment une forme différentielle de ξ^*.

Soit $f(\xi, y)$ une fonction numérique complexe, homogène en ξ, de degré $-n$ (n : entier >0, <0 ou $=0$), analytique en (ξ, y), en général multiforme et définie seulement sur une partie de $\Xi \times Y$. Nous dirons que *la projection* $\xi^*(\varphi, y)$ *uniformise la fonction* $f(\xi, y)$ (*ou la forme* $f(\xi, y)\omega^*(\xi)$) quand

$$f(\xi(\tilde{\varphi}, y), y) \quad (\text{ou } f(\xi(\tilde{\varphi}, y)\omega^*(\xi(\tilde{\varphi}, y)))$$

est une fonction (ou forme) de $(\tilde{\varphi}, y)$ holomorphe sur le domaine de $\tilde{\Psi} \times Y$ où $\xi(\tilde{\varphi}, y)$ est holomorphe et non nul; si $n = 0$ (ou $= l+1$), cette fonction (ou forme) dépend de $\xi^*(\tilde{\varphi}, y)$ et $f(\xi, y)$ sans dépendre du choix de $\xi(\tilde{\varphi}, y)$, qu'on fait au voisinage de chaque point.

Supposons que $\tilde{\Sigma}(y, y)$ contienne une sous-multiplicité $\tilde{\Sigma}_1$ de $\tilde{\Psi}$, indépendante de y; supposons $\tilde{\Sigma}(x,y)$ et $\tilde{\Sigma}_1$ en position générale pour $x \neq y$. Supposons donnée

si $n \leqslant l$, une classe d'homologie $h(\tilde{\Sigma}, \tilde{\Sigma}_1)$ de $\tilde{\Sigma}$ relativement à $\tilde{\Sigma}_1$, ayant la dimension $l-1$;

si $n > l$ une classe d'homologie $h(\tilde{\Psi}, \tilde{\Sigma} \cup \tilde{\Sigma}_1)$ de dimension l et notons alors $h(\tilde{\Sigma}, \tilde{\Sigma}_1)$ son bord dans $\tilde{\Sigma}$:

$$\partial h(\tilde{\Psi}, \tilde{\Sigma} \cup \tilde{\Sigma}_1) = h(\tilde{\Sigma}, \tilde{\Sigma}_1). \tag{1}$$

Supposons que ces classes dépendent continûment(¹) de (x, y).

Définissons la fonction $\mathfrak{L}[f]$ de (x, y), pour $y \in Y$ et $x \in Y - K(y)$, par l'intégrale:

$$\mathfrak{L}[f] = \frac{1}{(2\pi i)^{l-1}} \frac{1}{2} \int_{h(\tilde{\Psi},\, \tilde{\Sigma} \cup \tilde{\Sigma}_1)} \frac{(\xi \cdot x)^{n-l-1}}{(n-l-1)!} f(\xi, y)\, \omega^*(\xi) \quad \text{si } n > l, \tag{2.1}$$

$$= \frac{(-1)^{n-l-1}}{(2\pi i)^{l-1}} \frac{1}{2} \int_{h(\tilde{\Sigma}, \tilde{\Sigma}_1)} \frac{d^{l-n}[f(\xi \cdot y)\, \omega^*(\xi)]}{(d\xi \cdot x)^{1+l-n}} \quad \text{si } n \leqslant l, \tag{2.2}$$

où $\xi^* = \xi^*(\tilde{\varphi}, y)$; on suppose $f\omega^*$ uniformisé par cette projection, la seconde intégrale porte sur une classe-résidu (voir : [7], III).

La formule (2.1) généralise (1) puisque les projections de Σ (et $\tilde{\Sigma}_1$) sont dans x^* [et dans y^*]. Cette fonction $\mathfrak{L}[f]$ de (x, y) sera notée parfois $\mathfrak{L}[f]\,(x, y)$.

Voici ses *propriétés* :

Pour x et $y \in Y$, $x \notin K(y)$, $\mathfrak{L}[f]$ est une fonction *holomorphe* de (x, y).

Si les $c_i(y)$ sont des fonctions holomorphes de y et si une même projection uniformise les $f_i \omega^*$, alors

$$\sum_i c_i(y) \mathfrak{L}[f_i] = \mathfrak{L}[\sum_i c_i f_i]. \tag{3}$$

Quand $f\omega^*$ est uniformisé, alors

$$\frac{\partial}{\partial x_j} \mathfrak{L}[f] = \mathfrak{L}[\xi_j f]\,; \left[n - l - 1 - \sum_j x_j \frac{\partial}{\partial x_j}\right] \mathfrak{L}[f] = \mathfrak{L}[\xi_0 f]. \tag{4}$$

Quand f est uniformisé, alors

$$x_j \mathfrak{L}[f] = -\mathfrak{L}\left[\frac{\partial f}{\partial \xi_j}\right], \quad \mathfrak{L}[f] = -\mathfrak{L}\left[\frac{\partial f}{\partial \xi_0}\right] \tag{5}$$

$$\frac{\partial}{\partial y_j} \mathfrak{L}[f] = \mathfrak{L}\left[\frac{\partial f}{\partial y_j}\right]. \tag{6}$$

On a
$$\mathfrak{L}[f(\Theta\xi, y)]\,(x, y) = t^{l-n} \mathfrak{L}[f(\xi, y)]\,(Tx, y), \tag{7}$$

(¹) Quand (x,y) varie suffisamment peu, elles contiennent un cycle variant continûment.

T désignant l'homothétie de centre y, de rapport t :

$$T : x \to y + t(x-y)$$

et Θ la transformation, opérant sur Ξ et Ξ^* :

$$\Theta : \xi = (\xi_0, \xi_1, \ldots, \xi_l) \to (t\xi \cdot y - \xi_1 y_1 - \ldots - \xi_l y_l, \xi_1, \ldots, \xi_l),$$

qui vérifie $$\Theta\xi \cdot Tx = t\xi \cdot x.$$

Voici deux conséquences évidentes de ces propriétés :

Si $(\xi \cdot y)^{p+1} f(\xi)$ est uniformisé et si $f(\xi)$ est indépendant de ξ_0, alors

$$\mathfrak{L}\left[\frac{(-\xi \cdot y)^p}{p!} f(\xi)\right] = \mathfrak{L}\left[\frac{(-\xi \cdot y)^{p+1}}{(p+1)!} f(\xi)\right]$$

ne dépend que de $x-y$ et est homogène en $x-y$ de degré $n-l$. Si $f(\xi)$ est un polynome, alors la *fonction* $\mathfrak{L}[f]$ est nulle; cependant la *distribution* $\mathfrak{L}[f]$, que nous allons définir, ne le sera pas : elle aura pour support $x=y$.

2. *La distribution* $\mathfrak{L}[f]$ *pour* $x \neq y$. — Faisons maintenant les hypothèses suivantes :

$$y \in K(y) ;$$

pour $x \in K(y) - y$, $\tilde{\Sigma}(x, y)$ a un seul point singulier, $\tilde{\sigma}(x, y)$; $\tilde{\sigma}$ appartient à une partie compacte de $\tilde{\Psi}$; $\tilde{\sigma} \notin \tilde{\Sigma}_1$; $\tilde{\sigma}$ est un point double quadratique de $\tilde{\Sigma}$, c'est-à-dire, en notant Hess le Hessien :

$$\xi(\tilde{\varphi}, y) \cdot x = 0, \ \frac{\partial \xi(\tilde{\varphi}, y)}{\partial \tilde{\varphi}} \cdot x = 0, \ \mathrm{Hess}_{\tilde{\varphi}}\, [\xi(\tilde{\varphi}, y) \cdot x] \neq 0 \text{ pour } \tilde{\varphi} = \tilde{\sigma}(x, y).$$

Ces hypothèses ont les conséquences suivantes : $K(y) - y$ est une *sous-multiplicité* de Y, enveloppe des hyperplans de Y d'équation $\xi(\tilde{\varphi}, y)$. $x=0$ (y fixe, $\tilde{\varphi}$ variable); nous noterons l'équation locale ou cette sous-multiplicité :

$$K(y) - y : \ k(x, y) = 0.$$

La partie singuliére de $\mathfrak{L}[f]$, sur $K(y)$, est

$$H_0 k^{n-1-l/2} + H_1 \log k \quad (l \text{ pair} \geqslant 2n),$$

$$H_0 k^{n-1-l/2} \log k \quad (l \text{ pair} < 2n),$$

$$H_0 k^{n-1-l/2} \quad (l \text{ impair}),$$

où $H_i = H_i(x, y)$ est holomorphe; la restriction de H_0 à K se calcule sans quadrature.

L'étude de cette partie singulière définit sans ambiguité sur $X-y$ une *distribution* de x, fonction de y, $\mathfrak{L}[f]$; sa restriction à $X - K(y)$ est la fonction $\mathfrak{L}[f]$; cette distribution $\mathfrak{L}[f]$ possède les propriétés qu'énonce le n° 1.

3. *Le conoïde* $K(y)$. — Complétons l'hypothèse que $\tilde{\Sigma}(y, y)$ contient une sous-multiplicité $\tilde{\Sigma}_1$ de $\tilde{\Psi}$. Notons Γ l'ensemble des points singuliers de

$\dot{\Sigma}(y, y)$ appartenant à $\tilde{\Sigma}_1$; nous le supposerons non vide(¹) et $\neq \tilde{\Sigma}_1$; faisons alors les hypothèses les plus simples possibles :

1° $\tilde{\Sigma}_1$ *est l'espace projectif complexe*, de $\dim_c : l-1$;

2° $\tilde{\Sigma}_1 \cap \tilde{\Sigma}(x, y)$ *est un hyperplan de* $\tilde{\Sigma}_1$, arbitraire quand x décrit un voisinage de y; par contre $\tilde{\Gamma}$ n'est pas un hyperplan;

3° $\tilde{\Psi}$ est un voisinage de $\tilde{\Sigma}_1$ dans *un espace fibré* $\tilde{\Phi}$ *ayant* $\tilde{\Sigma}_1$ *pour section* et la droite complexe pour fibre.

D'après H. Grauert, son article [4] prouve, vu 1° et 2°, que 3° équivaut à

3°bis $\tilde{\Sigma}_1$ est *exceptionnelle* dans $\tilde{\Psi}$; c'est-à-dire : remplacer $\tilde{\Sigma}_1$ par un point, en conservant la structure analytique de $\tilde{\Psi}-\tilde{\Sigma}_1$, transforme $\tilde{\Psi}$ en un espace analytique.

Voici les *conséquences* de ces hypothèses :

D'après la théorie des espaces fibrés analytiques, on a

$$\tilde{\Phi} = (\Phi - \Phi_0)/\Lambda,$$

où : Φ est l'espace vectoriel complexe de $\dim_c : l+1$;

$(\tau, \eta_1, \ldots, \eta_l)$ désignera les coordonnées de $\varphi \in \Phi$;

Φ_0 est le sous-espace : $\eta_1 = \ldots = \eta_l = 0$ de Φ;

Λ est le groupe de transformations à un paramètre λ :

$\lambda : \varphi = (\tau, \eta_1, \ldots, \eta_l) \to \lambda\varphi = (\lambda^{1-m}\tau, \lambda\eta_1, \ldots, \lambda\eta_l)$ (m : entier).

La projection $$\xi^*(\tilde{\varphi}, y) : \tilde{\Psi} \times y \to \Xi^*$$

est induite par une application commutant avec λ :

$$\xi(\varphi, y) : \Psi \times Y \to \Xi \quad \text{où} \quad \Psi/\Lambda = \tilde{\Psi} ; \xi(\lambda\varphi, y) = \lambda\xi(\varphi, y).$$

Les équations de $\tilde{\Sigma}_1$ et $\tilde{\Sigma}(x, y) \cap \tilde{\Sigma}_1$ sont

$$\tilde{\Sigma}_1 : \tau = 0 ; \quad \tilde{\Sigma}(x, y) \cap \tilde{\Sigma}_1 : \tau = \eta_1(x_1 - y_1) + \ldots + \eta_l(x_l - y_l) = 0.$$

$\xi^*(\tilde{\varphi}, y)$ projette homéomorphiquement $\tilde{\Sigma}_1$ sur y^* : on peut faire en sorte que si $\varphi = (0, \eta_1, \ldots, \eta_l)$, alors $\xi(\varphi, y) = \eta = (\eta_0, \eta_1, \ldots, \eta_l)$ vérifiant $\eta \cdot y = 0$.

Le déterminant fonctionnel

$$\left|\frac{D(\xi_0(\varphi, y), \ldots, \xi_l(\varphi, y))}{D(\tau, \eta_1, \ldots, \eta_l)}\right|_{\tau=0} = g(\eta, y)$$

est un polynome en $\eta_1, \ldots, \eta_l$, homogène de degré m; on nomme g : polynome de la projection $\xi^*(\tilde{\varphi}, y)$; l'équation de $\tilde{\Gamma}$ est donc :

$$\tilde{\Gamma} : \tau = g(\eta, y) = 0 ;$$

donc $m > 1$ (ce qui permet de voir que 1°, 2° et 3° impliquent 3 bis). La projection de $\tilde{\Gamma}$ sur y^* est la variété algébrique :

(¹) Afin de pouvoir définir des classes d'homologie non nulles.

$$g^* : \eta \cdot y = g(\eta, y) = 0.$$

Nous supposerons $\quad$ Hess $g(\eta, y) \neq 0 \quad$ sur[1] Re g^*.

Alors : *$K(y)$ est un conoïde de sommet y*; ses plans tangents η^* en son point conique y sont les $\eta^* \in \mathrm{Re}\, g^*$.

Le support des singularités de $f(\xi, y)$ détermine $K(y)$, à l'exception de celles de ses nappes correspondant aux nappes de Re g^* sur lesquelles $f(\xi, y)$ est holomorphe.

4. *Choix*[2] *de* $h(\check{\Psi}, \tilde{\Sigma} \cup \tilde{\Sigma}_1)$.—Supposons la projection $\xi^*(\tilde{\varphi}, y)$ *hyperbolique*, c'est-à-dire le polynome $g(\eta, y)$ hyperbolique : quand $\eta_2, \ldots, \eta_l$ sont réels, non tous nuls, alors les η_1 annulant g sont réels et distincts.

Définissons d'abord une classe d'homologie $h(y^* - g^*, x^*)$:

si l est impair et $x_1 \geqslant y_1$, c'est celle de 2 Re y^*, muni d'une orientation qui change sur $x^* \cap y^*$, détourné[3] de g^*;

si l est pair et $x_1 \geqslant y_1$, c'est celle du cobord[4] de Re g^*, muni d'une orientation qui change sur $g^* \cap x^*$.

On constate que $\quad h(y^* - g^*, x^*) = 0 \quad$ pour $x_1 = y_1 \quad$ (8)

et on définit $\quad h(y^* - g^*, x^*) = 0 \quad$ pour $x_1 < y_1. \quad$ (9)

Note 4. Si l est *impair* on peut ne pas supposer la projection hyperbolique; on n'aura plus ni (8), ni (9); on choisira toujours $h(y^* - g^*, x^*)$ classe de 2 Re y^* détourné de g^*; ce qui suit vaudra à condition de remplacer $\frac{1}{2}$ par $\frac{1}{4}$ au second membre de (2).

L'homéomorphisme de y^* et $\tilde{\Sigma}_1$ transforme $h(y^* - g^*, x^*)$ en une classe $h(\tilde{\Sigma}_1 - \tilde{\Gamma}, \tilde{\Sigma})$; rappelons que $\tilde{\Sigma}_1 \cap \tilde{\Sigma}$ ne dépend que de la direction de $x - y$. On peut alors définir $h(\tilde{\Sigma}, \tilde{\Sigma}_1)$ par la condition de dépendre continûment de (x, y) pour $x \neq y$ et d'être voisin[5] de $h(\tilde{\Sigma}_1 - \tilde{\Gamma}, \tilde{\Sigma})$ quand x tend vers y, le long d'un segment non tangent à $K(y)$.

On définit enfin $h(\check{\Psi}, \tilde{\Sigma} \cup \tilde{\Sigma}_1)$ par la condition (1).

Ce choix a les conséquences suivantes :

$\mathfrak{L}[f] = 0$ *pour* $x_1 < y_1$, si g est hyperbolique;

$\mathfrak{L}[f] = 0$ pour $x \neq y$ signifie que f est un polynome;

$\mathfrak{L}[f]$ *ne dépend pas du choix de la projection $\xi^*(\tilde{\varphi}, y)$ qui uniformise f; si $1 + (l/2) < n$, $\mathfrak{L}[f]$ est une fonction $O(|x - y|^{n-l})$* pour $x - y$ petit.

5. *La distribution* $\mathfrak{L}[f]$. Supposons *$f(\xi, y)$ rationnellement uniformisable;* c'est-à-dire : quel que soit le polynome homogène $b(\xi)$ de $(\xi_1, \ldots, \xi_2)$, il existe une projection $\xi^*(\tilde{\varphi}, y)$ uniformisant $f(\xi, y)/b(\xi)$, ayant pour polynome $g(\eta, y) b(\eta)$, g étant indépendant de b. Supposons f hyperbolique, c'est-à-dire g hyperbolique (ou bien l impair : voir Note 4).

---

(1) Re...: partie réelle de ....

(2) L. Gårding a contribué à ce choix.

(3) *Détourner* d'une sous-variété analytique complexe des segments réels orientés la coupant, c'est remplacer chaque segment par la demi-somme de deux demi-circonférences, imaginaires conjuguées l'une de l'autre, ayant ce segment pour diamètre, ayant son orientation, ne rencontrant pas la sous-variété.

(4) Prendre *le cobord* de points d'une sous-variété analytique, c'est remplacer chaque point par une circonférence enlacée avec cette sous-variété.

(5) Contenir un cycle voisin d'un cycle choisi dans ....

Il existe alors une distribution $\mathfrak{L}[f]$ de $x \in Y$, fonction de $y \in Y$ et possédant toutes les propriétés énoncées ci-dessus : vu (4) on la définit par la formule

$$\mathfrak{L}[f] = b\left(\frac{\partial}{\partial x}\right) \mathfrak{L}[f(\xi, y)/b(\xi)],$$

où b est hyperbolique de degré $> 1 + (l/2) - n$, $\mathfrak{L}[f/b]$ est une fonction et $b\ \mathfrak{L}[f/b]$ est une distribution indépendante du choix de b.

$\mathfrak{L}[f] = 0$ implique $f = 0$; $\mathfrak{L}[1]$ est défini et vaut

$$\mathfrak{L}[1] = \delta(x-y) \quad (\delta : \text{mesure de Dirac}); \tag{10}$$

plus généralement, *quand $f(\xi)$ est une fonction rationnelle* homogène de $(\xi_1, \ldots, \xi_l)$ à dénominatueur hyperbolique, $\mathfrak{L}[f]$ est l'expression de sa transformée de Laplace classique qu'ont donnée Herglotz [5] et Petrowsky [8].

Note. Rappelons que Gel'fand, Shapiro, Shilov [3], Gårding [2] ont récemment étudié les transformées de Fourier des fonctions homogènes.

6. *La solution élémentaire* $E(x,y)$ d'un opérateur différentiel $a(y, \partial/\partial y)$, linéaire, d'ordre m et hyperbolique si l est pair est

$$E(x,y) = \mathfrak{L}[U(\xi,y)] \tag{11}$$

à condition que
$$a\left(y, \frac{\partial}{\partial y}\right) U(\xi, y) = 1, \tag{12}$$

$U(\xi,y)$ et ses dérivées en y d'ordres $< m$ soient rationnellement uniformisables. (13)

Or (13) implique que :

$$U(\xi,y) \text{ s'annule } m \text{ fois pour } \xi \cdot y = 0; \tag{14}$$

et la solution du problème de Cauchy (12), (14) possède la propriété (13); voir [7] II.

La formule (11), ainsi obtenue, permet d'analyser les singularités de $E(x,y)$ et aussi de calculer explicitement $E(x,y)$ dans le cas suivant : les coefficients de a d'ordres m, $m-1$ et $< m-1$ sont repectivement linéaires, constants et nuls [7]. F. John [6] a, dans le cas elliptique, déjà employé (11) (l impair) et pu traiter le cas : l pair.

7. *Un prolongement de la convolution par une transformée de Laplace*, analogue au prolongement $\mathfrak{L}$ de la transformation de Laplace, permet de résoudre de même le problème de Cauchy à données analytiques; voir [7] V; il se trouve, en partie, chez Fantappiè [1].

BIBLIOGRAPHIE

[1]. Fantappiè, L., L'indicatrice proiettiva dei funzionali lineari e i prodotti funzionali proiettivi. *Annali Mat. ser. IV*, 13 (1943).

—— *Second Colloque sur les Equations aux Dérivées Partielles* (Bruxelles, 1954), 95–128.

—— La théorie des équations aux dérivées partielles. *Colloque C.N.R.S.* (Nancy, 1956), 47–62.

[2]. GÅRDING, L., Transformation de Fourier des distributions homogènes. *Bull. Soc. Math. France*, 89 (1961).

[3]. GEL'FAND, I. M. & SHILOV, G. E., *La Théorie des Distributions Dunod*. 1961.

[4]. GRAUERT, H., Über Modifikationen und exceptionelle analytische Mengen. *Math. Ann.*, 146, (1962), 331–368.

[5]. HERGLOTZ, G., Über die Integration linearer, partieller Differentialgleichungen mit konstanten Koeffizienten. *Ber. Verh. Sächs. Akad. Wiss. Leipzig, Math. phys.*, 78 (1926), 92–126 et 287–318; 80 (1928), 68–114.

[6]. JOHN, F., The fundamental solution of linear elliptic differential equations with analytic coefficients. *Comm. Pure Appl. Math.*, 3 (1950), 213–304.

[7]. LERAY, J., Problème de Cauchy. I. *Bull. Soc. math. France*, 85 (1957), 389–429; II. *ibid.*, 86 (1958), 75–96; III. *ibid.*, 87 (1959), 81–180; IV. *ibid.*, 90 (1962), 39–156; V. *C.R. Acad. Sc. Paris*, 242 (1956), 953–959; VI. *Cahiers de physique*, 133 (1961), 373–381.

[8]. PETROWSKY, I., On the diffusion of waves and the lacunas for hyperbolic equations. *Mat. Sb.* (Recueil math.), 17 (1945), 289–370.

ANALYSE MATHÉMATIQUE. — *La théorie des résidus sur une variété analytique complexe.* Note de M. **Jean Leray.**

H. Poincaré a défini la *forme-résidu* d'une forme différentielle *fermée*, ayant une singularité polaire d'ordre 1 sur une sous-variété S. Dans l'anneau des formes régulières hors de S, holomorphes ou *non*, nulles sur S′ de toute forme *fermée* φ est cohomologue à des formes ayant sur S des singularités polaires d'ordre 1; leurs résidus constituent une classe de cohomologie de S rel. S′ : c'est la *classe-résidu* de φ.

1. Notations. — X est une variété analytique complexe, de dimension complexe l; $S_1, \ldots, S_m, S'_1, \ldots, S'_M, S''$ en sont des sous-variétés analytiques complexes, régulières, de codimension 1, en position générale

$$S = S_1 \cap \ldots \cap S_m; \qquad S' = S'_1 \cup \ldots \cup S'_M.$$

S_i a, près de $y \in S_i$, une équation locale

$$S_i: \quad s_i(x, y) = 0,$$

$s_i(x, y)$ étant une fonction de x, définie et holomorphe près de y; $\text{grd}_x s_i \neq 0$.

Une forme différentielle *régulière* sur X sera une forme extérieure $\varphi(x)$ des différentielles des parties réelles et imaginaires des coordonnées de $x \in X$; ses coefficients seront des fonctions numériques complexes, indéfiniment dérivables, de x. On dit $\varphi(x)$ *fermée* quand $d\varphi = 0$. On dit que $\varphi(x)$, régulière sur $X - S_1$, a une singularité *polaire* d'ordre p sur S_1 quand $s_1(x, y)^p \varphi(x)$ est régulière près de tout point y de S_1.

C. R., 1958, 2e *Semestre.* (T. 247, N° **25.**)

Reprinted from C. R. Acad. Sci. Paris, Vol. 247 (1958) pp. 2253–2257.

2. Forme-résidu. — ($m = 1$: $S = S_1$). Soit $\varphi(x)$ une forme *fermée* de $X - S$ ayant sur S *une singularité polaire d'ordre* 1. Alors en chaque point $y \in S$ existent des formes ψ et θ telles que

$$\varphi(x) = \frac{ds(x, y)}{s(x, y)} \wedge \psi(x, y) + \theta(x, y) \qquad \text{pour} \quad dy = 0;$$

la restriction de $\psi(x, y)$ à S ne dépend que de φ et S : c'est une forme *fermée* de S [Poincaré ([4]), de Rham ([5])]; notons-la

$$\text{rés}[\varphi] = \left.\frac{s\varphi}{ds}\right|_S.$$

Si $\varphi = 0$ sur S', alors rés$[\varphi] = 0$ sur S'. Si φ est holomorphe, alors rés$[\varphi]$ est holomorphe; c'est encore vrai, pour $d^0\varphi < l$, si l'on permet à S des points singuliers où $\text{grd}_x s = 0$, $\text{Hessien}_x(s) \neq 0$: rés$[\varphi]$ est la restriction à S de formes holomorphes en ces points : de Rham [[6]],

Exemple. — Si $\varphi(x) = [f(x, y)/s(x, y)]\, dx_1 \wedge \ldots \wedge dx_l$, $f(x, y)$ étant une fonction de x régulière près de y, alors

$$\text{rés}[\varphi] = f\frac{dx_2 \wedge \ldots \wedge dx_l}{s_{x_1}} = -f\frac{dx_1 \wedge dx_3 \wedge \ldots \wedge dx_l}{s_{x_2}} = \ldots.$$

3. Formule du résidu. — ($m = 1$). Notons $H_c(S, S')$ le groupe d'homologie, à supports compacts et à coefficients entiers, de S relativement à S'. Soit γ un cycle compact de S rel. S'. On peut construire un cycle $\delta\gamma$ de $X - S$ rel. S', fibré par des circonférences enlaçant une fois S, la base de cette fibration étant γ, un point de $\delta\gamma$ et sa projection sur γ appartenant à S' simultanément. δ induit un homomorphisme

$$\delta : \quad H_c(S, S') \to H_c(X - S, S')$$

qui sera nommé *cobord*; en effet, si S' est vide, $H_c(S, S') = H_c(S)$, $H_c(X - S)$ et δ s'identifient aux anneaux de cohomologie de S, de $X - S$ et à l'homomorphisme cobord de la cohomologie à supports compacts ([3]).

On a *la formule du résidu*, où $h(S, S') \in H_c(S, S')$, $\dim h = d^0\varphi - 1$:

$$\int_{\delta h(S, S')} \varphi = 2\pi i \int_{h(S, S')} \text{rés}[\varphi].$$

δ appartient à un triplet exact d'homomorphismes [si S' est vide, *voir* ([3])] :

$$\begin{array}{ccc} H_c(X, S') & \xleftarrow{\ \iota\ } & \\ \downarrow \pi & & \nwarrow \\ H_c(S, S') & \xrightarrow{\ \delta\ } & H_c(X - S, S') \end{array} \qquad \dim \pi = -2; \quad \dim \iota = 0; \quad \dim \delta = 1$$

ι est induit par l'application identique de $X - S$ dans X; π est induit par l'intersection par S.

$\delta^{-1}h(X - S, S')$ est l'ensemble des classes d'homologie $h(S, S')$ des intersections par S des chaînes de X dont le bord est dans $h(X - S, S')$: on peut donc énoncer la formule du résidu comme l'ont fait Poincaré et de Rham.

4. Définition de la classe-résidu. — ($m = 1$). Nommons forme de (X, S') toute forme différentielle *régulière* de X, nulle sur S'. L'anneau de cohomologie $H^*(X, S')$ des formes de (X, S'), est l'anneau de cohomologie, à coefficients numériques complexes et à supports arbitraires, de X relativement à S'.

Théorème. — *Soit* $\varphi(x)$ *une forme* fermée *de* $(X - S, S')$; *elle est cohomologue dans* $(X - S, S')$ *à des formes ayant sur* S *des singularités polaires d'ordre* 1; *leurs formes-résidus constituent une classe de cohomologie de* (S, S').

Cette classe est nommée *classe-résidu* de φ et est notée

$$\text{Rés}[\varphi] = \text{Rés}[h^*(X - S, S')],$$

h^* étant la classe de φ. Évidemment

$$\text{rés}[\varphi] \in \text{Rés}[\varphi] \qquad \text{si rés}[\varphi] \text{ existe};$$

$$\int_{\delta h(S, S')} \varphi = 2\pi i \int_{h(S, S')} \text{Rés}[\varphi].$$

Note. — Ce théorème serait faux si l'on remplaçait l'anneau des formes régulières par celui des formes holomorphes.

Note. — Si $l = 1$, nous nommons donc $\text{Rés}[f(x)\,dx]$ les nombres nommés habituellement résidus de f.

Preuve. — Vu les traités de Lefschetz (²) et de Rham (⁷), $H^*(X)$ est le dual de $H_c(X)$. De l'exactitude des triplets d'homologie et cohomologie relative (n° 5), résulte alors que $H^*(X, S')$ est le dual de $H_c(X, S')$. On peut donc définir le triplet δ^*, ι^*, π^* transposé de δ, ι, π :

$$\int_{\delta h} h^* = \int_h \delta^* h^*; \qquad \dots.$$

Son exactitude permet de construire π^*, puis δ^* par des opérations de calcul différentiel. Évidemment : $\delta^* = 2\pi i\,\text{Rés}$.

5. Propriétés de la classe-résidu. — ($m = 1$). $H^*(X - S, S')$ et $H^*(S, S')$ sont des algèbres sur $H^*(X)$ et Rés est un homomorphisme d'algèbre :

$$\text{Rés}[h^*(X - S, S').h^*(X)] = \text{Rés}[h^s(X - S, S')].h^*(X).$$

Notons p^*, i^*, ∂^* le triplet exact de la cohomologie relative : p^* est induit par l'application identique de l'anneau des formes de $(X - S, S'' \cup S')$ ou de $(S, S'' \cup S')$ dans celui de $(X - S, S')$ ou (S, S'); i^* est induit par la restriction à S'' ; ∂^* est le cobord. Dans le diagramme

$$\begin{array}{ccccc}
H^*(X-S, S') & & \xrightarrow{\text{Rés}} & & H^*(S, S') \\
\Big\downarrow i^* & \nwarrow p^* & & p^* \nearrow & \Big\downarrow i^* \\
 & H^*(X-S, S''\cup S') & \xrightarrow{\text{Rés}} & H^*(S, S''\cup S') & \\
 & \nearrow \partial^* & & \nwarrow \partial^* & \\
H^*(S''-S\cap S'', S') & & \xrightarrow{\text{Rés}} & & H^*(S\cap S'', S')
\end{array}$$

on a les règles de commutation :

$$\text{Rés}\, p^* = p^*\, \text{Rés}, \qquad \text{Rés}\, i^* = i^*\, \text{Rés}, \qquad \text{Rés}\, \partial^* = -\partial^*\, \text{Rés}.$$

6. Résidus composés. — $(m > 1)$. En composant les homomorphismes

$$H_c(S, S') \ldots \xrightarrow{\delta} H_c(S_1 \cap \ldots \cap S_i - S_{i+1} \cup \ldots \cup S_m, S') \xrightarrow{\delta} \ldots H_c(X - S_1 \cup \ldots \cup S_m, S')$$
$$H^*(X - S_1 \cup \ldots \cup S_m, S') \ldots \xrightarrow{\text{Rés}} H^*(S_1 \cap \ldots \cap S_i - S_{i+1} \cup \ldots \cup S_m, S') \xrightarrow{\text{Rés}} \ldots H^*(S, S')$$

on définit le cobord composé δ^m et le résidu composé Rés^m :

$$\delta^m : \quad H_c(S, S') \to H_c(X - S_1 \cup \ldots \cup S_m, S');$$
$$\text{Rés}^m : \quad H^*(X - S_1 \cup \ldots \cup S_m, S') \to H^*(S, S');$$
$$\int_{\delta^m h} \varphi = (2\pi i)^m \int_h \text{Rés}^m[\varphi];$$

on peut aussi définir rés^m. Ces compositions de δ, Rés et rés sont *associatives* et *anticommutatives* : permuter deux des S_i les change de signe.

7. Cas où $S_1, \ldots, S_m$ ont des équations globales. — Supposons que chaque S_i possède une équation globale près de S :

$$S_i : \quad s_i(x) = 0 \qquad [s_i(x) \text{ holomorphe près de } S].$$

Soit $\varphi(x)$ une forme régulière sur X, telle que

$$d\varphi = 0, \qquad ds_1 \wedge \varphi = 0, \qquad \ldots, \qquad ds_m \wedge \varphi = 0.$$

Alors $\text{Rés}^m(s_1^{-p} \ldots s_m^{-q} \varphi)$ est défini quels que soient les entiers positifs $p, \ldots, q$ et peut s'obtenir par *la construction* ([8]) *de Gelfand et Šilov* ([1]) que voici : il existe des formes $\varpi, \varpi_i, \varpi_{ij}, \ldots$ $(i, j_{=1}, \ldots, m)$ telles que

$$\varphi = ds_1 \wedge \ldots \wedge ds_m \wedge \varpi,$$
$$d\varpi = ds_1 \wedge \varpi_1 + \ldots + ds_m \wedge \varpi_m,$$
$$d\varpi_{i\ldots j} = ds_1 \wedge \varpi_{i\ldots j1} + \ldots + ds_m \wedge \varpi_{i\ldots jm};$$

de quelque façon qu'on choisisse ces formes, on a sur S :

$$\varpi_{p\ldots q}|s \in p! \ldots q!\, \text{Rés}^m \left[\frac{\varphi(x)}{s_1^{1+p}(x) \ldots s_m^{1+q}(x)} \right].$$

8. **Notation différentielle du résidu.** — La construction qui précède suggère la notation suivante, que nous utiliserons ultérieurement : Soit $\omega(x, y)$ une forme de x, définie et *régulière* près de chaque point y de S, *nulle* sur S′, telle que la forme

$$\varphi(x) = \frac{\omega(x, y)}{s_1^{1+p}(x, y) \ldots s_m^{1+q}(x, y)}$$

soit *indépendante* de y et *fermée*; Rés$^m[\varphi]$ est définie; c'est une *classe de cohomologie* de (S, S′); nous la noterons comme suit :

$$\left.\frac{d^{p+\ldots+q}[\omega(x, y)]}{ds_1(x, y)^{1+p} \wedge \ldots \wedge ds_m(x, y)^{1+q}}\right|_{(S, S')} = p! \ldots q! \,\text{Rés}^m\left[\frac{\omega(x, y)}{s_1(x, y)^{1+p} \ldots s_m(x, y)^{1+q}}\right].$$

Si $p = \ldots = q = 0$, rés$^m[\varphi]$ est définie; c'est une *forme* de S, *nulle* sur S′, appartenant à la classe précédente; nous noterons cette forme

$$\left.\frac{\omega(x, y)}{ds_1(x, y) \wedge \ldots \wedge ds_m(x, y)}\right|_{S} = \text{rés}^m\left[\frac{\omega(x, y)}{s_1(x, y) \ldots s_m(x, y)}\right].$$

(1) Gelfand et Šilov, *Les fonctions généralisées et leurs opérations*, I, 1958, chap. III, § 1, n° 5, p. 261.

(2) S. Lefschetz, *Algebraic Topology*, 1942.

(3) J. Leray, *J. Math. pures et appl.*, 24, 1945, chap. IV, § 1, p. 170; *Comm. Math. Helv.*, 20, 1947, p. 177.

(4) H. Poincaré, *Acta Mathematica*, 9, 1886, p. 321-380.

(5) G. de Rham, *Enseignement mathématique*, 35, 1936, p. 213-228.

(6) G. de Rham, *Comm. Math. Helv.*, 28, 1954, p. 346-352.

(7) G. de Rham, *Variétés différentiables*, 1955.

(8) J'avais moi-même utilisé cette construction, *Congrès math. canadien*, 1955 (Notes miméographiées).

ANALYSE MATHÉMATIQUE. — *Le calcul différentiel et intégral sur une variété analytique complexe.* Note (*) de M. **Jean Leray**.

Nous énonçons *les propriétés* des classes de cohomologie antérieurement notées $d^{p+\dots+r}[\omega]/ds_1^{1+p}\wedge\dots\wedge ds_m^{1+r}|_{(S,S')}$. Nous notons $\partial^{p+\dots+r}[\varpi]/\partial s_1^p\dots\partial s_m^r|_{(S,S')}$ certaines d'entre elles, qui ont *les propriétés formelles des dérivées partielles* des fonctions. Nous déterminons certaines autres par *la formule de Cauchy-Fantappiè.* Enfin nous les employons à l'étude des *intégrales, fonctions d'un paramètre.*

Nous conservons les notations et définitions d'une Note antérieure ([2]).

1. Les propriétés de $d^{p+\dots+r}[\omega]/ds_1^{1+p}\wedge\dots\wedge ds_j^{1+q}\wedge\dots\wedge ds_m^{1+r}|_{(S,S')}$ résultent immédiatement de cette Note : ce symbole est défini quand $\omega(x, y)$ est une forme différentielle de x, régulière près de $y\in S$ et que

$$\varphi(x)=\frac{\omega(x,y)}{s_1(x,y)^{1+p}\dots s_j(x,y)^{1+q}\dots s_m(x,y)^{1+r}}$$

est une forme de x, définie au voisinage de S, *fermée*, *nulle sur* S′, indépendante de y; ce symbole représente une classe de cohomologie, à supports arbitraires, à coefficients numériques complexes, de S rel. S′; cette classe ne dépend que de φ, S, S′; son degré est $d^0(\omega)-m$; elle est multipliée par -1 quand on permute deux des symboles ds_j^{1+q}; si $p\leqq P, \dots, r\leqq R$,

$$\frac{d^{p+\dots+r}[\omega]}{ds_1^{1+p}\wedge\dots\wedge ds_m^{1+r}}\bigg|_{(S,S')}=\frac{p!\dots r!}{P!\dots R!}\,\frac{d^{P+\dots+R}[s_1^{P-p}\dots s_m^{R-r}\omega]}{ds_1^{1+P}\wedge\dots\wedge ds_m^{1+R}}\bigg|_{(S,S')};$$

$$\frac{d^{p+\dots+q+r}[\omega]}{ds_1^{1+p}\wedge\dots\wedge ds_{m-1}^{1+q}\wedge ds_m^{1+r}}\bigg|_{(S,S')}=r!\,\text{Rés}\,\frac{d^{p+\dots+q}}{ds_1^{1+q}\wedge\dots\wedge ds_{m-1}^{1+\sigma}}\left(\frac{\omega}{s_m^{1+r}}\right)\bigg|_{(S_1\cap\dots\cap S_{m-1}-S_m,\,S')}.$$

Si $\psi(x)$ est une forme fermée régulière sur X et si $h^*(X)$ est sa classe de cohomologie, alors

$$\frac{d^{p+\dots+r}[\omega\wedge\psi]}{ds_1^{1+p}\wedge\dots\wedge ds_m^{1+r}}\bigg|_{(S,S')}=\frac{d^{p+\dots+r}[\omega]}{ds_1^{1+p}\wedge\dots\wedge ds_m^{1+r}}\bigg|_{(S,S')}.h^*(X).$$

Reprinted from C. R. Acad. Sci. Paris, Vol. 248 (1959), pp. 22–28.

Soit p^*, i^*, ∂^* le triplet exact de la cohomologie relative [(²), n° 5];

$$\begin{array}{ccc} H^*(S, S') & & \\ {\scriptstyle i^*}\downarrow & \nwarrow{\scriptstyle p^*} & \\ H^*(S\cap S'', S') & \xrightarrow{\partial^*} & H^*(S, S''\cup S') \end{array}$$

on a

$$p^* \frac{d^{p+\dots+r}[\omega]}{ds_1^{1+p}\wedge\dots\wedge ds_m^{1+r}}\Bigg|_{(S,S''\cup S')} = \frac{d^{p+\dots+r}[\omega]}{ds_1^{1+p}\wedge\dots\wedge ds_m^{1+r}}\Bigg|_{(S,S')};$$

$$i^* \frac{d^{p+\dots+r}[\omega]}{ds_1^{1+p}\wedge\dots\wedge ds_m^{1+r}}\Bigg|_{(S,S')} = \frac{d^{p+\dots+r}[\omega]}{ds_1^{1+p}\wedge\dots\wedge ds_m^{1+r}}\Bigg|_{(S\cap S'',S')};$$

$$(1)\quad \partial^* \frac{d^{p+\dots+r-m}[\varpi]}{ds_1^{p}\wedge\dots\wedge ds_m^{r}}\Bigg|_{(S\cap S'',S')}$$
$$= (-1)^m \frac{d^{p+\dots+r}}{ds_1^{1+p}\wedge\dots\wedge ds_m^{1+r}}\left[\frac{s_1\dots s_m}{p\dots r}\left(d\varpi - p\frac{ds_1}{s_1}\wedge\varpi - \dots - r\frac{ds_m}{s_m}\wedge\varpi\right)\right]_{(S,S''\cup S')}$$

si $\varpi(x, y)$ est une forme de x, régulière près de $y\in S$ et si $\varpi(x, y)/s_1(x, y)^p\dots s_m(x, y)^r$ est une forme indépendante de y, *fermée sur* S'', *nulle sur* S'. Enfin, si f est une application d'une autre variété analytique complexe X^* dans X et si f^* est l'application réciproque, qui transforme variétés, formes et classes de cohomologie de X en variétés, formes et classes de X^*, on a

$$f^*\left[\frac{d^{p+\dots+r}[\omega]}{ds_1^{1+p}\wedge\dots\wedge ds_m^{1+r}}\Bigg|_{(S,S')}\right] = \frac{d^{p+\dots+r}[f^*\omega]}{d(f^*s_1)^{1+p}\wedge\dots\wedge d(f^*s_m)^{1+r}}\Bigg|_{(f^*S,f^*S')}$$

si les f^*S_i, $f^*S'_j$ sont régulières, en position générale.

2. Cas où les S_i ont des équations globales. — Limitons-nous aux définitions et aux formules les plus simples, dont l'intérêt est de relier les définitions précédentes au calcul différentiel classique.

Cas $m=1$. — Soit $\varpi(x)$ une forme régulière sur X, telle que $ds(x)\wedge d\varpi(x)=0$; définissons

$$\frac{d^p\varpi}{ds^p}\Bigg|_{(S,S')} = \frac{d^p[ds\wedge\varpi]}{ds^{1+p}}\Bigg|_{(S,S')} = \frac{d^{p-1}[d\varpi]}{ds^p}\Bigg|_{(S,S')}$$

l'égalité des deux derniers termes résultant de la relation

$$p!\,s^{-1-p}\,ds\wedge\varpi \sim (p-1)!\,s^{-p}\,d\varpi.$$

Cas $m>1$. — Soit $\varpi(x)$ une forme régulière sur X, telle que

$$(2)\qquad ds_1(x)\wedge\dots\wedge ds_m(x)\wedge d\varpi(x)=0;$$

définissons de même, quels que soient $p, \dots, r\geq 0$, la classe de cohomologie, de degré $d^0(\varpi)$:

$$\frac{\partial^{p+\dots+r}[\varpi]}{\partial s_1^p\dots\partial s_m^r}\Bigg|_{(S,S')} = \frac{d^{p+\dots+r}[ds_1\wedge\dots\wedge ds_m\wedge\varpi]}{ds_1^{1+p}\wedge\dots\wedge ds_m^{1+r}}\Bigg|_{(S,S')};$$

cette classe ne change pas quand on permute $\partial s_1^p, \dots \partial s_m^r$.

Les propriétés de ces classes de cohomologie résultent aisément de la construction de Gelfand et Šilov. Si $\varpi(x)$ et $\pi(x)$ vérifient (2), on a *la formule de Leibnitz* :

$$\frac{\partial^{P+\ldots+R}[\varpi\wedge\pi]}{\partial s_1^P\ldots\partial s_m^R}\bigg|_{(S,S')} = \sum_{\substack{0\leq p\leq P\\ \ldots\\ 0\leq r\leq R}} \frac{P!}{p!(P-p)!}\cdots\frac{R!}{r!(R-r)!}\frac{\partial^{p+\ldots+r}[\varpi]}{\partial s_1^p\ldots\partial s_m^r}\bigg|_{(S,S')}\frac{\partial^{P-p+\ldots+R-r}[\pi]}{\partial s_1^{P-p}\ldots\partial s_m^{R-r}}\bigg|_{(S,S')}.$$

On a la formule du changement de variables : soient

$$t_1(s_1,\ldots,s_m),\quad\ldots,\quad t_m(s_1,\ldots,s_m)$$

m fonctions analytiques, telles que

$$\frac{D(t)}{D(s)}\neq 0\quad\text{pour}\quad s_1=\ldots s_m=0;$$

$$\frac{\partial^{P+\ldots+R}[\varpi]}{\partial s_1^P\ldots\partial s_m^R}\bigg|_{(S,S')} = \sum_{\substack{0<p+\ldots+r\leq\\ P+\ldots+R}} C^{p,\ldots,r}_{P,\ldots,R}\frac{\partial^{p+\ldots+r}[\varpi]}{\partial t_1^p\ldots\partial t_m^r}\bigg|_{(S,S')},$$

les nombres complexes $C^{p,\ldots,r}_{P,\ldots,R}$ ne dépendant que de l'allure des fonctions $t_i(s_1,\ldots,s_m)$ pour $s_j=0$: ce sont les coefficients de la formule analogue du calcul différentiel classique. En effet :

Cas où $\varpi(x)$ est de degré nul et S′ *vide.* — L'hypothèse (2) signifie que $\varpi(x)=F[s_1(x),\ldots,s_m(x)]$; on voit aisément que

$$\frac{\partial^{p+\ldots+r}[\varpi]}{\partial s_1^p\ldots\partial s_m^r}\bigg|_{(S)}\ \text{est le produit du nombre}\ \frac{\partial^{p+\ldots+r}F[s_1,\ldots,s_m]}{\partial s_1^p\ldots\partial s_m^r}\bigg|_{s_1=\ldots s_m=0}$$

par la classe de cohomologie unité de S.

3. LA FORMULE DE CAUCHY-FANTAPPIÈ ([3]) permet de calculer quelques résidus. Notons : X un domaine convexe d'un espace affin complexe de dimension l; Ξ l'espace vectoriel de ses fonctions linéaires, numériques complexes; Ξ^* l'espace de ses variétés planes de codimension 1 : Ξ^* est un espace projectif complexe de dimension l, image de Ξ. La valeur de $\xi\in\Xi$ en $x\in X$ est notée $\xi.x=\xi_0+\xi_1x_1+\ldots+\xi_lx_l$, $(x_1,\ldots,x_l)$ étant les coordonnés de x, $(\xi_0,\ldots,\xi_l)$ celles de ξ. Sur Ξ^*, près de ξ^*, image de $\xi=(\xi_0,\ldots,\xi_l)$, si $\xi_i\neq 0$, on utilise les coordonnées locales $(\xi_0/\xi_i,\ldots,\xi_l/\xi_i)$. Notons

$$\omega(x)=dx_1\wedge\ldots\wedge dx_l,$$

$$\omega^*(\xi)=\sum_{k=0}^{l}(-1)^k\xi_k\,d\xi_0\wedge\ldots\wedge d\xi_{k-1}\wedge d\xi_{k-1}\wedge\ldots\wedge d\xi_l;$$

puisque $\xi_0^{l+1}d(\xi_1/\xi_0)\wedge\ldots\wedge d(\xi_l/\xi_0)=\omega^*(\xi)$, $g(\xi)\omega^*(\xi)$ est une forme différentielle de Ξ^* si g est homogène de degré $-l-1$. Donnons-nous un point $y \in X$; considérons la variété plane P et la quadrique Q de $\Xi^* \times X$ que décrit le point (ξ^*, x) de $(\Xi^* \times X)$ quand

$$P : \ \xi . y = 0; \qquad Q : \ \xi . x = 0.$$

Quand ξ^* décrit la variété plane y^* de Ξ^* d'équation

$$y^* : \ \xi . y = 0,$$

munie de son orientation naturelle, alors (ξ^*, y) décrit un cycle compact de $P \cap Q$; nous noterons sa classe d'homologie $(-1)^{l(l-1)/2} h(P \cap Q)$; c'est une base du sous-groupe de $H_c(P \cap Q)$ de dimension $2l - 2$. Soit enfin $f(x)$ une fonction holomorphe sur X; la formule de Cauchy-Fantappiè et la formule du résidu donnent

$$f(y) = \frac{1}{(2\pi i)^{l-1}} \int_{h(P \cap Q)} \frac{d^{l-1}[f(x)\omega^*(\xi) \wedge \omega(x)]}{d(\xi . x) \wedge [d(\xi . y)]^l}; \tag{3}$$

la classe de cohomologie figurant sous le signe $\int$ est donc $(2\pi i)^{l-1} f(y)$ fois la classe de base.

Soient p et ∂ les deux homomorphismes, appartenant à deux triplets différents d'homologie relative

$$p : \ H_c(P \cap Q) \to H_c(P \cap Q, S); \qquad \partial : \ H_c(P, Q \cup S) \to H_c(P \cap Q, S);$$

de (1) et (3) résulte que, s'il existe $h(P, Q \cup S) \in H_c(P, Q \cup S)$ tel que

$$\partial h(P, Q \cap S) = p h(P \cap Q);$$

alors

$$f(y) = \frac{1}{(2\pi i)^{l-1}} \int_{h(P, Q \cup S)} \frac{d^l [f(x)\omega^*(\xi) \wedge \omega(x)]}{[d(\xi . z)]^{1+l}}. \tag{4}$$

4. Dérivation d'une intégrale, fonction d'un paramètre. — Reprenons les notations de ([2]), en supposant que $m = 1$ et que $S = S_1$ dépende d'un paramètre $t \in T$, dont S' et ω ne dépendent pas. Limitons-nous ici au cas où S appartient à une série *linéaire* de sous-variétés : son équation locale $s(x, y, t) = 0$ est linéaire en t; nous supposerons $s(x, y, t)/s(x, y, t')$ indépendant de y.

Soit $\omega(x, y)$ une forme régulière de $x \in X$, nulle sur S', telle que $\omega(x, y) s(x, y, t)^{-q}$ soit une forme *fermée* de $x \in X - S$, indépendante de y; $s^{-1} ds \wedge \omega$ est donc, pour $dy = dt = 0$, indépendante de t, si $q \neq 0$; nous supposons cela encore vrai pour $q = 0$.

Soient $h(X, S \cup S')$ et $h(S, S')$ des classes d'homologie à supports compacts, de $(X, S \cup S')$ et (S, S') variant continûment avec t;

$$\dim(X, S \cup S') = d^0(\omega); \qquad \partial : \ H_c(X, S \cup S') \to H_c(S, S'); \qquad \dim h(S, S') = d^0(\omega) - 1.$$

Soit P un polynome homogène de degré p. On a *les formules de dérivation*,

(5) $$P\left(\frac{\partial}{\partial t}\right)\int_{h(X,\,S\cup S')}\frac{[-s(x,y,t)]^{-q}}{(-q)!}\,\omega(x,y)$$
$$=\int_{h(X,\,S\cup S')}P(-s_t)\frac{[-s]^{-p-q}}{(-p-q)!}\,\omega \qquad \text{si } p\leqq -q;$$
$$=\int_{\partial h(X,\,S\cup S')}\frac{d^{p+q-1}[P(-s_t)\,\omega]}{ds^{p+q}} \qquad \text{si } 0<-q<p;$$

(6) $$P\left(\frac{\partial}{\partial t}\right)\int_{h(S,\,S')}\frac{d^{q-1}[\omega]}{ds^q}=\int_{h(S,\,S')}\frac{d^{p+q-1}[P(-s_t)\,\omega]}{ds^{p+q}} \qquad \text{si } 0<q.$$

5. Ramification d'une intégrale, fonction d'un paramètre. — Les points t tels que S ait un point singulier constituent l'enveloppe des sous-variétés planes de T d'équation $s(x, y, t)=0$. Soit K l'ensemble de ceux des points de cette enveloppe tels que S soit en position générale par rapport à S′ et ait un seul point singulier, y, où

$$s_x(x,y,t)=0, \qquad s_t(x,y,t)\neq 0, \qquad \text{Hessien}_x[s(x,y,t)]\neq 0 \qquad \text{pour } x=y.$$

K est une sous-variété analytique de T, de codimension 1, d'équation

$$K:\ k(t)=0 \qquad [k_t\neq 0;\ k_t \text{ est parallèle à } s_t(y,y,t)].$$

Nous nous proposons d'étudier *près de* K l'intégrale

(7) $$\begin{cases} J(t)=\displaystyle\int_{h(X,\,S\cup S')}\frac{[-s(x,y,t)]^{-q}}{(-q)!}\,\omega(x,y) & \text{si } q\leqq 0;\\[2ex] J(t)=\displaystyle\int_{h(S,\,S')}\frac{d^{q-1}[\omega]}{ds^q} & \text{si } q>0.\end{cases}$$

Théorème. — $J(t)$ *est* holomorphe *sur* K *si* $d^0(\omega)\neq l$.

Supposons $d^0(\omega)=l$. Les sous-groupes de $H_c(X,\ S\cup S')$ et $H_c(S, S')$ qui s'annulent pour $t\in K$ sont constitués par les multiples entiers des deux classes

$$e(X,\ S\cup S') \quad \text{et} \quad e(S,S')=\partial e(X,\ S\cup S')$$

images de deux classes d'homologie qu'E. Picard a nommées évanouissantes :

$$e(V,S) \quad \text{et} \quad e(V\cap S)=\partial e(V,S),$$

où V désigne un voisinage ouvert du point singulier qu'a S pour $t\in K$; ces classes ont les dimensions respectives l et $l-1$; quand t décrit un lacet autour de K, $h(X,\ S\cup S')$ et $h(S, S')$ deviennent $h(\ldots)+ne(\ldots)$, l'entier n étant l'indice de Kronecker de

$$h(S, S') \quad \text{et} \quad e(V\cap S) \qquad \text{(S. Lefschetz)}$$

Définissons, quand $P(k_t)\neq 0$; $p=d^0(P)$,

$$j_P(t)=\int_{e(V,S)}\frac{[-s]^{p-q}}{(p-q)!}\,\frac{\omega}{P(-s_t)} \quad (q\leqq p);$$
$$j_P(t)=\int_{e(V,S)}\frac{d^{l-p-1}}{ds^{q-p}}\,\frac{\omega}{P(-s_t)} \quad (p<q);$$

vu (5) et (6)

$$P\left(\frac{\partial}{\partial t}\right) j_{PQ}(t) = j_Q(t). \tag{8}$$

THÉORÈME. — *Si l est* impair

$$j_P(t)\, k(t)^{q-p-\frac{l}{2}} \quad \text{et} \quad J(t) - \frac{n}{2} j_1(t) \quad \textit{sont holomorphes sur } K.$$

Si l est pair, $j_P(t)$ *est* holomorphe *sur* K *et s'y annule* $p - q + (l/2)$ *fois;*

$$\textit{pour } q - \frac{l}{2} \leq p, \quad J(t) - \frac{n}{2\pi i} P\left(\frac{\partial}{\partial t}\right) [j_P(t) \log k(t)] \quad \textit{est holomorphe sur } K.$$

La preuve utilise les propriétés et la définition très simple de e dues à I. Fáry ([1]).

6. LA DISTRIBUTION QUE DÉFINIT UNE INTÉGRALE, FONCTION D'UN PARAMÈTRE. — Appliquons le théorème précédent aux hypothèses que voici :

T et K sont des variétés analytiques *réelles; h* varie *continûment* en fonction de t, *même aux points de* K.

Si l est *impair*, il existe un entier $n(t)$, constant de chaque côté de K, tel que

$$J(t) - \frac{1}{2} P\left(\frac{\partial}{\partial t}\right) [n(t) j_P(t)] \qquad \left(q - \frac{l+1}{2} \leq p\right) \tag{9}$$

soit holomorphe sur K; *la distribution* $P(\partial/\partial t)[n(t) j_P(t)]$ est indépendante du choix de P et est parfaitement déterminée par la donnée de h [vu (8) et l'inégalité imposée à p].

Si l est *pair*, il existe un entier constant n et un entier $N(t)$, constant de chaque côté de K, tels que

$$J(t) - \frac{n}{2\pi i} P\left(\frac{\partial}{\partial t}\right) [j_P(t) \log k(t)] - P\left(\frac{\partial}{\partial t}\right) [N(t) j_P(t)] \qquad \left(q - \frac{l}{2} \leq p\right) \tag{10}$$

soit holomorphe sur K; *la distribution*

$$\frac{n}{2\pi i} P\left(\frac{\partial}{\partial t}\right) [j_P(t) \log k(t)] + P\left(\frac{\partial}{\partial t}\right) [N(t) j_P(t)]$$

est indépendante du choix de P et déterminée par la donnée de h, à l'addition près d'une fonction holomorphe sur K. D'où :

THÉORÈME. — *Il existe, près de* K, *une* distribution *unique* $J(t)$, *égale hors de* K *à la fonction* $J(t)$ *et telle que la distribution* (9) (*l impair*) *ou* (10) (*l pair*) *soit une fonction holomorphe sur* K. [*La distribution* $J(t)$ *est la fonction* $J(t)$ *si* $2q \leq l + 1$]. *Convenons que l'intégrale* (7) *désigne cette distribution :* les formules de dérivation (5) *et* (6) *restent valables*.

Ces résultats permettent de poursuivre l'étude du problème de Cauchy ([4]).

(*) Séance du 22 décembre 1958.

(1) I. Fáry, *Ann. Math.*, 65, 1957, p. 35-37 et 47-53.

(2) J. Leray, *Comptes rendus*, 247, 1958, p. 2253.

(3) J. Leray, *Rend. Accad. Naz. Lincei*, 20, 1956, p. 589-590.

(4) J. Leray, *Bull. Soc. Math.*, 85, 1957, p. 389-429; 86, 1958, p. 75-96; *Comptes rendus*, 242, 1956, p. 953.

(avec Y. Hamada et C. Wagschal)

Systèmes d'équations aux dérivées partielles à caractéristiques multiples : probléme de Cauchy ramifié ; hyperbolicité partielle

J. Math. Pures Appl. 55 (1976) 297–352

INTRODUCTION

Cet article a pour objet l'étude du problème de Cauchy linéaire non caractéristique pour des opérateurs analytiques à caractéristiques multiples de multiplicité constante.

La première partie de cette étude est consacrée au problème de Cauchy ramifié; il s'agit du problème

$$(0.1)\qquad \begin{cases} a(x, \mathrm{D})\, u(x) = 0, \\ \mathrm{D}_0^h u(x)\,|_{\mathrm{S}} = w_h(x'), \qquad 0 \leqq h < m, \end{cases}$$

où l'opérateur $a(x, \mathrm{D})$ est un opérateur différentiel linéaire d'ordre m et à coefficients fonctions holomorphes de $x = (x^j)_{0 \leqq j \leqq n}$ au voisinage de l'origine de $\mathbf{C}^{n+1}$, où l'hyperplan $\mathrm{S} : x^0 = 0$ n'est pas caractéristique à l'origine pour cet opérateur et où les données $w_h(x')$, $x' = (x^j)_{1 \leqq j \leqq n}$, holomorphes sont supposées ramifiées autour de l'hyperplan de S :

$$\mathrm{T}: \quad x^0 = x^1 = 0.$$

Nous étudierons le problème (0.1) avec les hypothèses suivantes. Soit $g(x, \xi)$ le polynôme caractéristique de l'opérateur $a(x, \mathrm{D})$; considérons la décomposition de ce polynôme en facteurs irréductibles :

$$g(x, \xi) = \prod_s g_s(x, \xi)^{m_s}$$

et notons d le degré du polynôme réduit

$$g_0(x, \xi) = \prod_s g_s(x, \xi).$$

Nous supposerons que l'équation en $\xi_0 \in \mathbf{C}$:

$$g_0(0; \xi_0, 1, 0, \ldots, 0) = 0$$

admet d racines distinctes; cette hypothèse permet de construire d hypersurfaces caractéristiques issues de T (*cf.* § 1).

Sous les hypothèses précédentes, on a le résultat suivant (*cf.* th. 1.1) :

THÉORÈME 0.1. — *Le problème de Cauchy* (0.1) *admet une unique solution holomorphe ramifiée autour des hypersurfaces caractéristiques issues de* T.

Rappelons que ce théorème a d'abord été démontré par Y. Hamada lorsque les données de Cauchy (w_h) sont uniformes et présentent des singularités polaires ou essentielles : le cas des opérateurs à caractéristiques simples ($m_s = 1$ pour tout s) est traité dans [7], le cas des opérateurs à caractéristiques multiples vérifiant une condition de E. E. Lévi est traité dans [8], puis Y. Hamada [9] et L. Lamport [11] se sont affranchis de cette condition.

C. Wagschal [27] a ensuite montré que le théorème 0.1 subsistait sans aucune hypothèse sur la nature des singularités des données (w_h) pour des opérateurs à caractéristiques simples; l'objet essentiel de la première partie de ce travail est de montrer qu'il en est de même pour des opérateurs à caractéristiques multiples.

La méthode d'étude du problème (0.1) est celle qui a été développée dans [27]; elle consiste essentiellement à chercher la solution de (0.1) sous la forme

$$u(x) = \sum_{i=1}^{d} u^i(k^i(x), x), \tag{0.2}$$

où $k^i(x) = 0$ ($1 \leqq i \leqq d$) désigne les hypersurfaces caractéristiques distinctes issues de T; les inconnues $u^i(t, x)$ sont des fonctions de $x \in \mathbf{C}^{n+1}$ et d'une variable t décrivant le revêtement universel $\mathscr{R}$ d'un disque pointé centré à l'origine du plan complexe. Ceci conduit à des équations intégrodifférentielles [*cf.* les équations (3.17)] que nous résolvons par la méthode des approximations successives (*cf.* § 5) : les fonctions u^i sont alors données sous la forme de séries :

$$u^i(t, x) = \sum_{k \in \mathbf{N}} u^i_k(t, x), \tag{0.3}$$

où les fonctions $u^i_k : \mathscr{R} \times \Omega \to \mathbf{C}$ sont holomorphes, Ω désignant un voisinage ouvert de l'origine de $\mathbf{C}^{n+1}$. La principale difficulté de ce travail réside dans la démonstration de la convergence de ces séries; la méthode est celle qui avait été utilisée dans [27]; cette méthode décrite au paragraphe 5 consiste à composer les fonctions $t \to u^i_k(t, x)$ avec des chemins $\gamma : \mathrm{I} \to \mathscr{R}$, $\mathrm{I} = [0, 1]$, de classe $\mathscr{C}^\infty$ tracés sur le revêtement $\mathscr{R}$; on obtient ainsi des fonctions $x \to u^i_k(\gamma(.), x)$ holomorphes dans Ω à valeurs dans l'espace de Fréchet $\mathscr{C}^\infty(\mathrm{I}; \mathbf{C})$; on prouve alors la convergence (sous certaines conditions) des séries $\sum u^i_k(\gamma(s), x)$ dans l'espace $\mathscr{H}(\Omega; \mathscr{C}^\infty(\mathrm{I}; \mathbf{C}))$, d'où l'on déduit la convergence des séries (0.3).

La méthode précédente conduit en fait à la résolution d'un problème de Cauchy généralisé dans des espaces de fonctions holomorphes à valeurs dans des espaces du type de Gevrey (*cf.* th. 5.1). Cette remarque est fondamentale pour la seconde partie de notre travail : grâce à ce théorème 5.1, on peut en effet étudier le problème de Cauchy dans $\mathbf{R}^{n+1}$ pour des opérateurs analytiques partiellement hyperboliques.

Indiquons dans cette introduction le théorème essentiel auquel conduit cette étude. On considère au voisinage de l'origine de $\mathbf{R}^{n+1}$ le problème de Cauchy non caractéristique :

$$(0.4)\qquad \begin{cases} a(x, \mathrm{D})\,u(x) = v(x), \\ \mathrm{D}_0^h u(x)\,|_{\mathrm{S}} = w_h(x'), \qquad 0 \leqq h < m, \end{cases}$$

où $a(x, \mathrm{D})$ est maintenant un opérateur à coefficients $\mathbf{R}$-analytiques au voisinage de l'origine de $\mathbf{R}^{n+1}$ $(x = (x^j)_{0 \leqq j \leqq n})$. Etant donné une sous-variété linéaire T de S,

$$\mathrm{T}: \quad x^0 = x^1 = \ldots = x^q = 0 \qquad (1 \leqq q \leqslant n)$$

nous dirons, suivant la terminologie de J. Leray [18], que l'opérateur $a(x, \mathrm{D})$ est *partiellement hyperbolique relativement à* S *modulo* T, si pour tout $\eta = (\eta_1, \ldots, \eta_q) \in \mathbf{R}^q - \{0\}$, l'équation en ξ_0 :

$$g_0(0; \xi_0, \eta, 0, \ldots, 0) = 0$$

admet d racines réelles distinctes.

Posons $y = (x^1, \ldots, x^q)$, $z = (x^{q+1}, \ldots, x^n)$ et soit α un nombre $\geqq 1$. Nous supposons les données v et w_h de classe $\mathscr{C}^\infty$ au voisinage de l'origine et qu'il existe une constante $c \geqq 0$ telle que, pour tous les multi-indices de dérivation $p \in \mathbf{N}$, $\beta \in \mathbf{N}^q$, $\gamma \in \mathbf{N}^{n-q}$,

$$(0.5)\qquad \begin{cases} |\mathrm{D}_0^p \mathrm{D}_y^\beta \mathrm{D}_z^\gamma v(x^0, y, z)| \leqq c^{p+|\beta|+|\gamma|+1} (p!\,\beta!)^\alpha \gamma!, \\ |\mathrm{D}_y^\beta \mathrm{D}_z^\gamma w_h(y, z)| \leqq c^{|\beta|+|\gamma|+1} (\beta!)^\alpha \gamma!. \end{cases}$$

On a alors le théorème suivant (*cf.* th. 11.1) :

THÉORÈME 0.2. — *On suppose* $1 \leqq \alpha < m_0/(m_0 - 1)$, *où* $m_0 = \sup m_s$ *désigne l'ordre de multiplicité maximum des facteurs irréductibles du polynôme caractéristique de l'opérateur* $a(x, \mathrm{D})$. *Alors, si* $a(x, \mathrm{D})$ *est partiellement hyperbolique relativement à* S *modulo* T, *le problème de Cauchy* (0.4)-(0.5) *admet au voisinage de l'origine une unique solution de classe de Gevrey* α.

Ce théorème [que complètent, quand T est un hyperplan de S, la remarque 10.3 (domaine d'influence) et la proposition 10.1 (frontière du domaine d'analyticité lorsque $\alpha = 1$)] et en quelque sorte intermédiaire entre le théorème de Cauchy-Kowalewski où aucune hypothèse d'hyperbolicité n'est nécessaire et les théorèmes bien connus concernant des opérateurs hyperboliques non stricts (correspondant à T = S) tels que le théorème 5.7.3 de Hörmander [10], le théorème d'Ohya [20] et les théorèmes de Leray-Ohya ([16] et [17]).

(avec Y. Hamada et A. Takeuchi)

Prolongements analytiques de la solution du problème de Cauchy linéaire

J. Math. Pures Appl. 64 (1985) 257–319

Introduction

1. Le problème étudié. — Soit Ω une *variété analytique* complexe de dimension complexe n. Notons ω un point arbitraire de Ω et $x'=(x_1, \ldots, x_n)$ des coordonnées analytiques locales de ω. Supposons Ω connexe, paracompacte, non compacte. Soit $\bar{\Omega}$ le compactifié de Ω par adjonction d'un point $\partial\Omega$, appelé « point à l'infini »; les voisinages ouverts de $\partial\Omega$ sont les complémentaires dans $\bar{\Omega}$ des parties compactes de Ω.

Signalons ceci : l'hypothèse que la variété Ω est paracompacte équivaut à chacune des suivantes : elle est métrisable; elle peut être munie d'un ds^2 riemannien de classe C^∞. En effet : un théorème classique de Whitney prouve que la paracompacité implique l'existence d'un tel ds^2; tout espace métrisable est paracompact, d'après un théorème de A. H. Stone; *voir* [St] ou [K].

Soit Σ une *surface de Riemann*, non compacte, paracompacte et simplement connexe. Notons σ un point arbitraire de Σ et x_0 une coordonnée analytique locale de σ. Un point α de Σ est donné.

Notons :

$$x=(\sigma, \omega)\in X=\Sigma\times\Omega;$$

x a donc les coordonnées locales $(x_0, x')=(x_0, x_1, \ldots, x_n)$.

Notons :

$$\alpha\times\Omega=\{(\sigma, \omega)\in\Sigma\times\Omega;\ \sigma=\alpha\}.$$

Soit a un *opérateur différentiel* d'ordre m, holomorphe sur X au voisinage de $\alpha\times\Omega$, opérant sur les fonctions numériques holomorphes. Nous supposons qu'*aucune hypersurface $\sigma\times\Omega$ n'est, en aucun de ses points, caractéristique* pour l'opérateur a.

Soient v et w *deux fonctions* numériques holomorphes sur X au voisinage de $\alpha\times\Omega$.

Nous étudions dans le chapitre 1 et le chapitre 3 le problème de Cauchy : trouver une fonction numérique, holomorphe, u, telle que

(1) $$au=v; \qquad u-w \text{ s'annule } m \text{ fois sur } \alpha\times\Omega.$$

Notre but est *d'expliciter, aussi simplement que possible, des voisinages de* $\alpha\times\Omega$, *aussi grands que possible, sur lesquels le problème* (1) *possède une solution.*

Nous étendons nos conclusions dans le chapitre 4 au cas où les fonctions en jeu sont à valeurs dans $\mathbb{C}^N$ et où a est une $N\times N$ matrice, puis au cas où ces fonctions sont remplacées par des sections d'un espace fibré vectoriel complexe de base X.

2. Les principales définitions. — Rappelons que $T^*(Y)$ désigne le fibré cotangent d'une variété Y et $T^*_y(Y)$ sa fibre au-dessus du point y de Y.

Notons

ξ_0, ξ' et $\xi=(\xi_0, \xi')$ des covecteurs respectifs de Σ en σ, de Ω en ω et de x en X;

c'est-à-dire :

$$\xi_0\in T^*_\sigma(\Sigma), \qquad \xi'\in T^*_\omega(\Omega), \qquad \xi\in T^*_x(X);$$
$$(\sigma;\ \xi_0)\in T^*(\Sigma), \qquad (\omega;\ \xi')\in T^*(\Omega), \qquad (x;\ \xi)\in T^*(X).$$

Le polynôme caractéristique en $x\in X$ de l'opérateur a est un polynôme homogène de degré m, que nous notons

(2.1) $$g: \quad \xi\mapsto g(x;\xi)=\sum_{r=0}^{m} g_r(x_0, \omega; \xi')\,\xi_0^r;$$

donc g_r est un polynôme en ξ' homogène de degré $m-r$; il dépend du choix de la coordonnée locale x_0 de $\sigma\in\Sigma$. Nous supposons g *holomorphe sur* $T^*(X)$.

L'hypothèse que l'hypersurface $\sigma\times\Omega$ n'est caractéristique en aucun de ses points s'énonce :

$$(\forall(x_0, \omega)): \quad g_m(x_0, \omega)\neq 0.$$

Note. — Soit

$$\sum_{|\lambda|\leq m} a_\lambda(x)\,D^\lambda, \qquad \text{où } \lambda=(\lambda_0, \lambda_1, \ldots, \lambda_n)\in\mathbb{N}^{n+1}, \qquad |\lambda|=\lambda_0+\ldots+\lambda_n,$$
$$D^\lambda=D_0^{\lambda_0}\ldots D_n^{\lambda_n}, \qquad D_k=\partial/\partial x_k,$$

l'expression locale de l'opérateur a; par définition celle de son polynôme caractéristique g est

$$g(x;\xi)=\sum_{|\lambda|=m} a_\lambda(x)\,\xi^\lambda.$$

L'équation caractéristique sera, par définition, l'équation d'inconnue ξ_0 :

$$\sum_{r=0}^{m} g_r(x_0, \omega; \xi')\xi_0^r = 0;$$

ses racines sont les racines caractéristiques. Rappelons comment elles dépendent du choix de la coordonnée x_0 : pour tout $(\omega; \xi')$

(2.2) $\xi_0 \, dx_0$ est une forme différentielle de σ,

c'est-à-dire est indépendante du choix de la coordonnée locale x_0.

Les racines caractéristiques servent à l'étude de la propagation du support singulier : *voir*, en particulier, *D. Schiltz* [Si].

La majorante $\rho(x_0, \omega; \xi')$ *du module des racines caractéristiques*, définie comme suit, nous permettra de construire des domaines de X sur lesquels le problème (1) possède une solution holomorphe :

Si $g_r(x_0, \omega; \xi') = 0$ en $(x_0, \omega; \xi')$ pour $r = 0, \ldots, m-1$, alors $\rho(x_0, \omega; \xi') = 0$.

Sinon $\rho(x_0, \omega; \xi')$ est l'unique racine $\rho > 0$ de l'équation

$$\sum_{r=0}^{m-1} |g_r(x_0, \omega; \xi')| \rho^r = |g_m(x_0, \omega)| \rho^m. \tag{2.3}$$

C'est en mettant cette équation sous la forme

$$\sum_{r=0}^{m-1} |g_r(x_0, \omega; \xi')| (1/\rho)^{m-r} = |g_m(x_0, \omega)|$$

qu'on rend évidentes l'existence de ρ sur $\Sigma \times T^*(\Omega)$, son unicité et aussi sa continuité.

Évidemment, ρ est positivement homogène de degré 1 en ξ'; c'est-à-dire :

$$(\forall \theta \in \mathbb{C}) : \quad \rho(x_0, \omega; \theta\xi') = |\theta| . \rho(x_0, \omega; \xi'). \tag{2.4}$$

La section 5 établira les deux propriétés de ρ que voici :

Pour tout $(\omega; \xi') \in T^*(\Omega)$, la fonction

$$x_0 \mapsto \log \rho(x_0, \omega; \xi') \tag{2.5}$$

est sous-harmonique ou identique à $-\infty$.

Pour tout $(\omega; \xi') \in T^*(\Omega)$, l'expression

$$\rho(x_0, \omega; \xi') |dx_0| \tag{2.6}$$

est indépendante du choix de la coordonnée locale x_0.

Une partie fermée Φ^* de $T^*(\Omega)$ est dite *admissible* quand la fonction

$$x_0 \mapsto \rho_{x_0} = \sup_{(\omega;\, \xi') \in \Phi^*} \rho(x_0, \omega; \xi') \text{ est } \textit{localement bornée}. \tag{2.7}$$

La section 5 prouvera ceci :

Quand $(\omega; \xi') \in \Phi^*$, la fonction

(2.8) $$x_0 \mapsto \rho(x_0, \omega; \xi')$$

possède localement *un module de continuité indépendant de* $(\omega; \xi')$, mais dépendant de Φ^*, et la fonction

(2.9) $x_0 \mapsto 1/\rho(x_0, \omega; \xi')$, *tronquée* par une constante arbitraire,

vérifie localement *une condition de Lipschitz indépendante de* $(\omega; \xi')$, mais dépendant de Φ^* et de cette troncature.

(Tronquer une fonction par une constante c'est, là où sa valeur dépasse cette constante, remplacer cette valeur par cette constante.)

Commentaire. — La propriété (2.9) servira à établir la propriété (11.5), que la section 12 note (12.8) et emploie à prouver son lemme auxiliaire.

La fonction $x_0 \mapsto \rho_{x_0}$ est définie par (2.7); voici ses propriétés :

Cette fonction est *continue*, vu (2.8).

La fonction

(2.10) $x_0 \mapsto 1/\rho_{x_0}$, *tronquée* par une constante arbitraire, est *lipschitzienne*,

vu (2.9).

La fonction

(2.11) $x_0 \mapsto \log \rho_{x_0}$ est *sous-harmonique* ou identique à $-\infty$,

vu (2.5) et [R].

Vu (2.6), l'expression

(2.12) $ds = \rho_{x_0} |dx_0|$ est *indépendante du choix de la coordonnée locale* x_0 de $\sigma \in \Sigma$.

Par suite,

ds^2 est *riemannien, conforme à la structure analytique complexe* de Σ *et à courbure* $\leqq 0$ sur la partie de Σ où $\rho_{x_0} \neq 0$ et où la fonction $x_0 \mapsto \rho_{x_0}$ est de classe C^2 : *voir* [C], section 4.3, exercice 2, p. 237.

Commentaire. — Le chapitre 2 et donc la section 13 résultent de cette dernière propriété.

Définissons sur Σ

(2.13) $$dist(\sigma, \alpha) = \inf \int_\alpha^\sigma ds;$$

l'inégalité du triangle est vérifiée; mais *dist* $(\sigma, \alpha) = 0$ n'implique pas $\sigma = \alpha$.

Commentaire. — La « distance », ainsi définie sur Σ par la partie principale de l'opérateur a et le choix de Φ^* vérifiant (2.6), apparaît dans toutes nos conclusions.

3. LES RÉSULTATS PRINCIPAUX sont les théorèmes I, II et III que le chapitre 3 énonce dans le cas d'une équation et que le chapitre 4 étend au cas des systèmes. Citons ici le théorème III.

Notation. — Munissons Ω d'une métrique riemannienne; elle définit une métrique euclidienne sur chaque fibre $T^*(\Omega)$; supposons *admissible* le sous-espace fibré Φ^* de $T^*(\Omega)$ dont les fibres sont les sphères unitaires des $T^*_\omega(\Omega)$; c'est-à-dire :

$$\Phi^* = \{(\omega;\ \xi') \in T^*(\Omega);\ |\xi'| = 1\}.$$

Pour réaliser cette condition, quand elle ne l'est pas, il suffit de remplacer la métrique de Ω par une métrique conforme, décroissant à l'infini suffisamment vite. Notons $Dist(\omega, \partial\Omega)$ la distance dans cette métrique de $\omega \in \Omega$ au point à l'infini $\partial\Omega$ de Ω.

Le choix de Φ^* qui précède définit $dist(\sigma, \alpha)$ par (2.7), (2.11) et (2.12).

Théorème III. — *Le problème* (1) *possède une unique solution holomorphe sur le domaine* :

$$\Delta = \{(\sigma, \omega) \in X;\ dist(\sigma, \alpha) < Dist(\omega, \partial\Omega)\}$$

si les données a et v sont holomorphes sur Δ.

Note. — En particulier $\Delta = X$ dans chacun des deux cas :

$(\forall \sigma \in \Sigma)$: $dist(\sigma, \alpha) = 0$, c'est-à-dire : $a(x, \partial/\partial x)$ ne contient pas d'autre dérivation d'ordre m que $\partial^m/\partial x_0^m$.

$(\forall \omega \in \Omega)$: $Dist(\omega, \partial\Omega) = \infty$, condition qu'on peut satisfaire par exemple quand $\Omega = \mathbb{C}^n$; $\omega \mapsto g_r(x_0, \omega;\ \xi')$ est un polynôme de degré $\leqq m - r$.

La section 26 énonce deux résultats plus généraux : les exemples IV.1 et IV.2.

Commentaire. — Les résultats que n'impliquent pas les théorèmes sont nommés « propositions ». Les lemmes sont des résultats qu'englobent les théorèmes ou les propositions.

4. HISTORIQUE. — Cet article complète la note [HT], l'observation qui la suit et la note [HLT]. L'ensemble de ses résultats est présenté dans [L 2], [L 3] et [L 4].

La note [HT] et le présent article ont pour point de départ le lemme 9.1 de [L1], qui complète les précisions que J. Schauder [Sa] et I. Petrowsky [P] ont apportées au théorème de Cauchy-Kowalewski.

Nous n'employons pas les preuves de ce théorème données plus récemment par divers auteurs.

L'exposé [L4] *esquisse la théorie* qu'explicite le présent article.

La théorie des ondes, c'est-à-dire la théorie des opérateurs hyperboliques, a guidé non pas notre démonstration, que nous croyons originale, mais l'énoncé de notre théorème III : Dans cet énoncé, remplaçons la surface de Riemann Σ par l'axe du temps, la variété analytique complexe Ω par une variété réelle, jouant le rôle d'espace, $X = \Sigma \times \Omega$ devenant donc l'espace-temps; remplaçons l'opérateur holomorphe a par un opérateur hyperbolique de cet espace-temps et les hypothèses d'holomorphie par des hypothèses de régularité appropriées (Ω, g et a sont de classes C^2, C^1 et C^0; v et w sont localement de carrés sommables; u et ses dérivées d'ordres $<m$ sont localement de carrés sommables); alors le problème (1) possède encore une solution unique u sur le domaine Δ.

Certes, on sait que u existe et est unique sur un domaine plus grand que Δ dans le cas hyperbolique et aussi, localement, vu [Si], dans le cas analytique; certes nos théorèmes I, II et III ne donnent pas un prolongement analytique unique de la solution locale du problème (1), mais divers prolongements; certes nos théorèmes emploient et perfectionnent des méthodes dont nous n'avions d'abord tiré que les résultats restreints [HT], [HLT] et [L2]. On est donc tenté de croire que nos résultats se laissent englober dans un résultat plus général.

Pour l'obtenir, il faudrait effectuer un prolongement analytique plus efficace que celui dont les étapes sont les suivantes.